Sustainable Practices:
Concepts, Methodologies, Tools and Applications

Information Resources Management Association
USA

Volume III

Information Science REFERENCE
An Imprint of IGI Global

Managing Director:	Lindsay Johnston
Acquisitions Editor:	Kayla Wolfe
Production Manager:	Jennifer Yoder
Publishing Systems Analyst:	Adrienne Freeland
Development Editor:	Austin DeMarco
Multi-Volume Book Production Specialist:	Deanna Jo Zombro
Cover Design:	Jason Mull

Published in the United States of America by
Information Science Reference (an imprint of IGI Global)
701 E. Chocolate Avenue
Hershey PA 17033
Tel: 717-533-8845
Fax: 717-533-8661
E-mail: cust@igi-global.com
Web site: http://www.igi-global.com

Library of Congress Cataloging-in-Publication Data

Sustainable practices : concepts, methodologies, tools, and applications / Information Resources Management Association, editor.
volumes cm
ISBN 978-1-4666-4852-4 (hardcover) -- ISBN 978-1-4666-4853-1 (ebook) -- ISBN 978-1-4666-4854-8 (print & perpetual access) 1. Sustainable development. 2. Sustainability. 3. Environmentalism. I. Information Resources Management Association.
HC79.E5S86875 2014
338.9'27--dc23

2013033887

British Cataloguing in Publication Data
A Cataloguing in Publication record for this book is available from the British Library.

For electronic access to this publication, please contact: eresources@igi-global.com.

List of Contributors

Ahmadi, Arash / *Razi University, Iran* 335
Al-Bahadili, Hussein / *Petra University, Jordan* 429
Alhosseni, Sanaz Naghavi / *Amirkabir University of Technology, Iran* 517
Alimsi, Rene / *CERETETH, Greece* 1492
Allen, Julian / *University of Westminster, UK* 1297
Amanna, Ashwin / *Virginia Tech, USA* 495
Anayochukwu, Ani Vincent / *University of Nigeria, Nigeria* 413
Anpalagan, Alagan / *Ryerson University, Canada* 1331
Antonova, Albena / *Sofia University, Bulgaria* 384
Arora, Kalpana / *Indian Institute of Technology, India* 1519
Ates, Nuray / *Erciyes University, Turkey* 867
Aydogan, Emel Kizilkaya / *Erciyes University, Turkey* 867
Ayuk, Elias T. / *United Nations University Institute for Natural Resources in Africa, Ghana* 816
Badmasti, Bakhtiar / *Islamic Azad University—Marivan Branch, Iran* 1576
Baiocco, F. / *Istituto Superiore per la Protezione e la Ricerca Ambientale, Italy* 1137
Behzadan, Afshin / *Ryerson University, Canada* 1331
Bekić, Bojana / *Institute of Agricultural Economics, Belgrade, Serbia* 769
Bergeron, Susan J. / *Coastal Carolina University, USA* 537
Bevrani, Hassan / *University of Kurdistan, Iran* 1576
Bhagat, Someshwar / *National Centre for Integrated Pest Management, India* 566
Bhardwaj, Archana / *National Centre for Integrated Pest Management, India* 566
Bhutto, Arabella / *Mehran University Institute of Science & Technology Development, Pakistan* 693
Blackwood, David J. / *University of Abertay Dundee, UK* 581
Blin, Joël / *Centre de Coopération Internationale en Recherche Agronomique pour le Développement, Burkina Faso & 2iE, Institut International d'Ingénierie de l'Eau et de l'Environnement, Burkina Faso* 1380
Bochtis, Dionysis / *Aarhus University, Denmark* 786
Bose, Tamal / *Virginia Tech, USA* 495
Bošković, Olgica / *University of Belgrade, Serbia* 1203
Bradbury, Matthew / *Unitec Institute of Technology, New Zealand* 1683
Brini, S. / *Istituto Superiore per la Protezione e la Ricerca Ambientale, Italy* 1137
Browne, David / *Trinity College, The University of Dublin, Ireland* 1246

Browne, Michael / *University of Westminster, UK* 1297
Cassidy, Alice / *In View Education and Professional Development, Canada* 1454
Caulfield, Brian / *Trinity College, The University of Dublin, Ireland* 1246
Celesti, Antonio / *University of Messina, Italy* 279
Chen, Edward T. / *University of Massachusetts – Lowell, USA* 397
Chen, Xuetao / *Virginia Tech, USA* 495
Cherrett, Tom / *University of Southampton, UK* 1505
Ciocănea, Cristiana Maria / *University of Bucharest, Romania* 621
Comfort, Daphne / *University of Gloucestershire, UK* 748
Congedo, L. / *Istituto Superiore per la Protezione e la Ricerca Ambientale, Italy* 1137
Cotton, Matthew / *University of Leeds, UK* 1009
Cusick, John / *University of Hawai'i at Manoa Environmental Center, USA* 1443
Dabat, Marie-Hélène / *Centre de Coopération Internationale en Recherche Agronomique pour le Développement, Burkina Faso* 1380
Datla, Dinesh / *Virginia Tech, USA* 495
Dehghanian, Farzad / *Ferdowsi University of Mashhad, Iran* 944
D'Emilio, M. / *Institute of Methodologies for Environmental Analysis, Italy* 1400
Di Bella, Arturo / *University of Catania, Italy* 1121
Domazet, Ivana / *Institute of Economic Sciences, Serbia* 1640
Doukas, Haris / *National Technical University of Athens, Greece* 221
Drašković, Božo / *Institute of Economic Sciences, Serbia* 1640
Eid, Mohamed / *The British University in Egypt, Egypt* 1183
Ercan, Secil / *Istanbul Technical University, Turkey* 242
Ercoskun, Ozge Yalciner / *Gazi University, Turkey* 49
Evangelista, Pietro / *IRAT-CNR, Italy & University of Naples Federico II, Italy* 86
Falconer, Ruth E. / *University of Abertay Dundee, UK* 581
Filipović, Vladimir / *Institute "Tamis" Pancevo, Serbia* 769
Fleisch, Elgar / *ETH Zurich, Switzerland & University of St. Gallen, Switzerland* 731
Florou, Dimitra / *University of Athens, Greece* 988
Fokaides, Paris A. / *RD Hydraulis Ltd, Cyprus* 1742
Fonta, William M. / *United Nations University Institute for Natural Resources in Africa, Ghana* 816
Gilmour, Daniel / *University of Abertay Dundee, UK* 581
Gouscos, Dimitris / *University of Athens, Greece* 988
Grigoroudis, Evangelos / *Technical University of Crete, Greece* 158
Grosse-Dunker, Friedrich / *Dark Horse GmbH, Germany* 1106
Guo, Liping / *Northern Illinois University, USA* 318
Halasa, Azmi / *The Arab Academy for Banking and Financial Sciences, Jordan* *429*
Hanff, Elodie / *2iE, Institut International d'Ingénierie de l'Eau et de l'Environnement, Burkina Faso* 1380
Hansen, Erik G. / *Leupana University Lüneburg, Germany* 1106
Hasan, S. M. Shajedul / *Virginia Tech, USA* 495
He, An / *Virginia Tech, USA* 495
Heidmann, Olivier / *CERETETH, Greece* 1492

Higginson, Nancy / *University of Calgary, Canada* 713
Hillier, David / *University of Glamorgan, UK* 748
Hin, Leo Tan Wee / *Singapore National Academy of Science, Singapore & National University of Singapore, Singapore* 676
Hisarligil, Hakan / *Erciyes University, Turkey* 1027
Huemann, Martina / *WU Vienna University of Economics and Business, Austria* 201
Huge-Brodin, Maria / *Linköping University, Sweden* 86
Ieromonachou, Petros / *University of Greenwich, UK* 786
Ilic, Alexander / *ETH Zurich, Switzerland* 731
Imbrenda, V. / *University of Basilicata, Italy & National Research Council of Italy, Institute of Methodologies for Environmental Analysis, Italy* 1400
Iojă, Cristian / *University of Bucharest, Romania* 140
Isaacs, John P. / *University of Abertay Dundee, U K* 581
Isaksson, Karin / *Linköping University, Sweden* 86
Isik, Mine / *Istanbul Technical University, Turkey* 242
Jarvie-Eggart, Michelle Edith / *Barr Engineering, USA & University of Maryland University College, USA* 846
Jean-Vasile, Andrei / *Petroleum and Gas University of Ploiesti, Romania* 1701
Jones, Peter / *University of Gloucestershire, UK* 748
Kachris, Christoforos / *Athens Information Technology, Greece* *463*
Kajikawa, Yuya / *The University of Tokyo, Japan* 885
Kamel, Fouad / *University of Southern Queensland, Australia* 1614
Karaaslan, Sule / *Gazi University, Turkey* 1027
Karakosta, Charikleia / *National Technical University of Athens, Greece* 221
Karami, Saber / *Amirkabir University of Technology, Iran* 335
Kautz, Henry / *University of Rochester, USA* 640
Keeys, Lynn A. / *SKEMA Business School, France* 201
Kidd, Paul T. / *Cheshire Henbury, UK* 974
Kiertscher, Simon / *Potsdam Institute for Climate Impact Research, Germany* 658
Kim, Eun-Hee / *George Washington University, USA* 1089
Kouame, Euphrasie B. / *United Nations University Institute for Natural Resources in Africa, Ghana* 816
Kouikoglou, Vassilis S. / *Technical University of Crete, Greece* 158
Krasnoff, Andrea / *TenStep, Inc., USA* 121
Kumar, Ashwani / *Indian Institute of Technology, India* 1519
Kumar, Kamlesh / *National Centre for Integrated Pest Management, India* 566
Lanfredi, M. / *Institute of Methodologies for Environmental Analysis, Italy* 1400
Leff, Enrique / *Universidad Nacional Autónoma de México, Mexico* 1
Leonardi, Jacques / *University of Westminster, UK* 1297
Liberti, L. / *Istituto Superiore per la Protezione e la Ricerca Ambientale, Italy* 1137
Li, Guangming / *Shenzhen Tourism College of Jinan University, China* 1218
Maclennan, Barbara L. / *West Virginia University, USA* 537
Makarouni, Ioanna / *National Technical University of Athens, Greece* 221
Maltzman, Richard / *EarthPM, LLC, USA* 926

Manikas, Ioannis / *University of Greenwich, UK* 786
Maragkogianni, Alexandra / *Technical University of Crete, Greece* 65
Marinakis, Vangelis / *National Technical University of Athens, Greece* 221
Marwan, Marwan / *Queensland University of Technology, Australia* 1614
Mbzibain, Aurelian / *University of Wolverhampton Business School, UK* 1072
McGill, Carrie J. Boden / *University of Arkansas at Little Rock, USA* 1150
McKinnon, Alan / *Heriot-Watt University, UK* 1265
McLeod, Fraser / *University of Southampton, UK* 1505
Mehta, Neelam / *National Centre for Integrated Pest Management, India* 566
Merritt, Lauren / *University of Arkansas at Little Rock, USA* 1150
Michelini, Rinaldo C. / *PMAR Lab, Italy* 103
Mijajlović, Nada / *Institute of Agricultural Economics, Serbia* 18
Mochal, Tom / *TenStep, Inc., USA* 121
Moshnyaga, Vasily G. / *Fukuoka University, Japan* 479
Muga, Helen E. / *University of Mount Union, USA* 185
Munafò, M. / *Istituto Superiore per la Protezione e la Ricerca Ambientale, Italy* 1137
Naghshbandy, Ali Hessamy / *University of Kurdistan, Iran* 1576
Nakamura, Hiroko / *The University of Tokyo, Japan* 885
Nedeski, Snezana / *Maastricht University, The Netherlands* 903
Newman, Timothy R. / *University of Maryland, USA* 495
Niculae, Mihăiţă Iulian / *University of Bucharest, Romania* 621
Niroumand, Hamed / *National University of Malaysia, Malaysia* 517
Niţă, Mihai Răzvan / *University of Bucharest, Romania* 140, 621
Nyrose, Sarah / *Boucher Institute of Naturopathic Medicine, Canada* 1454
Oktay, Derya / *Eastern Mediterranean University, North Cyprus* 1723
Olla, Phillip / *Madonna University, USA* 303
O'Mahony, Margaret / *Trinity College, The University of Dublin, Ireland* 1246
Omer, Abdeen Mustafa / *Energy Research Institute, Sudan* 261
Ondieki, Christopher Misati / *Kenyatta University, Kenya* 798
Onose, Diana Andreea / *University of Bucharest, Romania* 621
Onwudinjo, Nkemdilim / *Madonna University, USA* 303
Ozansoy, Cagil / *Victoria University, Australia* 551
Ozbakir, Buket Ayşegul / *Yildiz Technical University, Turkey* 602
Oztaysi, Basar / *Istanbul Technical University, Turkey* 242
Paays, Jeffrey / *University of Groningen, The Netherlands* 1560
Papaefthimiou, Spiros / *Technical University of Crete, Greece* 65
Papajorgji, Petraq / *European University of Tirana, Albania* 867
Park, Sangho / *University of North Dakota, USA* 640
Pätäri, Satu / *Lapeenranta University of Technology, Finland* 1560
Pătroescu, Maria / *University of Bucharest, Romania* 621
Peters, Everson J. / *University of the West Indies, Trinidad and Tobago* 1657
Phillis, Yannis A. / *Technical University of Crete, Greece* 158
Pillai, Kamatchi / *Victoria University, Australia* 551
Platonova, Inna / *University of Calgary, Canada* 1047, 1591

Popović, Vesna / *Institute of Agricultural Economics, Serbia* 18
Privitera, Donatella / *University of Catania, Italy* 1279
Psarras, John / *National Technical University of Athens, Greece* 221
Puliafito, Antonio / *University of Messina, Italy* 279
Qureshi, S. M. / *Mehran University Institute of Science & Technology Development, Pakistan* 693
Ragosta, M. / *University of Basilicata, Italy* 1400
Razzoli, Roberto P. / *PMAR Lab, Italy* 103
Reed, Jeffery H. / *Virginia Tech, USA* 495
Renbi, Abdelghani / *Luleå University of Technology, Sweden* 447
Roljević, Svetlana / *Institute of Agricultural Economics, Belgrade, Serbia* 769
Roshani, Gholam Hossein / *Shahid Beheshti University, Iran* 335
Roshani, Sobhan / *Razi University, Iran* 335
Routray, Jayant K. / *Asian Institute of Technology, Thailand* 1414
Saroar, Md. Mustafa / *Khulna University, Bangladesh* 1414
Schipper, Ron / *Van Aetsveld Project Management and Change, The Netherlands* 903
Schnor, Bettina / *University of Potsdam, Germany* 658
Shah, A. A. / *Mehran University of Engineering and Technology, Pakistan* 693
Shah, Ambreen / *University of Sindh, Pakistan* 693
Shah, Asif A. / *Mehran University Institute of Science & Technology Development, Pakistan* 693
Shah, Wajiha / *Mehran University of Engineering and Technology, Pakistan* 693
Shama, Farzin / *Razi University, Iran* 335
Sharma, O P / *National Centre for Integrated Pest Management, India* 566
Sharma, Satyawati / *Indian Institute of Technology, India* 1519
Shibli, Murad / *Abu Dhabi Polytechnic, Institute of Applied Technologies, UAE* 348
Shirley, David / *EarthPM, LLC, USA* 926
Silvius, Gilbert / *HU University of Applied Sciences Utrecht, The Netherlands & Van Aetsveld, The Netherlands* 903
Simoniello, T. / *Institute of Methodologies for Environmental Analysis, Italy* 1400
Singh, Niranjan / *National Centre for Integrated Pest Management, India* 566
Sipos, Yona / *University of British Columbia, Canada* 1454
Staake, Thorsten / *ETH Zurich, Switzerland* 731
Stambouli, Amine Boudghene / *University of Sciences and Technology Mohamed Boudiaf, Algeria* 1357
Strohschen, Gabriele / *DePaul University, USA* 1478
Stupariu, Ileana Georgeta / *University of Bucharest, Romania* 140
Subramaniam, R. / *Singapore National Academy of Science, Singapore & Nanyang Technological University, Singapore* 676
Suzuki, Shinji / *The University of Tokyo, Japan* 885
Sweeney, Edward / *National Institute for Transport and Logistics, Dublin Institute of Technology, Ireland* 86
Talbot, Jude / *AMEC, Canada* 1313
Tavanti, Marco / *DePaul University, USA* 1164
Thomas, Ken D. / *Auburn University, USA* 185
Tomkos, Ioannis / *Athens Information Technology, Greece* 463

Tsalapata, Hariklia / *UTH, Greece* 1492
Turner, Rodney / *SKEMA Business School, France* 201
Tusa, Francesco / *University of Messina, Italy* 279
Uzal, Nigmet / *Nigde University, Turkey* 867
Vânău, Gabriel Ovidiu / *University of Bucharest, Romania* 621
van den Brink, Jasper C. / *HU University of Applied Sciences Utrecht, The Netherlands* 958
Venkataraman, Ray / *Penn State University, USA* 1313
Vennila, S / *National Centre for Integrated Pest Management, India* 566
Viardot, Eric / *EADA Business School, Spain* 1544
Villari, Massimo / *University of Messina, Italy* 279
Volos, Haris I. / *Virginia Tech, USA* 495
Vredenburg, Harrie / *University of Calgary, Canada* 713
Westerman, Wim / *University of Groningen, The Netherlands* 1560
Wong, Sam / *University of Liverpool, UK* 1232
Woodburn, Allan / *University of Westminster, UK* 1297
Zaharia, Catalin / *Polytechnic University of Bucharest, Romania* 365
Zain, M.F.M / *National University of Malaysia, Malaysia* 517
Zaral, Fulya / *Erciyes University, Turkey* 867
Zegrar, Mansour / *University of Sciences and Technology Mohamed Boudiaf, Algeria* 1357
Zerhouni, Fatima Zohra / *University of Sciences and Technology Mohamed Boudiaf, Algeria* ... 1357
Zerhouni, M'hammed Houari / *University of Sciences and Technology Mohamed Boudiaf, Algeria* 1357
Zinke, Jörg / *University of Potsdam, Germany* 658
Zopounidis, Constantin / *Technical University of Crete, Greece* 65
Zoran, Njegovan / *University of Novi Sad, Serbia* 1203
Zubović, Jovan / *Economics Institute, Serbia* 1640

Table of Contents

Preface xxv

Volume I

Section 1
Fundamental Concepts and Theories

This section introduces some of the principle topics currently being discussed in Sustainable Practices research. It is undeniable that the world is changing dramatically—socially, economically, and ecologically—and careful attention to these changes is imperative to weathering them effectively. The authors in this section evaluate the current status of ecology, industry, and urban development in order to support new strategies for adaptation and success. In the opening 15 chapters of this extensive reference source, readers will obtain a clear understanding of the fundamental concepts and theories integral to the field of Sustainable Practices.

Chapter 1
Environmental Rationality: Innovation in Thinking for Sustainability 1
Enrique Leff, Universidad Nacional Autónoma de México, Mexico

Chapter 2
Climate Change and Sustainable Development in Agriculture and Forestry 18
Vesna Popović, Institute of Agricultural Economics, Serbia
Nada Mijajlović, Institute of Agricultural Economics, Serbia

Chapter 3
A Paradigm Shift towards Urban Resilience 49
Ozge Yalciner Ercoskun, Gazi University, Turkey

Chapter 4
Emissions Trading Schemes in the Transportation Sector 65
Alexandra Maragkogianni, Technical University of Crete, Greece
Spiros Papaefthimiou, Technical University of Crete, Greece
Constantin Zopounidis, Technical University of Crete, Greece

Chapter 5
Purchasing Green Transport and Logistics Services: Implications from the Environmental Sustainability Attitude of 3PLs 86
Pietro Evangelista, IRAT-CNR, Italy & University of Naples Federico II, Italy
Maria Huge-Brodin, Linköping University, Sweden
Karin Isaksson, Linköping University, Sweden
Edward Sweeney, National Institute for Transport and Logistics, Ireland & Dublin Institute of Technology, Ireland

Chapter 6
The Ecological Industrialism Bet 103
Rinaldo C. Michelini, PMAR Lab, Italy
Roberto P. Razzoli, PMAR Lab, Italy

Chapter 7
GreenPM®: The Basic Principles for Applying an Environmental Dimension to Project Management 121
Tom Mochal, TenStep, Inc., USA
Andrea Krasnoff, TenStep, Inc., USA

Chapter 8
Resource Conservation: Key Elements in Sustainable Rural Development 140
Cristian Iojă, University of Bucharest, Romania
Mihai Răzvan Niță, University of Bucharest, Romania
Ileana Georgeta Stupariu, University of Bucharest, Romania

Chapter 9
Approaches for Measuring Sustainability 158
Evangelos Grigoroudis, Technical University of Crete, Greece
Vassilis S. Kouikoglou, Technical University of Crete, Greece
Yannis A. Phillis, Technical University of Crete, Greece

Chapter 10
Diffusion and Adoption of Innovations for Sustainability 185
Helen E. Muga, University of Mount Union, USA
Ken D. Thomas, Auburn University, USA

Chapter 11
Integrating Project Strategy for Sustainable Development: A Conceptual Framework 201
Lynn A. Keeys, SKEMA Business School, France
Martina Huemann, WU Vienna University of Economics and Business, Austria
Rodney Turner, SKEMA Business School, France

Chapter 12
EU—GCC Clean Energy Cooperation: From Concepts to Action 221
Haris Doukas, National Technical University of Athens, Greece
Ioanna Makarouni, National Technical University of Athens, Greece
Charikleia Karakosta, National Technical University of Athens, Greece
Vangelis Marinakis, National Technical University of Athens, Greece
John Psarras, National Technical University of Athens, Greece

Chapter 13
Multi-Criteria Decision Aid for Sustainable Energy Prioritization Using Fuzzy Axiomatic Design 242
Basar Oztaysi, Istanbul Technical University, Turkey
Mine Isik, Istanbul Technical University, Turkey
Secil Ercan, Istanbul Technical University, Turkey

Chapter 14
Cooling and Heating with Ground Source Energy 261
Abdeen Mustafa Omer, Energy Research Institute, Sudan

Chapter 15
Towards Energy Sustainability in Federated and Interoperable Clouds 279
Antonio Celesti, University of Messina, Italy
Antonio Puliafito, University of Messina, Italy
Francesco Tusa, University of Messina, Italy
Massimo Villari, University of Messina, Italy

Section 2
Tools and Technologies

This section describes emerging technologies in the field of Sustainable Practices that can be of use to conservationists and industrialists alike. A large part of innovation is developing the tools necessary to meet new challenges and explore new ideas. Particular topics discussed in this section include advances in harnessing solar energy efficiently, using energy resources more effectively, and constructing urban environments to minimize their impact on local ecosystems. With 18 chapters, this section offers a broad treatment of some of the many tools and technologies within Sustainable Practices.

Chapter 16
Productive Use of Renewable Energy (PURE) for Economic Development in Developing Countries 303
Phillip Olla, Madonna University, USA
Nkemdilim Onwudinjo, Madonna University, USA

Chapter 17
Control and Conversion of Solar Power 318
Liping Guo, Northern Illinois University, USA

Chapter 18
A Comparative Study on a Built Sun Tracker and Fixed Converter Panels 335
Farzin Shama, Razi University, Iran
Gholam Hossein Roshani, Shahid Beheshti University, Iran
Sobhan Roshani, Razi University, Iran
Arash Ahmadi, Razi University, Iran
Saber Karami, Amirkabir University of Technology, Iran

Chapter 19
Solar Photovoltaic Servo Tracking Controlled System 348
Murad Shibli, Abu Dhabi Polytechnic, Institute of Applied Technologies, UAE

Chapter 20
Polymer Solar Cells 365
Catalin Zaharia, Polytechnic University of Bucharest, Romania

Chapter 21
Green, Sustainable, or Clean: What Type of IT/IS Technologies Will we Need in the Future?......... 384
Albena Antonova, Sofia University, Bulgaria

Chapter 22
Issues and Technologies of Effective Energy Management 397
Edward T. Chen, University of Massachusetts – Lowell, USA

Chapter 23
Energy Optimization of Power Station for a Small Research Institute 413
Ani Vincent Anayochukwu, University of Nigeria, Nigeria

Chapter 24
A Location-Based Power Conservation Scheme for MANETs: A Step towards Green
Communications 429
Hussein Al-Bahadili, Petra University, Jordan
Azmi Halasa, The Arab Academy for Banking and Financial Sciences, Jordan

Chapter 25
Data-Stream-Driven Computers are Power and Energy Efficient 447
Abdelghani Renbi, Luleå University of Technology, Sweden

Chapter 26
Energy-Efficient Optical Interconnects in Cloud Computing Infrastructures 463
Christoforos Kachris, Athens Information Technology, Greece
Ioannis Tomkos, Athens Information Technology, Greece

Chapter 27
Display Energy Management based on Eye Tracking 479
Vasily G. Moshnyaga, Fukuoka University, Japan

Chapter 28
Green Communications: Realizing Environmentally Friendly, Cost Effective, and Energy Efficient Wireless Systems 495
Haris I. Volos, Virginia Tech, USA
Dinesh Datla, Virginia Tech, USA
Xuetao Chen, Virginia Tech, USA
An He, Virginia Tech, USA
Ashwin Amanna, Virginia Tech, USA
Timothy R. Newman, University of Maryland, USA
S. M. Shajedul Hasan, Virginia Tech, USA
Jeffery H. Reed, Virginia Tech, USA
Tamal Bose, Virginia Tech, USA

Chapter 29
Earth Building Materials, Production, and Construction Techniques 517
Hamed Niroumand, National University of Malaysia, Malaysia
M.F.M Zain, National University of Malaysia, Malaysia
Sanaz Naghavi Alhosseni, Amirkabir University of Technology, Iran

Chapter 30
3D Digital City Platforms as Collaborative and Decision-Making Tools for Small Municipalities and Rural Areas 537
Barbara L. Maclennan, West Virginia University, USA
Susan J. Bergeron, Coastal Carolina University, USA

Chapter 31
Web-Based Digital Habitat Ecosystems for Sustainable Built Environments 551
Kamatchi Pillai, Victoria University, Australia
Cagil Ozansoy, Victoria University, Australia

Chapter 32
ICT Based Pest Management System for Sustainable Pulse Production: A Case Study 566
O P Sharma, National Centre for Integrated Pest Management, India
Niranjan Singh, National Centre for Integrated Pest Management, India
Archana Bhardwaj, National Centre for Integrated Pest Management, India
S Vennila, National Centre for Integrated Pest Management, India
Someshwar Bhagat, National Centre for Integrated Pest Management, India
Neelam Mehta, National Centre for Integrated Pest Management, India
Kamlesh Kumar, National Centre for Integrated Pest Management, India

Volume II

Chapter 33
Real-Time Visual Simulation of Urban Sustainability 581
John P. Isaacs, University of Abertay Dundee, U K
David J. Blackwood, University of Abertay Dundee, UK
Daniel Gilmour, University of Abertay Dundee, UK
Ruth E. Falconer, University of Abertay Dundee, UK

Section 3
Utilization and Application

This section continues from the previous section to explore how new technologies can be integrated with real-world industries to enhance Sustainable Practices. As scientists discover new energy resources, develop more efficient engines and reactors, and discuss innovative methods of preventing climate change, leaders and professionals in a variety of industries must take advantage of these tools in their daily practices. Construction, agriculture, mining, and transportation are only some of the fields that can benefit immensely from advances in sustainable technologies. The 17 chapters in this section provide an in-depth examination of the utilization and application of the fundamental principles of Sustainable Practices.

Chapter 34
Urban Environmental Applications of GIScience: Challenges and New Trends 602
Buket Ayşegul Ozbakir, Yildiz Technical University, Turkey

Chapter 35
Natural Resources Conservation in the Influence Areas of Cities: A Case Study on Bucharest, Romania 621
Mihai Răzvan Niță, University of Bucharest, Romania
Mihăiță Iulian Niculae, University of Bucharest, Romania
Diana Andreea Onose, University of Bucharest, Romania
Maria Pătroescu, University of Bucharest, Romania
Gabriel Ovidiu Vânău, University of Bucharest, Romania
Cristiana Maria Ciocănea, University of Bucharest, Romania

Chapter 36
Energy-Aware Intelligence in Smart Spaces: A Case Study using Computer Vision and Machine Learning for User-Behavior Analysis 640
Sangho Park, University of North Dakota, USA
Henry Kautz, University of Rochester, USA

Chapter 37
Power Consumption Aware Cluster Resource Management 658
Simon Kiertscher, Potsdam Institute for Climate Impact Research, Germany
Bettina Schnor, University of Potsdam, Germany
Jörg Zinke, University of Potsdam, Germany

Chapter 38
Use of Policy Instruments to Promote Sustainable Energy Practices and Implications for the Environment: Experiences from Singapore 676
Leo Tan Wee Hin, Singapore National Academy of Science, Singapore & National University of Singapore, Singapore
R. Subramaniam, Singapore National Academy of Science, Singapore & Nanyang Technological University, Singapore

Chapter 39
Renewable Energy Scenario of Pakistan for Sustainable Development 693
Asif A. Shah, Mehran University Institute of Science & Technology Development, Pakistan
Arabella Bhutto, Mehran University Institute of Science & Technology Development, Pakistan
S. M. Qureshi, Mehran University Institute of Science & Technology Development, Pakistan
Ambreen Shah, University of Sindh, Pakistan
A. A. Shah, Mehran University of Engineering and Technology, Pakistan
Wajiha Shah, Mehran University of Engineering and Technology, Pakistan

Chapter 40
Finding the Sweet Spot of Sustainability in the Energy Sector: A Systems Approach to Managing the Canadian Oil Sands 713
Nancy Higginson, University of Calgary, Canada
Harrie Vredenburg, University of Calgary, Canada

Chapter 41
Using Sensor-Based Measurement Technology to Reduce the Environmental Impact of Perishable Goods 731
Alexander Ilic, ETH Zurich, Switzerland
Thorsten Staake, ETH Zurich, Switzerland
Elgar Fleisch, ETH Zurich, Switzerland & University of St. Gallen, Switzerland

Chapter 42
Sustainability and the UK's Major Food Retailers: Consumer Concentric Cause Marketing Writ Large 748
Peter Jones, University of Gloucestershire, UK
Daphne Comfort, University of Gloucestershire, UK
David Hillier, University of Glamorgan, UK

Chapter 43
Organic Production in Serbia: The Transition to Green Economy 769
Vladimir Filipović, Institute "Tamis" Pancevo, Serbia
Svetlana Roljević, Institute of Agricultural Economics, Belgrade, Serbia
Bojana Bekić, Institute of Agricultural Economics, Belgrade, Serbia

Chapter 44
Environmental Sustainability Initiatives in the Agrifood Supply Chain 786
Ioannis Manikas, University of Greenwich, UK
Petros Ieromonachou, University of Greenwich, UK
Dionysis Bochtis, Aarhus University, Denmark

Chapter 45
Hydrology and Integrated Water Resource Management for Sustainable Watershed Management in Kenya 798
Christopher Misati Ondieki, Kenyatta University, Kenya

Chapter 46
Application of Quantitative Methods in Natural Resource Management in Africa: A Review.......... 816
Elias T. Ayuk, United Nations University Institute for Natural Resources in Africa, Ghana
William M. Fonta, United Nations University Institute for Natural Resources in Africa, Ghana
Euphrasie B. Kouame, United Nations University Institute for Natural Resources in Africa, Ghana

Chapter 47
Towards Sustainable Mining: Diffusion of Sustainability Concepts into the Mining Industry within Canada.. 846
Michelle Edith Jarvie-Eggart, Barr Engineering, USA & University of Maryland University College, USA

Chapter 48
Sustainable Waste Management System and Reverse Logistic Network Design in Plastic Industry: A Case Study of Turkey 867
Emel Kizilkaya Aydogan, Erciyes University, Turkey
Nuray Ates, Erciyes University, Turkey
Nigmet Uzal, Nigde University, Turkey
Fulya Zaral, Erciyes University, Turkey
Petraq Papajorgji, European University of Tirana, Albania

Chapter 49
Innovation for Sustainability in Aviation: World Challenges and Visions... 885
Hiroko Nakamura, The University of Tokyo, Japan
Yuya Kajikawa, The University of Tokyo, Japan
Shinji Suzuki, The University of Tokyo, Japan

Chapter 50
Consideration of Sustainability in Projects and Project Management: An Empirical Study.............. 903
Gilbert Silvius, HU University of Applied Sciences Utrecht, The Netherlands & Van Aetsveld, The Netherlands
Ron Schipper, Van Aetsveld Project Management and Change, The Netherlands
Snezana Nedeski, Maastricht University, The Netherlands

Section 4
Organizational and Social Implications

This section discusses the human element of Sustainable Practices, encouraging citizens and leaders alike to take responsibility for solving the growing environmental crisis. As new technologies are implemented in organizational workflow, managers and leaders must train their constituents in the proper and efficient use of these new policies. In particular, issues of project management, citizen engagement, urban design, and environmental education are critical to the ultimate effectiveness of new technologies. In these 15 chapters, readers will find an in-depth discussion on some of the most pressing organizational and social implications of Sustainable Practices.

Chapter 51
Project Manager as a Pivot Point for Implementing Sustainability in an Enterprise 926
Richard Maltzman, EarthPM, LLC, USA
David Shirley, EarthPM, LLC, USA

Chapter 52
Sustainability in Supply Chain Management 944
Farzad Dehghanian, Ferdowsi University of Mashhad, Iran

Chapter 53
How Positive Psychology can Support Sustainable Project Management 958
Jasper C. van den Brink, HU University of Applied Sciences Utrecht, The Netherlands

Chapter 54
Social Networking Technologies as a Strategic Tool for the Development of Sustainable Production and Consumption: Applications to Foster the Agility Needed to Adapt Business Models in Response to the Challenges Posed by Climate Change 974
Paul T. Kidd, Cheshire Henbury, UK

Chapter 55
Social Media-based Communities of Practice for Education in Citizenship and Sustainability 988
Dimitra Florou, University of Athens, Greece
Dimitris Gouscos, University of Athens, Greece

Chapter 56
Community Opposition and Public Engagement with Wind Energy in the UK 1009
Matthew Cotton, University of Leeds, UK

Chapter 57
Energy Efficient Residential Block Design: The Case of Ankara 1027
Hakan Hisarligil, Erciyes University, Turkey
Sule Karaaslan, Gazi University, Turkey

Chapter 58
Diffusion of Renewable Energy Technologies in Rural Communities: Exploratory Study of Development Partnerships in Cajamarca, Peru.. 1047
Inna Platonova, University of Calgary, Canada

Chapter 59
The Effect of Farmer Capacities, Farm Business Resources and Perceived Support of Family, Friends and Associational Networks on Intentions to Invest in Renewable Energy Ventures in the UK 1072
Aurelian Mbzibain, University of Wolverhampton Business School, UK

Chapter 60
Government Policies to Promote Production and Consumption of Renewable Electricity in the US 1089
Eun-Hee Kim, George Washington University, USA

Chapter 61
Product-Service Systems as Enabler for Sustainability-Oriented Innovation: The Case of Osram's Off-Grid Lighting 1106
Friedrich Grosse-Dunker, Dark Horse GmbH, Germany
Erik G. Hansen, Leupana University Lüneburg, Germany

Chapter 62
Digital Urbanism in Southern Italy 1121
Arturo Di Bella, University of Catania, Italy

Chapter 63
Urban Environment Quality in the Italian Spatial Data Infrastructure 1137
L. Congedo, Istituto Superiore per la Protezione e la Ricerca Ambientale, Italy
F. Baiocco, Istituto Superiore per la Protezione e la Ricerca Ambientale, Italy
S. Brini, Istituto Superiore per la Protezione e la Ricerca Ambientale, Italy
L. Liberti, Istituto Superiore per la Protezione e la Ricerca Ambientale, Italy
M. Munafò, Istituto Superiore per la Protezione e la Ricerca Ambientale, Italy

Chapter 64
The Heifer International Model of Sustainable Adult Education 1150
Carrie J. Boden McGill, University of Arkansas at Little Rock, USA
Lauren Merritt, University of Arkansas at Little Rock, USA

Volume III

Chapter 65
Responsible Management Education in Practice: The Principles and Processes for Educating Socially Responsible and World Engaged Leaders 1164
Marco Tavanti, DePaul University, USA

Section 5
Critical Issues

This section takes an in-depth look at some of the problems facing Sustainable Practices in the twenty-first century in order to present potential solutions to future concerns. With any new technology, problem, or discovery comes uncertainty as to its impact on future practices, livelihoods, and the overall sustainability of a society; this uncertainty must be analyzed and addressed if any useful advances are to be made. As the global landscape continues to shift, questions of environmental change, economic reform, social policy, and industrial development are of paramount concern. In this section, 14 chapters explore some of the critical issues driving advances in Sustainable Practices.

Chapter 66
How Can Sustainable Development Redefine Project Management Processes? 1183
Mohamed Eid, The British University in Egypt, Egypt

Chapter 67
Conceptualization of the Base for Sustainable Rural Development Policy Implementation 1203
Njegovan Zoran, University of Novi Sad, Serbia
Olgica Bošković, University of Belgrade, Serbia

Chapter 68
A Paradigm of Constructing Industrial Symbiosis and Coupling in China's County-Region Economic Sustainable Development 1218
Guangming Li, Shenzhen Tourism College of Jinan University, China

Chapter 69
Mind the Gap Please!: Contrasting Renewable Energy Investment Strategies between the World Bank and Poor Customers in Developing Countries 1232
Sam Wong, University of Liverpool, UK

Chapter 70
Assessing the Barriers to Greener Fiscal Measures and Ecological Tax Reform in the Transport Sector 1246
David Browne, Trinity College, The University of Dublin, Ireland
Brian Caulfield, Trinity College, The University of Dublin, Ireland
Margaret O'Mahony, Trinity College, The University of Dublin, Ireland

Chapter 71
Improving the Sustainability of Road Freight Transport by Relaxing Truck Size and Weight Restrictions 1265
Alan McKinnon, Heriot-Watt University, UK

Chapter 72
Sustainable Transport and Quality of Life Analysis of Cycling Impact in Italy 1279
Donatella Privitera, University of Catania, Italy

Chapter 73
The Role of Urban Consolidation Centres in Sustainable Freight Transport 1297
Julian Allen, University of Westminster, UK
Michael Browne, University of Westminster, UK
Jacques Leonardi, University of Westminster, UK
Allan Woodburn, University of Westminster, UK.

Chapter 74
Evaluating Sustainability on Projects Using Indicators 1313
Jude Talbot, AMEC, Canada
Ray Venkataraman, Penn State University, USA

Chapter 75
Energy Efficient Data Query, Processing and Routing Techniques for Green Wireless Sensor Networks 1331
Afshin Behzadan, Ryerson University, Canada
Alagan Anpalagan, Ryerson University, Canada

Chapter 76
Study of Hybrid Sustainable Energy System Based on PEM Fuel Cells and Photovoltaic-Module Power Generator 1357
Fatima Zohra Zerhouni, University of Sciences and Technology Mohamed Boudiaf, Algeria
M'hammed Houari Zerhouni, University of Sciences and Technology Mohamed Boudiaf, Algeria
Mansour Zegrar, University of Sciences and Technology Mohamed Boudiaf, Algeria
Amine Boudghene Stambouli, University of Sciences and Technology Mohamed Boudiaf, Algeria

Chapter 77
Are Biofuels a Factor of Sustainable Development in a Food Insecurity Context in Africa? Case Study of Burkina Faso.. 1380
Marie-Hélène Dabat, Centre de Coopération Internationale en Recherche Agronomique pour le Développement, Burkina Faso
Joël Blin, Centre de Coopération Internationale en Recherche Agronomique pour le Développement, Burkina Faso & 2iE, Institut International d'Ingénierie de l'Eau et de l'Environnement, Burkina Faso
Elodie Hanff, 2iE, Institut International d'Ingénierie de l'Eau et de l'Environnement, Burkina Faso

Chapter 78
Indicators of Land Degradation Vulnerability due to Anthropic Factors: Tools for an Efficient Planning 1400
V. Imbrenda, University of Basilicata, Italy & National Research Council of Italy, Institute of Methodologies for Environmental Analysis, Italy
M. D'Emilio, Institute of Methodologies for Environmental Analysis, Italy
M. Lanfredi, Institute of Methodologies for Environmental Analysis, Italy
M. Ragosta, University of Basilicata, Italy
T. Simoniello, Institute of Methodologies for Environmental Analysis, Italy

Chapter 79
Desert in Bengal Delta-Changes in Landscape, Changes in Livelihood: Can Diffusion and Adoption of Sustainable Adaptation Make a Difference? 1414
Md. Mustafa Saroar, Khulna University, Bangladesh
Jayant K. Routray, Asian Institute of Technology, Thailand

Section 6
Emerging Trends

This section examines both future difficulties and innovative solutions for climate change and Sustainable Practices. Though the economic, ecological, and social troubles in the world today seem almost insurmountable, human ingenuity promises to offer ever more unique solutions to the problems faced by modern societies. Some notable advances explored in this section include environmental education, waste management and renewable energy, conservation of natural resources, and ecologically-friendly urban environments. The final 17 chapters of this extensive three-volume reference conclude with a detailed look at emerging trends in the field of Sustainable Practices.

Chapter 80
The University of Hawai'i at Manoa Sustainability Courtyard as a Center for Campus Engagement 1443
John Cusick, University of Hawai'i at Manoa Environmental Center, USA

Chapter 81
Supporting Sustainability Education and Leadership: Strategies for Students, Faculty, and the Planet 1454
Alice Cassidy, In View Education and Professional Development, Canada
Yona Sipos, University of British Columbia, Canada
Sarah Nyrose, Boucher Institute of Naturopathic Medicine, Canada

Chapter 82
Education Collaboration Development: A Blended Shore Education Approach to Sustainable Development 1478
Gabriele Strohschen, DePaul University, USA

Chapter 83
Environmental Education through Envkids Didactical Framework and ICT Tools 1492
Hariklia Tsalapata, UTH, Greece
Rene Alimsi, CERETETH, Greece
Olivier Heidmann, CERETETH, Greece

Chapter 84
Reverse Logistics for Sustainable Waste Management Processes 1505
Fraser McLeod, University of Southampton, UK
Tom Cherrett, University of Southampton, UK

Chapter 85
Energy from Waste: Present Scenario, Challenges, and Future Prospects towards Sustainable Development 1519
Kalpana Arora, Indian Institute of Technology, India
Ashwani Kumar, Indian Institute of Technology, India
Satyawati Sharma, Indian Institute of Technology, India

Chapter 86
Encouraging the Development of Renewable Energy: The Role of Cooperatives 1544
Eric Viardot, EADA Business School, Spain

Chapter 87
Value Creation with Wood-Based Energy Business Models 1560
Wim Westerman, University of Groningen, The Netherlands
Jeffrey Paays, University of Groningen, The Netherlands
Satu Pätäri, Lapeenranta University of Technology, Finland

Chapter 88
Impacts of High Wind Power Penetration on the Frequency Response Considering Wind Power Reserve 1576
Bakhtiar Badmasti, Islamic Azad University—Marivan Branch, Iran
Hassan Bevrani, University of Kurdistan, Iran
Ali Hessamy Naghshbandy, University of Kurdistan, Iran

Chapter 89
International Development Partnerships and Diffusion of Renewable Energy Powered Lighting Technologies in Off-Grid Communities in Developing Countries: Exploratory Study in Talamanca, Costa Rica 1591
Inna Platonova, University of Calgary, Canada

Chapter 90
Demand-Side Response Smart Grid Technique for Optimized Energy Use 1614
Fouad Kamel, University of Southern Queensland, Australia
Marwan Marwan, Queensland University of Technology, Australia

Chapter 91
Management and Utilisation of Natural Resources in Special Nature Reserves: A Case Study 1640
Božo Drašković, Institute of Economic Sciences, Serbia
Jovan Zubović, Economics Institute, Serbia
Ivana Domazet, Institute of Economic Sciences, Serbia

Chapter 92
Promoting Rainwater Harvesting (RWH) in Small Island Developing States (SIDS): A Case in the Grenadines 1657
Everson J. Peters, University of the West Indies, Trinidad and Tobago

Chapter 93
The Sustainable Waterfront 1683
Matthew Bradbury, Unitec Institute of Technology, New Zealand

Chapter 94
Multifunctional Agriculture and the Green Economy 1701
Andrei Jean-Vasile, Petroleum and Gas University of Ploiesti, Romania

Chapter 95
Sustainable Urbanism Revisited: A Holistic Framework Based on Tradition and Contemporary Orientations.. 1723
Derya Oktay, Eastern Mediterranean University, North Cyprus

Chapter 96
Towards Zero Energy Buildings (ZEB): The Role of Environmental Technologies.. 1742
Paris A. Fokaides, RD Hydraulis Ltd, Cyprus

Index..xxviii

Preface

The constantly changing landscape of Sustainable Practices challenges experts and practitioners to stay apprized of the field's most up-to-date research. That is why Information Science Reference is pleased to offer this three-volume reference collection that will empower students, researchers, and academicians with a strong understanding of critical issues within Sustainable Practices by providing both broad and detailed perspectives on cutting-edge theories and developments in the field. This collection is designed to act as a single reference source on conceptual, technical, and organizational issues, as well as provide insight into emerging trends and future opportunities within the discipline.

Sustainable Practices: Concepts, Methodologies, Tools, and Applications is organized into six distinct sections that provide comprehensive coverage of important topics. The sections are (1) Fundamental Concepts and Theories, (2) Tools and Technologies, (3) Utilization and Application, (4) Organizational and Social Implications, (5) Critical Issues, and (6) Emerging Trends. The following paragraphs provide a summary of what readers may expect from this invaluable reference tool.

Section 1, "Fundamental Concepts and Theories," introduces some of the principle topics currently being discussed in Sustainable Practices research. The section begins with chapters on the importance of sustainability in environmental conservation. In *Environmental Rationality*, Enrique Leff rethinks the current cultural perspective on natural resource management to encourage more ethical, sustainable principles, and *Climate Change and Sustainable Development in Agriculture and Forestry* by Vesna Popovic and Nada Mijajlovic presents similar arguments with a focus on local industries. The following chapters discuss environmentalism and sustainability in industry, including *Purchasing Green Transport and Logistics Services* by Pietro Evangelista, Maria Huge-Brodin, Karin Isaksson, and Edward Sweeney and *Integrating Project Strategy for Sustainable Development* by Lynn A. Keeys, Martina Huemann, and Rodney Turner. This section concludes with chapters on sustainable technologies and renewable energy, namely *Towards Energy Sustainability in Federated and Interoperable Clouds* by Antonio Celesti, Antonio Puliafito, Francesco Tusa, and Massimo Villari, an effective transition into the next section and its focus on the technologies of Sustainable Development.

Section 2, "Tools and Technologies," describes emerging technologies in the field of Sustainable Practices that can be of use to conservationists and industrialists alike. The opening chapters of this section focus on renewable energy technologies, primarily solar. Notable developments include a sun-tracking system, as described in *A Comparative Study on a Built Sun Tracker and Fixed Converter Panels* by Farzin Shama, Gholam Hossein Roshani, Sobhan Roshani, Arash Ahmadi, and Saber Karami, and Catalin Zaharia studies new materials for *Polymer Solar Cells*. Additional technologies focus on energy efficiency rather than generation, particularly *Energy-Efficient Optical Interconnects in Cloud Computing Infrastructures* by Christoforos Kachris and Ioannis Tomkos and *Energy Optimization of Power Station for a Small Research Institute* by Ani Vincent Anayochukwu. Finally, the section concludes with several chapters on sustainable urban planning, including *Earth Building Materials, Production, and Construction Techniques* by Hamed Niroumand, M.F.M Zain, and Sanaz Naghavi Alhosseni and *Web-Based Digital Habitat Ecosystems for Sustainable Built Environments* by Kamatchi Pillai and Cagil Ozansoy.

Section 3, "Utilization and Application," continues from the previous section to explore how new technologies can be integrated with real-world industries to enhance Sustainable Practices. The first chapters build off the end of Section Two by describing future urban environments in terms of renewable energy and natural resource conservation. Sangho Park and Henry Kautz, in their chapter *Energy-Aware Intelligence in Smart Spaces*, describe an automated sensor network for incorporation within human living environments that can adjust to human activity and avoid otherwise wasting valuable resources. As the world's population continues to grow exponentially, cities expand as arable land shrinks, making the future food supply a critical concern. Several chapters in this section address such concerns, including *Sustainability and the UK's Major Food Retailers* by Peter Jones, Daphne Comfort, and David Hillier and *Using Sensor-Based Measurement Technology to Reduce the Environmental Impact of Perishable Goods* by Alexander Ilic, Thorsten Staake, and Elgar Fleisch. The final chapter of this section, *Consideration of Sustainability in Projects and Project Management* by Gilbert Silvius, Ron Schipper, and Snezana Nedeski, leads into Section Four with an overview of practices that managers and leaders can adopt to further their own goals while also sustaining the resources they consume.

Section 4, "Organizational and Social Implications," discusses the human element of Sustainable Practices, encouraging citizens and leaders alike to take responsibility for solving the growing environmental crisis. Project management in business and industry serves as the primary topic of the first chapters in this section, the most notable being *How Positive Psychology can Support Sustainable Project Management* by Jasper C. van den Brink. This chapter also discusses elements of citizen engagement and personal responsibility, topics that are also covered in later chapters such as *Social Media-based Communities of Practice for Education in Citizenship and Sustainability* by Dimitra Florou and Dimitris Gouscos and *Government Policies to Promote Production and Consumption of Renewable Electricity in the US* by Eun-Hee Kim. This section concludes with chapters on enabling citizens to take charge of their own ecological footprint, either through new technologies, as in *Product-Service Systems as Enabler for Sustainability-Oriented Innovation* by Friedrich Grosse-Dunker and Erik G. Hansen, or through education, as in *Responsible Management Education in Practice* by Marco Tavanti.

Section 5, "Critical Issues," takes an in-depth look at some of the problems facing Sustainable Practices in the twenty-first century in order to present potential solutions to future concerns. This section begins with chapters on public policy, and how leaders can pave the way for a sustainable future. *Conceptualization of the Base for Sustainable Rural Development Policy Implementation* by Njegovan Zoran and Olgica Bošković examines socioeconomic inequalities in the development of sustainable societies in both rural and urban environments. Also of note in this section are chapters such as *Mind the Gap Please!* by Sam Wong, which discusses sustainable investments in an increasingly turbulent economy. Subsequent chapters explore the design of efficient urban infrastructure, such as *Improving the Sustainability of Road Freight Transport by Relaxing Truck Size and Weight Restrictions* by Alan McKinnon and *Energy Efficient Data Query, Processing and Routing Techniques for Green Wireless Sensor Networks* by Afshin Behzadan and Alagan Anpalagan. Finally, this section concludes with chapters on climate change and renewable resources to provide a smooth transition into the primary topic of the final section: future advances in the field.

Section 6, "Emerging Trends," examines both future difficulties and innovative solutions for climate change and Sustainable Practices. The global landscape is changing in ways that it never has before, but human ingenuity offers the possibility of a bright future. This section contains chapters on many unique ides for solving the global environmental crisis, including education and training, alternative energies, natural resource management, urban redevelopment, and others. *The University of Hawai'i at Manoa Sustainability Courtyard as a Center for Campus Engagement* by John Cusick describes a program in which students learn valuable lessons in sustainable solutions and environmental conservation. One such solution already in development is described in Kalpana Arora, Ashwani Kumar, and Satyawati Sharma's chapter, *Energy from Waste*. *Demand-Side Response Smart Grid Technique for Optimized Energy Use* by Fouad Kamel and Marwan Marwan, among other chapters, investigates methodologies and tools for effectively utilizing the energy generated on a daily basis to improve efficiency reduce wastefulness, and *The Sustainable Waterfront* by Matthew Bradbury outlines a plan for developing an efficient and ecologically healthy urban environment. The final chapter in this three-volume reference, *Towards Zero Energy Buildings (ZEB)* by Paris A. Fokaides, effectively illustrates the primary goal in sustainable urban development: a city that will serve its purpose without any negative impact on the surrounding environment, setting a positive example for future generations.

As a comprehensive collection of research on current findings related to the development of interdisciplinary technologies, *Sustainable Practices: Concepts, Methodologies, Tools, and Applications* provides researchers, administrators, and all audiences with a complete understanding of the latest advances, applications, and concepts in Sustainable Practices. Although the primary organization of the contents in this multi-volume work is based on its six sections, offering a progression of coverage on the important concepts, technologies, applications, social issues, critical concerns, and emerging trends, the reader can also identify specific content by utilizing the extensive indexing system found at the end of each volume. Given the vast number of issues concerning usage, successes and failures, policies, strategies, and applications of Sustainable Practices in countries around the world, *Sustainable Practices: Concepts, Methodologies, Tools, and Applications* addresses the demand for a resource that encompasses the most pertinent research on the technologies being employed to globally bolster the knowledge and implementation of Sustainable Practices.

Chapter 65

Responsible Management Education in Practice:
The Principles and Processes for Educating Socially Responsible and World Engaged Leaders

Marco Tavanti
DePaul University, USA

ABSTRACT

The Principles of Responsible Management Education (PRME) offer a platform for institutional commitment and leadership engagement toward business ethics and poverty alleviation. The author reviews the principles in light of a need for a renewed management education centered on social responsibility, sustainability and partnership values. The PRME present a possibility for universities to engage with United Nations Global Compact (UNGC), the Millennium Development Goals (MDGs) and other multi-stakeholder partnerships. The participating institutions share a commitment for a renewed management education where private organizations become positive agents for world benefit. The process of adopting and integrating the PRME into existing programs and curricula is instrumental for educating ethical, sustainable, and socially responsible leaders for the 21st century.

INTRODUCTION

Responsible management education is both a challenge and an opportunity. The time has come for business schools and public service management programs to center their education around global challenges of sustainability and poverty reduction. Social responsibility and ethics are no longer electives in the business of adequately preparing 21st century leaders (Gasparski, 2008). Social responsibility pertains to all stakeholders but it begins with a value-based commitment of management faculty and program administrators.

DOI: 10.4018/978-1-4666-4852-4.ch065

Academia can therefore provide opportunities to learn appropriate competencies for developing globally responsible leaders and promoting organizational practices for world benefits, actively contributing to poverty eradication, replenishing and restoring nature, and building foundations for peace (BAWB, 2006). These renewed management values are already priorities among the 363 academic institutions and programs (as of April 2011) who adopted the Principles of Responsible Management Education (abbreviated PRME and pronounced PRIME). Developed in 2007 by the academic institutions connected to the United Nations Global Compact (UNGC), the adoption of principles create opportunities for participating academic institutions to exchange best practices in teaching, learning and engaging for world benefit.

The idea of the PRME was introduced by the UNGC at the Global Forum "Business as an Agent of World Benefit" at Case Western Reserve University in October 2006. Inspired by the internationally accepted business values of the 10 principles of the United Nations Global Compact (UNGC) and the 8 Millennium Development Goals (MDGs), the PRME offer a boost for making management education an engaged, responsible and sustainable response to world poverty and inclusive development. The PRME were developed in July 2007 by an international task force consisting of 60 deans, university presidents and official representatives of leading business schools (PRME, 2007). At their official launch during the Global Compact Leaders Summit in Geneva, UN Secretary-General Ban Ki-moon said that 'The Principles for Responsible Management Education have the capacity to take the case for universal values and business into classrooms on every continent' (Forray & Leigh, 2010).

Today, PRME represents a growing movement of academic institutions and management programs committed to the promotion and integration of socially responsible principles and practices. They represent a platform for dialogue and implementation of social responsibility in education and for making management education relevant to local and global poverty reduction and sustainable development. Social responsibility and sustainability are not a fashion in management education (Christensen, Peirce, Hartman, Hoffman, & Carrier, 2007). They reflect fundamental shifts in our societies and economic systems that will surely develop in the years to come. The PRME offer an engagement model for management schools and academic institutions who want to stay "ahead of the curve" by integrating sustainability and social responsibility into their learning outcomes and programs (PRME, 2011b).

The PRME are part of a growing movement promoting education for global citizenship, sustainable development, social accountability and leadership for common good (Ball & Osborne, 2011; Godfrey & Grasso, 2000). The late-2000s financial crisis is also a 'good' opportunity to get serious about business ethics and placing social responsibility and sustainability at the core of management education (Rasche & Escudero, 2010). Dr. Ángel Cabrera, president of Thunderbird University and a strong promoter of the PRME, has made sustainability and social responsibility values the hallmarks of his academic institution, a leader in global management education. Commenting on PRME, he argued that 'the time is ripe for change' to professionalize business schools by starting a reform process towards accepting the broader responsibilities of management in society (Cabrera, 2009). As demonstrated by innovative leaders such as Mohammad Yunus (Yunus & Weber, 2010), profit maximization can be aligned with alleviating poverty and systemic solutions to societal problems (Wankel, 2008; Werhane, Kelley, Hartman, & Moberg, 2010). Dr. Cabrera explains well how the 'private' interests of business can be in line with a responsibility for the 'common' good. "A professional ideology of service to the greater good is not at odds with the principle of shareholder value creation. It actually grounds shareholder value morally and integrates it in a richer multidisciplinary context. It reaffirms the importance of shareholder value as both a source of societal prosperity in itself as well as an indicator of other forms of value. But it acknowledges that businesses create multiple forms of value and it attributes to managers responsibilities that go beyond profit maximization" (Cabrera, 2009).

This analysis of the PRME assumes that management education, through the right principles and processes, can be instrumental in advancing sustainable solutions to world problems. It offers a review of the principles in their values and practical application for educating managers and leaders engaged in world poverty reduction. After framing the connection between social responsibility of academia in relation to management education, the PRME is reviewed in the values and meanings of each principles. We will than make a case for adopting the PRME beyond 'moral' principles and in relation to the beneficial processes for engaging the entire management program and academic institution to center education around social responsibility and sustainability. The benefits of adopting PRME are reviewed in relation to practical possibility of engagement for poverty reduction and in relations to the United Nations Global Compact (UNGC) and the Millennium Development Goals (MDGs). The many global challenges, along with the past corporate scandals and the current economic crisis compel to revisit management education. The PRME offer a framework and an opportunity to make academia an agent for world benefit by educating socially responsible leaders and engaging in poverty reduction.

THE SOCIAL RESPONSIBILITY OF MANAGEMENT EDUCATION

Is management education at a tipping point for poverty reduction, social responsibility, sustainability, ethical leadership and common good? Foregoing the formulation of the PRME, the 2006 Global Forum *Business as an Agent of World Benefit* suggested that we are indeed at the tipping point (BAWB, 2006). Management guru Peter Drucker too recognized the urgency for socially responsible management when, shortly before his passing in 2005, 'he spoke passionately about the distressed state of the world with corporate meltdowns and a world where billions are condemned to lives of grinding poverty' (BAWB, 2006, p. 3). He invited all of us to aim higher as 'management is a matter of world affairs' and 'every single pressing social and global issue of our time is a business opportunity' (BAWB, 2006).

Dr. Jeffrey Sachs (2005) and UN Millennium Campaign in support of the MDGs emphasize that working toward the most pressing world problems, especially alleviating poverty, is the most pressing social responsibility of our generation. For the first time in human history, our generation has the knowledge, capacity and responsibility to halve extreme poverty by 2015 and end it by 2025 (Sachs, 2005, p. xiv). Therefore, the social responsibility toward poverty reduction is both an ethical challenge and an economic possibility for businesses to create 'inclusive capitalism' (Hart, 2005) and working for prosperity while serving humanity's most pressing needs (Yunus & Weber, 2010). As suggested by C. K. Prahalad's *The fortune at the bottom of the pyramid: Eradicating poverty through profit* (2005), the right mixture of innovation and entrepreneurship is a business opportunity of the 21st century.

Successful businessman and famous Harvard dropout Bill Gates challenged a university education that does not focus on world problems. During his remarks at the 2007 Harvard University commencement ceremony he confessed to having had a big regret:

I left Harvard with no real awareness of the awful inequities in the world – the appalling disparities of health, and wealth, and opportunity that condemn millions of people to lives of despair. I learned a lot here at Harvard about new ideas in economics and politics. I got great exposure to the advances being made in the sciences. But humanity's greatest advances are not in its discoveries – but in how those discoveries are applied to reduce inequity. Whether through democracy, strong public education, quality health care, or broad economic opportunity – reducing inequity is the highest human achievement. I left campus knowing little about the millions of young people cheated out of educational opportunities here in this country. And I knew nothing about the millions of people living in unspeakable poverty and disease in developing countries. It took me decades to find out (Gates, 2007).

The original purpose of management education to link economics, professionalism and universities has a new fundamental challenge today. David Cooperrider calls it a challenge of making business a positive (appreciative) agent for world benefit linking the power of transformation of organizations into ecology, peace and poverty (Cooperrider, Whitney, Stavros, & ebrary Inc., 2008). The real challenge and responsibility of management education is therefore not simply developing good MBAs but transformational agents and committed global leaders for a better world (Pless & Maak, 2009). Responsible management education is more than an ethical call for willing individuals – it is a systemic responsibility for administrators and faculty to direct their programs and institutions through their discernments, priorities and decision making. As Gates suggested to Harvard University's administrators and faculty in his remarks:

Let me make a request of the deans and the professors – the intellectual leaders here at Harvard: As you hire new faculty, award tenure, review curriculum, and determine degree requirements, please ask yourselves: Should our best minds be dedicated to solving our biggest problems? Should Harvard encourage its faculty to take on the world's worst inequities? Should Harvard students learn about the depth of global poverty ... the prevalence of world hunger ... the scarcity of clean water ...the girls kept out of school ... the children who die from diseases we can cure? (Gates, 2007).

Should management education share the responsibility to address world problems? The answer merging from this analysis of the PRME initiative is that the adoption of the principles and the processes connected with it, offer an opportunity for effectively engaging in addressing word major problems. As Schram and Harney (2003) highlight in their review of 2,300 research papers, most business and manager researchers focus on solving 'small technical problems' like product placement and supply chain, while failing to examine the larger social and political questions that could provide fundamental answers about how to create a better world. The PRME are about revising teaching and researching in management programs so that academic institutions and leaders engage in solving global social problems and serving the common good. (Godfrey & Grasso, 2000).

Principles and Processes of the PRME

The PRME include simple and straightforward values for orienting education toward societal and global responsibility. However, they do not include any specific technical insights of teaching and learning methods. The principles are not meant to be taken as a set of measurable standards for ranking classification or the like (Waddock, Rasche, Werhane, & Unruh, 2011, p. 14; Wolfe & Werhane, 2010, p. 146). Much like the UNGC 10 principles, the PRME six principles are more aspiration and voluntary than prescriptive and compulsory. Building on the Global Compact mission to inspire responsible management practices, the PRME aims to "inspire and champion responsible management education" (PRME, 2008, p. 2). The premise of both the UNGC and the PRME is that businesses and management education can be powerful forces to creating socially responsible leaders engaged to make the world a better place.

Under the guidance of Dr. Manuel Escudero, Head of Academic Initiatives of the UN Global Compact, the many institutional representatives convened in Geneva in July 2007 developed six key principles focusing on:

1. Creating the necessary capabilities among students so that they can become inclusive and sustainable value leaders;
2. Incorporating into the education programs the values of global social responsibility as illustrated by initiatives like the UNGC's 10 principles around human rights, labour rights, ecological sustainability, and anticorruption;
3. Creating methods, educational processes and material that enable effective and responsible leaders;
4. Engaging in research that advances understanding about the impacts of companies in creating sustainable social, environmental, and economic value;
5. Interacting in partnerships with managers and academics to meet environmental challenges and social responsibilities; and
6. Facilitating dialogue among stakeholders representing multiple sectors on critical social and sustainability issues.

Much more than a general organizational value statement, the formulation of these principles are more like a commitment for participating academic institutions to practice these values and share their practices (Table 1).

The PRME offer an opportunity to inspire educators, administrators and professionals to embed the values of sustainability and corporate responsibility into the core mission of manage-

Table 1. The values and commitments of the PRME (source: Adapted from www.unprme.org)

Participating Commitment: As institutions of higher education involved in the development of current and future managers we declare our willingness to progress in the implementation, within our institution, of the following principles, starting with those that are more relevant to our capacities and mission. We will report on progress to all our stakeholders and exchange effective practices related to these principles with other academic institutions.
Principle 1-Purpose: We will develop the capabilities of students to be future generators of sustainable value for business and society at large and to work for an inclusive and sustainable global economy.
Principle 2-Values: We will incorporate into our academic activities and curricula the values of global social responsibility as portrayed in international initiatives such as the United Nations Global Compact.
Principle 3-Method: We will create educational frameworks, material, processes, and environments that enable effective learning experiences for responsible leadership.
Principle 4-Research: We will engage in conceptual and empirical research that advances our understanding about the role, dynamics, and impact of corporations in the creation of sustainable social, environmental and economic value.
Principle 5-Partnership: We will interact with managers of business corporations to extend our knowledge of their challenges in meeting social and environmental responsibilities and to explore jointly effective approaches to meeting these challenges.
Principle 6-Dialogue: We will facilitate and support dialogue and debate among educators, business, government, consumers, media, civil society organizations, and other interested groups and stakeholders on critical issues related to global social responsibility and sustainability.
Organizational Practices: We understand that our own organizational practices should serve as an example of the values and attitudes we convey to our students.

ment education programs. They offer a timely global call to business schools and management programs to gradually but systemically adopt values and practices for sustainability, social responsibility, human rights, labor rights, environmental rights and anti-corruption mechanisms. Beyond the adoption of the principles, the process of engaging the whole institution or program toward a commitment for socially responsible education can be transformative. The PRME offers a framework for engagement of management related academic institutions at three levels:

1. **Continuous Improvement:** Any school that is willing to engage in a gradual but systemic manner is welcome to join the initiative. Implementation of the Principles should be understood as a long-term process of continuous performance improvement and the PRME can provide a framework of general principles through which to engage faculty and staff, and build institutional support.
2. **A Learning Network:** The PRME initiative also functions as a learning network. By collecting and channeling good practices, it will facilitate an exchange of existing and state-of-the-art experiences within the PRME network.
3. **Report to Stakeholders:** Adopting the PRME implies that the signatory school is willing to report regularly - annually - on progress to all stakeholders. Public reporting is the best way to ensure the credibility of the initiative and allows giving recognition to good performances (PRME, 2011a)

The process could also generate innovative ways of integrating sustainability and ethics across the curricula and in the learning outcomes of the program(s) (Matchett, 2008; Swanson & Fisher, 2011). Business ethics, social responsibility and corporate sustainability have entered into the teaching of most management programs, however, they have not become an integral part of the strategic core of most management education colleges and universities. Social entrepreneurship, business ethics, green business and sustainable development management are increasingly present topics in management courses and curricula worldwide, but they are often relegated to electives and not placed at the center of the core courses and competencies of the programs (Net Impact, 2008). The adoption of the PRME can enhance these processes in academic institutions who want to adequately respond to the growing demands of a new generation of students more concerned on sustainability and global citizenship values. The voluntary adoption of the Principles along with the institutional commitment to engage in the processes of PRME can provide dynamic opportunities for preparing world engaged, socially responsible and sustainable business leaders for the 21st century.

In tackling sustainable and inclusive solutions to complex world problems like world poverty and exploitative and unsustainable business practices, the PRME recognizes the importance of involving multiple stakeholders and sectors. The representatives of key institutions in the steering committee and larger academic network point out to the need of multi-stakeholder dialogue and worldwide collaboration. Currently the PRME Secretariat works in coordination with PRME steering committee representing the United Nations Global Compact (UNGC), the Association to Advance Collegiate Schools of Business (AACSB International), the European Foundation for Management Development (EFMD), the Aspen Institute's Business and Society Program, Globally Responsible Leadership Initiative (GRLI), the European Academy of Business in Society (EABIS), Association of MBAs, Central and East European Management Development Association (CEEMAN), Latin American Council of Management Schools (CLADEA) and Net Impact, a student organization with more than 13,000 members (PRME, 2011c). The effective education of future socially responsible leaders requires the commitment of current higher education leaders, administrators and scholars of key organizations and institutions (Khurana, 2007; Nemerowicz & Rosi, 1997).

Meanings and Values of the PRME

The principles, which appear to be very general at first, provide undersigning institutions with a holistic perspective on responsible management education. Adopting these principles gives institutions the possibility to integrate ethics, social responsibility and sustainability into the curricula; promote and engage cutting edge socially responsible research and establish dialogues and collaborations between academics and practitioners. Institutions of higher learning involved in the education of current and future managers cannot find in the six principles 'what to do'. Rather it is the responsibility of the many, diverse and worldwide institutions to 'fill' these six principles with contextualized meaning based on reflections and discussions (Rasche & Escudero, 2010). In his analysis of German Universities, Andrew Rasche (2010b), recognizes how the PRME are a practical call to action. The principles have an emphasis on 'what' are the values of social responsibility and sustainability; on 'how' to implement these principles and 'within' what institutional and social sectors they can best be put into practice (p. 9). The meaning and values of the six Principles are therefore a framework for implementing a long term process academic social responsibility (PRME, 2007, p. 4).

Principle 1 Purpose: We will develop the capabilities of students to be future generators of sustainable value for business and society at large and to work for an inclusive and sustainable global economy. This principle summarizes the primary purpose of PRME to provide academic conditions for developing sustainable and engaged leaders. The most important aspect here is to engage all parts of the academic program (faculty, staff and administrators) in a process to integrate the principles in the curricula and educational support services. This engagement process is critical especially for those disciplines that traditionally do not include ethical issues in their courses (e.g., finance, accounting and marketing). The principle can be integrated into the course by including reflections on the school's mission in relation to the PRME.

Principle 2 Values: We will incorporate into our academic activities and curricula the values of global social responsibility as portrayed in international initiatives such as the United Nations Global Compact. This second principle explicitly recognizes the UN Global Compact values as foundation for socially responsible education. The UNGC principles along with the values of human rights, labor issues, environmental issues, and anti-corruption act like a 'moral compass' for concerned institutions. These values are not exclusive. Other value-based practices like fair trade, sustainable development and peace could also well integrate the purpose of the 'values' principle.

Principle 3 Method: We will create educational frameworks, materials, processes and environments that enable effective learning experiences for responsible leadership. Teaching responsible management and sustainable ethics can take many shapes and forms. Most top-ranked MBA programs worldwide teach ethics and social responsibility as an integrated unit across the curricula (Christensen, et al., 2007). Many professors resist this integration as they find it burdensome to include ethics systematically in their area of expertise (Swanson & Fisher, 2011). PRME can not only help to initiate a process to integrate ethics across curricula but it also helps create collaborations among institutions for travel courses, international development projects and other innovative teaching methods in the field of management (Rasche, 2010b).

Principle 4 Research: We will engage in conceptual and empirical research that advances our understanding about the role, dynamics, and impact of corporations in the creation of sustain-

able social, environmental and economic value. The practical application of teaching through real project is inherent to management education (Wankel & DeFillippi, 2005). Yet, integration of research that benefits not only organizations but adequately responds to communities in developing countries, for example, needs to be coordinated by competent and collaborative institutes and centers. Unfortunately the coordination between business ethics centers with other institutes working on human rights, intellectual property or community development is often limited by the sectoral fragmentation of the disciplines. Effective engagement in complex and multifaceted problems like poverty reduction requires coordinated multidisciplinary and multi-stakeholder efforts across universities and in collaboration with businesses, government agencies, international nongovernmental organizations and local nonprofits. PRME with the United Nations support has the possibility to assist academic institutions to act as agents for change through their engaged research across disciplines.

Principle 5 Partnership: We will interact with managers of business corporations to extend our knowledge of their challenges in meeting social and environmental responsibilities and to explore jointly effective approaches to meeting these challenges. Although partnerships are important for the necessary practical and hands-on purpose of good management education, it cannot be limited to business corporations. The UNGC, although primarily targeting for-profit organizations attempts to engage them in a larger dialogue with NGOs, academic institutions and government and inter-governmental agencies. The United Nation's broad experience offers numerous opportunities to engage in multi-sector partnerships.

Principle 6 Dialogue: We will facilitate and support dialogue and debate among educators, business, government, consumers, media, civil society organizations and other interested groups and stakeholders on critical issues related to global social responsibility and sustainability. The most important value added by PRME is that of engaging institutions and leaders in dialogue toward cutting edge education trends in sustainability and social responsibility. The recently concluded 2nd Global Forum for the Responsible Management Education is an example of the benefits produced by these intersectoral partnerships and international dialogues. This is accompanied by the growing importance of the exchange of best strategies shared in the required self-reporting of participating institutors.

Adoption and Recognition of the PRME

PRME can also serve as a framework for systemic change for business schools and management-related institutions on the basis of three distinctive characteristics of the initiative. First, PRME can provide schools and universities a gradual but systemic trajectory for a long term process for continuing performance improvement. Second, PRME can function as a learning network collecting and exchanging good practices and effective strategies across institutions, programs and national education systems. Third, PRME can engage educational institutions and communities of stakeholders through the required public annual reporting. Although self-reporting, PRME public reporting is the best way to monitor institutional commitment and progress over time. It also gives institutions opportunities to highlight their social, ethical and environmental stewardship practices through curricula, research, service and administrative commitments (PRME, 2011a).

Dr. Ángel Cabrera recognizes the importance of adopting PRME not just on the values and meaning of the 'principles' as much as they are a useful framework to engage a university in the 'process' of educating socially responsible leaders: "The process of adopting the Principles can be as important as the adoption itself. The PRME framework offers a unique opportunity to ask some fundamental questions about our educational mission, to test some of the core assumptions that currently dominate our curriculum and research, and to generate a multi-stakeholder conversation to drive change" (PRME, 2008, p. 4).

The benefits of a school adopting PRME resemble the strategic advantages that a business has to adopt socially responsible and sustainable business model (Olson, 2010; Tan & Geh, 2008). First, they can become leading institutions ahead of competition, in line with important trends and along internationally recognized standards (Rasche, 2010a). Second, business schools and management-related institutions can increase their visibility and effectiveness in engaging with stakeholders and local and global communities they serve. In the same way that most schools strive to balance sound scholarship with excellent practice, PRME offers a context for appreciating research that is relevant to world problems (either in the pure or applied research modes). Third, PRME gives scholars, schools, students and practitioners the platform to seriously invest in socially responsible and sustainable development educational practices. Taking part of the PRME networks and publically reporting on the progress forces participating institutions to go beyond superficial 'green-washing' in reference to sustainability or 'blue-washing' in reference to the relation with the United Nations. Taken into account that PRME is at its beginning stages, it already represents and will surely stimulate new processes for enhancing, measuring, teaching, researching and practicing socially responsible management in schools and society.

Although accreditation bodies like the Association to Advance Collegiate Schools of Business (AACSB), the Association of MBAs (AMBA) and the European Foundation for Management Development (EFMD) have endorsed the principles, the PRME is not a certification/verification standard. While the provision of general principles act as a catalyst for the formation of innovative education practices championed by leading institutions and scholars, this strength can also be considered a weakness of PRME. It is understandable that the PRME seeks to act more as a vehicle for reflection and action rather than a measuring stick to enforce practices. However, the principles need to be accompanied and reinforced by a process (not necessarily headed by PRME) that stimulates measurement, accountability and criteria for assessing achievements (Waddock, et al., 2011). Indeed, the PRME will acquire more practical relevance as clearer assessment criterion are developed for measuring progress on faculty development, curricula integrations, research accomplishments and community engagement (Wolfe & Werhane, 2010).

The PRME do not intend to create criteria for program accreditation and ranking. However, with the adoption of these values, along with the participation in their global networks, management educational programs and institutions could enhance their academic quality while refocusing on their ethical mission. Swanson & Fisher's (2008) comprehensive analysis of administrators of AACSB accredited business programs, finds that 80% of administrators believe that business schools should place more emphasis on ethics education. They conclude that accreditation agencies have contributed to management schools' lack of integration (or superficial adoption) of business ethics, social responsibility and sustainability in their curricula. For example, the Association to Advancement College Schools of Business (AACSB) in their attempt to become the primary accrediting body of business schools internationally did not emphasize ethics and social responsibility in their standards (Swanson & Fisher, 2008). The National Association of Schools of Public Affairs and Administration (NASPAA), the main accreditation body for public service programs, appears to face similar challenges in their internationalization of accreditation standards (McFarland, 2007). In the shift from 'mission-driven" accreditation to "quality standards" even public service management programs risk undermining the ethical foundation and service orientation of its educational mission. By adopting the PRME in both the principles and the processes connected to it, business and public service management curricula could receive a bust for integrating ethics, sustainability and social responsibility across the curricula. Martell and Castiñeira (2011) suggest that the use of PRME could be instrumental in directing the qualifications of a management program more toward the alternative ranking of ethics and social responsibility as measured by the Aspen Institute's Beyond Grey Pinstripe (p. 110).

PRME is gaining momentum by offering participating institutions the possibility to excel in socially responsible management education through global forums, shared best strategies and worldwide cutting edge trends. The growing interest in socially responsible themes is also clearly manifested in the growing participation of member institutions associated with PRME (Wolfe & Werhane, 2010). In less than 4 years, the initiative witnessed a tremendous growth going from 40 to more than 360 participating institutions (PRME, 2011b). Most of the participating academic institutions and management programs are from Europe (38%) and United States (23%), but there is a large representation also from Latin American and South-East Asia. Of the 364 participants, 138 (38%) have completed an annual public report on their progress showing some good practices in implementing the PRME across the curricula and in the service of communities (PRME, 2011b). The growing relevance of the PRME in management education is also reflected by the high participation rate in the first and second Global Forum for Responsible Management Education held at UN headquarters in December 2008 and 2010, respectively.

The PRME reflects also the new generation of management students' growing interest and demand for social responsibility, business ethics and sustainability education. The 2008 Net Impact survey of more than 2,980 MBA students across 95 programs showed that 60% of students indicated a strong interest to learn more about concepts such as social entrepreneurship, environmental sustainability and Corporate Social Responsibility (CSR) (Net Impact, 2008). Academic institution sensitive to these trends will likely to participate in important forums and networks created by the PRME and their corresponding contexts of the UN Global Compact and the UN Millennium Goals.

ACADEMIC SOCIAL RESPONSIBILITY FOR POVERTY REDUCTION

The PRME go beyond merely adopting general ethical principles. They attempt to inspire academic institutions to rediscover their educational mission in the context of their social responsibility for poverty reduction. Academic social responsibility (ASR) is therefore a practical challenge for universities in general and management programs in particular to create the appropriate educational opportunities to engage with other sectors of society in solving social problems. Although deferring to the decision and adaptation of each individual institution, the PRME invite participants to avoid the temptation to engage in shallow sustainability PR (the so-called 'green-washing') or superficial international relations – what some have called 'blue-washing' in reference to the UN blue flag (Bruno & Karliner, 2000).

To fully understand and integrate responsible management values into education we need to understand the contexts and challenges represented by the UN Global Compact and Millennium Development Goals (MDGs). In addition to responding to the call of management scholars to renew management education and integrate it into more sustainable practices for world benefit, the PRME represent an entry step into the larger contexts of multi-sector partnerships (represented by the UNGC) and the sustainable development challenge (represented by the MDGs). These contexts are reinforced by and connected with the growing trends in management education advocating for

Figure 1. The context of responsible management education (source: Adapted from www.un.org/millenniumgoals, www.unglobalcompact.org, and www.unprme.org)

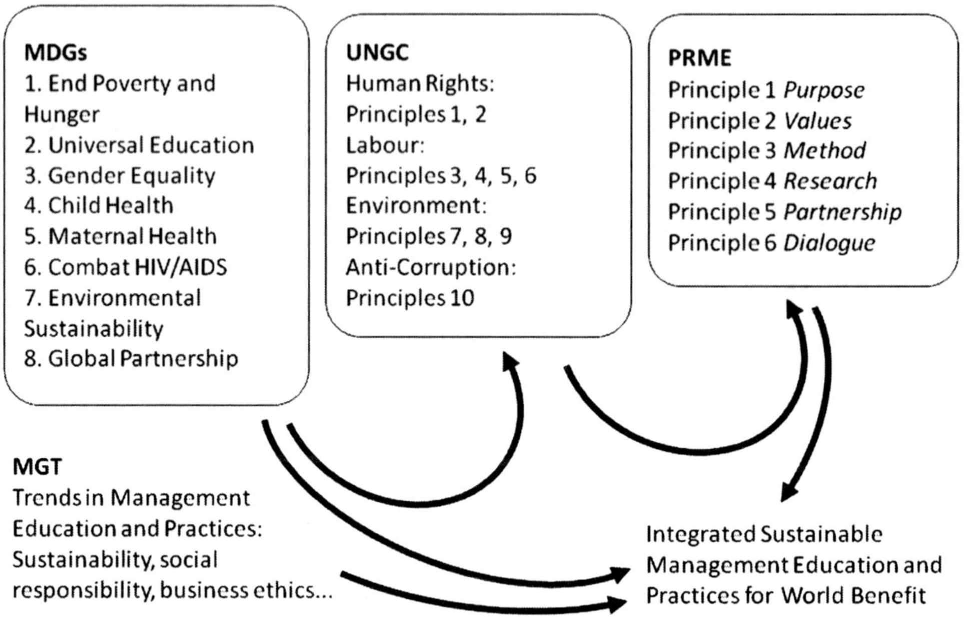

programs in line with sustainability, business ethics and social responsibility (Figure 1).

The UNGC is a strategic policy initiative to engage businesses committed to fulfill the 10 universally accepted principles in the areas of human rights, labor justice, environmental sustainability and anti-corruption. By doing so, the United Nations recognizes the importance of the private sector to collectively achieve a more humane, sustainable and responsible globalization. They do so by engaging corporations and business organizations in a global platform of dialogue and collaboration with various inter-governmental agencies of the United Nations, nongovernmental organizations (NGOs) and academic institutions. Global Compact recognizes how academia adds critical dimensions to their objective and advances the UNGC agenda at the local and global level. They can do so through commitment of academic leadership, socially responsible research, shared educational infrastructure, and educational programs in line with the Global Compact's 10 principles on human-labor-environmental rights and anti-corruption. In addition, academia plays an important role in shaping future business leaders and educating them on the importance of responsible citizenship. The participation in the Global Compact Academic Network is open to all those universities, colleges, and business schools; university departments, research departments willing to collaborate as strategic partners and committed to implementing the 10 principles.

Facing the enormous challenges of global markets and increasing global inequalities, former UN Secretary General Kofi Annan launched the challenge of the global compact at the 1999 World Social Forum in Davos inviting business leader to commit to human rights, labors standard and environmental practices:

I believe what I am proposing to you is a genuine compact, because neither side of it can succeed without the other. Without your active commitment and support, there is a danger that universal values will remain little more than fine words -- documents whose anniversaries we can celebrate and make speeches about, but with limited impact on the lives of ordinary people. And unless those values are really seen to be taking hold, I fear we may find it increasingly difficult to make a persuasive case for the open global market (UNSG, 1999).

After 10 years, the words of Kofi Annan continue to inspire true world engagement welcoming these principles not through superficial undersigning but through corporate commitment and systemic practices. Since its foundation on July 26, 2000, the UNGC has grown exponentially representing six UN agencies at its core and engaging more than 8,000 participants, including more than 5,300 businesses in 130 countries around the world. Participating academic institutions have numerous engagement possibilities going from exchange of best strategies in the Communication of Progress to influencing global trends in the 2010 UN Global Compact Leaders Summit (Rasche & Kell, 2010).

The challenge of poverty reduction, education access, global health and sustainable societies can only be reached through global commitments. Academic institutions play a vital role not only for exposing students to the realities and possibilities of international sustainable development and global social entrepreneurship, but also by offering practical opportunities for experiential learning and transformative world engagement initiatives. Although not as formalized as in the PRME and UNGC, management schools can engage in practical partnership initiatives for poverty reduction like those of the United Nations Development Programme promoting inclusive markets and growing sustainable businesses (UNDP, 2011).

In 2010, the PRME Secretariat joined CEEMAN, the international development association and accreditation agency, in sponsoring a Survey on Management Education and Poverty Reduction. The study builds on the first global survey to explore the relationship between poverty alleviation and business education conducted by CEEMAN in 2008. The major findings in that first study which surveyed 154 individuals from 33 countries were that:

- The issue of global poverty was a legitimate topic for management education;
- Courses on corporate social responsibility and business ethics were said to be the most logical "homes" for this discussion; yet
- Innovations in teaching about global poverty were occurring in business courses outside of the ethics and corporate social responsibility course (CEMANN & PRME, 2010).

At all teaching levels of management (undergraduate, MBAs, Executive MBAs and Ph.D.s) students expressed a clear and strong preference to increase the number of study travel trips that have a business and poverty focus (respectively 34%, 45%, 42%, 41%) and increase the number of student consulting projects to NGOs/governments/corporations that focus on poverty (respectively 23%, 23%, 22%, 22%). They also expressed interests in adding new assignments that deal with poverty, new cases that address the issue of poverty and articles and/or books that discuss poverty. Some other suggestions that came from the surveyed management students included the importance of added service learning in an organization that deals with poverty, increase the number of sessions taught by outside experts in poverty and increase the number of sessions taught by faculty experts on poverty (pp. 21-23).

CONCLUSION: TOWARDS A WORLD ENGAGED EDUCATION

Beyond diverse economic theories and systemic change methods, the fundamental management education question remain: *what kind of person should an MBA or MPA graduate be if she/he is ultimately going to help shape the direction of our world's most powerful organizations and institutions?* Unfortunately, most business school discourse today emphasizes a narrow technical vision over a broader overview of the principles and social values behind their career (Parkin, 2010, p. 123). Courses in related fields to business ethics, social entrepreneurship, sustainable business or socially responsible investing still occupy a marginal, elective role in most MBAs. Even with their emphasis on public service, most MPAs still experience the challenge of effectively integrating ethics and social responsibility in their curricula (Bowman & Menzel, 1998; Plant & Ran, 2009). In addition, the promises of higher salaries, quicker promotions or personal decisions to in a degree programs for business administration (MBAs) and public administration (MPAs). Henry Mintzberg, in his critique of management education, invites educators to rethink the wrong approaches and make room for a more engaged type of managers and leaders:

The trouble with "management" education is that it is business education, and leaves a distorted impression of management. Management is a practice that has to blend a good deal of craft (experience) with a certain amount of art (insight) and some science (analysis). An education that overemphasizes the science encourages a style of managing I call "calculating" or, if the graduates believe themselves to be artists, as increasing numbers now do, a related style I call 'heroic.' Enough of them, enough of that. We don't need heroes in positions of influence any more than technocrats. We need balanced, dedicated people who practice a style of managing that can be called "engaging." Such people believe that their purpose is to leave behind stronger organizations, not just higher share prices. They do not display hubris in the name of leadership. The development of such managers will require another approach to management education, likewise engaging, that encourages practicing managers to learn from their own experience. In other words, we need to build the craft and the art of managing into management education and thereby bring these back into the practice of managing (Mintzberg, 2005, p. ix).

Increasing complexity and interdependence requires new approaches in management education (PRME, 2007, p. 4). The complexity of world problems requires the integration of environmental, social, and governance responsibility with strategic thinking and hands-on educational opportunities both locally and globally. With new emerging human development inequities, the threats of climate change, energy and food shortages and the international financial markets crisis, we cannot continue to do business as usual. The creation of sustainable and responsible societies with appropriate principles, practices and institutions begins in academia (Global University Network for Innovation., 2009). Management programs in particular have the possibility and responsibility to shape the values, attitudes and behavior of managers and leaders that can generate a new wave of positive change worldwide. With the integration of socially responsible principles and practices into their curricula and experiential learning opportunities can engage leaders for world benefits. Recent analysis of worldwide higher education indicate the need for programs to be more clearly directed toward human and social development, civic engagement and sustainable development (Global University Network for Innovation., 2009, pp. 55-56).

World engaged and competent leaders for the 21st century require more than the usual managerial skills. Twenty-first century leaders need to develop values, skills and attitudes capable of effectively engaging in complex international environments, multi-sector collaborations and multicultural contexts. 'They require talented and ethical leaders who can not only advance organizational goals and fulfill legal and fiduciary obligations to shareholders, but who are also prepared to deal with the broader impact and potential of business as a positive global force in society' (PRME, 2007, p. 4).

Management programs and business schools have the primary responsibility to revisit the effectiveness of their education programs in developing socially responsible managers and engaged value leaders. They have the responsibility to adequately prepare students for a more integrated and interdependent world. As the corporate scandals at the beginning of the 21st century suggested a prioritization of ethics in management curricula, the current crisis should remind us that the interdependence of our global economy and the necessary integrated look at the economic factor in relation to social and environmental consequences. Business scandals and economic crises should be strong reminders of the importance of teaching integrated systemic analysis of ethical and practical managerial education centered in 'social responsibility' and 'sustainability' principles.

Social responsibility and sustainability are not just business ethical buzzwords. They are an opportunity to transform our management programs into laboratories for engaged world leaders capable of promoting sustainable business and societies. The growing awareness and interest in sustainability of young leaders and students entering management programs in business and public service should suggest a stakeholder approach for adopting PRME in academic institutions. Following the example of Thunderbird School of Global Management and other best practices in adopting PRME, management programs and institutions can implement university-wide processes for integrating sustainable management and socially responsible leadership into their management education. The European Foundation for Management Development (2006) has recognized the crucial role of academia in making the idea of responsibility a cornerstone for global leadership development:

> *Globally responsible leaders at all organizational levels face four key challenges. First, they should think and act in a global context. Second, they should broaden their corporate purpose to reflect accountability to society around the globe. Third, they should put ethics at the centre of their thoughts, words and deeds. Fourth, they - and all business schools and centres for leadership learning - should transform their business education to give corporate global responsibility the centrality it deserves (European Foundation for Management & Global, 2006, p. 2).*

Academic institutions and management programs have the social responsibility to develop engaged world leaders, globally responsible and sustainability leaders. This requires the right vision, oriented by sound universal principles and values as expressed in PRME, the UNGC and the MDGs. The pursuit of experiential and principles-based leadership development programs oriented toward economic and societal progress and sustainable development is based on the recognition of prioritizing management practice in a global context. The foundation of capable, globally responsible, engaged world leaders must be value-based. Hence the process toward the integration of socially responsible practices in management education must be principle-based. This overview of the PRME in their values, meaning and practical application in academic engagement for poverty reduction exemplify a trajectory toward world engaged management education. Working for the common good is at the core of public service careers, but also the foundation for renewed socially responsible managers and leaders. Young leaders have the right to receive an education that is not just based on skill-management trainings but also values, principles and engaging experiential learning capable of inspiring them toward the creation of economic and societal progress in a globally responsible and sustainable way.

REFERENCES

Ball, A., & Osborne, S. P. (2011). *Social accounting and public management: Accountability for the common good.* New York: Routledge..

BAWB. (2006). Business as an agent of world benefit: Management knowledge leading positive change. *Forum Overview.* Retrieved from: http://www.bawbglobalforum.org

Bowman, J. S., & Menzel, D. C. (1998). *Teaching ethics and values in public administration programs: Innovations, strategies, and issues.* Albany, NY: State University of New York Press..

Bruno, K., & Karliner, J. (2000). *Tangled up in blue: Corporate partnerships at the United Nations.* San Francisco, CA: CorpWatch..

Cabrera, Á. (2009). Let's professionalize management. *Harvard Business Review Blog – how to fix business schools.* Retrieved from: http://blogs.hbr.org/how-to-fix-business-schools/2009/04/a-hippocratic-oath-for-future.html

CEMANN, & PRME. (2010). *Final report 2010 Survey on Global Poverty as a Challenge for Management Education.* Retrieved from: http://www.unprme.org/resource-docs/PovertySurvey-FinalReportOct2010.pdf

Christensen, L., Peirce, E., Hartman, L., Hoffman, W., & Carrier, J. (2007). Ethics, CSR, and sustainability education in the Financial Times top 50 global business schools: Baseline data and future research directions. *Journal of Business Ethics, 73*(4), 347-368.

Cooperrider, D. L., Whitney, D. K., Stavros, J. M., & ebrary Inc. (2008). *Appreciative inquiry handbook for leaders of change.* Retrieved from: http://www.columbia.edu/cgi-bin/cul/resolve?clio7609938

European Foundation for Management. D., & Global, C. (2006). *Globally responsible leadership a call for engagement.* Retrieved from: http://www.efmd.org/index.php/component/efmd/?cmsid=040929yuch&pub=060614xvqa

Forray, J., & Leigh, J. (2010). Special issue: Principles of responsible management education (PRME). *Journal of Management Education, 34*(2), 327–328. doi:10.1177/1052562910364759.

Gasparski, W. (2008). Responsible management education. Warsaw: Academic and Professional Press; Kozminski Business School..

Gates, B. (2007). *Harvard University commencement PM exercises Bill Gates.* Chelsea, MA: Commencement Video..

Global University Network for Innovation. (2009). *Higher education at a time of transformation: new dynamics for social responsibility.* New York: GUNI/Palgrave Macmillan..

Godfrey, P. C., & Grasso, E. T. (2000). *Working for the common good: Concepts and models for service-learning in management.* Washington, D.C.: American Association for Higher Education..

Hart, S. L. (2005). *Capitalism at the crossroads: The unlimited business opportunities in solving the world's most difficult problems.* Upper Saddle River, N.J.: Wharton School..

Khurana, R. (2007). *From higher aims to hired hands: The social transformation of American business schools and the unfulfilled promise of management as a profession.* Princeton: Princeton University Press..

Martell, J., & Castiñeira, Á. (2011). Assessing what it takes to earn a beyond grey pinstripes ranking. In D. L. Swanson, & D. G. Fisher (Eds.), *Toward assessing business ethics education* (pp. 101–132). Charlotte, N.C.: Information Age Pub.

Matchett, N. (2008). Ethics Across the Curriculum. *New Directions for Higher Education*, (142): 25–38. doi:10.1002/he.301.

McFarland, L. (2007). *International Accreditation Issues: NASPAA White Paper*. Retrieved from: http://www.naspaa.org/accreditation/document/NASPAAWhitePaperonInternationalAccreditationIssuesbyLaurel.pdf

MDGs. U. N. (2010). *A gateway to the UN system's work on the MDGs*. Retrieved from: http://www.un.org/millenniumgoals

Mintzberg, H. (2005). *Managers not MBAs: A hard look at the soft practice of managing and management development*. San Francisco: Berrett-koehler..

Nemerowicz, G. M., & Rosi, E. (1997). *Education for leadership and social responsibility*. Washington, DC: Falmer Press..

Net Impact. (2008). *New Leaders, New Perspectives: A Survey of MBA Student Opinions on the Relationship between Business and Social/Environmental Issues*. San Francisco, CA: The Aspen Institute..

Olson, E. G. (2010). *Better green business: Handbook for environmentally responsible and profitable business practices*. Upper Saddle River, N.J.: Wharton School Pub.

Parkin, S. (2010). *The positive deviant: Sustainability leadership in a perverse world*. Washington, DC: Earthscan..

Plant, J., & Ran, B. (2009). Education for ethics and human resource management. *Public Integrity*, *11*(3), 221–238. doi:10.2753/PIN1099-9922110302.

Pless, N., & Maak, T. (2009). Responsible leaders as agents of world benefit: Learnings from "Project Ulysses.." *Journal of Business Ethics*, *85*(Supplement), 59–71. doi:10.1007/s10551-008-9947-1.

Prahalad, C. K. (2005). *The fortune at the bottom of the pyramid*. Upper Saddle River, NJ: Wharton School Pub.

PRME. (2007). *The Principles of Responsible Management Education*. New York: United Nations Global Compact..

PRME. (2008). *Principles for Responsibe Management Education: A global initative, a global agenda*. New York: United Nations Global Compact..

PRME. U. N. (2011). *Principles for responsible management education (PRME) official website*. Retrieved from: http://www.unprme.org

PRME. (2011a). *Engagement model for PRME schools & academic institutions*. Retrieved from http://www.unprme.org/the-6-principles/engagement-model.php

PRME. (2011b). *PRME academic institutions*. Retrieved from: http://www.unprme.org/participants/index.php

PRME. (2011c). *PRME steering committee*. Retrieved from: http://www.unprme.org/participants/coconvening-organisations-steering-committee/index.php

Rasche, A. (2010a). The limits of corporate responsibility standards. *Business Ethics (Oxford, England)*, *19*(3), 280–291. doi:10.1111/j.1467-8608.2010.01592.x.

Rasche, A. (2010b). The principles for responsible management education (PRME) – A 'call for action' for German universities. In M. Haase, S. Mirkovic, & O. J. Schumann (Eds.), *Stand und Perspektiven der unternehmens- und wirtschaftsethischen Ausbildung in Deutschland* (pp. 119–136). Mering, Germany: Hampp..

Rasche, A., & Escudero, M. (2010). Leading change – The role of the principles for responsible management education. *Journal for Business and Economic Ethics*, *10*(2), 244–250.

Rasche, A., & Kell, G. (2010). *The United Nations global compact: Achievements, trends and challenges*. New York: Cambridge University Press..

Sachs, J. (2005). *The end of poverty: Economic possibilities for our time*. London: Allen Lane..

Schram, S. F., & Harney, S. (2003). Review of state work: Public administration and mass intellectuality. *American Journal of Sociology, 109*(2). doi:10.1086/381631.

Swanson, D. L., & Fisher, D. G. (2008). *Advancing business ethics education*. Charlotte, NC: Information Age Pub.

Swanson, D. L., & Fisher, D. G. (2011). *Toward assessing business ethics education*. Charlotte, NC: Information Age Pub.

Tan, G., & Geh, E. (2008). *Developing business advantages throughout corporate social responsibility: A conceptual framework*. Singapore: Lee Kong Chian School of Business, Singapore Management University..

UNDP. (2011). *Inclusive market development (IMD)*. Retrieved from: http://www.undp.org/partners/business/IMD.shtml

UNGC. U.N. (2011). *United Nations Global Compact (UNGC) official website*. Retrieved from: http://www.unglobalcompact.org

UNSG. (1999). Secretary-general proposes global compact on human rights, labour, environment, in address to world economic forum in Davos. *United Nations Press Release* (February 1). Retrieved from: http://www.un.org/News/Press/docs/1999/19990201.sgsm6881.html

Waddock, S., Rasche, A., Werhane, H. P., & Unruh, G. (2011). The principles of responsible management education: Implications for implementation and assessment. In D. L. Swanson, & D. G. Fisher (Eds.), *Toward assessing business ethics education* (pp. 13–28). Charlotte, N.C.: Information Age Pub.

Wankel, C. (2008). *Alleviating poverty through business strategy*. New York: Palgrave Macmillan. doi:10.1057/9780230612068.

Wankel, C., & DeFillippi, B. (2005). *Educating managers through real world projects*. Greenwich, CT: Information Age Pub.

Werhane, P. H., Kelley, S. P., Hartman, L. P., & Moberg, D. J. (2010). *Alleviating poverty through profitable partnerships: Globalization, markets and economic well-being*. New York: Routledge..

Wolfe, R. W., & Werhane, P. H. (2010). Academic insitutions in the United Nations global compact: The principles for responsble management education. In A. Rasche, & G. Kell (Eds.), *The United Nations global compact: Achievements, trends and challenges* (pp. 144–160). New York: Cambridge University Press..

Yunus, M., & Weber, K. (2010). *Building social business: The new kind of capitalism that serves humanity's most pressing needs*. New York: Public Affairs.

ADDITIONAL READING

Aspen, I. (2008). *The Aspen Institute guide to socially responsible MBA programs, 2008-2009*. San Francisco: Berrett-Koehler..

Aspen Institute Center for Business. E. (2009). The sustainable MBA: the 2010-2011 guide to business schools that are making a difference. New York: Aspen Institute Center for Business Education..

Wankel, C., & Stoner, J. A. F. (2009). *Management education for global sustainability*. Charlotte, N.C: Information Age Pub.

KEY TERMS AND DEFINITIONS

Academic Social Responsibility: A concept extending the educational mission of academic institutions into actively engaging for the benefits of society though teaching, research, service and partnerships.

Millennium Development Goals (MDGs): Include eight international development goals and 21 targets that all 192 United Nations Member States have agreed to achieve by the year 2015. The goals include eradicating extreme poverty, reducing child mortality rates, fighting disease epidemics such as AIDS, and developing a global partnership for achieving sustainable development worldwide.

Poverty Reduction: Also called poverty alleviation, is a process which seeks to reduce economic and non-economic poverty levels in groups of people, communities or countries. Poverty reduction strategies may include program in education, health, entrepreneurship, technology, income redistribution and various forms of economic development.

Principles for Responsible Management Education (PRME): They are about purpose, values, method, research, partnership and dialogue to implement socially responsible organizational practices as models for students.

Social Engagement: A concept referring to the individual, collective or institutional relations or involvements with some elements of society particularly communities and social service organizations.

Social Responsibility: An ethical theory that an individual, organization, or institution (including a university) has the obligation to act positively act to benefit society at large. Although some interpret it simply as passive value, avoiding engaging in socially harmful acts, it includes an active obligation to perform activities that directly advance social goals.

Sustainable Education: A concept that involves active academic participation to create economic, social and environmental programs improving life standards, generating empowerment and respecting interdependence.

United Nations Global Compact (UNGC): Conveys businesses, academic institutions, non-governmental organizations and inter-governmental agencies and programmes of the United Nations around ten principles the areas of human rights, labour, the environment and anti-corruption.

World Benefit: It implies leadership and commitment toward producing positive impact to the planet, people and the economy and promoting initiatives for poverty alleviation, peace building and the promotion of human rights.

This chapter was previously published in Handbook of Research on Teaching Ethics in Business and Management Education edited by Charles Wankel and Agata Stachowicz-Stanusch, pages 546-563, copyright 2012 by Information Science Reference (an imprint of IGI Global).

Section 5
Critical Issues

This section takes an in-depth look at some of the problems facing Sustainable Practices in the twenty-first century in order to present potential solutions to future concerns. With any new technology, problem, or discovery comes uncertainty as to its impact on future practices, livelihoods, and the overall sustainability of a society; this uncertainty must be analyzed and addressed if any useful advances are to be made. As the global landscape continues to shift, questions of environmental change, economic reform, social policy, and industrial development are of paramount concern. In this section, 14 chapters explore some of the critical issues driving advances in Sustainable Practices.

Chapter 66
How Can Sustainable Development Redefine Project Management Processes?

Mohamed Eid
The British University in Egypt, Egypt

ABSTRACT

The relationship between Project Management (PM) and Sustainable Development (SD) is tested in this chapter through its application on the construction industry. When PM is embedded in construction projects, it has the capacity to be a significant leverage point of great influence, and it becomes one of the cornerstones for rethinking the relationships between PM, SD, and the Construction Industry. The work presented discusses the need for integrating sustainable development into project management processes to ensure a better outcome for the construction industry, which is directly related to the degradation of our quality of life on the economic, social, and environmental levels. The author explores the origins and philosophies behind sustainability, the core of project management processes, the strategic implications of the construction industry practices, and puts forward the "systems thinking and points of leverage" approach to facilitate an efficient environment of integration. Thinking and acting sustainably requires not only incremental change but also a revolution in approach, a shift of perspective; sustainable project management processes are possible to achieve.

INTRODUCTION

The major issues facing the world are mostly from the effect of human development activities. Economic instability, widespread social exclusion and inequalities and global environmental degradation mean the world is fighting for a better quality of life for all present and future generations. In order to start facing up to these challenges, the global community has to realize that these problems are no longer specific to certain countries at a national level, but in fact, they have now risen to a global scale. When tackling such huge issues, everyone has a significant role to play. Enhancing our quality of life will not be achieved unless the world's communities rise to

DOI: 10.4018/978-1-4666-4852-4.ch066

the challenge and contribute individually as well as collectively to a better quality of life, each within their field of expertise. Acknowledging that there are always better ways of practice, and that the wheel of development never stops, this thesis embarks on researching and studying the construction industry's practices and indeed challenging the impacts of global development on our quality of life. The research tackles the need for changing the strategies, policies, and standards which normally control the practices and guide the projects from inception to completion (Eid, 2009, pp.1-2).

Sustainable development is emerging as an important agenda for better practice in the construction industry; an industry that in conjuction with the built environment—which constitutes more than one half of the real capital—they are the main consumers of resources in energy and materials (CIB, 1999). The relationship between construction and sustainability has created new performance agendas driven by sustainable construction guidelines.

With projects increasing in complexity, needs and standards to be fulfilled, project management has grown to be an essential and vital tool for managing and delivering those projects, meeting if not exceeding expectations. This underpins the major contributions of project management to our quality of life which justifies its increasing recognition as a profession and practice. The work presented herein describes a suggested relationship between sustainability, project management, and construction. The author puts forward a rich environment encircling this relationship based on fundamental criteria that embrace the opportunities for introducing "change"; this milieu is the systems thinking approach. This is implemented by identifying the places within this system (leverage points) where the earlier the change is introduced to its early fundamental stages, the better and more efficient is the impact expected to derive in terms of the overall performance of the system (Eid, 2009)[1].

The research reported in this chapter examines the potential of a tri-dimensional integration to explore the way forward for construction, project management, and sustainable development to contribute effectively to a better quality of life. This is investigated to unveil whether or not it would have fundamental impact on the three professions in general and in turn, enhance their contributions to people's quality of life.

BACKGROUND: DEFINING THE THREE DIMENSIONS OF THE RELATIONSHIP AND THE ENCIRCLING MILIEU

Figure 1 represents the first part of the methodology, the research uses a Venn diagram on three-sets to exhibit the existing and proposed relationships between the three dimensions of the relationship in question; the construction industry, sustainable development and project management, while the encircling milieu is the systems thinking theories and leverage points.

Figure 1. Establishing the relationships between the four pillars of the research

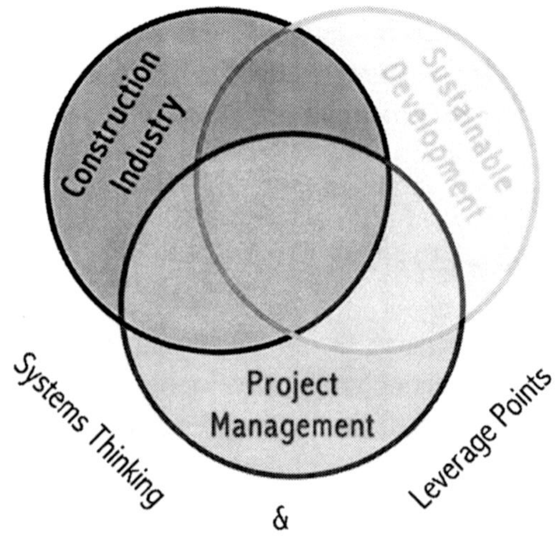

THE CONSTRUCTION INDUSTRY

The construction industry is a major constituent of the European Union's economy (The Commission of the European Parliament, 2002). The construction sector is an active and dynamic industry throughout Europe, but one that contains a large number of hazards and risks. It plays a very important role in the economic well-being of the European Union, representing approximately 11% of Community GDP (Gross Domestic Product) in 1996 (The European Union Online; Europa). In spite of a slow-down in the rate of increase of productivity, the construction industry remains the largest industrial sector in the community, ahead of the foodstuffs and chemicals industries and it is crucial in producing investment goods. It accounted for 10% of GDP and 48.5% of Gross Fixed Capital Formation in 2000 (European Construction Industry Federation (FIEC). It is also the largest sector in terms of employment, providing jobs for 8.8 million people (7% of the working population), and gives rise to 2.5 million construction related jobs, and 14.3 million other service jobs, acquiring 28.5% of industrial employment. In the EU, 26 million workers depend, directly or indirectly, on the construction sector making it the biggest industrial employer in Europe. It is a fact that the construction industry and the built environment are the main consumers of resources (i.e. energy and materials) within the European Union where buildings are estimated to consume approximately 40% of total energy. Buildings are also responsible for some 30% of CO2 emissions – and generate approximately 40% of all man-made waste (CIB, 1999). The destruction of the resource base and the whole scale pollution of the life support systems upon which continued human development depends are among the consequences of operating economic limits beyond ecological limits (Royal Commission on Environmental Pollution (RCEP), 2000).

As addressed in the Egan Report, leading companies in other sectors (and a very small number of leading UK construction companies) are already measuring, and taking systematic steps to manage, their sustainability impacts. These companies recognize that a reorientation towards sustainable development is essential for ensuring their long-term viability. However, most companies within the UK construction industry appear to be missing opportunities for taking a more pro-active role, within their broader sphere of influence, in helping to provide a sustainable built environment (The Construction Task Force, 1998).

An urgent need—and a considerable opportunity—exists to include sustainable development as an explicit objective into existing and new initiatives. This could be facilitated by including indicators of sustainability (such as those being developed by CIRIA, BSRIA and others) into the Key Performance Indicators (KPIs) framework put forward in "Rethinking Construction."

The construction industry has successfully passed the milestone of recognizing the importance of integrating sustainable development into the core of the industry's practices. Creating sustainable agendas for the industry is a vital step forward towards the implementation of change in the culture of the industry and attitudes towards a sustainable future. The scale of the UK construction industry, its immediate environmental and resource use impacts, coupled with the influence of its products on the nation's quality of life, make it a key player in delivering sustainable development (Eid, 2000).

The implementation of sustainable development agendas and the business case behind the integration demonstrate that the business case for sustainable construction has many opportunities to create environmental and social improvements whilst increasing profits and market shares which can create new sources of employment and ensure better business benefits and sustainable economic base and beneficial impacts. The industry is facing up to its weaknesses and building upon its strengths, and, it has not yet used its full potential for change and doing things differently. The construction industry was part of the problem of delivering a better quality of life for people and there was a persisting need to make the industry part of the solution. Integrating sustainable development into the construction industry practices and indeed into the culture of the industry was, and still is, the biggest challenge facing the industry (Eid, 2009).

SUSTAINABLE DEVELOPMENT

Sustainable development goes beyond environmental awareness; it tackles social and economic aspects as well as the environment, hence, this section discusses the ways in which sustainable development touches upon our daily activities. Sustainable development represents a vital element of the integration proposed in order to attain sustainable construction practice and more specifically in the management of projects. European and British policies for sustainable development are quite similar in their commitment to the implications which sustainable development might introduce within social, environmental and economic trends, but whatever the policies are, sustainable development faces significant challenges in defining its "Business Case."

In order to achieve an effective implementation of sustainable development and face up to the challenge, putting these strategies into action demands the response of different sectors in society. More sector-specific initiatives are needed to allow the sustainable development framework to touch on the everyday aspects of life. They must provide environments in which people live and work enjoyably and efficiently and which encourage working and social communities to flourish.

The construction industry stands a great potential for change because "Sustainable Construction" is about much more than the fabric of the built environment. Housing and the social, commercial and transport infrastructures around them must all be built in ways that are sustainable in environmental and economic as well as in social terms. They must add value to the quality of life for the individual and the community (Eid, 2009).

The "Business Case" for sustainable development stands as a major challenge for businesses to accept and embrace sustainability into their financial planning and economic development (WBCSD, 2002). In general, sustainable development calls for more efficiency, higher standards of quality in delivering social progress, protecting the environment and ensuring economic growth. The new paradigm of sustainable progress and acceptance of businesses to the concept of more involvement of sustainable development into their financial planning and core commitment has provided a significant opportunity for government sectors to follow suit. But this has proven yet as challenging as it was for general businesses to concur. For the construction industry, the challenge for construction businesses to embrace sustainable practices is still in its early phase and more needs to be done. From the previous analysis of the interconnections between business and sustainability, it emerges that the concept of sustainable development is closely linked to efficiency and competitiveness (The UK Government, 2003).

Sustainable development is not incompatible – and can be mutually supportive- with competitiveness. Business has many "win-win" opportunities to create and/or social improvement whilst increasing profits and market share (James, 2000).

Sustainable business practice is the way forward and this will reflect on the different business sectors including the construction industry and its sustainable approach. Sustainable Development has established itself as the road to a better future and a better quality of life. With all the contributing factors to its current popularity, it highlights for governments, businesses, and society the importance of not only future development on the economic level but also on the social and environmental levels. The reasons behind the emergence of sustainable development, its importance to and its impact on our lives, demonstrate that although future and continuous development is the aim of all nations, it cannot follow the same strategies of the past with no regard for the environmental, and social impacts. It has to be generated by the need for a change of attitude. National economies can no longer envisage development without considering the social and environmental aspects. Businesses have to integrate the sustainable development triple bottom line into the core of their financial plans.

The construction industry plays a major role in any nation's development and therefore, the potential it holds for sustainable approaches to a better performance is immense. Achieving more sustainability into the construction processes is possible by breaking down the existing system without losing perspective of the holistic nature of the construction processes. By breaking down the system, it is possible to identify where the sustainability gaps exist and work on better implementation for these gaps. Within the construction industry, project management stands as a significant stage of all projects and it is the role of the following chapter to introduce the functions of project management in general and more specifically within construction practices (Eid, 2009).

PROJECT MANAGEMENT

In the construction industry as well as in all other industries, project management applications have proved indispensable for the continuous progress of better performance along with the undeniable need for its existence among the early phases of projects. The effective involvement of project management processes in projects is shown as an essential factor of success in delivering/exceeding stakeholders'/clients' expectations and needs. The Project Management Institute (PMI) created a project management reference that is undeniably one of the most comprehensive bodies of knowledge for PM; the Guide to the Project Management Body of Knowledge (PMBOK).

The purpose of project management described for the construction industry is to add significant and specific value to the process of delivering construction projects. PM adds significant opportunities for better and more efficient completion. Whether on small or large scale projects, they both share the same properties of "projects" described earlier, but the contribution of PM to the successful completion does not vary relatively to their sizes. For construction and all other fields, projects are not only confined to major undertakings or large scale projects but they come in all sizes, varying in time plans, resources, goals and areas of applications.

While it is true that major projects have provided the impetus for identifying and studying the principles of effective project management, these principles are nevertheless just as applicable to small projects as to larger projects (Wideman, 1995, p.72).

The importance of the project management's contribution remains the same whatever the scale or area of application. This contribution is underpinned by a core of knowledge and guidelines which generally remain the same but only vary in their applications according to the field of incorporation.

Due to the increasing demand on PM and since the skills required to manage projects successfully have multiple dimensions, models of the project management body of knowledge needed to be identified. Models which guide practitioners or academics working or researching on PM to identify the core basics of PM processes, stages and enable them to apply its basics to all fields and areas of application given that the management team is always responsible for determining what is appropriate for any specific project.

During the late 1980s and early 1990s, a significant increase of interest in knowledge management originated from a convergence of at least four interrelated trends: the first as described by Argyris and Schön (1978) is the development of organizational learning, the second is an emphasis on re-engineering and re-invention of business processes as expressed by Davenport (1993) and Hammer and Champy (1993). The third trend is the total amount of advances in information technology such as computational speed and Web applications as explained by Snider and Nissen (2003) and the fourth trend being the associated development of information systems management theory and practice. These four trends have resulted, together, in an awareness of and attention to knowledge, particularly in the private sector, as an organizational asset or resource that may be leveraged for competitive advantage as illustrated in the work of Grant (1996) and Spender (1996).

The bodies of knowledge, in the case of project management, represent generally accepted sets of skills and tools that all certified professionals should possess (Sneider & Nissen, 2003). However, the studies mentioned before, deal with knowledge as something that is capable of being transferred, shared, or transformed in the context of any specific business sector (Berger & Luckman, 1967).

This fluid view of knowledge is at odds with the more static view implied by the bodies of knowledge of professional fields. This is tempered by recognition that none of the several project management bodies of knowledge describe themselves as complete and exhaustive repositories of all project management knowledge (Wirth & Tryloff, 1995).

The main reasons behind choosing The Project Management Institute PMBOK for this research study are (Eid, 2009):

- In comparison to the previously mentioned BOKs of project management, the PMBOK® guide excels in its breakdown of application areas of project management and processes. The level of details provided in the PMBOK is not matched by any other BOK, specifically in the illustration of interactions between processes through the corresponding lists of inputs, tools & techniques and outputs.
- PMBOK allows a simple application to the circumstances of the construction industry and therefore an easy path to the successful completion of construction projects following the core knowledge of project management described in PMBOK.
- The new Construction Extension to PMBOK, is another demonstration of the important partnership between PM and construction practices, hence the new document details the additional knowledge areas which are more specific to construction.
- The practices and the project management of construction projects were one of the resources on which the original PMBOK was based. The involvement of construction from the early beginnings of PMBOK as an important body of PM knowledge makes the final document even more relevant to this research and touches on many sides of the industry.
- Unlike any of the international PM bodies of knowledge, no other PM initiatives had ventured in creating a PM document specific to construction as PMI pioneered in creating the Construction Extension to PMBOK.
- Due to the international exposure of PMBOK, it covers the practices of project management in many countries over the world including the UK, which makes its body of knowledge more applicable to more countries and industries than any other PM body of knowledge.

The essence of project management has been identified by PMBOK; the project management knowledge areas, their component processes along with their inputs, tools and techniques and outputs.

These component processes and lists represent the initial information which project management processes deal with, to incorporate them into the PM life cycle. The construction industry holds within its practices significant opportunities that allow project management processes to contribute more sustainably into its framework. But after displaying all the knowledge areas from the PMBOK and adding the new areas from the Construction Extension, the combined knowledge areas from both publications do not portray any serious commitment to sustainable construction.

Integrating sustainability strategies into the PM life cycle has to start from within the core of project management processes, following the theory of systems thinking. The sooner sustainability changes are integrated into project management, the more significant and stronger the impact we can expect. The following chapter examines "Systems Thinking" as the tool which will facilitate the integration as the vehicle to assist the integration of sustainable development into the project management processes for construction projects (Eid, 2009).

SYSTEMS THINKING AND LEVERAGE POINTS

General systems theory was introduced in the 1940s by Ludwig Von Berttalanffy, but has been vastly expanded since its inception (Cummings, 1980). It developed as a response to rapid technological complexities that confronted engineering and science. It was a radical departure from traditional science which dealt with cause and effect explanations (Gardner & Demello, 1993).

Systems thinking viewed an organization and its respective environment, as a complex whole of interrelating, interdependent parts. It stressed the relationships and the processes that make up the organizational contact, rather than the separate entities or the sum of the parts (Cummings, 1980).

Leverage points allow individuals to identify the opportunities in existing systems/organizations where "change" could be implemented to lead the system to the long planned goals and objectives. Senge describes them as places to pursue the goals in a way that takes advantage of, instead of working against, the systemic structures around them (Senge, 1999).

Leverage points are seen as opportunities in existing systems where embedded actions and changes can lead to significant, enduring improvements. Donella Meadows describes Leverage points as:

Places within a complex system where a small shift in one thing can produce big changes in everything (Meadows, 1999).

The reason behind the identification of leverage points and places to intervene in a system, is mostly because individuals and participants in systems need to study where changes can be implemented and the manner in which these changes should be introduced as well as the nature of these changes. Senge and Meadows, both argue that small changes and well-focused actions can produce significant, enduring improvements if they are introduced to the system in the right place, with the right value needed and at the right time (Senge, 1999).

The lead taken in sustainable development has changed the global reality and commitment to a better future, and has also affected the commitment of private and public sectors in pushing forward the application of a smooth transition from environmental management systems to sustainable management systems. The initiation of development of such management systems highlights the potential role and contribution of conventional business and public sector organizations in corporate and public environmental, social and economic management for all practices. The conversion of environmental management systems to sustainable management systems, represent a smooth transition to face the new global challenge, embracing sustainable business thinking, tools and techniques to identify, evaluate and improve social, economic and environmental organizational commitment.

It is not only about environmental concerns, but it is about a better future on the three parallel strands of economic, social and environmental concerns. Sustainable approach to management systems is proven to be a valid contribution to sustainable performance which aligns with the arguments for implementing sustainability into the project management process, from the early phases of any project, as advised by Meadows. Achieving a more effective impact is possible when the change is implemented from the early stages using the right leverage points. The next stage of this thesis illustrates the interaction between the three pillars of this research (i.e. Project Management, Sustainable Development and Construction Industry's Practices) within a tri-dimensional framework to achieve a more sustainable project management process for construction practices starting from the earliest phases of projects. Systems thinking and its theory's basics and the identification of leverage points will act as the "milieu" which will allow the tri-dimensional integration to take place and will fortify their existing relationships (Eid, 2009).

RETHINKING THE RELATIONSHIPS: THE TRI-DIMENSIONAL INTEGRATION

Figure 2 represents a mirrored image of the same Venn diagram; the image on the left entails the intersections between the three dimensions of the relationship while the image on the right explains how these intersections are manifested by the enclosed case studies or initiatives.

The body of the chapter relies on three distinctive case studies as a fundamental part of the methodology of the research. The first practical application portrays the Egan Agenda's Drivers from a systems thinking point of view; when integrating SD into construction practices, it generates a sustainable construction agenda.

The second case study is the identification of leverage points within one of the most important phases of all construction projects; the procurement stage. This is demonstrated as an example to the integration between PM and construction practices. Project lifecycles and phases are demonstrated through the procurement stage to attain a more sustainable application to the project phase in question (i.e. procurement) through the identification of sustainability gaps within an existing Government model which should have

Figure 2. The tri-dimensional integration

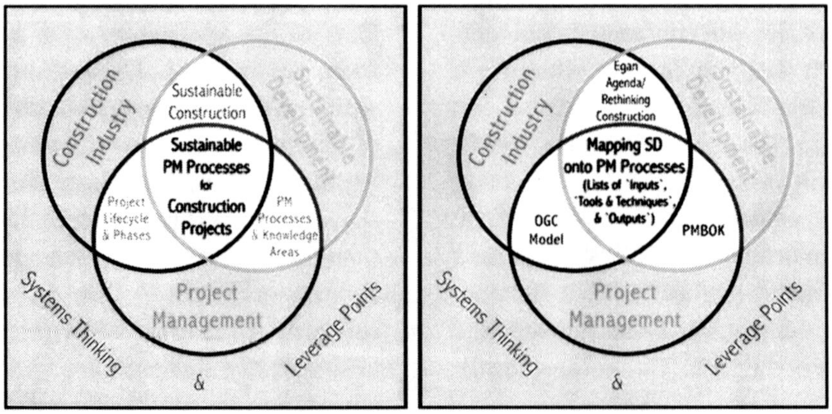

been following Government's sustainable initiatives (OGC Model).

The final case study application withdraws the research into its fundamental conclusion; integrating SD into PM processes for construction projects. The case study examines the notions of a sustainable development approach to the PMBOK processes and knowledge areas and describes the way forward to implement the desired change into the core knowledge of PM to envisage an efficient implementation of SD guidelines into the PM standards of the PMBOK. The integration explores the potential of a holistic approach to strategic project management and show how a systems thinking approach to PM processes would enhance PM's contribution to construction projects.

As a governing milieu, systems thinking theories would allow the Integration of sustainable development guidelines more easily and clearly into construction processes starting with the core of its activities which are represented by the project management knowledge areas and processes. This would highlight the connections between and among sustainable development, construction and project management.

Integrating SD into PM Processes

Following the guidelines of systems thinking, introducing sustainable development to the existing framework of project management has to take place through the identification of leverage points within the system. This takes the analysis back to the components of project management as a functioning system, to the fundamental parts out of which it is formed; the processes and their interactions which rely on their inputs and outputs. Implementing a significant change such as "sustainable development" into project management has to take place from within its core of knowledge or paradigm out of which the project management process arises.

This research relies on changing the interactions between the lists of inputs and the lists of outputs through embedding sustainable development guidelines, indicators and measuring tools into the list of tools and techniques of all the PM processes and their knowledge areas listed in the PMBOK. Leverage point 2 (the mindset or paradigm out of which the system arises) is a very high point on Meadows list in terms of the effectiveness of the implementation of the desired change. This point of leverage allows the change to take place from within the sources of the system. However small this change could be, its effectiveness is expected to be of very high significance because the change is introduced to the core of the system. This will mean therefore that all subsequent activities, knowledge areas, process groups, project processes, project management practices and ultimately, the construction projects would be affected by the impact of this change.

The PMBOK was criticized in this research for its lack of commitment to sustainable development and failure to address sustainability issues more seriously and direct practitioners' interest towards the global pressure for a better sustainable future. PMBOK holds, within its knowledge areas, significant opportunities to embrace sustainability; the following analysis represents the main hypothesis for integrating sustainable development within project management processes. The hypothesis will be demonstrated by introducing a new sustainable approach to the interactions between the groups of process and their lists of inputs and outputs as the origin of the continuity of the whole PM functioning system.

As shown in Figure 3, when analyzing each process, the PMBOK displays the interactions between the list of inputs and outputs by listing the suggested tools & techniques which facilitate the transformation of inputs to outputs. For all combined fifty two processes, PMBOK uses the format shown in Figure 3 to display the linear interconnections among the lists of inputs, tools and techniques and outputs for each process. The example shown in the figure is for the first process of "Initiation" which is part of the knowledge area

Figure 3. The proposed phases of integrating SD into PM processes

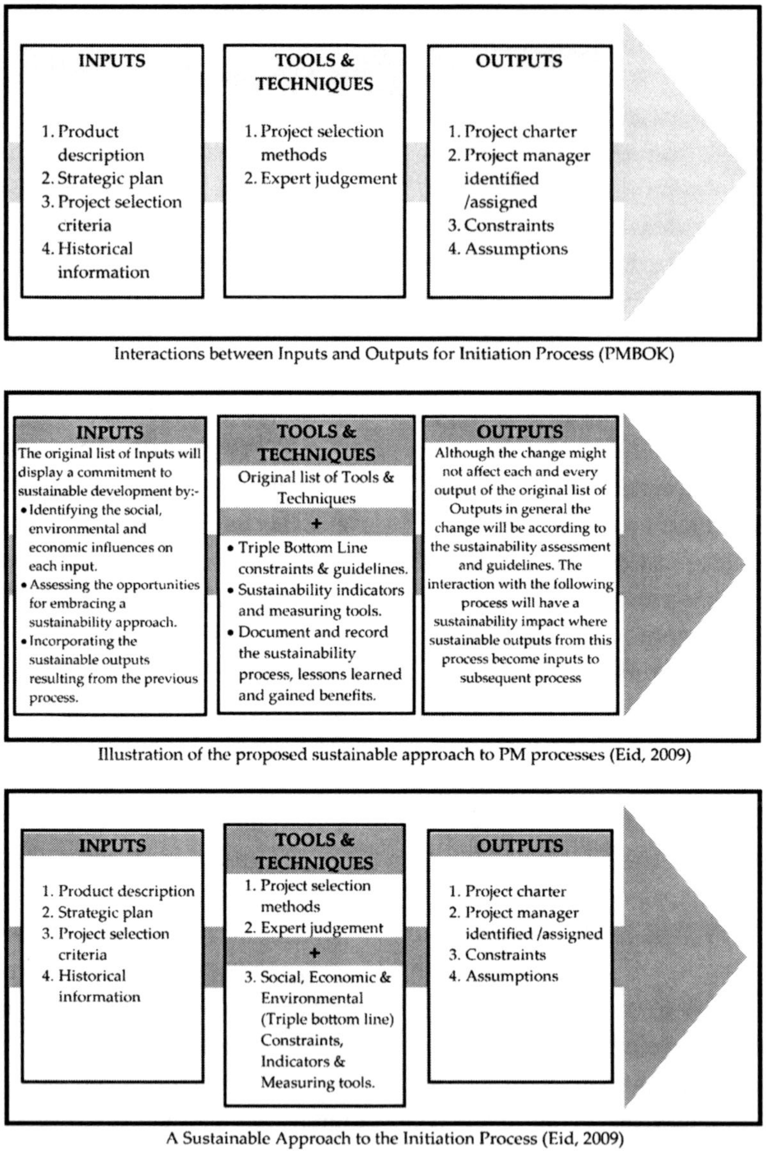

of "Project Scope Management" and the Process Group of "Initiation."

The argument is based on adding sustainable development guidelines, indicators and measuring tools to the original list of tools and techniques for each corresponding process. Clearly, it is not an attempt to substitute the existing list but to add sustainability drivers to its content to enhance and direct all of the interactions towards sustainable development.

Figure 3 displays the transformation from the PMBOK format at the top, to an illustration of the proposed sustainable approach to the PM process at the middle and then finally at the bottom of the figure an illustration of the new sustainable approach to the Initiation PM process. From the figure, it is now obvious that the approach is applicable to the processes and relies on adding sustainable development guidelines, indicators and tools to the lists of tools and techniques and also assesses the opportunities for change starting from the inputs.

In order to move forward, it is vital at this stage to map the opportunities for integrating sustainable development to the general as well as the specific elementary origin of construction PM processes; the knowledge areas.

Mapping the Opportunities for SD to PM Processes and Knowledge Areas

The knowledge areas, as the origin of PM processes, hold within them the core catalyst for change in performance. This also represents a very efficient leverage point on Meadows' efficiency scale to introduce sustainable development to the whole process and guarantee significant positive impact.

The matrix shown in Table 1 demonstrates that the relevance of each knowledge area to sustainable development is not necessarily the same among its own processes (e.g. for 6- Project Time Management; "6.1 Activity Definition," "6.4 Schedule Development" and "6.8 Progress Curve Development") nor is the relevance constant among the three elements of sustainability when compared for one process (e.g. "7.2 Cost Estimating": relevance of "* *" "medium relevance" to the Social element, relevance of "* * *" "high relevance" to the Economic element and finally, relevance of "*" "low relevance" to the Environmental element). This, of course, does not rule out the possibility of similarities in relevance among the processes nor does it rule out the "None" relevant option (e.g. Claim Identification). The table is not meant to be exclusive, but to demonstrate how the PM processes can be integrated with sustainability and to identify the potential areas of knowledge, which lead to a more sustainable approach.

The matrix lays out the knowledge areas and their processes in terms of any project's main five process groups (Initiating to Closing).

In line with the theories of systems thinking and leverage points, it is apparent that the highest relevance and potential for integrating sustainable development exists within the Initiating and Planning Process Groups. This is the most efficient leverage point where the project is still in its early phases of creating the scope, generating the project's plans, identifying the risks and producing the rules. As explained earlier, these process groups keep repeating the cycle from Initiating processes to Closing processes for every stage of the project, hence the appearance of high relevance and potential points during the Executing and the Controlling Processes (e.g. "14.2 Environmental Assurance" for the Environmental bottom line, "5.5 Scope Change Control" for the three bottom lines and "7.4 Cost Control" for the Economic bottom line). The high potential for integration at this stage is underpinned by the significance of the process itself and its contribution to and authority over the following process.

Although the matrix presents the possibilities of integrating sustainable development into project management process provided in the PMBOK (2000 Edition), a chance was provided for PM practitioners, professionals, academics and researchers to introduce their own professional opinions on this integration's hypothesis in an online questionnaire.

Table 1. Matrix of mapping sustainability opportunities to PM processes and knowledge areas

	SOCIAL		ECONOMIC		ENVIRONMENTAL	
	PM Processes	Relevance	PM Processes	Relevance	PM Processes	Relevance
1- INITIATING	5.1 Initiation	* * *	5.1 Initiation	* * *	5.1 Initiation	* * *

continued on following page

Table 1. Continued

	SOCIAL		ECONOMIC		ENVIRONMENTAL	
	PM Processes	Relevance	PM Processes	Relevance	PM Processes	Relevance
2- PLANNING	4.1 Project Plan Development	* * *	4.1 Project Plan Development	* * *	4.1 Project Plan Development	* * *
	5.2 Scope Planning	* * *	5.2 Scope Planning	* * *	5.2 Scope Planning	* * *
	5.3 Scope Definition	* * *	5.3 Scope Definition	* * *	5.3 Scope Definition	* * *
	6.1 Activity Definition	* * *	6.1 Activity Definition	* * *	6.1 Activity Definition	* * *
	6.2 Activity Sequencing	* *	6.2 Activity Sequencing	* * *	6.2 Activity Sequencing	* *
	6.3 Activity Duration Estimating	* *	6.3 Activity Duration Estimating	* *	6.3 Activity Duration Estimating	* *
	6.4 Schedule Development	* *	6.4 Schedule Development	* *	6.4 Schedule Development	* *
	7.1 Resource Planning	* * *	7.1 Resource Planning	* * *	7.1 Resource Planning	* * *
	7.2 Cost Estimating	* *	7.2 Cost Estimating	* * *	7.2 Cost Estimating	*
	7.3 Cost Budgeting	None	7.3 Cost Budgeting	* * *	7.3 Cost Budgeting	None
	8.1 Quality Planning	* *	8.1 Quality Planning	* *	8.1 Quality Planning	* *
	9.1 Organisational Planning	* * *	9.1 Organisational Planning	* *	9.1 Organisational Planning	*
	9.2 Staff Acquisition	* * *	9.2 Staff Acquisition	* *	9.2 Staff Acquisition	*
	10.1 Communication Planning	* *	10.1 Communication Planning	*	10.1 Communication Planning	*
	11.1 Risk Management Planning	* * *	11.1 Risk Management Planning	* * *	11.1 Risk Management Planning	* * *
	11.2 Risk Identification	* * *	11.2 Risk Identification	* * *	11.2 Risk Identification	* * *
	11.3 Qualitative Risk Analysis	* *	11.3 Qualitative Risk Analysis	* * *	11.3 Qualitative Risk Analysis	* *
	11.4 Quantitative Risk Analysis	* *	11.4 Quantitative Risk Analysis	* * *	11.4 Quantitative Risk Analysis	* *
	11.5 Risk Response Planning	* *	11.5 Risk Response Planning	* * *	11.5 Risk Response Planning	* *
	12.1 Procurement Planning	* *	12.1 Procurement Planning	* * *	12.1 Procurement Planning	* *
	13.1 Safety Plan Development	* * *	13.1 Safety Plan Development	* *	13.1 Safety Plan Development	* * *
	14.1 Environmental Planning	* *	14.1 Environmental Planning	* *	14.1 Environmental Planning	* * *
	15.1 Financial Planning	*	15.1 Financial Planning	* * *	15.1 Financial Planning	*
	16.1 Claim Identification	None	16.1 Claim Identification	* *	16.1 Claim Identification	None
	16.2 Claim Quantification	None	16.2 Claim Quantification	* *	16.2 Claim Quantification	None

continued on following page

Table 1. Continued

	SOCIAL		ECONOMIC		ENVIRONMENTAL	
	PM Processes	**Relevance**	**PM Processes**	**Relevance**	**PM Processes**	**Relevance**
3- EXECUTING	4.2 Project Plan Execution	* * *	4.2 Project Plan Execution	* * *	4.2 Project Plan Execution	* * *
	6.6 Activity Weights Definition	* *	6.6 Activity Weights Definition	* *	6.6 Activity Weights Definition	* *
	8.2 Quality Assurance	* *	8.2 Quality Assurance	* *	8.2 Quality Assurance	* *
	9.3 Team Development	* * *	9.3 Team Development	None	9.3 Team Development	* *
	10.2 Information Distribution	*	10.2 Information Distribution	* *	10.2 Information Distribution	*
	12.4 Source Selection	* *	12.4 Source Selection	* * *	12.4 Source Selection	* *
	13.2 Safety Plan Execution	* * *	13.2 Safety Plan Execution	* *	13.2 Safety Plan Execution	* *
	14.2 Environmental Assurance	*	14.2 Environmental Assurance	*	14.2 Environmental Assurance	* * *
4- CONTROLLING	4.3 Integrated Change Control	*	4.3 Integrated Change Control	*	4.3 Integrated Change Control	*
	5.4 Scope Verification	* * *	5.4 Scope Verification	* * *	5.4 Scope Verification	* * *
	5.5 Scope Change Control	* * *	5.5 Scope Change Control	* * *	5.5 Scope Change Control	* * *
	6.5 Schedule Control	* *	6.5 Schedule Control	* *	6.5 Schedule Control	* *
	6.7 Progress Curve Development	None	6.7 Progress Curve Development	None	6.7 Progress Curve Development	None
	6.8 Progress Monitoring	None	6.8 Progress Monitoring	None	6.8 Progress Monitoring	None
	7.4 Cost Control	*	7.4 Cost Control	* * *	7.4 Cost Control	*
	8.3 Quality Control	* *	8.3 Quality Control	* *	8.3 Quality Control	* *
	10.3 Performance Reporting	*	10.3 Performance Reporting	* *	10.3 Performance Reporting	*
	11.6 Risk Monitoring & Control	* *	11.6 Risk Monitoring & Control	* *	11.6 Risk Monitoring & Control	* *
	14.3 Environmental Control	*	14.3 Environmental Control	*	14.3 Environmental Control	* * *
	15.2 Financial Control	*	15.2 Financial Control	* * *	15.2 Financial Control	*
	16.3 Claim Prevention	*	16.3 Claim Prevention	*	16.3 Claim Prevention	*
5- CLOSING	Relevance of Sustainability Approach to this Process Group is significantly minor not only because this stage consists mainly of the legal aspects of the project but also because it is too late to incorporate sustainability at this phase of administrative closure and contract closeout (ineffective leverage points). The only relevance for future projects is to document and record all lessons learned including gained sustainability benefits.					

The Questionnaire

There is a persisting need for the project management community to act upon the conviction that a sustainable approach is possible and as shown, it creates a whole new and significantly more powerful role for project management; to guide the industry to even better performance.

In order to examine further these possibilities and opportunities for sustainable project management processes, the research now moves forward to examine the results of an online survey made accessible to PM practitioners, professionals and academics which attempts to discover the potential and identification of a sustainable approach. An online questionnaire was created discussing the proposed set of arguments as an invitation for all project management users to agree and/or disagree with the hypothesis. The questionnaire is also an opportunity to share the views of the international audience on the sustainable approach to an international PM document publication such as the PMI's PMBOK.

The overall objective of the questionnaire was to investigate the views of the respondents on the possibilities and the opportunities for integrating sustainable development into project management processes listed in the PMBOK (2000 Edition). The broad as well sophisticated nature of the area covered by the questionnaire meant that the targeted audience should represent a specific class of respondents; PM practitioners, professionals, academics and/or researchers. The questionnaire examines specific aspects of project management with specific interest in the PMBOK processes; this implied that the respondents should be familiar of the contents of the PMBOK and preferably well into practicing its guidelines.

There were two versions of the questionnaire, a detailed version entailing all fifty two numerous processes from the PMBOK which demanded the inputs and full dedication of certain type of respondents to complete it. The second version was a much less detailed document which took about 20 minutes to complete (this version can be found in the only annex to this chapter). Both versions followed the same design, methodology, format and answering techniques. The primary difference is that the final version does not contain the detailed illustration of the interactions of each process from "Inputs" to "Outputs" through the involvement of the lists of "Tools and Techniques" but rather a general outlook on the process without its lists of interactions.

Questionnaire Findings

Table 2 summarizes the answers retrieved from the first part of section 5 of the questionnaire. To illustrate all the answers, the same matrix used before in Table 1 has been re-introduced to map the opportunities and the relevancy of sustainable development to each of the PM processes. In this section, but with minor changes although following the same framework, the responses were compiled to calculate the percentage of respondents who chose any of the possible selections (A, B, C).

The outcome of the compilation demonstrates that "B" is the most popular choice. Hence, for 79.5% of the listed processes, the respondents thought that when sustainable development guidelines would be added to the list of "tools and techniques" of these processes, there will be considerable changes to their lists of outputs and therefore guide the overall performance towards sustainability. This significant majority supports the argument that endorses the integration of SD into PM processes. The analysis of the responses clearly shows that the respondents agree that the processes hold within them efficient leverage points for introducing SD to PM practices.

From Table 3, another conclusion is apparent; among the other two selections (A, C); the respondents seem more leaning towards choosing (A) rather than (C). In fact, only two processes out of the thirty nine had a majority choice of (C) denying them the opportunity for SD enhancement (i.e. Activity Duration Estimating and Schedule

Table 2. Matrix of the respondents' selections concerning mapping the opportunities and the impact of integrating SD into PM processes

	SELECTION A		SELECTION B		SELECTION C	
	PM Processes	**%**	**PM Processes**	**%**	**PM Processes**	**%**
1- INITIATING	5.1 Initiation	**82.4**	5.1 Initiation	**11.8**	5.1 Initiation	**5.9**
2- PLANNING	4.1 Project Plan Development	**76.5**	4.1 Project Plan Development	**20.6**	4.1 Project Plan Development	**2.9**
	5.2 Scope Planning	**82.4**	5.2 Scope Planning	**17.6**	5.2 Scope Planning	**0**
	5.3 Scope Definition	**67.6**	5.3 Scope Definition	**32.4**	5.3 Scope Definition	**0**
	6.1 Activity Definition	**41.2**	6.1 Activity Definition	**41.2**	6.1 Activity Definition	**17.6**
	6.2 Activity Sequencing	**20.6**	6.2 Activity Sequencing	**52.9**	6.2 Activity Sequencing	**26.5**
	6.3 Activity Duration Estimating	**11.8**	6.3 Activity Duration Estimating	**44.1**	6.3 Activity Duration Estimating	**44.1**
	6.4 Schedule Development	**20.6**	6.4 Schedule Development	**50**	6.4 Schedule Development	**29.4**
	7.1 Resource Planning	**55.9**	7.1 Resource Planning	**32.4**	7.1 Resource Planning	**11.7**
	7.2 Cost Estimating	**44.1**	7.2 Cost Estimating	**44.1**	7.2 Cost Estimating	**11.8**
	7.3 Cost Budgeting	**17.7**	7.3 Cost Budgeting	**64.7**	7.3 Cost Budgeting	**17.7**
	8.1 Quality Planning	**59.4**	8.1 Quality Planning	**37.5**	8.1 Quality Planning	**3.1**
	9.1 Organisational Planning	**23.5**	9.1 Organisational Planning	**47.1**	9.1 Organisational Planning	**29.4**
	9.2 Staff Acquisition	**20.6**	9.2 Staff Acquisition	**52.9**	9.2 Staff Acquisition	**26.5**
	10.1 Communication Planning	**35.3**	10.1 Communication Planning	**50**	10.1 Communication Planning	**14.7**
	11.1 Risk Management Planning	**61.8**	11.1 Risk Management Planning	**38.2**	11.1 Risk Management Planning	**0**
	11.2 Risk Identification	**67.6**	11.2 Risk Identification	**29.4**	11.2 Risk Identification	**3**
	11.3 Qualitative Risk Analysis	**26.5**	11.3 Qualitative Risk Analysis	**64.7**	11.3 Qualitative Risk Analysis	**8.8**
	11.4 Quantitative Risk Analysis	**20.6**	11.4 Quantitative Risk Analysis	**61.8**	11.4 Quantitative Risk Analysis	**17.6**
	11.5 Risk Response Planning	**35.3**	11.5 Risk Response Planning	**55.9**	11.5 Risk Response Planning	**8.8**
	12.1 Procurement Planning	**48.5**	12.1 Procurement Planning	**48.5**	12.1 Procurement Planning	**3**
	12.2 Solicitation Planning	**33.3**	12.2 Solicitation Planning	**60.6**	12.2 Solicitation Planning	**6.1**
3- EXECUTING	4.2 Project Plan Execution	**26.5**	4.2 Project Plan Execution	**67.6**	4.2 Project Plan Execution	**5.9**
	8.2 Quality Assurance	**30.3**	8.2 Quality Assurance	**63.6**	8.2 Quality Assurance	**6.1**
	9.3 Team Development	**20.6**	9.3 Team Development	**52.9**	9.3 Team Development	**26.5**
	10.2 Information Distribution	**32.6**	10.2 Information Distribution	**50**	10.2 Information Distribution	**17.6**
	12.3 Solicitation	**15.2**	12.3 Solicitation	**57.6**	12.3 Solicitation	**27.2**
	12.4 Source Selection	**33.3**	12.4 Source Selection	**51.5**	12.4 Source Selection	**15.2**
	12.5 Contract Administration	**21.2**	12.5 Contract Administration	**57.6**	12.5 Contract Administration	**21.2**

continued on following page

Table 2. Continued

	SELECTION A		SELECTION B		SELECTION C	
	PM Processes	%	PM Processes	%	PM Processes	%
4- CONTROLLING	4.3 Integrated Change Control	23.5	4.3 Integrated Change Control	70.6	4.3 Integrated Change Control	5.9
	5.4 Scope Verification	35.3	5.4 Scope Verification	55.9	5.4 Scope Verification	8.8
	5.5 Scope Change Control	29.4	5.5 Scope Change Control	70.6	5.5 Scope Change Control	0
	6.5 Schedule Control	11.8	6.5 Schedule Control	44.1	6.5 Schedule Control	44.1
	7.4 Cost Control	23.5	7.4 Cost Control	50	7.4 Cost Control	26.5
	8.3 Quality Control	30.3	8.3 Quality Control	45.5	8.3 Quality Control	24.2
	10.3 Performance Reporting	11.7	10.3 Performance Reporting	55.9	10.3 Performance Reporting	32.4
	11.6 Risk Monitoring & Control	15.2	11.6 Risk Monitoring & Control	66.7	11.6 Risk Monitoring & Control	18.1
5- CLOSING	10.4 Administrative Closure	5.8	10.4 Administrative Closure	47.1	10.4 Administrative Closure	47.1
	12.6 Contract Closeout	3	12.6 Contract Closeout	60.6	12.6 Contract Closeout	36.4

Control). In this case, the majority of choice was equally divided between (B) and (C) and never for (C) alone. This analysis underpins the research's findings of the great potential that exists within the processes to embrace sustainable development guidelines and therefore, enhance the overall performance of project management.

The respondents' feedback on each section of the questionnaire was very informative in many ways. This enhanced the outcome of the analysis of the responses. The responses were analyzed with a clear approach in mind which allowed the statistical analysis to represent, in an honest and fair way, the participants' point of views and professional feedback.

From the statistical analysis of the responses, it is evident that the respondents acknowledge the possibilities for integrating sustainable development into project management processes and support the positive impact this integration would have on the overall performance of PM within construction in specific and in all fields in general.

Significant majority of the participants identified several valid opportunities within PM processes listed in the PMBOK which have been identified earlier within the approach of this research. The outcome of the questionnaire's analysis has strengthened this research's hypothesis and identified the ways in which project management, as significant profession in the current modern age, should contribute to the global commitment for sustainable development within all fields. Therefore, since the PMBOK is considered as an international source of project management standards, a way forward for enhancing the PM pledge to embrace a sustainable approach to its practices, is to have sustainable criteria implications on the contents of the PMBOK. These implications have to be transferred and applied to the lists of PM processes and knowledge area. This would serve the whole PM profession with a document of standards that embraces sustainable development. In this case, changing PM practices to sustainable PM practices would have originated from the source out of which PM practices arise; the sustainable approach to PM standards.

Table 3. Summarizes the results taken from the respondents' answered questionnaires with a rating scale based on: A= significant sustainable outputs which enhance sustainable performance; B = considerable changes to outputs and guiding performance towards sustainability; C = resulting outputs have no modification or enhancement of sustainability

PM Knowledge Areas	PM Component Processes	Majority of Respondents chose: -	%
Project Integration Management	Project Plan Development	A	76.5
	Project Plan Execution	B	67.6
	Integrated Change Control	B	70.6
Project Scope Management	Initiation	A	82.4
	Scope Planning	A	82.4
	Scope Definition	A	67.6
	Scope Verification	B	55.9
	Scope Change Control	B	70.6
Project Time Management	Activity Definition	A/B	41.2
	Activity Sequencing	B	52.9
	Activity Duration Estimating	B/C	44.1
	Schedule Development	B	50
	Schedule Control	B/C	44.1
Project Cost Management	Resource Planning	A	55.9
	Cost Estimating	A/B	44.1
	Cost Budgeting	B	64.7
	Cost Control	B	50
Project Quality Management	Quality Planning	A	59.4
	Quality Assurance	B	63.6
	Quality Control	B	45.5
Project Human Resource Management	Organisational Planning	B	47.1
	Staff Acquisition	B	52.9
	Team Development	B	52.9
Project Communications Management	Communications Planning	B	50
	Information Distribution	B	50
	Performance Reporting	B	55.9
	Administrative Closure	B/C	47.1
Project Risk Management	Risk Management Planning	A	61.8
	Risk Identification	A	67.6
	Qualitative Risk Analysis	B	64.7
	Quantitative Risk Analysis	B	61.8
	Risk Response Planning	B	55.9
	Risk Monitoring & Control	B	66.7

continued on following page

Table 3. Continued

PM Knowledge Areas	PM Component Processes	Majority of Respondents chose: -	%
Project Procurement Management	Procurement Planning	A/B	48.5
	Solicitation Planning	B	60.6
	Solicitation	B	57.6
	Source Selection	B	51.5
	Contract Administration	B	57.6
	Contract Closeout	B	60.6

FUTURE RESEARCH DIRECTIONS

The built environment could benefit from the involvement of several disciplines in its creation. The construction industry is the main executer of the elements which define the built environment, but the involvement of sustainable development guidelines should not be limited to construction practices or the contribution of project management processes into construction projects, but future work should tackle the other elements which contribute to the built environment such as architecture, commerce, and manufacturing. Further research should look into the progress of these disciplines along the years to reveal different ways of improving their performances. This would ultimately result in a better quality of life which does not only rely on the impact of construction but an overall response from the built environment.

In the practice of construction projects, new case studies should be identified to demonstrate the potential benefits which exist in the implementation of sustainable PM guidelines. This research introduced and tested the desired change of SD guidelines into the standards of PM. This change was intended to be introduced to the policies and strategies of practice before heading to the real life projects. The research makes the initial step towards the implementation phase but it is no longer about enforcing new initiatives on the construction sector, and expecting stakeholders to justify their business cases within these regulations. Further research should be done to examine the practical implementation of this research's recommendation on real life case studies. Future work should concentrate on the identification of stakeholders/clients who would be interested in implementation such recommendations on different scales of projects (i.e. large, medium or small scale projects). This would add an additional dimension to the findings to reveal whether or not the scale of the impact of such change in policies would differ in the case of large scale projects from its impact in the case of small scale projects.

For developing countries, in the preliminary implementation of sustainable regulatory frameworks, there is a greater opportunity and a more efficient leverage point than in the case of developed economies. For the latter, it is a chance to reshape existing frameworks and disseminating such knowledge to the developing countries. The developing world, paradoxically, has greater prospects for effective change in the application of existing expertise through new frameworks, conceived with sustainability as a key integral component. This contrasts with the difficulties the western hemisphere has in grafting sustainable policy to entrenched methodologies.

CONCLUSION

This chapter has discussed the decision making level of current construction projects including project management practice. The involvement of project management in construction projects is fundamental, although not exclusive, to the successful completion of projects. When embraced from the initial phases of a project, project management standards ensure a systematic breakdown of all activities to fit into a comprehensive, detailed plan for the project as a whole, from initiation to completion and closure.

This research has demonstrated that project management, when embedded in construction projects, has the capacity to be a significant leverage point of such great influence that it is the cornerstone for rethinking the relationships between the three main elements of the integration.

The study has defined and examined the proposal of a tri-dimensional integration between construction, project management and sustainable development when approached from a systems thinking point of view. The main focus of the research is not only to explain the potential and the benefits of such integration. It also tested the criteria on which the integration is based and offers guidance concerning the practical aspects of construction, and project management practices, in the form of illustrative examples taken from current initiatives and practice standards.

REFERENCES

Argyris, C., & Schon, D. (1978). *Organizational learning: A theory of action perspective*. Reading, MA: Addison-Wesley Publication..

Berger, P., & Luckman, T. (1967). *The social construction of reality*. Hoboken, NJ: Anchor Publications..

CIB. (1999). *Agenda 21 on sustainable construction*. CIB Report Publication 237.

Cummings, T. (1980). *Systems theory of organizational development*. New York: Wiley Publications Ltd.

Davenport, T. (1993). *Process innovation: Reengineering work through information technology*. Boston, MA: Harvard Publications..

Eid, M. (2000). A review of project management & sustainable development for construction projects. *EAR Journal, 27*.

Eid, M. (2009). *Sustainable development & project management: Rethinking relationships in the construction industry, integrating sustainable development into project management processes*. Koln, Germany: Lambert Academic Publishing..

European Construction Industry Federation (FIEC). (n.d.). Retrieved June 2002, from http://www.fiec.org/main.html

Gardner, B., & Demello, S. (1993). Systems thinking in action. *The Healthcare Forum Journal, 36*(4). PMID:10127050.

Grant, R. (1996). Toward a knowledge based theory of the firm. *Strategic Management Journal*, 17.

Hammer, M., & Champy, J. (1993). *Reengineering the corporation: A manifesto for business revolution*. New York: Harper Business School Press. doi:10.1016/S0007-6813(05)80064-3.

James, P. (2000). *Business, eco-efficiency and sustainable development*. Lisbon: The European Commission..

Meadows, D. (1999). *Leverage points, places to intervene in a system*. New York: The Sustainability Institute..

Royal Commission on Environmental Pollution (RCEP). (2000). *Energy - The changing climate*. London: RCEP..

Senge, P. (1999). *The fifth discipline field book, strategies and tools for building a learning organization*. London: Nicholas Braeley Publishing Limited..

Senge, P. (1999). *The fifth discipline, the art & practice of the learning organization*. London: Randome House Business Books..

Sneider, K. F., & Nissen, M. E. (2003). Beyond the body of knowledge: Knowledge-flow approach to project management theory and practice. *Project Management Journal, 34*(2).

Spender, J. (1996). Making knowledge the basis of a dynamic theory of the firm. *Strategic Management Journal*, 17.

The U. K. Government. (2003). Sustainable development fact sheets: The UK government's policy for SD for UK businesses. London: Her Majesty's Stationery Office (DETR).

The Commission of the European Parliament. (2002). *The competitiveness of the construction industry*. Brussels: European Union..

The Construction Task Force. (1998). *Rethinking construction*. London: DETR..

The European Union Online. Europa. (n.d.). *The European commission*. Retrieved June 2002, from http://europa.eu.int/comm/enterprise/construction/it/connet/connet.htm

Wideman, R. M. (1995). Criteria for a project management body of knowledge. *International Journal of Project Management*, *13*(2), 72. doi:10.1016/0263-7863(94)00020-D.

Wirth, I., & Tryloff, D. (1995). Preliminary comparison of six efforts to document the project management body of knowledge. *International Journal of Project Management*, *13*, 109–118. doi:10.1016/0263-7863(95)00003-9.

World Business Council for Sustainable Development. (2002). *The business case for sustainable development, making a difference toward the Johannesburg summit 2002 and beyond*. Switzerland: WBCSD..

World Business Council for Sustainable Development (WBCSD). (2002). *The business case for sustainable development, making a difference toward the Johannesburg summit 2002 and beyond*. Switzerland: WBCSD..

ENDNOTES

1 This chapter is based on an existing published PhD Thesis by the same author; the Thesis was concluded in 2004 at the University of Edinburgh, Scotland, UK. The book was published in 2009 by Lambert Academic Publishing AG & Co. KG and licensors. ISBN: 978-3-8383-1578-2

This chapter was previously published in Sustainability Integration for Effective Project Management edited by Gilbert Silvius and Jennifer Tharp, pages 76-94, copyright 2013 by Business Science Reference (an imprint of IGI Global).

Chapter 67
Conceptualization of the Base for Sustainable Rural Development Policy Implementation

Njegovan Zoran
University of Novi Sad, Serbia

Olgica Bošković
University of Belgrade, Serbia

ABSTRACT

Looking in wider perspective, the problems of inequality have emerged relatively early even from the period of establishment of economy as a scientific discipline. However, those problems are also the subject of different socio-humanistic scientific research. That process lasts up to date, and it could be said that nowadays it is much more important than before. The main reason for that is that current inequalities are significant and radicalised nowadays as a cause of non-adequate development process.

NONBALLANCED DEVELOPMENT: GLOBAL CONSTRAINTS

Rural Development theory represents the reaction on unequally and poverty that was caused by intensive development at the global level. But, as any other theory, it must be related to the fact that "Scientific paradigm sometimes might isolate the society from important problems that could be reduced in parts of the whole scientific theory mosaic, because it may be expressed through conceptual and instrumental meanings of the paradigm itself" (paraphrased from Kuhn, 1962). As a result from that usually appears disorientation in development and often negative effects that create poor conditions for perspective rising of its own performances.

DOI: 10.4018/978-1-4666-4852-4.ch067

Some authors like Sir Arthur Lewis who has formulated so called dual-sector model put the pressure on that problem in the second half of XX Century (1954 and 1978). In the model, he focused attention on surpluses of agricultural labour force and on permanent growth of food producer's productivity as well domestic demand. In continuation, author Mellor, J., pointed out the inferior position of rural sector (1976). He was writing about and trying to specify factors that influence such situation in agriculture and that are connected with that process. Also he tried to specify the role of the factors in the process of creation employment and income of rural population. Later on, author Meier, G. (1989) has tried to formulate the common opinion that agriculture is not only the source of supplying the industry with its surplus of labour, but that it is a dynamic factor of growth, employment and better distribution of income. In addition, Michael Lipton (1977) wrote about rural bias in the economic development process and like some other authors has described the territorial inequalities particularly in poverty i.e. wellbeing between urban and rural areas. The main argument he took is the conflict that stands in theory between capital and labour as well as domestic and international interest in that field.

The importance of rural development and rural development needs as a part of overall economic processes continues to be the main topic of different authors also during the last decade of XX Century. Many of them like Duncan and Howell (1992) as well Sahn at all (1997) had tried to articulate its research on potential effects of programs for adapting the development policies in order to lower the rural poverty caused by terms of trade. But at the same time Sahn, D. et al (2003) reminds that promoting export oriented agriculture in case of less developed countries, may result with lowering of its food security. Further on, Brayden at all (2001-2004) were explaining the importance and the role of particular segments like intersectoral connections and stressed the need for interaction of public and private partnership. In connection with that Porter, M. (1990) has pointed out the importance of social and cultural environment as a factor of improvement of competition. Besides that, many other authors and/or international organizations had tried to improve the theory and practice in the field of rural development and to make its own contribution in the field of rural development and poverty reduction.

And while the discussions on rural development and its role in overall economic development theory are expanding, it must be said that there is a huge number of opinions and actions that put great pressure and concern on the permanent growth of poverty and malnutrition also in urban zones (Haddad et al, 1999). They are usually focused on rural urban gaps in the cases when urban poverty exceeds (von Brown, 1993), as well as on urban poverty and malnutrition (Maxwell, 1998). Those are very important problems, but our opinion is that launching such problems as a main ones we are refocusing ourselves from rural problems toward overall poverty problems confirming in that way the inability to solve rural problems in less developed countries.

Rural development represents alternative development concept itroduced to solve inequality caused by the process of world globalization and huge polarization that happens also at the local level. Promising XX century doesn't help much to the removing of unequality. In spite of that the unequality of the countries has expanded immensly in the last few decades. At the beginning of the XXI century it could be recorded that 5% of the whealthies population is in the possession of the one third of world income like 80% of the poorest population (Milanović, 2006). Such inequality of the world poor is standing against neoliberal opinion that is expressing fast growth of the poor countries (Washington Consensus) influencing further progress of globalization process in the conditions of nonadequat redistribution of income. It means that it is not possible to register any significant global development on the sustainable basis using relatively poor and inadequate adoptions to the existing conditions.

RURAL DEVELOPMENT OBSTACLES IN THE REPUBLIC OF SERBIA

The problems of regional and rural differences in the Republic of Serbia and AP Vojvodina are present for more than fifty years (Zakić & Njegovan, 1995; Bogdanov, 2007; Nikolić, & Živanović, 2008; Njegovan, & Pejanović, 2009; etc.). The general feature and the focus of different authors were mainly articulated toward examination of actual situation and problems on one side and monitoring of the theory and practice of developed and developing countries on other. The majority of the work was also directed toward creation of as efficient as possible rural development policy, but all the time there was a lack for some general preconditions. First of all there were no government orientation toward real decentralization of the country and improvement of territorial base that is out of political frame and that will promote economic units similar to EU NUTS classification. (EU, 2003). The practice of OECD, EU, IBRD, and FAO as well as domestic sources and literature, are directing the activities toward definition of rural areas and territorial differentiation. This territorial differentiation in the Republic of Serbia is recent. But actual districts still doesn't serve for rural development needs. That causes another problem of lack of formal and informal institutions as a tool for further and much more efficient rural development. Current problems could be described as structural by nature, with immense institutional imperfection, demographic retrograde processes and different material limitations.

In general, districts in the Republic of Serbia were introduced by the political power that was manifested in the actual conditions within the country. That way the districts were introduced as a way for cooperation and control between centre and periphery. Thus the situation in the Republic of Serbia could be seen as a sort of abstraction that is a part of political willingness to introduce a territorial vs. central sectoral concept of the society. The huge number of regional and rural problems is created politically. The results of such approach were poor so new concept of integrated and sustainable rural policy must be created and introduced to promote so called socio-economic model (Table 1).

The base and precondition for such approach relates to the existing territorial differentiation of Serbia. That is going in hand with the latest promoted orientation of the country to start the process of decentralization and implementation of rural policies based upon the theory and practice of EU as well some OECD countries. Existing international methodologies are representing the possible way for implementation of the best practices and the creation of the paradigm for the focusing of the further efforts that needs to be put in rural development policy creation and implementation.

EXPERIENCES IN CREATING METHODOLOGYFOR MEASURING RURAL DEVELOPMENT

The term rural is used to describe areas of land with a relatively low population density or with specific socio-economic characteristics. OECD methodology provides the only internationally recognized definition of rural areas, based on

Table 1. New rural policy frame

Rural Policy Levels	**Policy Instruments and Measures**			
	General		**Specific**	
Internationsl	↓	↓	↓	↓
National	↓	↓	↓	↓
Regional	↓	↓	↓	↓
Municipal	↓	↓	↓	↓
Community Level	↓	↓	↓	↓
Types of Regions	Extreme Rural	Rural	Partly Rural	Urban

Source: Authors.

two hierarchical levels of geographic units: local community and regions.

Territorial network at the local community level is detailed and consists of basic administrative or statistical units. Rural analysis is normally based on these local units (usually municipalities), as it applies to characteristics of "homogenous" areas that can either be classified as rural or urban.

Territorial network at the regional level is less detailed and consists of larger administrative units or functional areas, such as districts (administrative counties) or regions. At this level, rural analysis is aimed at functional relationships and a broader context of the rural development. Issues at this level are usually managed by State Administrations and International Organizations. Regions can normally be identified only as more or less rural.

Taking the two aforementioned geographic units as a starting point, OECD methodology defines the term rural in two steps. First step - territorial units are defined as rural if their population density is under 150 inhabitants per square kilometre. Second step - territorial units are classified in one of the 3 categories: distinctly (predominantly) rural area (if more than 50% of the population is living in rural local units), partially (intermediately) rural area (if 15% to 50% of the population is living in rural local units) and distinctly (predominantly) urban area (if less than 15% of the population is living in rural local units). In addition to the main criterion (regional rural population share), since 2005, a secondary criterion – size of an urban centre has been in use. If there is an urban centre in the region with more than 200.000 inhabitants, representing no less than 25% of the regional population it is classified as intermediate (partially) rural region. If there is an urban centre in the region with more than 500.000 inhabitants, representing no less than 25% of regional population it is classified as predominantly urban region. An urban centre is defined as a local unit with a population density above 150 inhabitants per square kilometre and total population above 200.000 inhabitants.

European Union is using Eurostat practice (European Union Statistic Institute) where all regional data refers to NUTS (Nomenclature of territorial units for statistics), which is a coherent system for division of territory for the purpose of gathering, compiling and disseminating harmonised statistics at a regional level. This nomenclature has divided the EU territory into territorial units, using single criteria.

According to the NUTS criteria, average size of region (measured by size of population) in one country at the NUTS1 level should be between 3.000.000 and 7.000.000 inhabitants, at the NUTS2 level – between 800.000 and 3.000.000 inhabitants and at the NUTS3 level – between 150.000 and 800.000 inhabitants. If a country does not have appropriately sized administrative units, in accordance with stated criteria, smaller administrative units will be merged to form the NUTS sized region.

It has been suggested in previous Eurostat working papers, that the NUTS3 level regions, whose population density is under 200 inhabitants per square kilometre, should be considered rural. More recently, for example in the Rural Development Program 2007-2013 and the rural development report for 2007, due to the lack of a single internationally recognized definition, European Union used the OECD definition of rural areas.

National practice for defining and establishing rural areas is pretty diverse. In many countries, the official definition of rural does not exist, and where they do exist, definitions defer significantly.

Principle territorial units that are used in national definitions are pronouncedly different in size, population and area size. For example, the communes in France (the smallest administrative units in Europe), have, on average, 1.500 inhabitants and 15 square kilometres of area size, while the districts in England have, on average, 118.000 inhabitants and more than 500 square kilometres in area size.

Analysis of various national practices lead to three major conclusions:

1. Rural cannot be defined as an urban residual,
2. Number of inhabitants isn't a sufficient criterion to describe the rural, and
3. Statistics that didn't originate from commonly accepted definitions tend to produce very inconsistent results.

Definition and use of indicators has been initiated by the increased interest in sustainable development that put those indicators into focus of the scientific and professional public. Apart from the general dimensions of sustainable development (demographic, economic, social, and physical environment), attention has been directed at specific area indicators, such as rural development. There are several different approaches in defining the term indicator: The First approach, the widest and the most common approach is that indicators are simply the combination of two or more statistical data (results), through an established procedure (defined algorithm), to form a new derived measure. The Second approach, a somewhat more narrow approach to the term indicator, uses a normative interpretation, with the possibility to establish rankings or league tables if applied to comparison of regions, countries or other appropriate units for the same period. The third approach concerns those indicators that cover only a part of the target concept, which remains unmeasured within the framework of official statistics. The fourth approach concerns the so-called synthetic indicators. These are composite measures of individual indicators designed to give an easy overview of a complex array of indicators or statistics. Different methods of weighting can be used.

ESTABLISHING OF THE METHODOLOGY FOR THE RURAL AREA OF SERBIA DEFINITION

It should be pointed out that it is very difficult in practice to summarily and comprehensively assess the level of economic and geographical development of the Serbian territory and to encompass it with single standardized methodology. This is affected by both natural and socio-economic allocation factors. Over time, their influence in certain cases caused regionalisation that has elements of a more static approach. This is why the former classifications were limited to selecting features that are equal in all points of the observed area and creating homogenous agricultural – rural regions. However, in this case, the selection of basic rural regions was based on characterization of territory from the aspect of function and existing relations that were established between the features in the selected area. It provides far dynamic aspect of the rural – agricultural region and is more applicable from the aspect of social and economic objectives.

In a technical sense, as noted by the proposed methodology in previous chapters, this approach represents an attempt at synthesis of ecological, technical/technological, economic and social characteristics of a designated territory and fully takes into account the significance of agriculture as a dominant sector and economic activity that affected the lifestyle of the population in rural areas for a long time. We also took care that the said approach for area classification upholds the integral rural development principle.

At the same time, we cannot forget the fact that the definition of the new concept of regional – rural development is tied to the objective that is to be achieved - a successful sustainable rural development policy. This is why the proposed methodology framework for regional classification of rural economic areas indicates the possibility of demarcation, which uses an analytic or a synthetic method.

PROBLEMS WITH TERRITORIAL ORGANISATION IN THE REPUBLIC OF SERBIA

When discussing rural areas and rural development policy, it is important to mention that EU and OECD practice is mainly related to defining rural areas. That is certainly a prerequisite for adequate territorial demarcation and for implementation of their methodology.

To this effect, the Nomenclature of territorial units for statistics was adopted in July 2010, via the Decree of the Serbian Government, according to which, the Republic of Serbia is divided into two units on the NUTS1 level: Northern Serbia (Vojvodina region and Belgrade region) and Southern Serbia (Sumadija and Western Serbia region, Southern and Eastern Serbia region and Kosovo and Metohija region). NUTS2 level consists of five regions which constitute Northern and Southern Serbia. At the NUTS3 level, Serbia is divided into 30 districts – the so called administrative counties. Belgrade region consists of the Belgrade district (city of Belgrade). Vojvodina region has seven districts: Zapadnobacka, Juznobanatska, Juznobacka, Severnobanatska, Severnobacka, Srednjebanatska and Sremska district. Sumadija and Western Serbia region consists of Zlatiborska, Kolubarska, Macvanska, Moravicka, Pomoravska, Rasinska, Raska and Sumadijska districts. Southern and Eastern Serbia region consists of: Borska, Branicevska, Zajecarska, Jablanicka, Nisavska, Pirotska, Podunavska, Pcinjska and Toplicka districts. Kosovo and Metohija region consists of Kosovska, Kosovsko-mitrovacka, Kosovsko-pomoravska, Pecka and Prizrenska districts. Further division into lower NUTS levels is at the country's discretion, hence the NUTS 4 level has 194 municipalities and the NUTS 5 level consists of 6169 settlements.

Since nomenclature of territorial units for statistics has been established at the NUTS 3 level, districts (administrative counties) act as regions at this level in the Republic of Serbia.

In order to enable comparison, not only between rural areas, but also between rural and urban areas, it is recommended for the system of indicators to cover not only rural areas, but the entire territory. Hence, the NUTS 2 and NUTS 3 classification level indicators will be presented here.

IDENTIFICATION OF THE INDICATORS FOR CONDUCTING OF RURAL POLICY

A list of indicators, that can be used for initial analysis of rural areas in the Republic of Serbia and later monitoring of the development process, has been made, based on available statistics. These indicators have been classified, by subjects, into three groups. These are socio-economic indicators, agriculture sector indicators and quality of life and well-being indicators.

What follows is a review of the most important indicators of rural development on the level of geographical regions (NUTS 2) in the Republic of Serbia, as shown in Tables 2, 3, and 4.

From the aspect of socio-economic indicators in Serbia, the Belgrade region has a 4 ranking for 5 of the 6 indicators and it is the only urban region with 509 inhabitants per km^2, while based on the same indicators, the Southern and Eastern Serbia region is the most rural region in Serbia (and has a 1 ranking for all indicators).

From the aspect of agriculture indicators, the Belgrade region has a 4 ranking for 3 of the 4 indicators, while based on the same indicators, the Southern and Eastern Serbia region is the most rural region in Serbia (has a 1 ranking for 2 indicators, and a 2 ranking for the other 2) when compared to the other regions.

From the aspect of quality of life and well-being indicators, the Belgrade region has a 4 ranking for 3 out of 4 indicators, while the other regions have approximately even rankings, on average.

CREATING OF A SYNTHETIC INDICATOR

There are two questions posed when defining synthetic indicators: 1) selection of indicators that will make up the synthetic indicator and 2) how to aggregate the chosen indicators into one com-

Table 2. Values and rankings by the regional socio-economic indicators

Region/ Indicator	Vojvodina		Belgrade		Šumadija and Western Serbia		Southern and Eastern Serbia	
		Rank		Rank		Rank		Rank
GDP per capita(in 000)	366	3	690	4	275	2	243	1
Population Density	92	3	509	4	78	2	64	1
Index of Dependency	50,2	3	39,2	4	54,2	2	61,2	1
Indexof Maturity	102,7	4	113,3	2	107,1	3	120,1	1
No of population educated in secondary and high schools as well faculty - in %	53,4	3	70,2	4	46,9	3	40,0	1
Average salary (total in RSD)	31203	3	39862	4	26601	3	26085	1

Source: authors.

Table 3. Values and rankings by the agriculture indicators

Region/ Indicator	Vojvodina		Belgrade		Šumadija and Western Serbia		Southern and Eastern Serbia	
		Rank		Rank		Rank		Rank
Population per km^2 agricultural land	111	1	748	4	129	3	111	2
Portion of agricultural population in total working contingent	7,3	3	1,8	4	15,7	1	12,8	2
Maturity index of agricultural population	1,13	4	1,32	2	1,27	3	1,33	1
Average salary in agricultural sector	27727	3	37958	4	22207	3	20056	1

Source: Authors.

Table 4. Values and rankings by the quality of life and well-being indicators

Region/Indicator	Vojvodina		Belgrade		Šumadija and Western Serbia		Southern and Eastern Serbia	
		Rank		Rank		Rank		Rank
Lengths of roads - km at km^2	0,275	1	1,811	4	0,697	3	0,509	2
No of phone payers per 1000 inhabitants	393	3	504	4	376	2	349	1
No of beds in tourism per 1000 inhabitants	6,7	1	9.9	2	19,3	4	18,1	3
No of inhabitants per one doctor	410	3	272	4	415	1	365	2

Source: Authors.

posite measuring unit. Thirteen aforementioned indicators were selected for the attempt to create a synthetic indicator of rural area development in the Republic of Serbia. However, the bigger problem is how to aggregate the chosen indicators into one composite measure of rural development. Obviously, the selected indicators are not expressed in the same measurement units; hence, a simple method of aggregation does not exist. On the other hand, it is necessary to ensure that the synthetic indicator is expressed as a measure that will be understandable and suitable for further analysis.

This is why, instead of the nominal value of individual chosen indicators, the average ranking of regional rural development for each chosen indicator has been selected as the starting point to create a composite measure for rural development.

Considering that Serbia has 4 regions, they will be ranked from 1 to 4 and an average regional ranking will be 2.5. Calculation will provide the results shown in the following tables. However, the assumption that all selected indicators have the same significance in defining the aggregate measure is the weakness of this calculation of average ranking of the municipal rural development. Therefore, a synthetic indicator based on the weighted arithmetic mean of chosen indicators would better reflect the overall level of rural development. The weights for each indicator determine its significance (involvement) in forming the synthetic indicator. This is the approach taken in calculating the synthetic indicator – called the composite indicator for relative ranking of rural development of municipalities in Vojvodina.

Since there are no empiric weights to determine the significance for each of the chosen indicators, the weights have been defined subjectively by the author's expertise. Overall synthetic indicator is calculated using the following formula:

$$I = \sum_{k=1}^{n} w_k i_k$$

where $\sum w_k = 1$, w_k are the so-called weights (impact of each indicator on the composite indicator), and i_k are the selected indicators for the calculation of the composite indicator, k=1,2,3,... n.

Value of the composite indicator for the relative ranking of rural development of regions and areas in the Republic of Serbia can be between 1 and 4 for the regions, or between 1 and 25 for the districts. This type of calculation provides somewhat different and more realistic results when compared to the average unweight ranking of rural development. Table 5 contains the values for the composite indicator for relative ranking of regions and districts (administrative counties).

Southern and Eastern Serbia region has the lowest composite ranking - 1.45 and is considered to be the most rural region in the Republic of Serbia. Only the city of Belgrade with a 3.70 composite ranking is considered an urban region. The other two regions are considered rural.

As for the districts (administrative counties), shown in Table 6, Zajecarska, Branicevska, Pirotska, Toplicka and Borska districts have the lowest composite indicators. The city of Belgrade and Juznobacka district have the biggest average ranking and are considered to be developed rural areas. The span of the composite indicator for relative ranking is 16.55, and for the average ranking it is 13.70; Severnobanatska, Srednjebanatska, Juznobanatska and Zapadnobacka districts are below the average ranking of 4 and they are considered to be more rural than the Sremska and Severnobacka districts which have a ranking between 4 and 5 and are considered rural.

On the basis of analysis and calculation, the following can be concluded:

1. Based on criteria for determining how rural the region is, only the City of Belgrade is considered non-rural, while all other regions

Table 5. Composite indicator of the rural development relative rankings

Region	Vojvodina	Belgrade	Šumadija and Western Serbia	Southern and Eastern Serbia
Composite rankings	2,55	3,70	2,30	1,45
Rankings of regions by the composite indicator	3	4	2	1

Source: Authors.

Table 6. Composite Indicator of the rural development relative rankings

Area – County	Composite Rankings	Rank
Zaječarski	6,25	1
Braničevski	7,00	2
Pirotski	8,05	3
Toplički	8,15	4
Borski	8,75	5
Kolubarski	9,60	6
Pčinjski	10,60	7
Zlatiborski	10,90	8
Jablanicki	11,15	9
Severnobanatski	11,95	10
Srednjebanatski	12,10	11
Pomoravski	12,35	13
Rasinski	12,90	14
Južnobanatski	13,20	15
Mačvanski	13,35	16
Zapadnobački	14,95	17
Sremski	15,40	18
Raški	15,50	19
Severnobački	15,75	20
Podunavski	16,55	21
Nišavski	17,05	22
Šumadijski	17,20	23
Južnobački	21,15	24
Grad Beograd	22,80	25

Source: Authors.

(Vojvodina, Sumadija and Western Serbia and Southern and Eastern Serbia) are considered rural.

2. As for the districts, when the population density criteria is applied, only three districts are non-rural (the City of Belgrade, Podunavska and Juznobacka district), while the other ones (administrative counties) are rural.
3. Based on available county level statistics for certain areas (demographic, employment and salary statistics as well as agricultural data), rural development indicators have been selected and grouped together as socio-economic indicators, agricultural sector indicators and quality of life and well-being indicators.
   The calculated figures of the most significant selected indicators are shown for the regional and district levels and ranked according to their values.
4. Rural development indicators provide an overall picture of the state of rural development of districts in the Republic of Serbia and represent primary elements for conducting rural development policy.

PRINCIPLES OF RURAL POLICY

Modern socio-economic development is full of examples where terms of exchange between agriculture and non-agricultural sectors became very complex. As a consequence, one side was impoverished and the other experienced tremendous growth, which lead to a critical point in global and national politics. These types of events usually lead to redistribution of revenue and affect both growth and development. Since agriculture is by definition tied to rural areas, this attitude towards rural areas gave them serious trouble and limited further development. This mostly refers to the decrease of both quantity and quality of the workforce. As a result, instead of having a policy of balanced rural development, creators of economic policy now have an additional task to decrease poverty and, kind of paradoxically, to eliminate hunger in rural areas. These kinds of activities require more focus on rural economy, above all the following aspects:

- Narrow field of development and improvement to the agricultural sector.
- Involving the State in promoting rural development as part of a new development concept.

- Insertion into wider international community and establishing new development strategies.
- Introduction of issues that have long been ignored, such as land reform, labour issues, gender equality, policy of human well-being, environmental policy etc.

In this sense, many authors are pointing out that old issues are being raised in a new way (Njegovan, & Pejanovic, 2009). It is a fact that there's no place in contemporary society for the approach that has been used for decades, where the development of agriculture as a one-dimensional process used to represent a synonym for development of villages and rural community – the rural development.

We should not have to emphasize that rural development is a complex field that includes, besides the agricultural development policy, other sector policies, structural and social policies, as well as all types of local initiatives that contribute to the development of rural community. That is why it's necessary to frame a sensible, efficient and, above all, sustainable concept of rural development that has to be in concordance with its own development capacities, regardless of whether national resources or foreign financial resources are used.

Rural economy should represent an integral, rounded economy consisting of an ensemble of inter-connected activities. New living needs, new habits, new world views and a new social life are more and more forcing the inhabitants of rural areas to adopt new ways of living and working. They are in a situation where they can participate in the development process and gradually find economic and even political means of struggle, although it is important to note that in these processes, the role of the State is crucial, whether it is on the centralized, regional, or local level.

THE BASE FOR SUSTAINABLE RURAL DEVELOPMENTPOLICY FORMULATION

Postulates for building an efficient and sustainable concept of rural development and implementation of appropriate policy, can be grouped into several basic ones as follows:

1. The approach to the rural development has its own limitations in the current development which is based on a sectoral instead of an integral concept, which is obvious from numerous uncoordinated activities of the competent Ministries and other state institutions.
2. Legal interregnum–single regulations that will regulate the issues of rural development still have not been enacted.
3. Lack of a clear, complete, and legally based institutional framework for conducting rural policy.
4. The decentralization process, which is tied to regional policy and implementation of the subsidiarity principle, is late.

Inadequate monitoring and valuation of investments into rural development, directing the state stimulus packages to undeveloped areas, insufficient transparency and lack of coordination, all represent additional limitation to measuring the effects of state activities (through investments by the Development Funds, competent Ministries and the secretariat, other funds and Directorates). Therefore, the formulation and implementation of the rural development policy must connect with the accumulated problems that emerged as a consequence of long-term implementation of inadequate development policies (Pejanović,

Njegovan, & Tica, 2007). Consequently, it is to be expected that rural development policy will not be able to solve these accrued problems in a short period of time.

For a long time, rural development policy represented the sectoral policy with agriculture as the main focus. However, rural policy represents a territorial concept, and a regionalization of rural areas is needed on the basis of which individual policies or integrated policy would be differentiated.

The formulated options for the reform of the rural development management and planning system contain many options. The ones that have been elaborated the most and include research, at least in its rudimentary form, represent the connection between alternative models and various "future scenarios."

The backbone of the favourable scenarios consists of:

- Building a modern and democratic ("developing") post-socialist state.
- Creating democratic regional and local coalitions.
- A developed private sector and cooperation ("partnership") between all forms of ownership.
- Developed institutions and social market economy mechanisms.

On the other hand, unfavourable scenarios consist of:

- Prolonged existence of the "pre-modern" state.
- Weak regional coalitions.
- Growing of the "wild/untamed" market.
- Conducting one-sided and unfair privatizations.

Instruments of rural policy to achieve goals could be systemized as follows:

- Regulatory
    - Drafting and adopting a new Constitution, which will pay special attention to decentralization and minority rights.
    - Enacting a Law on rural regionalization and rural development – with clear separation of development responsibilities on individual levels of governance, a Law on public finances, a Law on investment as well as accompanying Laws and regulations that would address the new situation.
    - Enact regulations that will deal with issues of rural development on all levels (national, regional, county level, municipalities, etc.).
    - Adjust and harmonize all strategic documents on all levels.
- Institutional
    - Reform of the Provincial government and local self-government.
    - Give more unequivocal significance to the Ministry of agriculture in relation to rural development and rural decentralization and accompanying institutions at all levels.
    - Form development institutions and development agencies on all levels.
- Economic/Financial
    - Reform or better structuring of the Ministry of finance to go hand in hand with the degree of centralization/decentralization of public revenue and expenditure, with the needs of rural development in mind;
    - Reform the system of financing and investment;
    - Strengthen existing and form new development and financial institutions on all levels.
- Education and public information
    - Improve the education and training system.

- ◦ Improve the system of technical assistance, advisory departments, etc.
- ◦ Improve the information system.

Rural development policy must take into account the fact that rural areas are not only a problem. They also provide the opportunity to contribute to the overall competitiveness of the entire economy through e.g. growth of micro companies and women's companies. Providing infrastructure (transportation, telecommunication, water supply and sewer, energy and gas, etc.) and stimulating the development of human – social capital will ensure positive effects for rural development. Initiative for rural development doesn't have to come exclusively from the central level. Central level should ensure more flexible arrangements for rural development, but the basic focus should, above all, be on the investments into human and social capital, diversification of economic activities in entrepreneurial segments, innovation, environment, research and development as well as specialized regional institutions and centres.

In the process of building a concept of rural development and implementing mapped out goals, it is necessary to define rural area as well as indicators of rurality at various levels (national, regional and municipal). After that, it is necessary to monitor the development dynamics and establish differentiated development policies related to individual problems that appear in the development.

Basic areas of intervention for rural development policy are as follows:

1. Population and employment – rural development motor.
2. Decentralization trends – movement for rural decentralization.
3. Entrepreneurship and creation of jobs in rural areas.
4. Are manufacturing and services the pillars of rural development?
5. Business cooperation with the industrial sector.
6. Relationship between industrial structures and characteristics of rural areas and urban centres.
7. Sectoral mix and territorial dynamics.
8. Education and employment in rural areas.
9. Role of tourism and agriculture.
10. Significance of communication.
11. Significance of information technology for development.
12. Service standards in rural areas.

It is especially important to note that interregional cooperation in all aspects from legal and institutional to economic and environmental represents a new direction in rural development. That is why it is so important to develop coordinating policy models. Such orientation is supported by the EU and OECD through various programs. Prominent measures for less favoured areas that were proposed and adopted on the basis of joint actions of countries and regions are: afforestation, training, other forestry measures, investments in agricultural farms and households, early retirement, improvement in processing and marketing of agricultural products, diversification of agricultural activities. Less successful measures are financial engineering and renewal of agricultural potential.

Basic measures of development and research have the following share:

- Agro-ecological payments.
- Modernization of agricultural farms.
- Encouraging young farmers to become independent.
- Creating added value in agricultural and forestry products.
- Payments to farmers in handicapped areas.
- Training and information activities.
- Implementation of local development strategies.

- Farmer participation in projects for improvement of food quality.
- Technical assistance.
- Diversification into non-agricultural activities.
- Conservation and improvement of rural heritage.
- Development and creation of new businesse.
- Using advisory service.
- Payments for Natura 2000 program and other Directive related ones.

This kind of layout is very important in order to get familiar with EU policies that we have to follow and based on which we ensure our participation in projects and programs as well as pre-accession programs. On the other hand, it is instructive enough and ensures and provides a basis that can potentially be exploited, bearing in mind the state of rural development as well as the problems, limitations and chances for a more efficient rural development. Of course, this base is also significant for the second and further stages of this project, considering that in the upcoming years, rural development policy will be created on the basis of regionalization of rural areas by various indicators.

After analysing the rural development management process, we can say that it represents a complex and multidimensional process. As a proactive process that doesn't relativize the consequences of the former development, it includes the identification of opportunities and resources that can change the unfavourable present image of the development of less favoured areas. It is especially important to emphasize the significance of three very important aspects: obtain/create favourable development opportunities, ensure the process of innovation in comparison to the existing state and finally, ensure economic growth.

Creators and implementers of rural policy must bear in mind, when conceptualizing, designing and implementing rural development policy, that the State will be the one to create the needed environment and promote orientation to a certain development concept. Regional and local participation is crucial. However, a certain level of decentralization is required to do that, as well as setting the development policy at the regional and local level. In that process, the State will be the one to prescribe the rules, prepare lower levels of governance, which never had such powers, to quickly and efficiently organise in order to create and set the tempo for their own development. They have to seek change, react to it – monitor it, evaluate and use it.

It can be concluded that, above all, success lies in improving the foundation for the development breakthrough. Also, we can conclude that unwanted processes might occur and that there is a need to monitor and alleviate potential failures.

REFERENCES

Atkinson, G. et al. (1997). *Measuring sustainable development – Macroeconomics and the environment*. Cheltenham, UK: Edward Elgar..

Bogdanov, N. (2007). *Small rural husbandries in Serbia and rural non-agriculturaleconomy. Belgrade, Serbia: Ministry of Agriculture*. Belgrade: Forestry and Water-management of Republic of Serbia..

Bryden, J. (2001). Rural development. In *Landsis g.e.i.e*. Luxembourg: Proposal on Agri-Environmental Indicators PAIS..

Bryden, J. M., Copus, A., & MacLeod, M. (2002). Rural development indicators. Report of the PAIS project, Phase 1. Report for EUROSTAT with LANDSIS. Luxembourg..

Duncan, A., & Howell, J. (1992). *Structural adjustment and the African farmer*. Portsmouth, NH: Heinemann Educational Books..

EU. (2003). *Nomenclature of units for territorial statistics*. Brussels, Belgium: Local Administrative Units, European Office for Statistics..

EU Directorate-General for Agriculture and Rural Development. (2007). *Rural development in the EU*. Brussels, Belgium: Statistical and Economic Information, Brussels..

European Commission. (2002). *Guidelines for the mid-term evaluation of rural development programs 2000-2006*. Supported from the European Agricultural Guidance and Guarantee Fund..

Filipović, M., & Vujošević, M. (2008). New generation of documents in sustainable development and lessons for Serbia. Belgrade, Serbia: Faculty of Economics..

Haddad, L. et al. (1999). Are urban poverty and under-nutrition growing? Some newly assembled evidence. *World Development*, *27*(11), 1891–1904. doi:10.1016/S0305-750X(99)00093-5.

Heideman, C. (1992). *Regional planning: A reader*. Karlsruhe, Germany: University of Karlsruhe..

Janković, D. (2007). Importance of social capital in local rural communities' development. In *Proceedings of Matica srpskafor social sciences*. Novi Sad: Serbia..

Kinnear, P., & Gray, C. D. (1994). *SPSS for windows made simple. Hove, UK*. UK: Lawrence Erlbaum Associates Publishers..

Kuhn, T. (1962). *The structure of scientific revolutions* (3rd ed.). Chicago: University of Chicago Press..

Lewis, A. (1954). *Economic development with unlimited supplies of labour*. Manchester, UK: The Manchester School of Economics and Social Sciences. doi:10.1111/j.1467-9957.1954.tb00021.x.

Lipton, M. (1977). *Why poor people stay poor: Urban bias in world development*. Cambridge, MA: Harvard University Press..

Maxwell, D. (1998). *The political economy in sub-Saharan Africa*. FCND Discussion Paper 41. Washington, DC: International Food Policy Research Institute.

Meier, G. (1989). *Leading issues in economics development* (5th ed.). New York: Oxford University Press..

Mellor, J. (1976). *The new economics of growth: A strategy for India and the developing world*. Ithaca, NY: Cornell University Press..

Milanović, B. (2006). *Separated worlds – Measurement of international and global inequality*. Belgrade, Serbia: UNDP.

Nikolić, M., & Živanović, J. (2008). *Indicators of rurality and developmental chances of local communities*. Belgrade, Serbia: IAUS.

Njegovan, Z., & Pejanović, R. (2009). *Rural regionalization of AP Vojvodina. Novia Sad, Serbia*. Faculty of Agriculture..

Njegovan, Z., & Zarić, V. (1998). *Contribution for education and regions delineation*. Paper presented at Growth, structural changes and functioning of Serbian economy. Kragujevac, Serbia.

OECD. (1996). *Creating rural indicators for shaping territorial policy*. Paris: OECD..

OECD. (2007). *Rural households livelihood and well-being, statistics on rural development and agriculture household income*. New York: the Wye Group Handbook, UN.

Pejanović, R., Njegovan, Z., & Tica, N. (2007). *Transition – Rural development and agricultural policy*. Novi Sad and Economics Institute..

Porter, M. E. (1990). *The competitive advantage of nations*. London: MacMillan Press..

Republic Office for Development. (1996). *Road to balanced regional development – Policy, priorities, incentives*. Belgrade, Serbia: ROD.

Sahn, D., Dorosh, P., & Younger, S. (1997). *Structural adjustment reconsidered: Economic policy and poverty in Africa*. New York: Cambridge University Press..

Stojiljković, D., & Bošković, O. (2008). Methodological remarks relating to identification of rural areas and determination of indicators for rurality level measuring. *Journal Agroekonomika, 37-38*.

Von Braun, J. et al. (1993). *Urban food insecurity and malnutrition in developing countries: Trends, policies and research implications*. Washington, DC: International Food Policy Research Institute..

WB – IBRD. (2001). *World development indicators*. Washington, DC: WB-IBRD..

Zakić, Z., & Njegovan, Z. (1995). *Base for establishment of strategy, policy and management to rural development. Union of engineersand technicians of Yugoslavia, Žabljak*. FRY..

This chapter was previously published in Sustainable Technologies, Policies, and Constraints in the Green Economy edited by Dorel Dusmanescu, Andrei Jean-Vasile, Turek Rahoveanu Adrian, and Jonel Subic, pages 189-203, copyright 2013 by Information Science Reference (an imprint of IGI Global).

Chapter 68
A Paradigm of Constructing Industrial Symbiosis and Coupling in China's County-Region Economic Sustainable Development

Guangming Li
Shenzhen Tourism College of Jinan University, China

ABSTRACT

This chapter introduces the sustainable development of county-region economy under the constraint of natural resources and environment in China when parts of classic industries along the south-eastern coast are shifting to the poorer mountain areas and middle-western areas. It argues that the county regional economic development at the stage should focus on the agricultural growing and breeding their processing sectors, natural resource exploitation and its processing, manufacturing, tourism, and ecological industrial (shift) park, according to the resource endowment. The local government should guide the scientific planning of county-region industries' layout, structure, symbiosis, and coupling relationships. By means of environmental cost internalization, region industries' symbiosis and coupling size, roles of social network, cultivation of innovative culture, guidance of government subsidy, pressure from public monitoring, the market mechanism, and economic incentives will play roles in resource allocation. In practice, the local government or industrial authority can design the "lack in" value chain parts and package them into some feasible and profitable projects open to the market investors, encourage firms to participate regional industrial symbiosis and coupling, and construct a complete industrial chain or network, in order to realize the integration of closed-circuit industry, higher value-added ecological agriculture, tourism and related service sectors, creative economic industrial park, urbanization, and ecological environment. This will avoid the heavy-damage from traditional industrialization and urbanization on county-regional ecological environment, and then realize the harmony development of county-regional recycle economy, society, and environment.

DOI: 10.4018/978-1-4666-4852-4.ch068

INTRODUCTION

County-regional economy belongs to regional economy scope in which market allocates resource based on county-level administrative district boundary, taking county's city as the centre, town as the bond, and rural area as the hinterland (Gong, Shen, & Li, 2005). It is also the fundamental parts of national economy (Lu, 2008). County-regional economy largely belongs to small sized regional economy in which the infrastructure is not complete, industries are not developed, and agricultural economy is dominant and constrained by county-regional administrative district. But it is in China a kind of area which 800 millions of villagers reside in and live on. It is related to the national strategy by which a wide range of villagers jointly increase their wealth. It is strategically important for China's new county construction, national economy's coordinative and harmonious development. The main problems at China's current stage county-regional economic development include deficiency of basic production factors, insufficiency of county's common goods, big population of unemployed labor force with lower skills and difficulty of reemployment, insufficient fine and deep processing with the primary natural resources and agricultural products, unreasonable industrial structure, lagging tertiary industry development, etc(Lu, 2008). Part of China's county-regional economy fall into borderline regional economy between different provinces, and have their specific development rule (Luo, 2007).

The key of regional recycle economy is the cultivation and construction of good symbiosis relationships between the regional industries, to realize the "reducing, reusing and recycling" of materials and energy (Feng, 2004; Chertow, 2000; McManus & Gibbs, 2008). The key of creative economy is increasing the value-added and core competitiveness of goods by product innovation and industrial chain extension (Cao, 2007). Recycle economy and creative industries can, to some extent, deal with the existing problems in county-regional economy. To systematically investigate the county-regional economy, recycle economy, creative economy and their relationships between each other, to study how recycle economy assists county-regional economy solving the hard lock of environment deterioration, to study how creative economy assists solving the problems such as lower value-added of products, unreasonable industrial structure, and big unemployment by extending and improving industrial chain and constructing good symbiosis relationships between industries, to realize harmonious development of economy, society and environment, have important theoretical and practical significance.

The industrial transfers from the coastal well-developed provinces in China have raised severe challenges to county-regional economy's sustainable development. In order to avoid the old model of "first development and later rehabilitation" during the preliminary stage of industrialization and to protect the natural resources and ecological environment in mountain areas, China should construct its modern industrial economics system under the guidance of recycle and creative economic theories. For industrial transferring, urbanization and agricultural economic industrialization, China should insist, under the recycle economic frame, taking the good industrial symbiosis design and its cultivation mechanism as the breaking through gate, to realize the harmonious economic development of well-developed central cities, county-regions and natural environment. Under the frame of creative industries theory, China should construct modern ecological agriculture, modern tourism and tertiary industries to particularly increase the product value-added of featured agriculture and tourism industries by extending industrial chain and building the more-widely industrial symbiosis. This chapter will, taking Deqing county of Guangdong as an example, preliminarily investigate the industrial symbiosis and coupling of China's county-regional economic sustainable development under the situation of industrial and labor dual shifting.

Current Trends of China County-Regional Economic Sustainable Development

Existing literatures show that the research of China county-regional economy mainly still focuses on the traditional development strategic questions of industrial structure, featured economy, model innovation, government function, county district layout, infrastructure and investment environment improvements from the view of industrialization and agglomeration(Gong, Shen, & Li, 2005; Lu, 2008; Ji, 2008; Wei & Yu, 2005; Huang, 2008; Zhan & Han, 2005; Wu & Wang, 2008; Liang, 2008). About the county-regional recycle economy development, researchers raised different paradigms and countermeasures, such as industrial park modification, lean production model construction, ecological industrial park propulsion, construction and improvement of ensuring mechanism and green technical supporting system for recycle economy development (Chen & Xu, 2007); recycle economic model dominating by agriculture industrialization, or ecological tourism leading model (Jiang, Tao, & Liu, 2008); recycle economy embedded in county-regional economic development layout, and industrial layout or structure modification (An, Tian, & Zhang, 2008); green investment (Yang, Zhang, Li, & Dong, 2007); reduced natural resource input and increased natural resource recycling and utilizing efficiency (Luo, Wu, & Qiu, 2007); building the whole resource recycling system within internal and external firms, villages and towns of the county (Zhan & Zhang, 2007). The non-harmonious development of economy, society and environment is a consensus along the country. That is, researchers have explored the possible paradigms of local county-region recycle economy development, considering the natural resource endowments of different regions, market, location conditions, development level, government strategic positioning, existing policies and humanistic environment. Unfortunately, these works have not yet been in implementation stage.

According to the author's observations, the county-regional economies in most of China's areas (as central and western areas) currently are still at the stage of preliminary stage of development. Their leading industries are mainly in great agriculture, natural resource exploitation and handicraft industries. But quite a few of them have already in the deeper stage of development, and the underlying leading industries are in deep processing of mining, agriculture, forest, herd, fishing products, textile, mechanical manufacturing, and so on, such as in the well-developed counties in central areas, and backward counties in south China). Part of county regions may already in the comprehensive development stage with advanced manufacturing and assembling industries, or with complete industrial scale and structure, or even with industrial clusters, such as in fast developed county regions of eastern and southern coastal provinces. Furthermore, thanks to the paid attention on tourism and featured industrial development by the governments in recent years, and to the promotion effects from the socialism new county construction in China, many county-regions (some from mountain and backward districts) take advantage of their good ecology, agriculture, forest, fruit, and humanistic resource, greatly develop tourism industry and have achieved satisfied results.

However, China's county-regional economic development most takes the regional level administrative district boundary as the border without regional economic cooperation to encourage the reasonable mobility of production factors cross regions. There are no integrating layouts of county-regional recycle economy under greater regional economic development than county level. These layouts are not coordinated with other industry, city and commercial layouts, without systematical and overall consideration, and far from implementation. These layouts are reflected in investment plan, industrial layout and design. There are no coordination between governments and no innocuous cultivating mechanism for recycle economy paradigm. Although some researchers brought forward innovative models for county-regional economic development, the application of creative economy paradigm in China are still only in the fields of cultural creative industry and e-commerce, but not in county–regional economy.

A long with the deep development of industrialization and world economic globalization, land and natural resource scarcity and ecological environment damage have become more and more serious. Until now, most areas of China are still not considering sufficiently the negative impacts from industrialization and urbanization on natural resources and environment. Without prospective plan and prediction, unilaterally pursuing economic GDP growth had paid too much ecological environmental costs. Taking this lesson, China should systematically explore the industrial symbiosis and coupling, their relationships cultivation mechanism under the frame of county-regional recycle economy, and benefits of creative economy to industrial structure improvement, extension, optimization or product value-added increase. China should also pay attention to how to achieve the organic integration of recycle, creative and county-regional economy, put forward the county-regional economic development measures in the situation of newly initiated industrial transfer, structure optimization and modification, embed recycle and creative economy paradigm into county-regional economy layout, investment plan and industrial structure design, to realize the scientific, harmonious and sustainable development of its county-regional economy.

Features of Future Modern Industrial System and its Impacts on County-Regional Economy

Next stage of scientific development of China's coastal well-developed areas will require systematical industrial transformation, structural optimization. This will need the building of industrial core with high-tech, high value-added, low energy consumption, low pollution and self-innovative capability. This also needs the support from highly efficient industrial supplement system with tech, human talents, capital and information. Based on the industrial development environment with natural environment beauty, complete infrastructure, reliable social security system and good-ordered market, new type of industrial system should be built with features of creativity, openness, amalgamation, agglomeration and sustainability (Anonymous, 2008).

To meet the requirements of China's economic and industrial structural optimization and regional economic balance development, the transfer of partial traditional industries from southern and eastern coastal to poor mountain and central or western areas has been on the way. This will bring in new opportunities and challenges to China's county-regional economic development. In the future years, China's county-regional economic development should follow the requirements of its industrial system positioning and construction, that is, to build sustainable, scientific and harmonious county-regional economic structure. Along with the continuous improvement of central city's transportation and infrastructure in most of provincial or municipal districts neighboring the coastal areas, most of these districts will integrate into the "1-5 hours economic circle" of well developed regions. The related county-regional economic should actively engage in the division and collaboration of modern industrial system of developed districts, in addition to fostering endogenous industrial system, featured economy, ecological agriculture and tourism economy, in order to form the collaboration development trends with these developed areas, through the industrial cores, industrial shifting park, recycle economy system, processing trade transformation demonstrating zones, and creative industrial parks.

However, most of China's lower developed mountain areas belong to ecologically weak, sensitive or protective water head districts. China's central and western area economic development is relatively backward and the local ecological environment is relatively protected, but these areas have to some extent been spoiled by industrialization. They all belong to important protective waterhead, natural protective regions, agricultural production base and green ecological barriers, and then cannot bear any more damage similar to coastal areas. In these areas, economics and social development is constrained much by ecological environment, environmental capacity is limited, and natural factors account for important or even dominant roles in regional imbalance development (Hu, 2008).

China's county-level economic development status quo shows that a number of county governments do not sufficiently understand the importance of collaborative development between population, natural resources, environment, economy and society. To achieve the higher economic growth rate, these county governments usually run into the myths of "development first, treatment later." Not even for pre-schematizing ecological industrial park, entrance criteria will be depressed to attract sufficient firms in, and the requirements of ecological industrial park construction can not been met. As a result, the county-level economy still features as high input, lower output, and much emission. The transfer of processing trade oriented and labor intensive industries to river's upstream and green ecological barrier areas will necessarily bring in seriously negative impacts on ecological environment, if county-regional economy and industries are not scientifically designed according to the requirements of recycle economy development model, in case of over-development of land or natural resources, and environment-unfriendly investment.

CONSTRUCTING MECHANISM OF INDUSTRIAL SYMBIOSIS AND COUPLING OF COUNTY-REGIONAL ECONOMY

China's market and industry environment have changed a lot during the past 30 years. Particularly in coastal developed areas, it is a right time of industrial transformation. On one side, how to motivate local firms to increase their technical input, product value-added, and international competitiveness of industries through tech innovation, self-owned brand cultivation or relocation, is an important task facing coastal developed areas. On the other side, economic development of China, particularly coastal areas' industrialization have resulted in too much consumption of natural resources and environment capacity, industrial structure optimization should be initiated to relieve the pressure. Although industrial transfer brings in good opportunities to China county-regional economic development, how to make first, secondary and tertiary industries collaboratively develop, how to reduce natural resource consumption, how to decrease pollution emission, and how to keep better ecological environment must also be paid much attention.

China's county-regional economic development at the current stage pursues the main goals of agricultural industrialization, county-regional industrialization and urbanization taking county town as core to radiate villages and small towns. This developing progress, or even the tourism which itself belongs to modern service industry, is reflecting the questions of fast resource consumption and serious environment damage. To realize county-regional economic sustainable development, it is necessary to apply recycle economic development model. Meanwhile, well industrialized county-regions in China have relatively shaped leading or non-leading industries compatible with the natural resource endowment and production factor features, such as agricultural growing and its deep processing, mineral resource exploitation and its processing, construction materials, handcrafts, mechanical manufacturing and tourism. Therefore, part of China's county-regions has already met the pre-requirements to initiate developing recycle economy.

As per the industrial layout, the county level government should play the role of guidance, design the industrial layout, and construct ecological parks, taking county town centre as development poles. These development poles can radiate and drive surrounding areas such as villages or small towns, to form industrial bands or clusters, or specialty town which particularly specialize on somewhat industrial production and service. There have already been relatively shaped industrial systems in China's developed county regions. The distances between the county town and villages or small towns are roughly about 10 km, and the cross transportation is not so busy yet. This will reduce the transport cost of transaction of county regional by-products or discarded substances. To take the county-level administrative districts as units, cultivation of symbiosis and coupling relationships between industries should be paid much attention during investment and development, and it is reasonable and feasible to systematically consider the construction of regional recycle economic industrial system. The external symbiosis between industries can be dealt with by implementing virtual ecological industrial network paradigm (Chertow, 2000; McManus & Gibbs, 2008).

As per the industrial structure, the county level government should scientifically design county-regional industrial composition and select leading industry according to its natural resource endowment characteristics, applying industry ecology, recycle economy and sustainable development theories. Other counter measures include constructing ecological industrial park based on the principles of industrial symbiosis and coupling, managing firms by clean production philosophy, popularizing production and consumption paradigm of natural resource saving and environment friendly, developing environment protection, comprehensively resource utilization and reusing industries, to build ecological modern industrial system.

Therefore, the whole frame of China county-regional industrial symbiosis and coupling relationship building currently should be to give guiding play to county level government, to systematically consider in existence county-regional industries or sectors, future possibly taking over industries or sectors, future possibly participating division and collaboration into well-developed central cities and industrial systems, and to integrate and design county-regional industries' composition and layout. To ecologically integrate in existence industries, county-regional economy should take agricultural, planting, breeding sectors and their processing, mineral exploitation and its processing, mechanical and equipment manufacturing, processing industry, tourism, creative sector and urbanization as the core, and systematically design and integrate the symbiosis and coupling relationships between industries and sectors, to guarantee the technical and economic feasibility. Then the key links of industrial symbiosis and coupling should be designed and packaged into profitable investment projects that both county level government and firms can input on.

As per the relationship building of regional industrial symbiosis or coupling, and the green or ecological development of regional industrial system, how to give play to market mechanism for resource allocation and how to motivate firms to participate in by economic incentives are the keys, besides the planning and leading roles of government.

The main benefits of firms participating industrial symbiosis and coupling in the regional industrial system are described in the following.

Valuable Component Benefit (S): Industrial civilization consumes huge quantity of natural resources including energy, environment capacity. This has made the natural resources in the world market more and more scarce, current reserve smaller and smaller, resource goods price higher and higher, the valuable components (oil, coal, nonferrous metals, iron ore) recovered from discarded wastages more and more important for national economy, and their market values become more and more obvious. This benefit from recovery mainly depends on the world price and processing technology of wastages.

Environment Benefit (E): Because firms, which participate in regional ecological industrial symbiosis treat the wastages, reduce emission and environmental damage (including the harm to human body, damage to environmental ecology, air quality deterioration, water quality exacerbation, detrimental impacts on agriculture production, and unfavorable effects on climate change, etc by heavy metals, effluent water or solid scrap, BOD and COD), environment is protected. The corresponding environmental cost saving should theoretically be in the reduction of environmental emission fee and purchasing fee for emission permits, or government subsidy to participating firms in industrial symbiosis.

Firm's Intangible Asset Value Increment (Br): Firms participating in ecological industrial symbiosis have fulfilled the social responsibility if environment protection and resource saving. This will greatly improve public relationships, increase social recognition, credibility or popularity, and then increase their intangible assets value which reflects in firm's brand value added.

Technical Innovation Compensation (T): Facing with the pressure of environmental regulation from environmental damage and external cost internalization, firms will actively participate in regional industrial symbiosis. If firms build processing facility to recycle or reuse the waste emission, more inputs are needed and firms will face with the pressure from technical innovation and investment failure. Firms actively participating in industrial symbiosis will greatly promote technical innovation and favor firms to receive various innovation compensation benefits.

Agglomeration Benefits (J): Large number of firms participating in regional ecological industrial symbiosis will form a kind of industrial agglomeration in which these firms will share the "cluster surplus" (Chen, 2003). Compared with external firms, these inside firms have more types of benefits. These benefits are mainly in learning effects, innovative pressure effects, knowledge broadcasting and diffusion effects, and innovation success effects. Firms' economies of scale and scope (mainly from collaborative effects between firms) may also be inclusive.

The main costs of firms participating in regional ecological symbiosis are described in the following.

Transportation Cost (TC): The emissions of discarded wastages from firms in regional ecological industrial symbiosis network are relatively dispersed. They need to be concentrated to one or few firms to be processed, ensuring the continuity of the technology, economy of scale and scope or agglomeration. Considering the features of ecological industrial symbiosis, the transportation cost from emission collection process is one of the important costs.

Site Dump or Warehouse Storage Cost (WC): To ensure the production continuous or stable, firms need building some storage or dump sites, to store some quantity of emission waste.

Waste's Sorting and Processing Cost (RC): This cost reflects in the production cost incurred from waste processing.

Depreciation of Innovation Input and Fixed Assets (D): The cost is the abrasion cost of machine or facility.

Transaction Cost between Firms (DC): The "producer" and "decomposer" of discarded waste need search information, screening, communication, and negotiation to determine the bargain and transactions with each other, which will make regional ecological symbiosis and coupling formed and sustainable. Information search, credibility between firms, people's social network, cultural difference, contract implementation and monitoring will incur various transaction costs.

Secondary Pollution Costs (Ds): The processing and treatment of waste discarded are restrained by technology so that there will be some materials cannot be reused and discharged into the environment or changed to new pollutant harmful to environment. It is reasonable that Ds will mainly rely on technical progress.

Therefore, the net benefit of firms participating in regional ecological industrial symbiosis should be equal to above mentioned total benefit minus total costs. That is

$$\pi = TB - TC = (S + E + Br + T + J) - (TC + WC + RC + D + DC + Ds)$$

As per the firms which optimize net benefit, only would positive net benefits encourage them to participate in. Government administrative departments should guide and motivate firms to participate in regional ecological industrial symbiosis and coupling through issuing environmental regulations and economic policies, which include environmental cost internalization, proper choice of regional industrial symbiosis size, role of social network, cultivation of innovative culture, leading of government subsidy and formation of social pressure, etc.

CASE STUDY

According to the "Eleventh Five Plan" of Zhaoqing regional economic development, "Twelve Five Plan" of Zhaoqing national economics ans social development, the subordinate Deqing County has been positioned as mountain block. This will require Deqing County promotes the industrial pillars with local resource endowment features, in order to accelerate its economic development to form an integrative forest, pulp and paper unity, forest and chemical engineering unity, building material (cement), general equipment manufacturing, auto parts, metallic fabricating industries with some scale of agglomeration (Anonymous, 2006; Anonymous, 2010). The tourism industry will be featured by the "Kilo KM tourism corridor." Agriculture will focus on forest, featured fruits for planting and feeding sectors with some scale. Recently Deqing County is cooperating with Guangdong Geology and Exploration Bureau to exploit the local gold, silver, ceramic, and mineral spring resources, and to develop downstream industries.

The forest chemical engineering, building material or cement, and general equipment manufacturing will still be the main parts of Deqing county-regional secondary industries. The transfer of traditional industries to mountain counties will necessarily bring in negative impacts on ecological environment. If local mineral resources are overexploited, the negative effects will be some serious on natural resources and ecological environment. The essence of recycle economy is to realize the "reducing, reusing and recycling" of substance material through industrial symbiosis, clean production, recycling and saving society model in order to achieve the maximum economic output consuming the least natural resources and energy. Reusing of natural resources or energy and recycling of wastage will decrease pollution or even achieve zero-emission to minimize environment impacts. If recycle economy paradigm can be applied to dominate the industrial economic development (particularly the industrial transferring park and mineral resource exploitation) of Zhaoqing in the short run, above-mentioned questions can be solved effectively.

Effective agriculture and tourism with local features will compose the important principal part of primary and secondary industries. They will have strategic significance for Deqing County regional economic sustainable development and industrial symbiosis. The updated world economic development showed that recycle and creative economy can extend and create some additional effective parts of agriculture and tourism value chains that will bring in innovative products, add value, and optimize industrial structure. Therefore, recycle and creative economies dominating Deqing county's regional effective agriculture and tourism with local features will take advantages of fruits, Chinese medicine, forest, local products and tourism resources to realize the coordinative development of the three industrial sectors in the region.

The county regional economic layout including industrial park's location, guidance or cultivation of market and industrial clusters, policy or regulation constitution, should follow the technical principles of recycle and creative economies. The coordination between regional recycle industrial system, Deqing county town, Zhaoqing central city, proximal Hezhou city of Guangxi province, or integration between city and countryside must also be designed to make different regional economies cooperatively, co-prospectively and collaboratively developing.

Based on Deqing county region's resource endowment and the development status quo of Guangdong or China, the short run economic development objectives in this region can be designed as: to apply the recycle and creative economic paradigms for economic and industrial structure upgrading; to enhance technical innovation and application; to accelerate new type of industrialization march; to actively foster ecological agriculture, industries with low energy and material consumption or emission, to eliminate backward technology, techniques, and equipments; to seriously limit the industries with high consumption of energy, water and with heavy pollution; to make county-regional economy turn to the growth model with low input, low consumption, low emission and high efficiency. In the medium or long run, it is reasonable to make Deqing County become national or provincial recycle economic selected units; to motivate Deqing firms to become the demonstration units of national clean production, water saving, energy saving, material saving, land saving and resource comprehensive; to actively encourage the region to become provincial industrial cluster demonstration zone or specialty small town (district) to form the provincial recycle economic ecological industries park; to strive for the selected provincial saving city to promote the ecological transform with Deqing county town and to strive for the national environmental protection model city.

Designing the frame of recycle economic industrial chain should follow the constructing principle of ecological industrial network (Casavant & Jeffrey, 2004). According to the County's current industrial status quo, natural environment and regional characteristics, Deqing should concentrate on building and improving these ecological industrial chains:

- Mineral exploitation-metallic extraction and processing-general equipment manufacturing-automobile accessory-metallic fabricating-environmental protection equipment-resource recycling chain;
- Integrative forest, pulp and paper-forest chemical engineering-chemical fertilizers-effluent utilization chain;
- Mineral exploitation-building material (cement)-residual heat power generation-acid production and chemical engineering-industrial tourism chain;
- Stock raising sector-methane making-fruit planting-ecological agriculture-tourism chain;
- Township production and sanitary sewage-solid scrap-resource recovery-recycle chain.

It is through these recycle industrial chains and their symbiosis relationships that greater circular economic system within the industrial, agricultural production and their connection with township and countryside is built. The design of underlying recycle industrial chain should, under investigating local natural environment and industrial characteristics, base on the technical principle of recycle economy. The systematical integration and nodes between recycling industrial chains should be paid much attention, and particularly design of the coordinating mechanism of relationships between firm's clean production, industrial symbiosis, and the greater economy-society circulation (Figure 1). Agricultural growing-breeding-processing in-

Figure 1. Diagram of symbiosis and coupling system of principal industries of Deqing County of Guangdong province[1]

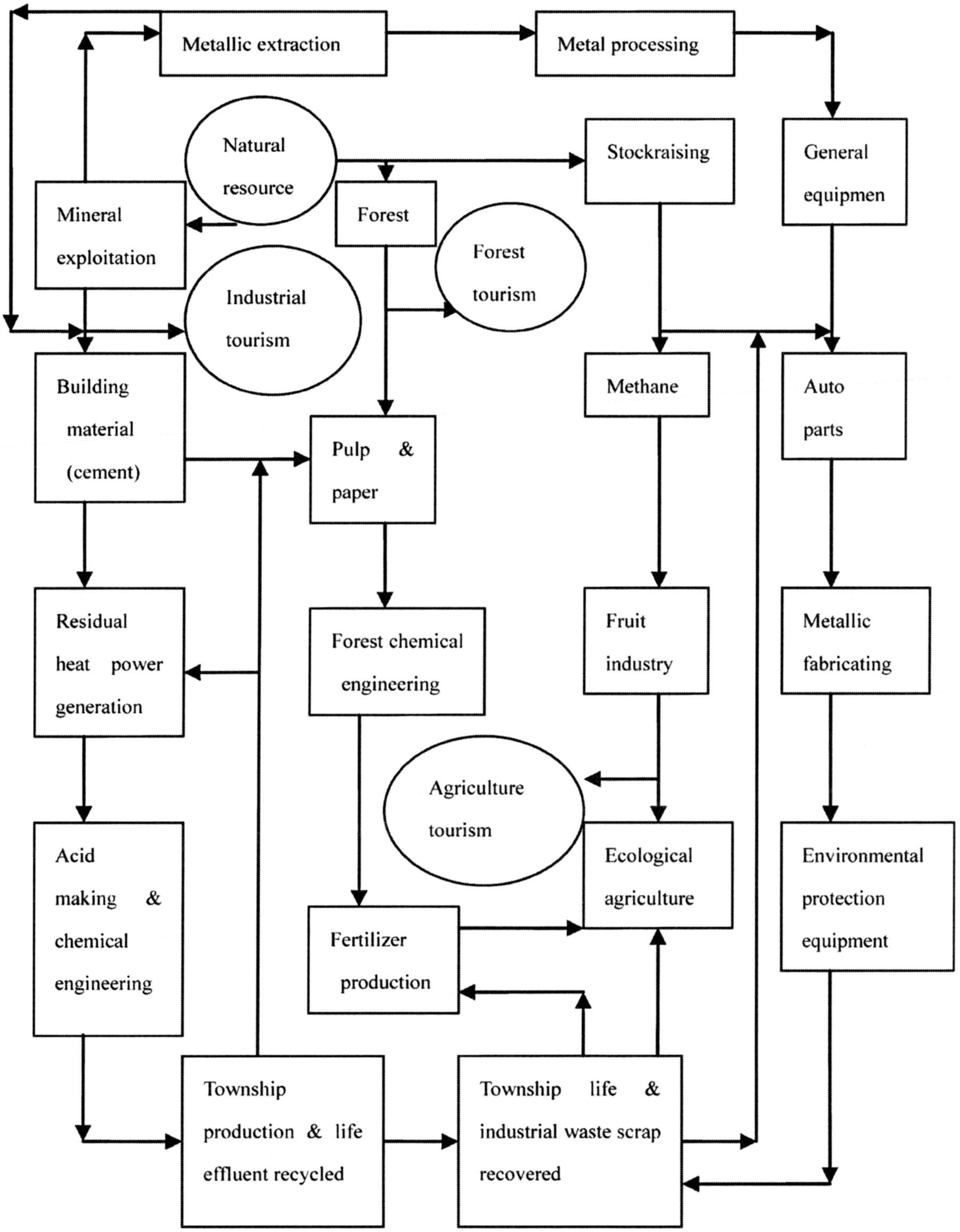

tegration, development of the connections of agriculture with other sectors and divisions, extension of industrial chain, all will pull the development of agricultural processing and tourism industries, to realize the collaboration between production, life and ecology. It is also necessary to foster local leading or dominant firms in the above-mentioned recycle economic industrial chains to drive industrial agglomeration development.

COUNTERMEASURES-TAKING DEQING COUNTY AS AN EXAMPLE

Plan and Layout

By means of weaving county-level recycle economic plan, it is expected to effectively design industrial system and recycling industrial chain or network for the whole county region, and also the symbiosis relationships between industries, to form closed circulation with lower consumption and effluent emission. If it is not too late to consider circular economy only after the environment has already been damaged and various problems have been obvious, prospective thinking will avoid the old path of "developing first and rehabilitating later."

Transformation of Industrial Parks

Along with the "dual shifts" strategy in Guangdong province, it is expected to carrying out the circular economy paradigm, to consummate and to upgrade the "dual shifts" industrial parks. Applying the technical principles of circular economy (such as substance metabolism analysis, life cycle assessment, flexibility analysis, systematical integration, and clean production) to actively patch the existing industrial chains in the county region, each industrial chain's imperfect nodes can be packaged into profitable investment projects to weave ecological industrial chain. Transforming shifting industrial parks into ecological ones will make firms agglomerative and integrative, and form the symbiosis and coupling relationships between industries. When industrial economy is greatly developed, meanwhile, ecological environment and natural resources are effectively protected. Keeping the mountains green and water clean with good surrounding for living will be supportive for tourism industrial development.

Goods Value Increment

Development of effective agriculture with local features is strategically significant for Deqing county's economic growth, local villager's income increase, regional featured brand's building, and tourism industrial chain improvement. "Local features" require combining local natural resource and environment endowment like climate, soil and water, relying on tribute orange, Chinese medicine herbals, turpentine, silkworm, cassava, forest, mountain chicken, and local produce, taking use of concepts and technical principles of creative economy to design series of featured agriculture. For example, around the flower appreciation, scientific education, fruit tasting, personal experience of tribute orange, a kind of orchard tourist spot can be created. By cultivating ecological agriculture sightsee park mainly for melon and fruit appreciation, an agriculture sightsee tourist spot can also be created. Around silkworm growing, processing, and appreciating, a specific scenic spot can be built. Fostering countryside new type of economic organization, innovating agriculture producing model, constructing high-tech agriculture, ecological and featured agriculture parks, producing quality, environment protecting, profitable goods, and incrementing their technical contents and value added, all will contribute to "effective" agriculture.

Industrial Chain Perfect

Deqing's tourism industry has already been the good foundation of Zhaoqing's "Kilo KM Tourism Corridor." The main question is imperfect tourism

"6 elements"[2] composition, especially in foods, shopping and entertainment to be well developed. By means of cuisine competition, Deqing's local foods and dishes, or snacks can be discovered to let tourists enjoy themselves. Deeply processing Deqing's local goods or tourism souvenirs, and packaging them creatively will make them attractive. In healthcare, hospitality, sports experience, and cultural leisure, Deqing can build entertainment features and let tourists satisfied. Above mentioned "effective" and "local featured" agriculture development can exactly offset the shortage of foods, shopping and entertaining in tourism industries. Additionally, construction and improvement of tourism industrial chain need to be systematically guided by the technical principle of creative economy. Improvements of "6 elements" will also require coordination with socialism new countryside construction, township design and building.

Government Subsidy and Policy

Taking Deqing's main tourism routes and townships or villages with featured agriculture bases as the core, will make it easier to weave the virtual creative economic industrial parks' development plan. It is necessary to introduce and internally foster creative subjects and backbone firms, and on this base, to build marketable technical innovation system. Local government should invest leading capital, and actively apply for the modern industrial development guidance capital which is totally 1 billion RMB annually, and other various specific funds for energy sawing, effluent reducing, dual shifting and supportive independent innovation and industrial development. Other measures include weaving creative economic industrial parks' plan, constituting policies, intensifying guidance and collaboration, building human capital introducing mechanism, properly developing modern professional service sectors for production, speeding up the development of tertiary sector leading by tourism industry, improving the environmental protection regulation and criteria of ecological industrial parks for project admission according to ecological principles, and constitutions for clean production or scrap recycling and reusing, etc.

Ecological Brand Building

Developing Deqing's circular and creative economy will compose the circumstance of collaborative development of economy, society and environment, or harmonious or scientific developments, in order to construct this county region as the "no.1 Mountainous City" in Guangdong province suitable for living with green environment and clean water. On this base, continuous extraction and propaganda will form Deqing's regional brand, which will accelerate the agglomeration of production inputs to the county and further drive Deqing's regional economic development.

REFERENCES

An, J. F., Tian, C. Y., & Zhang, A. Z. (2008). Theory and practice of county regional circular economy. *Market Modernization*, *532*(3), 280–281.

Anonymous,. (2006). *11th five years plan of Zhaoqing's regional economic development.* Bureau of Development and Reform of Zhaoqing City of China..

Anonymous,. (2008). *Decisions on accelerating modern industrial system building*. Guangdong Province Government..

Anonymous (2010). *Information on the planning layout of Zhaoqing city's 12th five years plan of national economic and social development.* Zhaoqing city documents No.16.

Cao, C. Y. (2007). Secret of profits from creative economy. *New Capital*, *2*, 87.

Casavant, T., & Jeffrey, M. (2004). Industrial ecology and eco-industrial networking backgrounder. The guide to eco-industrial networking for greater Vancouver municipalities, Canada..

Chen, X. M. (2003). *Theory and practice of medium and small size firms agglomerations*. Beijing, China: Economic Science Press..

Chen, Y. P., & Xu, X. (2007). Study on county regional circular economic development in Heilongjiang province of China. *Heilongjiang Foreign Trade and Economics*, *160*(10), 17–19.

Chertow, M. R. (2000). Industrial symbiosis: Literature and taxonomy. *Annual Review of Energy and the Environment*, *25*, 313–337. doi:10.1146/annurev.energy.25.1.313.

Feng, Z. J. (2004). *Introduction to circular economy*. Beijing, China: Renmin Press..

Gong, T., Shen, H., & Li, X. P. (2005). Exploration of model and direction of county regional economic development-taking Yaoan County as an example. *Agricultural Economics*, *10*, 49–53.

Hu, Z. Y. (2008). New ideas to solve the imbalance regional development of Guangdong province of China. *Fast News on Think Tank, 159*(4).

Huang, X. M. (2008). Industries-driver of regional economic development. *Entrepreneur Forum*, *2*, 30–31.

Ji, M. (2008). Understanding a number of important questions' significance on China's county regional economic development. *Specific Zone Economy,* 271-272.

Jiang, L., Tao, X. S., & Liu, Z. L. (2008). Model selection of resource-saving and environment-friendly county regional economic development in Hunan province of China. *Journal of Hubei Economic College*, *5*(2), 57–58.

Liang, T. (2008). Successful experience and implications of Rongcheng City's economic development – "Shandong No.1" county. *Entrepreneur Forum*, *2*, 180–181.

Lu, K. (2008). County regional economic development of central part rising abruptly. *Journal of the Social Sciences*, *175*(2), 111–113.

Luo, B., Wu, C., & Qiu, Z. L. (2007). Study on county regional agriculture circular economy development-taking Liuyang city of Hunan province of China as an example. *Agriculture Environment and Development*, *4*, 25–30.

Luo, Z. L. (2007). *Regional border economy*. Changsha, China: Human People Press..

McManus, P., & Gibbs, D. (2008). Industrial ecosystems? The use of tropes in the literature of industrial ecology and eco-industrial parks. *Progress in Human Geography*, *32*(4), 525–540. doi:10.1177/0309132507088118.

Wei, X. F., & Yu, Z. P. (2005). Development model of China county regional economy. *Agricultural Business and Management*, *3*, 47–48.

Wu, Y. M., & Wang, X. L. (2008). Effective path to develop county regional economy. *Firms Standardization*, *2*(1), 45–46.

Yang, Y. Z., Zhang, J. P., Li, S. J., & Dong, B. (2007). Sustainable development of county regional economy and green investment-taking Henan province of China as an example. *District Study and Development*, *26*(5), 38–42.

Zhan, G., & Zhang, M. J. (2007). Circular economy theory and county regional economic development. *Anhui Agriculture Science*, *35*(3), 916–917.

Zhan, Z. L., & Han, L. (2005). Model classification and equilibrium selection of innovation oriented county regional economy. *Agricultural Economics*, *6*, 66–71.

ENDNOTES

1 In ecological industrial network, the symbiosis between industries and townships or treatment facilities is also referred as a kind of industrial symbiosis (Casavant & Jeffrey, 2004).

2 6 elements normally mean foods, accommodation, transportation, sightseeing, shopping and entertainments.

This chapter was previously published in Green Technologies and Business Practices edited by Patricia Ordóñez de Pablos, pages 1-14, copyright 2013 by Information Science Reference (an imprint of IGI Global).

Chapter 69
Mind the Gap Please!
Contrasting Renewable Energy Investment Strategies between the World Bank and Poor Customers in Developing Countries

Sam Wong
University of Liverpool, UK

ABSTRACT

This chapter scrutinises the World Bank's nine guiding principles for investment strategies on renewable energy in developing countries. Drawing on two World Bank-funded solar lighting projects in Bangladesh and India as examples, it demonstrates a wide gap in investment strategies between the Bank and local people. It suggests that a rigid distinction of renewable and non-renewable options risks restricting poor people to adopt an energy-mix approach to cope with poverty. The economic assumptions of the strategic choice for renewable energy investment pay inadequate attention to the cultural norms that shape people's preferences for energy sharing. A lack of participation of NGOs and local communities in shaping the Bank's investment strategies also undermines the effectiveness of its renewable energy policies in the long term. This chapter suggests that the World Bank re-conceptualises the complex relationships between energy and poverty and seeks a better understanding of local people's daily energy consumption practices.

INTRODUCTION

The World Bank plays a multiple and indispensible role in promoting renewable energy in developing countries. It is a big investor—US$11 billion has been spent on renewable energy since 1990 (2007: ix). It has approved 366 renewable energy and energy efficiency projects in 90 countries between 2004 and 2009 (2009). It has spent billion dollars on research and development in order to find suitable renewable technologies that match with the needs of poor people. It is also an ambitious campaigner—the 'Lighting Africa' campaign alone, launched in 2008, was intended to provide

DOI: 10.4018/978-1-4666-4852-4.ch069

solar lighting to more than 250 million Africans (Wamukonya, 2005).

What are the driving forces underlying the World Bank's very considerable interest in renewable energy? Sustainable development, concerns over climate change, and the Clean Development Mechanisms have provided the Bank with a continuous flow of funding and legitimate reasons for interventions. What is equally important is that renewable energy fits into the Bank's overall plan of poverty reduction (World Bank, 2000). From the Bank's perspective, solving energy-poverty by renewable energy is crucial to providing solutions to other poverty-related problems, such as economic stagnation, environmental degradation and gender inequalities.

Based on this complex rationality, the World Bank has developed nine strategic investment principles for renewable energy in developing countries (World Bank, 2007a, 2004, 2002). They are:

1. Use localised renewable energy supplies;
2. Apply sustainable technologies to provide renewable energy;
3. Give heavy subsidies to promote renewable energy market;
4. Recognise that poor people are potential customers for renewable energy;
5. Ensure that charges for renewable energy should not be as high as non-renewable options;
6. Insist that customers pay for services, no matter how poor they are;
7. Create different services for different customers;
8. Decentralise energy supplies at local levels; and
9. Work with NGOs and private companies to provide renewable energy services.

How are these strategic investment principles actually translated on the ground? How effective are they? What impact have they made on poor people's lives, and how will people respond to the changes? To provide answers to these questions, we will draw on two World Bank-funded solar lighting projects in South Asia: an individual solar home system in South Bangladesh and a solar lantern project in Rajasthan, India. We will compare the World Bank's renewable energy strategies with local people's energy consumption practices.

In this chapter, we will expose a wide gap in the investment strategies between the World Bank, as the funder and project initiator, and local people, as users and customers. Firstly, it will suggest, with evidence, that, in contrast to the World Bank's rigid distinction between renewable and non-renewable energy options, poor people take a pragmatic approach to energy use. Rather than considering renewable and non-renewable energy as either-or, they adopt a flexible energy-mix strategy which enables them to cope with their daily struggle against poverty.

Secondly, it will suggest that, most of the World Bank's guiding principles for renewable energy are over-economic and pay inadequate attention to socio-political factors, such as cultural norms of sharing collective resources and the right-to-free-electricity, that shape their preferences for energy uses. Thirdly, it will argue that the World Bank considering NGOs as purely an implementer of its renewable energy policies has missed the opportunity to hear the voice of local people.

This chapter will suggest that the widening gaps raise doubt over the effectiveness of the Bank's investment in renewable energy on the ground. In addressing the problems, it urges the Bank to reconceptualise the 'energy-poverty' relationships, to seek a better understanding of poor people's energy consumption practices, and to encourage NGOs to be more critical of its own energy policies.

This structure of this chapter is as follows: it will first discuss the nature of the World Bank's nine guiding principles of strategic investment in renewable energy in developing countries. It will then provide contextual backgrounds of the two case studies and examine local customers' actual investment strategies in solar lighting. By exposing the gaps in investment strategies between the World Bank and local people, it will bring out the problems that may undermine the effectiveness of the Bank's renewable energy policies. It will conclude by offering solutions to overcoming the barriers.

WORLD BANK'S COMPLEX INVESTMENT STRATEGIES IN RENEWABLE ENERGY

The nine principles (see Table 1), guiding the World Bank's investment strategies in renewable energy, are derived from its 60 years' engagement in developing countries. They are the lessons that the Bank has learnt from its success, as well as mistakes and failure, of its interventions and development reforms.

From the Bank's perspective, the multiple socio-economic-political-environmental problems that have baffled the developing world are the results of under-development. Securing constant, reliable, self-sufficient, and the recent idea of green, energy supplies have, therefore, been regarded as a solution to energy poverty. In the 1950's and 60's, driven by the agenda of modernisation, the Bank had introduced, predominantly, non-renewable energy projects, such as coal-based power stations, as the engine for economic growth. Some renewable energy-related programmes, such as building dams for hydroelectricity, had also been promoted.

The outcomes of the interventions, however, were not all satisfactory (Wong, 2009a). Firstly, electricity, produced by coal-burning and hydro-driven power stations, was distributed via the national grids. The coverage of the grids was limited because of the high costs of construction. As a result, the grids failed to reach the majority of the population living rural areas. Secondly, the big energy projects were implemented without paying sufficient attention to environmental protection and costs to human lives, such as massive scale of deforestation and forced migration. At the same time, the 'small is beautiful' campaigns, led by Ernst Schumacher in the West in the 1970s, had placed the World Bank's development programmes under a close scrutiny (Bortman, 2005).

Table 1. 9 principles guiding the World Bank's investment strategies in renewable energy in developing countries

	Sources of energy and role of technology
Principle 1	Use localised renewable energy supplies
Principle 2	Apply sustainable technologies to provide renewable energy
	Characteristics of the market and customers
Principle 3	Heavy subsidies needed to promote renewable energy market
Principle 4	Poor people are potential customers for renewable energy
Principle 5	Charges for renewable energy should not be as high as non-renewable
Principle 6	Customers pay for services, no matter how poor they are and how small a contribution they make
Principle 7	Create different services for different customers
	Service delivery and governance
Principle 8	Decentralise energy supplies at local levels
Principle 9	Work with NGOs and private companies to provide services

(sources: author's own table, inspired by World Bank, 2007b, 2004, 2000; Martinot et al., 2001)

The 1980's and the early 1990's were considered as the 'lost decade' for many developing countries. Poverty worsened; many states went into bankruptcy. To cope with energy poverty, not-so-poor people relied on kerosene and oil lamps for lighting and cooking while the very poor used animal waste and firewood for survival. These coping strategies had, however, resulted in undesirable environmental and health problems. Cutting down trees in order to collect firewood resulted in deforestation and soil erosion, and land degradation led to declining farm productivity (Asian Development Bank, 2000). Burning animal waste and firewood created black smoke and affected people's health. Huge human resources were also wasted since the dim lighting by kerosene had forced people to stop working in the evening.

The emergence of the notion of sustainable development in the late 1980's had raised awareness of the balance between social well-being, environmental conservation and economic growth. Concerns over the impact of the climate change, such as rising sea level and higher frequencies of extreme weather, in the mid-1990's, have highlighted the role of renewable energy and sustainable technology in mitigating, and adapting to, changing climate, both in developed and developing countries (Paavola & Adger, 2002).

Amid these socio-political-environmental changes, the World Bank pondered: how was it possible to help poor people obtain energy to improve their well-being, without compromising the long-term environment and social development? Since the late 1990s, the World Bank has integrated the energy-poverty relationships into its development framework. Renewable energy strategies are highlighted as key 'pro-poor' interventions to address wider problems (Chaurey *et al.*, 2004). Firstly, renewable energy creates economic growth. Small-scale irrigation and mechanisation is possible, thus improving rural productivity. New job opportunities are created; technicians are trained to provide maintenance support. Secondly, renewable energy enhances well-being. Smoke-free environment improves health. Artificial lighting raises literacy, and improving communication by radio and television makes the flow of information easy. Thirdly, renewable energy empowers women. Lighting provides women and girls with confidence, thus increasing their mobility.

Nine Guiding Principles

Table 1 summarises the nine principles that shape the World Bank's investment strategies in renewable energy in developing countries.

Principles 1 and 2 touch on the sources of renewable energy and the role of sustainable technology in generating energy. The first principle highlights self-sufficiency of obtaining renewable energy from localised sources, rather than from outside. This principle is based on two observations: firstly, discussion of energy poverty in developing countries in the past, focused too much on what poor countries lacked, and too little on what they had. From the investment perspective, developing countries have huge potential for developing renewable energy since many of them own abundant and unlimited supplies of natural resources, such as sunlight and wind, and they have large populations to support the market. Secondly, these clean and renewable sources are resources and assets that have not been fully tapped or utilised for development before (World Bank, 2000).

Principle 2 ascertains the role of sustainable technology in generating renewable energy and in poverty reduction (Willis *et al.* 2006). Different types of renewable innovations have been adopted and applied in developing countries, such as reverse osmosis plants to provide clean water and solar mirrors for cooking. Favourable conditions, such as the Clean Development Mechanism and the agreed North-South technology transfer in the Kyoto Protocol, have made the introduction of sustainable technology in developing countries easier and bigger in scale.

Principles 3 to 7 summarises the World Bank's understanding of the characteristics and features of renewable energy markets in developing countries (World Bank, 2008; IEA International Energy, 2008). Principle 3 identifies an obvious fact that free market of provision of renewable energy, based on demand and supply, does not yet exist in poor countries (Robins, 2006). Hands-on policies, such as the provision of heavy subsidies, are considered necessary to incentivise both the suppliers and the users of the renewable energy sectors. Principle 4 may sound ironic because customers, by definition, are those who can afford for goods or services. Poor people in developing countries are usually not able to pay. From the World Bank's perspective, however, people, suffering from poverty, still pay for non-renewable energy for living. The obvious option is kerosene. In light of this, Principle 4 asserts that each person, no matter how poor, should be treated as a potential customer for renewable energy. In order to tap into the big pool of potential customers, how much the renewable energy providers should charge for their services should depend on the customers' affordability. Principle 5 suggests that the service providers can compare the price of the renewable energy with that of non-renewable options. In order to maintain the competitiveness and attractiveness of the renewable options, it should be set at the same, or even lower, price level of the non-renewable alternatives.

Principle 6 is based on the user-pays principle that each customer should make some contributions for obtaining the renewable energy (Redwood & Eerikainen, 2009). The forms of contributions can be in cash or in kind, such as crops and animals. The World Bank argues that providing services free-of-charge to poor people would cause more harm than good. Making contributions, no matter how small, empowers poor customers. It creates a sense of ownership by generating demand for better services and monitoring the quality of services.

Principle 7 deconstructs the potential customers into three groups: the not-so-poor, the poor, and the-very-poor. Each group is different, in terms of energy consumption patterns, affordability and needs. In order to tap into the potential of each group, renewable energy sectors, the World Bank argues, should provide tailor-made services for each group.

The last two principles touch on service delivery and governance (World Bank, 2004; Havet *et al.* 2009). The World Bank considers the central and regional governments in developing countries weak, inefficient and corrupt. Principle 8 builds on the Bank's good governance agenda, adopting a decentralised approach to provide funding and services directly to local communities. Principle 9 highlights the importance of building partners with local NGOs, grassroots community groups and the private sector. The assumptions are that NGOs have long been working with local people, so the sense of trust helps promote renewable energy. Driven by profit-making, the private sector is responsive to changing needs of customers and they should provide better services than civil servants.

CASE STUDIES AND RESEARCH METHODS

How effective are the World Bank's investment strategies in renewable energy at local levels? How have these principles been put into practice? What impact have they made on poor people's lives and how do people respond to the changes? In answering these questions, seeking a deeper understanding of local people's investment decisions in energy use is crucial. By comparing and contrasting the investment strategies of renewable energy between the World Bank, as the donor and initiator, and local people living in small communities, as users and customers, it will be useful to examine the effectiveness of the World Bank's renewable energy intervention on the ground.

We draw on two World Bank-related solar lighting projects in South Asia as case studies: an individual solar home system in Char Kajal, South Bangladesh[1] and a solar lantern system in Rajasthan, India. To promote individual solar home systems in Bangladesh, the World Bank assisted the Bangladeshi Central Government to set up a Dhaka-based company, called the Infrastructure Development Company. It oversaw all solar home projects in Bangladesh. It sub-contracted the projects to 15 NGOs. These NGOs were responsible for recruiting customers. They offered their clients three options: 50 watts, 40 watts and 20 watts. While the down-payment of the first option cost 5,200 taka[2] which supported four bulbs, one TV and one mobile phone charger, the down-payment of the last option was 1,000 taka which supported two bulbs and one mobile phone charger. In addition, the owners of the first option needed to pay 1,000 taka for monthly subscription fees while the owners paid 400 taka for the third option.

The down-payments might sound a lot to many poor households, yet the World Bank had already subsidised US$100 to each solar home system subscriber. The NGOs would provide ten years maintenance service support for the PV panels, five years for batteries and three years for electrical circuits. They went to each household to collect payment monthly. In order to incentivise NGOs in promoting the solar home systems, they would obtain US$80 for each transaction. In order to achieve the goal of building 200,000 units of solar home systems in Bangladesh by the end of 2009, the Infrastructure Development Company deliberately arranged a number of NGOs to work in the same areas in order to create a sense of competition.

We conducted our research in Char Kajal in January 2007. Char Kajal had 35,000 population, comprising 5,300 households. There were six NGOs selling the systems in that area. Through our personal networks, we successfully contacted two NGOs working in Char Kajal. There were no figures showing how many units of systems had been sold, but according to our interviews with the NGOs, totally 1,000 households in Char Kajal had bought the systems. This covered nearly one-fifth of the total population in the area. We found out that those without the solar home systems were largely poor or very poor households. Some of them had never been contacted by the NGOs. In their words, the NGOs only approached those families who could afford the 'expensive' systems. Considering the solar home systems as a 'golden goose', the NGOs were highly motivated to promote the systems in order to raise their revenues. However, the keen competition had caused bitterness amongst the NGOs. In addition, some NGO workers expressed their reservations about the success of the projects. They were worried that, when all free maintenance services expired in ten years, many solar home owners would not be financially capable of getting them replaced or repaired.

Our second case study was solar lantern projects in East Rajasthan, India[3]. Partially funded by the World Bank, the Energy and Resource Institute (TERI), a Delhi-based energy research organisation, was responsible for implementing the solar lantern projects in Rajasthan. It then contracted the projects out to a Rajasthan-based NGO, called Humana People to People (HPTP). HPTP received funding from TERI and served nine communities, with each community being offered 50 free solar lanterns. Apart from building solar battery-recharging facilities, HPTP was responsible for recruiting an entrepreneur from each community. The entrepreneur would assist HPTP to provide services, such as running the battery-recharging shops and offering technical maintenance, for their local users. Regarding the payment, HPTP told the entrepreneurs that they could not charge more than 60 rupees[4] for monthly subscribers. This amount of payment was based on estimation that each household paid 2 rupees for kerosene a day. For non-regular users, entrepreneurs were allowed to charge flexibly. The entrepreneurs could also expand the market by serving customers living in other communities. According to HPTP's calculation, if all 50 lanterns were successfully rented out, each entrepreneur could make more than 3,000 rupees profit a month. In order to ensure financial viability of the projects, each entrepreneur was requested to set up a maintenance fund by reserving 1,000 rupees of their incomes (or one thirds of their profits) into a bank at the end of each month.

We went to two villages, Banganga and Chhota Kantrana, to examine the impact of the solar lantern systems in April 2009. The former village was managed by a female entrepreneur while the latter by a male. The former system was regarded by HPTP as a 'failure' because of low subscription rate (only 15 out of 50 solar lanterns were regularly rented out). The latter, in contrast, had a 75% subscription rate. HPTP explained that the former received an older-generation of the solar lanterns which had suffered from many technical problems. The hands-off policy by HPTP worsened the situations. Assuming that the entrepreneurs and local users would develop a self-regulatory mechanism, HPTP did not provide immediate technical support to the entrepreneur in Banganga. The high break-down rate had first reduced the confidence of local users, and then undermined the profitability of the business. As a result, the vicious cycle had eroded the effectiveness of the solar lantern system in Banganga.

During our fieldwork, we adopted qualitative research methods. Based on the stakeholder analysis, we targeted different community groups, including women, the elderly, NGO leaders, government officials and entrepreneurs. We raised questions as to what renewable and non-renewable energy options were available in their villages, and what factors affected their choices between the renewable and non-renewable energy. We interviewed in total thirteen males and twelve females in our two case studies.

GAPS IN INVESTMENT STRATEGIES

Based on the nature of the nine design principles, we divide this section into three parts: sources of energy and role of technology, characteristics of the market and customers, and service delivery and governance. Our analysis will explore the potential gaps in investment strategies between the World Bank and local people over renewable energy.

Sources of Energy and Role of Technology

The principle of self-sufficiency by obtaining clean energy from local sources is highly appreciated. However, while localised renewable energy is utilised, equipment to make renewable energy technology often comes from outside. Many components of sustainable innovations are built and transported from another side of the countries concerned or even from abroad. In the case studies of Bangladesh and India, while sunshine was tapped into for producing solar lighting on site, many parts of the solar home systems came from Germany. Nearly all solar lanterns were assembled and transported to Rajasthan from Delhi. The Chief Executive of the Infrastructure Development Company explained, in interviews, that it was much cheaper to buy in bulk components from Germany than produce them locally (interview, 10/01/08) because they lacked the right kind of technology and expertise in Bangladesh.

Using sustainable technology to produce renewable energy gains support from the local population, and that helps promote public acceptance for clean energy. Villagers, from our case studies, considered solar home and solar lantern systems as a sign of modernisation and a source of community pride. Suffering from smoke and coughing by burning animal waste and firewood, they appreciated the smoke-free cooking environment. In the fishing community of Char Kajal, the solar lighting made it possible for women to make money by repairing fishing nets in the evening (Wong, 2009b). Children were also able to do their homework. Villagers exclaimed that this was not possible in the past because kerosene produced dim light, and they were forced to go to bed after 9pm. People in Rajasthan also considered the solar lanterns very useful. Because of the size and weight, they could take them out for outdoor activities, such as weddings, funerals, or harvesting in the field.

A new 'hierarchy of inferiority', built along the use of different sources of lighting, has also created favourable conditions for promoting renewable energy in poor communities. In the case study of the solar lantern systems, at the bottom, the poorest could not afford anything and they lived in the darkness. A layer above was those who relied on clay lamps and kerosene. The next layer was solar lantern users, and at the top were the villagers connected to national grids. Our research reveals that the solar lanterns had provided the users with new social status. That said, they still considered the national grids more powerful since it provided more functions and supported irrigation and TVs. The preferences for energy use had shown the users' pragmatic attitudes towards renewable and non-renewable energy. This point will be further elaborated on in the next section.

Characteristics of the Market and Customers

These five principles could be found in our case studies. In the individual solar home systems, the World Bank offered US$100 subsidies for each household. Each NGO would also receive US$80 as a bonus for each installation. In the case study of Rajasthan, solar lanterns were supplied free-of-charge; the constructing costs of the battery-recharging facilities were paid by the Bank and other donors.

Setting the prices of renewable energy at the same, or lower, level of non-renewable options had been implemented in this solar lantern case study. HPTP had made estimation that each household, on average, spent 2 rupees a day for kerosene. In order to make solar lanterns competitive, HPTP set the rental fees at 2 rupees a day. To meet different demands for the solar home systems in Char Kajal, customers could choose between 50 watts, 40 watts and 20 watts. Those who could afford the down-payment of 5,200 taka and monthly subscription fees of 1,000 taka chose the first option.

How effective are these principles in promoting and sustaining the renewable energy services? Do they match with the customers' actual investment strategies? We make our analysis from four aspects: pragmatic decision-making, affordability, service reliability, and unintended social impact.

Pragmatism

Villagers in our case studies showed a pragmatic approach to energy use. They considered energy as a means to achieve other economic and social goals. Whether the energy being produced was renewable or not did not bother them. Contrasting to the World Bank's division of energy as renewable and non-renewable, poor people in our case studies did not regard these artificial divisions useful. Neither did they consider renewable energy more desirable than the non-renewable. What they were more concerned were the availability of the energy options as well as the suitability of the options to meet the needs (c.c. Practical Action, 2009).

In the case study of Rajasthan, poor villagers had developed special energy-mix strategies. While solar lanterns were used for lighting, kerosene was still in demand for cooking and firewood was collected for creating warmth in winter. When harvests were good, farmers made use of the light and mobile design of the solar lanterns and took them to the fields to do extra work in the evening. Solar lanterns were also popular on occasions of weddings and funerals when external lighting was needed.

For the not-so-poor customers, gaining access to various sources of energy supplies was desirable. Successful connections to the national grids might boost their social status in their villages, but the unreliability of the grids in electricity supplies had created inconvenience. So they got connected to the grids and also rented the solar lanterns if needed in order to maximise energy security.

Both Indian and Bangladeshi governments have long been providing subsidies for using kerosene since it is still the most popular source of energy consumption. This policy may be changed because, according to a senior governmental official in Bangladesh, the government might consider reducing the subsidies on kerosene and divert the resources to support solar energy (interview, 12/01/07). Changes in the subsidy policies would only benefit the not-so-poor customers who could afford solar lighting. They would penalise the very poor population who still heavily rely on kerosene. This would push them to use more animal waste and firewood to cope with their daily needs.

Affordability

The three options for the solar home systems (50, 30 and 20 watts) offered in Char Kajal seemed to offer sufficient choices to meet different demands. Yet, the cheapest option still required 1,000 taka for down-payment and 400 taka for monthly subscription fees. The capital costs were still too high for many poor Bangladeshi households. The Infrastructure Development Company had considered introducing the 10 watts systems to capture the unmet market. The NGO workers realised that heavy down-payments could reduce the attractiveness of the solar home systems, but they were reluctant to introduce micro-credit schemes to deal with financial exclusion. Theoretically, micro-credit schemes help by pulling the needy together and provide them with some money to pay for the down-payments. In practice, however, the NGOs in Char Kajal considered the solar home systems as consumptive, rather than productive, goods. Although the solar lighting provided more income-generating opportunities for the poor, the NGOs were worried that, unlike using the credits to buy fertilisers or to install irrigation, the solar home systems would not be able to help poor families raise sufficient money to repay the debts.

How to pay for monthly subscription fees is another concern to the poor. In the case study of the solar lantern systems in Rajasthan, two rupees a day were charged in order to compete with kerosene. Yet, these two-rupee charges were only applicable for monthly subscribers. Each monthly subscriber was requested to pay 60 rupees in a lump sum. If villagers wanted to rent a lantern on a daily basis, they actually needed to pay up to five rupees. The NGO defended this policy, explaining that this would encourage more monthly subscription, and that would provide a better financial foundation for the systems in the long-run.

Our research, however, shows that many households could not afford 60 rupees in a lump sum. Villagers explained that farming is a seasonal business and income generation was highly fluctuating. Even though some of them were able to afford the lump sum, they were reluctant to do so because they feared they might need the money for unanticipated events, such as health problems.

Charging solar lighting at the same or lower prices of non-renewable options does not always make the renewable energy more attractive. Apart from pricing, customers' investment decision lies in how energy is actually kept and saved. Take solar lantern and kerosene as examples for comparison. Solar lanterns are generally able to provide five to six hours lighting for each household. Once the family pays the rents, the fees would become the 'sunk cost' because no matter the household uses the lantern or not, they have already paid for the services. They cannot save the solar energy for future use since they need to return the lantern to the shopkeeper the day after. Kerosene, in contrast, is kept in liquid form. Villagers can control when, and how much, kerosene is used and for what purposes. Unlike solar lighting, the extra kerosene can be kept for future use.

The practice of stealing electricity from the national grids also poses a threat to the demand for renewable energy. Some poor families, in interviews, suggested that this was their 'human right' to obtain free electricity from the government in order to alleviate poverty. Showing us how they connected to the national grids illegally, they defended their actions by saying that it was extremely unfair that the grids were built for the rich who were heavily subsidised by the government. The NGO workers believed that the 'right-to-electricity' thinking was inspired by the 'right-to-water' campaigns by other NGOs in the same region (c.c. Tully, 2006). Despite a heavy penalty being imposed to deter illegal connection, villagers suggested that corrupted government officials were willing to take bribes. In addition, villagers helped one another by setting up an early warning system in case of inspection.

Service Reliability

The success of renewable energy also relies on efficient and robust technical support. Unlike non-renewable energy options, local users find renewable energy innovations interesting, but technologically incomprehensible. This does not necessarily create barriers for adoption of renewable energy as long as the users receive quick service support in case of technical breakdown. The efficiency of obtaining support is, therefore, crucial to building public confidence and acceptance of the innovations. In the case study of Rajasthan, one village was still using the first-generation solar lantern system, and technical problems occurred frequently. Worse still, the slow responses of the NGO and the system manufacturers had made the situation worse. 15, out of 50, lanterns were broken. This high break-down rate and inefficient maintenance support created a bad reputation, thus reducing the demand for the services. The entrepreneur running the business was disincentivised as profits dropped. Once this vicious cycle was developed, it made the promotion of renewable energy difficult.

Unanticipated Social Impact

Poor people tend to help one another to cope with the daily struggle against poverty. Yet the forms of mutual help and cooperation are different between renewable and non-renewable energy options. Before the introduction of solar lighting, villages in our case studies relied on kerosene for lighting. Kept in liquid form, kerosene plays a special social function: it can be consumed-individually but transferred easily from one family to another. This means that the needy families can obtain kerosene from other villagers and use it whenever, and wherever, they like. Sharing animal waste and firewood has similar facility.

Solar lighting may produce better lighting as it covers a wider area, but sharing solar lighting is more geographically- and temporally-bound than kerosene. Unlike kerosene, people need to be close together at the same time and in the same place in order to enjoy the solar lighting. For example, households enjoying free lighting provided by their neighbours are those living next to ones who could afford solar lighting.

Solar lighting would also add a new layer of social divide in the communities, between those who could afford the solar lighting and those couldn't (c.c. Wong, 2009b). This inequality between the 'haves' and the 'have-nots' is built on other social differences, such as caste and land ownership. As mentioned before, the villagers had created a 'hierarchy of inferiority' with reference to different means of lighting. This psychological impact might offer strong incentives for poor people to improve their lives, but inequality could be further widened when renewable energy innovations become more technically advanced and provide more functions, such as radio and mobile phone charging. The poorest would feel they were further lagging behind.

Applying the pay-for-service principle in renewable energy also risks creating tension in the villages, and that could affect cooperation amongst villagers in the long-run. The entrepreneurs in the solar lantern case study could earn more than 3,000 rupees a month if all 50 lanterns were successfully rented out. This could easily stir up jealousy and conflict because many poor villagers regarded the solar lanterns as a gift from the donors to the whole community. Perceived as communal, rather than private, goods, villagers condemned the entrepreneurs for making personal gains out of the communal resources. Secondly, they felt exploited by the entrepreneurs when they were in great need. On special occasions, such as their children preparing for exams, community members hired the solar lanterns for one day, and were charged 5 rupees a day. This meant that non-monthly subscribers paid 150% higher than the daily rate of their monthly counterparts.

Service Delivery and Governance

NGOs play a strategic role in achieving the World Bank's renewable energy policies in developing countries (World Bank, 2008; Bank Information Centre, 2006). Ideally, NGOs act as a bridge between the World Bank and local people. They work closely with local communities and understand poor people's energy consumption patterns and habits. They then reflect their observations and concerns to the World Bank, which helps the Bank design more poverty-sensitive investment strategies in renewable energy.

The reality, however, could be very different. The reasons for a lack of engagement of the NGOs in our case studies, which will be illustrated below, were that the intervention model of the World Bank remained top-down in nature. It took an instrumental perspective and used NGOs to gain legitimacy for its renewable energy policies on the ground. By considering NGOs as an implementer, rather than a partner, the Bank designed the whole investment strategies and set the targets, such as area coverage and number of solar panels being sold, and then asked the NGOs to achieve this. The problems of this approach were that: firstly, NGOs failed to reflect the voices of local communities to the Bank. Secondly, NGOs internalised their role as purely executors. They became less critical of the Bank's energy policies, and that would further widen the gaps in investment strategies in renewable energy between the Bank and local people.

NGOs as Executors

NGOs are one of the winners of recent trends for stronger renewable energy development. The continuous flow of funding has resulted in income generation and new job opportunities. New innovations and external expertise have also widened their scope of services. There is a wide range of NGOs in developing countries: campaigning for social and ecological justice from one side and simply following policies set by the donors from the other. Our two case studies belong to the latter. They considered working with the World Bank as a way to raise revenues. The financial reward was obvious in the Bangladeshi case study in which the NGO would earn US$80 for each transaction. HPTP in the solar lantern systems in Rajasthan was only involved in recruiting and training entrepreneurs. How the entrepreneurs worked with local people, and what level of quality of service had been achieved, were not their prime concerns.

Using financial rewards to promote renewable energy policies could create pervasive incentives because only not-so-poor customers were served. The NGOs in the Bangladeshi case study targeted not-so-poor households because they knew they could afford the systems. The poor and very poor people were largely excluded in the promotion campaigns of the solar home systems. The conscious act of raising competition by putting six NGOs in a tiny area had made the situations worse. While the deliberate selection of customers had violated the World Bank's strategic principle that poor people were potential customers for renewable energy, the NGOs defended their actions, arguing that they needed to target the audience in order to maximise the impact of their limited manpower.

The hands-off attitudes of the NGO in Rajasthan resulted in a lack of vigilant monitoring of the quality of the service being delivered on the ground. The entrepreneurs kept changing the rules, without consulting HPTP and the customers. In Chhota Kantrana, for example, the entrepreneur imposed a new rule that no refund was allowed if the users did not report any technical faults within two hours after collecting the solar lanterns. The solar lantern users found the new rule unreasonable because they did not often check the batteries till they used the lanterns in the evening. To ensure the financial sustainability of the project, each entrepreneur was expected to set up a maintenance fund. The entrepreneur in our case study, however, admitted that she had not set it up even though the project had been operated for more than one year.

CONCLUSION

Framing renewable energy from a wider developmentalist perspective, the World Bank has avoided a narrow focus on energy poverty. Instead, it has been successful in integrating renewable energy policies into its global poverty-reduction strategies. In designing the guiding principles for the investment strategies on renewable energy in developing countries, the Bank has also drawn lessons from its previous mistakes, such as the over-reliance on the national grids to reach the poor. Recent ideas of maximising local natural resources to generate energy and decentralising the decision-making process have achieved some success on the ground.

That said, the Bank's investment strategies remain top-down in nature. The assumptions underlying the guiding principles are too economic. The use of the NGOs in implementing its renewable energy policies is too instrumental. Based on our detailed analysis of local people's energy consumption practices in two of the World Bank-funded solar lighting projects in Bangladesh and India, our chapter has exposed a wide gap in investment strategies between the World Bank and local people.

Poor people have adopted a pragmatic energy-mix in their daily lives because various sources of energy serve different purposes at different times. The distinctions between the renewable and non-renewable energy options and the Bank's explicit preferences for renewable energy risk reducing poor people's flexibility in coping with poverty. The guiding principles concerning payment stress the importance of making renewable energy more competitive than non-renewable energy options. This assumes that local customers making decisions on energy consumption are based on rational choice. We have provided evidence to suggest that poor people's decision-making process is socially-shaped. The right-to-free-electricity and the cultural norms of sharing collective resources are a few socio-political factors affecting their choice of energy use.

The Bank's instrumental use of the NGOs as an executor, rather than a genuine working partner, in its renewable energy policies is also problematic. Without the NGOs' critical reflections of the needs of local people, our study has shown that the Bank has missed the opportunity to improve its investment strategies on renewable energy by making them more poverty-sensitive.

In addressing these limitations, this chapter suggests that the World Bank should have a better understanding of the 'energy-poverty' relationships as well as poor people's daily energy consumption practices. While renewable energy may help lessen poverty, poverty can undermine the actual impact of the renewable energy policies on the ground. People living in chronic poverty, for instance, may sell the components of the solar home systems in return for short-term benefits (Practical Action, 2009). The Bank should also make the design of its strategic investment policies more participatory and reflective (Bank Information Centre, 2006). Paying more attention to the voice of local people and encouraging the NGOs to be more critical of its energy policies could achieve better outcomes of its interventions on renewable energy in developing countries.

REFERENCES

Asian Development Bank. (2000). *Review of the Energy Policy*. Manila. *Energy*, 2000.

Bank Information Centre & Bretton Woods Project. (2006). *How the World Bank's Energy Framework Sells the Climate and Poor People Short. A Civil Society Response to the World Bank's Investment Framework for Clean Energy and Development*. (available at: www.bicusa.org/proxy/Document.9515.aspx; accessed on 10 May 2009)

Bortman, J. (2005). Small Is Beautiful – Technology as if People Mattered. *The Journal of Technology Transfer*, *2*, 77–82. doi:10.1007/BF02221571.

Chaurey, A., Ranganathan, M., & Mohanty, P. (2004). Electricity Access for Geographically Disadvantaged Rural Communities – Technology and Policy Insights. *Energy Policy*, *32*, 1693–1705. doi:10.1016/S0301-4215(03)00160-5.

Havet, I., Chowdhury, S., Takada, M., & Cantano, A. (2009). *Energy in National Decentralization Policies*. United Nations Development Programs..

IEA International Energy. (2008). *Renewable Energy Services for Developing Countries. In support of the Millennium Development Goals: Recommended Practice & Key Lessons* (available at www.ieapvps.org/products/download/REEEP%20RESDC% 20%20MDGs%20v9%202.pdf; accessed on 7 July 2009)

Infrastructure Development Company Ltd. (2007). *Annual Report 2006-7*. Kawranbazar, Dhaka.

Martinot, E., Cabraal, A., & Mathur, S. (2001). World Bank/GEF Solar Home System Projects: Experiences and Lessons Learned 1993-2000. *Renewable & Sustainable Energy Reviews*, *5*, 39–57. doi:10.1016/S1364-0321(00)00007-1.

Paavola, J., & Adger, N. (2002). *Justice and Adaptation to Climate Change*. Tyndall Centre for Climate Change Research. Working Paper 23.

Practical Action. (2009). Energy Poverty: the Hidden Energy Crisis. London..

Redwood, J., & Eerikainen, J. (2009). *Climate Change and the World Bank Group: Phase I: An Evaluation of World Bank Win-Win Energy Policy Reforms*. Washington, D.C: World Bank..

Robins, B. (2006). Subsidised Solar Lighting: The Only Option for 1 Billion People. *Refocus*, *7*, 36–39. doi:10.1016/S1471-0846(06)70571-6.

Tully, S. (2006). The Human Right to Access Electricity. *The Electricity Journal*, *19*, 30–39. doi:10.1016/j.tej.2006.02.003.

Wamukonya, N. (2005). Solar Home System Electrification as a Viable Technology Option for Africa's Development. *Energy Policy*, *35*, 6–14. doi:10.1016/j.enpol.2005.08.019.

Willis, M., Wilder, M., & Curnow, P. (2006), *The Clean Development Mechanism: Special Considerations for Renewable Energy Projects*. Renewable Energy and International Law Project (available at: http://www.reilproject.org/about.htm; accessed on 7 July 2009)

Wong, S. (2009a). Linking Climate Change and Sustainable Energy Technologies to Develop a 'People-Centred' Intervention Framework. In W. Filho (ed.) Interdisciplinary Aspects of Climate Change, 449-468. New York: Peter Lang Scientific Publishers..

Wong, S. (2009b). Climate Change and Sustainable Technology: Re-linking Nature, Governance and Gender. *Gender and Development*, *17*, 95–108. doi:10.1080/13552070802696953.

Wong, S. (forthcoming). Overcoming Obstacles Against Effective Solar Lighting Interventions in South Asia. *Energy Policy*.

World Bank. (2000). *Energy and Development Report*. Washington: Washington DC..

World Bank. (2004). *World Development Report 2004: Making Services Work for Poor People*. New York: Oxford University Press..

World Bank. (2007a). *Technical and Economic Assessment of Off-grid, Mini-grid and Grid Electrification Technologies*. Energy Sector Management Assistance Program (ESMAP) Technical Paper 121/07. World Bank: Washington D.C.

World Bank. (2007b). *Catalyzing Private Investment for a Low-Carbon Economy - World Bank Group Progress on Renewable Energy and Energy Efficiency in Fiscal 2007*. Washington, D.C.: World Bank..

World Bank. (2008). *Issues Note of the REToolkit: A Resource for Renewable Energy Development*. Washington, D.C.: World Bank..

World Bank. (2009). *Renewable Energy & Energy Efficiency Financing by World Bank Group Hits All Time High*. Press release on September 10, 2009 (available at: http://beta.worldbank.org/climatechange/news/renewable-energy-energy-efficiency-financing -world-bank-group-hits-all-time-high; accessed on 15 Sept 2009)

ENDNOTES

1. This project was funded by the British Council (2006-7).
2. Currency rate: £1 = 114.9 taka (on 10/11/09).
3. This project was funded by the British Academy (2007-09).
4. Currency rate: £1 = 77.1 rupees (on 10/11/09).

This chapter was previously published in Advanced Analytics for Green and Sustainable Economic Development edited by Zongwei Luo, pages 50-63, copyright 2012 by Business Science Reference (an imprint of IGI Global).

Chapter 70
Assessing the Barriers to Greener Fiscal Measures and Ecological Tax Reform in the Transport Sector

David Browne
Trinity College, The University of Dublin, Ireland

Brian Caulfield
Trinity College, The University of Dublin, Ireland

Margaret O'Mahony
Trinity College, The University of Dublin, Ireland

ABSTRACT

The objective of this chapter is to: (1) appraise the purpose of fiscal measures in the transport sector; (2) provide an overview of potential fiscal measures and their application; and (3) examine potential barriers to the introduction of more efficient pricing and ecological tax reform (ETR) in the transport sector. It is concluded that, although vehicle and fuel taxation measures vary between jurisdictions, ultimately policy-makers should consider calibrating vehicle and fuel taxation to ensure that externalities are internalised, costs are shifted from vehicle ownership to use and that marginal variable costs are transparent to the network user.

However, it is possible that achieving ETR through full internalisation of external costs could be contentious due to political concerns over the potental impact on lower socio-economic groups and rural communities as well as commercial concerns over the competitiveness of the freight sector and macro-economic impacts.

DOI: 10.4018/978-1-4666-4852-4.ch070

INTRODUCTION

Transport causes a significant number of economic, social and environmental costs, resulting from, *inter alia*, congestion, opportunity costs associated with travel times, social exclusion, accidents and fatalities, air pollution, greenhouse gas (GHG) emissions, noise, waste, water pollution, loss of biodiversity and land fragmentation. Delucchi (2003) estimates that environmental externalities account for 36-64% of all external costs of motor vehicle use and 3-16% of the social costs of motor vehicle use. However, these externalities are often under-priced or not internalised at all (Calthrop and Proost, 1998; Proost et al., 2002; Litman, 2009).

Thus, users are not aware of the marginal external costs of their activities and do not factor this into their purchasing patterns or consumer behaviour, unless such costs are immediately transparent and/or vary with vehicle use. Lakshmanan et al. (2001) argue that transport users do not accurately perceive variable costs or the full monetary costs of a trip or journey and only consider direct costs such as parking or time costs. Furthermore, evidence suggests that consumers are more likely to consider the cost of vehicle purchases and do not consider long-run fuel or operating costs over the lifetime of the vehicle.

The objectives of the chapter are to:

- Provide an overview of how efficient pricing and ecological tax reform (ETR) could deliver a more sustainable transport system and minimise the negative impacts of transport;
- Present an overview of fiscal measures that are applicable in the transport sector;
- Identify potential barriers to ETR; and
- Provides selected examples of ETR in the European Union (EU).

The chapter concludes by suggesting that greater focus should be made on internalising external costs in a variable usage taxation system.

EFFICIENT PRICING AND ECOLOGICAL TAX REFORM

Pricing of transport network use and vehicle ownership has a number of purposes, including, *inter alia*:

- Payment for the cost of transport infrastructure provision;
- Provision of local authority and metropolitan funding;
- Contribution to general Exchequer taxation;
- Travel demand reduction;
- Traffic management in urban areas;
- Ensuring that the 'polluter pays' principle is enshrined in transport policy; and
- Internalisation of external costs.

However, although pricing has a number of simultaneous and complementary purposes, these may not be transparent or apparent to the network user and may not be explicitly stated or considered when designing, implementing and adjusting fiscal measures in the transport sector. The key is to ensure that the pricing system is efficient, effective, transparent and progressive and that it incentivises sustainable travel patterns and purchasing decisions through appropriate pricing mechanisms.

Environmental or Pigouvian taxes may be used to internalise the external costs of transport and encourage more efficient behaviour under the principle of fair and efficient pricing (Banister, 2007). Estimates vary as to the extent that such external costs are already currently internalised through excise rates, infrastructure charges and vehicle taxation (Proost et al., 2002). Many of the external environmental and social costs of transport are already internalised through fuel excise and vehicle taxation. However, these are seen more as revenue-raising measures and the external cost element is not transparent.

Delucchi (2003) argues that, in order to ensure the maximum gain in social welfare, an environmental tax should be equal to the cost of damage and should be levied on the immediate source of damage rather than on some proxy or correlate measure. Thus, damages from motor vehicle air pollution could be addressed by taxes that vary with the pollutant emitted as well as the time and location of emissions.

Greener fiscal measures, which can ensure more efficient pricing of the transport network, can encourage more sustainable travel and transport by:

1. Encouraging a reduction in demand for travel and transport;
2. Increasing the relative price attractiveness of alternative travel options;
3. Changing travel behaviour patterns;
4. Incentivising the purchase and use of fuel-efficient vehicle models; and
5. Raising awareness by sending market signals to consumers and network users about the external costs of transport.

Dresner et al. (2006a) argue that ETR achieves revenue neutrality by lowering other taxes, predominantly on labour. Revenues may be recycled through:

- Tax reform or revenue neutrality, whereby revenues from an environmental tax are used to reduce other taxes and taxation is shifted from economic 'goods' such as work, income or employment to environmental 'bads' like pollution;
- Revenues can be earmarked or hypothecated for specific projects; and
- Revenues can be used to fund 'compensation measures', e.g. redistribution to correct for some of the negative impacts on lower-income groups (Baranzini et al., 2000).

In addition, depending on the way revenues are recycled, environmental taxes may generate additional benefits such as the 'double dividend', where recycling revenues by reducing distortionary taxes may have positive impacts on economic growth and employment. However, tax reforms have met with opposition, even where they were combined with reductions in taxation on employment (Dresner et al., 2006a).

Types of Fiscal Measures

A number of fiscal measures exist and these are broadly classified under the following categories:

1. Vehicle taxation, including vehicle registration tax and annual motor or circulation tax;
2. Road pricing and infrastructure charging;
3. Congestion charging in urban areas;
4. Parking pricing, including on-street parking and workplace parking levies;
5. Fuel excise and carbon levies; and
6. Trading mechanisms, including personal carbon trading, tradable permits and emissions trading.

Vehicle Taxation

Vehicle-related taxes include:

- Vehicle purchase or registration taxes and
- Annual ownership or circulation taxes (Hayashi et al., 2001).

Tax incentives to encourage the purchase of more fuel-efficient vehicles have the potential to increase the overall fuel efficiency of the vehicle fleet and "reward" consumers who purchase more fuel-efficient vehicles. This can be done through structuring the vehicle taxation system so that tax, which is payable at the point of vehicle registration, and annual circulation tax (ACT) are based on the environmental impacts of motor use, e.g. CO_2 emissions or tailpipe emissions, rather than ownership or engine size (Gerard and Lave, 2003;

Austin and Dinan, 2005; Greene et al., 2005; Ryan et al., 2009).

A number of studies have looked at the effectiveness of this measure and concluded that vehicle taxation and, in particular, usage or circulation tax could incentivize more fuel-efficient purchasing patterns (Hayashi et al., 2001; Ryan et al., 2009). Johnstone and Karasakis (1999) argue that a vehicle characteristics tax, when used in conjunction with a fuel tax, can lead to economically efficient outcomes.

Road Pricing and Infrastructure Charging

Road pricing involves the levying of fees for the use of certain road sections and may be used to either:

- Finance infrastructure;
- Reduce travel demand;
- Limit the environmental impacts of transport; or
- Regulate or ration road use, for example in cases where road capacity is scarce and the network becomes overloaded (Lindberg, 1995).

It may refer to different schemes such as toll rings, differentiated charges or zone-based systems (Odeck & Brathen, 1997; Hug et al., 1997; Fournier & Monsigny, 1998; Odeck & Brathen, 2002). Nationwide road pricing schemes include measures where car users are either charged a flat fee or charges are differentiated based on, for instance, time of day, period of use, level of congestion, type of vehicle or distance travelled (Johansson & Mattsson, 1995; Bonsall et al., 2007). Infrastructure charging essentially involves charging transport users to compensate for the use of particular infrastructure or a part of the road network and using the revenue to fully or partially pay the marginal costs of infrastructure use through tolls at particular points along the network. It can be differentiated from road pricing in that it generally relates to the use of linear sections of the network rather than the whole road network and is levied at particular points on the network through tolls rather than on the basis of distance.

Urban Congestion Charging

Congestion is essentially a relative phenomenon and is linked to the difference between the roadway system performance that users expect and how the system actually performs. Congestion is endemic in all transport systems as demand for road space occasionally exceeds supply, even in well-maintained systems. It results in longer queuing, slower speeds and increased travel times, which impose costs on the economy (Taylor, 1997). The economic solution is that individuals should bear a certain proportion of the marginal external cost of their actions, i.e. the cost that their decision to drive imposes upon others.

Making individuals pay some of the marginal costs imposed on others could lead to the efficient use of road space, although making motorists pay the full marginal cost is not necessarily optimal as there may be some costs imposed upon society, in terms of lost productivity/sales, by reducing the volume of traffic (Clinch & Kelly, 2001). Congestion charges can be used as a form of travel demand management in urban areas and ensure that users of the road network partially or fully pay the marginal costs imposed by their contribution to congestion. Cordon and area charging schemes focus on the time and place of a journey but can be limited in scale and do not distinguish between lengths of journey or multi-trip journeys.

Parking Pricing

Parking pricing mechanisms include both workplace parking levies and on-street parking pricing (Calthrop & Proost, 1998; Johansson-Stenman, 1999). The availability of free workplace parking is generally considered to have a significant influence on car use for work trips (Willson, 1992;

Shoup & Willson, 1992). In addition, low-priced or free employer-provided parking at the workplace may discourage use of public transport (Merriman, 1998). Workplace parking levies arguably provide an incentive for occupiers of property to reduce the total number of parking spaces (Ison & Wall, 2000; 2002).

Managing workplace parking has, therefore, been examined as a demand management measure to see what its potential is for reducing car trips to the workplace (Feeney, 1989). Options may include:

- Parking charges, which can be applied (or increased);
- The number of parking spaces can be reduced or some reserved for car sharers;
- A benefit-in-kind tax can be applied to the provision of spaces;
- A workplace parking levy can be applied by local authorities; or
- Employees can be offered a cash value in lieu of a parking space, typically referred to as a cash-out payment (Watters et al., 2006).

On-street parking pricing also offers potential for incentivising modal shift and trip reduction (Shiftan, 2002). As the price of parking is often the largest component of an urban vehicle journey, its influence can be significant at higher tariff levels (Button & Verhoef, 1998). Its effectiveness depends not only on the responsiveness of demand to price but also the effect of price changes on the profile of the parking population in a given area. Low-cost parking has been shown to be a significant influencing factor with regard to trip generation decisions and modal shift (Higgins, 1992; Thomson & Richardson, 1998).

Fuel Excise and Carbon Levies

There are three purposes of a road fuel tax, including:

- Behavioral change;
- Reduction in environmental impacts; and
- Revenue-raising.

Most vehicle and fuel tax regimes have been designed with revenue generation rather than environmental benefits as a primary motive (Sterner, 2007). This has resulted in vehicle taxes, which differ greatly among jurisdictions by the amount charged and the method by which they are calculated and politics can often determine the level of fuel tax and the way in which it is applied. For example, some countries use fuel taxes to favour diesel fuel due to its higher energy efficiency, while others prefer to incentivize petrol, since there are lower particulate matter emissions associated with petrol combustion (Rietveld & Van Woudenberg, 2005).

One possible way of internalising the environmental impacts specifically related to GHG emissions is through a carbon tax or levy. This is a tax levied on the specific carbon content of fuels (Parry, 2007). The desirable level of the carbon tax is a complicated issue. Some would argue for cost-benefit analysis and, thus, set the tax equal to the social cost of carbon, if that can be estimated (Tol, 2005). Others argue that the tax should be set at a level that is sufficiently high to meet the emission target with reasonable certainty (Den Elzen et al., 2007). Li et al. (2004) argue that the carbon tax should not exceed the level that is acceptable to the electorate.

The impact of an increase in fuel tax depends on whether one adopts a short-term perspective or long-term perspective. A driver's response to higher prices is largely composed of a reduction in the amount of driving, purchases of more fuel-efficient vehicles and an increase in the fuel efficiency of driving through, for example, improved vehicle maintenance or changes in driving behaviour such as slower acceleration or reduced vehicle speed. Long-term changes may include shifts in demand for structural travel through changes in commuting patterns.

Carbon Trading

Carbon trading is based on the equitable allocation of GHG emissions among, for example the adult population in a jurisdiction, and includes *inter alia:*

- Tradable energy quota (TEQ) schemes;
- Domestic tradable quota (DTQ) schemes;
- Personal carbon allowances (PCAs);
- Personal carbon rationing;
- Cap and share;
- Rate all products and services (RAPS);
- The Ayres scheme; and
- The Sky Trust or cap and dividend.

It is argued that such schemes are economically efficient, fiscally progressive, morally egalitarian and provide theoretical certainty, compared with, for example, carbon taxes or fuel excise rates (Ayres, 1996; Starkey and Anderson, 2005; Roberts and Thumin, 2006; Johnson et al., 2008).

Transferable permit instruments such as tradable permits and emissions trading have a number of criteria in common, including:

1. The setting of quantified physical constraints in the form of obligations, permits, credits or rights allocated to target groups and
2. The permission granted to agents to transfer quotas between activities, products or places (offsetting), periods of time (banking) or to other agents (trading) (Raux, 2004).

In cases where a given environmental performance must be achieved in a context of uncertainty, a permit system is more likely to achieve a quantitative objective than taxation (Harwatt, 2008). In cases where agents are more sensitive to quantitative signals than price signals, particularly if the price-elasticity of demand is low in the short or medium-term, a permit system is more appropriate (OECD, 2001).

Comparative Analysis

The measures outlined above vary on the basis of a number of criteria, including:

1. Spatial dimension, i.e. whether they are applied on a local or national basis;
2. Temporal dimension, i.e. whether they apply at a particular point-in-time or are based on continuous vehicle use;
3. Functional dimension, i.e. the purpose of the fiscal measure;
4. To what extent external environmental and/or social costs are included;
5. Impact on modal shift;
6. Impact on travel demand and energy consumption;
7. Other environmental impacts;
8. Distributional and social equity impacts;
9. Impact on rural communities; and
10. Macroeconomic impacts, i.e. inflation, consumer price, tourism and competitiveness.

Table 1 shows a comparative analysis of potential fiscal measures under a number of relevant criteria.

Barrier Evaluation

Banister (2005) argues that barriers to sustainable transport can be divided into seven main categories, including:

1. Financial barriers;
2. Technical barriers;
3. Institutional and administrative barriers;
4. Public acceptability;
5. Legal or regulatory barriers;
6. Policy failures and unintended outcomes; and
7. Physical barriers.

Table 1. Comparative analysis of fiscal measures

Type of Fiscal Measure	Spatial Application	Temporal Dimension	Impact on Modal Shift	Direct Impact on Congestion	Impact on Rural Communities
Vehicle Taxation	National	Purchasing stage/annual	No	No	No
Nationwide Road Pricing	National	Varies with use	Yes	Yes	Yes
Infrastructure Charging	Local	Varies with road network	Yes	No	No
Urban Congestion Charging	Local	Varies with location	Yes	Yes	No
Parking Pricing	Local	Varies with location	Yes	Yes	No
Fuel Excise and Carbon Levies	National	Varies with use	Yes	No	Yes
Personal Carbon Trading	National	Varies with use	Yes	No	Yes
Emissions Trading	National	Varies with use	Yes	No	Yes

Financial Barriers

The potential financial barriers to ETR include:

1. Increased costs to consumers and network users. The specific costs will vary depending on the scheme and impacts will vary depending on willingness to pay (WTP) and availability of alternatives;
2. Loss of revenue to central Government or local authorities, e.g. loss of revenue from vehicle taxation, fuel excise or fuel tourism where fuel is purchased within a particular jurisdiction by residents outside the state;
3. Indirect macroeconomic impacts, i.e. impacts on Gross Domestic Product (GDP) and employment, increase in consumer price index (CPI) and inflation, competitiveness, impact on tourism, etc.
4. Costs of implementing the measure, including capital investment costs, costs of back-office equipment associated with road pricing and congestion charging, costs of administration associated with carbon trading or tradable permits and costs of enforcement. Investment in charging systems is a sunk or fixed cost and may require a significant payback period. Introduction of fiscal measures may also require a simultaneous investment in public transport services or park and ride.

Technical Barriers

Technical barriers associated with system design include (Jones, 1998; Eliasson, 2000; Rye & Ison, 2005; Bottrill, 2006; Rich & Nielsen, 2007; Holguín-Veras & Cetin, 2009):

1. It may be difficult to estimate the optimal fuel excise rate, parking charge, congestion charge or road toll due to different values of time, different externalities and WTP values;
2. Design and boundary issues, i.e. population coverage, system boundary, administration of the scheme, inclusion of heavy goods traffic, which is registered outside the state, and whether exemptions should be granted for domestic users of the network;
3. Interaction with other transport policy instruments;
4. Lack of empirical evidence supporting the benefits and costs of the measure;
5. Road pricing may face implementation problems, e.g. global positioning system

(GPS) technology may not be available or reliable; and
6. Difficulties with workplace parking levies, e.g. ensuring all required spaces are registered and difficulties compiling an inventory of spaces.

Another potential barrier may be a lack of alternatives or 'pull measures', i.e. where there is no viable alternative such as public transport or there is crowding on existing services (Stradling et al., 2000; Jansson, 2008; Kottenhoff & Brundell Freij, 2009). Behaviour may be more inelastic if consumers are prevented from changing their lifestyles due to entrenched structural barriers such as a lack of local employment opportunities, rigidity of the housing market or a lack of supply-side alternatives. Furthermore, the household structure may impact on the efficacy of the scheme, i.e. where household members are dependent on automobile transport, for example elderly, children, mobility-impaired, infirm, etc.

Institutional and Administrative Barriers

Institutional barriers may include (Newbery, 2001; De Moor, 2002; Santos & Rojey, 2004; Mackie, 2005; Prud'homme & Bocarejo, 2005; Button, 2006; Tight et al., 2007; Gudmundsson et al., 2009):

1. Political and institutional resistance to market reforms, in part due to various interests that benefit directly from automobile dependency or fuel consumption;
2. Reluctance to impose common pricing strategies or tax harmonisation at supra-national level in order to avoid antagonising national interests;
3. Concerns over the effectiveness or benefits of a new toll or charge;
4. Charges may be difficult to set at optimal rates due to institutional and administrative divisions;
5. Lock-in mechanisms such as rent-seeking, i.e. once a subsidy is in place, the beneficiaries receive the concentrated benefits of the subsidy payments and have a vested interest in seeing subsidies continue. In contrast, because the costs of the subsidy support are generally diffused throughout society, there is little incentive for taking political action to end subsidies;
6. Failure of institutional arrangements and communication strategies to ensure the credibility and legitimacy of information for decision-makers; and
7. Many formal and informal policy positions remain entrenched or embedded within organisations,

Political factors may include (Oberholzer-Gee & Weck-Hannemann, 2002; McQuaid & Grieco; 2005; Parry & Small, 2005; King et al., 2007; Isaksson & Richardson, 2009):

1. Lack of political courage or commitment to implement unpalatable demand restraint measures, particularly within a short-term electoral cycle;
2. Lack of a political champion;
3. Perverse subsidies may be maintained for political reasons in order to satisfy business interests, e.g. automobile industry, oil industry, car lobbies, etc.;
4. Fears over public protests and reaction; and
5. National governments may be unwilling to unilaterally introduce potentially uncompetitive fiscal measures and may adopt a 'wait and see' approach.

Administrative factors may include:

1. Long lead-in times may be required to gather public support;
2. Logistical difficulties of exempting certain drivers or network users from paying charges;

3. Administrative burden on Government departments, local authorities or commercial organisations; and
4. Complexity, e.g. number of operators in carbon trading schemes.

There have been a number of recent studies, which evaluated the business response to the implementation of fiscal measures. It was found that many employers are concerned that localised measures, such as parking charges, devalue the fringe benefits for their employees or may have repercussions in terms of punctuality and attendance (Van der Maas, 1998; Wang & Sharples, 1999; Van der Mede & Visser, 1999; Van Vuren & Johnson, 2000; Gerrard et al., 2001; Rye & Ison, 2005). Other commercial barriers associated with localised measures such as parking charges and congestion charging include concerns over impacts on commercial activity or retail revenue and 'retail flight' from the urban centre to the wider suburban hinterland (Eliasson and Mattsson, 2001; Windsor-Cundell, 2003; Quddus et al., 2006; Daunfeldt et al., 2009).

Public Acceptability

Public acceptability is a critical issue in implementation and it is generally acknowledged that acceptability, particularly in relation to equity, is one of the major barriers to successful implementation of more efficient pricing measures (Dresner & Ekins, 2004; Karlstrom & Franklin, 2009).

Public resistance might include (Baron & Jurney, 1993; Baron, 1995; Jones, 1998; Fransson & Garling, 1999; Ison, 2000; Harrington et al., 2001; Oberholzer-Gee & Weck-Hannemann, 2002; Johnson et al., 2008):

1. Electorate may oppose attempts to introduce full social cost pricing measures and progressive taxation;
2. Exclusion of modes of transport or exemptions for certain population cohorts could damage acceptability, particularly among those who may object on the grounds of double standards;
3. Perceived infringement on freedom and social norms;
4. Perception that roads are a public good and existing motor taxes more than cover the costs of their provision, i.e. the public are generally unwilling to pay for something that was previously 'free';
5. Perception of socially-adverse distributional effects or negative net welfare impact;
6. Concerns over the privacy of data collected from road pricing as well as the potential for fraud;
7. Concerns over impact on rural communities;
8. Opposition to coerced reforms and perception that negative effects are caused by others rather than by oneself; and
9. Public support for trading schemes may be tempered by widespread scepticism about the trading market itself.

Concerns over use of revenue include:

- A perception that the scheme is designed to raise revenue and
- A perception that revenue is not being used fairly.

In general, consumers are less averse to tax increases if the revenue is ring-fenced or hypothecated, e.g. improvements in the frequency and reliability of public transport, cheaper public transport fares, the creation of a comprehensive network of safe cycle and walking routes and the introduction or expansion of park and ride sites. The way potential revenue is used is very important in determining whether the measures have a positive or negative impact on the economy, e.g. if additional revenues are used to fund a reduction in income tax or social insurance contributions, it could produce a small increase in output and employment in the medium term, i.e. a "double dividend" (Smith, 2000; Schade & Schlag, 2000; Mayeres & Proost, 2001).

Lack of prior public support may affect the success of a scheme, although acceptability tends to increase after introduction of schemes where benefits are visible and more widely understood. Factors that might affect this include:

- Lack of public consultation or a public referendum and
- Lack of a trial period (King et al., 2007; Eliasson et al., 2009).

Public acceptability may also be affected by:

- Insufficient knowledge of the benefits or purpose of the scheme, which may be compounded by lack of effective advertising and communication, and
- The complexity of the scheme.

The scheme should be sufficiently easy and convenient for the public to use and there should be certainty over impacts (Klandermans, 1997; Sturmer & Simon, 2004; Schade & Baum, 2007; Bonsall et al., 2007; Johnson et al., 2008).

Regulatory or Legal Barriers

Regulatory barriers might include (Santos & Rojey, 2004; Rye & Ison, 2005):

1. Local authorities or central government may not have extant legislative authority to introduce measures; and
2. Possible legal challenges by stakeholders who perceive a threat to competitiveness.

Policy Failures and Unintended Outcomes

Policy failures may include (Borck & Wrede, 2005; Banister, 2007; Eliasson, 2009; Karlstrom & Franklin, 2009):

1. Increasing charges may not be effective after a certain threshold;
2. An ill-designed charging system may actually induce more congestion on non-charged roads than what is alleviated on the section of the network that is charged;
3. Believing that others will reduce car use may also lead to "free-riding," i.e. people decide not to reduce car use since the beneficial effects will materialise anyway;
4. Some public transport users may find car use more attractive than their current mode of transport and, therefore, switch to car in order to increase their welfare;
5. Compromise on fiscal measures may result in discounts and exemptions, which reduce the effectiveness of the policy;
6. Peak spreading may occur, in particular towards a later departure time;
7. Introducing carbon levies on fuel, which does not take into account the biofuel content, may not incentivise renewable energy producers;
8. Unavailability of alternatives or resistance to change may result in minimal behavioural change with a significant impact on prices and competitiveness; and
9. If parking is incorrectly priced and spaces are allowed to reach complete saturation, this may result in vehicles 'cruising' while searching for a space, resulting in 'stacking' with other vehicles.

Unintended outcomes may include (Schipper et al., 2002; Newbery, 2005; Eliasson, 2009; Karlstrom & Franklin, 2009):

1. Rebalancing taxation systems may distort the second-hand car market or impact on the sales of new vehicles;
2. Economic instruments may encourage more fuel-efficient vehicles and create a 'rebound effect';
3. Restructuring of the taxation system on the basis of fuel economy has the potential to

incentivize diesel vehicles with higher fuel economy but which emit higher proportions of carcinogenic particulate matter;

4. Existing public transport users might experience the effects of increased patronage;
5. Congestion charges and workplace parking levies may result in 'retail flight' and traffic being diverted from urban centres to suburban areas; and
6. Congestion charges may lead to "rat running" or driving in un-priced areas.

Physical Barriers

Physical barriers might include (Parry & Small, 2005; Knoflacher, 2006):

1. Degree of low-density residential development has implications for the responsiveness of an increase in fuel prices;
2. Lack of closer employment opportunities or supply-side alternatives; and
3. Long walking distances to public transport stops.

CASE-STUDIES OF ECOLOGICAL TAX REFORM IN THE EUROPEAN UNION

The objective of this section is to provide an overview of some examples of where ETR has been applied in European Union (EU) Member States. A number of European countries have introduced ETR, including, *inter alia*, Finland, Denmark, Sweden, the Netherlands, the UK and Germany. For example, the German approach consists of increases in existing energy taxes. Revenues from the taxes have been used to decrease social security contributions, with the reductions divided evenly between employees and employers.

The approach to ETR in the UK consisted of the introduction of the Climate Change Levy (CCL), which is a tax on commercial and industrial energy consumption. Revenues from the CCL have been recycled back to business by decreasing employers' tax contributions (Agnolucci, 2009). Dresner et al. (2006b) have assessed the impact of the CCL in the UK and found that the problem that the ETR faces in terms of public acceptability is not so much outright hostility but rather conceptual problems with the design, including a lack of trust about the use of revenues and difficulty in understanding the purpose of a tax shift. The precursor to the CCL in the UK was the fuel duty escalator, which was introduced in 1993. Between 1993 and 1997, the escalator increased road fuel duties by 5% in real terms each year. In 1997, this was increased to 6%. However, in 1999, it was announced that this duty would be abandoned.

The UK Green Fiscal Commission (2009a, b) argued that, if ETR is to succeed, there is a need to:

- Motivate people about the importance of climate change,
- Overcome scepticism that taxes would be reduced and
- Communicate the financial implications for households and institutions.

Deroubaix and Leveque (2006) argue that the failure of ETR in France has been due to:

- The fact that decision-makers were badly informed about business and general public perceptions;
- Is difficult to reach a balance between social acceptability and political feasibility; and
- Failures at administrative level.

DISCUSSION AND CONCLUSION

This chapter sought to critically examine a range of fiscal measures that could potentially lead to more efficient pricing in the transport sector. There is extensive academic literature on optimal transport pricing, marginal social cost pricing and Pigouvian taxes, which varies in its objectives but generally aims to put a case for the internalisation of external costs and to calculate the optimal fuel excise or tax. In addition, there is extensive debate on how such external costs should be internalised although there is generally broad consensus that this should be done by 'getting the prices right'.

However, moves towards more efficient pricing are often opposed on the basis that transport demand is inelastic, particularly in the short-term, due to lack of alternatives. Furthermore, in many jurisdictions, increasing mobility and dispersal of land use have become self-reinforcing so that car use has become more of a necessity than a choice. Therefore, consumers are more likely to change purchasing patterns rather than trip patterns. This raises an important point about the short-term and long-term implications of increasing fuel taxes or imposing other fiscal measures. Transport demand can be differentiated between structural and discretionary demand. Structural demand is long-term and generally fairly inelastic as it is dictated by residential settlement and commuting patterns, which are not easy to alter in the short-term. Discretionary demand for leisure and recreational travel, however, is more elastic and can be influenced through taxation and price changes.

Thus, a fiscal measure, which has a significant impact in urban areas or in areas with high density and compact spatial distribution, may have a negligible impact in rural areas with long-distance commuting. This negative impact is exacerbated where it captures low-income earners, who are 'structurally locked-in' to their existing commuting patterns. Furthermore, the long-term and cumulative impact of fiscal measures needs to be evaluated and understood. This is because, as consumers adapt to increased price changes through behavioural change and purchasing patterns, there will be a threshold beyond which consumer choice could be completely inelastic.

Transport users already pay a significant amount but the price they pay often bears little connection to the real costs on society of their choices. Price signals may not be effective if the market does not offer realistic alternatives such as cleaner vehicles at an affordable price or an appropriate level of service in another mode of transport. These alternatives may not always exist, in particular where there are market failures or where institutional or technological 'lock-in' favours a particular technological approach or solution.

In terms of design of a fiscal measure or a suite of fiscal instruments for the transport sector, it is necessary that such measures are efficient, effective and socially progressive. In order to ensure that fiscal measures do not result in an onerous burden on the economy, it has been argued that any rebalancing of fiscal measures should be accompanied by concomitant reductions in income tax and/or social insurance. ETR can be achieved where the burden shifts from economic goods such as labour (personal income tax), capital (corporate income tax), and consumption (value-added tax (VAT) and other indirect taxes) to activities that lead to environmental pressures and natural resource use. It may also result in the accrual of a 'double dividend' as a tax shift from the relatively abundant factor of labour leads to positive effects on employment and GDP.

Thus, changes to fuel taxation can avoid being socially regressive by having a balancing mechanism to compensate for those in lower socio-economic groups, who may be disproportionately burdened. One way to compensate against this would be to hypothecate or ring-fence revenues for improvements in public transport services or for deductions in income tax or social insurance. User charges should be applied as closely as possible to the source of the externality to optimize economic efficiency and no single mechanism can capture all external costs due to their diverse nature. Tax and price increases should be gradual and predictable to allow individuals and firms to adjust when making long-term decisions.

However, support for ETR tends to be conditional. Therefore, if a programme of green fiscal reform is to be advanced, there is a need to:

- Motivate people about the importance of climate change and the health and economic benefits of sustainable transport;
- Overcome scepticism about taxation measures;
- Effectively communicate the financial implications for households and commercial operators; and
- Facilitate the transition to low-carbon lifestyles.

Consideration should also be given to those residing in rural areas, who do not have access to public transport or cannot walk or cycle for a variety of reasons, e.g. distance, safety, lack of facilities, etc. If green taxes are to be both politically attractive and economically effective, they must be clearly distinguished from other taxes or charges; set at levels determined by acceptable methods and applied uniformly to all sources of the same damage.

All the measures analysed in this chapter have particular strengths and weaknesses and vary in terms of their point of application, administration, ease of implementation, distributive effects, macroeconomic impacts and efficiency. It may be that no one single measure is the optimal solution and that a plurality of policy options is needed, depending on:

- The objective, e.g. revenue-raising, internalisation of external costs, sending appropriate market signals, etc.;
- The mode of transport, e.g. passenger and freight transport; or
- Geographical focus, e.g. urban parking charges versus nationwide road pricing.

Some measures may be preferable in the short-term, while others could be phased in over a longer time period.

Ideally, all costs of vehicle use would be internalised as marginal variable costs, which would send a constant market signal to consumers and incentivise more efficient behaviour. Where the fixed cost of car ownership is more predominant than the variable costs of use, it is possible that vehicle owners may feel entitled or incentivised to use their car due to the initial expenditure and ancillary costs of insurance and motor tax. One of the ways to address this would be adjust vehicle taxation and replace it with a taxation system, which levies taxation according to network use rather than car ownership and internalises all fixed and variable costs in a marginal pricing system.

REFERENCES

Agnolucci, P. (2009). The effect of the German and British environmental taxation reforms: A simple assessment. *Energy Policy*, *37*(8), 3043–3051. doi:10.1016/j.enpol.2009.03.052.

Austin, D., & Dinan, T. (2005). Clearing the air: The costs and consequences of higher CAFE standards and increased gasoline taxes. *Journal of Environmental Economics and Management*, *50*(3), 562–582. doi:10.1016/j.jeem.2005.05.001.

Ayres, R. U. (1996). *Environmental market failures: Are there any local market-based corrective mechanisms for global problems?* France: INSEAD..

Banister, D. (2005). Overcoming barriers to the implementation of sustainable transport. In P. Rietveld, & R. R. Stough (Eds.), *Barriers to sustainable transport: institutions, regulation and sustainability*. London: Spon Press..

Banister, D. (2007). Sustainable transport: Challenges and opportunities. *Transportmetrica*, *3*(2), 91–106. doi:10.1080/18128600708685668.

Baranzini, A. B., Goldemberg, J., & Speck, S. (2000). A future for carbon taxes. *Ecological Economics*, *32*(3), 395–412. doi:10.1016/S0921-8009(99)00122-6.

Baron, J. (1995). Blind justice: Fairness to groups and the do-no-harm principle. *Journal of Behavioral Decision Making*, *8*(2), 71–83. doi:10.1002/bdm.3960080202.

Baron, J., & Jurney, J. (1993). Norms against voting for coerced reform. *Journal of Personality and Social Psychology*, *64*(3), 347–355. doi:10.1037/0022-3514.64.3.347.

Bonsall, P., Shires, J., Maule, J., Matthews, B., & Beale, J. (2007). Responses to complex pricing signals: Theory, evidence and implications for road pricing. *Transportation Research Part A, Policy and Practice*, *41*(7), 672–683. doi:10.1016/j.tra.2006.06.001.

Borck, R., & Wrede, M. (2005). Political economy of commuting subsidies. *Journal of Urban Economics*, *57*(3), 478–499. doi:10.1016/j.jue.2004.12.003.

Bottrill, C. (2006). *Personal Carbon Trading: Excluding Public Transport*. Retrieved from: http://www.ukerc.ac.uk/Downloads/PDF/1/12062006_PCT_transport_cb.pdf

Button, K. J. (2006). The political economy of parking charges in "first" and "second-best" worlds. *Transport Policy*, *13*(6), 470–478. doi:10.1016/j.tranpol.2006.05.004.

Button, K. J., & Verhoef, E. T. (1998). *Road pricing, traffic congestion and the environment*. Cheltenham, Northampton: Edward Elgar..

Calthrop, E., & Proost, S. (1998). Road transport externalities: Interaction between theory and empirical research. *Environmental and Resource Economics*, *11*(3-4), 335–348. doi:10.1023/A:1008267917001.

Clinch, J. P., & Kelly, J. A. (2001). *Economics of traffic congestion. University College Dublin*. Retrieved from: http://www.ucd.ie/gpep/gpepinfo/downloads/index.html

Daunfeldt, S.-O., Rudholm, N., & Ramme, U. (2009). Congestion charges and retail revenues: Results from the Stockholm road pricing trial. *Transportation Research Part A, Policy and Practice*, *43*(3), 306–309. doi:10.1016/j.tra.2008.09.005.

De Moor, A. (2002). *Building a grand deal to phase out harmful energy subsidies*, Presentation to the INFORSE-Europe Seminar, 29 November 2002, Brussels. Retrieved from: http://www.inforse.dk/europe/ppt_docs/Subsidies-Andre.ppt

Delucchi, M. A. (2003). Environmental externalities of motor vehicle use. In D. A. Hensher, & K. J. Button (Eds.), *Handbook of transport and the environment* (Vol. 4). Elsevier..

Den Elzen, M. J. G., Meinshausen, M., & Van Vuuren, D. P. (2007). Multi-gas emission envelopes to meet greenhouse gas concentration targets: Costs versus certainty to limiting temperature tncrease. *Global Environmental Change*, *17*(2), 260–280. doi:10.1016/j.gloenvcha.2006.10.003.

Deroubaix, J.-F., & Leveque, F. (2006). The rise and fall of French ecological tax reform: social acceptability versus political feasibility in the energy tax implementation process. *Energy Policy*, *34*(8), 940–949. doi:10.1016/j.enpol.2004.08.047.

Dresner, S., Dunne, L., Clinch, P., & Beuermann, C. (2006a). Social and political responses to ecological tax reform in Europe: an introduction to the special issue. *Energy Policy*, *34*(8), 895–904. doi:10.1016/j.enpol.2004.08.043.

Dresner, S., & Ekins, P. (2004). *The social impacts of environmental taxes: Removing regressivity*. Research Discussion Paper 19, Policy Studies Institute (PSI).

Dresner, S., Jackson, T., & Gilbert, N. (2006b). History and social responses to environmental tax reform in the United Kingdom. *Energy Policy*, *34*(8), 930–939. doi:10.1016/j.enpol.2004.08.046.

Eliasson, J. (2000). *The use of average values of time in road pricing. A Note on a Common Misconception. Department of Infrastructure and Planning*. KTH..

Eliasson, J. (2009). A cost-benefit analysis of the Stockholm congestion charging system. *Transportation Research Part A, Policy and Practice*, *43*(4), 468–480. doi:10.1016/j.tra.2008.11.014.

Eliasson, J., Hultkrantz, L., Nerhagen, L., & Smidfelt-Rosqvist, L. (2009). The Stockholm congestion-charging trial 2006: Overview of effects. *Transportation Research Part A, Policy and Practice*, *43*(3), 240–250. doi:10.1016/j.tra.2008.09.007.

Eliasson, J., & Mattsson, L.-G. (2001). Transport and location effects of road pricing. *Journal of Transport Economics and Policy*, *35*(3), 417–456.

Feeney, B. P. (1989). A review of the impact of parking policy measures on travel demand. *Transportation Planning and Technology*, *13*(4), 229–244. doi:10.1080/03081068908717403.

Fournier, P., & Monsigny, M. (1998). *Road pricing via tolling*. Paper presented at the 8th World Conference on Transport Research (WCTR), Antwerp, 12–17 July 1998.

Fransson, N., & Garling, T. (1999). Environmental concern: Conceptual definitions, measurement methods and research findings. *Journal of Environmental Psychology*, *19*(4), 369–382. doi:10.1006/jevp.1999.0141.

Gerard, D., & Lave, L. B. (2003). *The economics of CAFE reconsidered: A response to CAFE critics and a case for fuel economy standards*. AEI-Brookings Joint Centre of Regulatory Studies..

Gerrard, B., Still, B., & Jopson, A. (2001). The impact of road pricing and workplace parking levies on the urban economy: Results from a survey of business attitudes. *Environment & Planning A*, *33*(11), 1985–2002. doi:10.1068/a3412.

Green Fiscal Commission (2009a). *Lessons from two green tax shifts in the United Kingdom – Briefing Paper One.*

Green Fiscal Commission (2009b). *Public opinion on a green tax shift – Briefing Paper Three.*

Greene, D. L., Patterson, P. D., Singh, M., & Li, J. (2005). Feebates, rebates and gas-guzzler taxes: A study of incentives for increased fuel economy. *Energy Policy*, *33*(6), 757–775. doi:10.1016/j.enpol.2003.10.003.

Gudmundsson, H., Ericsson, E., Beser Hugosson, M., & Smidfeldt Rosqvist, L. (2009). Framing the role of decision support in the case of the Stockholm congestion charging trial. *Transportation Research Part A, Policy and Practice*, *43*(3), 258–268. doi:10.1016/j.tra.2008.09.003.

Harrington, W., Krupnick, A. J., & Alberini, A. (2001). Overcoming public aversion to congestion pricing. *Transportation Research A. Policy & Practice*, *35*(2), 87–105.

Harwatt, H. M. (2008). *Reducing carbon emissions from personal transport through the application of a tradable permit scheme: Empirical findings and policy implications*, Paper presented at the International Transport Forum, Leipzig.

Hayashi, Y., Kato, H., & Teodoro, R. V. R. (2001). A model system for the assessment of the effects of car and fuel green taxes on CO_2 emission. *Transportation Research Part D, Transport and Environment*, *6*(2), 123–139. doi:10.1016/S1361-9209(00)00021-3.

Higgins, D. (1992). Parking taxes: Effectiveness, legality and implementation, some general considerations. *Transportation*, *19*(3), 221–230. doi:10.1007/BF01099978.

Holguin-Veras, J., & Cetin, M. (2009). Optimal tolls for multi-class traffic: analytical formulations and policy implications. *Transportation Research Part A, Policy and Practice*, *43*(4), 445–467. doi:10.1016/j.tra.2008.11.012.

Hug, K., Mock-Hecker, R., & Wurtenberger, J. (1997). *Transport demand management by electronic fee collection in a zone-based pricing scheme: The Stuttgart MobilPASS field trial.* Report from the Research Institute for Applied Knowledge Processing..

Isaksson, K., & Richardson, T. (2009). Building legitimacy for risky policies: The cost of avoiding conflict in Stockholm. *Transportation Research Part A, Policy and Practice*, *43*(3), 251–257. doi:10.1016/j.tra.2008.09.002.

Ison, S. (2000). Local authority and academic attitudes to urban road pricing: A UK perspective. *Transport Policy*, *7*(4), 269–277. doi:10.1016/S0967-070X(00)00019-6.

Ison, S., & Wall, S. (2002). Attitudes to traffic-related issues in urban areas of the UK and the role of workplace parking charges. *Journal of Transport Geography*, *10*(1), 21–28. doi:10.1016/S0966-6923(01)00035-7.

Ison, S. G., & Wall, S. (2000). An efficient transport system to support a competitive economy: The role of workplace parking charges. In L. Lloyd-Reason, & D. Walls (Eds.), *Dimensions of competitiveness: Issues and policies*. Cheltenham, UK: Edward Elgar..

Jansson, J. O. (2008). Public transport policy for central-city travel in the light of recent experiences of congestion charging. *Research in Transportation Economics*, *22*(1), 179–187. doi:10.1016/j.retrec.2008.05.027.

Johansson, B., & Mattson, L.-G. (Eds.). (1995). *Road pricing: theory, empirical assessment and policy*. Boston: Kluwer..

Johansson-Stenman, O. (1999). Regulating road transport externalities: Pricing versus command and control. In T. Sterner (Ed.), *The market and the environment*. Cheltenham, UK: Edward Elgar..

Johnson, M., Harfoot, M., Musser, C., Wiley, T., Pollitt, H., Chewpreecha, U., & Tarafdar, J. (2008). *A study in personal carbon allocation: Cap and share, Comhar Sustainable Development Council*. SDC..

Johnstone, N., & Karousakis, K. (1999). Economic incentives to reduce pollution from road transport: The case for vehicle characteristics taxes. *Transport Policy*, *6*(2), 99–108. doi:10.1016/S0967-070X(99)00011-6.

Jones, P. (1998). Urban road pricing: Public acceptability and barriers to implementation. In K. J. Button, & E. T. Verhoef (Eds.), *Road pricing, traffic congestion and the environment: Issues of efficiency and social feasibility*. Cheltenham, UK: Edward Elgar..

Karlstrom, A., & Franklin, J. P. (2009). Behavioral adjustments and equity effects of congestion pricing: Analysis of morning commutes during the Stockholm trial. *Transportation Research Part A, Policy and Practice*, *43*(3), 283–296. doi:10.1016/j.tra.2008.09.008.

King, D., Manville, M., & Shoup, D. (2007). The political calculus of congestion pricing. *Transport Policy*, *14*(2), 111–123. doi:10.1016/j.tranpol.2006.11.002.

Klandermans, B. (1997). *The social psychology of protest*. Oxford, UK: Blackwell..

Knoflacher, H. (2006). A new way to organize parking: The key to a successful sustainable transport system for the future. *Environment and Urbanization*, *18*(2), 387–400. doi:10.1177/0956247806069621.

Kottenhoff, K., & Brundell Freij, K. (2009). The role of public transport for feasibility and acceptability of congestion charging – The case of Stockholm. *Transportation Research Part A, Policy and Practice*, *43*(3), 297–305. doi:10.1016/j.tra.2008.09.004.

Lakshmanan, T. R., Nijkamp, P., Rietveld, P., & Verhoef, E. T. (2001). Benefits and costs of transport: Classification, methodologies and policies. *Papers in Regional Science*, *80*(2), 139–164. doi:10.1007/PL00013618.

Li, H., Berrens, R. P., Bohara, A. K., Jenkins-Smith, H. C., Silva, C. L., & Weimer, D. L. (2004). Would developing country commitments affect US households' support for a modified Kyoto Protocol? *Ecological Economics*, *48*(3), 329–343. doi:10.1016/j.ecolecon.2003.10.010.

Lindberg, G. (1995). Road pricing: Policy and options for the future. In B. Johansson, & L.-G. Mattson (Eds.), *Road pricing: Theory, empirical assessment and policy*. Boston: Kluwer. doi:10.1007/978-94-011-0980-2_12.

Litman, T. A. (2009). *Transportation cost and benefit analysis: Techniques, Estimates and implications, Victoria Transport Policy Institute*. VTPI..

Mackie, P. (2005). The London congestion charge: A tentative economic appraisal. A comment on the paper by Prud'homme and Bocajero. *Transport Policy*, *12*(3), 288–290. doi:10.1016/j.tranpol.2005.04.001.

Mayeres, I., & Proost, S. (2001). Marginal tax reform, externalities and income distribution. *Journal of Public Economics*, *79*(2), 343–363. doi:10.1016/S0047-2727(99)00100-0.

McQuaid, R., & Grieco, M. (2005). Edinburgh and the politics of congestion charging: Negotiating road user charging with affected publics. *Transport Policy*, *12*(5), 475–476. doi:10.1016/j.tranpol.2005.06.005.

Merriman, D. (1998). How many parking spaces does it take to create one additional transit passenger? *Regional Science and Urban Economics*, *28*(5), 565–584. doi:10.1016/S0166-0462(98)00018-0.

Newbery, D. M. (2001). *Harmonizing energy taxes in the EU*. Retrieved from: http://www.econ.cam.ac.uk/faculty/newbery/files/ROTTERDM.PDF

Newbery, D. M. (2005). *Why tax energy? Towards a more rational energy policy*. CMI Working Paper 72.

Oberholzer-Gee, F., & Weck-Hannemann, H. (2002). Pricing road use: Politico-economic and fairness considerations. *Transportation Research Part D, Transport and Environment*, *7*(5), 357–371. doi:10.1016/S1361-9209(02)00005-6.

Odeck, J., & Brathen, S. (1997). On public attitudes toward implementation of toll roads — The case of Oslo toll ring. *Transport Policy*, *4*(2), 78–83. doi:10.1016/S0967-070X(97)00008-5.

Odeck, J., & Brathen, S. (2002). Toll financing in Norway: The success, the failures and perspectives for the future. *Transport Policy*, *9*(3), 253–260. doi:10.1016/S0967-070X(02)00030-6.

Organisation for Economic Cooperation and Development. (2001). *Domestic transferable permits for environmental management: Design and implementation*. Paris: OECD..

Parry, I. W. H. (2007). Are costs of reducing greenhouse gases from passenger vehicles negative? *Journal of Urban Economics*, *62*(2), 273–293. doi:10.1016/j.jue.2006.12.003.

Parry, I. W. H., & Small, K. A. (2005). Does Britain or the United States have the right gasoline tax? *The American Economic Review*, *95*(4), 1276–1289. doi:10.1257/0002828054825510.

Proost, S., Van Dender, K., Courcelle, C., De Borger, B., Peirson, H., & Sharp, D. et al. (2002). How large is the gap between present and efficient transport prices in Europe? *Transport Policy*, *9*(1), 41–57. doi:10.1016/S0967-070X(01)00029-4.

Prud'homme, R., & Bocajero, J. P. (2005). The London congestion charge: A tentative economic appraisal. *Transport Policy*, *12*(3), 279–287. doi:10.1016/j.tranpol.2005.03.001.

Quddus, M. A., Carmel, A., & Bell, M. G. H. (2006). The impact of the London congestion charge on retail: The London experience. *Journal of Transport Economics and Policy*, *41*(1), 113–133.

Raux, C. (2004). The use of transferable permits in transport policy. *Transportation Research Part D, Transport and Environment*, *9*(3), 185–197. doi:10.1016/j.trd.2004.01.001.

Rich, J., & Nielsen, O. A. (2007). A socio-economic assessment of proposed road user charging schemes in Copenhagen. *Transport Policy*, *14*(4), 330–345. doi:10.1016/j.tranpol.2007.03.003.

Rietveld, P., & Van Woudenberg, S. (2005). Why fuel prices differ. *Energy Economics*, *27*(1), 79–92. doi:10.1016/j.eneco.2004.10.002.

Roberts, S., & Thumin, J. (2006). *A rough guide to individual carbon trading: The ideas, the issues and the next steps*. DEFRA..

Ryan, L., Ferreira, S., & Convery, F. (2009). The impact of fiscal and other measures on new passenger car sales and CO_2 emissions intensity: Evidence from Europe. *Energy Economics*, *31*(3), 365–374. doi:10.1016/j.eneco.2008.11.011.

Rye, T., & Ison, S. (2005). Overcoming barriers to the implementation of car parking charges at UK workplaces. *Transport Policy*, *12*(1), 57–64. doi:10.1016/j.tranpol.2004.11.002.

Santos, G., & Rojey, L. (2004). Distributional impacts of road pricing: The truth behind the myth. *Transportation*, *31*(1), 21–42. doi:10.1023/B:PORT.0000007234.98158.6b.

Schade, J., & Baum, M. (2007). Reactance or acceptance? Reactions towards the introduction of road pricing. *Transportation Research Part A, Policy and Practice*, *41*(1), 41–48. doi:10.1016/j.tra.2006.05.008.

Schade, J., & Schlag, B. (2000). Acceptability of urban transport pricing. VATT Research Report 72, Helsinki..

Schipper, L., Marie-Lilliu, C., & Fulton, L. (2002). Diesels in Europe: Analysis of characteristics, usage patterns, energy savings and CO_2 implications. *Journal of Transport Economics and Policy*, *36*(2), 305–340.

Shiftan, Y. (2002). The effects of parking pricing and supply on travel patterns to a major business district. In E. Stern, I. Salomon, & P. H. L. Bovy (Eds.), *Travel Behaviour*. Cheltenham: Edward Elgar..

Shoup, D. C., & Willson, R. (1992). *Commuting, congestion, and pollution: The employer-paid parking connection.* Papers Presented at the Congestion Pricing Symposium, US Department of Transportation, June 10–12.

Smith, Z. (2000). *The petrol tax debate*, The Institute for Fiscal Studies, Briefing Note No.8.

Starkey, R., & Anderson, K. (2005). *Domestic tradable quotas: A policy instrument for reducing greenhouse gas emissions from energy use*, Tyndall Centre for Climate Change Research, Technical Report 39.

Sterner, T. (2007). Fuel taxes: an important instrument for climate policy. *Energy Policy*, *35*(6), 3194–3202. doi:10.1016/j.enpol.2006.10.025.

Stradling, S. G., Meadows, M. L., & Beatty, S. (2000). Helping drivers out of their cars: Integrating transport policy and social psychology for sustainable change. *Transport Policy*, *7*(3), 207–215. doi:10.1016/S0967-070X(00)00026-3.

Sturmer, S., & Simon, B. (2004). Collective action: towards a dual-pathway model. *European Review of Social Psychology*, *15*(1), 59–99. doi:10.1080/10463280340000117.

Taylor, M. A. P. (1997). *The effects of lower urban speed limits on mobility, accessibility, energy and the environment: Trade-offs with increased safety*. Canberra, Australia: Federal Office of Road Safety..

Thomson, R. G., & Richardson, A. J. (1998). A parking search model. *Transportation Research Part A, Policy and Practice*, *32*(3), 159–170. doi:10.1016/S0965-8564(97)00005-0.

Tight, M. R., Vicat, A., Bristow, A. L., Pridmore, A., & May, A. D. (2007). An exploration of household response to personal travel carbon reduction targets. *International Journal of Sustainable Transportation*, *1*(3), 143–159. doi:10.1080/15568310601113769.

Tol, R. S. J. (2005). The marginal damage costs of carbon dioxide emissions: An assessment of the uncertainties. *Energy Policy*, *33*(16), 2064–2074. doi:10.1016/j.enpol.2004.04.002.

Van der Maas, C. (1998). *The use of disincentives in mobility management*. Rotterdam, Netherlands: Report by Ministry of Transport..

Van der Mede, P., & Visser, J. (1999). Effects of parking measures on traffic congestion. In J. D. De Ortuzar (Ed.), *Stated preference modelling techniques. Perspectives 4*. London: PTRC Education and Research Services Ltd.

Van Vuren, T., & Johnson, P. (2000). *Assessing possible impacts of workplace charging and investments on traffic levels in the West Midlands*, presented at the European Transport Conference, Seminar on Transport Modelling, Cambridge, England.

Wang, T., & Sharples, J. (1999). *Workplace parking levy. TRL Report 399*. Crowthorne..

Watters, P., O'Mahony, M., & Caulfield, B. (2006). Response to cash outs for workplace parking and workplace parking charges. *Transport Policy*, *13*(6), 503–510. doi:10.1016/j.tranpol.2006.05.008.

Willson, R. W. (1992). Estimating the travel and parking demand effects of employer-paid parking. *Regional Science and Urban Economics*, *22*(1), 133–145. doi:10.1016/0166-0462(92)90029-Z.

Windsor-Cundell, E. (2003). *The retail survey: The impact of the congestion charge on the retail sector*. London Chamber of Commerce and Industry..

KEY TERMS AND DEFINITIONS

Ecological Tax Reform: Rebalancing the taxation system from vehicle ownership to vehicle use.

Efficient Pricing: Internalisation of external costs in the price of network use.

Fiscal Measures: Structure of vehicle and fuel taxation.

Transport Externalities: Non-user costs and benefits which result from private network use.

This chapter was previously published in Sustainable Systems and Energy Management at the Regional Level edited by Marco Tortora, pages 19-37, copyright 2012 by Information Science Reference (an imprint of IGI Global).

Chapter 71
Improving the Sustainability of Road Freight Transport by Relaxing Truck Size and Weight Restrictions

Alan McKinnon
Heriot-Watt University, UK

ABSTRACT

Increasing legal limits on the size and weight of trucks allows companies to achieve a higher degree of load consolidation. This reduces the total number of vehicle-kilometres required to distribute a given quantity of goods, saving money and reducing environmental impacts. Proposals to legalise longer and heavier vehicles, (LHVs) have, nevertheless, generated intense debate, particularly in Europe where they are strongly resisted by railway and environmental organisations. This chapter reviews recent studies on this subject, presents an analytical framework and focuses on three critical issues: the extent to which loads can be consolidated in LHVs, their effect on the freight modal split and the possibility that the resulting reduction in road freight costs will stimulate additional traffic growth. Most of the recent studies support the development of LHVs, particularly those based on actual experience of their use in countries such as Australia, Sweden and the United States.

INTRODUCTION

Consolidating loads in larger and heavier vehicles is one of most effective ways of improving the efficiency of road freight transport and reducing its impact on the environment. Governments around the world have therefore been under pressure from industry to relax truck size and weight restrictions and thereby reduce the number of trips required to move a given quantity of freight. There are, nevertheless, physical limits to the size and weight of trucks that can be accommodated within the existing road infrastructure and representatives of

DOI: 10.4018/978-1-4666-4852-4.ch071

competing transport modes have strongly opposed the introduction of larger and heavier vehicles. This has made the maximum weight and dimensions of trucks one of the most controversial public policy issues to arise in the field of logistics for many years. The contradictory positions on this issue are well illustrated by the European Shippers Council (2007) and German Environment Ministry (Umwelt Bundes Amt, 2007). The former argued that LHVs will *reduce fuel consumption and pollution, lower transport costs, contribute towards an increase in the available capacity utilisation of the primary road network, reduce the number of lorry trips per annum, and increase the competitiveness and sustainability of EU industry*. The latter asserts that *megatrucks do not contribute towards sustainable development of freight transport. Their use relieves neither the environment nor road infrastructure*.

This chapter reviews the evidence that has accumulated on this issue from both desk-based studies and operational experience in countries where larger and heavier trucks are allowed to operate. It examines the problems that researchers experience when trying to assess the net effects of allowing these vehicles onto the road. The discussion relates mainly to EU countries though makes reference to the use of these trucks in other parts of the world. It focuses on operational, economic and environmental issues. For detailed discussion of the implications for road safety and transport infrastructure readers should consult Knight et al (2008), TML et al (2008) and Leduc (2009).

This chapter is mainly concerned with a combined relaxation of vehicle size and weight restrictions in trucks that are both longer and heavier. It is, of course, possible for length and weight limits to be increased independently. In the UK in 2001, for example, the government raised the maximum weight of an articulated lorry (running on six axles) from 41 to 44 tonnes. It is currently considering an extension to the length limit within current weight restrictions. To maximise the economic and environmental benefit of relaxing legal limits on truck capacity, however, it is often desirable to increase weight and size limits simultaneously.

The semantics of truck enlargement also requires some comment at the outset. Various terms have been used to describe bigger trucks. Some, such as 'megatruck', 'monster truck' and 'juggernaut', tend to be used by opponents of these vehicles and are fairly emotive. Others such as 'high productivity vehicle', 'high capacity vehicle', 'super truck' or 'load efficient vehicle' have much more positive connotations. Other expressions, such as the Australian 'road train', the US 'long combination vehicle' (LCV), the German 'gigaliner', European 'Eurocombi' and the British 'longer and heavier vehicle' (LHV) simply describe the vehicles. The last of these terms is now in common use and will be adopted in this chapter.

A BRIEF HISTORY OF LHVs IN EUROPE

LHVs with lengths in excess of 20 metres and gross weights of over 50 tonnes have been operating in Scandinavia for over 20 years. Sweden, for example, set a 24 metre length limit for trucks in 1968 and allowed them to run at weights up to 51.4 tonnes in 1974 (Vierth, 2008). At the time of Sweden and Finland's accession to the EU in 1996, special provision had to be made to allow these countries to continue to operate these vehicles which were much bigger and heavier than those permitted elsewhere in Europe. An EU Directive (96/53) was approved which granted all EU member states the right to operate longer vehicles so long as they conformed to the standard modular dimensions in existence at that time. This has since become known as the European Modular System (EMS). The main objectives of the directive were to avoid a proliferation of vehicle dimensions and promote the harmonisation of modules that could be transferred between modes. Although the 96/53 Directive allowed member states other than Sweden and Finland to legalise LHVs for use within their national borders, no country did so until recently.

Outside Scandinavia, the demand from hauliers and their clients for LHVs has arguably been greatest in the Netherlands. The Dutch government has responded by permitting two trials of LHVs, between 1999 and 2003 and 2004 and 2006. The results of the more recent trial were considered sufficiently positive to justify the general legalisation of LHVs in the Netherlands up to 25.25 metres long with maximum gross weights of 60 tonnes from 2008. Following a more limited trial in several German states (or 'lander') between 2006 and 2007, the federal government rejected the case for allowing LHVs to operate on German roads, ostensibly on safety grounds, though it is reckoned that this decision was also strongly influenced by plans to privatise half the country's rail system. A limited trial of LHVs was launched in Denmark in 2008. Following a desk-based study of seven regulatory options for LHVs published in 2008 (Knight et al, 2008), the UK government decided that only one option might be permissible involving the use of a semi-trailer 2 metres longer than the current maximum length of 13.5 metres. At the time of writing the government has still to make its final decision on this issue. In response to the growth of interest in LHVs at the national level, the European Commission has funded three studies of LHVs between 2007 and 2010 by TML et al (2008), Christidis and Leduc (2009) and one currently underway.

Despite all this research, wide consultation on the subject and the inclusion of changes in truck size and weight regulations in the EU's Logistics Action Plan (European Commission, 2007), no decisions have yet been made on modifications to the 96/53 Directive which would alter the maximum size and weight of trucks making international journeys across the continent. In the meantime, the Joint Transport Research Centre of the OECD and International Transport Forum have published the results of a global review of opportunities to *move freight in better trucks*, which is largely concerned with increases in vehicle carrying capacity (OECD/ITF, 2010).

Before examining the conclusions of these various studies it is important to identify the factors influencing public policy on truck size and weight and present an analytical framework within which the relationship between these factors can be mapped.

ANALYTICAL FRAMEWORK

Figure 1 shows the key elements in a cost and benefit analysis of an increase in truck carrying capacity. At its centre is the consolidation of loads in fewer vehicles, the source of all LHV benefits. This yields two streams of benefits: (i) a reduction in vehicle operating costs which are internalised by the carrier and/or shipper and (ii) a reduction in lorry traffic levels which helps to alleviate environmental impacts and congestion, cutting the external costs of road transport borne by the community at large. Against these benefits must be set a series of four potentially offsetting factors:

1. Possible shift of freight from less-environmentally-damaging modes (mainly rail and waterborne services) to LHVs to take advantage of the lower road freight costs they offer. As LHVs have higher external costs per tonne-km than these alternative modes, the environmental benefits from load consolidation must be scaled-down accordingly.
2. Possible generation of new freight traffic if companies respond to the reduction in freight costs by making their logistics operations more transport-intensive, by, for example, sourcing products from more distant suppliers or centralising their inventory. This can offset some of the vehicle-km, fuel and emission savings accruing from load consolidation

Figure 1. Inter-relationships in the cost-benefit analysis of LHVs

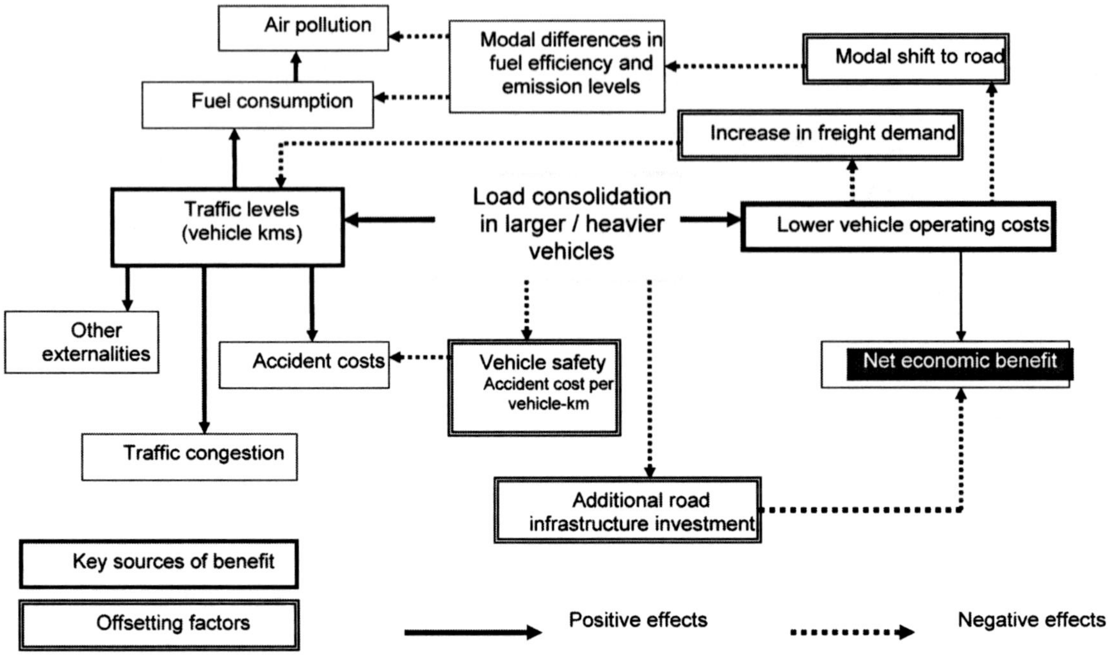

3. Possible increase in the severity of accidents as a result of the greater weight and size of LHVs. Accident costs per vehicle-km may therefore rise, though there will be an offsetting reduction in total vehicle-kms. Operational experience and simulation modelling, nevertheless, suggests the LHVs have a positive effect on safety. The OECD/ ITF (2010) review concludes that *the various studies and experiences from recent years agree that a transfer of goods to trucks with higher cargo capacity should result in a reduction in casualties per unit of goods moved* (p.23).
4. Possible increase in expenditure on road infrastructure to accommodate LHVs. As it is normally assumed that maximum axle weights will remain unchanged and that any additional payload weight will be spread across more axles, 'wear and tear' on the road pavement is unlikely to increase. On the contrary, the reduction in truck-kms will be likely to reduce maintenance costs. The extra infrastructural cost is likely to be capital expenditure on modifications to road layouts and bridge strengthening.

The solid black lines in the diagram represent the positive/beneficial effects of LHVs while the broken lines show the offsetting negative effects. In assessing the case for permitting LHVs, researchers are therefore presented with a classic cost-benefit analysis problem whose solution depends critically on the weighting factors attached to the various positive and adverse effects and their monetary valuation.

In later sections of the paper we will examine the analytical problems that one encounters in trying to quantify three of the key variables in the cost-benefit analysis: degree of load consolidation, freight modal shift and induced traffic growth. The analysis is complicated by differences in the maximum length and weight proposed for LHVs and the diversity of LHV designs. These differences are considered in the next section.

DIFFERENT TYPES OF LHV

There are many types of LHV in commercial operation and in prototype form around the world, varying in design, size, weight and ability to meet infrastructural and safety requirements. In this chapter we are more interested in differences in the carrying capacity of LHVs than in their technical design features. A recent study compared the performance of 18 types LHVs in operation in nine countries (OECD/ITF, 2010). Thirteen of these LHVs combined a long trailer with a shorter one (up to a maximum combined length of 30 metres) and were described as 'higher capacity' vehicles. Five comprised either two full length trailers or three trailers and were classified as 'very high capacity' (with a combined length in excess of 30 metres). The 'higher capacity' vehicles typically offer 50% more cube than the standard 'workhorse' articulated truck, while the 'very high capacity' vehicles can increase the cube by 100% of more.

In Scandinavia the dominant type of LHV is a rigid vehicle hauling a 13.6 metre trailer with a maximum length of 25.25 metres and gross weight of 60 tonnes. Other European countries that have recently examined the case for LHVs have explored other configurations but focused mainly on this 25.25 metre maximum length (falling within the OECD/ITF's 'higher capacity' category). In the Netherlands 162 LHVs were trialled conforming to five LHV configurations, two-thirds of which operated at a maximum weight of 50 tonnes with the remainder at a 60 tonne limit. The sample of LHVs trialled in Germany (Lower Saxony) were 25.25 metres long but subject to the same weight limit as existing articulated trucks (40 tonnes). In the desk-based study conducted in the UK, seven LHV scenarios were tested (Figure 2). These were essentially regulatory scenarios relating to maximum lengths and weights. Each scenario could accommodate differing vehicle designs. The scenarios ranged from the *status quo* to a 34 metre road train with gross weight of 82 tonnes, which would be deemed 'very high capacity' by

Figure 2. Seven LHV scenarios tested the UK study (source: Knight et al, 2008)
** relative to standard articulated truck with 13.6 m trailer and 44 tonne gross weight limit*

Scenario	Design	Length limit	Weight limit	Axles	Capacity change*
1	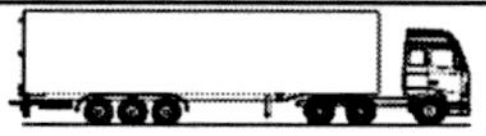	18.75m	44 tonnes	6	Increased cube Less weight
2	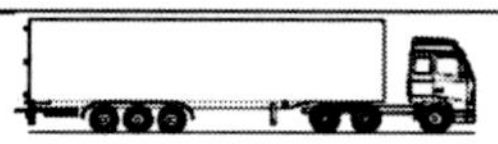	18.75m	45.8 tonnes (payload neutral)	6	Increased cube Same weight
3	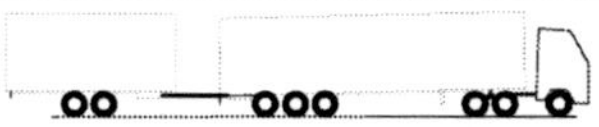	25.25m	44 tonnes	8	Increase cube Less weight
4	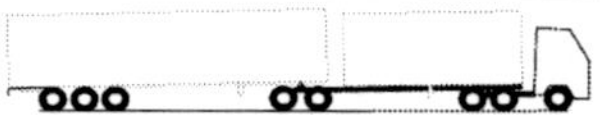	25.25m	49.5 tonnes (payload neutral)	8	Increased cube Same weight
5	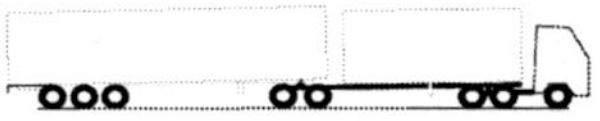	25.25m	60 tonnes	8	Increased cube Increased weight
6	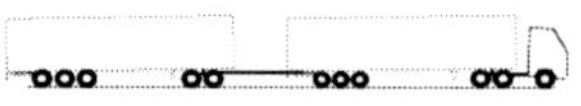	34m	63.2 tonnes (payload neutral)	11	Increased cube Same weight as 5.
7	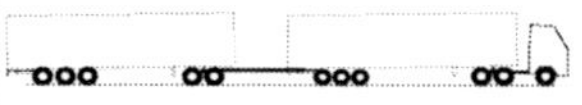	34m	82 tonnes	11	Increased cube Increased weight

the OECD/ITF study. Three of the scenarios, for the longer semi-trailer and the 25.25 metre vehicle, introduced the concept of *payload neutrality*. This would have involved increasing the maximum gross weight of the vehicle in line with its tare (i.e. empty) weight but not permitting any increase in payload weight. It would allow companies moving low density products which currently *cube out* to take advantage of the additional space in LHVs without suffering a loss of weight-carrying capacity.

Several countries, such as Australia, Canada and South Africa, adopt a more flexible, 'performance-based' approach to regulating vehicle parameters. This involves defining the infrastructural, environmental and safety outcomes that must be achieved rather than quantitative size and weight limits. In other words it relates '*not to what the vehicle looks like*' but *what the vehicle can do* (Sullivan, 2010). In Australia, for example, *performance based standards have been used to authorise access to suitable parts of the road network for vehicles that do not conform to prescriptive limits on mass or dimensions* (OECD/ITF, 2010, p.8). This allows truck designers and manufacturers to be more innovative in developing vehicles which match particular traffic flows and the specific environmental and safety requirements of the areas through which they pass.

Reference is often made to the LHV as a single distinct type of vehicle. In practice it represents a heterogenous category of vehicles varying both in their maximum weights and dimensions and in their design. As the carrying capacity, operating costs, energy efficiency and emissions all vary with regulatory specification and design, it is difficult to generalise about the net economic and environmental effects of LHVs.

DEGREE OF LOAD CONSOLIDATION

The use of LHVs in those countries where their operation has been legalised indicates the actual degree of load consolidation that can be achieved. On the basis of Swedish and Finnish data, TFK (2002) estimated that the use of LHVs cut vehicle trip numbers by 32% and road transport costs by 23%. More recent research has reversed this analysis by assessing the likely impact of the removal of LHVs from Swedish roads and imposition of current EU regulations on truck size and weight (i.e. 18.75 metre and 40 tonne maxima). This indicates that vehicle numbers would have to increase by 37% and road freight costs by 24%, broadly corroborating the results of the TFK study. A study by Woodroofe and Ash (2001) in Alberta concluded that the use of single 'semi-trailer configurations' rather than 'long combination vehicles' would increase truck-kms by 80%, transport costs by 40%, fuel/CO_2 emissions by 32% and *pavement wear on highways infrastructure* by 40%. The Dutch trial, discussed earlier, indicated that *depending on the level of the preconditions* permitting LHVs could cut truck numbers by between 2000 and 5000 and raise fuel efficiency (expressed relative to tonne-kms carried) by around third (Aarts and Feddes, 2008). A study based on this trial assessed the wider impact of the legalisation of LHVs on traffic conditions and road transport costs in the Netherlands. It concluded that road congestion would be reduced by 0.7-1.4% and road transport costs by 1.8-3.4% (Arcadis, 2006).

Given the international differences that exist in the nature of freight markets, transport infrastructure and regulatory frameworks, one must exercise caution in geographically extrapolating on the basis of only a few countries' experience. Governments contemplating a relaxation of truck size and weight limits must therefore assess the likely impact of this regulatory change on their particular country's transport system. Increased load consolidation is a key element in this assessment.

It is possible to predict the extent to which loads will be consolidated in LHVs by:

1. Measuring the proportion of loads that are capacity-constrained in trucks operating at

the current maximum size and weight limits, and
2. Estimating the proportion of these loads that will migrate to LHVs

Unfortunately very few countries compile the statistics required to make the first of these assessments. A recent review of available freight road data in the EU revealed a serious lack of statistics on vehicle loading (McKinnon, 2010). The UK government is exceptional in having collected data on truck load constraints since 1998. This differentiates the proportions of loads that are weight- or volume-constrained and has revealed that in the UK a much larger proportion of road tonne-kms are constrained by volume than by weight (Department for Transport, 2009). From this we can conclude that expanding the deck area/cubic capacity of trucks will offer greater load consolidation opportunities than increasing their weight limit. It is generally acknowledged too that the average density of freight is declining, partly as a result of the substitution of lighter materials for heavier ones, but also because of increased packaging. Unfortunately, however, no official data exists on the density of loads carried by road. If this information were available, one could analyse the statistical distribution of freight densities across all the loads surveyed. This would permit quantitative modelling of the migration of loads of varying density to LHVs with differing maximum weights and dimensions.

In an earlier UK study of the consolidation of loads in vehicles that were heavier but no larger (maximum weight raised from 41 tonnes to 44 tonnes), various percentage migration rates were applied to test the sensitivity of economic and environmental benefits to this variable (McKinnon, 2005). In the case of LHVs, this sensitivity analysis is much more complicated as vehicle dimensions can also be varied independently of the maximum weight. In the desk-based LHV study conducted in the UK by Knight et al (2008) for the Department for Transport, three maximum length and three maximum weight options were assessed. The three length options met the requirements of the EMS in conforming to existing modular lengths. The maximum weight options were

- Retaining the *status quo*,
- Allowing a payload-neutral increase in gross weight and
- Permitting an increase in gross weight commensurate with the increased number of axles[1].

In theory, the optimum gross weight for a particular modular length, offering the greatest load consolidation opportunities, could lie at an intermediate position between these three benchmark weights. Insufficient data exists, however, to model this empirically.

The load consolidation issue is further complicated by the likelihood that at least some categories of LHV would have their movements confined to particular classes of road. In Sweden and Finland, for example, where LHVs have operated for many years, they are denied access to most urban roads. The imposition of route restrictions on LHVs can be justified on several grounds, mainly their inability to negotiate local road layouts, the overtaking problems they create for other categories of traffic on single carriageways and public opposition in environmentally sensitive neighbourhoods. The tighter the network restrictions on LHVs, the less flexible will be their use and the lower the rate of load migration from existing trucks. In the course of the UK LHV study an attempt was made to model the effects of differing degrees of network restriction on the likely uptake of LHVs using trip-specific data from the main annual survey of road freight movements (Knight et al, 2008). This analysis was underpinned by numerous assumptions relating to vehicle loading, the compatibility of loads for consolidation and the coupling/decoupling of trailers at strategic locations on the road network. Table 1 summarises the results for one category of LHV, a 60 tonne

Table 1. Effects on distance travelled, transport cost and CO_2 emissions of restricting LHVs to particular road networks (% savings relative to delivery by non-LHV, standard articulated truck)

Permitted Categories of Road	Trips	Vehicle-kms	Cost	CO_2
motorways + roads within 20 kms	-10.4	-10.3	-6.5	-2.7
motorways and dual carriageways + roads within 20 kms	-24.8	-29.8	-19.3	-7.6
all roads	-34.6	-34.6	-22.9	-8.8

25 metre truck, and three levels of road network restriction. It shows the % reductions in traffic levels, cost and CO_2 emissions from operating this LHV rather than a standard articulated vehicle with a 13.6 metre trailer and gross weight of 44 tonnes. It can be seen how confining LHVs to particular categories of road significantly reduces these potential benefits.

Not all journeys made by LHVs would move loads that are too large or too heavy to be transported in trucks meeting current regulations. Some will be run empty, while others will be only partially laden. In assessing the degree of consolidation, therefore, allowance must be made for the likely utilisation of LHV capacity. This too is very difficult to quantify in a desk-based study. It can simply be assumed that the average levels of empty running and lading will be the same as those achieved by trucks operating at the current maximum size and weight limits. This may underestimate the level of LHV utilisation, however, because, as these vehicles require additional capital investment, companies are likely to deploy them mainly on journeys which take advantage of their greater carrying capacity. This was a view strongly expressed by logistics managers consulted during the UK study (Knight et al, 2008 Appendix F). Another way of addressing this issue is to determine how high the vehicle utilisation factor would have to be for LHVs to yield a net economic and/or environmental benefit. A study conducted by the German Ministry of the Environment, for example, estimated that the average load factor[2] on LHVs 25.25 metres in length would have to be at least 77% for them to yield a net environmental benefit, well above the average achieved by trucks within current regulations (64%) (Umwelt Bundes Amt, 2007).

FREIGHT MODAL SPLIT

LHV impact studies have been criticised for not taking enough account of the effects of truck size and weight increases on the division of freight traffic between modes (Schulman, 2003). Railway trade bodies and their railfreight member companies have, nevertheless, commissioned several studies on the subject and strongly campaigned against wider introduction of LHVs in EU countries. They see LHVs as a major threat not only to the future growth of railfreight traffic but also to the existing railfreight market. UIC et al (2006), for example, argue that the *advent of mega-trucks* would result in the transfer of 7 bn tonne-kms per annum from road to rail in Germany (i.e around 16% of railfreight tonne-kms carried by rail in 2005). In the UK, the main railfreight operator, EWS (2007) has claimed that *taking both the vulnerable bulk market sectors and the high-value intermodal sector, roundly half of all the freight moved by train in the UK is at risk from moving to road if LHVs were introduced.* On the other hand, it is worth noting that in Sweden where LHVs have been operating for several decades, rail's share of total domestic tonne-kms in 2006 was roughly twice the EU25 average (35% as opposed to 17%) (Eurostat, 2009).

As discussed earlier, the displacement of freight from rail (and to a lesser extent water-borne modes) to road is considered to be one of the major disbenefits of LHVs. Some modal shift may be acceptable, however, so long as this disbenefit is more than offset by economic and environmental benefits accruing from load consolidation on the road network. As the Commission for Integrated Transport (2000) in the UK stated, *encouraging freight consignors to use rail, canal or sea transport should be regarded as a means towards the end of environmental benefit, not as an end in itself.* For many potential users of LHVs, transporting their goods by other modes is not an option. The nature of their products, the geography of their distribution operations and/or their delivery schedules often require the use of road transport. Such companies legitimately ask why they should be denied the opportunity of using LHVs to rationalise their road freight movements when the modal split issue has little or no relevance to them. As the OECD/ITF (2010) report explains, the impact of LHVs *is limited by the fact that many freight markets are not contestable between modes*' (p.19).

Predicting the impact of LHVs on freight modal split is fraught with difficulty. This is partly because of a lack accurate, reliable and up-to-date modal cross-elasticity values. These values measure the extent to which changes in the price of one transport mode affect the demand for another. The LHV analysis undertaken in 2007 by Oxera for the UK railfreight operator EWS employed price elasticity values for freight transport quoted in a 1990 paper and largely based on North American experience (Oum et al, 1990). Christidis and Leduc (2009) reviewed road-rail cross-elasticity values cited in more recent studies and found that they ranged from *0.3 to 2 depending on the trip length and commodity type* (p.11). This suggests that a 10% reduction in the cost of road haulage could cut railfreight demand by between 3% and 20%. Dings (2007) quotes cross-elasticity values for rail of +1.8 to 3.0 which were derived by the Netherlands Economic Institute and CE Delft. On the assumption that the use of LHVs would reduce road freight rates by 20% he calculates that between 36% and 60% of railfreight would divert to LHVs. Given the broad spectrum of cross-elasticity values it is very difficult to generalise about the impact on modal split of the cost savings that would result from the use of even a single type of LHV.

To address this problem, cross-elasticity values should be derived for those sectors of the freight market within which rail and LHVs would be in direct competition. The nature of this competition partly depends on the maximum gross weight of the LHVs. As the railways tend to carry heavier, higher-density freight, competition from LHVs is particularly sensitive to their gross weight limits. If vehicle length were increased and maximum weight left unchanged, there is likely to be limited traffic erosion from rail. The main exception to this general statement would be containerised traffic. Although many commodities moved by rail in containers are relatively light, LHVs can still pose a major threat to the rail container market. 25.25 or 34 metre long LHVs, after all, permit the carriage of multiples of 40 ft and 20 ft containers and thus effect a step-change in the cost of moving containers by road. Research in the UK has suggested a price elasticity value of 2.5 for deep-sea container traffic moving by rail[3] (MDS Transmodal, 2006). If, as has been suggested by a UK railfreight operator, the use of a 25.25 metre LHV reduced the average cost of transporting containers by road by 15%, one might expect 37.5% of rail deep-sea container traffic to transfer to LHVs in the UK (Knight et al, 2008).

The Dutch trial suggested that the impact of LHVs on modal split was very limited with only 1.4-2.7% of railfreight diverting and 0.2-0.3% of inland waterway traffic switching (Aarts and Feddes, 2008). Given the wide differences in national freight markets it would be very unwise to extrapolate these modal shift estimates to other countries or to cross-border transport in Europe (Debauche and Decock, 2007). It is worth noting, however, that two studies undertaken for the European Commission of EU-wide LHV impacts yielded modal shift estimates of a similar magnitude. One of them, involving a Monte Carlo simulation analysis of the impact of allowing 60 tonne 25 metre LHVs to operate across the EU indicated that by 2020 railfreight traffic would be only 2.1% lower that would otherwise be the case (Christidis and Leduc, 2009). The other by TML et al (2008) which modelled the effects of allowing this type of LHV to run on EU motorways suggested that *rail and waterway volumes would respectively decrease by 3.8% and 2.9%* (p.72). On the other hand, a study commissioned by the rail sector (CER), using much higher cross-elasticity factors, estimated that within the EU rail tonne-kms would decline by, respectively, 11.6% and 12.9% if 60 tonne 25 metre LHVs cut trucking costs by 18% and 20% (Doll et al, 2008). This clearly illustrates the sensitivity of the modelling to the calibration of cross-elasticity values.

The modal split analyses conducted in the course of LHV impact studies have been largely confined to econometric modelling within a narrow set of parameters. Numerous studies have shown, however, that the mode choice decision is multi-dimensional, with cost being only one of a range of factors affecting the relative attractiveness of different transport services. Improvements in the marketing and quality of railfreight services could help to offset some of the loss in price competitiveness that would result from a wider use of LHVs. Opportunities also exist to increase the carrying capacity of freight vehicles on the railfreight network, for example by strengthening track to support heavier axle weights, expanding the loading gauge and lengthening sidings to accommodate longer freight trains, though exploiting them usually carries a high capital cost.

INDUCED ROAD FREIGHT TRAFFIC GROWTH

An argument regularly used against increases in vehicle capacity limits is that it will induce a *rebound effect* with the resulting transport cost reductions stimulating an increase in the demand for freight movement. It has been claimed that, in the case of LHVs in Europe, *increased demand for road transport would cancel out any environmental gains from the increased efficiency* (European Federation of Transport and Environment, 2007). Attempts to quantify this traffic generation effect are again frustrated by a lack of accurate and reliable price elasticity values, in this case solely for road freight transport[4]. Graham and Glaister (2004) and Christidis and Leduc (2009) reviewed the published estimates and found that most lie within the range -0.5 to -1.5, suggesting that a 20% reduction in road freight costs from the use of 60 tonne 25 metre LHV would generate 10-30% more freight movement.

A study conducted for the main UK railfreight company, Oxera attempted to estimate the strength of the *market generating* effect of LHVs (EWS, 2007). It predicted that the use of LHVs in the UK would cut road freight costs per tonne-km by 14.2-15.3% and that this would cause a surge in road tonne-kms. It gave no indication, however, of the likely sources of this large amount of additional freight traffic. Its econometric analysis was detached from the real world of production and logistics management. It merely asserts that *over time the structure of supply chains will adapt to lower road haulage costs* and that *it would be naïve not to expect a substantial increase in the amount of freight transport (i.e. tonne-kilometres) in the UK as a result of the introduction of LHVs* (p.16).

There is strong counter evidence to suggest that any freight traffic generation effect is likely to be weak. A retrospective analysis of the impact of the increase in maximum truck weight in the UK from

41 tonnes to 44 tonnes in 2001 could find little statistical evidence of this measure reinforcing the rate of freight traffic growth (McKinnon, 2005). Focus group discussions with logistics managers in the UK also challenged the premise that road transport savings arising from the introduction of LHVs would be likely to unleash a new wave of logistical restructuring (Knight et al, 2008). This concurs with the view of the OECD/ITF (2010) report that *induced demand effects are likely to be small* (p.9) The analysis undertaken by TML et al (2008) for the European Commission quantified this effect and found that it would be marginal. In its 60 tonne 25 metre LHV scenario *road volumes are expected to increase by 0.99%* (p.72).

NET ENVIRONMENTAL IMPACT

Several studies have conducted wide-ranging cost benefit analyses of the introduction of LHVs, taking account of the factors discussed above, plus others such as safety and the upgrading of road infrastructure, and applying monetary values to the various transport-induced externalities. Some of the results are not directly comparable as they differ in the range of factors included in the calculation, the definition of the LHV scenarios, the underlying assumptions and valuation of externalities. If one focuses on 60 tonne 25 metre LHVs, a consensus emerges in several of the main international studies that the use of these vehicles yields a net welfare benefit. For example, Christidis and Leduc (2009) place a '*mean value*' of 700 million Euro on the net welfare gain in 2020 from having 3.2% of EU road freight carried in these LHVs. Increasing this level of LHV useage raises the net benefit. Taking a global view of the use of LHVs, the OECD/ITF (2010) conclude that, *Higher capacity vehicles have been operated extensively for a variety of freight tasks in some areas of the world without adverse impacts. The evidence available indicates significant safety, sustainability and productivity improvements (p.10).*

FUTURE RESEARCH DIRECTIONS

One hesitates to recommend further research in a field that has already been extensively studied around the world. There are, nevertheless, several issues that merit more detailed investigation.

Route-specific restrictions on LHVs: it is generally acknowledged that it is desirable, for environmental, safety and infrastructural capacity reasons, to deny LHVs access to particular roads and districts. Advances in telematics now make it possible to monitor the movement of LHVs (and other trucks), providing route guidance to drivers and helping to enforce route restrictions. The costs and benefits of this application of telematics need to be more fully explored.

Price-elasticity of demand for freight transport: As discussed earlier, analyses of the impact of LHVs have been inhibited by the lack of an agreed set of price elasticity values for freight movement within a single mode, in this case road, and between modes. A recent review of freight elasticities in European road freight transport by De Jong et al (2010) revealed a '*substantial*' variation in published elasticity values, only some of which they were able to explain. It would be a worthy research aspiration to derive, on a consistent and transparent basis, a broader suite of elasticity values for different freight markets. This would certainly help to make the modelling of LHV impacts more analytically robust.

Refining public policy on LHVs: The aim of public policy in this arena is to maximise the economic, environmental and safety benefits of LHVs while minimising the disbenefits of modal shift and higher infrastructural investment. For some countries to meet this aim, public policy on this issue will have to be more finely tuned, striking a more delicate balance between conflicting objectives. This can be done, for example, by using the 'performance-based approach' to LHV regulation applied in Australia, Canada and South Africa, by adopting the principle of 'payload neutrality' where appropriate and finding more innovative ways of promoting the use of LHVs within rail- and water-based intermodal systems. The development of such policy refinements will be another fertile area for future research.

CONCLUSION

A large body of evidence now exists to confirm that increasing the maximum length and weight of trucks yield significant economic, environmental and safety benefits. On the basis of long experience of LHV useage in countries such as Sweden, Canada and Australia and the results of numerous trials and desk-based studies it is possible to build a persuasive case for relaxing current length and weight limits on trucks across the EU and elsewhere. Most of the evidence to the contrary is found in studies commissioned by railway organisations. They are naturally keen to defend their share of the freight market against the more intense competition that LHV would inevitably bring. Rather than allow the rail lobby to obstruct the consolidation of road freight in higher capacity vehicles, policy makers could use a combination of taxation and regulation to achieve a modal compromise that would maximise the net benefit to the economy, society and environment.

This chapter has focused on three issues at the heart of the current debate in Europe over the merits of LHVs: load consolidation, modal split and traffic generation. Space limitations have precluded detailed discussion of other important issues such as safety, energy efficiency, emissions, road investment and enforcement. This review of recent literature has, nevertheless, highlighted the difficulties of conducting research on this controversial subject and drawing general conclusions. First there is the lack of survey data on truck utilisation by weight and volume, the density of loads and the price elasticity of demand for freight services both within the road sector and cross-modally. Second, there is the diversity of LHV designs and capacities, each with a different cost-benefit ratio to be modelled. The third complication is the large number of operational, economic, environmental variables that must be factored into the analysis, many of which, as noted earlier, cannot be empirically calibrated. And finally, there is the need to 'customise' the analysis to the particular region, country or continent across which LHVs will operate, taking account of differences in its geography, transport infrastructure, industrial mix and freight market.

As the LHV issue combines analytical complexity and political sensitivity, there is a danger that it will never be resolved and that politicians will continually 'kick it into the long grass' of yet another study. At some point, however, policymakers at both EU and national levels will accept that few other transport measures offer such large productivity and environmental benefits as increasing the maximum weights and dimensions of trucks.

REFERENCES

Aarts, L., & Feddes, G. (2010). *Experience with Longer and Heavier Vehicles in the Netherlands.* Presentation to the 10th International Symposium on Heavy Vehicle Transport Technology, Paris.

*Arcadis (2006).*Monitoringsonderzoek Vervolgproef LZV: Resultaten van de Vervolgproef met Langere of Langere en Zwaardere Voertuigcombinaties op de Nederlandse Wegen. *Arnhem.*

Christidis, P., & Leduc, G. (2009). *Introducing Mega-Trucks: A Review for Policy Makers.* Institute for Prospective Technological Studies, Seville..

Commission for Integrated Transport. (2000). *Interim Report: Permitting 44 tonne Lorries for General Use in the UK.* London: Commission for Integrated Transport..

De Jong, G., Schroten, A., van Essen, H., Otten, M., & Bucci, P. (2010). *Price Sensitivity of European Road Freight Transport – Towards a Better Understanding of Existing Results*. Report 9012-1, Significance/CE Delft.

Debauche, W., & Decock, D. (2007). *Working Group on Longer and Heavier Vehicles: A Multidisciplinary Approach to the Issue*. Brussels: Belgian Road Research Centre..

(2007). *Department for Transport*. London: Road Freight Statistics..

Dings, J. (2008). *LHVs and the Environment – Addressing the Big Forgotten Issue*. presentation to LHV Stakeholder Meeting 4th March, Brussels: European Commission.

Doll, C., Fiorello, D., Pastori, E., Reynaud, C., Klaus, P., Lückmann, P., et al. (2008). Long-Term Climate Impacts of the Introduction of Mega-Trucks. Study for the Community of European Railways and Infrastructure Companies (CER) co-ordinated by Fraunhofer ISI, Karlsruhe.

(2007). *European Commission*. Brussels: Freight Transport Logistics Action Plan..

European Federation of Transport and Environment. (2007). Longer and Heavier Lorries (LHLs) and the Environment. Brussels..

European Shippers Council. (2007). Road Transport Reduction through the European Modular System: the Challenges for European Transport Markets. Brussels.

Eurostat (2007). *Panorama of Transport*. Luxembourg.

EWS. (2007). *Longer, Heavier Road Vehicles Study by TRL/Heriot-Watt University: A Response by English Welsh & Scottish Railway*. Retrieved from http://www.ews-railway.co.uk/news/downloads/LHVs%20research% 20by%20EWS.pdf

Graham, D. J., & Glaister, S. (2004). Road Traffic Demand Elasticity Estimates: A Review. *Transport Reviews*, *24*(3), 261–274. doi:10.1080/0144164032000101193.

Knight, I., Newton, W., & McKinnon, A. C. et al. (2008). *Longer and/or Longer and Heavier Goods Vehicles (LHVs): a Study of the Likely Effects if Permitted in the UK. Report prepared for the Department for Transport*. TRL, Crowthorne..

Leduc, G. (2009). *Longer and Heavier Vehicles: an Overview of Technical Aspects*. Institute for Prospective Technological Studies, Seville..

McKinnon, A. C. (2005). The Economic and Environmental Benefits of Increasing Maximum Truck Weight: The British Experience. *Transportation Research Part D, Transport and Environment*, *10*(1), 77–95. doi:10.1016/j.trd.2004.09.006.

McKinnon, A. C. (2010). *European Freight Statistics: Limitations, Misinterpretations and Aspirations*. Brussels: ACEA..

OECD/ITF. (2010). *Moving Freight with Better Trucks*. Paris: OECD..

Oum, T., Waters, W., & Jong, Y. (1990). *A Survey of Recent Estimates of Price Elasticities of Demand for Transport*. Working Paper, World Bank, Washington DC.

Schulman, J. F. (2003). *Heavy Truck Weight and Dimension Limits in Canada*. Ottawa: Railway Association of Canada..

Sullivan, P. (2010). *The PBS Model in Australia*. Paper presented at the 11th Heavy Vehicle Transport Technology conference, Melbourne, March.

TFK. (2002). *Improved Performance of European Long Haulage Transport*. Stockholm..

TML et al. (2008). Effects of Adapting the Rules on Weights and Dimensions of Heavy Commercial Vehicles as Established with Directive 96/53/EC. Report for the European Commission, Brussels..

Transmodal, M. D. S. (2006). *Impact of Track Access Charges on Rail Freight Traffic*. Chester..

UIC et al. (2007). *Mega-trucks versus Rail Freight*. Paris..

Umwelt Bundes Amt. (2007). *Longer and Heavier on German Roads: Do Megatrucks Contribute towards Sustainable Transport*. Dessau..

Vierth, I. et al. (2008). *The Effects of Long and Heavy Trucks on the Transport System*. Stockholm: VTI..

Woodrooffe, J., & Ash, L. (2001). *Economic Efficiency of Long Combination Transport Vehicles in Alberta*. Woodrooffe and Associates..

ENDNOTES

1. Most of the European studies have assumed that maximum axle weight will remain unchanged and that heavier weights will be distributed across more axles.
2. Measured in terms of loaded pallets per trip.
3. This elasticity value is expressed with respect to rail freight rates, though as road is the only competing mode and container volumes are relatively insensitive to inland transport rates, it can be applied to a relative reduction in road freight costs.
4. As there is an inverse relationship between road freight costs and the demand for haulage services, the elasticity values are negative. The cross-elasticity values discussed earlier were positive to reflect the direct relationship between road freight costs and the volume of freight carried by alternative modes.

This chapter was previously published in Supply Chain Innovation for Competing in Highly Dynamic Markets edited by Pietro Evangelista, Alan McKinnon, Edward Sweeney, and Emilio Esposito, pages 185-198, copyright 2012 by Business Science Reference (an imprint of IGI Global).

Chapter 72
Sustainable Transport and Quality of Life Analysis of Cycling Impact in Italy

Donatella Privitera
University of Catania, Italy

ABSTRACT

In the last 20 years cities have undergone considerable changes. The current phase of expansion, which took place in the absence of demographic pressures, has diluted urban space into functional areas, scattered randomly throughout the area. The authors therefore identify new mobility needs, met mainly by private means, with implications in terms of congestion and air pollution. Sustainable mobility is a priority intervention for the EU. This chapter studies, after a discussion of the importance of sustainable mobility, cycling in Italy from an economic and strategic perspective. Promoting cycling is important for individual health, environmental sustainability, and transport demand management. In Italy, very few people use a bicycle on a regular basis. The analysis relies on national aggregate data as well as case studies of large and small cities in an Italian region.

INTRODUCTION

Over the past decade, environmental problems have no longer been caused simply by local effects from local activities. The globalization of ecological change has been accompanied by the localization of human populations and settlements (Rees, 2003). Cities, of course, have undergone changes and their shape is changing dramatically. They presents the typical features of the current process of urban transformation "characterized by growth and the differentiation of urban activities which are resulting in new metropolitan spaces that are enlarged and distantiated in terms of scale and scope" (Salet, 2007, p. 6). In fact, the European Commission recognizes urban sprawl ("the low-density expansion or leapfrog development of large urban areas into the surrounding rural land")

DOI: 10.4018/978-1-4666-4852-4.ch072

as the most urgent of the urban design issues as it leads to green-space consumption, expensive infrastructure and energy, increased social segregation and land use functional divisions, reinforcing the need to travel and sharpen dependence on a private motorized transport model, leading in turn to increased traffic congestion, energy consumption and polluting emissions (CEC, 2004; EEA, 2006; OECD, 2000). Transport is associated with many positive connotations, but the impacts and the benefits have a high price (Geerlings, Shiftan & Stead, 2012).

It is important to identify new mobility needs. Sustainable mobility is a priority for action for the EU, although the issue is limited resources. At the same time, it has been found that tourism plays a significant role in the economies of cities, with investment in leisure activities alongside the processes of urban regeneration. In this context, the urban supply system, more oriented towards sustainable patterns, is enhanced by services aimed towards new ways of enjoying the city.

The role reserved for the urban quality of life and livability of the cities is significant and is reflected in numerous studies in national and foreign literature. Urban transport, the excesses of motor traffic, population density and traffic congestion, the concentration of industrial activities and not least the many spatial elements and the weather are all factors contaminating the quality of urban life with the consequences of not livable cities (Cori, 1988, 1997). The power of the automobile has reduced the alternatives offered by other transport modes since, besides the psychological attachment to the private comfort of speed and seamlessness offered by the car (Sheller & Urry, 2000; Urry, 2006), it has inevitably shaped the physical form of urban mobility so that other modes of mobility do not fit in: cyclists and pedestrians have to fight against the symbol of cars, and space for bicycle paths or pavements has to be "stolen" from the "natural." The importance of urban policies and their effectiveness is closely linked to the ability to exploit the land in all its components by focusing on quality of life and meeting the needs of residents. A quality of life and more specifically, quality of urban life, as "the state of social wellbeing of an individual or group, either perceived or as identified by observable indicators" (Pacione, 2005, p. 673) oriented also to the principles of sustainability. Urban quality, based on strong elements of identity and specificity, has thus become the determining factor in a process of continuous economic development in an integrated way that involves public and private resources (Travisi & Camagni, 2005).

In this context, strong planning policies that favour non-auto modes of transport assume increasing importance (Kenworthy & Laube, 1997). The efficiency and effectiveness of the transport system, in fact, help to qualify the supply of city services, also it cannot ignore the important role in shaping the image on tourism (Page, 2009). Consequently there are numerous decision to make on the mobility system – including all modes of transportation – concerning policies and actions to revitalize and enhance the local area.

In this perspective, addressing the issues concerning sustainable mobility (meaning "to move people and goods in an environmentally friendly, ergonomic, safe, economical and timely manner") would make a positive contribution to the environmental, social and economic sustainability of the communities they serve, where the aim is also to contribute in an integrated network of actors involved to the development of competitive cities.

Cycling in particular is a sustainable mode of transport; it is ecological while at the same time offering a tourist attraction role as a possible niche to exploit and promote among the different ways of enjoying the city (Gatersleben & Haddad, 2010; Lumsdon, 2000). Increased cycling should contribute to reducing noise and air pollution and it has important implications for both personal health, having been linked to lower rates of overweight and obesity (Wen & Rissel, 2008), as well as livability and environmental resources.

The bicycle is more energy-efficient than walking (Isfort, 2011). Moreover, its daily use significantly increases levels of physical activity and well-being, generating environmental benefits (ECF, 2009). Its use, if properly supported, also improves the overall level of road safety, reducing the social costs related to traffic and accidents. No less important is the economic impact in terms of the income generated in the local economies and in relation to infrastructure (e.g. bike lanes), networks, and marketing activities and business. In this context, the success and growth in the demand for sustainable tourism that led to the intervention goals of sustainability have spurred destinations to compete on the environmental and social quality of tourism offered. Despite these positive features, cycling does have negative aspects related to the lack of inadequacy of road and parking infrastructure, cyclists' safety and security, weather conditions and also poor intermodality (Krizek & Stonebraker, 2011), but there are also success cases in some countries (the Netherlands, Denmark and Germany) where policies have reduced these risks (Pucher & Buehler, 2008). There have been many initiatives to promote cycling but a long-term strategy is important to see how a change may be achieved (Tight et al., 2011).

This chapter investigates the development and the importance of sustainable mobility in urban areas, with particular reference to economic and strategic cycling, and then attempts to influence the improvement of urban use as a sustainable destination. Furthermore, it shows a comparative analysis of different Italian cities in terms of development, sustainability and infrastructures to promote cycling as a sustainable mode of transport.

BACKGROUND: SUSTAINABLE MOBILITY

Mobility is a multifaceted term when used in different contexts and meanings (Adey, 2010; Kellerman, 2012). In social sciences, mobility refers to two main areas of research, social mobility and spatial mobility, but they are interrelated (Urry, 2007). On one side, social mobility – or "intergenerational mobility," as economists prefer to call it – corresponds to those studies connected with social stratification and measures the degree to which people's social status changes between generations (Breen, 2004). On the other side, spatial mobility (or motility) is the objective of those researchers who analyze movements, the activity of people within space, thus considering a spatial or a territorial perspective (Kaufmann et al., 2004). Spatial mobility is also the ability to move between activity sites (Hanson, 1995), a meaningful condition, implying modernity, freedom, speed and extensibility (Cresswell, 2006). Mobility is a complex term, since it involves aspects connected to transportation, travel, journey and movement, depending on the focus of attention. Mobility ranges from the analysis of the tools and modes of transport utilized by people, such as cars, bicycles and buses, to the "direction" of and the reasons for movement, thus where people move to and why, as in tourism or home-work commuting.

Transport is one of the sectors of the economy with the most challenging tasks facing decision makers (UNECE, 2008). Transportation refers to the study of how the need for mobility is connected with the size and the organization of the city. By assuming a rational-choice approach, transportation analysis has been addressed in particular by urban economists, who try to determine how cities are dependent upon transportation needs and costs (O'Sullivan, 2003).

Sustainable mobility is an evolving concept and there is no universally accepted definition (Zito & Salvo, 2011). The first appearance of the term (CEC, 1992) refers to strict environmental standards for all modes of transport in connection with the reduction of transport volume. Then attention was paid to the development needs of individuals, businesses and society in a manner consistent with human and ecosystem health (OECD, 2004), and then to the reduction of traffic, particularly in urban areas. A modern transport system must be sustainable from an economic and social as well as an environmental viewpoint, and able to adapt to the requirements of sustainable development (CEC, 2001).

In recent times, the policies for sustainable mobility and transport planning have attracted increasing interest from the European Union (CEC, 2007, 2011) and therefore the priority of policies being expressed as binding rules of law. Only France and England seem to have made strict choices with regard to strategic planning of transport in terms of sustainability. Among other countries, such as Italy, Germany and Spain, there are regulations that are binding on the realities of sustainable urban planning, but only as not mandatory instruments: the urban mobility plane, introduced by the Italian legislation in 2000 (decree of 27/3/98 on the integration of "sustainable mobility in urban areas") for cities with over 100,000 inhabitants, has no binding force and is a prerequisite for access to national funds in case of investments (Socco, 2010). In particular, according to White Paper (CEC, 2011), with a broad strategy and the long-time horizon until 2050, transport will also help sustainable competitive policies of urban and regional planning. The European Commission adopted a road map of 40 concrete initiatives for the next decade to build a competitive transport system that will increase mobility and remove major barriers in key areas including increasing fuel prices and unemployment. At the same time, the proposals will dramatically reduce Europe's dependence on imported oil and cut carbon emissions from transport by 60% by 2050.The central point is that the transport system should meet society's economic, social and environmental needs.

Mobility is a central problem for the city. In 2010, more than 72% (United Nations, 2012) of the European population lived in urban areas, which are key to growth and employment. Cities need efficient transport systems to support their economy and the welfare of their inhabitants.

In many EU countries, especially in Italy, the emergence of a mobility model focused mainly on the car, and the lack of proper planning resulted in excessive land consumption and processing functions in the city with new residential, tertiary and trade shapes. Moreover, the predominance of speculation and lack of recovery of the existing buildings have certainly not helped a mobility model centred on urban sustainability.

In urban areas, the mobility model is relevant from the point of view of the quality of life of those directly involved (time devoted to the movement, chronic congestion, costs, etc.), in terms of environmental pressure exerted by automobile mobility. The statement is explained by car use in situations characterized by poor quality of public transport. In fact, the car is the best means to manage a mobility that has taken on a space-time profile that is increasingly fragmented (Hochschild, 1997) due to the progressive changes that have occurred in the styles of mobility of the actors who, in turn, refer to most common changes of urban societies (Colleoni, 2008). The centrality of the automobile and its increased use is only one aspect of the general mobility, although the dependence on the use of private transport is the principal factor that helps to characterize the profile in the sense of its lack of sustainability, not only environmental and economic but also social.

Many studies have investigated how the externalities of the transport system (non-renewable fuel use and greenhouse gas emissions, traffic congestion, low and unequal mobility, pollution, accidents and fatalities, degradation of ecosystems) may be minimized, while the benefits of mobility are maximized and shared equally through sustainable urban and transport development means (Banister et al., 2000; Kenworthy et al., 1996; Shore, 2006). Unfortunately, global emissions of carbon dioxide (CO_2) – the main cause of global warming – increased by 3% in 2011, reaching an all-time high of 34 billion tonnes[1] (Olivier, Janssens-Maenhout & Peters, 2012) and continue to increase with time (primarily due to electricity generation and road transportation). It is worth considering that in many countries CO_2 emissions decreased – in the European Union by 3%, in the United States by 2% and in Japan by 2% – mainly due to weak economic conditions in many countries, mild winter weather in several countries and high oil prices. However, it is significant that CO_2 emissions from OECD countries now account for only one-third of global emissions – the same share as that of China and India, where emissions increased by 9% and 6%, respectively, in 2011 (United Nations, 2012).

In particular, among all the sources of origin of the total CO_2 emissions generated by Italian citizens, gas is the most significant (37.7%), but the incidence of private mobility is significant with a value of 31.2%, given that the city is just one of the actors helping to determine the overall environmental impact of the metropolitan area (Cittalia Report, 2010).

Clean and energy-efficient vehicles aim to reduce vehicle- and fuel-related greenhouse gas and exhaust gas emissions, fuel consumption and fossil fuel dependency, through the use of measures including exhaust gas treatment systems, alternative drive trains and the use of alternative fuels and energy storage systems.

VEHICLES, PEDESTRIANS AND CYCLISTS IN ITALY

Automobile dependence has been studied with many different motivations – social, cultural, economic, climatic, and political – but without doubt it is interconnected with planning and the contemporary urban landscape (Gatersleben & Uzzel, 2002; Gorham, 2002; Newman & Kenworthy, 1999, 2000; Urry, 2004). In fact, the city is defined by the forms and structures of its network of mobility and hence in accordance with user requirements and management of the car (Thrift, 2004). Indeed, cities have been shaped by the needs of cars: parking places, garages, huge mobility infrastructures etc. have been created to serve automobile use. One major impact the motor car has had is on the geography of roadspace. In fact, motor roads have shaped our experience of space and place, and their design, inhabitation and regulation, but they have also facilitated social and cultural relations with individual freedom of movement (Merriman, 2009).

In Italy, the global economic crisis (in recent years) has affected the lifestyles of the population and, consequently, the mobility of people and goods (Table 1).

From Isfort's (2011) survey, some elements emphasize the peculiarities of urban areas and mobility: some people are abandoning cars for bikes (at least for some trips), others are abandoning bikes for cars (often for all trips). But even within a specific town, at any particular time, there are, of course, major mobility inequalities. Both levels of mobility and attitudes tend to vary, often quite dramatically, according to gender, class, ethnicity and age. In 2011, public transport gained space, but without sudden accelerations, and increasingly assumed a role for popularity, a perspective of sustainable urban mobility, especially in the face of the structural decline of non-motorized transport. In the previous two years a decrease in trips had been recorded that was more compatible with environmental needs (pedestrians), but clearly the dependence on the car continued to prevail, although there was some variation in user behaviour. Travelling for reasons of work and study was the main reason for mobility; also, between 2010 and 2011, the distribution of displacements recorded significant changes: the sharing of trips for study (from 4.3% in 2010 to 5.8% in 2011) and family (3.8%), along with a decrease in the number of journeys to work and, especially, leisure (-4.1% and 28.7%). In 2011 this reason for travelling lost the leadership in favour of the family (Isfort, 2011).

In Italy, the development of cycling as a sustainable mode of transport is still far from that of other northern European countries. There are 0.44

Table 1. Modal split of travel in Italy (%)

	2000		2005		2009	
	Trips	Passenger/km	Trips	Passenger/km	Trips	Passenger/km
Walk	21.4	3.1	20.5	2.9	17.3	2.2
Cycle	3.8	1.1	3.9	1.2	3.7	1.1
Motorcycle	5.4	3.2	4.7	3	4.3	2.8
Car driver	50.5	59.1	54.2	62.6	57.7	65.6
Car passenger	8.8	10.7	7.2	7.8	7.6	8.8
Bus/Tram/Metro	5.3	4.2	4.7	3.5	4	2.1
Work vehicle	1.4	4.8	1.9	6.2	1	2.4
Rail	0.9	4.6	0.9	4	0.6	2.4
Transit Integration	1.8	6.9	1.9	8.4	3.6	11.3
Other	0.7	2.3	0.1	0.4	0.3	1.3
Total	100	100	100	100	100	100

Source: Isfort, Observatory "Audimob" on Italian mobility, 2000-2009

bicycles per capita in the Netherlands, more than one for every Dutch inhabitant, and they cover on average 1,019 km a year. Furthermore, the cycle paths in countries such as Denmark, Germany and the Netherlands are convenient, safe and attractive. In these countries, cycling is not only recognized as an important means of daily transport but it remains a symbol of national identity and the strong bicycle culture contributes to sustainable urban behaviours (Carstensen & Ebert, 2012).

We note that sustainable mobility fails to adopted n in urban areas (the same is happening in the extra-urban), disregarding the expectations vested in the middle of the last decade. One reason is the lack of space in Italy and bike lanes where the cycling index (m_eq/100 inhabitants) is equal to 34.86 (metres per 100 inhabitants) in Reggio Emilia, a city that boasts 155.5 km the longest network in absolute terms and of 0.0 for L'Aquila and Nuoro (Urban Ecosystem, 2011), although it must be said that some cities lend themselves more easily to cycling, even for morphological and configuration issues and the quality of services offered. In big cities (with over 250,000 inhabitants), pedestrian mobility numbers in 2011 were slightly below the overall average (22.9% vs. 23.2%), with a decided downturn compared to 2010 (Table 2). The bicycle has a traditional lack of penetration, with just 2.3% of trips served (a slight increase from 2010). In the complex sustainable mobility is in the big cities of unsatisfactory levels. In the midsize towns (100,000-250,000 inhabitants) confirm a important role. Conversely, the increased use of "two wheels" to replace a portion of the pedestrian mobility in urban areas reaches 20% of intermediate demand; the smaller towns favour cycling and walking instead (5.1%, a share slightly above the overall average).

Table 2. Total trips of non-motorized urban mobility with amplitude of the cities

	Towns with > 250,000 inhabitants		Towns with 100,000-250,000 inhabitants		Town < 250,000 inhabitants	
	2010	2011	2010	2011	2010	2011
Cycle	2.1	2.3	7.9	8.4	6.0	5.1
Walk	27.5	22.9	19.5	18.2	27.0	24.5
Total	29.6	25.2	27.4	26.6	33.0	29.6

Source: Isfort, Observatory "Audimob" on Italian mobility, 2011

Isfort's survey (2011) shows how cycling has a typological relief in part similar to that of walking as the choice is derived from motivations which tend to be diversified by social references. The reasons are to be found in the economic and ecological characteristics but in sum the cycling is not preferable to a quick and more comfortable car.

All over the country the share of non-motorized travel (walking or cycling) is equal to 21.6% of the total (Isfort, 2011). Focusing on urban centres, the percentage increases: in the city, one-third of journeys are made on foot, but despite the numbers, the attention given to pedestrians (and cyclists) by national and local administrators is relatively low.

The systematic displacement cycle have a minority of those generated by the need for family and leisure, with a marked characteristic of regularity – 53.7% of the mileage is repeated every day, or almost – confirming that choice is always less determined by contingent causes more and more, but rather, connected to a structured model of a search for modal and sustainable solutions in response to daily transport needs.

Wide, however, the difference in cycling in a part of North-European cities: the mobility bike, properly protected and encouraged, has become a significant part of the management of urban mobility (Pucher & Buehler, 2008). The available data show that the spread of cycle paths is a primarily cultural factor, independent of climatic conditions: they are all in small and medium-sized cities like Helsinki and Aarhus – those in which the network appear than a metre per inhabitant. Medium to large areas such as Stockholm and Copenhagen also have a good supply per capita (Ambiente Italia, 2007). Dutch, Danish and German cities have reinforced the safety, convenience and attractiveness of their excellent cycling rights of way with extensive bike parking, integration with public transport, making their cities people-friendly rather than car-friendly and thus more livable and more sustainable than American and British cities (Pucher & Buehler, 2008).

In many parts of Italy the emergence of modern lifestyles and a new mobility model, which recognizes the important role of cycling, faces difficulties, but in some places (and times) the practice of cycling is easier than in other places (and times). As the contributions which follow also demonstrate, different cycling practices are either encouraged or discouraged by variously favourable and unfavourable conditions. Within any city, cyclists tend to favour some routes over others. Within a region, some cities are considered more cycle-friendly than others, which may nonetheless still be identified with cycles and cycling, through, for example, association with long-established cycle races (Krizek, 2012).

Since the Italian economic growing years, after World War II, also a political choice, Italian cities have grown around a mobility essentially based around the private car and not responded to the application of the principles of sustainability instead of less polluting modes of transport. Especially in urban areas, "car regime" vision (Holtz, Brugnach & Pahl-Wostl,, 2008), which still continues, has changed the landscape. Infact "(...) it is important to recognise the car as part of a system (…) which we have called the 'automobility paradigm'. It is often not appreciated to what extent our modern culture is integrated with the car and its systems: we have literally built our world around the car in its current form (…)." (Nieuwenhuis, Wells, & Vergragt, 2004, p. 10) and has gradually transformed the urban environment (Marletto, 2011). The environment is occupied with increased traffic congestion, air pollution and noise, a lack of safety, accidents as well as difficulty with lifetime management and excessive individual and social costs (Urry, 2007).

The analysis of indicators of sustainable urban mobility (Ispra, 2011) reveals a growing trend on average (2000-2009), but it is not yet sufficient to guarantee adequate levels of sustainability in transport and in the movement of Italian citizens. One cause is related to public transport since the possibility of easy access to quality public transport contributes to the spread of forms of sustainable mobility. There is little evidence of growth in the number of trams, trolley buses and subways other than for buses and railways. In Europe the main approach to increasing cycling and making it safer has beeen the provision of more bike paths and lanes (Pucher et al., 2010). In Italy there is a spatial inhomogeneity (Table 3). In general, the use of these lanes, as well as those used for recreation and sports for systematic trips of a short distance, is more common in the cities of North Italy. The recent focus has been on expanding bike lanes, off-street bike paths are often the most heavily used and highest profile cycling facilities but tend to serve mainly recreational purposes. In 2009, the provincial capitals had 13.7 km of cycle paths for the municipal area of 100 km^2, which has more than doubled in the last decade.

Table 3. Trend in bike paths and lanes (km/100 km^2 land) in Italian cities

	2000	2005	2009
Torino	49,9	77,6	91,8
Vercelli	1,9	12,0	44,8
Milano	21,9	40,9	41,1
Brescia	27,6	107,0	132,3
Cremona	38,4	54,4	75,3
Mantova	29,1	63,5	72,2
Bolzano	42,0	72,6	95,5
Verona	2,7	7,2	38,2
Padova	35,5	81,9	148,6
Trieste	4,7	4,7	27,8
Imperia	0,9	0,9	0,9
Modena	41,0	60,0	76,0
Bologna	22,0	45,5	61,1
Ferrara	8,2	15,3	22,2
Rimini	17,1	44,9	54,3
Firenze	17,6	35,2	67,9
Reggio Emilia	24,0	43,6	67,2
Roma	2,3	5,2	8,8
Campobasso	0,9	0,9	11,5
Bari	4,7	4,7	6,5
Cosenza	10,7	10,7	32,2
Reggio Calabria	-	0,6	0,6
Catanzaro	-	1,3	8,5
Palermo	-	3,8	13,3
Agrigento	0,4	2,0	2,0
Ragusa	-	-	0,2
Italia	5,3	9,3	13,7

Source: Ispra, 2011

There are 25 cities in Italy who have not made their paths (e.g. Napoli, Taranto, Sassari and Genova). Among others, Padova (148.6 km), Brescia (132.3), Bolzano (95.5) and Torino (91.8) have the highest density and Imperia (0.9 km per 100 km2 land), Reggio Calabria (0.6) and Ragusa (0.2) are characterized by the lowest. On average, in large cities such as Milano - still dominated by the car (covering 42.4% of trips) rather than the bike (which accounts for 3.8% of the total) – the network of cycle tracks has been redesigned to more precisely defined links, with "direct" paths from the centre connecting with the periphery in order to improve mobility between home and work and home and school. This is in line with the objectives of the municipality to "reinvent" Milan, and make it a more green and sustainable city (Dell'Agnese & Anzoise, 2011). Separate facilities are not sufficient but they are certainly necessary to ensure that cycling is possible for a broad spectrum of the population (Garrard, Rose & Lo, 2008).

It therefore seems essential to regulate mobility in urban centres by encouraging cycling as well as the use of public transport. Looking, for example, at data related to the space for pedestrians, in 2010, the different urban realities had, on average, for every 100 inhabitants, 34 square metres of area closed to motorized traffic (Legambiente, 2010). The municipalities of Venezia, Verbania, Cremona and Terni exceed 100 square metres for every 100 inhabitants, but the big cities present more surface area for pedestrians: Turin (which has 469,239 square metres), Roma (396,195) and

Milano (362,410). Eight cities say they do not have traffic islands within their territory (e.g. Ascoli Piceno and Agrigento).

But who is the typical cyclist? In Italy, and also in other countries, very few people use a bike on a regular basis. Urbanczyk (2010) has identified different types of cyclists, which coincide or differ depending on the routine, the frequency and the purpose of their cycling (Figure 1). It should be noted that studies of this kind are numerous and similar (Davies et al., 2001; Jensen, 1999), and identify distinctions mostly related to the motivations for and frequency of use (commuting, health, shopping, leisure), or at least the attitude towards individual sustainability and lifestyle. Social identities are important to the promotion of cycling because in marketing they are often used to promote products (Gatersleben & Haddad, 2010).

The goal, in fact, is to encourage people to use bicycles on holiday, at leisure, but transferring this "good practice" to everyday use becomes a tool for disseminating sustainable behaviour within an urban mobility mainly characterized by the use of the automobile (Sener, Eluru & Bhat, 2009).

It therefore appears to be a social phenomenon – a different leisure culture – and particularly in urban areas related to the deployment of bike sharing[2], which is playing an important role in urban mobility services in many cities worldwide. These are interesting recent developments in cycling for recreation, but at present in Italy it appears to be a niche business. The cyclists using the bicycle as their main mode of transport in the activity, although not exclusively recreational (Ritchie, 1998) depending on whether the cyclists "excited" or "occasional," in reference to the parameters of time and purpose of the holiday (Simonsen & Jorgenson, 1998). The bicycle is used by a wide variety of individuals and offers potential market segments in terms of sustainable development in tourism (Lumsdon, 2000); it may refer to either a circuit to a journey of at least one day as an integral part of a trip or vacation, for both short and long recreational sports but not (Ritchie, 1998; Sustrans, 1999), although it also includes participation in cycling events (Faulks et al., 2008; Lamont, 2009). It is still experience an activity that involves both recreation and transportation as the relationship between tourism and transport is inseparable (Chang & Chang, 2003).

CASE STUDY: CITIES OF SICILY, ITALY

It was intended to analyze the performance of provinces in Sicily as a statistical indicator reference and other measures, considering like a limit the extension of the municipal employees who, for historical reasons and administrative governance, vary considerably from one city to

Figure 1. Principal groups of cyclists (Urbanczyk, 2010; Gatersleben & Haddad, 2002)

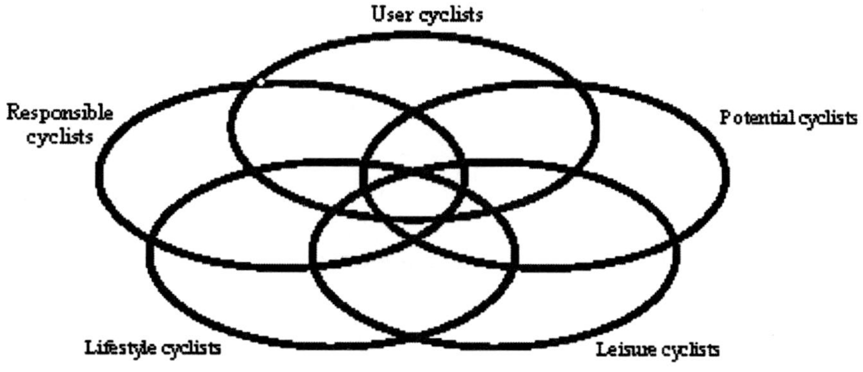

another. In fact, the land of some cities in some cases is 10-15 times smaller than others, implying a different density of population (Figure 2). It's worthy of note that specific studies demonstrate that moderate and high-density populations both correlate well with less motor traffic and create more attractive conditions for cycling (Dumbaugh & Li, 2010). But high density city populations may also intensify exposure to pollution, less safe travel, etc., so generally smaller cities are more cyclefriendly than large ones (Krizek, 2011). Furthermore, the urban structure of some cities (Trapani, Siracusa, Messina, Palermo, Catania) is founded on their relationship with the sea and the use of the seafront. They alternate high-density to low-density areas, with a high level of urbanization, in particular by the seaside. That widely increases the distance between different points inside the cities (e.g. Palermo, Catania), with a consequent increase in the average travel distance.

Data produced are therefore not comparable and sometimes rase quesions over to the effective provision of services to residents. It was also not possible to find specific data because the local authorities show little propensity due to the lack of specify data

The overall view of the Sicilian provinces has a high attention to projects not defined "by environmental value," especially in relation to sustainable mobility (urban railways, tunnels, trams, exchange parking, cycle lanes), although there is a need for an improvement in the quality of life in cities and the development of local economies (Tables 4 and 5). The mobility system of the main Sicilian cities is mostly based on private means. Reorganization of the urban mobility system, in addition to urban environmental safety and reorganization of services inside the city are among the most important themes in local government policies.

As long as car use remains frequent and urban transportation policy keeps aiming at motor transport systems, bicycles will be used primarily for recreation and not for daily or systematic trips. The resistance shown to such means of transport, today accepted in other Italian urban contexts as being extremely useful transport alternatives – both for occupational and recreational trips - stem from certain factors that are characteristic of Mediterranean cities, namely orographic, urbanistic (urban structure with concentrated and multifunctional historic inner city) and climatic factors (high humidity rate and high temperatures) but, above all, cultural and sociological factors. Often users show resistance to using collective means of transport because these are not efficient and punctual, and environmental sensibility is a minor consideration.

The total size of the Limited Traffic Zone (LTZ) in relation to the municipal area is an interesting indicator of the intentions of the local government

Figure 2. Population and land of the cities of Sicily: 2009

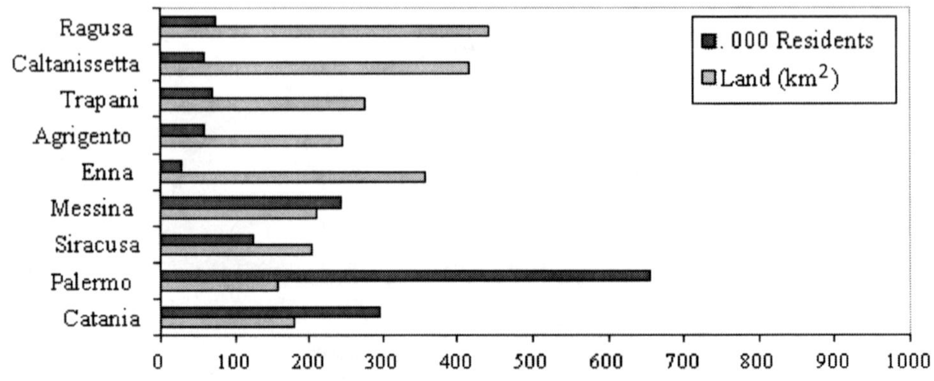

Table 4. Some indicators for sustainable mobility – 2009

	Bike Paths km/100km² land	LTZ km²/100 km²	Urban pedestrian areas m²/100 ab.	Urban Traffic plan	Cars/100 residents	Bike sharing	Events ***	Cycling indicator**
Catania	1,05*	n.a.	7,9	-	72	x	2	5,1
Palermo	13,3	4,8	8,9	-	61		4	7,7
Siracusa	7,5*	0,2	4,5	-	65	x	2	20,5
Messina	1,2	2,1	4,5	X	60	-	3	-
Enna	-	-	-	-	65	-	-	-
Agrigento	2,0	0,6	-	X	68	-	-	n.d.
Trapani	-	0,7	-	X	58	-	-	7,7
Caltanissetta	-	0,6	5,3	X	69	-	-	-
Ragusa	0,2	0,2	14,6	X	69	-	-	5,1

* Common data, 2011.

** This indicator is composed from the presence of: bike lanes; rent shops; bicycle paths; service bikes; traffic signs; bike parking; bike sharing (Urban Ecosystem, *op. cit.*).

*** Cycling days ; Kids cycle days; Live in town; Bike in town (Euromobility, 2011).

Source: ISTAT, 2011

Table 5. Some indicators for sustainable mobility – 2009

	Railways km/100km² land	Road bus km/100km² land	Subways km/100km² land	Tram 10,000 inhab.	Buses x 10,000 inhab.	Change parking x 1,000 cars	Car parking x 1,000
Catania	16	152	2.1	-	11.4	8.1	38.4
Palermo	23.4	214.6	-	-	8.5	7	46.,6
Siracusa	12.4	59.1	-	-	3.7	-	15
Messina	14.9	187.7	-	0.4	3	8.2	30.4
Enna	8	28.8	-	-	5.4	-	16.7
Agrigento	1.6	90.2	-	-	3.4	4.8	5.6
Trapani	16.5	46	-	-	8.5	12.3	14.8
Caltanissetta	12.5	18.5	-	-	2.7	7.3	-
Ragusa	13.1	51.7	-	-	3.4	-	17.9

Source: ISTAT, 2011

to make a real change to the prevailing patterns of movement, trying to limit the role of the private vehicle. It is worth noting Messina (2.1) which rates between the cities of medium size, although the quality of this city's measures remains to be verified – their daily and weekly timetables, what special permits are available – as these are relevant in terms of the selectivity and efficacy of measures in terms of the containment of private traffic and the real advantage (more space on the road, driving speed, reliability, punctuality of buses) assured for public transport.

The synthetic index of cycling (with values from 0 to 100), calculated by Legambiente (Urban Ecosystem, 2011), measures the ability of cities to activate a range of tools that promote cycling.

Specifically, among the cities of Sicily, Siracusa has the highest score (20.5) for the presence of bike sharing, established in 2009, which represents a widely available service. The bike-sharing stations are within 15 points of the city, with the distinction of being "mixed," offering both traditional bicycles and electric ones (5 bicycles and 59 users per 10,000 residents and subscribers almost doubled from 400 to 730; Euromobility, 2011).

With regard to bike paths, used for recreational and sports but also systematic short-distance trips, Palermo has the highest availability while four provinces do not detect any value in the knowledge that we are very far from the values of the cities of North Italy. In some cases events are organized to promote the widespread use of bicycles despite the presence of non-cycling zones and limitations of the traffic area. In 2009, the Italian government, like many other international governments, actively examined methods to promote cycling and other sustainable modes of transport with a travel sustainable plan.

These policies were driven by the desire to reduce greenhouse gas emissions and decrease congestion, particularly the policies introduced to promote cycling by reducing the cost of purchasing a bicycle. In many cities the success was significant, but no data demonstrate that citizens have changed their perceptions of cycling and improved accessibility to the bicycles and cycling after using this scheme.

Discussion and Solutions

Cycling continues to increase in popularity and garner attention for its ability to achieve various environmental, health and congestion-mitigating benefits for local communities. However, very few people use a bicycle on a regular basis. If cycling is to be a viable mode of transportation with extreme variation in the experience of its users, it must have appropriate facilities.

The diffusion of cycling currently has limited added value for the study area but if properly supported could meet the requirements of sustainability by many operators, public authorities and users in particular in urban areas. From synthetic and fragmented a small fraction, even more so in urban areas in Sicily, but potentially growing.

The image is one of cities still characterized by the proliferation of cars, public transport and little-used eco-friendly mobility. The use of the bicycle is still limited; big cities with more than 10% of the local population using sustainable mobility include Milan and Bologna. Residents of the Sicilian cities are car-dependent with costs for the quality of life and health that are likely to remain high. While implementation of innovative facilities such as bicycle boulevards and coloured bike lanes is low, surveys have indicated that there are projects currently being implemented or in the process of approval. The number of cities that report having implemented innovative facilities has increased significantly in recent years.

To implement a system of bicycle paths (from different regions) means promoting and implementing soft mobility, which is the first step towards a sustainable development perspective. One likely development to be maintained and enhanced over time is the awareness of communities and local governments. The sustainable vision of urban mobility is present, at least in its potential, and the idea of cycle paths should be communicated and disseminated with its territorial, economic and social implications. Informal programmes and policies (e.g. marketing programmes) and several independent bicycle organizations and events may also help create a "bicycle culture" that will likely influence cycling behaviour. Cycling may be combined with other modes of transportation and be used by adults to meet the recommendations for daily physical activity, but a network of different types of infrastructure is necessary to attract new people to cycling (Dill, 2009).

The solution to a complex problem to ensure a sustainable urban mobility requires a strong commitment on innovation as well as methods and content of urban policies. Sustainable urban mobility, like cycling, requires good use of transport infrastructures, coordination between different transport modes and the promotion of strategies of less polluting modes. Today, space, especially in Italian cities, is unfortunately a limited resource. And its distribution appears consequential to government policy. On the other hand, the development prospects for encouraging more cycling for urban travel are strongly affected by the policy choices of mobility, urban and environmental planning, protection of landscape and local governance.

FUTURE RESEARCH DIRECTIONS AND CONCLUSION

This chapter investigates the issues related to the development cycling in Italy and considers the nexus between cycling, environment, and economics. Cycling is a clean and efficient mode of transportation especially suited for short to moderate distances. Cycling provides numerous benefits compared to motorized vehicles, including healthy exercise, does not use fossil fuels or create air or noise pollution, and helps reduce traffic congestion.

The bicycle should not be considered only as a two-wheeled means of transport, nor a sustainable transport symbolic but the preferred mode of transport for most short distance. It's important to promote the "pedal power" (Horton & Parkin, 2012). Cycling needs a well defined system in which operate where the role of urban planners is fundamental because that system needs to be closely allied to the needs of society with the environment.

With increasing pressure for cities to reduce their carbon footprint and to enhance their part in providing for healthy living, there is a greater need for more research focusing on the economic, social and environmental benefits of cycling for quality of life in urban areas, including leisure and cycling tourism specifically. Considerable research seeks improved models of travel behaviour but it is important to investigate reciprocal correspondences and relationship with environmental outcomes. The results presented in this chapter also demonstrate the potential success of government policies to encourage cycling and how these types of intervention can achieve a modal shift. Furthermore, research should focus on a more detailed close examination between proportion of bicycle traffic and index for sustainable urban development so to include the possibility of different transport modes.

REFERENCES

Adey, P. (2010). *Mobility*. Abingdon, UK: Routledge..

Banister, D., Watson, S., & Wood, C. (1997). Sustainable cities: Transport, energy, and urban form. *Environment and Planning. B, Planning & Design*, *24*, 125–143. doi:10.1068/b240125.

Breen, R. (2004). The comparative study of social mobility. In R. Breen (Ed.), *Social Mobility in Europe* (pp. 37–75). Oxford, UK: Oxford University Press. doi:10.1093/0199258457.003.0003.

Carstensen, T. A., & Ebert, A. (2012). Cycling cultures in Northern Europe: From golden age to renaissance. In J. Parkin (Ed.), *Cycling and sustainability* (pp. 23–58). Bingley, UK: Emerald Group Publishing Limited. doi:10.1108/S2044-9941(2012)0000001004.

Chang, H.W., & Chang, H. L. (2003). A strategic study of bicycle tourism in Taiwan. *Journal of the Eastern Asia Society for Transportations Studies, 5*.

Colleoni, M. (Ed.). (2008). *La ricerca sociale sulla mobilità urbana: Metodo e risultati di indagine*. Cortina, Italy: Libreria Milano..

Commission of the European Communities. (1997). *Towards a urban agenda in the European Union. COM 197*. Bruxell..

Commission of the European Communities. (1998). *Sustainable urban development in the European Union: A framework for action. COM*. Bruxell..

Commission of the European Communities. (1999). *Green paper on the urban environment. COM 218*. Bruxell..

Commission of the European Communities. (2001a). *A sustainable strategy for a better world: A European strategy for sustainable development. COM 264*. Bruxell..

Commission of the European Communities. (2001b). *European transport policy for 2010: Time to decide. COM 370*. Bruxell..

Commission of the European Communities. (2004). *Towards a thematic strategy on the urban environment. COM 60 final*. Bruxell..

Cori, B. (Ed.). (1988). Traffico urbano e qualità della vita in Italia. Bologna, Italy: Pàtron editore.

Cori, B. (Ed.). (1997). La città invivibile: Nuove ricerche sul traffico urbano. Bologna, Italy: Pàtron editore..

Cresswell, T. (2006). The right to mobility: The production of mobility in the courtroom. *Antipode*, *38*(4), 735–754. doi:10.1111/j.1467-8330.2006.00474.x.

Dell'Agnese, E., & Anzoise, V. (2011). Milan, the unthinking metropolis. *International Planning Studies*, *16*(3), 217–235. doi:10.1080/13563475.2011.591143.

Dematteis, G. (Ed.). (2012). *Le grandi città italiane: Società e territori da ricomporre*. Milano, Italy: Marsilio Editore..

Dill, J. (2009). Bicycling for transportation and health: The role of infrastructure. *Journal of Public Health Policy*, *30*, 95–110. doi:10.1057/jphp.2008.56 PMID:19190585.

Ecosystem, U. Legambiente. (2011). *XV report about environmental quality of Italian provinces*. Retrieved from http:/www.legambiente.eu

European Cyclists' Federation. (2009). *The future of transport*. Retrieved June 10, 2012, from www.ecf.com

European Environment Agency. (2006). *Urban sprawl in Europe: The ignored challenge*. Retrieved from http://www.eea.europa.eu/publications/eea_report_2006_10

Garrard, J., Rose, G., & Lo, S. (2008). Promoting transportation cycling for women: The role of bicycle infrastructure. *Preventive Medicine*, *46*(1), 55–59. doi:10.1016/j.ypmed.2007.07.010 PMID:17698185.

Gatersleben, B., & Haddad, H. (2010). Who is the typical bicyclist? *Transportation Research Part F: Traffic Psychology and Behaviour*, *13*, 41–48. doi:10.1016/j.trf.2009.10.003.

Gatersleben, B., & Uzzell, D. (2002). Sustainable transport and quality of life. In W. R. Black, & P. Nijkamp (Eds.), *Social Change and Sustainable Transport*. Bloomington, IN: Indiana University Press..

Geerlings, H., Shiftan, Y., & Stead, D. (2012). The complex challenge of transitions towards sustainable mobility - An introduction. In H. Geerlings, Y. Shiftan, & D. Stead (Eds.), *Transition towards Sustainable Mobility* (pp. 1–9). Aldershot, UK: Gower Publishing..

Gorham, R. (2002). Car dependence as a social problem: A critical essay on the existing literature and future needs. In W. R. Black, & P. Nijkamp (Eds.), *Social Change and Sustainable Transport*. Bloomington, IN: Indiana University Press..

Hanson, S. (Ed.). (1995). *The geography of urban transportation*. New York: Guilford Press..

Hochschild, A. R. (1997). *The time bind: When works becomes home and home becomes work*. New York: Metropolitan Books..

Holtz, G., Brugnach, M., & Pahl-Wostl, C. (2008). Specifying regime - A framework for defining and describing regimes in transition research. *Technological Forecasting and Social Change*, *75*(5), 623–643. doi:10.1016/j.techfore.2007.02.010.

Horton, D., & Parkin, J. (2012). Conclusion: Towards a revolution in cycling. In J. Parkin (Ed.), *Cycling and sustainability* (pp. 303–325). Bingley, UK: Emerald Group Publishing Limited. doi:10.1108/S2044-9941(2012)0000001014.

Isfort, Observatory Audimob on Italian mobility. (2011). *Rapporto su stili e comportamenti di mobilità degli Italiani*. Retrieved October 18, 2011, from http://www.isfort.it

Ispra, Istituto Superiore per la Protezione e la Ricerca Ambientale. (2011). *Qualità dell'ambiente urbano: VII rapporto*. Retrieved from June 14, 2011, from http://www.isprambiente.it

ISTAT. Istituto Nazionale di Statistica. (2011). *Indagine sui trasporti*. Retrieved from September 22, 2011, http://www.istat.it

Italia, A. (2007). *Urban ecosystem Europe: An integrated assessment on the sustainability of 32 European cities*. Retrieved March 10, 2012, from http://www.ambienteitalia.it

Kaufmann, V., Bergman, M. M., & Joice, D. (2004). Motility: Mobility as capital. *International Journal of Urban and Regional Research*, *28*(4), 745–756. doi:10.1111/j.0309-1317.2004.00549.x.

Kellerman, A. (2012). *Daily spatial mobilities: Physical and virtual*. Surrey, UK: Ashgate Publishing Group..

Kenworthy, J., & Laube, F. (1996). Automobile dependence in cities: An international comparison of urban transport and land use patterns with implication for sustainability. *Environmental Impact Assessment Review*, *16*, 279–308. doi:10.1016/S0195-9255(96)00023-6.

Krizek, K. J. (2012). Cycling, urban form and cities: What do we know and how should we respond? In J. Parkin (Ed.), *Cycling and sustainability* (pp. 23–58). Bingley, UK: Emerald Group Publishing Limited. doi:10.1108/S2044-9941(2012)0000001007.

Krizek, K. J., & Stonebraker, E. W. (2011). Assessing option to enhance bicycle and transit integration. *Journal of Transportation Research Board*, *2217*, 162–167. doi:10.3141/2217-20.

Lamont, M. J. (2009). Reinventing the wheel: A definitional discussion of bicycle tourism. *Journal of Sport & Tourism*, *14*(1), 5–23. doi:10.1080/14775080902847363.

Lumsdon, L. (2000). Transport and tourism: Cycle tourism - A model for sustainable development? *Journal of Sustainable Tourism*, *8*(5), 361–377. doi:10.1080/09669580008667373.

Marletto, G. (2011). Structure, agency and change in the car regime: A review of literature. *European Transport*, *47*, 71–88.

Merriman, P. (2009). Auto mobility and the geographies of the car. *Geography Compass*, *3*(5), 586–599. doi:10.1111/j.1749-8198.2009.00219.x.

Newman, P. W. G., & Kenworthy, J. R. (1999). *Sustainability and cities: Overcoming automobile dependence*. Washington, DC: Island Press..

Nieuwenhuis, P., Wells, P., & Vergragt, P. J. (2004). Introduction to the special issue on technological change and regulation in the car industry. *Greener Management International*, *47*, 5–11.

O'Sullivan, M., & Watling, D. (2011). Visions for a walking and cycling focused urban transport system. *Journal of Transport Geography*, *19*, 1580–1589. doi:10.1016/j.jtrangeo.2011.03.011.

Olivier, J. G. J., Janssens-Maenhout, G., & Peters, J. A. H. W. (2012). Trends in global CO2 emissions: 2012 report. The Hague: PBL Netherlands Environmental Assessment Agency, Ispra: Joint Research Centre..

Organisation for Economic Cooperation and Development. (2000). *Managing urban growth*. Paris: DT/Tdpc..

Organisation for Economic Cooperation and Development. (2004). *Implementing sustainable urban travel policies: Moving ahead: National policies to promote cycling*. Paris: OECD..

Page, S. J. (2009). *Transport and tourism: Global perspectives*. Harlow: Prentice Hall..

Pucher, J., & Buehler, R. (2008). Making cycling irresistible: lessons from the Netherlands, Denmark and Germany. *Transport Reviews*, *28*(4), 495–528. doi:10.1080/01441640701806612.

Pucher, J., & Buehler, R. (2009). Integrating bicycle and public transport in North America. *Journal of Public Transportation*, *12*(3), 79–104.

Rees, W. (2003). Ecological footprint and urban transportation. In R. Tolley (Ed.), *Sustainable Transport: Planning for walking and cycling in urban environments* (pp. 3–19). London: Woodhead Publishing Limited..

Ritchie, B. W. (1998). Bicycle tourism in the south island of New Zealand: Planning and management issue. *Tourism Management*, *19*(6), 567–582. doi:10.1016/S0261-5177(98)00063-6.

Salet, W. (2007). New metropolitan spaces and metropolitan strategies in the face of modernisation. *Metropoles, 2*, 1-26. Retrieved August 22, 2012, from http://metropoles.revues.org/262

Sener, I., Eluru, N., & Bhat, C. R. (2009). Who are bicyclists? Why and how much are they bicycling? *Transportation Research Record*, *2134*(1), 63–72. doi:10.3141/2134-08.

Sheller, M., & Urry, J. (2000). The city and the car. *International Journal of Urban and Regional Research*, *24*(4), 737–757. doi:10.1111/1468-2427.00276.

Shore, W. B. (2006). Land-use, transportation and sustainability. *Technology in Society*, *28*, 27–43. doi:10.1016/j.techsoc.2005.10.014.

Simonsen, P. S., & Jørgensen, B. (1998). Cycle tourism: An economic and environmental sustainable form of tourism? In Simonsen, Jørgensen, & Robbins (Ed.), *Cycling tourism*. Retrieved from www.crt.dk/media/Cycling_tourism.pdf

Socco, C. (2010). *Il piano urbano di mobilità sostenibile*. Retrieved March 12, 2012, from http://www.ocs.polito.it/casi/dwd/pianourb_annex.pdf

Thrift, N. (2004). Driving in the city. *Theory, Culture & Society*, *21*(4-5), 41–59. doi:10.1177/0263276404046060.

Tight, M., Timms, P., Banister, D., Bowmaker, J., Copas, J., & Day, A. et al. (2011). Visions for a walking and cycling focused urban transport system. *Journal of Transport Geography*, *19*, 1580–1589. doi:10.1016/j.jtrangeo.2011.03.011.

Travisi, C., & Camagni, R. (2005). *Sustainability of urban sprawl: Environmental-economic indicators for the analysis of mobility impact in Italy*. Retrieved from http://www.feem.it/Feem/Pub/Publications/WPapers/default.htm

United Nations. (2012). *World urbanization prospects: The 2007 revision population database*. Retrieved July 20, 2012 from http://www.esa.un.org

United Nations Economic Commission For Europe. (2008). *Working together for sustainable and healthy transport: Guidance on supportive institutional conditions for policy integration of transport, health and environment.* Retrieved June 10, 2012, from www.unece.org

Urbanczyk, R. (2010). *PRESTO cycling policy guide: Promotion of cycling.* Retrieved June 22, 2012, from www.presto-cycling.eu

Urry, J. (2004). The system of auto mobility. *Theory, Culture & Society, 21*(4-5), 25–29. doi:10.1177/0263276404046059.

Urry, J. (2006). Inhabiting the car. *The Sociological Review, 54,* 7–31. doi:10.1111/j.1467-954X.2006.00635.x.

Urry, J. (Ed.). (2007). *Mobilites.* Cambridge, UK: Polity Press..

Wen, L. M., & Rissel, C. (2008). Inverse associations between cycling to work, public transport, and overweight and obesity: Findings from a population based study in Australia. *Preventive Medicine, 46,* 29–32. doi:10.1016/j.ypmed.2007.08.009 PMID:17904210.

Zito, P., & Salvo, G. (2011). Toward an urban transport sustainability index: A European comparison. *European Transport Research Review, 3,* 179–195. doi:10.1007/s12544-011-0059-0.

ADDITIONAL READING

Aldred, R. (2010). On the outside: Constructing cycling citizenship. *Social & Cultural Geography, 11*(1), 35–52.

Banister, D. (2005). *Unsustainable transport: City transport in the new century.* London, UK: Routledge..

Banister, D. (2008). The sustainable mobility paradigm. *Transport Policy, 15,* 73–80. doi:10.1016/j.tranpol.2007.10.005.

Blickstein, S., & Hanson, S. (2010). Critical mass: Forging a politics of sustainable mobility in the information age. *Transportation, 28*(4), 347–362. doi:10.1023/A:1011829701914.

Brebbia, C. A. (Ed.). (2011). *The sustainable world.* Southampton, UK: Wit Press..

Cervero, R. (2002). Built environments and mode choice: Toward a normative framework. *Transportation Research Part D, Transport and Environment, 7,* 265–284. doi:10.1016/S1361-9209(01)00024-4.

Holden, E. (2007). *Achieving sustainable mobility: Everyday and leisure-time travel in the EU.* Aldershot, UK: Ashgate Publishing Limited..

Metcalfe, R., & Dolan, P. (2012). Behavioural economics and its implications for transport. *Journal of Transport Geography, 24,* 503–511. doi:10.1016/j.jtrangeo.2012.01.019.

Nicolopoulou-Stamati, P., Hens, L., & Howard, C. V. (Eds.). (2005). *Environmental health: Impacts of transport and mobility.* Amsterdam, The Netherlands: Springer. doi:10.1007/1-4020-4307-4.

Parkin, J. (Ed.). (2012). *Cycling and sustainability.* Bingley, UK: Emerald Group Publishing Limited..

Rietveld, P., & Daniel, V. (2004). Determinants of bicycle use: Do municipal policies matter? *Transportation Research Part A, Policy and Practice, 38,* 531–550. doi:10.1016/j.tra.2004.05.003.

Rubulotta, E., Ignaccolo, M., Inturri, G., & Rofè, Y. (2012). Accessibility and centrality for sustainable mobility: A regional planning case study. *Journal Urban Planning and Development.* Retrieved October 20, 2012 from http://ascelibrary.org/loi/jupddm

Spinney, M. (2007). Cycling the city: Non place and the sensory construction of meaning in a mobile practice. In D. Horton, P. Rosen, & P. Cox (Eds.), *Cycling and Society* (pp. 25–45). Aldershot, UK: Ashgate Publishing Limited..

Vega, A. (2012). Using place rank to measure sustainable accessibility. *Journal of Transport Geography*, *24*, 411–418. doi:10.1016/j.jtrangeo.2012.04.008.

KEY TERMS AND DEFINITIONS

Accessibility: The ability of people to reach the destinations that they need to visit in order to meet their needs. It's different than mobility that means the ability to move.

Bike Lanes: A striped lane on a roadway.

Cycling: For society one of the more sustainable modes of transport for people and goods; for individual is a healthy and cheap mode of transport because is a physical activity and physical activity is beneficial for health.

Sustainable Development: Is the development that meets the needs of the present generations without compromising the ability of future generations to meet their own needs. The definition is the integration of economic efficiency, ecological stability, and social equity.

Travel Sustainable Plan: Is a package of measures put in place to by an public organisation to encourage and support more sustainable travel patterns amongst staff and visitors and reduce the demand for unnecessary car use.

ENDNOTES

1. Includes not only fossil fuel combustion on which the BP reports are based, but also incorporates all other relevant CO_2 emissions sources including flaring of waste gas during oil production, cement clinker production and other limestone uses, feedstock and other non-energy uses of fuels, and several other small sources (Olivier et al., 2012).
2. It is a bike rental service in the short term, self-service, organized and made available to the network of public spaces, lets you leave your bike in a location other than sampling and is designed for different types of users. In Italy is currently active in 160 cities (OBIS, 2011; www.obisproject.com).

This chapter was previously published in Sustainable Technologies, Policies, and Constraints in the Green Economy edited by Dorel Dusmanescu, Andrei Jean-Vasile, Turek Rahoveanu Adrian, and Jonel Subic, pages 251-268, copyright 2013 by Information Science Reference (an imprint of IGI Global).

Chapter 73
The Role of Urban Consolidation Centres in Sustainable Freight Transport

Julian Allen
University of Westminster, UK

Jacques Leonardi
University of Westminster, UK

Michael Browne
University of Westminster, UK

Allan Woodburn
University of Westminster, UK.

ABSTRACT

The chapter considers the role that urban consolidation centres (UCCs) can play in helping to reduce goods vehicle traffic and its environmental impacts in urban areas. A UCC is a logistics facility that is situated in relatively close proximity to the area that it serves. Goods destined for this area are dropped off at the UCC, and are sorted and consolidated onto goods vehicles for delivery to their final destinations, often using environmentally friendly vehicles such as electric and gas-powered goods vehicles, and electrically-assisted cycles. The development of UCCs since their first appearance in the 1970s is reviewed and the various types of UCC categorised. The freight transport and logistics advantages offered by UCCs are discussed. Case studies of four recent UCC trials are included. The objectives, operational and financial aspects and impacts of these UCCs are compared and their critical success factors identified.

INTRODUCTION

This chapter considers the role played by urban consolidation centres (UCCs) in rationalising the movement of freight in urban areas to cut delivery costs and reduce environmental impacts. They not only permit greater consolidation of loads, but also permit greater use of electric and other alternatively-fuelled goods vehicles which emit less CO_2 and other, noxious gases.

The introductory sections provide an overview of the development of UCCs and the range of benefits that they offer. Case studies of four recent UCC trials and operations in the UK are then presented: the London Construction Consolidation Centre, the London Heathrow Retail UCC, the Bristol UCC and an urban 'micro-consolidation' Centre based in the City of London. The final two

DOI: 10.4018/978-1-4666-4852-4.ch073

sections draw on the experience of these UCCs to highlight key issues and identify critical success factors and future research needs.

OVERVIEW OF UCCs

A UCC is a logistics facility that is situated in relatively close proximity to the urban area that it serves be that a city centre, an entire town or a specific site such as a shopping centre, airport, hospital or major construction site. Goods destined for these locations are dropped off at the UCC. The UCC operator sorts and consolidates these loads dropped off by logistics companies and makes deliveries to the final destinations, often using environmentally-friendly vehicles such as electric and gas-powered goods vehicles, and electrically-assisted cycles (Browne et al, 2005).

By improving the lading factor of goods vehicles making final deliveries in congested locations, UCCs reduce the total distance travelled by delivery vehicles in urban areas, as well as reducing greenhouse gas emissions and local air quality pollutants associated with these journeys (both through reductions in the total distance travelled, and through the use of low emission vehicles) (Browne et al., 2007). In addition the total kerbside time and space occupied by vehicles making on-street deliveries can be reduced through consolidation further reducing the impact of freight operations on traffic congestion. Other social and environmental advantages can include noise reductions through the use of quieter vehicles, reductions in conflicts between goods vehicles and other road users and greater pedestrian safety (Gonzalez-Feliu and Morana, 2008; WSP, 2008).

The logistics companies dropping their loads at the UCC benefit by avoiding the need to enter congested urban areas and thereby saving time and costs. Those receiving goods from the UCC benefit in terms of delivery reliability. In addition to consolidation and final delivery, a range of other value-added logistics and retail services can also be provided at the UCC.

Initial research into UCCs as an urban freight initiative commenced in the early 1970s and has continued ever since with levels of interest in this approach increasing during the last decade (see for example Browne et al., 2005; Institut für Seeverkehrswirtschaft und Logistik, 2005; McKinnon, 1998a; McKinnon, 1998b; van Duin et al., 2010).

An international literature review of UCCs carried out in 2005 identified 67 UCCs in 15 countries worldwide that had either been the focus of a feasibility study, trial or fully operational scheme. The earliest was a UCC study that took place in Columbus, Ohio in the early 1970s (McKinnon, 1998a). The majority of these UCCs had taken place in Europe, with France, Germany, Italy, the Netherlands and the UK accounting for 46 of the 67 identified (Browne et al., 2005).

The review indicated that 26 of the 67 UCCs had been the subject of a research study/feasibility project only, 13 had proceeded to a trial, and 28 had become fully operational schemes (Browne et al., 2005). At the time of the review (2005) 27 UCC schemes were understood to still be in operation, 14 of these dated from 2000 or later showing the growth in interest in setting up UCC operations in the last decade. Much of this had been driven by private sector involvement (including logistics service providers, property developers and landlords). Three major categories of UCC can be defined.

UCCs serving all or part of an urban area: These UCCs are usually associated with the supply of retail products, but are also used for the supply of office products, and occasionally food supplies for restaurants and cafes. These UCC schemes are often used to serve locations with features such as narrow streets, historic layouts, and limited unloading facilities. The introduction of this type of UCC is usually initially suggested by the local authority on traffic and environmental grounds.

Existing examples of such UCCs include Reglog in Regensburg, Cityporto in Padua, La Rochelle UCC, Nijmegen UCC and Bristol UCC.

UCCs serving large sites with a single landlord: These UCCs are most commonly associated with the supply of retail products and food supplies for restaurants and cafes. There are also examples of them being used for supplying hospital products. The types of large sites served by these UCCs include airports, shopping centres and hospitals. In some instances they serve only one large site (for example London Heathrow airport retail UCC and Meadowhall shopping centre UCC in Sheffield) and in other cases they serve several large sites (such as the Hospital Logistics Centre in London which delivers to several major hospitals). Although these UCCs only serve one or a few large sites, within the sites many different outlets may be served (such as many different retailers in a shopping centre or airport, or many different departments within a hospital). The interest in UCCs among developers and owners of airports and shopping centres usually stems from the desire to maximise retail space by minimising on-site storage and the need for multiple delivery bays (WSP, 2008). In the case of a hospital the interest is more commonly in reducing on-site stock levels and storage space. These UCCs can also be insisted on by the planning authority through the planning permission process in the case of new sites.

Construction project UCCs: These are UCCs that are used for consolidating construction materials for major building projects including housing, office blocks and hospitals. Examples include UCCs established at London Heathrow airport during major development work and in Hammerby in Stockholm for a major housing project. This type of UCC can serve either a single major building project or several. This type of UCC can either exist only for the lifetime of a building project or can be on-going serving new major building projects as they are established, but experience to date suggests that the former is more common. Either the site developer or the main construction contractor would decide to make use of a UCC, or it could be made mandatory through the planning permission process.

Each of these three types of UCC can offer either relatively basic consolidation and delivery services or offer a wider range of value-added logistics activities. In the 2005 literature review previously discussed 59 of the 67 UCCs identified served part or all of an urban area (type 1), five served large sites with a single landlord (type 2), and 3 were construction project UCCs (type 3) (Browne et al., 2005). This illustrates the predominance of UCCs serving part or all of an urban area over the other types.

Many additional UCC feasibility studies, trials and operations have been established since the 2005 literature review, especially in Italy, the Netherlands and the UK (see for example Gonzalez-Feliu and Morana, 2010; JMP and TTR, 2010; van Duin et al., 2010; van Rooijen and Quak, 2010; Ville et al., 2010). A recent report lists ten UCCs that are operational in the UK alone in 2010, all of which opened since 2000 (TTR, 2010). All of these UCCs are operated by the private sector and several receive no public funding. Using the above categorisation approach, three are UCCs serving a town/city (Bristol, Regent Street in central London, and Snetterton which serves Norwich city centre), five are retail UCCs on single sites with a private landlord (three serve airports: Manchester, East Midlands and Heathrow, and two serve shopping centres: Meadowhall and Bluewater), and two of these are construction project UCCs in London. In addition, another UCC serving a town/city commenced operations in the City of London in 2009. This UCC handles office products and is the subject of the detailed case study in the next section. This indicates the current level of interest and activity in this city logistics innovation.

FREIGHT TRANSPORT AND LOGISTICS ADVANTAGES OFFERED BY UCCs

Impacts on Transport Operations

One of the primary reasons for considering the implementation of a consolidation centre is the potential to reduce freight transport activity and its associated social and environmental impacts. This is achieved through the improvements in vehicle load factor that can be attained through load consolidation at the UCC, together with the potential to use alternatively-fuelled vehicles (electricity or gas) for final delivery from the UCC. In addition some UCCs have made use of electrically-assisted tricycles (Attlassy and the University of Westminster, 2005; Transport for London, 2009). They are proving to be important test sites for these innovative vehicle technologies. UCC trials and operations that had been subject to evaluation show that reductions in vehicle kilometres can be up to 80 per cent, improvements in vehicle load factors range from 15-100 per cent, and reductions in greenhouse gas emissions range from 25-80 per cent as a result of the UCC.

In addition to the reduction in the total distance travelled per delivery in the urban area as a result of using a UCC, these schemes also offer an important operational advantage to freight transport operators using them. They allow a driver to drop their load at the UCC without having to make the final delivery in a congested urban area, and thereby save time and hence operating costs. In the case of a shopping centre, the use of a UCC removes the need for the driver to transport goods from shopping centre loading bays to receivers located long distances away as this is taken care of by UCC personnel (some of whom may be based in the shopping centre). Also, in the case of deliveries to construction sites, the use of a UCC means that the driver does not have to spend time searching for the correct location or person to deliver to.

UCCs also improve delivery reliability, as they are located in close proximity to the customers. Longer direct deliveries are more likely to experience delays, especially within urban areas. UCCs can also reduce the total kerbside time and space occupied in urban areas by vehicles making deliveries as they increase the quantity of goods delivered per stop.

Impacts on Other Logistics Activities

UCCs can also help to achieve other supply chain and logistics benefits. UCCs can offer a range of value-added logistics services including off-site stockholding, consignment unpacking, preparation of products for display and price labelling, goods return and waste collection services. These can benefit receivers by reducing their on-site space requirements, saving time by reducing the number of tasks that have to be performed on-site, and enhancing productivity and sales in core activities.

UCCs can also be used to offer community collection and delivery point facilities (for other consumer and business products), while home delivery operations can also be operated from the UCC.

Figure 1 illustrates the range of logistics and pre-retail activities that can be carried out by a UCC and the potential benefits of these activities.

Inventory Control and Associated Activities

A UCC can be used to offer stockholding services to users, subject to available capacity and appropriate storage conditions for the products involved. In the main, only short-term storage tends to be envisaged, providing a useful local buffer stock that can be called off quickly when needed, thus reducing delivery lead times and improving product availability and customer service. For maximum benefit this depends on the UCC offering the flexibility for quick call-off of product as required. For certain products or customers,

Figure 1. Range of potential logistics and pre-retail activities at UCC and possible benefits

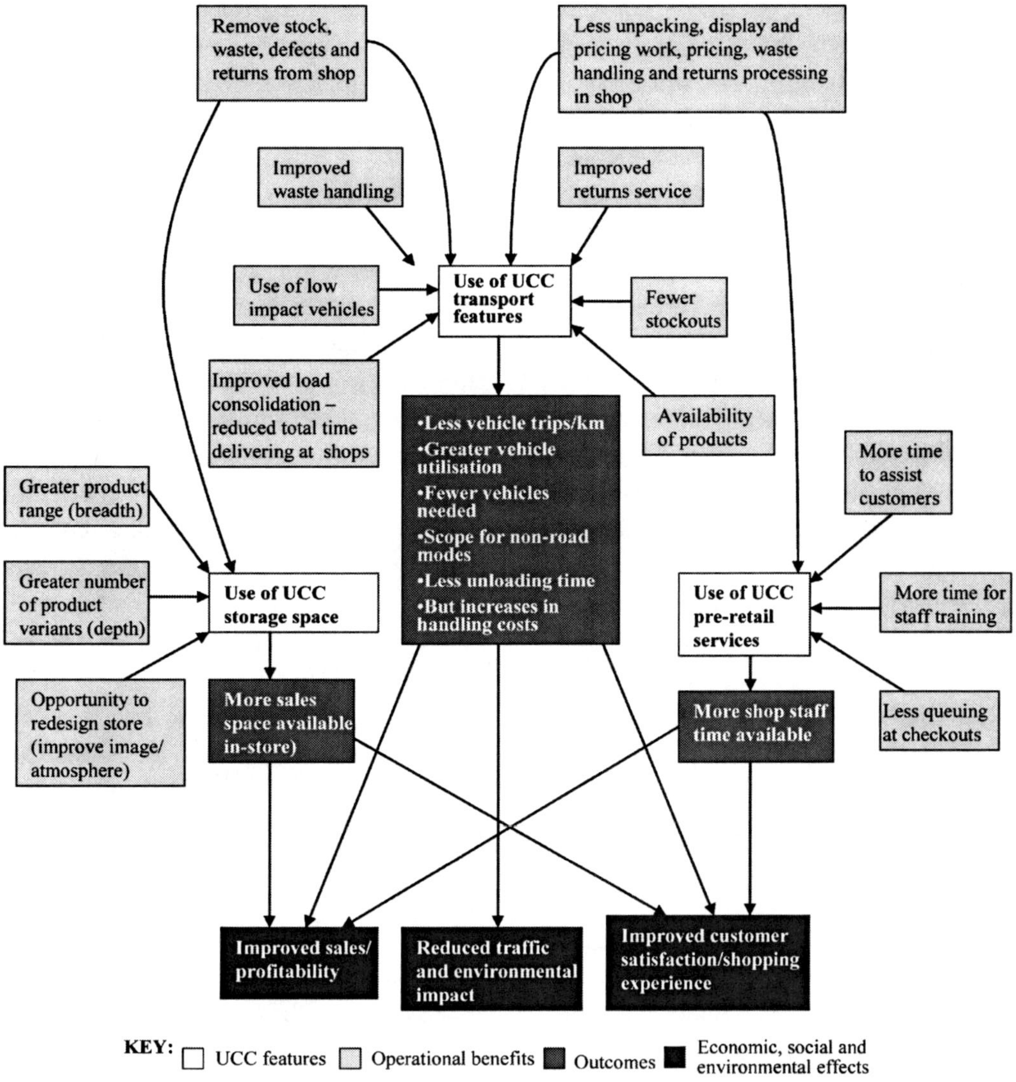

seasonal or peak storage can sensibly be provided by a consolidation centre so as to provide valuable additional space.

Other services can be offered, particularly by retail UCCs but perhaps also by other types of centre. Inventory monitoring and information collection and analysis, linked to in-store systems, can be provided. This can increase the visibility of the supply chain, again leading to better product availability and service levels, as well as reducing loss of stock. Product quality and quantity checking can be carried out upon consignments' arrival at the centre, giving advance notice to the customer of any problems with supplies. Various pre-retailing activities, such as consignment unpacking, preparation of products for display and price labelling, can also be carried out at the consolidation centre to reduce time and space requirements at the delivery point.

Impacts at the Delivery Location

As a result of the additional logistics services that can be offered by a UCC, space at the receiving establishment can be freed up for other activities that are more productive or profitable, such as retailing floor space or construction sites, which is likely to be particularly beneficial when space is at a premium or expansion is desirable. This can prove especially attractive to developers of shopping centres as the removal of storage space allows them to fit more retail space and units into the prime development.

If the UCC offers an enhanced delivery service, with more flexible and reliable delivery times, higher product or component availability may be achieved and, ultimately, sales volume or site productivity may be increased. Fewer deliveries to the destination may be required as a result of the load consolidation undertaken at the centre, thereby reducing the disruption and labour requirements associated with receiving multiple deliveries, leading to improvements in staff planning and productivity at delivery locations.

Off-site storage can also lead to reductions in damage and theft of stock, especially on construction sites, and thereby further reductions in development costs. The same benefit is true of the use of UCCs for hospital medical and equipment supplies.

Due to the reduction in deliveries associated with the use of UCCs it is also possible for developers of shopping centres and other major retailing centres to reduce the delivery bay requirements and thereby free-up this space for other, more profitable activities.

Impacts for Suppliers and Logistics Companies

UCCs may be open at more flexible times than city centre delivery points. Staff will be available to receive the consignment and the delivery company can avoid problems of congested loading bays found in many urban areas.

In some cases, the number of locations that a supplier has to deliver to may be reduced, since a particular UCC could be serving more than one customer location. In a broader sense, there may be some potential to ease skills shortages within the transport and related sectors (e.g. construction) through more efficient methods of working.

Product Flow

UCCs have a role in the handling of return and recycling flows, including product returns and the coordination of waste and packaging collection for reuse or recycling, instead of individual customers having to deal with this. As packaging waste regulations tighten, such coordinated approaches may offer greater benefits over time. UCCs may also offer a new distribution channel which, as a result of serving multiple customers, may offer opportunities to smaller suppliers to access new markets that are currently beyond their reach. Finally, UCCs may be in a position to offer B2B (business-to-business) and B2C (business-to-customer) services within their catchment area, perhaps including inter-store transfers, home delivery or customer collection of products purchased in town centres.

INTRODUCTION TO THE CASE STUDIES

This section contains case studies of four UCC schemes established in the UK in recent years. This includes two UCCs that serve part of an urban area, one UCC that serves a major site with a single landlord (retail outlets in an airport), and one construction project UCC. For each UCC, the objectives, operational characteristics, financial aspects and impacts are examined.

Table 1 shows basic details of the four UCC schemes. The sample comprises a range of UCC types, a mix of sectors and examples of optional and compulsory scheme participation.

Table 1. Key characteristics of case study UCCs

Centre	Location	UCC Type	Product Type	Trial or Fully Operational	Use of Electric Vehicles?	Current Status	Terms of Use
London Construction Consolidation Centre	London, UK	Construction	Construction	Trial	No	Closed *	Compulsory **
Bristol UCC	Bristol, UK	Part of urban area	Retail	Trial then fully operational	Partial	Active	Voluntary
City of London Micro-Consolidation Centre	London, UK	Part of urban area	Office products	Trial then fully operational	Yes	Active	Compulsory
London Heathrow UCC	London, UK	Airport	Retail	Trial then fully operational	Partial	Active	Compulsory

Notes: * - concept now applied to new site; ** - with certain exceptions (see case study)

Case Study: London Construction Consolidation Centre

Background and Objectives

A two-year trial project referred to as the London Construction Consolidation Centre (LCCC) operated from 2005 to 2007. The LCCC was a 5,000 sq. m. facility located in South Bermondsey, approximately five kilometres south of the City of London. Its objective was to serve four major construction sites in the City of London. It was intended for a rapid flow, on a just-in-time basis, of material from suppliers to site with storage time limited to ten days. The LCCC was intended to reduce the number of deliveries going directly to the construction sites and thereby reducing traffic congestion and vehicle emissions. The main reason for carrying out this LCCC project was to assess the potential benefits that such a scheme could bring (Transport for London et al., 2008).

Operational Aspects

Contractors working on the four construction sites placed orders for their material requirements with their suppliers in the normal way, but instructed that the delivery was made to the LCCC, not the construction site. Contractors then placed a delivery order with the LCCC for the materials they required. This was assembled at the LCCC and delivered to the sites. Approximately a sixth of all the deliveries requested from the LCCC by the construction sites were needed within less than 24 hours notice (i.e. on a just-in-time basis). This level of service would probably have been difficult to achieve if deliveries were made directly to site by suppliers. The delivery from the LCCC to the construction sites consolidated numerous contractors' orders onto each vehicle.

As well as delivering construction materials to the sites, the vehicles operating from the LCCC also collected recyclable packaging and unused materials and brought them back to the LCCC. This was then recycled, returned through the supply chain for re-use or collected by a waste operator. It is important to note that some supplies continued to be sent directly to the construction sites rather than via the LCCC. These included aggregates, structural steel, ready-mix concrete, escalators and furniture, all of which were usually delivered as a full vehicle load.

The LCCC employed 16 staff and had a fleet of six delivery vehicles. This comprised four flatbed rigid vehicles (ranging from 18 to 26 tonnes gross weight), one 7.5 tonne gross weight curtain-sider rigid vehicle and one 3.5 tonne gross weight van.

Financial Aspects

The project cost of £3.2 million was met by a partnership of private developers/construction companies and Transport for London (TfL), the public organisation responsible London's transport system. TfL funded £1.85 million, while the developers and construction companies funded the other £1.35 million. The companies participating in the LCCC project voluntarily agreed to use the LCCC for the four construction sites. Some of the costs of the LCCC were passed on to the individual building contractors (Transport for London et al., 2008).

Impact Assessment and Outcome

Operational targets were set for the performance of the LCCC. These are shown in Table 2 alongside the operational performance actually achieved. The performance targets were equalled or surpassed..

For those materials delivered via the LCCC, it is estimated there was a 60-70 per cent reduction in the number of vehicles delivering to the four construction sites. It is estimated that this resulted in a 70-80 per cent reduction in CO_2 emissions for these flows. Taking account of all deliveries from suppliers, there was a reduction of approximately 40 per cent in total vehicle deliveries to the Unilever House building site. This was considered the best managed site in that it had the lowest proportion of deliveries direct (Transport for London et al., 2008).

Table 2. LCCC targets and achievements

Key Performance Indicator	Target	Achieved
Reduction in freight journeys	40%	60-70% of journeys via LCCC 40% of journeys to construction site
Reduction in journey time of supplier deliveries to contractors	30-60 minutes	120 minutes
Delivery reliability	97%	97%

Source: Transport for London et al., 2008.

The vast majority of deliveries from the LCCC to construction sites were made in rigid goods vehicles. Consolidating goods at the LCCC eliminated the use of articulated goods vehicles for site delivery, and significantly reduced the use of vans. Two hours average reduction in supplier journey times were achieved by delivering to the LCCC rather than driving directly to the sites in the City of London (this included unloading time).

Deliveries from the LCCC to the sites achieved 97 per cent delivery reliability (i.e. 97 per cent materials of the correct type and quantity were delivered within 15 minutes of the scheduled time). In addition, productivity of the labour force on the construction sites increased by up to 25 minutes per person day as a result of the delivery reliability from the LCCC (Transport for London et al., 2008).

Negative impacts included some work scheduling problems at the construction sites caused by a lengthening of the order lead times for products channelled through the LCCC. There were also a few instances of incorrect components being sent from the LCCC to the sites and items being misplaced at the LCCC.

The LCCC trial was therefore judged to have been a success. While the original centre closed down at the end of the trial period, a 7,500 m^2 replacement construction UCC opened shortly afterwards in east London. This new UCC is operated by Wilson James, the operator of the trial UCC, on a commercial basis, highlighting the viability of the concept. It is serving a number of major construction projects, such as the redevelopment of Barts and the Royal London Hospital (Wilson James Ltd, 2008).

Case Study: London Heathrow Retail Consolidation Centre

Background and Objectives

Retail development within Heathrow airport increased dramatically over the decade up to 2000, but the capacity of road and unloading infrastructures was not expanded accordingly. Problems included traffic congestion on airport roads and at loading bays, and a lengthy and unpredictable delivery times. These issues, together with growing environmental pressures and the potential requirements of the planned Terminal 5, led to a re-examination of the supply system serving retail and catering outlets in Terminals 1-4.

A study reviewed the various retail delivery options. It concluded that the specific conditions would be best met by an urban consolidation centre through which all deliveries should pass with the exception of newspaper deliveries, cash and valuable items which should continue to be delivered directly to retail outlets.. A four-month trial was established in March 2000 in which the Heathrow Airport Limited (HAL) (the owner of Heathrow airport) contracted the logistics provider DHL Supply Chain to operate a consolidation centre, vehicle fleet and related goods reception and inventory management systems. Eight retailers with a total of 40 retail outlets across the four terminal buildings participated. Results showed a 66 per cent reduction in vehicle deliveries. Operations started in May 2001 (Department for Transport, 2002).

Operational Aspects

A 25,000 sq ft warehouse with cold store was established at Hatton Cross (2.5 km from Terminal 4) employing 20 operational and clerical staff. Deliveries by suppliers to the centre were subject to security checks, and then put into roll cages and sealed ready for dispatch to the retail outlets. By 2002 the delivery fleet consisted of three tractor units and three urban rear-steer 11 metre box van trailers with tail-lifts. Two trailers had dual compartments and were fitted with refrigeration to cater for chilled and frozen food deliveries. The delivery vehicles are also used to collect and remove empty roll cages, recyclable packaging waste, such as cardboard and plastic. In 2007 DHL introduced electric vehicles to the consolidation centre fleet (Avison, 2007a).

A new consolidation centre was developed in 2006 to cope with the growing space demands, higher product throughput and planning requirements for Terminal 5. This new centre is also based approximately 2.5 km from the airport and operates 24 hours a day, 7 days a week. Incoming deliveries are cross-docked onto the centre's vehicles and deliveries made to airside and landside stores from which other consolidation centre staff deliver orders to each retail outlet. These staff can also even stock shop shelves and carry out pre-retailing services such as pricing and ticketing.

By 2009 the consolidation centre was handling deliveries to 323 retail and catering outlets, pubs and restaurants based in the airport (DHL, 2009). The consolidation centre operation employs approximately 100 people (Avison, 2007b).

All retail and catering concessions operating at Heathrow airport use the consolidation centre as part of their rental agreement. It must therefore accommodate ambient, chilled and frozen goods. HCC operations have been effectively streamlined, increasing average accuracy to 99.9 per cent on all retail deliveries. Retail consolidation was a prerequisite for the planning permission to build Terminal 5 and is part of the future development of the airport including the renovation of Terminals 1 and 2.

Financial Aspects

The consolidation centre contract has a value to the operator DHL of approximately £9 million per year (DHL, 2009). By 2009 all retail and catering outlets based in the airport used the consolidation centre as part of their contractual agreement with HAL. Half of the income of the Heathrow consolidation centre is met by retailers and other outlets in the airport paying for the consolidated delivery service. 10 per cent of the income is generated from charges made for value added services such as pre-retailing and off-site storage. The remaining 40 per cent if income is met by the airport operator HAL (Hapgood, 2009).

Impact Assessment and Outcome

The Heathrow retail consolidation centre reduces vehicle deliveries to outlets in the airport by approximately 80 per cent (Bastien, 2007). It has been estimated that in 2008 these consolidated deliveries resulted in a saving of 218,000 km (DHL, 2009). This reduction in total distance travelled together with the use of an electric vehicle has been estimated to result in a 158,000 kg reduction in CO_2 emissions (DHL, 2009).

Case Study: Bristol Urban Consolidation Centre

Background and Objectives

The urban consolidation centre in Bristol was originally established as a trial in an EU-funded CIVITAS project. The objectives of the trial were to reduce goods vehicle activity in the central retailing area (Broadmead), thereby helping to relieve traffic congestion, improve air quality and minimise conflict between vehicles at loading areas/delivery bays. The trial was also intended to provide suppliers and retailers with improved logistics services, removing the need for suppliers to send their vehicles into central Bristol, increasing delivery reliability and offering a range of value-added services for such as pre-retailing, remote storage, and packaging and waste collection. The consolidation centre and delivery operations have been operated by DHL since it opened in 2004.

Operational Aspects

The trial commenced in May 2004. It has since become a permanent urban consolidation centre scheme. Retailer participation in the consolidation centre scheme is voluntary. The area served has been extended from Broadmead to the new Cabot Circus Centre retail development which opened in 2008 adding 40 per cent to the retail floorspace in the area. Also in 2008, new pedestrian areas were introduced and tighter loading time restrictions imposed (0500-0800 for the main pedestrian area of Broadmead). This has encouraged more retailers to make use of the consolidation centre (START, 2008). The catchment area comprises over 400 retail outlets approximately 70 of which currently use the consolidation scheme (Minihane, 2009).

The consolidation centre has a floorspace of 500 sq. m. and is located on an established industrial estate on the north western edge of Bristol, close to both the M4 and M5 motorways, 15 kilometres (25 minutes travel time) from Broadmead district, (Minihane, 2009).

Two delivery vehicles are operated from the centre, an 18 tonne diesel-powered vehicle with a Euro IV engine and a 9 tonne electric goods vehicle (Hapgood, 2009). The centre handles retail goods only and operates six days per week (Bastien, 2007). The local transport authority allows UCC vehicles to access a bus lane to and from the city centre on a trial basis to cut delivery times and improve reliability (Hapgood, 2009).

In addition to the consolidation and delivery services, other value-added services are provided including collection of waste cardboard and plastic for recycling, and pre-retailing activities.

Financial Aspects

The development and trial phase of the UCC from 2002-2006 was funded through the EC CIVITAS VIVALDI project. Some subsequent financial support has been provided by Bristol City Council. In 2007/08, the total UCC operating cost was £459,000, of which £285,000 came from Bristol City Council against income from retail users of £174,000. This represents a cost recovery rate of almost 40 per cent from users. In 2008/09, Bristol City Council contributed £225,000 (Hapgood, 2008). DHL Supply Chain is tasked with further increasing the cost recovery from retailers as public funding is planned to reduce progressively over time. This is intended to be achieved through signing up new retailers to use the service, and selling more value-added services to users. In addition, the UCC is to be offered to non-retail end users, and the scope of UCC services can also be expanded to include council deliveries (Minihane, 2009). However, despite the reduction in the proportion of total costs met from public funds it is likely that public support will be required on a continuing basis, especially as there is currently no mechanism for enforcing participation by retailers.

Impacts Assessment and Outcome

The UCC has resulted in a 77 per cent reduction in vehicle deliveries to participating retailers. This is equivalent to a total reduction of approximately 10,000 vehicle deliveries per annum, with a saving of 250,000 vehicle kilometres. It is estimated that the UCC scheme yields annual savings of approximately 30 tonnes of CO_2, 1,000 kg of NOx, and 30 kg of PM10. In addition 20 tonnes of cardboard/plastic are collected and recycled (Hapgood, 2009; Minihane, 2009).

Feedback from retailers has been positive, with 94 per cent stating that they would recommend the UCC service to another retailer. There have been no reports of losses or damage to stock, and a majority of retailers report that they are saving more than 20 minutes per delivery. As a consequence, 38 per cent say that this enables their staff to spend more time with customers, and 45 per cent state that staff morale has improved and stress levels have reduced (Minihane, 2009). Retailers generally seem to appreciate the improved service and cost reduction opportunities offered by channelling deliveries through the UCC.

While the scheme has been successful in operational terms, the main challenge is to improve the commercial viability of the scheme and to find further sources of income from potential users.

Case Study: City of London Micro-Consolidation Centre

Background and Objectives

Office Depot is a major supplier of stationary and other office products to businesses in the UK. In 2009 Office Depot made the decision to trial a new urban delivery system in the City of London in order to reduce the environmental impacts of their delivery operation. This decision was taken as part of the company's corporate social responsibility and environmental efforts (Office Depot, 2010a). However, it was also necessary for the new scheme to meet customer needs cost-effectively if it was to be adopted on a permanent basis (Office Depot, 2010b).

The City of London lies within the area of the Clear Zone Partnership serves as a testing ground for researching, trialling, monitoring and setting best practice for new transport technologies, innovations and physical measures. Since 1999 the London Borough of Camden, the City of London and the City of Westminster have worked jointly to achieve this.

Operational Aspects

The trial involved the use of an urban micro-consolidation centre located within the City of London together with electrically-assisted tricycles and electric vans. The customers to whom deliveries were made were all located in the City of London which is the historic core of London with an area of 2.9km^2. It is an area comprising a mix of very narrow streets and wider roads with double red lines on which no stopping is allowed, and have very heavy traffic flows in the morning peak. The urban micro-consolidation centre and the deliveries made from it were operated by Gnewt Cargo, a new company specialising in green urban freight deliveries.

Prior to the trial Office Depot operated a distribution system for deliveries in the City of London that involved using 3.5 tonne gross weight diesel-powered vans that were dispatched from a warehouse located approximately 30 kilometres away in the London suburbs. A total of approximately 1,000 parcels were delivered per day, all to businesses addresses.

During the trial the micro-consolidation centre in the City of London was used as a transhipment facility for the transfer of parcels from the suburban depot onto electric vans and tricycles and for overnight vehicle storage. An 18-tonne goods vehicle was used to transport parcels overnight from the Office Depot warehouse in the suburbs of London to the micro-consolidation centre. Electrically-assisted tricycles and electric vans were used to make parcel deliveries from the micro-consolidation centre to customers in the City of London. Neither of these vehicles resulted in any fossil fuel consumption or CO_2 emissions as the electricity they used was produced from renewable sources.

Financial Aspects

The London Borough of Camden made a small financial contribution towards the costs of the trial as the City of London delivery area and much of the route taken from the suburban depot to the City of London was within the Clear Zone Partnership area. All the other costs were met by Office Depot, who expected the new delivery system to have similar costs to their existing system. The system was trialled in the City of London as it comprises a high density of Office depot delivery locations. This results in a relatively short travel distance between delivery locations and was therefore felt to be an appropriate setting in which to test the cost-effectiveness of using electrically-assisted tricycles for freight delivery (Office Depot, 2010b).

The trial proved successful from Office Depot's perspective in transport, environmental and financial terms. Office Depot has calculated that the new system with its electric vehicles and micro-consolidation centre has the same operating costs as the previous system using diesel vans dispatched the suburban London depot (Office Depot, 2010b).

Impacts Assessment and Outcome

The results show that the use of the micro-consolidation centre together with the complete replacement of the diesel van fleet by electrically-assited tricycles and electric vans led to a reduction of 20 per cent in the total distance driven per parcel delivered between the suburban depot and the customer delivery locations. Total CO_2 equivalent emissions per parcel delivered was 54 per cent lower in May 2010 than in October 2009 before the trial. This was due to the reduction in the total distance travelled per parcel and the use of electric vehicles whose batteries are charged with electricity from carbon-free sources.

The distance travelled between the suburban depot and the City of London per parcel delivered fell by 82 per cent due to the use of a single truck to transport goods between the suburban depot and the micro-consolidation centre in the City of London. However, within the City of London the total distance travelled per parcel delivered increased by 349 per cent by May 2010. This was due to the lower carrying capacity of the electric vans and tricycles compared to the diesel vans, thereby resulting in the need for more delivery activity per day. CO_2 emissions fell by 49 per cent, per parcel delivered between the suburban depot and the City of London, and by 83 per cent per parcel delivered within the City of London compared with the situation before the trial.

The results therefore reflect the trade-off between total distance travelled and CO_2 emissions associated with the use of electric vehicles replacing diesel vehicles with greater carrying capacity. The trial was of particular interest as the existing Office Depot delivery system prior to the trial already involved a high degree of product consolidation and hence high vehicle lading factors. It did not therefore typify the types of operation in which UCCs are usually considered beneficial – namely systems with poor load consolidation.

Due to the environmental and financial success of the trial, Office Depot decided to continue the operation beyond the end of the trial period. The scheme has now been made permanent (Office Depot, 2010b). Office Depot intends to implement the same delivery system in other cities, and is also considering the potential use of electrically-assisted tricycles in other environmentally sensitive locations or where noise pollution is an issue such as university campuses (Office Depot, 2010b).

FUTURE RESEARCH DIRECTIONS

A range of organisational, operational and financial issues have to be addressed in assessing the performance of UCC schemes. These include: the potentially high set up costs (and sometimes high operating costs) that need to be recovered; contractual issues concerning legal responsibility for goods; the loss of the direct interface between suppliers and customers possibly deterring some users; the relatively low rates of product throughput in the early stages; and the potential for local authorities to create monopolistic UCC situations, thus eliminating competition and perhaps leading to legal action.

Two factors appear to be decisive in the success of a UCC scheme. The first is the level of demand. Sufficient UCC users and product throughput is required to drive down the costs per unit handled thereby making the UCC highly competitive with traditional urban distribution systems. The second is the sharing of UCC costs and benefits between the various parties involved in the scheme. Although the UCC may result in total benefits that outweigh total costs (taking into account commercial, traffic, social and environmental impacts), each party must be prepared to share these costs and benefits fairly if the scheme is to succeed (Browne et al., 2005; TTR, 2007).

These two issues are less problematic in the case of UCCs that serve major sites with a single landlord and construction project UCCs as in these cases a single commercial party is typically responsible for financing the UCC. It either enjoys all the benefits or has leverage to recover some of the costs from the beneficiaries of the scheme.

The situation for UCCs that serve all or part of an urban area is more problematic. In these cases the decision of users to participate in the UCC is typically a voluntary one and there is no single private body responsible for financing the UCC or exploiting the benefits. Instead there are typically many business users, as well as a local authority that is keen to implement the UCC in its efforts to reduce traffic and environmental impacts. The financial costs in this type of UCC scheme tends to be largely borne by the local authority at least in the initial stages. This level of public subsidy is usually unsustainable in the longer term and therefore efforts to increase product throughput, and reduce unit costs become critical.

In the short-term the local authority can implement accompanying policy measures that encourage greater use of the UCC, such as allowing UCC vehicles access to bus lanes, or imposing tighter vehicle access or loading restrictions in the UCC catchment area. As new users are attracted to use the UCC and throughput increases, the level of public subsidy can be gradually reduced, until the UCC becomes financially self-supporting. It is important that mechanisms are put in place so that all users (carriers and receivers) make financial contributions in return for the commercial gains that the UCC offers them. The provision and marketing of value-added logistics services generates an additional stream of revenue and broadens the role of the UCC within the city logistics system.

There is a need for further research into the economic and environmental costs and benefits that arise from different types of UCC and the means by which these costs can be minimised and the benefits maximised. In addition, further consideration is required of how best to apportion the economic costs associated with the UCC between the various parties involved in order to ensure that the scheme is equitable and therefore attractive to all. In addition, it would be useful to understand the most promising and realistic actions that could be taken by various local authorities if they wished to encourage the use of a UCC approach.

CONCLUSION

This chapter has considered the role that can be played by UCCs in rationalising the movement of freight in urban areas. By increasing the consolidation of loads UCCs can assist in reducing delivery costs and environmental impacts. They also permit greater use of electric and other alternatively-fuelled goods vehicles which emit less CO_2 and other, noxious gases and help to reduce vehicle noise. UCCs should therefore be viewed as a potentially important innovation in helping to achieve greater sustainability in urban freight transport in both economic and environmental terms.

The case studies presented in the chapter have helped to demonstrate the various types of UCC that can be established and their attributes. As discussed the two key factors in the success of a UCC are the obtaining sufficient product throughput, and determining how the costs and benefits that arise from the use of a UCC will be shared fairly between the various parties involved in the scheme. UCCs can benefit from the public and private sectors working closely together to attempt to ensure their success. This is particularly necessary in the case of UCCs that serve all or part of an urban area, as operating incentives and some degree of public subsidy may well be required for a UCC to achieve viability, especially during its establishment.

REFERENCES

Attlassy, M. & University of Westminster (2005). *The tricycle experimentation in Paris "la Petite Reine."* University of Westminster.

Avison, R. (2007a). *DHL hopes to expand airport consolidation centre, RoadTransport.com, 24 July 2007.* Retrieved December 18, 2010, from http://www.roadtransport.com/Articles/2007/07/24/127981/DHL-hopes-to-expand-airport -consolidation-centre.htm

Avison, R. (2007b). *Making consolidation centres work.* RoadTransport.com, 27 July 2007. Retrieved December 18, 2010, from http://www.roadtransport.com/Articles/2007/07/27/128022/Making-consolidation-centres-work.htm

Bastien, O. (2007). *An Introduction to Retail Consolidation Centres.* presentation to the West London Freight Quality Partnership, 10th January 2007.

Browne, M., Woodburn, A., & Allen, J. (2007). Evaluating the potential for urban consolidation centres. *European Transport, 35*, 46–63.

Browne, M., Woodburn, A., Sweet, M., & Allen, J. (2005). *Urban Freight Consolidation Centres.* Report for Department for Transport, London: University of Westminster. Retrieved December 20, 2010, from http://www.freightbestpractice.org.uk/urban-freight-consolidation-centre-report

Department for Transport. (2002). *Heathrow Airport Retail Consolidation Centre.* Good Practice Case Study 402, Energy Efficiency Best Practice Programme. Retrieved December 18, 2010, from http://www.freightbestpractice.org.uk/heathrow-airport-retail- consolidation-centre

DHL. (2009). *DHL wins major contract extension for Heathrow Consolidation Center*. press release 16 November 2009, DHL. Retrieved January 5, 2011, from http://www.dp-dhl.com/en/media_relations/press_releases/2009/dhl_wins_contract_extension_for_heathrow_ consolidation_center.html

Gonzalez-Feliu, J., & Morana, J. (2010). Are City Logistics Solutions Sustainable? The Cityporto case. *Territorio Mobilità e Ambiente*, *3*(2), 55–64.

Hapgood, T. (2008). *Broadmead Freight Consolidation Scheme*. presentation made 28 July 2008 (provided by T. Hapgood).

Hapgood, T. (2009). *The Importance of Freight Management and the Role of Urban Consolidation Centres*. Paper presented at Seventh Transport Practitioners Meeting: Ideas and Reality, 13-15 July, University of Reading.

JMP & TTR. (2010). *Tactran Freight Consolidation Feasibility Study*. Draft Feasibility Report, Prepared for Tactran, JMP and TTR. Retrieved December 18, 2010, from http://www.tactran.gov.uk/documents/TactranFreightConsolidationDraft FeasibilityReport280410.pdf

McKinnon, A. (1998a). *International Review of Urban Transhipment Studies and Initiatives*, Report prepared for the Retail and Distribution Panel of the UK Government's Foresight Programme, Edinburgh: Heriot Watt University. Retrieved December 20, 2010, from http://www.sml.hw.ac.uk/logistics/downloads/UKTranshipINT.pdf

McKinnon, A. (1998b). *Urban Transhipment: Review of Previous Work in the UK*. Report prepared for the Retail and Distribution Panel of the UK Government's Foresight Programme, Edinburgh: Heriot Watt University. Retrieved December 20, 2010, from http://www.sml.hw.ac.uk/logistics/downloads/UKTranshipUK.pdf

Minihane, E. (2009). *Bristol Consolidation Centre*. Presentation made at Chartered Institute of Logistics and Transport Seminar 14th October 2009. Retrieved January 5, 2011, from http://www.ciltuk.org.uk/pages downloadfile?d=52F89921-7B0D-4B0E-A8A7- 30462F93EFC7&a=stream

Office Depot. (2010a). *We all contribute: 2010 Corporate Citizenship Report*. Florida: Office Depot. Retrieved December 20, 2010, from http://www.officedepot.cc/environment t/downloads/OD-Corporate-Citizenship-Report-2010.pdf

Office Depot. (2010b). *Launch of Cargo Cycles*. Office Depot Briefing Document, Office Depot. Retrieved December 20, 2010, from http://www.officedepot.co.uk/OD_content/data/documents/press-kit/Cargo%20Cycle%20%20 briefing%20doc1.pdf

START. (2008). *City of Bristol: Access Management and Priority Measures*. START project factsheet. Retrieved January 4, 2011, from http://www.start-project.org/download/fact%20sheets/START%20 Bristol%20WP3%20Dec08.pdf

Transport for London. (2009). *Cycle Freight in London – A Scoping Study*. London: Transport for London. Retrieved December 20, 2010 from http://www.tfl.gov.uk/assets/downloads/businessandpartners/cycle-as-freight-may-2009.pdf

Transport for London, Bovis Lend Lease, Constructing Excellence, Stanhope, Wilson James and the University of Westminster. (2008). *London Construction Consolidation Centre: Final Report*, London: Transport for London. Retrieved December 20, 2010, from http://www.tfl.gov.uk/microsites/freight/documents/publications/LCCC_final_report_July_2009.pdf

TTR. (2007). *South London Freight Consolidation Centre Feasibility Study: Final Report*. Report prepared on behalf of South London Freight Quality Partnership, TTR. Retrieved December 18, 2010, from http://www.londonsfqps.co.uk/LinkClick.aspx?fileticket=1FSau5i7vTI%3d& tabid=188 &mid=520&forcedownload=true

TTR. (2010). *Freight Consolidation Centre Study: Main Report*. Prepared for Department for Transport, TTR. Retrieved December 18, 2010, from http://www.dft.gov.uk/pgr/freight/research/freightreport/pdf/mainreport.pdf

van Duin, R., Quak, H., & Muñuzuri, J. (2010). New challenges for urban consolidation centres: A case study in The Hague, The Sixth International Conference on City Logistics. *Procedia Social and Behavioral Sciences*, 2, 6177–6188.

van Rooijen, T., & Quak, H. (2010). Local Impacts Of A New Urban Consolidation Centre – The Case of Binnenstadservice.nl, The Sixth International Conference on City Logistics, *Procedia Social and Behavioral Sciences,* 2, 5967–5979.

Ville, S., Gonzalez-Feliu, J., & Dablanc, L. (2010). *The limits of public policy intervention in urban logistics: The case of Vicenza (Italy) and lessons for other European cities*. At the 12th World Conference on Transport Research, 11-15th July, Lisbon, Portugal.

Wilson James Ltd. (2008). Personal communication from representative of Wilson James Ltd, July 2008.

WSP. (2008). *Freight Consolidation and Remote Storage*, BCSC Educational Trust. Retrieved December 18, 2010, from http://www.bcsc.org.uk/media/downloads/FreightConsolidation2008.pdf

This chapter was previously published in Supply Chain Innovation for Competing in Highly Dynamic Markets edited by Pietro Evangelista, Alan McKinnon, Edward Sweeney, and Emilio Esposito, pages 199-214, copyright 2012 by Business Science Reference (an imprint of IGI Global).

Chapter 74
Evaluating Sustainability on Projects Using Indicators

Jude Talbot
AMEC, Canada

Ray Venkataraman
Penn State University, USA

ABSTRACT

The concept of balancing people, planet, and profit to maximize the absolute value of an enterprise is known as sustainability. It is concerned with the economic, social, and environmental effects of an enterprise in the long term. However, in practice, this definition does not provide companies with a meaningful framework to integrate sustainability into their projects, which by definition are one-off endeavors. Given this divide between the long-term nature of sustainability and the temporary nature of projects, companies have found it difficult to incorporate relevant sustainability indicators into project baselines. In this chapter, the authors examine a methodology for integrating sustainability into project baselines for consultants in the industrial and resource extraction fields. The methodology is comprised of an indicator set and a procedure for using the indicator set. This chapter's goal is to help standardize the sustainability process, making it easier to implement and more mainstream.

The objectives of this chapter are: (1) identify different sustainability indicator sets and their strengths and weaknesses; (2) explain what a multi-level analytical hierarchy project is and why it is important to integrating sustainability into such projects; and (3) state the steps in a procedure to integrate sustainability into project baselines.

DOI: 10.4018/978-1-4666-4852-4.ch074

INTRODUCTION

The 1987 World Commission on Economic Development, the Brundtland Commission (Brundtland, 1987), provided the most common working definition of sustainability as the balancing of people, planet, and profit to maximize the absolute value of an undertaking. This definition forms the foundation of sustainability programs at the executive level in many corporations. A majority of major corporations now provide a public sustainability report, or a sustainability section in their annual report, that is organized around the Bruntland definition and outlines how the company incorporates sustainability in their operations.

One problem with the Bruntland definition is that it does not provide good day to day guidance for attempting to integrate sustainability into a company's operations and/or projects. One definition of a project is a temporary undertaking that has a specific objective and a definite beginning and end, with the key focus often being the creation of a unique product or service (Labuschagne & Brent, 2005). Sustainability, however, has a long term orientation and this direct contrast with the short term nature of projects creates an imbalance that makes it challenging to incorporate sustainability into projects in a meaningful and measureable way.

It is becoming more common for project organizations to drive sustainability deep into their organizations by conducting projects such as building new facilities or designing new production methods to achieve internal sustainability objectives or targets. These projects are usually very company-specific and are not applicable outside the organization, except in a general sense. There are few, if any, large consulting companies in the resource extraction or industrial sphere that have driven sustainability deeply into their processes and, therefore, can offer their clients a range of sustainability services within their project execution philosophies. It is also very rare to find a large engineering company with sustainability offerings that contain documented processes, procedures, indicators, and methodologies.

This chapter examines sustainability indicators through the lens of one global engineering, consulting, and project management services organization. This organization is one of the leaders in its peer group in regards to sustainability and for a number of years has been the world sector leader on the Dow Jones Sustainability World Index. This organization works on world scale projects for many global resource companies, which are very advanced organizations when it comes to corporate sustainability and therefore demand high sustainability acumen from their consultants. In addition to these highly advanced sustainability clients, this consulting organization also has clients that are not yet mature in the sustainability field. These clients are looking to their consultants to guide them in improving their sustainability practices.

Given the global landscape, there is an opportunity for this consulting firm to be a leader in driving sustainability deeper, and perhaps fully, into its processes and project execution. Doing this would differentiate this firm from its competitors when pursuing new projects, both from a reputational standpoint and, for some clients, from a material standpoint.

The goal of this chapter is to assist in standardizing the sustainability process, making it easier to implement and more mainstream. Along with an examination of a number of current indicator sets in use, this chapter will show a proposed indicator set for use by consulting engineers and their resource extraction and industrial clients on projects in industries such as oil and gas, mining, manufacturing, and pulp and paper. The indicator sets examined will include those used by the International Federation of Consulting Engineers (PSM), the Global Reporting Initiative (G3), the US Green Building Council (LEED), the International Finance Corporation (IFC), the Ceres and Tellus Institute (FRP), and the Mining Association of Canada (TSM). The proposed indicator set is a sub-set of indicators that are used by existing sustainability indicator sets and is tailored to be applicable to industrial or resource extraction projects.

The chapter will conclude by evaluating the implementation of the indicator set and implementation process against a current design build project in the transportation industry.

Because each project is, by definition, unique, the proposed indicator has been developed to be broader in scope than the majority of projects would require. Project dimensions such as industry type, stakeholder requirements, location, and the owner's sustainability culture will determine the exact indicator set for each project. One of the first steps in using such an indicator set is for the engineering consultant to work with the client to determine which indicators are applicable to the project in question and whether new indicators will be needed. For a project to be successful, this step must be done up front at the beginning of a project.

SUSTAINABILITY IN PROJECTS

The traditional top-down and bottom-up approaches to incorporate sustainability within the organization, to a large extent, have not been effective. In order to integrate sustainability into the evaluation process, practical tools are needed to align business methodologies with the principles of sustainable development (Gladwin, Kennelly, & Krause 1995). Project management methodologies, which are core business methodologies for most companies are not excluded from this requirement (Labuschagne & Brent, 2005). In addition, there is a definite need to develop indicators that can be used in decision-making processes to ensure that projects are managed according to practices that will contribute to sustainable development (Warhurst, 2003).

There is a growing body of research literature that examines the interface between project management and sustainability. However, this body of work is still small and there has not been a lot of research done into integrating sustainability into project management processes and specifically into a framework that is useful for consultants involved in resource extraction or industrial projects. According to the United Nations Environment Programme 2002 report, industry needs to develop a set of sustainability indices, against which it can benchmark performance towards increased sustainability. An indicator is defined as an operational representation of an attribute (quality, characteristic, property) of a system (Moldan & Billharz, 1997). An index of sustainability indicators is defined as an amalgam of more than one indicator (Liverman, Hanson, Brown, & Meredith, 1988).

Existing Sustainability Indicator Frameworks

One of the more useful starting points for building a framework that can be used for selecting sustainability indicators on projects was produced by Koo and Ariaratnam. Their example was specific to a water main replacement project but their concepts are generally applicable. There are four steps in the process.

1. Define the project details.
2. Determine the spatial and temporal project boundaries.
3. Identify the indicators (using existing indicator lists as a basis).
4. Analyze the indicator attributes for assessment.

Using the analytical hierarchy process, originally developed by Saaty (1980), on the indicator set we get the matrix represented in Figure 1.

When there are options that are well defined, like in some municipal construction projects, the analytical hierarchy process can work very well. However, in most natural resource extraction and industrial consulting projects, the options are not

Figure 1. Project sustainability requirements matrix for qualitative indicator analysis

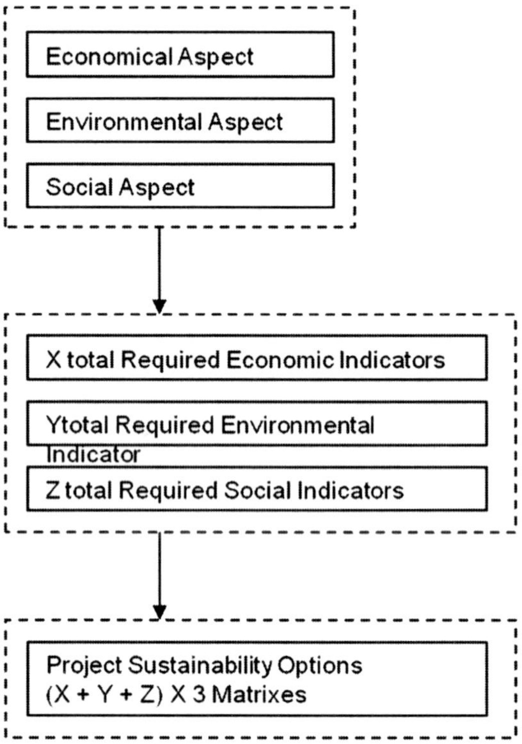

known with any degree of certainty during the initial definition or feasibility study phases because the variables are not defined in enough detail. In other words, it is difficult to use the analytical hierarchy process in most resource extraction and industrial projects because it is during the feasibility study phase, or perhaps even in earlier phases, that many of the significant decisions regarding project sustainability indicators are made.

Accounting of Sustainability Indicators on Projects

While the Global Reporting Initiative (GRI) is the accepted global standard for corporate sustainability reporting, there is currently no such standard for project sustainability reporting or management. In most organizations, a bespoke process is built from the ground up to handle each project's sustainability aspirations, with little learning and/or data used from past projects. Sustainability data collection is difficult, which leads to wasted resources throughout the project lifecycle.

Casella Stanger, Forum for the Future, and Carillion plc (Stranger, Forum for the Future, & Carillion plc, 2002) provide a framework for the final step of the process outlined in the previous section, which they have called sustainability accounting. The framework was developed for construction projects but is generally applicable. There are three steps in this framework.

1. Categorizing the indicators into economic, environmental, and social categories.
2. Creating sub categories for direct and indirect accounts within each of the categories in step one. Direct accounts show benefits/expenditures that are internal to the organization undertaking the project. Indirect accounts show externalities or benefits/expenditures that are external to the organization undertaking the project.
3. Determine what information is already collected via the firm's existing project accounting methods and can be restated in a project sustainability report and determine what additional information requires collection.

Sustainability in Project Management in the Resource Sector

Labuschagne, Brent, and van Erck (2007) have developed a framework that can be used to organize indicators into categories. This methodology gives us a very good starting point for developing a framework for evaluating sustainability for consultants in natural resource extraction and industrial projects at the project level. The key to this methodology is to break the three pillars of sustainability: economic, environmental, and social, into sub categories containing groups of indicators. For example, social sustainability indicators could be broken up into the sub-categories of:

- Internal Human Resources
- External Populations
- Land Resources
- Macro Social Performance

Most indicator sets that have been published use this multi-level grouping approach. One key point that is often overlooked when setting up an indicator set however, is that the level at which indicators are tracked cannot contain too many indicators or the indicator set will be too difficult to track in a project setting. This point is made implicitly in Figure 1. None of the indicator sets discussed next contain a manageable number of indicators on the tracking level, and hence are very difficult to track at the project level. An unmanageable number of indicators to track is seen as a leading cause of poor uptake of the indicator sets on projects by consulting organizations. Labuschagne, Brent, and van Erck (2007) help solve this problem by identifying related groups of indicators at the lowest level, resulting in a small enough data set for accurate and transparent decision making process on whether sustainability has been achieved for the project. The sustainability indicators section of this chapter shows a framework that can be used by consultants on natural resource extraction and industrial projects.

SUSTAINABILITY INDICATOR SETS

The Global Reporting Initiative (GRI) is considered the global standard for corporate sustainability reports (APEG BC, 2005; Labuschagne, Brent, & van Erck, 2007). However, at the beginning of 2012 there were over 100 recognized indicator sets for measuring sustainability. Some of these indicator sets cut across industry boundaries, while others are specific to a single industry/geography. This multitude of indicator sets shows that there is currently no standardization of sustainability indicators, accounting practices, and measurement tools for consultant companies to use on resource extraction and manufacturing projects.

Representative indicator sets that have some applicability to natural resource extraction or industrial industries are reviewed in this section to give an overview of the tools that are currently available for consultants and their clients to work with. Each indicator set is introduced and a table describes the strengths and weaknesses for each.

- **Global Reporting Initiative (GRI):** G3 Reporting Guidelines

In 1997, the United Nations Environment Programme, along with the United States non-governmental organization, created the GRI as a Coalition for Environmentally responsible Economics. The mission of the GRI is to make economic, environmental, and social impacts transparent and a fundamental component in effective stakeholder relations by providing a trusted and credible framework for sustainability reporting that can be used by organizations of any size, sector, or location (Global Reporting Initiative, 2002).

- **Ceres and the Tellus Institute:** Facilities Reporting Project

In order to develop consistent, comparable, and credible economic, environmental and social reporting guidance for individual facilities in the United States the Facility Reporting Project (FRP) was launched as a multi-stakeholder effort by Ceres and the Tellus Institute in 2003. (Facilities Reporting Project, 2004). The structure, indicators, and guidance contained in the FRP have been designed to match up with the GRI sustainability reporting indicators and guidance so that facilities in an organization can report against the FRP and roll up to a corporate GRI report.

- **International Finance Corporation:** Performance Standards for Social and Environmental Sustainability

The International Finance Corporation (IFC) is an organization that is part of the World Bank. The IFC's vision and purpose are that people should have the opportunity to escape poverty and improve their lives. The IFC performance standards apply to all the projects it finances. There are eight performance standards.

1. Social and Environmental Assessment and Management System
2. Labour and Working Conditions
3. Pollution Prevention and Abatement
4. Community Health, Safety and Security
5. Land Acquisition and Involuntary Resettlement
6. Biodiversity Conservation and Sustainable Natural Resource Management
7. Indigenous Peoples
8. Cultural Heritage (International Finance Corporation, 2006)

- **International Federation of Consulting Engineers:** Project Sustainability Management Guidelines

The International Federation of Consulting Engineers (FIDIC) has published a Project Sustainability Management (PSM) process, indicator set, and guidance as the basis for the development of a system to integrate sustainability into projects. In the PSM process, the project owner and engineer balance the owner's project vision against cost and available alternatives, by working together to select appropriate project goals and indicators for sustainable development which are linked back to higher level goals. Objectives for sustainable development are therefore addressed in much the same way as other project objectives are addressed in the project's quality plan.

PSM also provides a methodology for benchmarking sustainable development project performance, and for ensuring that advances in one dimension of sustainable development are not accomplished at the expense of another (International Federation of Consulting Engineers, 2004).

- **Towards Sustainable Mining:** Performance Indicators

Towards sustainable mining (TSM) is an initiative of the Mining Association of Canada. It has come up with a set of indicators that is geared towards the issues that its members, the mining industry within Canada, face on their projects and operations. The focus areas for these indicators are:

1. Tailings Management
2. Energy use and Greenhouse Gas Emissions Management
3. Aboriginal and Community Outreach
4. Crisis Management Planning
5. Biodiversity Conservation Management
6. Safety and Health

- **Green Building Council:** Leadership in Energy and Environmental Design Indicators

The Leadership in Energy and Environmental Design (LEED) indicator set is a green building rating system that encourages and accelerates global adoption of sustainable green building and development practices through the creation and implementation of universally understood and accepted tools and performance criteria in occupied buildings (Canadian Green Building Council – LEED, 2009). This indicator set is not designed to be applicable to natural resource extraction and industrial projects, however, LEED is a very well-known indicator set in North America and it is often one of the first things that clients reference for integrating sustainability on a project.

- Envision Rating System for Sustainable Infrastructure

Envision provides a holistic framework for evaluating and rating the community, environmental, and economic benefits of all types and sizes of infrastructure projects. The Envision rating system evaluates, grades, and gives recognition to infrastructure projects that use transformational, collaborative approaches to assess the sustainability indicators over the course of the project's life cycle (Sustainable Infrastructure, 2012).

The Envision system is very similar to the LEED in that it is a voluntary program containing a prescriptive set of indicators, and projects are given a rating based on how many indicators they are able to achieve on the project. Therefore, from the perspective of this chapter, it has the same strengths and weaknesses as the LEED rating system.

INDICATOR SET'S STRENGTHS AND WEAKNESSES

Each of these indicator sets has advantages that can be used when managing natural resources extraction and industrial projects and each set also has weaknesses that reduce the effectiveness of these indicator sets when managing sustainability on these projects. Table 1 outlines these strengths and weaknesses.

SUSTAINABILITY INDICATOR SET SELECTION

There are three main criteria for choosing which indicators from the indicator sets discussed in the previous section to include in the indicator set in this chapter. These criteria are as follows.

Table 1. Indicator set strengths and weaknesses

Indicator Set	Strengths	Weaknesses
GRI/FRP	• Role as the global standard for corporate sustainability reports • Reporting project sustainability against the GRI may lead to alignment between corporate and project sustainability outcomes.	• Some indicators are difficult to report against, especially in a project setting and • The guidance provided by the GRI tends to be too general.
IFC Performance Standards	• The standards are complete and there is a framework presented that can be used to track achievement • They are specifically designed for World Bank Projects, and often apply to industrial and resource extraction projects.	• The guidance does not contain suggestions for tracking measures to implement on projects • The guidelines are written in a way that makes it difficult to track achievement using simple data sets.
FIDIC PSM Guidelines	• Conceptually, the guidelines were designed to cover effects of the project from a whole society perspective, which could help prevent a project from focussing on one area such as water pollution while neglecting other areas of sustainability • FIDIC's analysis starts from envisioning integrating sustainability into the project management processes and ensures that as many of the indicators as possible is associated with measurable success criteria.	• Many of the measurable success criteria are too broad and/or difficult to measure at the project level. • While the whole society perspective of the guidelines and the focus on project management processes is an excellent way to approach the problem of integrating sustainability into projects, the end result is that the tracking indicators set is not conducive to efficient project management.
TSM	• Excellent baseline for mining projects • Indicators mostly well written for project use.	Much of the indicator set is very specific to the mining industry.
LEED/Envision	• Success criteria are easily measureable • Success criteria are generally applicable.	Much of the indicator set specific to occupied buildings and infrastructure.

- The indicators can be measured in a project setting.
- The indicators are relevant to natural resources extraction and industrial projects.
- The indicators have applicability to a wide range of natural resources extraction and industrial settings.

In many cases indicators were duplicated between the different indicators sets. When this occurred, the indicator that was most clearly written and simplest to track on a consulting project was chosen. To make this indicator set applicable to as wide a range of projects as possible, the indicators are not filtered with a view for what type of project a consultant may encounter. They can be customized for each project depending on the project location, the client, and other factors.

Economic Indicators

This set includes only economic indicators that focus on the project's economic impact outside of the project, e.g. items that have an economic impact on external stakeholders. Economic indicators that are part of the project normally have well defined processes that do not need duplication. In addition, the evaluation of performance in the area of community capital is of utmost importance in evaluating the social sustainability of a project as it has been implied that the core of sustainability is how people feel about their surroundings and their way of life. These perceptions can directly influence stakeholder participation initiatives (Labuschagne, Brent, & van Erck, 2007).

Environmental Indicators

Environmental sustainability is not a new concept to most resource extraction and industrial organizations. Therefore, this is the most developed dimension of sustainability in the indicator sets studied. In fact, indicator sets such as LEED have been designed to be almost exclusively environmentally focused. The number and breadth of environmental indicators made it necessary to only focus on those indicators that were applicable to resource extraction or industrial projects.

Social Indicators

Compared to environmental sustainability, less work has been done on social sustainability as it applies to business (Edum-Fotwe & Price, 2008), which means that there is little or no consensus on what social sustainability means and how to best apply it in a project setting. A consequence of this is a multitude of social sustainability indicators that have a different focus for different indicator sets. One benefit of this excess of indicators and focuses is that it is unlikely that any one project will address all the indicators and the list created for this chapter can be customized and shortened for most projects.

Another consequence of the lack of consensus around social sustainability is that there are usually few, if any, well-defined processes to apply social sustainability in a project setting. Therefore, unlike the economic dimension indicators, we look at social business sustainability from both an internal and external perspectives. The internal focus concerns the health and well being of employees, disciplinary practices, and equity and human rights aspects in employee sourcing. Training and development opportunities for employees are also included. The external focus concerns the impact of the project on three different levels of society: local, regional, and national communities, as applicable (Labuschange, Brent, & van Erck, 2007).

Indicator Organization

The total set of indicators for tracking sustainability in natural resource extraction and industrial projects consists of approximately 300 unique indicators. No project could effectively manage this task without an inordinate amount of resources. Simply sorting through this number of indicators and selecting the appropriate ones provides project management a challenge that hinders sustainability implementation and measurement on projects. To make this process feasible, another higher level, called high level sustainability themes, on the indicator hierarchy is required. Sustainability themes are groups of indicators that cover a portion of the project, such as minimizing or eliminating air emissions. When a theme is to be addressed in a project, the indicators within this theme are reviewed to determine which ones apply on the project.

The list of sustainability themes in each dimension is given in Table 2. Due to space considerations, an example of one sustainability indicator that applies to each indicator is presented in Table 3. In total, there are approximately 300 indicators in the indicator set, which is considerably more comprehensive than the 26 given in Table 3. The indicators presented were included because they apply to a wide range of resource extraction and manufacturing projects; however each indicator within the theme is unique and covers a different item that could be part of the project. For example there are fourteen indicators within the Minimize or Eliminate Air emissions including indicators on counting, auditing, and minimizing greenhouse gases and on other gasses such as ozone depleting substances.

Table 2. High level sustainability themes

Social	Environmental	Economic
Maintain community health, safety, and security	Manage natural resources in a sustainable manner to promote conservation and maintain biodiversity	Provide economic opportunities for the local community - employment
Maintain a crisis communication plan	Comply with all applicable environmental regulations	Provide economic opportunities for the local community - supply chain
Preserve cultural heritage	Minimize or eliminate air emissions	Determine the local taxation and infrastructure benefits and impacts of the project
Create a social and environmental management system to IFC standards	Minimize or eliminate energy use and procure energy from renewable sources	Design the project to maximize its lifespan with a minimum level of care and maintenance
Maintain effective communication with local stakeholders	Water pollution: discharges, effluents, and spills	General economic considerations
Maintain human rights standards	Fresh water withdrawal and recycling	
Enable positive outcomes for indigenous peoples	Waste and hazardous waste control, minimization, or elimination	
Ensure decent work and labour practices are adhered to	Transportation to and from site	
Eliminate, minimize, or fairly compensate for land acquisition and involuntary settlement	Indoor Environmental Quality - occupied buildings only	
Take responsibility for impacts of the projects outputs (products and services)		
Contribute to local/national housing and educational system		

Table 3. Sustainability indicator list

Dimension	Theme	Possible Project Indicators	Source
Social	Maintain community health, safety, and security	The client will prevent or minimize the potential for community exposure to water-borne, water based, water-related, vector-borne disease, and other diseases that could result from project activities.	IFC
Social	Maintain a crisis communication plan	The President and CEO has endorsed and demonstrated support for crisis management planning.	TSM
Social	Preserve cultural heritage	The client is responsible for siting and designing a project to avoid significant damage to cultural heritage. The client will not disturb any chance finds further until an assessment by a competent specialist is made and actions are identified.	IFC
Social	Create a social and environmental management system to IFC standards	The management program will be accompanied by an action plan where specific mitigation measures and actions necessary to comply with applicable laws and regulations.	IFC
Social	Maintain effective communication with local stakeholders	The communication and engagement process includes formal agreements with stakeholders respecting the process for dialogue.	TSM
Social	Maintain human rights standards	Operations identified as having significant risk for incidents of child labour, and measures taken to contribute to the elimination of child labour.	GRI
Social	Enable positive outcomes for indigenous peoples	The client will identify all communities of Indigenous Peoples who may be affected by the project within the project's area of influence, as well as the nature and degree of the expected social, cultural, and environmental impacts on them, and avoid adverse impacts whenever feasible.	IFC
Social	Ensure decent work and labour practices are adhered to	Ratio of basic salary of men to women by employee category.	GRI
Social	Eliminate, minimize, or fairly compensate for land acquisition and involuntary settlement	When displacement cannot be avoided, the client will offer displaced persons and communities compensation for loss of assets at full replacement cost and other assistance to help them improve or restore their standards of living or livelihoods.	IFC
Social	Take responsibility for impacts of the projects outputs (products and services)	Monetary value of significant fines and total number of non-monetary with laws and regulations concerning the provision and use of products and product services.	GRI
Social	Contribute to local/national housing and educational system	Contribution to K-12 education by building facilities and incorporating auxiliary programmes in the project.	FIDIC
Environmental	Manage natural resources in a sustainable manner to promote conservation and maintain biodiversity	Effect of the project on arable and permanent crop land area.	FIDIC
Environmental	Comply with all applicable environmental regulations	List categories of environmental permits in effect at the project and their issuing authorities. Describe the scope of these permits and how stakeholders can access them.	FRP
Environmental	Minimize or eliminate air emissions	The organizational level of accountability for GHG emissions management.	TSM
Environmental	Minimize or eliminate energy use and procure energy from renewable sources	In addition to fundamental building systems commissioning, hire an independent agent for the commissioning, conduct reviews of the contract documents prior to issuing and the contractor submittals, provide the owner with a single manual for recommissioning the building, and have a system in place for dealing with follow up commissioning items and indoor air quality issues.	LEED
Environmental	Water pollution: discharges, effluents, and spills	Total number and volume of significant spills.	GRI

continued on following page

Table 3. Continued

Dimension	Theme	Possible Project Indicators	Source
Environmental	Fresh water withdrawal and recycling	Concentration of faecal coliform in freshwater.	FIDIC
Environmental	Waste and hazardous waste control, minimization, or elimination	Weight of transported, imported, exported, or treated waste deemed hazardous under the terms of the Basel Convention Annex I, II, II, and VIII, and percentage of transported waste shipped internationally.	GRI
Environmental	Transportation to and from site	Provide showers for 5% and secure bicycle storage for 15% of building occupants.	LEED
Environmental	Indoor Environmental Quality - occupied buildings only	Install a permanent carbon dioxide monitoring system that provides feedback on space ventilation performance in a format that affords operational adjustments. Follow CO2 differential guidelines in ASHRAE 62-2001, Appendix C.	LEED
Environmental	General environmental considerations	Initiatives to mitigate environmental impacts of products and services, and extent of impact mitigation.	GRI
Economic	Provide economic opportunities for the local community - employment	Effect of project on GDP per capita.	FIDIC
Economics	Provide economic opportunities for the local community - supply chain	Cost and percentage of all goods, materials, and services purchased locally.	FRP
Economics	Determine the local taxation and infrastructure benefits and impacts of the project	Development and impact of infrastructure investments and services provided primarily for public benefit through commercial, in-kind, or pro bono engagement.	GRI
Economic	Design the project to maximize its lifespan with a minimum level of care and maintenance	Durability of the constructed works.	FIDIC
Economic	General economic considerations	Financial implications and other risks and opportunities for the organization's activities due to climate change.	GRI

HOW TO USE THE INDICATOR SET

Procedure for Using the Indicator Set

Natural resources and industrial projects can implement this indicator set by following a simple procedure.

1. Determine the project's stakeholders and their requirements.
2. Sort these requirements into the indicator set's high level sustainability themes, creating new themes if necessary.
3. For each theme that is applicable to the project, determine which of the theme's indicators are best suited for tracking the project's sustainability achievement. Create new indicators if necessary.
4. Assign ownership for each indicator and define when in the project the indicator will be completed.
5. Create success criteria for each selected indicator and determine the information required to measure achievement.
6. Collect project information and track the project's progress in completing the indicators.
7. Report on the indicator set at regular intervals, preferably as part of the regular project report.

Key aspects of this procedure are explained next.

Determining Stakeholders

As a general rule, engineering consultants do not own the assets that they design in their projects. They also don't often operate these facilities without a separate business group taking part under different arrangements. It is even rarer that they are associated with the marketing, handling, and development of the products that are developed as a result of their projects. However, it is....the asset life cycle resulting from the project, and the subsequent product life cycle resulting from the asset that have economic, social, and environmental consequences, which are in turn associated with an implemented project (Labuschagne & Brent, 2005). Therefore, in all dimensions of sustainability on a project, it is important for a project to define [the project's] stakeholders. Treating stakeholders in an ethically and socially responsible manner has been seen as the core of CSR (corporate sustainability) (Hopkins 2002). Additionally, stakeholder management has been regarded as the tool to connect strategy to social and ethical issues (Wartick, 1998). Because of this, a complete indicator set for sustainability cannot be achieved when the project stakeholders, along with their needs and desires, are not all identified. Stakeholders are people or groups that have, or believe they have, legitimate claims against substantive aspects of the project. A stake is an interest or share or claim in a project; it can range from informal interest in the undertaking at one extreme, to a legal claim of ownership at the other extreme (Cleland 1998).

The methodology for selecting project stakeholders is covered elsewhere including in Winch 2004 (Winch, 2004) and is not the focus of this chapter. However it is imperative that only after the project stakeholders have been determined that indicator selection for the project can proceed.

Assigning Indicator Responsibility

From the perspective of the consultant's role on an industrial or natural resources extraction project, all indicators in a project need to contain project information such as how to measure the indicator, when in the project to address the indicator, and where the responsibility is for the indicator's achievement. Two examples of indicators with responsibilities assigned and with project lifecycle stages for measurement identified are presented in Figure 2.

A natural resources extraction or industrial project is often broken down into the following project phases.

- **Idea Generation:** Generate the idea for a new project and prepare the initial proposal that describes the business need. This phase does not require a formal project plan.
- **Pre-Feasibility:** Evaluate the existing proposal in terms of financial, operational, and technical viability as well as against the company's strategy. Overlap or synergy with other projects should also be checked out.
- **Feasibility:** Identify and define the optimum solution to address the business need. All areas of the solution must be analyzed and assessed to determine killer concerns and risks.
- **Development and Execution or Engineering and Construction:** Design, develop, create, and build the chosen solution. Also develop the supporting system manuals, business processes, and training for the solution during this phase.
- **Commissioning:** Test the solution in an operational environment to validate the acceptance and capabilities of the solution.

Figure 2. Examples of sustainability indicators organization

	Example 1	Example 2
Dimension	Environmental	Social
High Level Theme	Comply with all applicable environmental regulations	Maintain a crisis communication plan
Indicator	List categories of environmental permits in effect at the project and their issuing authorities. Describe the scope of these permits and how stakeholders can access them.	Establish a local crisis management team, with defined roles and responsibilities.
Owner	Consultant or project owner	Project owner, with consultant involvement.
Project Lifecycle Phase for Measurement	Pre-Feasability with a review at every stage until commissioning.	Every stage where there is a site presence. Review plan at every stage.

- **Launch or Start-Up:** Hand over to the customer (internal or external) and thus release to the operational environment. This phase also marks the beginning of operational support.
- **Post Implementation Review or Project Follow-Up Review:** Assess the project after sufficient time (9-15 months), to determine if the benefits were delivered and what the impact of the project was on the business. Capture lessons learned for future reference.

Two examples of sustainability indicators organization in a project setting with responsibility assigned and lifecycle stage for measurement identified are shown in Figure 2.

PRACTICAL SIGNIFICANCE OF SUSTAINABILITY INDICATORS

An example of the robustness and usefulness of the indicator set can be seen when checking the use of the set and high-level themes by a design build project in the transportation sector. This project includes the design, procurement, and construction, by a consortium of engineering and construction companies, of a road through a medium sized Canadian city. Its purpose is to strengthen transportation and trade links, reduce road congestion, and foster economic growth. The road's strategic importance for economic growth, urban location, and unique ecological context necessitate strong design and planning principles to guide infrastructure development. It is to be a state-of-the-art facility within a contextually sensitive landscape setting that has ecological integrity, builds physical and cultural connections, and establishes a sustainable network of amenities that can be enjoyed by present and future generations. The legacy of the project will not only be its significant contribution as a trade and transportation route, but also include the establishment of a contiguous and sustainable green space system that contributes to the quality of life in the community and supports the re-establishment of an ecologically rich landscape.

High Level Themes and Specific Indicators

Table 4 shows that many of the high level themes applied to the project.

For each of the applicable themes, the specific objective that is applicable to the project has been recorded. In addition a SMART (Specific, Measurable, Achievable, Relevant, and Time-bound) indicator with a specific target, who in the project is responsible for the indicator, and the frequency of measurement are included. For example:

Table 4. High level themes applicability to the project

Number	Social	Environmental	Economic
1	**Applicable**: Maintain community health, safety, and security	**Applicable**: Manage natural resources in a sustainable manner to promote conservation and maintain biodiversity	Not Applicable: Provide economic opportunities for the local community - employment
2	**Applicable**: Maintain a crisis communication plan	**Applicable**: Comply with all applicable environmental regulations	**Applicable**: Provide economic opportunities for the local community - supply chain
3	**Applicable**: Preserve cultural heritage	**Applicable**: Minimize or eliminate air emissions	Not Applicable: Determine the local taxation and infrastructure benefits and impacts of the project
4	**Applicable**: Create a social and environmental management system to IFC standards	**Applicable**: Minimize or eliminate energy use and procure energy from renewable sources	**Applicable**: Design the project to maximize its lifespan with a minimum level of care and maintenance
5	**Applicable**: Maintain effective communication with local stakeholders	**Applicable**: Water pollution: discharges, effluents, and spills	Not Applicable: General economic considerations
6	**Implicitly Applicable**: Maintain human rights standards	Not Applicable: Fresh water withdrawal and recycling	
7	Not Applicable: Enable positive outcomes for indigenous peoples	**Applicable**: Waste and hazardous waste control, minimization, or elimination	
8	**Implicitly Applicable**: Ensure decent work and labour practices are adhered to	Not Applicable: Transportation to and from site	
9	Not Applicable: Eliminate, minimize, or fairly compensate for land acquisition and involuntary settlement	Not Applicable: Indoor environmental quality - occupied buildings only	
10	**Applicable**: Take responsibility for impacts of the projects outputs (products and services)		
11	Not Applicable: Contribute to local/ national housing and educational system		

- **Sustainability Aspect:** Environmental
- **High Level Theme:** Minimize or eliminate energy use and procure energy from renewable sources
- **Specific Indicator:** Reduce overall energy use of the road and associated green space
- **SMART Indicator:** Electrical use in KWhr/year
- **Target:** 35,533,480 KWhr/year
- **Owner:** OM&R Manager
- **Frequency:** Annual

Review of the Project Indicators

Using the methodology and indicator set described in this chapter, and incorporating the Green Roads indicator set as well as the indicator set in this chapter, the project determined that there were a total of 33 indicators for the 14 applicable high level themes. Most of the specific indicators used on the project, such as the previous example maps directly to economic indicator 4.17, are taken straight from the indicator set in the appendix. In other cases new indicators have been written within a theme to meet the specific requirements of the project. This is to be expected in any large scale project as when this indicator set and implementation procedure are used as the indicator set will normally have applicable themes and indicators for most, but normally not all of the project stakeholder needs.

Note that two social themes, maintaining human right standards and ensure decent work and labour practices are adhered to were implicitly applicable because this project is being conducted in Canada where there are already strong baseline laws, standards, and culture that ensure that virtually all projects meet the requirements for these themes. Additionally, two of the not applicable economic indicators, provide economic opportunities for the local community – employment and determine the local taxation and infrastructure benefits and impacts of the project were both applicable to the project but did not have specific indicators written for them. For the first of these indicators, there was a target for youth employment, but no specification that it needed to be local. For the second indicator, it was inherent in the project's purpose and baseline assumptions so it is not being specifically tracked as part of the project.

FUTURE DIRECTIONS

One possible area for future study can be found in the procedure presented in the last section, namely creating a list of sustainability indicator success criteria that can be used on projects. The unique nature of each project means that this list cannot be all-inclusive, but building the list will highlight what information is needed for tracking and achieving sustainability, and will give projects a common starting point. Efforts could then be directed towards investigating integration of sustainability reporting into project reports to streamline information collection in projects and ease project management.

Another possible area of further study is determining the utility of this sustainability indicator set and methodology both in more resource extraction and industrial projects and in other industries. There may be common sustainability indicators shared by most industries and discovering these similarities and leveraging the material presented in this chapter will assist in making project sustainability management a defined and accepted aspect of project management.

CHAPTER SUMMARY

There is an imbalance between the long term nature of sustainability and the unique, finite nature of projects, which makes it difficult to incorporate meaningful sustainability indicators into project execution plans. Specifically there is not yet a standardized method of measuring sustainability on consulting projects in resource extraction and industrial industries.

This chapter has examined such a method centered on an indicator set based on existing sustainability indicators sets. The indicator set used set contains approximately 300 indicators, too many to manage in a project setting, so the indicators are grouped as high level themes organized around the three aspects of sustainability, social, environmental, and economic. Projects can use a sub-set of these indicators to measure and report the attainment of sustainability on the project.

This chapter also outlines, with examples, a process to use the indicator set to create project specific SMART objectives that have specific owners and are addressed at specific points in the project lifecycle.

After reading this chapter you should be able to:

- Identify different sustainability indicator sets and their strengths and weaknesses.
- Explain what a multi-level analytical hierarchy project is and why it is important to integrating sustainability into such projects.
- Develop a procedure to integrate the sustainability into project baselines.

REFERENCES

Association of Professional Engineering and Geoscience. B.C. (2004). Part 3e: Practice specific module - Consulting engineering and geosciences. In Sustainability in Professional Engineering and Geoscience: A Primer. Academic Press..

Brundtland, G. E. (1987). *Our common future: The world commission on environment and development*. Oxford, UK: Oxford University Press..

Canadian Green Building Council - LEED. (2009). Retrieved from http://www.cagbc.org/AM/Template.cfm?Section=LEED

Cleland, D. I. (1998). Stakeholder management. In *Project Management Handbook*. San Francisco, CA: Jossey Bass..

Facilities Reporting Project. (2004). Retrieved from https://www.ceres.org/resources/reports/facility-reporting-project-sustainability-reporting-guidance/view

Gladwin, T. N., Kennelly, J. J., & Krause, T.-S. (1995). Shifting paradigms for sustainable development: Implications for management theory and research. *Academy of Management Review*, *20*(4), 874–907.

Global Reporting Initiative. (2002). *GRI sustainability reporting guidelines*. Boston: GRI..

Hopkins, M. J. D. (2002). Sustainability in the internal operations of companies. *Corporate Environmental Strategy*, *9*(4), 398–408. doi:10.1016/S1066-7938(02)00121-5.

International Federation of Consulting Engineers. (2004). Retrieved from http://www1.fidic.org/bookshop/account/reading_pdf.asp?sesskey=0&productcode=FI-EN-T-AA-10&range=0&f=IENTAA&p=2&n=Project+Sustainability+Management+Guidelines%2C+2004

International Finance Corporation. (2006). Retrieved from http://www1.ifc.org/wps/wcm/connect/5159190048855a4f85b4d76a6515bb18/SustainabilityPolicy.pdf?MOD=AJPERES&CACHEID=5159190048855a4f85b4d76a6515bb18

Labuschagne, C., & Brent, A. C. (2005). Sustainable project life cycle management: The need to integrate life cycles in the manufacturing sector. *International Journal of Project Management*, *23*(2), 159–168. doi:10.1016/j.ijproman.2004.06.003.

Labuschagne, C., Brent, A.C., & van Erck Ron, P.G. (2007). *Assessing performance of industries*. Pretoria, South Africa: University of Petroria openUP.

Liverman, D. M., Hanson, M. E., Brown, B. J., & Meredith, R. W. Jr. (1988). Global sustainability: Toward measurement. *Environmental Management, 20*(2), 133–143. doi:10.1007/BF01873382.

Moldan, & Billharz, S. (Eds.). (1997). Indicators and their use: Information for decision making sustainability indicators: Report on the project on indicators for sustainable development. Chichester, UK: Academic Press.

Saaty, T. T. (1990). *The analytic hierarchy process*. New York: McGraw-Hill..

Stranger, C. (2002). *Sustainability accounting in the construction industry*. London: Ciria Publishing Services..

Sunter, W. V. A. C. (2002). *Beyond reasonable greed: Why sustainable business is a much better idea*. Cape Town, South Africa: Human & Rousseau Tafelberg..

Sustainable Infrastructure. (2012). Retrieved from http://www.sustainableinfrastructure.org/rating/index.cfm

Warhurst, A. (2003). *Sustainability indicators and sustainability performance management*. Mining, Minerals and Sustainable Development..

Wartick, S. L. W. D. J. (Ed.). (1998). International business and society. Malden, MA: Blackwell..

Winch, G. M. (2004). *Managing project stakeholders*. Hoboken, NJ: John Wiley & Sons..

ADDITIONAL READING

Abood, K. A. (2007). *Sustainable and green ports: Application of sustainability principles to port development and operation*. Paper presented at Ports 2007 30 Years of Sharing Ideas. New York, NY.

Ashley, R., Blackwood, D., Butler, D., Jowitt, P., Davies, J., & Smith, H. et al. (2008). Making asset investment decisions for wastewater systems that include sustainability. *Journal of Environmental Engineering, 134*(3). doi:10.1061/(ASCE)0733-9372(2008)134:3(200).

Beheiry, S. M. A., Chong, W. K., & Haas, C. T. (2006). Examining the business impact of owner commitment to sustainability. *Journal of Construction Engineering and Management, 132*(4). doi:10.1061/(ASCE)0733-9364(2006)132:4(384).

Brundtland, G. E. (1987). *Our common future: The world commission on environment and development*. Oxford, UK: Oxford University Press..

Cleland, D. I. (1998). Stakeholder management. In *Project Management Handbook*. San Francisco, CA: Jossey Bass..

Dandy, G., Roberts, A., Hewiston, C., & Chrystie, P. (2006). *Sustainability objectives for the optimization of water distribution networks*. Paper presented at 8th Annual Water Distribution Systems Analysis Symposium. Cincinnati, OH.

Gladwin, T. N., Kennelly, J. J., & Krause, T.-S. (1995). Shifting paradigms for sustainable development: Implications for management theory and research. *Academy of Management Review, 20*(4), 874–907.

Hopkins, M. J. D. (2002). Sustainability in the internal operations of companies. *Corporate Environmental Strategy, 9*(4), 398–408. doi:10.1016/S1066-7938(02)00121-5.

Koo, D. H., & Ariaratnam, S. T. (2008). Application of a sustainability model for assessing water main replacement options. *Journal of Construction Engineering and Management, 134*(8). doi:10.1061/(ASCE)0733-9364(2008)134:8(563).

Labuschagne, C., & Brent, A. C. (2005). Sustainable project life cycle management: The need to integrate life cycles in the manufacturing sector. *International Journal of Project Management*, *23*(2), 159–168. doi:10.1016/j.ijproman.2004.06.003.

Labuschagne, C., Brent, A.C., & van Erck Ron, P.G. (2007). *Assessing performance of industries*. Pretoria, South Africa: University of Petroria openUP.

Saaty, T. T. (1990). *The analytic hierarchy process*. New York: McGraw-Hill..

Sunter, W. V. C. (2002). *Beyond reasonable greed: Why sustainable business is a much better idea*. Cape Town, South Africa: Human & Rousseau Tafelberg..

Talbot, J., & Venkataraman, R. R. (2011). Integration of sustainability principles into project baselines using a comprehensive indicator set. *International Business & Economics Research Journal*, *10*(9), 29–40.

KEY TERMS AND DEFINITIONS

Global Reporting Initiative (GRI): A non-profit organization that promotes economic, environmental, and social sustainability. GRI provides all companies and organizations with a comprehensive sustainability reporting framework that is widely used around the world.

Indicator: An operational representation of an attribute (quality, characteristic, property) of a system.

Indicator Set: An amalgam of more than one indicator.

SMART Objective: An objective that is Specific, Measurable, Achievable, Relevant, and Time-bound.

Stakeholder: People or groups that have, or believe they have, legitimate claims against substantive aspects of the project. A stake is an interest or share or claim in a project; it can range from informal interest in the undertaking at one extreme, to a legal claim of ownership at the other extreme.

Success Criteria: A method of measuring achievement of an indicator on a project. To be most effective, a success criterion should contain a SMART objective.

Sustainability: The concept of balancing people, planet, and profit to maximize the absolute value of an enterprise.

Sustainability Dimension: An aspect of sustainability. Traditionally sustainability is separated into three dimensions, social, environmental, and economic. However, alternate schemes exist with the most common including an operational dimension in addition to the usual three.

This chapter was previously published in Sustainability Integration for Effective Project Management edited by Gilbert Silvius and Jennifer Tharp, pages 194-211, copyright 2013 by Business Science Reference (an imprint of IGI Global).

Chapter 75
Energy Efficient Data Query, Processing and Routing Techniques for Green Wireless Sensor Networks

Afshin Behzadan
Ryerson University, Canada

Alagan Anpalagan
Ryerson University, Canada

ABSTRACT

While wireless sensor networking plays a critical role in many important applications, it also contributes to the energy footprint - which continues to increase with the proliferation of wireless devices and networks worldwide. Energy-efficiency becomes a major concern in the development of next generation sensor systems and networks. This chapter discusses data management techniques from energy efficiency point of view for green wireless sensor networks.

INTRODUCTION

Wireless sensor networks (WSNs) usually contain power-constraint nodes that require energy-efficient (a.k.a. green) transmission techniques, resource sharing algorithms and networking protocols (see an example in Figure 1). Each WNS node consists of one or more sensors for sensing the surrounding environment. Sensors are small and usually inexpensive, and have limited processing resources. Sensors sense and gather information from the environment, based on the decision guidelines provided by users in forms of queries. Acquired data are transmitted, via embedded radio in nodes, among nodes in a multi-hop way and finally reach the root, i.e., the base station (Estrin, Girod, Pottie, & Srivastava, May 2001).

DOI: 10.4018/978-1-4666-4852-4.ch075

Figure 1. A wireless sensor network scheme

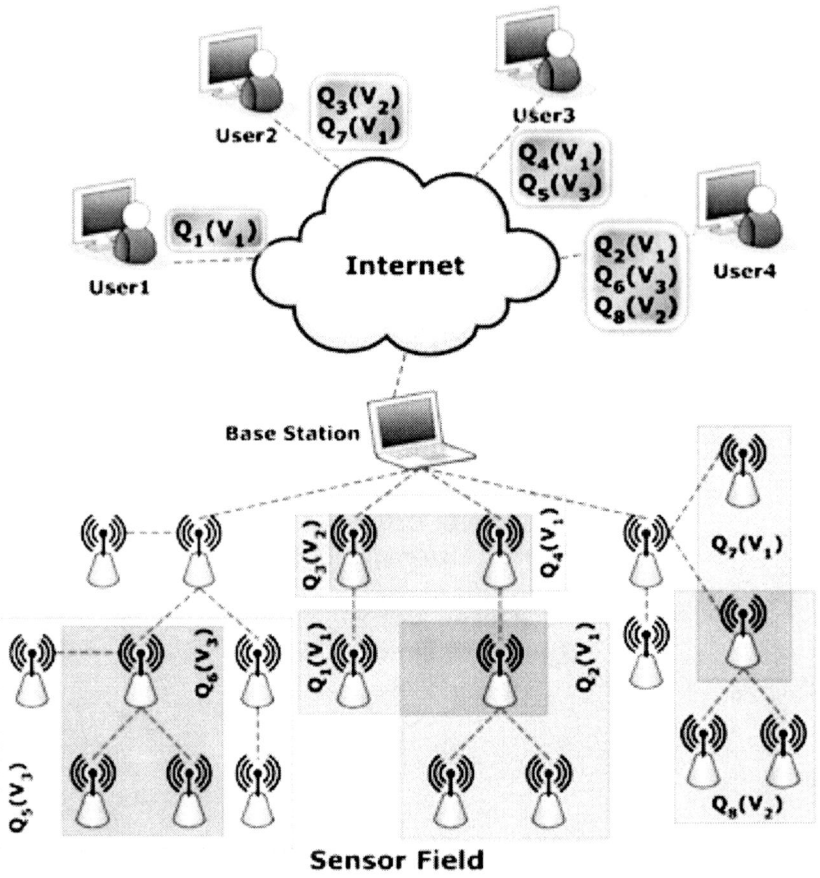

Usually, a WSN has no infrastructure and has an ad-hoc topology. This topology is due to the fact that sensor networks are typically deployed in hard-to-access environment. Obstructions in the environment can limit the wireless communication between nodes, affecting the network connectivity. These networks have many applications, such as military target tracking and surveillance, natural disaster relief, biomedical health monitoring and hazardous environment exploration and sensing.

Data Query, Processing, and Routing

Data sensing and transmitting consume a considerable portion of nodes' energy which is limited and vital (Potdar, Sharif, & Chang, May 2009). The main goal of *query processing* is to answer queries posed by users, while decreasing the energy consumption and prolonging network lifetime are also considered (Gehrke & Madden, March 2004). Queries are declared by users to the network and the network returns the required data by using the query processing engine (Madden, Franklin, Hellerstein, & Hong, Tinydb, An acquisitional query processing system for sensor networks, March 2005). A single query processing scheme assumes one active query in the network. However, in reality, different users may connect to the base station and query various data. In many cases, although different users may have different requests, their requests are somewhat similar. Thus, assigning a network to a single query and running queries sequentially not only lead to undesirable delays in responding to user requests, but also results in a vast energy wasting by doing redundant operations corresponding to different queries. Therefore, *multi-query processing* has been considered in which multiple queries run in the network simultaneously (Trigoni, Yao, Demers, Gehrke, & Rajaraman, June 2005).

Severe energy constraints and consequently communication and processing limitations of sensor nodes make centralized methods not suitable for these networks, as their execution involves noticeable processing overhead and needs a global view of the network. Generating this global view especially when the network is large will lead to a quick energy drain of nodes. Therefore, distributed and lightweight methods which are executable by low-powered sensor nodes and that need only local information to run properly, are required (Akyildiz, Su, Sankarasubramaniam, & Cayirci, March 2002).

Multi-hop data routing, driven by the above mentioned limitations, is typically used in a tree topology for data delivery, where intermediate nodes (as parents) receive data from their beneath nodes (i.e., child nodes), apply aggregation functions on them and their local data, and send the aggregated data to the next hop towards the base station (Yao & Gehrke, January 2003). This topology is formed by issuing a "flood" from the base station towards the network, through which nodes are assigned to different levels. Due to flooding and the location of nodes, each node might receive multiple flooding messages from some other nodes which can be its candidate parents. Thus, each node can have some candidate parents. However, at the end of flooding process, each node selects a parent in one level lower than its level, to which it will send the acquired data. Higher amount of traffic leads to faster energy depletion and the nodes suffering this problem have a shorter lifetime compared to nodes dealing with a less amount of traffic. Thus, uneven distribution of children among parents implies uneven drain of energy and lifetimes for nodes. As the nature of sensor networks is ad-hoc and randomly distributed, the connectivity of the network highly depends on the lifetime of nodes.

STATE-OF-THE-ART: FROM ENERGY EFFICIENCY POINT OF VIEW

Load balancing can be used to reduce the overall energy consumption of the wireless sensor networks. Generally, the load balancing problem in wireless sensor network can be considered from the routing and query processing point of view, since one of the main purposes of load balancing in such networks is the data delivery in an energy efficient way. The main areas for this purpose are routing and query processing. Routing aims to route the data in the network with the least cost, mostly based on the energy and least number of hops. On the other side, query processing tries to decrease the amount of data which travels through the network. Both of the mentioned areas try to decrease the energy consumption of individual nodes, and have many related concepts associated with each other.

Load balancing also tries to decrease the energy consumption not only for individual nodes, but also for the whole network. Approaches which rely on load balancing adjust the load. The load is usually defined from the energy point of view and as the energy consumption due to data communication among nodes. However, the inseparable part of load balancing is routing, since it benefits from the routing concepts and techniques, but in a balanced way. Therefore, three areas should be reviewed and investigated for proposing a solution for the load balancing problem.

When the concept of "utility" is considered for nodes in sensor networks, utility based approaches will be a relevant area. Also, sensor nodes can be considered as independent agents as they have local information. A utility based approach which models nodes as independent agents having their own local knowledge, implies using a game theoretical approach with nodes as players. Game theory approaches have recently received attention in different areas of sensor networks such as routing, load balancing and security, as the nature of these networks encourages this tendency. As our work considers a utility-based approach based on game theory concepts, a section of this chapter is dedicated to reviewing the use of game theory in sensor networks.

Query Processing in Sensor Networks

Many works have been proposed in the area of query processing and energy conservation techniques in wireless sensor networks. The authors in (Rosemark, Lee, & Urgaonkar, May 2007) present some fundamentals for processing of queries and discuss some optimization techniques for data gathering. TAG (Madden, Franklin, Hellerstein, & Hong, December 2002) introduces in-network aggregation to conserve energy.

The following works propose various techniques to increase energy efficiency. The authors in (Gehrke & Madden, March 2004) moot some fundamentals for query processing systems and mention some optimization techniques for query plan generation. The proposed work in (Munteanu, Beaver, Labrinidis, & Chrysanthis, February 2005) partitions the network among multiple routing trees to run a single query, to increase performance and fault tolerance. In (Muller & Alonso, October 2006), the authors propose some multi-query processing fundamentals and use a framework to support it. In their work, base station (BS) batches and aggregates queries together and injects a new query resulted from them to the network. In the run time if any new query arrives, it will be aggregated by the previous queries, but the new resulted query will be sent as the second query to the network. This is a transient state in which two combined queries run in the network. The transient state continues until the first query finishes, and then only the second query continues to run. However, the proposed solution aggregates all the received queries together and, in cases that queries are not compatible with each other their aggregation will not be beneficial.

In (Trigoni, Yao, Demers, Gehrke, & Rajaraman, June 2005) the idea of dividing the network into multiple zones and processing of results inside each zone is investigated. Results from each zone are collected and aggregated by nodes inside the zone, and finally only one aggregated data will be sent out. This way, the amount of outgoing traffic from each zone decreases, which leads to a considerable energy conservation. The authors in (Trigoni, Guitton, & Skordylis, 2006) propose some routing techniques for optimum result transition to BS in multi-query environments. The authors in (Lee, Lee, Zheng, & Winter, April 2006) use some pre-computed partial results, called materialized in-network views, to decrease energy consumption and share network among multiple queries. In their work, some partial aggregation results are computed before running of queries and they are kept in nodes. When the queries run, if a node has the desired value, it returns its partial aggregation value and does not send the query further to its beneath nodes. This way a, decrease in communication and computation overhead can be gained. However, computation of materialized views at first can impose a considerable energy overhead, and also in time dependent applications, after a while views are spoiled and cannot be used anymore. The proposed work in (Zhang, Yu, & Chen, September 2005) supports concurrent run of queries; however in that work each node has to aggregate received queries and run aggregated query which can impose considerable computation overhead on it and energy consumption.

Load Balancing in Sensor Networks

Several authors studied balanced data gathering in sensor networks. Harvey et al. (Harvey, Ladner, Lovasz, & Tamir, July 2003) propose an optimum centralized method for making an optimal semi-matching which minimizes the variance of the load among nodes. Considering a bipartite graph, a semi-matching is a balanced assignment of vertices in one part to vertices in another part. The graph can be represented as a set of tasks in one part and a set of machines in another part, each is able to process a subset of tasks. Then, an edge exists between a node from tasks part and another node in machines part, if and only if that machine is able to perform the task. Loads on machines are determined by the number of tasks they run. Thus, the goal is to assign tasks to machines in a fair way, so at the end, machines run as equal number of tasks as possible. Their method produces optimum load balanced assignments in all cases. To avoid unbalanced assignments, formation of a kind of path, called *alternating path* in the graph is prevented, as it is a sign of unbalancing.

Due to the limitations of sensor nodes and their autonomous mode, most of the proposed works in the literature tend to be distributed. Sadagopan et al. in (Sadagopan, Singh, & Krishnamachari, 2006) propose a decentralized utility based method for making the balanced data gathering tree. They use a game theoretic approach, where nodes are the selfish players of a tree construction game. Their method is suitable where in-network aggregation can be applied as they consider nodes with similar amount of data. In-network aggregation gives the opportunity to interpret the whole game as a number of games per level of tree. Thus in each level, the game is the balanced joining of children to possible parents in one level higher. However, this technique may cause a number of switchings by children among parents when different parents increase their bids resulting in considerable communication overhead.

In (Sadagopan, Singh, & Krishnamachari, 2006) the authors also discussed the construction of a spanning tree where no in-network aggregation is applied. They assume each sensor node samples and transmits one unit of data per epoch. With this situation, each leaf node transmits one unit of data, while every intermediate node transmits its local data plus data from nodes in the sub-tree for which it is the root. Assuming each data unit transmission consumes one unit of energy, each leaf node dissipates one unit of data and each intermediate node dissipates energy proportional to size of its sub-tree. The goal is to construct a spanning tree such that all the sensor nodes are fairly utilized on the average. If s^i shows the size of sub-tree rooted at node *i*, and e^i shows its current energy, then the fractional remaining energy of node *i* is defined as $\frac{e^i - s^i}{e^i}$. To ensure fair utilization of energy resources the goal is the construction of a spanning tree which maximizes the summation of the fractional remaining energy of all nodes. If the network is modeled by graph *G=(V,E)*, the strategy space of s_k^i of each node *i* in iteration *k* is

$$s_k^i = \left\{ j \mid (i,j) \in E \right\} \tag{1}$$

Then, the utility function of each node is defined as follows:

$$u_k^i = \min_{j \in s_k^i} \left\{ \frac{1}{e^i} + Q_k^j \right\} = \min_{j = s_k^i} Q_k^j \ , \tag{2}$$

where Q_k^j is the length of the shortest path from sensor node *j* to the BS at iteration *k*. $\forall_k : Q_{BS}=0$. Thus, at each iteration, a sensor node *i* chooses a node *j* that offers it the shortest path to BS, by using $\frac{1}{e^z}$ as the metric of edge length for any edge $(z,w) \in E$, where $z, w \in V$. This mechanism can be implemented by a distributed distance vector algorithm. Thus, the energy balanced tree construction terminates when the distance vector algorithm terminates.

As mentioned earlier, different characteristics can be considered for classification of the literature in the context of load balancing and energy conservation. First, the proposed works are either centralized and decentralized (or distributed). The latter is preferred for sensor environment since it is scalable and provides light weight solutions. Next, existing works can be investigated from the heterogeneity aspect. Heterogeneity includes some criteria such as different power, communication range, the amount of traffic from each node, transmission range and etc. Considering the current literature, most of works can be divided into two major groups. The first group considers homogenous nodes, and the second one considers two different types of nodes. One type includes more powerful nodes having more capabilities which are deployed in a fewer number. The other group includes a higher number of nodes with lower power and fewer capabilities. While consideration of homogenous nodes suggests flat network structure, heterogeneity in the mentioned form suggests a hierarchy for the network structure formed by clustering in which and powerful nodes act as cluster heads and the rest constitutes nodes inside clusters. In the following, based on this brief discussion, some related works are briefly discussed.

Chu et al. (Chu, et al., October 2009) proposed a centralized approach for dynamic monitoring of nodes' energy to determine if changing the parent of a node is necessary. The process includes two procedures in the gateway. In the first procedure, priority of each node is determined. It starts from the node with heaviest nodes and repeats for every node. The number of parents is used for breaking ties, meaning that the node with the least parents is the winner. The energy of selected node's parent is compared to the energy of other nodes from lower layer in its communication range. The link among node and its parent is remains intact unless the energy of parent is less than the other nodes from the lower layer in its communication range. In that case, the second procedure starts and a new parent with the least load is selected. The least number of neighbours is used for breaking ties in this procedure.

Application of Game Theory in Sensor Networks

Game theory is a discipline aiming to model situations in which decision makers have to make specific actions that have mutual, possibly conflicting, consequences (Fudenberg & Tirole, 1991). It has been used primarily in economics, in order to model the competition between companies. In the area of wireless networks, game theory is often used as a tool for making cooperation between nodes, terminals or various authorities. Game theory is applied to help solving problems in routing, resource allocation and power management (Niyato & Hossain, 2007).

Game theory is related to the actions of decision makers who know that their actions affect each other. A game includes a set of players $i=\{1, 2, ..., N\}$, each of which selects a strategy $s \in S_i$, where S_i is a strategy space including all possible strategies for i. Player i has the objective of maximizing its utility u_i. A game can be modeled by a set of players, a set of available resources in the game, all possible choices for each player made possible from the set of resources, and the payoffs assigned to each player after choosing a specific resource.

Two types of games can be considered: in *non-cooperative* games, each player selects strategies without coordination with others. On the other hand, in a *cooperative* game, the players have the choice and try to cooperate with each other, so that they can gain a maximum benefit which is higher than what they could have obtained by playing the game without cooperation (Uliman, Pomalaza, Oppernann, & Lehtomaki, August 2004). The objective is to allocate the resources so that the total utility is maximized. In wireless networks, the formation of coalitions involves the sharing of certain resources. However, when the costs of such resource sharing are more than the benefits gained by the nodes, they are less likely to participate.

Nash equilibrium: The equilibrium strategies are chosen by the players in order to maximize their individual payoffs. In game theory, the Nash equilibrium is a solution concept of a game involving two or more players, in which no player has anything to gain by changing only his own strategy unilaterally (Charilas & Panagopoulos, 2010). If each player has chosen a strategy and no player can benefit by changing his strategy while the other players keep their selected strategy unchanged, then the current set of strategy choices and the corresponding payoffs constitute a Nash equilibrium.

Repeated games: In strategic games, the players make their decisions simultaneously at the beginning of the game. On the contrary, the model of an extensive game defines the possible orders of the events. In repeated games, players can make decisions during the game and react to other players' decisions. In this case, a game is played many times and the players can check the outcome of the previous game before playing in the next repetition.

A game theoretic approach is suitable for this purpose, since it considers independent players with local knowledge. This feature supports the nature of sensor environments, in that nodes have only local information due to their limitations. In proposing such an approach, nodes are considered as players of the game trying to reach a higher benefit according to their individual utility functions. On the top level, a global utility function also should be defined in a way that when each player tries to gain more individual utility, it implicitly helps in maximizing the global benefit defined by the global utility function. As each game should have a termination at the end, Nash equilibrium can be used to show that the game will stop at some point. However, Nash equilibrium proves the termination of the game, and not necessarily the optimum outcome resulted by playing the game.

When a rational interaction is considered between nodes, the forwarding nodes can be encouraged to cooperate in data routing. Then, based on a price-based scheme nodes try to benefit other nodes to receive more benefits from them in a corporation scheme. In this section, the discussion of game theory in wireless sensor networks is provided by surveying recent work on routing among nodes, which relies on price-based or utility-based approaches and use concepts of the game theory.

Non-Cooperative Games: The authors in (Han & Liu, 2007) use game theory to analyze the outcome of a game, in which the deployed sensors belong to different authorities and can receive incentives for cooperative forwarding for routing, data storage and aggregation. When sensors request a service from another sensor belonging to a different authority, the other sensor may choose to cooperate or decline based on its resources. In this game, where none of nodes from different domains decline to cooperate with nodes from another domain, the Nash equilibrium results in non-cooperation of nodes belonging to different sponsors. To avoid this situation, the authors propose the use of tokens as incentives to encourage cooperation between sensors which belong to different sponsors.

In (Suris, DaSilva, Han, & MacKenzie, 2007), the authors consider the problem of packet forwarding in wireless sensor networks using game theory. In classical game theory, players choose a particular strategy in response to strategies of other players and this strategy does not change over time. However, in (Suris, DaSilva, Han, & MacKenzie, 2007) the frequency with which a player chooses a given strategy varies over the time in response to the strategies chosen by other players. This allows players to choose from a set of actions and strategies and use of only local information. The authors assume a heterogeneous sensor network from different classes, where any two non-neighbouring classes communicate via multi-hop routing. Nodes can be selfish. They consider inter-class relaying, when a class can cooperate and forward packets or decline to relay. They model the game as that of non-cooperative repeated N-player game between classes of nodes, where nodes participate repeatedly in games with other nodes. In repeated games, a node's action in a given round is influenced by the actions of other nodes and corresponding payoffs in previous rounds. Thus, a repeated game offers ways to punish nodes that do not cooperate by decreasing their payoffs at the end of the game. This can be done by making bad reputation or decrease in incentives resulting in reduced payoffs at the end of the game. Cooperation is similarly rewarded, by examining the payoffs after repeated rounds of the game. Nodes with richer history of cooperation have better reputation, accumulate incentives faster and are included in routes. In transmitting or forwarding a packet, classes spend battery energy b and gain an incentive c. If classes refuse to retransmit, they gain nothing and there is no cost to them. They show that for packet forwarding between stationary classes, Nash equilibrium is achieved if each player plays the Patient Grim strategy (Suris, DaSilva, Han, & MacKenzie, 2007) and the discount factor is approximately close to unity.

The proposed work in (Frattasi, Fathi, & Fitzek, 2006) can be also considered in the set of approaches that use the game theory in sensor networks, as it models the load balancing problem by using techniques from game theory to make a routing tree in sensor network by a decentralized way. The authors design the utility functions of individual nodes such that the network objectives are met when the sensors maximize their individual utility functions. The problem here is to construct a routing tree rooted at the base station. Every node has a level in the network which is the number of hops from the sink node. A node must find and attach to a parent with fewer children than the current one. The decisions taken by a node at every level are independent of the decisions taken by nodes located at other levels. They describe a distributed algorithm to design the utility functions of individual nodes, such that when these utility functions are optimized by the sensor nodes, the overall objective of the network is met.

Cooperative Games: A reliable query routing scheme is proposed in (Wang, Xue, & Schulz, 2006), where it is suggested that the number of sensors working simultaneously to collaborate on aggregation should be chosen such that global objectives are achieved. The global objectives are defined as increasing network utilization, communication efficiency and energy consumption. For this purpose the authors use a game-theoretic approach. In this approach, sensors are modeled as intelligent agents cooperating to find optimal network architectures that maximize their payoffs. Payoff for a sensor is defined as benefits resulted from its action minus its individual costs. The problem is modeled as a reliable routing, in which a set of sensors are the players of the routing game. When the base station sends a query to the nodes, it is checked for a match with the attributes of the data sensed by the node. They model this idea by a value v_i that represents the closeness of the match. If v_i=0, it implies that the query does not match any attributes. Data is routed to the sink node through an optimally chosen set of sensors. They call this game as the reliable query routing (RQR) game. Each sensor node is modeled as relaying a received data packet to only one neighbour and hence forms only one link between any pair of source and destination nodes. The strategy space of a sensor node is modeled in the form of a binary vector, $\{l_{i1}, l_{i2}, \ldots, l_{in}\}$, where l_{ij} = 1/0 represents decision of a sensor node s_i to send or not to send a packet to sensor node s_j. Each node's payoff is a function of the reliability of the path between it and the base station and the expected value of information at that node. This results in a routing tree that is optimal, since if a sensor node decides to choose a different neighbour on another tree, it results in reduced payoffs to/from other nodes. Hence, this also forms the Nash equilibrium for PQR game. Since the network is unreliable, they use a path metric called the *path weakness* to evaluate various suboptimal paths. The path weakness determines how much the node would have gained by switching from its current path to an optimal one. A negative switching suggests that a node s_i is benefiting more from its given strategy, but at the expense of some other sensor. A positive switching indicates that the sensor node could have performed better.

Directions to Node Heterogeneity

In general, the investigation of node heterogeneity in the literature is mostly limited to two aspects such as nodes with different load, or two kinds of deployed nodes in the network with different capabilities like different communication ranges for each type. Considering two different types of sensor nodes in the network is a big help towards network clustering and localization of data processing and communicating in most of nodes. However, uniform distribution of powerful nodes acting as cluster heads among other normal nodes is a challenging problem in practice, especially in situations where human interference in not possible. In such cases some cluster heads may have larger number of nodes in their clusters, while other have less, and the existing load balancing algorithms cannot achieve a good performance. Even, it is possible that a collection of nodes cannot find a nearby cluster head, and have to connect in multi hop way to the closest cluster head. Increasing number of cluster heads relative to number of normal nodes is a solution, but it inevitably ends up with more expensive outcome. Thus, considering load balancing among all nodes looks important. This is also true in large clusters which may be considered as small networks.

Also, it looks very interesting if an algorithm can adapt itself according to different application needs. While some applications are more sensitive to delay, for some others more quality even with the cost of more delay is preferred. Thus, just consideration of prolonging the lifetime of the network is not enough and a proper solution should consider application needs too. In the next section, a decentralized utility-based approach for the load balanced data routing is presented. The main goal of the algorithm is to prolong the network lifetime by balancing load among nodes.

LOAD BALANCING ALGORITHM: GAME THEORETICAL APPROACH

In this section the algorithm for balanced data routing in wireless sensor networks is explained in details. The approach is based on game theoretic concepts by defining of utility functions for nodes as players of the games. Nodes' actions are based on utility function so that they gain more benefits according to utility function. The most important of advantage of using game theory, is that nodes are viewed as independent entities having their local knowledge, which facilitates the distributed implementations of algorithms.

Network Model and Problem Formulation

System Model: Wireless sensor networks are usually modeled via a graph $\boldsymbol{G = (V,E)}$ with set of nodes (G) and their adjacency (E). As mentioned before, each node has a level equal to its distance to the base station in terms of number of hops. Data routing tree is a tree topology $\boldsymbol{T=(V_T,E_T)}$, $\boldsymbol{V_T \subseteq E_T}$ and $\boldsymbol{E_T \subseteq E}$ built on a graph G. Using this tree, the acquired data from the network are streamed to the base station. The routing tree is formed when each node selects a closer node to the base station as its parent to send its data. Different policies can be considered for parent selection such as random selection, selection based on distance, or selection based on the quality of the link. However, as the network becomes denser, consideration of load on nodes looks more important, since nodes with higher load have a faster energy depletion and failure. Here, load is defined as the required energy for receiving, processing and transmitting of data.

Data Aggregation: One of the advantages of sensor nodes is their processing ability (Madden, Franklin, Hellerstein, & Hong, March 2005). Although this ability is limited, it gives the opportunity of moving some of the data processing operations into the network. As the energy resources are restricted in sensor nodes, the main goal of in-network data processing is to decrease the amount of communication overhead among nodes. In-network aggregation (Madden, Franklin, Hellerstein, & Hong, December 2002) is the most common case of such processing, where each node aggregates its acquired data with the received data from other nodes, and sends the aggregated data to its parent in the routing tree. Aggregation functions can be divided into two main groups: perfect and imperfect (Sadagopan, Singh, & Krishnamachari, 2006). For perfect aggregations, the number of output data items is constant and independent of the number of input data items. Examples of these kinds of aggregations are maximum and average functions. The former causes one data item to be sent out from each node, and the latter leads to two output data items from each node, i.e. the sum and the number of values. Perfect aggregations can be completely performed in the network. On the other hand, the number of output data items for imperfect aggregation is variable and depends on the number of input data items. Median function is an example of such kind of aggregations. For the computation of a median function, all data should be available. This requirement forces nodes to send all raw data to the base station, where the result can be determined, after these data has been gathered.

While in-network aggregations decrease processing and communication overheads on nodes, load balancing techniques try to distribute them evenly among nodes (Bouabdallah, Bouabdallah, & Boutaba, 2009). Load balancing tends to distribute excessive loads on nodes among other possible nodes in a fair manner to prolong their lifetime (Bouabdallah, Bouabdallah, & Boutaba, 2009) and maintain network connectivity as a global objective. Most of the proposed works in the context of load balancing are accompanied by aggregation techniques. Existing works in the literature which use aggregation along with load balancing, can be divided into two groups. The first group considers perfect aggregation (Sadagopan, Singh, & Krishnamachari, 2006). The main feature of this consideration is the equal load sent out from each node. In the second group, different loads from nodes are assumed (Sadagopan, Singh, & Krishnamachari, 2006). This assumption makes the load balancing problem more complicated when dealing with the second group compared to the first group, but it is more practical. In this case, in addition to perfect aggregations, imperfect aggregations can be included, where the load sent out from each node can be different due to the fact that complete in-network aggregations cannot be performed. A perfect solution for heterogeneous case should also work for homogenous case as load homogeneity is a specific case of load heterogeneity when the output loads of nodes are equal.

Node heterogeneity is not limited to various amounts of produced data. Nodes may also differ in the amount of power supply, total amount of data which they can transmit in a time slot, or the total bandwidth that they can allocate for data transmission to their children. Besides the energy issue, time and bandwidth are two important factors in the quality of data delivery. Thus, the importance of load balancing can mainly be discussed from three aspects: nodes energy consumption, delay and link quality in terms of available bandwidth, all of which are critical for applications in sensor networks, and it would be more realistic if they are considered in the proposed solutions.

Terminology

For a better description of the algorithms, some definitions and system model are provided in this section.

- $G = (V,E)$ is a graph which models the network and includes a set of nodes V and their adjacency (set E).
- $T=(V_T,E_T)$ is a tree that models the routing tree, where $V_T \subseteq V$ is the set of its constitutive nodes with their adjacency set (E_T) where $E_T \subseteq E$.

- **Node Level:** Level of node $v \in V$ in graph G is its distance to the root in number of hops.
- P_v^p is the set of potential parents for node v in graph G, including nodes from which v accepts flooding message during the construction process of graph G.
- P_v^c is the set of potential children for node v in graph G, including nodes that receive flooding message from v during the construction process of graph G.
- N_v^u is the set of nodes which are located in the same level as v and are two hops away from it with the intermediate node u between v and them.
- N_v is the set N_v^u $\forall u \in P_v^c$, i.e. all two hops away nodes from v which are located in the same level as of v.
- E_V is the set of children for node v in tree T, called v's existing children.
- $P(v)$ is v's parent in tree T.

The construction of tree T is implemented on graph G. The construction of graph G itself is explained in the next section. By the end of construction of graph G each node v knows sets P_v^p and P_v^c.

Nodes are considered heterogeneous in terms of their initial energy, data transmission rate, the amount of their produced data, and the total bandwidth which they can provide their children with, to transmit their data. The presented solution to the mentioned problems consider Stackelberg model for the game. With this model nodes playing parent role are leaders of the game and nodes with child role are followers. Parents have cooperative behaviour, while children have selfish behaviour. Leaders make decisions before followers, since they can predict decision of followers. The behaviour of parents influences the behaviour of their children, so that even with selfish actions of children as followers, they still help to the global benefit of the game.

A wireless sensor network can be modeled as a graph $G=(V,E)$ where sets V and E correspond to nodes in the network and the connections between them respectively. An edge $e_{ij} \in E$ exists between nodes $v_i, v_j \in V$, if they are located within each other communication range. Due to similar communication ranges, edge e_{ij} is considered as a bidirectional link which can be used for both transmitting and receiving purposes. Streaming of data from nodes towards the base station is supported by a routing infrastructure made by flooding mechanism starting from the base station. During this flooding, each node may accept several flooding messages from other forwarding nodes in its adjacency. However, based on an adopted parent selection policy, it finally selects only one of the sender nodes as its parent. At the end of the parent selection process a topology is formed which can usually be stated by a spanning tree $T=(V,E_T)$ where $E_T \subseteq E$ denotes parent-child} relationship among nodes. The root of this tree is the base station. There is a variety of policies for parent selection in the literature. However, taking the energy constraint of nodes into account is highly critical for network longevity. From the energy consumption point of view, as sensor networks are deployed in an ad hoc manner, their connectivity is highly dependent on node lifetimes.

As mentioned before, using aggregation functions is highly considered in sensor networks as it decreases the data traffic in the network. Two major groups of aggregation functions are perfect and imperfect aggregations. A high percentage of applications rely on perfect aggregations. However, in some other applications, queries impose the use of imperfect aggregations, where in-network aggregation cannot be applied on the acquired data.

Heterogeneous Load Balancing Formulation

Since for the imperfect aggregation case, different amounts of outgoing traffic from nodes are considered, heterogeneity on load would cover perfect aggregation case which implies homogeneity of load. Thus, in the following explanations load variety of nodes is assumed. Let **E(v)** show the initial energy of a node **v**. A portion of this energy is consumed for sensing and processing purposes and the rest will be used for communication. Again, as the energy required to sense and process data is negligible comparing to communication energy consumption, it is omitted in the computations. A simple and well-known model is used to model radio communication between sensor nodes. Let $\mathbf{E}_{\text{elec}}$ is the energy needed to operate transmitter and receiver circuit, and ε_{amp} is the required energy to amplify transmission for achieving an acceptable signal to noise ratio. Then, energy required for sending **k** bits to distance **d**, and the energy required to receive these **k** bit can be determined through (3) and (4).

$$E_{Tx}(k,d) = E_{elec}k + \varepsilon_{amp}kd^2 \tag{3}$$

$$E_{Rx}(k) = E_{elec}k \tag{4}$$

To find out the value of k, consider C_v as the set of v's children with $|C_v|$ members, and each node $u \in C_v$ sends $L(u)$ bits. Then, $\varepsilon_{R(v)}$ shows the consumed energy for reception of all data units by v from its children, and can be determined by (5).

$$\varepsilon_{R(v)} = E_{elec} \sum_{i=0}^{|C_v|-1} L(u_i). \tag{5}$$

After receiving data, v performs aggregation function on its local data and the received data. Aggregation functions on k, bits of data can be modeled by

$$Agg(\lambda_1,\lambda_2) = \lambda_1 k + \lambda_2, \tag{6}$$

where $0 \leq \lambda_1 \leq 1$ is the *compression factor* and $\lambda_2 \geq 0$ is the *overhead factor*. Compression factor shows the average amount of compression obtained by applying the aggregation function, and the overhead factor shows the amount of data imposed by applying aggregation function, regardless of the input data amount. For example maximum or minimum functions can be shown by $Agg(0,1)$, average by $Agg(0,2)$, and median by $Agg(0,1)$.

Now, if v's local sensed data is $S(v)$ bits, and it has $|C_v|$ children, the amount of data after applying aggregation function, $L(v)$, which will be sent by v is

$$L(v) = \lambda_1 (S(v) + \sum_{i=0}^{|C_v|-1} L(u_i)) + \lambda_2 \tag{7}$$

Thus, the energy required to transmit the received data along with its local data is shown by $\varepsilon_{S(v)}$ and defined as:

$$\varepsilon_{S(v)} = E_{Tx}(L(v), r) = E_{elec} L(v) + \varepsilon_{amp} L(v) r^2, \tag{8}$$

where r is the v's communication range. Considering (5) and (8), the energy that v spends to receive, process and transmit data is determined by (9):

$$\varepsilon(v) = \varepsilon_{R(v)} + \varepsilon_{S(v)} \tag{9}$$

Then, the remaining energy of v is

$$\varepsilon'(v) = E(v) - \varepsilon(v) \tag{10}$$

Also, v's lifetime can be defined as

$$\Delta(v) = \frac{E(v)}{\varepsilon(v)} \tag{11}$$

Network lifetime is defined as the time to the first node failure. By this definition, the longer node lifetimes are, the longer network lifetime is.

From the delay point of view, if node v can receive and transmit with rate $R(v)$ data units, the time required to receive data from its children is shown by $T_{R(v)}$ and computed by

$$T_{R(v)} = \frac{1}{R(v)} \sum_{i=0}^{|C_v|-1} L(u_i) \tag{12}$$

Then, v processes the received data and aggregates them with its local data. Assuming that the processing time is negligible, the time in terms of epochs, required to transmit v's data can be determined by

$$T_{S(v)} = \frac{L(v)}{R(v)} \tag{13}$$

Thus, the total delay that v incurs to receive, process and transmit all data is defined by:

$$T(v) = T_{R(v)} + T_{S(v)} \tag{14}$$

The delay amount should be specifically considered for applications that are real-time and delay is important for them, such as safety or security monitoring systems. A special case can be obtained when each node v has the communication rate of one data unit, $R(v)=1$, and produces one data unit per time slot, $L(v)=1$, and the aggregation function is perfect, with function $Agg(0,1)$. In this case $T_{R(v)} = |C_v|$ and $T_{S(v)}=1$ leading to the total delay $T(v) = |C_v| + 1$ imposed by v.

On the other hand, when the reliability is more important and some delays can be tolerated by the application, then a proper utilization of bandwidth should be considered. The most common way suggests the division of available bandwidth among children relative to the amount of their sent data. Thus, the available bandwidth for data transmission gained by node u with parent v, shown by $B_{T(v)}$ is:

$$B_T(u) = L(u) \frac{B_R(v)}{\sum_{i=0}^{|C_v|} L(u_i)}, \tag{15}$$

where $B_R(v)$ is the total bandwidth that v can allocate to its children for data transmission. In the special case, as mentioned before, and when $B_R(v)$ is considered with unit of one, $B_T(v)$ equals to $|C_v|^{-1}$.

A combination of delay and bandwidth functions forms the utility function for nodes when they play child role. This combination is able to support different kinds of application needs for delay and reliability where response time and bandwidth have specific importance based on the application. Utility function for child u with parent v is defined through

$$\Gamma(v, \varepsilon'(v), |C_v|) = U_C(u) = \varepsilon'(v) \left[F_t \frac{R(v)}{\lambda_1 (L(v) + \sum_{i=0}^{|C_v|} L(u_i)) + \lambda_2} + F_f \frac{B_R(v)}{\sum_{i=0}^{|C_v|} L(u_i)} \right],$$

$$0 \le F_t \le 1,\ 0 \le F_f \le 1, \tag{16}$$

where F_t and F_f are time and frequency coefficients correspondingly and indicate the importance of them. Γ shows the bid generated and sent by the parent v, with remaining energy $\varepsilon'(v)$ and $|C_v|$ children. Each child in a selfish manner tends to join a parent by which it obtains the highest possible utility.

Load factor corresponding to an arbitrary set S consisting of $|S|$ nodes is defined as the standard deviation of loads on nodes $v \in S$, and formulated as follows:

$$\sigma_S = \sqrt{\frac{\sum_{i=0}^{|S|} (\varepsilon(v_i) - \varepsilon(v_i))^2}{|S|}}, \tag{17}$$

where $|S|$ is the number of $v \in |S|$ and $\varepsilon(v_i)$ is the average of loads on them. For a node v when it plays parent role, the utility function can be defined as

$$U_P(v) = \frac{1}{\sigma_{N_v}}, \tag{18}$$

where N_v is the set of nodes two hops away in the same level as of v in the graph G including v, or in other words, the set of connected parents to v including v. Minimizing σ_{N_v} leads to maximizing the output of $U_{P(v)}$ which is the individual benefit for node v in parent role.

As nodes may have both child and parent roles based on their location in the network, the utility function for an arbitrary node v is defined by

$$U(v) = U_C(v) + U_P(v). \tag{19}$$

At a top level, minimizing load factor with a set equal to the set of all nodes in the network can be considered as the global goal in construction of routing tree. In this case, no extra load is imposed on any node and all of them have as closest as possible lifetime to each other. Although in this case the lifetime for all nodes is not the maximum possible lifetime, no node suffers from extra load imposed by other nodes. Because the maximum life time for a node can be achieved when it behaves selfishly and does not accept children to join itself. However, this behaviour has the cost of joining its possible children to other parents resulting in an extra load on them and shortening their and network lifetimes. For simplicity of explanation V_T is considered equal to V, though it does not limit the generality of the solution. Thus, the game is to create a routing tree for which U_V, defined by (20), is maximum.

$$U_V = \frac{1}{\sigma_V} \tag{20}$$

The difference between various nodes in terms of their lifetime, imposed delay and offered bandwidth to their children arises from the number of children. If some parents have extra number of children or loads and some others have less, then based on the mentioned definitions, the delay imposed by them in data transmission and the bandwidth offered by them to their children for data transmission are also uneven. This unbalancing leads to a longer delay and poorer quality for children of nodes with higher load, but less delay and richer quality and even unused resources like energy or bandwidth for children of nodes with lighter load. Thus, adjusting loads among parents is also a critical problem for quality of service guarantee.

Construction of Communication Infrastructure

As mentioned in the previous section, it is assumed that the graph **G** = (**V**,**E**) exists. Thus, before going through the details of algorithms, the construction process of this graph or in other words infrastructure of both algorithms is explained in this section. Algorithm 1 shows the construction process of communication infrastructure which is modeled by graph **G** = (**V**,**E**).

The construction is based on the node distances to the base station, and starts with broadcasting a flooding message from this point as the *root* in level 0 of the tree. Receiving nodes of this message set their level to 1, their potential parent to the root, inform the root about their setting and re-broadcast the flooding message. Upon receiving the flooding message by other nodes in the network, each receiver node goes through a series of investigations. Firstly, it accepts the flooding message if its level is not determined yet. By accepting this message, the node sets its level one more than the sender's level, adds the sender to its potential parents set, and informs the sender of this new setting. When a sender of flooding message receives a related setting message from another node, it adds the sender of the setting message to its potential children set. If a node accepts the flooding message for the first time, it re-broadcasts that message. Due to flooding nature, it is generally possible that a node receives multiple flooding messages from different nodes. Thus, a node may receive a flooding message after determination of its level raising one of the following cases:

Algorithm 1. Graph G construction, run in each node $v \in \boldsymbol{V}$

```
if v is root then
   v.level ← 0
      broadcast floodingMsg
else
   if v.level = null and receive(floodingMsg) then
      v.level ← floodingMsg.sender.level + 1
                                       P_v^p ← floodingMsg.sender
      send stateAck
      broadcast floodingMsg
   else if v.level ≠ null $ and receive(floodingMsg) then
      if floodingMsg.sender.level ≥ v.level then
         discard floodingMsg
      else if floodingMsg.sender.level = v.level − 1
         P_v^p ← floodingMsg.sender
         send stateAck
      else if floodingMsg.sender.level < v.level − 1
         remove all members in P_v^p
         send updateAck
         v.level ← floodingMsg.sender.level + 1
                                       P_v^p ← floodingMsg.sender
         send stateAck
         broadcast floodingMsg
      end if
   end if
end if
if receive(stateAck) then
   P_v^c ← stateAck.sender
end if
if receive(updateAck)
   remove stateAck.sender from P_v^c
end if
```

- The sender has a higher level than the level of node's potential parents. In this case, receiving node ignores the received message.
- The sender has a level equal to the level of node's potential parents. In this case, receiving node accepts the sender as another potential parent for itself and notifies it by sending a state acknowledgment message.
- The sender has a lower level than the level corresponding to potential parents of the node. In this case, receiving node removes all previous parents from its potential parents set and informs them about this action, so they remove the node's ID from their potential children set. Then, it adds the sender ID to its potential parents set as

its new potential parent, sets its level one more than the sender's level, informs the sender of this new setting, and re-broadcasts the flooding message.

The flooding process continues until all nodes receive this request and do not broadcast it anymore. Eventually, each node $v \in \boldsymbol{V}$ has a set of potential parents $\boldsymbol{P}_v^p$ and a set of potential children $\boldsymbol{P}_v^c$.

Heterogeneous Balanced Data Routing Algorithm

In this section, details of heterogeneous balanced data routing (HBDR) is explained. The presented algorithm is not only suited for the case that imperfect aggregation functions are used and nodes are heterogonous, but also works for homogenous case as well where perfect aggregation functions are used. HBDR algorithm includes a series of games, each is played in one level of graph G. These games are played in a sequential order. The first game is the game in the highest level among leaf nodes and their parents. As soon as this game terminates, the second game starts in one level lower, and after its termination, next game corresponding to one level lower can start. This procedure continues until the termination of game in the lowest level, among nodes in the first level and the root that is base station. In fact, playing order follows a bottom-up model from leaves towards root.

Each game includes a number of rounds. In each round some children join parents. The first round of the games is called *initial round*. Generally, in this round, nodes gather information about their one-hop neighbours. In the following rounds, referred as *subsequent rounds*, children gradually join parents in a load balanced way. Algorithm 2 and Algorithm 3 show the process of tree construction.

The first game starts from leaves. Based on the information obtained through flooding, each node knows the set of its potential parent and potential children. If a node has just one potential parent, it immediately joins that parent, since the joined parent is its only option. On the other hand, if a node has some potential parents, it sends a join request (*joinReq*) message indicating it is a common children among some potential parents. Each *joinReq* message contains three areas. First field indicates the data amount which was sent by the sender of message. Two other fields, which will be discussed later in this section, do not have any value or in other words are *null* in the first round. Therefore, the *joinReq* message in the initial round is referred as *null joinReq*, and is used to inform potential parents about their potential children status.

Each parent $v \in \boldsymbol{V}$, updates its load ($\varepsilon(v)$) and potential load ($\varepsilon_p(v)$) based on joining nodes and received *joinReq* messages respectively, using (9). In the computation of $\boldsymbol{C}_v = \boldsymbol{E}_v$, and in the computation of $\varepsilon_p(v)$, $\boldsymbol{C}_v = \boldsymbol{E}_v \bigcup \boldsymbol{P}_v^c$. In fact, $\varepsilon_p(v)$ shows the load on v when all its potential children join it, in the worst case. After determining the remaining energy and potentially remaining energy, v computes its bid, i.e. $bid(v) = \Gamma(v, \varepsilon(v), \boldsymbol{E}_v)$ and potential bid, i.e. $bid_p(v) = \Gamma(v, \varepsilon_p(v), \boldsymbol{E}_v \bigcup \boldsymbol{P}_v^c)$, which it can offer to its children. These bids are broadcasted to its potential children in the form of $\{bid(v), bid_p(v)\}$ pairs, called *bid pair*.

Each child node can receive some bid pairs according to the number of its potential parents. By receiving these pairs, child node $u \in \boldsymbol{V}$ selects parent $v \in \boldsymbol{P}_u^p$ which offers the highest benefit or bid. Based on the formula for bid, the highest bid from a node implies better quality of service in longer time meaning more stability of the offer than others. If for some parents the offered bids are the same, their potential bids are used for breaking ties. This means from parents with equal offered bids, those are chosen which have the

Algorithm 2. Node $v \in V$ in parent role

while $\left|P_v^c\right| > 0$ do the following steps
if initial round **then**
  wait for all $u \in P_v^c$
  update $\varepsilon(v)$

$$\varepsilon'(v) \leftarrow E(v) - \varepsilon(v)$$

  if receive $(joinReq)$
    update $(\varepsilon_p(v))$

$$\varepsilon_p'(v) \leftarrow E(v) - \varepsilon_p(v)$$

  end if
  generate $bid(v)$ and $bid_p(v)$

$$bidPair(v) \leftarrow bid(v), bid_P(v)$$

  broadcast $bidPair(v)$
else
  wait for $joinReq\, timeout$
  if receive $(joinReq)$ **then**
    $nextbid(v) \leftarrow bid(v)$
    $oldMinParent \leftarrow null$
    boolean $skipFlag \leftarrow$ **false**
    while $nextbid(v) > bid(w)$, $\forall w$ of received bid s **and** $skipFlag \leftarrow$ **false**
      $minBid \leftarrow$ **minimum** $(received\, bid_{Min} \in joinReq)$
      $selectedChild \leftarrow joinReq.sender$
      $minParent \leftarrow$ the parent which sent $minBid$ to $selectedChild$
      if $minParent = oldMinParent$ **then**
        $skipFlag \leftarrow$ **true**
      else
        $oldMinParent \leftarrow minParent$
        compute $Nextbid(v)$ based on new load
        $acceptMsg(v, selectedChild) \leftarrow bid(v)$
       send $acceptMsg(v, selectedChild)$ to $selectedChild$
      end if
    end while
    send $rejectMsg$ to the rest requesters
    wait for $joinAck$ timeout
    for all received $joinAck(u, w)$ **do**
      $P_v^c \leftarrow P_v^c - \{u\}$
      if $v = w$ **then**
        update $\varepsilon(v), \varepsilon_p(v), \varepsilon'(v), \varepsilon_p'(v)$
      else
        update $\varepsilon_p(v), \varepsilon_p'(v)$
      end if

Algorithm 3. Node $u \in \boldsymbol{V}$ in child role

while $p(u) = \varnothing$ do the following steps
if *initial round* and v is *leaf* **then**
  if $\boldsymbol{P}_v$ |= 1
| **then**
    join $(P(v))$
  else if $\boldsymbol{P}_v$ |> 1
| **then**
    send $\left(joinReq, \boldsymbol{P}_v^p\right)$
  end if
else if receive $(bidPair)$ **then**
  wait for all $v \in \boldsymbol{P}_v^p$
  $BestBids$ set $\leftarrow \varnothing$
  $BestParents$ set $\leftarrow \varnothing$
  $BestBids$ $\leftarrow$ received bids with maximum bid
  $BestBids \leftarrow bids \in BestBids$ set with maximum bid_p
  $BestParents \leftarrow bid.sender$, $\forall bid \in BestBids$ set
  for all $w \in BestParents$ set **do**
    $bid_{max} \leftarrow \textbf{maximum}\left(bid(v)\right), \forall v \in \boldsymbol{P}_v^p - \{w\}$
    $bid_{min} \leftarrow \textbf{minimum}\left(bid(v)\right), \forall v \in \boldsymbol{P}_v^p - \{w\}$
    $joinReq(u,w) \leftarrow \{L(u), bid_{max}, bid_{min}\}$
    send $(joinReq(u,w), w)$
  end for
else if receive $(acceptMsg)$
  wait for all involved parents
  $bestChoice \leftarrow \textbf{minimum}\,(bid(v))$ in all received $acceptMsg$ s
  $bestParent \leftarrow bestChoice.sender$
  join $bestParent$
() **broadcast** $(JoinAck(u, bestParent))$
else if $receive(rejectMsg)$
  wait until next round
end if

maximum the potential bid. Thus, in the worst case when all potential children join that parent it still has better benefit for u. The chosen parents after these two steps investigation are the best choice for children. Child u shows its willingness for joining those parents by sending *joinReq* messages to them. As mentioned before, a *joinReq* from node u to node v contains three fields. The first field includes the amount of data which u will send to v in each time unit. The second and third fields indicate the maximum and minimum bids respectively among the received bids by u from its parents except v.

A *joinReq* from child u to parent v is an indication about the load on other parent in $\boldsymbol{N}_v^u$. Based on the third field in the received *joinReq* messages, v chooses a child which has a parent with minimum offered bid indicating a combination of high load and short lifetime for that parent. This cooperative behaviour relaxes the load on the second parent and helps it to have a longer lifetime. After choosing u, based on its *joinReq* message, v adds the amount of data indicated as the first field in *joinReq* to its current amount data, and computes the benefit based on this new amount of data showing the benefit in the case that u finally joins v. This predicted bid is called $nextBid$. Also, v sends an acceptance message to u to inform it. Each accept message contains the $nextBid$ value. Then, v compares *nextBid* to the second fields of received *joinReq* messages which show maximum bids of other potential parents. If the *nextBid* is larger than all other maximum bids, v continues with child acceptance in a same way as before and re-computes the *nextBid*. Child acceptance continues unless one maximum bid larger than *nextBid* can be found. In that case, v stops child acceptance and sends the reject message to the rest of potential children which request joining.

A child may receive some acceptance messages. By receiving acceptance messages, child u chooses a parent with maximum included *nextBid* in its acceptance message. When u joins a parent, it broadcasts a join acknowledge message to not only inform the joined parent, but also to inform other parents about leaving them.

Let child u joins parent v. Then, u is added to $\boldsymbol{E}_v$ and also removed from $\boldsymbol{P}^c_w, \forall w \in \boldsymbol{P}^p_u$. By reception of every *joinAck* message which is destined for v, v re-computes its remaining energy, bid and potential bid, and re-broadcasts the bid pair. From this point and by broadcasting the new bid pair to the its potential children, the next round begins with similar steps mentioned above.

When a child joins a parent the game terminates for it. For a parent v, when $|\boldsymbol{P}^c_v| = 0$ the games ends. As mentioned before, games are played in a sequence from leaf nodes towards the base station. An intermediate node has two roles: parent for its children in one level higher and child role for its potential parents in one level lower. Thus, it plays two games. Because the sequence of games is from higher level to lower levels, such nodes first play a game in parent role with nodes in one level higher, and after its termination they start playing another game in child role with nodes in one level lower. This way may affect synchronization among nodes in different levels of the graph, as some nodes may finish their first game faster than others, and start the second game sooner. Thus, in the initial round of each game a node in parent role waits to receive null *joinReq* from all its potential children and then each child node waits to receive *bidPair* from all its potential parents. Considering such scheduling helps to synchronization of related nodes to each other and guarantees when a node continues the game it has received all necessary data. Let node v_1 waits for v_2. To make sure that v_2 has not failed and is waiting for reception of data from some other nodes, v_1 and v_2 can exchange a Hello messages on some certain intervals to differentiate between node failure and waiting status.

Algorithm terminates when the last games ends, i.e. when nodes in the second level join the parent in the first level.

SIMULATION RESULTS

In this section the performance of the algorithm, referred to as HBDR, is investigated and discussed. A network with size $500m \times 500m$ is considered. The performance of the algorithm is compared to the proposed algorithm in the (Sadagopan, Singh, & Krishnamachari, 2006) which we refer to it as cumulative algorithm, since it considers cumulative cost based the residual energy of nodes on the path from each node to the base station.

Table 1 includes default values of various parameters in simulations. Scenario-specific changes in this default values is mentioned in the corresponding section of each scenario. The following are explanations of different scenarios.

Table 1. Node u ∈ V in child role

Parameter	Value
Network Dimension	500(m)×500(m)
Number of Nodes	150
Node Communication Range	150(m)
Aggregation Function (λ_1,λ_2)	(1,0)
(Time Coefficient, Frequency Coefficient)	(1,1)
E_{elec}	50 nj
E_{Amp}	100 pj
Initial Energy	10 joule
Data Amount	128 bits
Data Rate	3 kbps
Bandwidth	30 KBuad (BPSK modulation is used)

Network Lifetime

Network lifetime is defined as the time from the start of running the network until the first node failure. Network lifetime is an important factor, since as mentioned earlier, sensor networks have ad-hoc topology and consequently network connectivity depends on aliveness and data forwarding of nodes. In this scenario the effect of changes in number of nodes and node communication ranges on network lifetime is investigated. Generally, increase in number of nodes or node communication range leads to higher adjacency among them which gives more chance of load balancing.

In one scenario, number of nodes increase from 50 nodes to 300 nodes in steps of 50 and the communication ranges is fixed on 150 meters. In the other, communication range increases from 50*m* to 350 *m* in the steps of 50 *m* and the number of nodes is fixed on 150 nodes. Figure 2 and Figure 3 depict the results of these two scenarios respectively.

As shown in the figures, HBDR method has a better performance than cumulative method in both cases. The reason is that cumulative method run proactive distance vector algorithm updating and regenerating routing tree periodically. Since it is implemented in a distributed way, and the flooding mechanism is used and during this flooding nodes choose a parent with minimum cost to join. Thus, it is probable that a bunch of potential children joins a specific potential parent because it has the best cost among others. Although the algorithms is able to update, the routing tree joining of number of nodes to one parent because of

Figure 2. Number of nodes vs. network lifetime

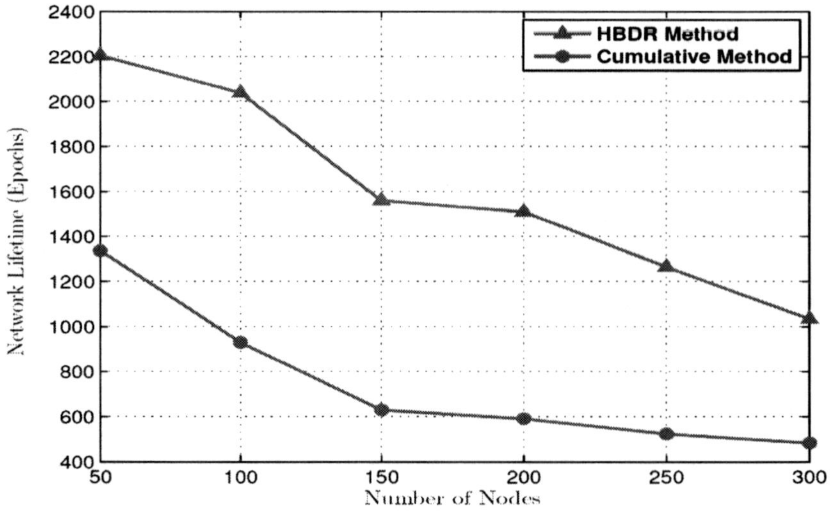

Figure 3. Node communication range vs. network lifetime

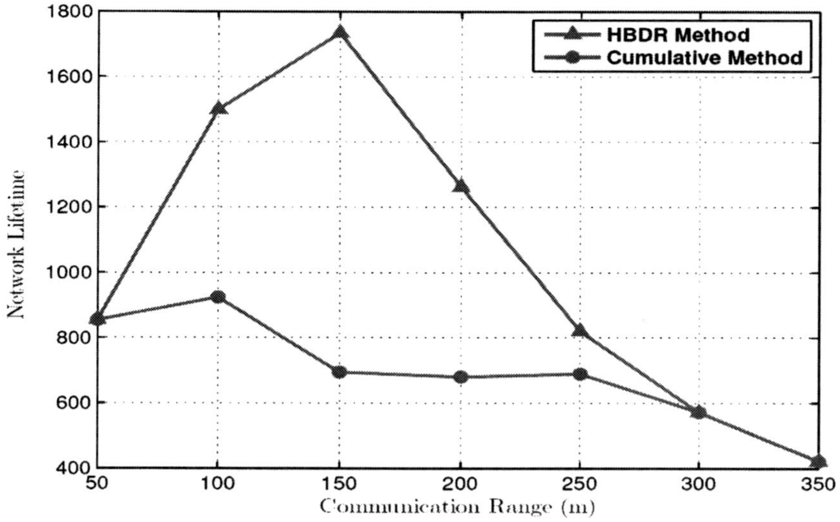

its appropriate offer results in faster energy drain of that node. This fast energy drain forces a sooner regenerating process of routing tree. However, for HBDR the balanced assignment of potential children based on their load and also the cost of the parent is considered.

For the scenario in which communication range is considered variable when the communication range is short because of less amount of adjacency among nodes, the strategy space of nodes for parent selection is limited. Thus, two methods have similar performance. While the communication range increases, the chance of selection among different parents also increases. This increases the chance of more load balancing specifically for HDBR algorithm, resulting in longer network lifetime. When the communication range becomes larger than 200 meters, since the network dimension is $500m \times 500m$, more nodes can directly connect to base station, and this decreases the impact of load balancing, resulting in the same performance for both methods.

Standard Deviation of Node Lifetime

As mentioned before, standard deviation from the average node lifetime is an indication of load balance efficiency of the methods. Higher variation to the average implies shorter lifetime for some nodes leading to shorter network lifetime. Figure 4 and Figure 5 shows the standard deviation of network lifetimes, when number of nodes and node communication range are variable respectively. All the parameters for simulation are the same as the previous scenario.

As shown in the figures, HBDR method has less standard deviation compared to cumulative method. This means that nodes that run HDBR algorithm have closer lifetime to each other, resulting from more balanced assignment of loads to nodes. For the case in which communication range is variable, when communication range relative to network dimension increases, more nodes are able to attach base station directly. As a result, both methods have close node lifetime leading to similar standard deviation of node lifetime.

Updating Factor

Although the total number of message exchanges is important, it is not a precise factor for communication overhead of algorithms. For more precise investigation, the frequency of exchanges

Figure 4. Number of nodes vs. standard deviation node lifetime

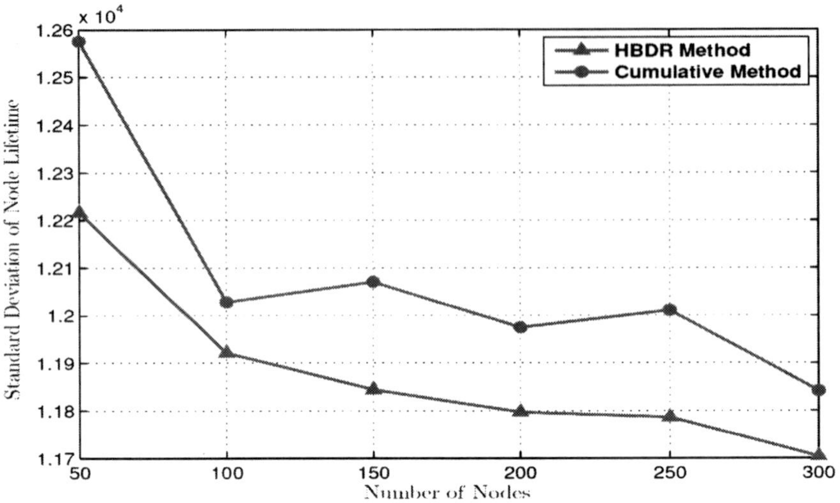

Figure 5. Node communication range vs. standard deviation node lifetime

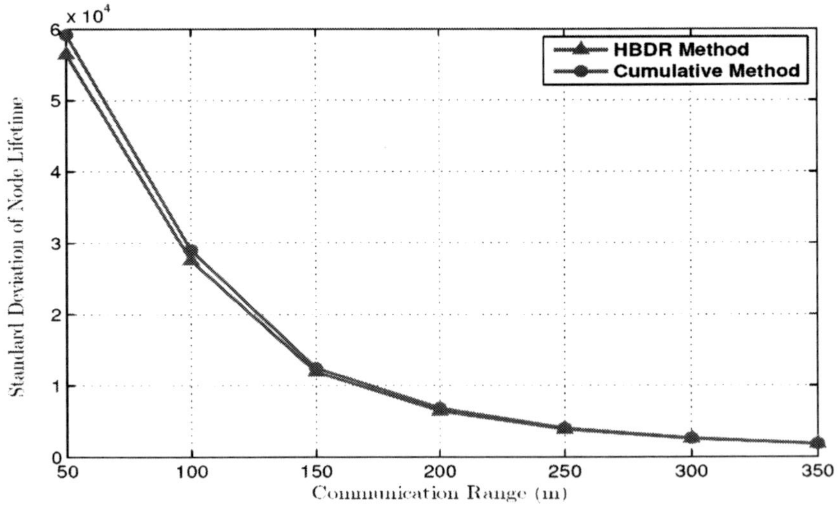

should also be investigated. The reason is that it is probable a large number of message exchanges occurs but not frequently, and on the other hand less number of message exchanges occurs more frequently. The latter case may impose larger communication overhead, specifically when the difference of message exchanges is not high. The mentioned point is a good reason to define *updating factor* as:

$$\text{updating factor}=\frac{\text{total number of message exchange}}{\text{network lifetime}} \tag{21}$$

Network lifetime is a good indication of how fast it is needed that the tree is regenerated. Thus, updating factor is an indication of what amount of message exchanges is required in each epoch to keep the tree up-to-date and network connected.

The less the updating factor is, the less communication overhead per epoch is imposed. Again, simulation results are reflected under the condition of different number of nodes, and different node communication range respectively.

As it can be seen in Figures 6 and 7, HDBR method mostly outperforms cumulative method. This is because, although cumulative method imposes fewer amounts of message exchange, it is not a more balanced algorithm than HDBR algorithm. Thus, over-used nodes have faster energy depletion and updating of the tree have to occur more often than HBDR method, resulting in a higher updating factor than HBDR method. However, as indicated in Figure 7, when the communication range is variable and it becomes close to the network dimension, more nodes can directly connect to base station. This direct

Figure 6. Number of nodes vs. updating factor

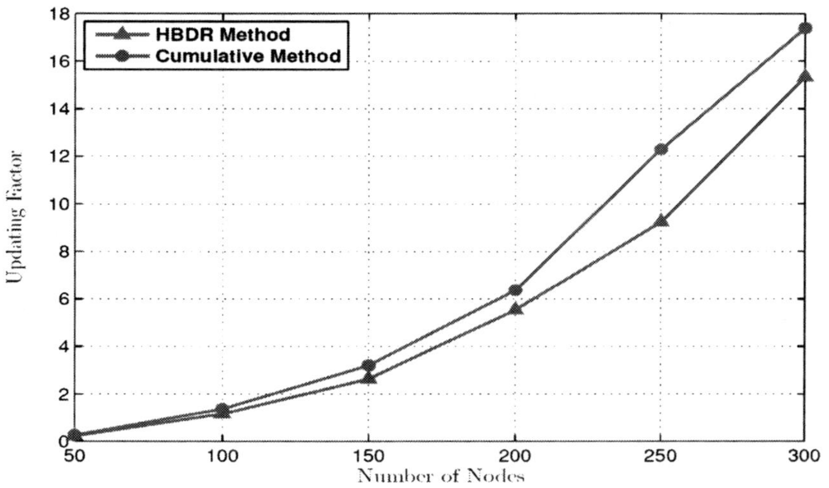

Figure 7. Node communication range vs. updating factor

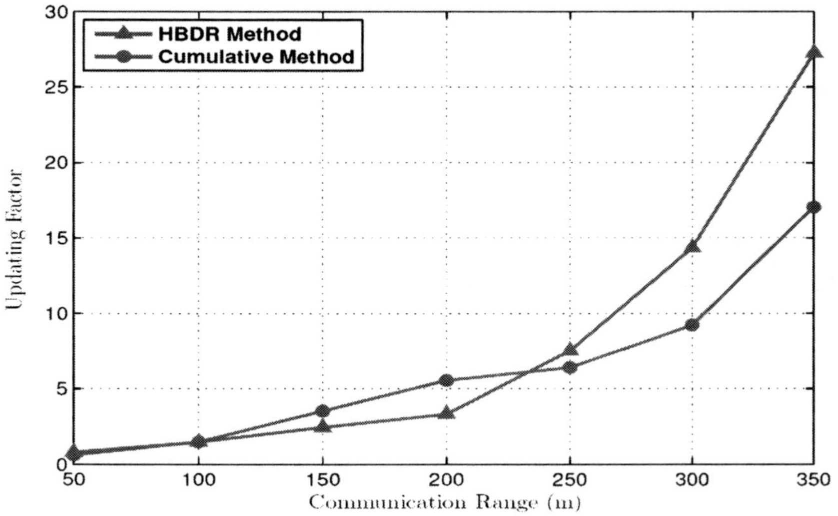

connection causes fewer loads on nodes in both approaches. Thus, both approaches update trees less frequently. However, because the number of message exchanges needed in each construction of tree using cumulative method is less than the number of message exchanges required in each construction of tree using HBDR method, it outperforms HBDR method in this case.

CONCLUSION AND FUTURE RESEARCH DIRECTIONS

This chapter introduced an algorithm based on game theoretic approach for load balanced data routing in wireless sensor networks. Load is defined as the amount of energy required to receive, process and transmit information by nodes. Load balancing is critical for sensor networks as the only energy resource for sensor nodes is their limited battery supplies resulting in computation and communication limitations. The presented algorithm specifically aimed to solve the problem of constructing a load balanced tree for data routing. The game theoretic approach implies designing of games in that, nodes as players have two roles as child and parent. In a child role, nodes are selfish players trying to gain more possible available bandwidth, while in parent role they act as cooperative nodes trying to decrease the load on their connected parents which are two-hop away nodes in the same level. Each parent is connected to its connected parents via nodes in one lower level.

Considering load heterogeneity of nodes opens a challenging area. Accordingly, taking heterogeneous nodes in terms of the amount of sending data into account is a solution for applications for which in-network aggregation is not possible. As this case is more general than the homogenous case, solving this problem will lead to a solution for all kinds of queries. Considering a few mobile nodes in the network will add some challenging dynamics to the problem, where some nodes move from one communication area to another. This will cause different load for both nodes which are located in the communication range of the mobile node and mobile node itself, when it goes through communication range of different nodes.

Although the presented algorithm is discussed for construction of routing tree, they can easily be adapted to support updating a portion of trees, i.e. sub-trees during runtime. Because in a long duration of query running, it is probable from a certain time, initial conditions based on which the construction is conducted change and consequently the current tree topology becomes non-optimum. However, reconstruction of the whole tree may not be beneficial, and fixing the problem in a sub-tree of the whole tree solves the problem. Also, nodes can have different communication ranges as another heterogeneity criterion. This difference may rise from different levels of residual energy, or deploying nodes with different cases. In both cases, importing different range of communication in the formulas looks practical.

REFERENCES

Akyildiz, I., Su, W., Sankarasubramaniam, Y., & Cayirci, E. (2002). Wireless sensor networks: A survey. *Computer Networks*, *38*, 393–422. doi:10.1016/S1389-1286(01)00302-4.

Bouabdallah, F., Bouabdallah, N., & Boutaba, R. (2009). On balancing energy consumption in wireless sensor networks. *IEEE Transactions on Vehicular Technology*, 58.

Charilas, D., & Panagopoulos, A. (2010). A survey on game theory applications in wireless networks. *Computer Networks*, *54*(18). doi:10.1016/j.comnet.2010.06.020.

Chu, Y., Tseng, C., Hung, C., Liao, K., Ouyang, C., Yen, C., et al. (2009). Application of load-balanced tree routing algorithm with dynamic modification to centralized wireless sensor networks. *IEEE Sensors*, (pp. 1392–1395). New Zealand.

Estrin, D., Girod, L., Pottie, G., & Srivastava, M. (2001). Instrumenting the world with wireless sensor networks. *International Conference on Acoustics, Speech, and Signal Processing (ICASSP)*, (pp. 2033–2036).

Frattasi, S., Fathi, H., & Fitzek, F. (2006). 4g: A user-centric system. *Special Issue on Advances in Wireless Communications: Enabling Technologies for 4G.*

Fudenberg, D., & Tirole, J. (1991). *Game theory*. MIT Press..

Gehrke, J., & Madden, S. (2004). Query processing in sensor networks. *Pervasive Computing, 3.*

Han, Z., & Liu, R. (2007). Non-cooperative resource competition game by virtual referee in multi-cell OFDMA networks. *IEEE Journal on Selected Areas in Communications*, *25*, 1079–1090. doi:10.1109/JSAC.2007.070803.

Harvey, N., Ladner, R., Lovasz, L., & Tamir, T. (2003). *Semi-matchings for bipartite graphs and load balancing*. Workshop on Algorithms and Data Structures (WADS 2003), Canada.

Lee, K., Lee, W., Zheng, B., & Winter, J. (2006). *Processing multiple aggregation queries in geo-sensor networks*. The 11th International Conference on Database Systems for Advanced Applications (DAS-FAA06). Singapore.

Madden, S., Franklin, M., Hellerstein, J., & Hong, W. (2002). Tag a tiny aggregation service for ad-hoc sensor networks. 5th Symposium on Operating Systems Design and Implementation. Boston, Massachusetts, USA.

Madden, S., Franklin, M., Hellerstein, J., & Hong, W. (2005). Tinydb, An acquisitional query processing system for sensor networks. *ACM Transactions on Database Systems*, *30*, 122–173. doi:10.1145/1061318.1061322.

Muller, R., & Alonso, G. (2006). *Efficient sharing of sensor networks*. IEEE International Conference on Mobile Adhoc and Sensor Systems. Vancouver, BC, Canada.

Munteanu, A., Beaver, J., Labrinidis, A., & Chrysanthis, P. (2005). Multiple query routing trees in sensor networks. *The IASTED International Conference on Databases and Applications (DBA05)*, (pp. 145–150). Innsbruck, Austria.

Niyato, D., & Hossain, E. (2007). Radio resource management games in wireless networks: An approach to bandwidth allocation and admission control for polling service in IEEE 802.16. *Wireless Communications*, *14*, 27–35. doi:10.1109/MWC.2007.314548.

Potdar, V., Sharif, A., & Chang, E. (2009). Wireless sensor networks: A survey. *WAINA*, *09*, 636–641.

Rosemark, R., Lee, W., & Urgaonkar, B. (2007). *Optimizing energy-efficient query processing in wireless sensor networks*. International Conference on Mobile Data Management, Mannheim, Germany.

Sadagopan, N., Singh, M., & Krishnamachari, B. (2006). Decentralized utility-based design of sensor networks. *Mobile Networks and Applications*, *11*, 341–350. doi:10.1007/s11036-006-5187-8.

Suris, J., DaSilva, L., Han, Z., & MacKenzie, A. (2007). *Cooperative game theory for distributed spectrum sharing*. IEEE International Conference on Communications.

Trigoni, N., Guitton, A., & Skordylis, A. (2006). Routing and processing multiple aggregate queries in sensor networks. *The 4th International Conference on Embedded Networked Sensor Systems*, (pp. 391–392). Colorado, USA.

Trigoni, N., Yao, Y., Demers, A., Gehrke, J., & Rajaraman, R. (2005). *Multi-query optimization for sensor networks*. The First IEEE International Conference on Distributed Computing in Sensor Systems (DCOSS 2005). Marina del Rey, CA: IEEE.

Uliman, I., Pomalaza, R. C., Oppernann, I., & Lehtomaki, J. (2004). *Radio resource allocation in heterogeneous wireless networks using cooperative games*. Nordic Radio Symposium 2004/ Finnish Wireless.

Wang, L., Xue, Y., & Schulz, E. (2006). *Resource allocation in multi-cell ofdm systems based on non-cooperative game*. 17th Annual IEEE International Symposium on Personal, Indoor and Mobile Radio Communications.

Yao, Y., & Gehrke, J. (January 2003). *Query processing in sensor networks*. The First Biennial Conference on Innovative Data Systems Research (CIDR). Asilomar, California.

Zhang, X., Yu, X., & Chen, X. (2005). *Inter-query data aggregation in wireless sensor networks*. International Conference on Wireless Communications, Networking and Mobile Computing (WCNM 2005). Wuhan, China.

KEY TERMS AND DEFINITIONS

Data Aggregation: Data aggregation is any process in which information is gathered and expressed in a summary form, for purposes such as statistical analysis.

Load Balancing: A technique to distribute network or computational load among multiple resources.

Network Lifetime: The time duration for a network to be operational.

Query Processing: A method to execute query on a dataset.

This chapter was previously published in Network and Traffic Engineering in Emerging Distributed Computing Applications edited by Jemal H. Abawajy, Al-Sakib Khan Pathan, Mukaddim Pathan, Mustafizur Rahman, and Mustafa Mat Deris, pages 275-301, copyright 2013 by Information Science Reference (an imprint of IGI Global).

Chapter 76
Study of Hybrid Sustainable Energy System Based on PEM Fuel Cells and Photovoltaic-Module Power Generator

Fatima Zohra Zerhouni
University of Sciences and Technology Mohamed Boudiaf, Algeria

Mansour Zegrar
University of Sciences and Technology Mohamed Boudiaf, Algeria

M'hammed Houari Zerhouni
University of Sciences and Technology Mohamed Boudiaf, Algeria

Amine Boudghene Stambouli
University of Sciences and Technology Mohamed Boudiaf, Algeria

ABSTRACT

The computer is the greatest innovation of the 20th century. It has changed our lives. It executes tasks with precision. There is no limit with what we can do with software. Computers are seductive. Companies and students cannot work without them. They help students to perform mathematical computations. It is very important that mathematical ideas are expressed in computer programs in order to have theoretical results and to verify them practically. Nowadays, the development of new and non-polluting energy producing and energy-storage systems is a great challenge for scientists. An alternative to the nuclear and fossil fuel power is renewable energy technologies. Due to ever-increasing energy consumption, rising public awareness of environmental protection, and steady progress in power deregulation, alternative (i.e., renewable and fuel cell based) distributed generation systems have attracted increased interest. There is an accelerating world demand for environmentally friendly power. Among the renewable energy sources, the Photovoltaic (PV) energy is the most promising candidate for research and development for large scale users. Fuel cells have been receiving a lot of attention lately due to their potential of becoming a new energy source with a large range of applications. Fuel cells can be incorporated with other components to create high efficiency industrial power plants. Fuel cells permit clean and efficient energy production. The purpose of the work is to optimize the system's operation. The main reason to build described system is to supply stand-alone systems using renewable energy sources. Therefore, the power plant has to produce energy independent of any weather fluctuations. Integrating photovoltaic energy

DOI: 10.4018/978-1-4666-4852-4.ch076

sources with fuel cells, as a storage device replacing the conventional lead-acid batteries, leads to a non-polluting reliable energy source. In this chapter, an energy system comprising different energy sources, namely PV and fuel cells, is proposed. Photovoltaic cells coupled with electrolytic devices can be used to produce hydrogen and oxygen in a sustainable manner. With the produced hydrogen from the electrolysis process, it is possible to generate electricity through fuel cells. Photovoltaic panels in particular can provide a good source of producing green electricity. It is autonomous, its operation does not pollute the atmosphere, and it is an inexhaustible and renewable source with great reliability. The simulation program developed also allows the exportation of different configurations. The experimental system described has permitted the validation of the proposed method.

INTRODUCTION

Computers are great resources to own. They are everywhere and society depends on them to manage our world. The computers have unlimited applications. Many of these applications are benefiting society directly or indirectly. There exist other functions of these computers, such as their use for running simulations and calculating large, complex data sets. In this present work, a hybrid system, particularly a photovoltaic fuel cell one, in Algeria, is studied. The fuel cell power supply is a very attractive option to be used with an intermittent power generation source like PV power because the fuel cell power system is characterized with many attractive features such as efficiency, fast load-response, modular production and fuel flexibility.

Photovoltaic (PV) has emerged in last decades since it has the previously mentioned advantages and less maintenance. Photovoltaic panels have received increasing attention over the last decade as one response to the challenges of global warming, increasing demand for energy, high fuel costs, and local pollution. Solar modules have the disadvantage of high initial installation costs. So, different methods have been proposed in the literature to increase the photovoltaic generator (SCA: Solar Cells Array) efficiency. The method, proposed here is based on selecting in real time, the SCA optimum configuration for a given load and given working conditions. This method is supposed to be particularly convenient for all direct coupling between a photovoltaic generator and a load system such as a pumping system. Experimental results strongly confirm the effectiveness of the proposed method. The system is an environmentally friendly solution since it tries maximising the use of a renewable energy source.

GENERAL SYSTEM DESCRIPTION

The alternative energy system that we propose to study in Algeria, more precisely in Oran city, is represented in Figure 1.

The overall system includes:

- Photovoltaic field containing photovoltaic polycrystalline silicon cells;
- Fuel cells stack;
- Gas storage equipment;
- Electrolyser;
- Stand-by battery; and
- Power Management and Control device (PMU).

The load is fed by a photovoltaic field. Electricity excess supplies thus the electrolyser producing hydrogen and the oxygen. The gases are stored for a length or a short term. In the event of insufficient energy, the PV generator is adopted to feed the load. A battery is added to the system in the transitory moments where fast load's variations occur. The energy management (PMU) equipment is used to control the power between each compo-

Figure 1. Overall system

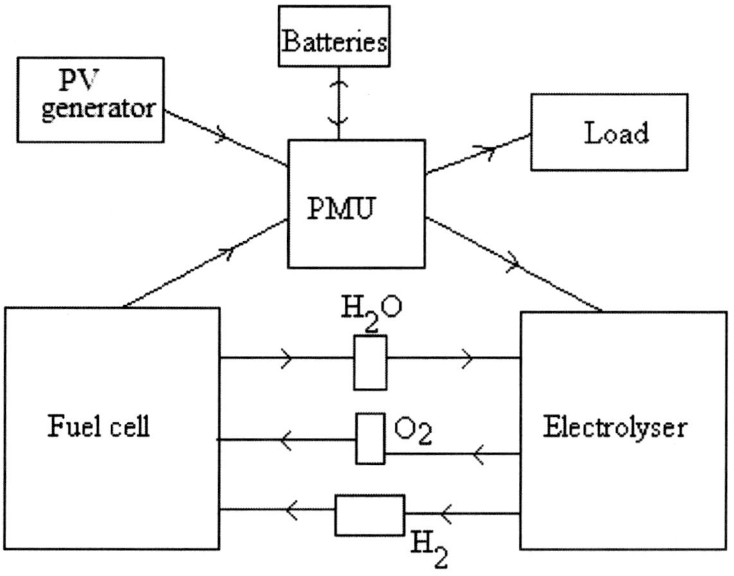

nent. A Proton Exchange Membrane (PEM) fuel cell is used to keep the system's reliability at the same level as for the conventional system while decreasing the environmental impact of the whole system. The PEM fuel cell consumes gases which are produced by an electrolyser to meet the user load demand when the PV generator energy is deficient, so that it works as an auxiliary generator (Busquet, et al., 2004). The experimental system used a microcontroller. It is robust when undergoing environment changes and load variations. The performance of the methods is verified through simulations and experiments. The experimental system used a microcontroller. It is robust to environment changes and load variations.

PHOTOVOLTAIC SOURCE

Introduction

The fact that oil is running out fast, alternatives should be adopted. Renewable energy is one of the most promising solution to the above problems (Boudghéne, 2011a). In particular, photovoltaic panels can provide a good source of producing *clean* electricity (Boudghéne, 2011b, 2011c). The most attractive features of photovoltaic panels are the non-existence of movable parts, the very slow degradation of the sealed photovoltaic cells, the flexibility in the association of modules, and the extreme simplicity of its use and maintenance. In addition, photovoltaic energy is a very relevant source. It is autonomous, its operation does not pollute the atmosphere, and it is an inexhaustible and renewable source with great reliability. To deal with these problems have been proposed several methods (Salamah, et al., 1990a; Applbaum, 1989; Zerhouni, 1996; Weidong, et al., 2007; Zerhouni, et al., 2008). Sun trackers have been utilized to face the PV system to the direction of the sun so that maximized power at any point in time is achieved. The Maximum Power Point Tracker (MPPT) method can increase the efficiency and the performance of the PVG (Photovoltaic Generator) (Enrique, et al., 2007; Zerhouni, et al., 2010; Mutoh, et al., 2007; Zerhouni, et al., 2005; Kwon, et al., 2006; Zerhouni, 2006a).

The methods vary in complexity, convergence speed, and cost (Zerhouni, et al., 2006b; Salameh, et al., 1990b; Benmessaoud, 2010).

The method presented in this chapter is extended to any load and based on reconfiguring on line the PVG.

Fundamentals

The equivalent electrical circuit of a single cell is represented in Figure 2 (Busquet, 2003; Appelbaum, 1989; Zerhouni, 1996).

In Figure 2, we can find the following parameters:

- The photocurrent generator I_{ph} which is compared to the short circuit current I_{cc}. Its value is proportional to the solar radiation
- The diode D symbolizing the junction PN.
- The serial resistance Rs which takes into account contact's and connection's resistances.
- The parallel resistance Rsh representing the leakage current at the junction.
- The diode's voltage V_d
- The diode's forward current I_d

The simple equivalent circuit is sufficient for most applications. Kirchhoff's first law provides the current I–voltage V characteristics of the simple photovoltaic cell equivalent circuit. From Figure 1, we can drive Equation 1 which gives the cell current I function of the cell voltage V.

Figure 2. Photovoltaic cell's equivalent model

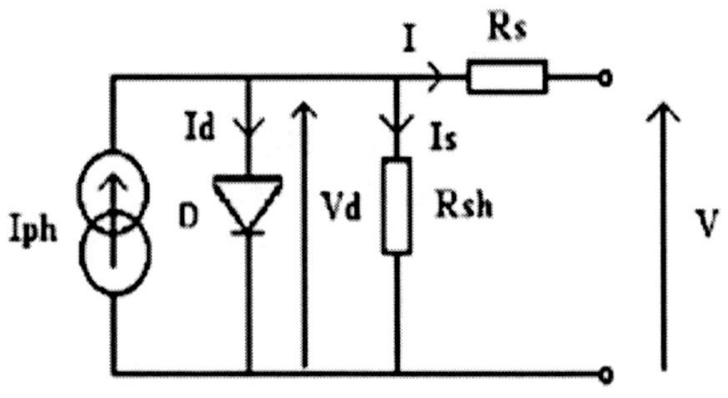

$$I = I_{cc} - I_o\left[\exp\frac{(V + R_s I)}{A.U_T} - 1\right] - \left(\frac{V + R_s I}{R_{sh}}\right) \tag{1}$$

where I_o is the saturation current, which varies with the temperature; A is the ideality factor. It depends on the semiconductor used and its nature (single-crystal, polycrystallin, amorphous...); and U_T is thermal potential.

$$U_T = \frac{K_B T}{q} \tag{2}$$

where K_B is Boltzmann constant; T is junction temperature in Kelvin degrees; and q is electron charge.

If we neglect Rsh which presents an important value, Equation (1) becomes:

$$I = I_{cc} - I_o\left[\exp\frac{(V + R_s I)}{A.U_T} - 1\right] \tag{3}$$

The Characteristic of a Photovoltaic Module

The effective solar radiation comprises the spectral components such as wavelength of the solar spectrum, which the photovoltaic cells convert into electrical energy. A photovoltaic system is an uncontrolled power source whose behavior is quite different from conventional power supplies. Solar cells naturally exhibit a nonlinear characteristic, which vary with the solar irradiation and cell temperature. In the climatic conditions of solar radiation (1000W/m^2 corresponding to E_s=100%), temperature T=25°C, the characteristic of the current according to the voltage I=f(V) of a photovoltaic module, as given by Equation (1) is illustrated by Figure 3.

From Figure 3, we can highlight three relevant parts (Zerhouni, 1996):

Figure 3. I-V characteristics of a photovoltaic module

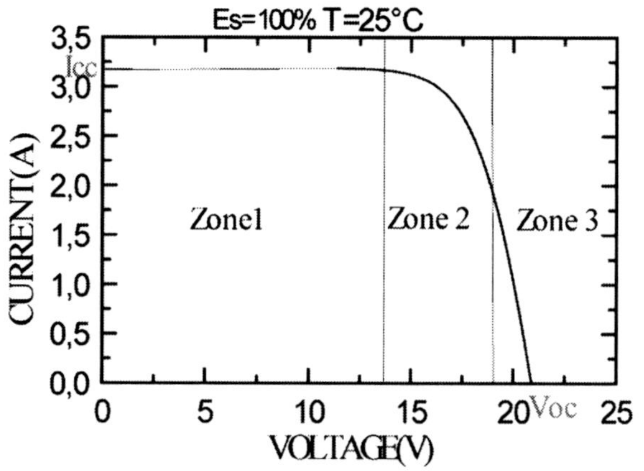

- Part 1 where the current remains almost constant for voltage variations. In this area, the photovoltaic module is comparable to a current generator.
- Part 2 which is characterized by a current fluctuation corresponding to an appreciably constant tension. In this area, the photovoltaic module is a voltage source.
- Finally, part 3 has the feature of being to be an intermediate zone between an operation of the current generator and the voltage generator. It is affirmed by an evolution, very significant of the output variables (I, V). This area includes a point M of co-ordinates $I_{opt,}\ V_{opt}$ delivering the optimum power P_{opt}.

The PVG electrical characteristics are essentially affected by the solar radiation and the temperature (Zerhouni, 1996, 2006c).

The Connection PVG-Load

It is the simplest connection without any system between the PVG and the load. This system is represented by the Figure 4, which directly associates the SCA to a load (Busquet, et al., 2004; Applbaum, 1989; Zerhouni, 1996).

In a directly coupled system it is important to match the load to the photovoltaic generator otherwise the power loss resulting from this mismatch would lead to a lower efficiency of the overall system. To overcome this problem different methods have been proposed (Zerhouni, 2009).

PV arrays are built up with combined series/parallel combinations of PV solar modules (Salameh, et al., 1990a, 1990b). The number of modules required to build a SCA depends on the power required and also on the modules types used and the operating conditions (photovoltaic radiation, temperature, etc).

The method presented in this chapter is extended to any load and based on reconfiguring on line the PVG by changing the connections between different modules in order to minimise these losses. We determine the current Icom and the voltage Vcom corresponding to the commutation point against the photovoltaic radiation Es and/or the temperature T change. Consequently, the global system performance is improved.

Commutation Factors

We consider the case where a load is connected to a PVG. The commutation from a configuration to another can occur when the load change or when the climatic conditions change. In Figure 5, in the case of two configurations $N_{p1}\times N_{s1}$ noted "parallel configuration" and $N_{p2}\times N_{s2}$ noted "series configuration" we assume a constant load of characteristics

Figure 4. The direct connection PVG-load

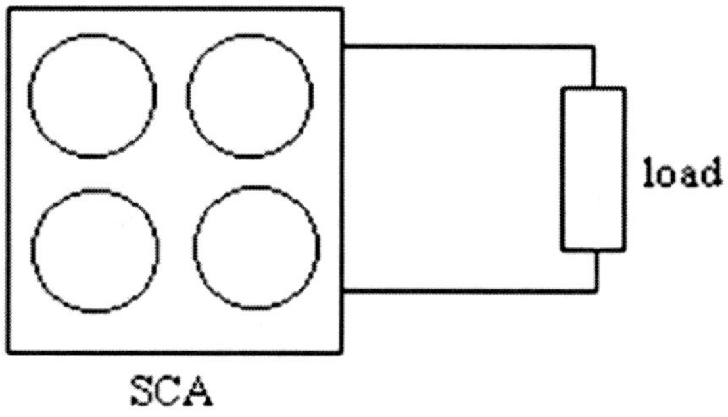

P-V noted 'Z,' a constant temperature and a change of solar radiation (insulation) E_s equals to 10% which corresponds to 100W/m², 50% (500W/m²), 100% (1000W/m²). For example, at E_s=10%, if the parallel configuration is adopted the operating power point is P_b, if the series one is adopted the operating power point is P_a, but $P_b>P_a$ then it is necessary to adopt the configuration parallel because the power extracted by the load is the best. If the solar radiation varies, the operating point for example at E_s=100% is P_c because the configuration is parallel. P_d is the power if the series one is adopted. In this case $P_d>P_c$ then it is necessary to switch to the series one for a good exploitation of the PVG (Zerhouni, 1996, 2009).

The other case is for the constant climatic conditions, a load variation. As illustrated in Figure 5b, the load is noted Z_1. If the parallel configuration is adopted the load Z_1 has an operating point of power P_A. P_B is the power extracted by Z_1 if the series configuration is adopted, or $P_A>P_B$ so the configuration parallel is the best: it is adopted. A load variation suddenly occurs, the load becomes Z_2, and the operating point is then P_c because the parallel configuration is maintained. The series configuration provide to the load P_D with $P_D>P_c$. Therefore, a switch from the parallel configuration to the series one is necessary to provide a best power to the load.

If N_T is the total photovoltaic modules used:

$$N_T=N_s \times N_p \tag{4}$$

where N_p is the number of branches in parallel; and N_s is the number of modules in series in a branch.

The points P_1, P_2, P_3 in Figure 5a and *P'* in Figure 5b are the intersect points of the P_V characteristics for series and parallel configurations under the given climatic conditions. In these points for example *P'* these configurations have equal power. These points are the reference. So in Figure 5b as example, for any load with operating point situated at the right of *P*, configuration 2 ($N_{p2} \times N_{s2}$) will be used. On the other hand, for any load with operating point situated at the left of *P'*, configuration 1 ($N_{p1} \times N_{s1}$) will be used.

In our work, N_T=6 with the configurations: parallel N_{p1}=2 N_{s1}=3 and Series N_p=1 N_s=6. At T=25°C E_s=100%, the *I-V* characteristic (current-voltage) is in series N_{p2}=1 N_{s2}=6

$$I = 3.192 - 1.1295e^{-7}\left[\exp\frac{(V+2.4I)}{2.842e^2.U_T} - 1\right] \tag{5}$$

parallel N_{p1}=2 N_{s1}=3

$$I = 6.384 - 2.25e^{-7}\left[\exp\frac{(V+0.6I)}{1.421e^2.U_T} - 1\right] \tag{6}$$

The idea to improve the PVG operation, extended to any load is based on the following principle: with a number of modules N_T is constant

Figure 5. The commutation factors

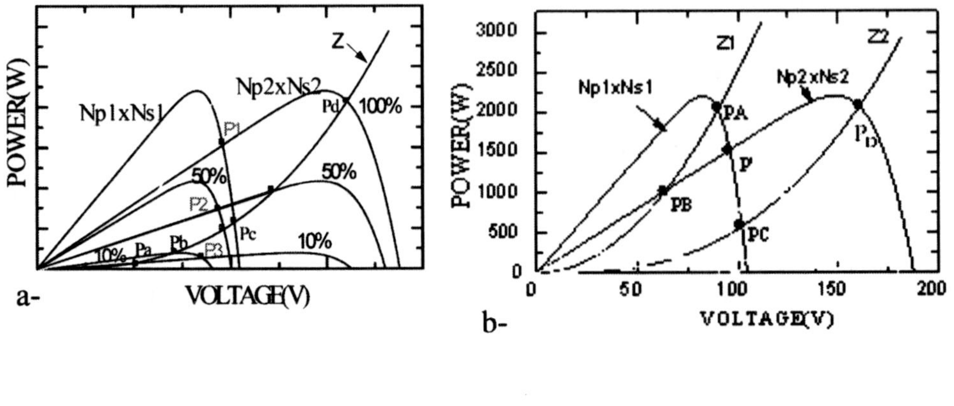

which satisfy the load demand, one sets either two or three configurations.

A simulation program developed calculates at each photovoltaic radiation the co-ordinates of the commutation points P_{com} (I_{com}, V_{com}) of the characteristics current-voltage $I=f(V)$ or power-voltage $P=f(V)$ of the various arrangements.

The points 1-5 of curve noted 'C1' on Figure 6a or P-V noted 'C2' on Figure 6b are calculated at each solar radiation Es and temperature T. In Figure 6 the solar radiation varies from 10% to 100% at T=25°C. They are independent of the constraint load but are a function of the climatic conditions and the selected configurations. They are the marks of commutation towards the suitable configuration.

The commutation mechanism from one configuration to the other, independently of the load constraint will use the commutation points in curve 'C1' or 'C2' for which the two configurations have an equality in power for the given conditions of temperature and the photovoltaic radiation. Each point on the curve C1 is defined by its current Icom and its voltage Vcom. It will be a matter of comparing the load voltage Vfct and/or current Ifct to Vcom and/or Icom and switch the PVG onto appropriate configuration (Zerhouni, 1996, 2009).

The Complexity of the Switches Number

To allow the commutation passage from one configuration to the other one, we use switches. By control the closing of some and the opening of others, one can obtain the desired configuration. In the practical case, it was observed that to allow only 2 changes of the system configuration is sufficient. The number of switches, for a specific number N_T of photovoltaic modules follows especially the configurations considered. These last are determinated according to the particular needs for the equipment to supply and the modules total number. If two modules are adopted, there will be 2 configurations: $N_p \times N_s$=1×2: the series configuration and the parallel one $N_p \times N_s$=2×1. As shown in Figure 7, this choice leds us to the use of 3 switches: Rels allows the series configuration. Relp, Relp' allow the parallel one.

As for N_T=4, one can have 3 configurations (parallel $N_p \times N_s$=4×1, mixed $N_p \times N_s$=2×2, series $N_p \times N_s$=1×4) as shown in Figure 8.

For this case, for the choice of:

- Parallel and series configurations, one needs 9 switches,
- Parallel and mixed configurations, one needs 6 switches,

Figure 6. Different characteristics

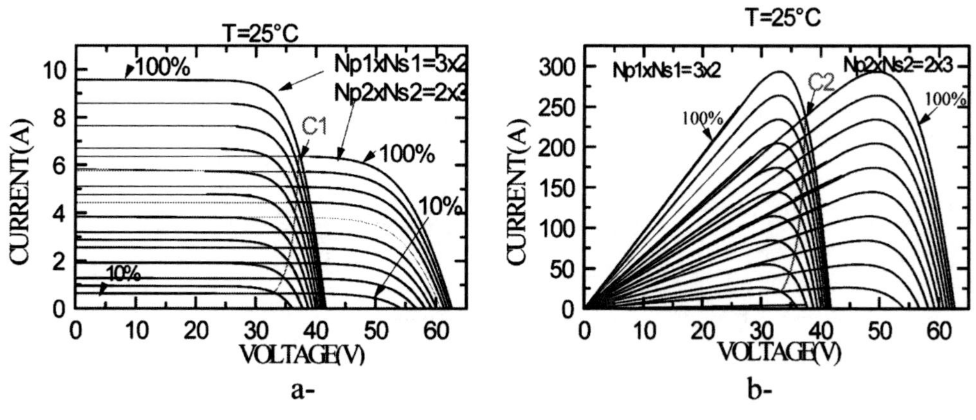

Figure 7. Case of N_T=2

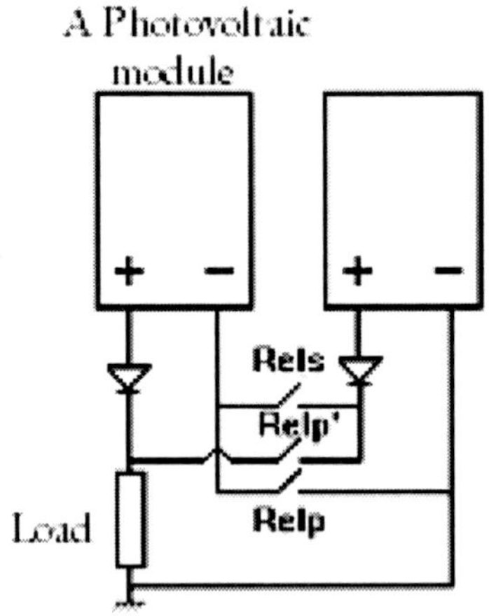

- Series and mixed configurations, one needs 4 switches,
- Parallel, mixed, and series configurations, one needs 9 switches.

These results show that for the photovoltaic modules same total number, the switches number to be auctioned differs according to the selected configurations and not only to the total number of modules used.

Commutation with many configurations can be characterized by a switches significant number. For example, 9 switches for the three configurations of the 4 photovoltaic modules. If we tolerate a change between several configurations, we must make a compromise between the number of switches used and the power's gain obtained. In the case of our study, we will tolerate only the change of 2 configurations with N_T=6.

To allow the opening or closing of the switches used in order to have the required commutation, one must have a reference mark which can be either the photovoltaic radiation (Zerhouni, 2009) or another reference. The photovoltaic radiation reference mark was already studied (Zerhouni, 2009). The drawback of this method is that we must change this reference when the load varies. It is not applicable on real time and for any load. The object of the next study is the our reference mark utilization which is applicable for any load and climatic conditions. This mark is the points on curve C1 in Figure 6a (Zerhouni, 2009).

The Variations of the Commutation Points

The commutation points (I_{com}-V_{com}) vary with E_s and T as it is illustrated in Figure 9. In Figure 9, one represents these points variation for Es=10% to 100% at T=25°C and extreme temperatures of 70°C and -5°C. As observed, these points are not constant and varie if the climatic conditions change. The important deduction is that the current Icom remains constant in spite of the temperature changes but it is affected by the solar radiation variations. V_{com} is affected by both E_s and T variations. Computer simulations were made to investigate this method. A simulation program developed calculates at each photovoltaic radiation

Figure 8. Case of N_t=4

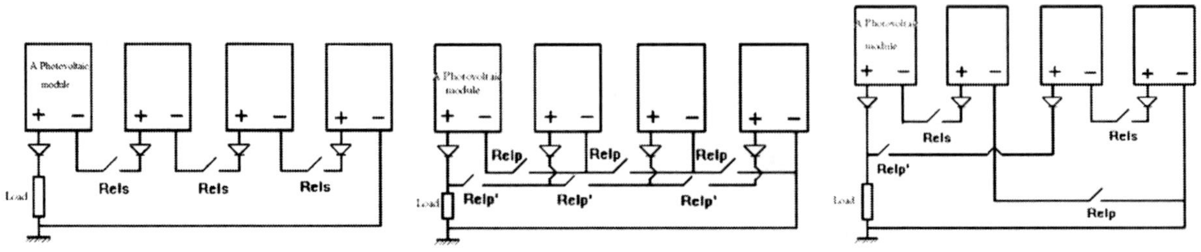

a- b- c-

Figure 9. Variation of the commutation points

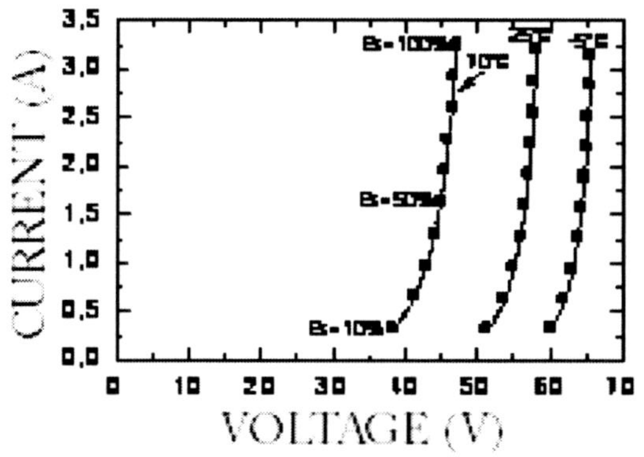

the commutation points co-ordinates P_{com} (I_{com}, V_{com}) of the current-voltage characteristics $I=f(V)$ of various arrangements (Zerhouni, 2009, 2010).

The New Approach using the Commutation Current I_{com}

To concretize the idea of a configuration commutation from one towards the other without any influence of the load, our attention is particularly focused on the current I_{com}. The case considered 6 modules in $N_p \times N_s = 2\times3$ named parallel configuration and $N_p \times N_s = 1\times6$ named series one. The results can be applied to another number of modules. We are interested in the Icom variations according to climatic conditions. A simulation program has been developed. The I_{com} layout according to the photovoltaic radiation for a constant temperature of 25°C is represented in the Figure 10.

It is noticed from Figure 10 that Icom is linear according to the photovoltaic radiation. It can thus be written:

$$I_{com}=K_i.E_s \tag{7}$$

where K_i is constant which depends on the adopted configurations and the modules selected number. For our example: K_i=0.032 A.m²/w.

Figure 10 also shows the simulation results of I_{com} versus T for the configurations chosen of the 6 modules. From Figure 9, it is clear that the commutation current varies linearly with the photovoltaic radiation and hardly affected by the temperature. The simulation program has been run for other different configurations and the results led to the same conclusion.

To work with I_{com} linearly proportional to the photovoltaic radiation and slightly depend on the temperature will simplify the practical realization according to the relation:

$$I_{com}=K_i.E_s, \forall T \tag{8}$$

E_s is deduced from I_{cc} which varies linearly with it.

For a load supplied by a photovoltaic generator, one is interested in the operation current position *Ifct* compared to I_{com}. The control system has been implemented. A variable load has been used in order to cover different operating conditions. The practical realization has been based on the following flowchart in Figure 11.

The experimental characteristics for E_s=24%, T=27°C are illustrated in the Figure 12. The Figure 12a is related to the commutation from parallel to series configuration. The commutation occurs at '*A*.' The Figure 12b illustrates the commutation from series to parallel configuration, but the major drawback was summarized in the switch from the series towards the parallel configuration never obtained. The commutation does

Figure 10. I_{com} variation against Es

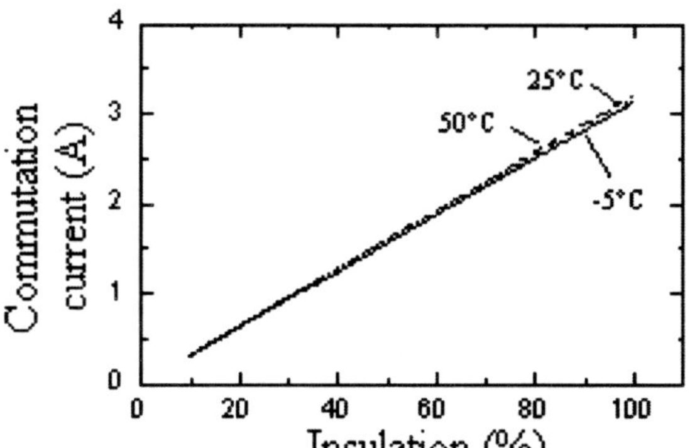

Figure 11. The implementation flowchart follows using I_{com}

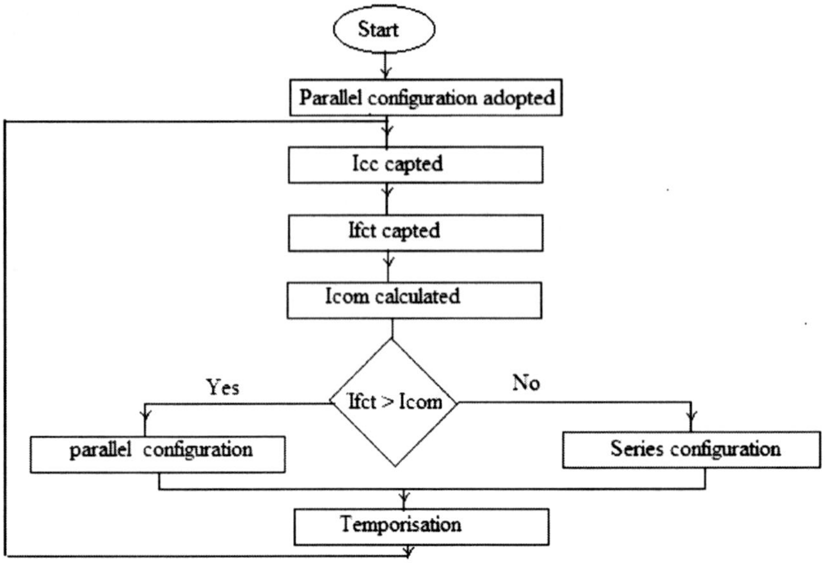

not occur where it should be or later. Indeed, the difference between *Ifct* and I_{com} is very small in the flat region of the (I, V) characteristic of the PVG (current generator). Therefore, the commutation using Icom is not appropriate.

The New Approach using the Commutation Voltage V_{com}

Instead of considering the currents, one is interested in the voltages. For a load supplied by photovoltaic way, one is interested in the operation voltage position *Vfct* compared to V_{com}. Computer simulations were made to investigate this idea. The variation of V_{com} as a function of the photovoltaic radiation for different temperatures has been determined and is illustrated in Figure 13a. As it can be seen the characteristic is non linear at low photovoltaic radiation level, so a simple linear procedure has been applied. As a consequence for a given temperature the commutation voltage V_{com} could be written as follows:

$$V_{com}=a_i.E_s+b_i \tag{9}$$

where, a_i and b_i are constants, which are derived from the linear procedure at 25°C. The Figure 13a also shows that the commutation voltage V_{com} is significantly affected by the temperature T. As shown in Figure 13, as T increases V_{com} dimin-

Figure 12. Experimental results using I_{com}

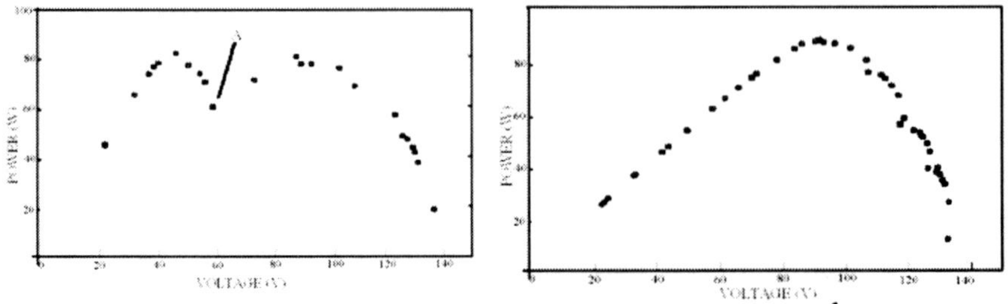

a- b-

Figure 13. Vcom variation against climatic conditions

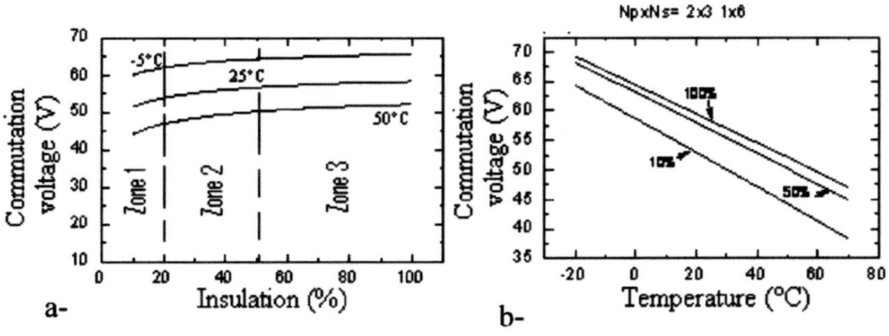

ishes linearly. To implement the reconfiguration method using the commutation voltage V_{com} as reference, (9) should be corrected in order to take into account the linear variation of V_{com} as a function of the temperature. After modification (9) becomes:

$$V_{com}=a_i.E_s+b_i-c_i.(25-T) \quad (10)$$

where a_i and b_i are as defined previously and c_i is the slope of the curve $V_{com}=f(T)$ at 25°C.

The control system based on Equation (10) has been implemented following the flowchart in Figure 14 and an example of the practical results is given in Figure 14. A variable load has been used in order to cover different operating conditions. One distributes the characteristic $I=f(V)$, $P=f(V)$ in three noted zones Z1, Z2, Z3, as the Figure 13a.

The calculation of the coefficients for the example of six modules led to the following values:

E_s€[0,20%]; a_i=0.23V.m²/W; b_i= 49.19V; c_i=-0.288V/°C

E_s€[20%,50%]; a_i= 0.09V.m²/W; b_i= 52.21V; c_i= -0.27V/°C

E_s€[50%,100%]; a_i=0.03 V.m²/W; b_i= 55.21V; c_i=-0.253V/°C

Figure 14. The implementation flowchart using V_{com}

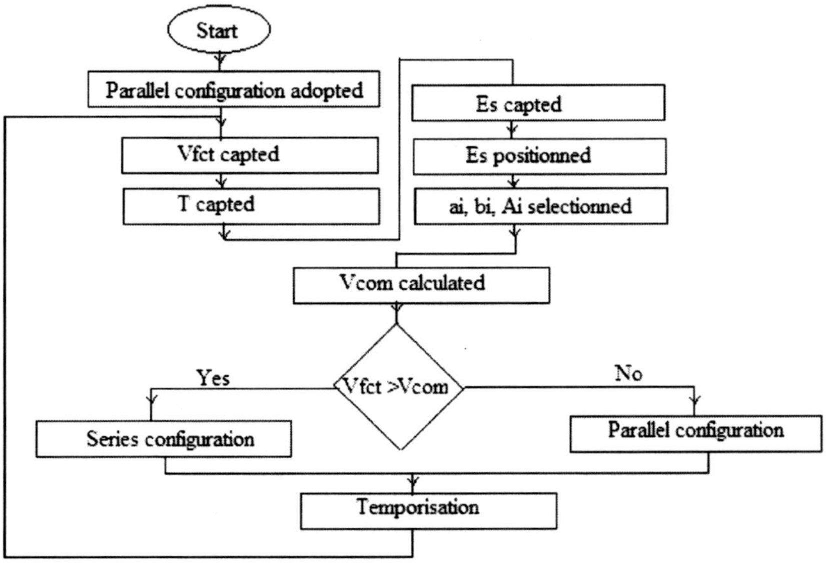

The implementation follows the flowchart given in Figure 14.

The experimental switching characteristics for E_s=24%, T=27°C are illustrated by the Figure 15. The Figure 15a is related to the commutation from parallel to series configuration. The commutation occurs at '*Pps*.' The Figure 15b illustrates the commutation from series to parallel configuration. The commutation occurs at '*Psp*.' But, the major handicap is that '*Pps*' differs from '*Psp*.' The zone of descent of the curve *I* - *V* (current - voltage) of the parallel configuration is similar to a generator of voltage that presents a stiff slope. The difference in voltage between the voltages in this area is weak and the detection of the threshold of commutation is delicate. Thus, the commutation will be erroneous during the passage from the parallel configuration to the series one.

The New Approach using I_{com} and V_{com}

In order to avoid the cited problems, the solution is to use both I_{com} and V_{com}. An experimental hardware has been implemented in the laboratory as shown in Figure 16.

Figure 15. Experimental results using V_{com}

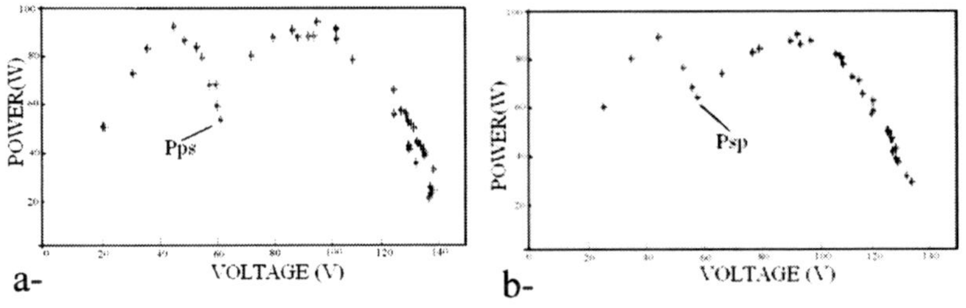

Figure 16. Experimental system synoptic

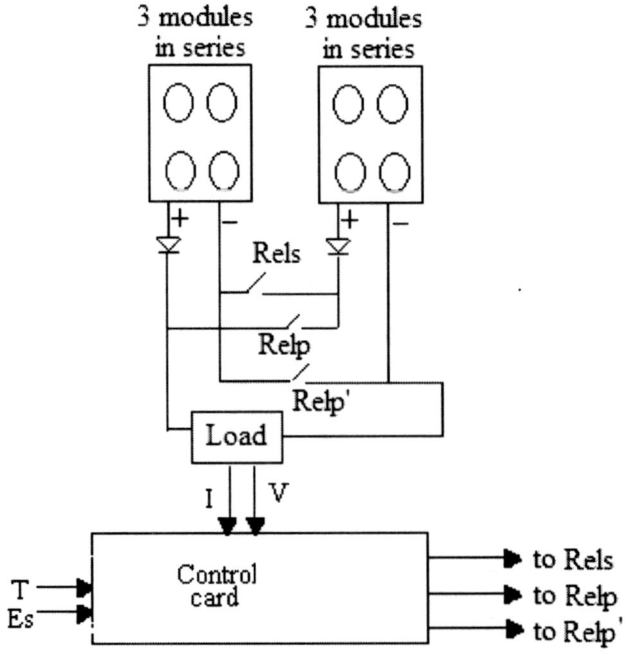

The panels are composed of 3 modules in series. To allow the commutation passage from one configuration to the other one, we use switches Rels, Relp, Relp'. By control the closing of some and the opening of others, one can obtain the desired configuration automatically, in real time. The control card is based on microcontroller. The control card permits the insulation Es and the temperature T measure, to determine then V_{com} or I_{com} using Equations 8 and 10. If the SCA is in the parallel configuration then the current flowing into the load is measured and compared to I_{com}. If it is higher, the SCA remains in its actual configuration otherwise it is reconfigured into the series configuration (Relp, Relp' opened and Rels closed). If however the SCA is in the series configuration, then the load's voltage is measured and compared to V_{com}. If it is greater, then the SCA remains in its actual configuration otherwise it is switched onto the parallel configuration (Relp, Relp' closed and Rels opened). As a result, the SCA is always switched onto its best configuration for any load and any weather conditions, on line. A variable load has been used in order to cover a large currents and voltages range.

The flowchart in Figure 17 shows the principle of the method.

The experimental results for E_s=24%, T=27°C are illustrated in Figure 18. Figure 18a shows the commutation at '*Pps*' from the parallel to the series configuration. The Figure 18b shows the commutation at '*Psp*' from the series to the parallel one. The commutation occurs precisely at the same point. This point coincides too with

Figure 17. Implementation flowchart using V_{com} and I_{com}

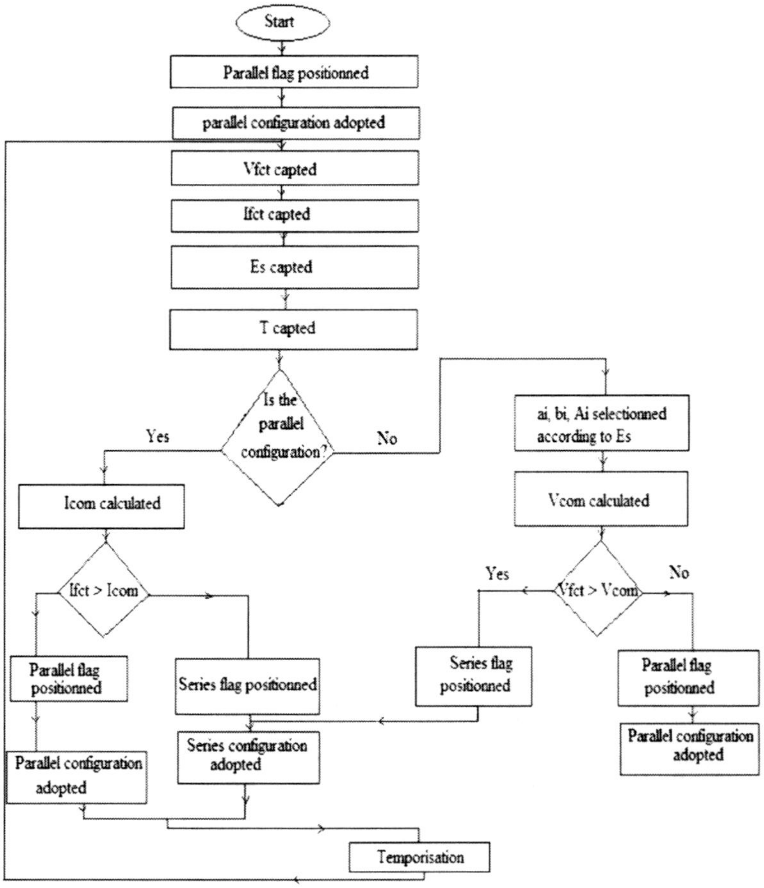

Figure 18. Experimental results using I_{com} and V_{com}

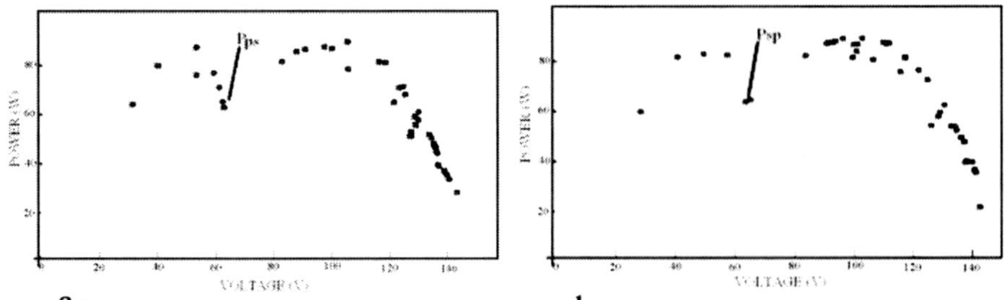

a- b-

'P_{com}' that obtained by the intersection of the layout of the characteristic series alone and that parallel in Figure 19. There is no hysteresis phenomenon observed. Consequently, this idea is considered adequate. The PVG is always switched onto the best configuration for any load and any weather conditions.

The Uncertainty Analysis

Figure 19 illustrates the voltage commutation real curves and the approximate one for the chosen configurations (N_pxN_s=1×6 and 2×3) at solar radiation from 10% to 100%. Figure 20a is for T=25°C. Figure 20b is for T=-5°C. Figure 19c

Figure 19. Series and parallel experimental results

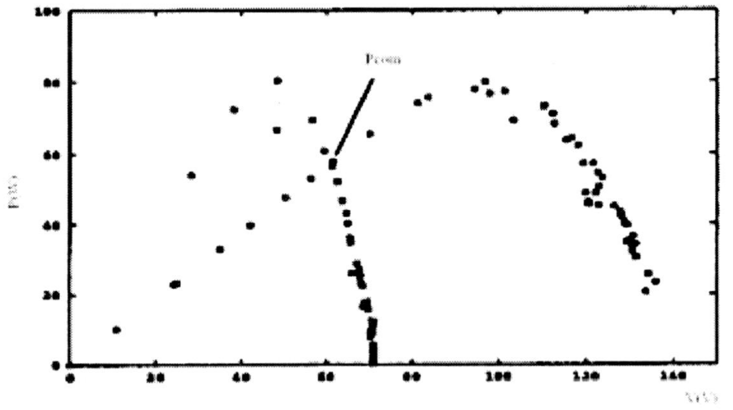

Figure 20. Commutation voltage curves

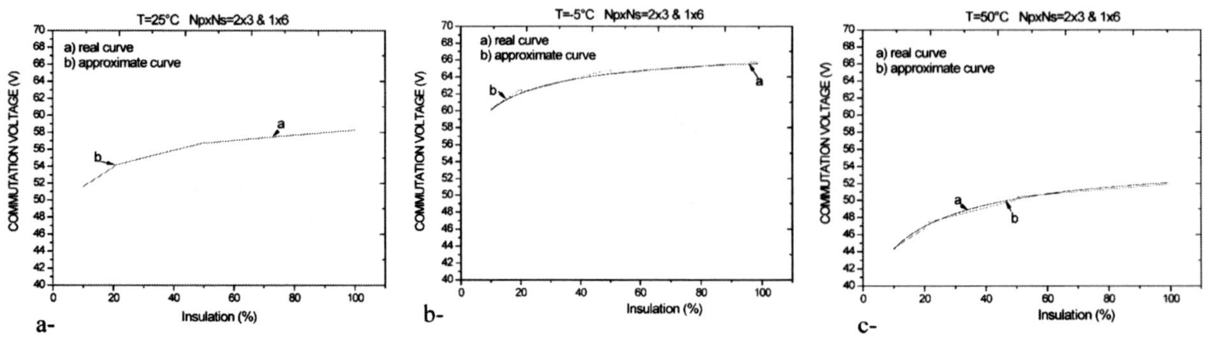

a- b- c-is for T=50°C. The curves noted 'b' in Figure 20 used the relationship (10). As seen in Figure 20, the curves are overlaped each other. The mean square error EQM calculus obtained is:

EQM=0.007 at 25°C.
EQM=0.027 at -5°C.
EQM=0.047 at 50°C.

Figure 21 represents the current real and approximate curves obtained for the adopted configurations at several solar radiations from 10% to 100% and temperatures of -5°C, 25°C and 50°C. In Figure 21, the approximate curve using Equation (8) is noted "approximate curve." As seen in Figure 21, the curves are overlapped each other whatever is the temperature and these curves are closed to the approximate one.

The mean square error EQM calculus for this case is:

For -5°C, EQM=0.004;
For 50°C, EQM=0.002;
For 25°C, EQM=0.001.

Uncertainty analysis has been made on some other experimental works for different modules number and configurations and was assessed to be around 2.5% for the commutation voltage and 0.002% for the commutation current.

Figure 21. Commutation current curves

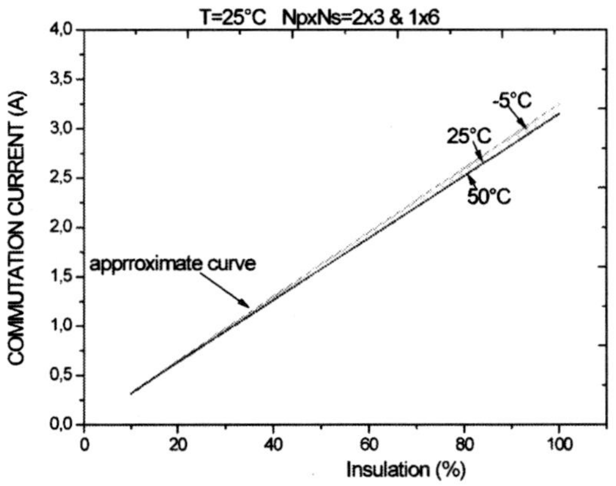

Conclusion

From this combined study, the use at the same time of Icom and Vcom is the best solution. Icom is a reference to allow the commutation from the parallel configuration towards the series one; Vcom is to allow the reverse commutation from series configuration towards the parallel one. This method is in real time, on line independently of the load and climatic conditions. This technique has been concretized in our laboratory. It was tested on a resistive load and a DC motor. The method of configuration change on line, in real time is a promising method since it guarantees the most power to the load. Thus, advantage of power is transmitted to the motor, more is the rise of the fluid through pump, and most important is the depth of water pumping reached. Another advantage of this configuration change is that a system motor-pump can sometimes begin its pumping earlier for such configuration contrary to another according to the current request. One can commutate towards another configuration if this last provides more power. To tolerate a configuration change will allow a PVG good exploitation and improve its operation. The technique based on the change of configurations is a simple technique, at moderate price.

FUEL CELLS

Fuel cells, although they have existed for more than 50 years, are just starting to capture the general public interest. Fuel cells promise clean and efficient operation along with high energy densities (Boudghéne, 2011a). While there are many different fuel cell types (Zerhouni, 2009; Boudghéne, et al., 2002a), the proton exchange membrane fuel cell stands out as one of the most promising for transportation and small stationary applications (Boudghéne, et al., 2002b).

The Proton Exchange Membrane or PEM fuel cell is a compact type of hydrogen fuel cell that has a relatively fast electrical response time and a low operating temperature and pressure (Busquet, 2003).

Polymer Electrolyte Membrane (PEM) fuel cells have received much attention, in scientific and industrial research (Boudghéne, 2011a). Proton Exchange Membrane (PEM) fuel cells take hydrogen and air and produce power and water. The mechanism is exactly opposite to that of electrolysis (Boudghéne, 2003).

$H_2 \Rightarrow 2H^+ + 2e^-$ Anodic half reaction

$\frac{1}{2}O_2 + 2H^+ + 2e^- \Rightarrow H_2O$ Cathodic half reaction

$$H_2 + \frac{1}{2}O_2 \Rightarrow H_2O \text{ The overall reaction} \quad (11)$$

Due to the increasing and highly promising use of fuel cells as an important electrical energy source, its modeling has been a major research issue, trying to find mathematical models to accurately predict how a cell will behave in a real system. The overall goal is the verification of fuel cell characteristics to provide a diagnostic tool to ensure the safe operation of the system. The model is based on the empirical electric model developed by NR Albrieux (Busquet, 2003). The used model is illustrated in Figure 22.

The characteristic equation is given by:

$$Vcell = E_{oc} + \frac{b}{Ln\left(\frac{J}{J_d \times e^2}\right)} + \left(\frac{b}{4 \times J_d} - \Delta\right) \times J \quad (12)$$

Vcell is the cell's potential, J the current density, E_{oc} is the cell's open circuit voltage, J_d and Δ are model's parameters. This equation takes into account the diffusion problems which exist for strong current densities, but the zone where the diffusion problems are dominating is not modelled by this model.

Characterization of the fuel cell polarization curves based on pressures, temperature allows analysis and simulation of the fuel cell power generation. The parameters Eoc, B, Jd, and Δ depend on the temperature T and on the partial oxygen pressure Po2. Each parameter is dissoci-

Figure 22. Fuel cells simulink model

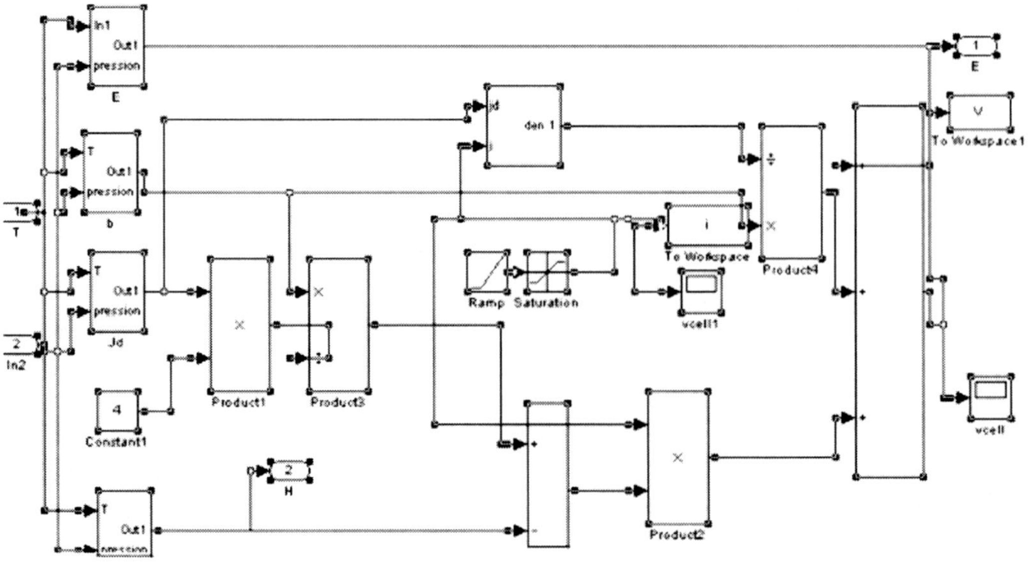

ated in three components according to the following equation:

$$\begin{bmatrix} E \\ Jd \\ b \\ \Delta \end{bmatrix} = \begin{bmatrix} E1 & E2 & E3 \\ Jd1 & Jd2 & Jd3 \\ b1 & b2 & b3 \\ \Delta 1 & \Delta 2 & \Delta 3 \end{bmatrix} \times \begin{bmatrix} 1 \\ T \\ T \times \ln\left(P_{O2}\right) \end{bmatrix} \qquad (13)$$

The pressure Po2 is maintained constant at 1.4 Bar (Figure 23a). For various temperatures, the temperature improves the fuel cell performances as seen in the figure. In addition, the temperature is maintained constant at T=35°C with various pressures. The simulation results are described in the Figure 23b. The partial oxygen pressure increase increases the cell's voltage V for the case of the strong densities J. The temperature's influence is comparable to that of the partial oxygen pressure,. It improves the fuel cell's performances (Zerhouni, 2009; Busquet, 2003).

ELECTROLYSER

Electrolysis is the process of decomposition of a substance into its chemical components by an electric current. An electrolyser produces both hydrogen and oxygen gas for use in a fuel cell by the process of electrolysis. In the process of electrolysis, the atoms that comprise a water molecule, hydrogen and oxygen, are disassociated from each other. This is the only method of hydrogen production with no requirements for fossil fuels. In addition, with polymer membrane technology it is possible to have an electrolyser with a solid polymer membrane. In a solar hydrogen system, the electrolyser is the component, which changes electrical energy from the photovoltaic panels into hydrogen gas. Actually, if properly designed, the same PEM unit can be operated as an electrolyser to produce hydrogen, as well as hydrogen driven fuel cell to produce electricity. Hydrogen storage is chosen for its potential to make a clean energy option and low cost in construction.

The basic difference between the electrolyser and PEM fuel cell is that the hydrogen and oxygen reactions on the electrodes are opposite. Therefore, the electrolyser produces H2 at the cathode and O2 at the anode, while the fuel cell consumes H2 at the anode and O2 at the cathode. Also, the water's decomposition into H2 and O2 in the electrolyser is non-spontaneous and requires an electric power for the reaction to occur, while the opposite fuel cell reaction is spontaneous and produces an electric power. In electrolysis, an electric current is run through water and the water reacts to form hydrogen and oxygen.

Figure 23. Simulation results

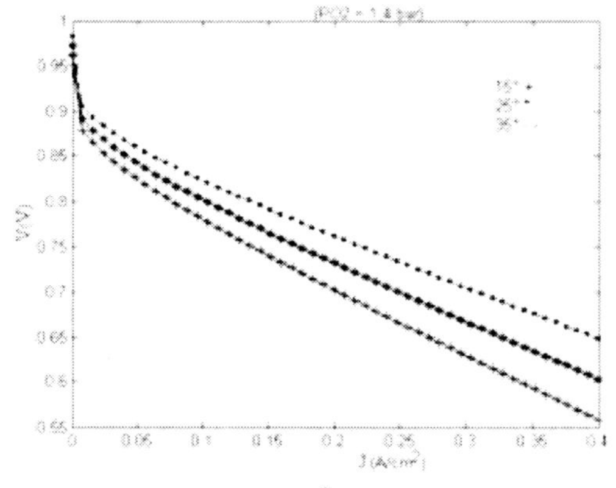

a-

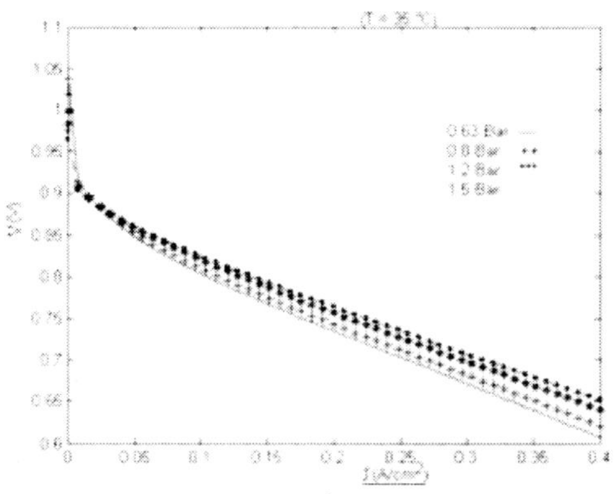

b-

The Simulink model developed is the same illustrated in Figure 22 with other constants values. The parameter pressure is kept constant (Po_2 = 9.6 Bar). For various temperatures (Figure 24a), it is noticed that the tension decreases when the temperature increases. The temperature's influence increases with the intensity. The electrolyser's performances are thus a temperature function. And when the temperature is maintained at a constant value of T=53°C, for example, with various pressures (Figure 24b), it is noticed that the pressure increase improves the electrolyser's performances (Busquet, 2003; Zerhouni, 2009).

CONCLUSION

Nowadays, the development of new and non-polluting energy producing and energy-storage systems is a great challenge for scientists. Hybrid power systems are based on renewable energy sources and especially on photovoltaic systems. Among the renewable energy sources, the Photovoltaic (PV) energy has been widely utilized in low power applications. The electricity generated from solar cells is one of the best options for the sustainable future energy requirements of the world. It is also the most promising candidate for research and development for large-scale users as the fabrication of low cost PV devices becomes a reality.

Fuel cells have been receiving a lot of attention lately due to their potential as becoming a new energy source with a large range of applications. Fuel cells can be incorporated with other components to create high efficiency industrial power plants. Fuel cells permit clean and efficient energy production for stationary and transportation applications. Considering economical and ecological aspects, low temperature fuel cells like Polymer Electrolyte Fuel Cells (PEMFC) seem to represent well-suited systems. Fuel cells can help in contributing in environment by maximizing the use of renewables in the energy production and using hydrogen produced with no pollutant emission. A simple and accurate fuel-cell model is studied for green power-system. The results of the model are used to predict the output voltage, and power of FC as a function of pressure and temperature. Circuit model parameters of many series-connected cells can be acquired using the results of a single stack simulation. Electrolysis, which is the simplest process to generate hydrogen, is powered by the DC voltage from the photovoltaic cells. To allow for simultaneous utilisation of emission-free renewable energy sources, the hydrogen shall be produced by implementation of electrolysis of water. The application of fuel cell technologies to advanced power generation systems implies the most significant advancement in energy conservation and environmental protection for the next years.

Figure 24. Simulation results

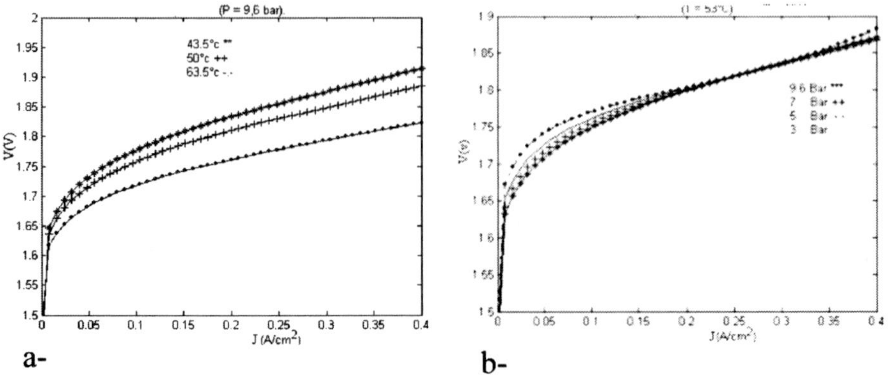

a-   b-

The purpose of the work is to optimize the system's operation. The main reason to build described system is to supply stand alone system using renewable energy sources. Therefore, the power plant has to produce energy independent of any weather fluctuations. In order to compensate for their cost, Photovoltaic (PV) systems need to be operated in a manner that ensures the optimum use of the available photovoltaic energy. In this chapter, the concentration is on photovoltaic systems. A simple and efficient method to optimise the SCA's power output has been presented. This method applies a dynamical electrical array reconfiguration strategy on the Photovoltaic (PV) generator, in order to improve its energy production. The illustrated method allows a continuous SCA reconfiguration in order to insure a good load matching. The principle of the proposed method consists of determining on line and in real time for a fixed number of photovoltaic modules, which configuration is the best for a given load under given working condition and the switching the SCA into that configuration. Unfortunately, the reference point is not fixed but varies with weather conditions (solar radiation, temperature, etc.). Positions are found according to the variation of the already mentioned conditions. A better way of calculating the current and voltage is therefore determined.. The reconfiguration could occur during the insulation variation or when a load's change occurs. By rearranging connections between different modules, the matching between the load and PVG is improved. This enables the system to react to changes in loads and environmental conditions. The simulation program developed also allows to explore different configurations and similar results were obtained. The experimental system described has permitted to validate the proposed method. To avoid the untimely commutations if the operating point situates in the intersection zone of I - V(or P - V) of the two configurations, we have to tolerate a certain time between two succeed commutations: an hysteresis.

This method, prevent the decrease of the output power when environmental changes occured. The advantage of the method we presented appears for an engine suing a hydraulic pump for example. If advantage of power is transmitted to the engine, more will be the climbing of the fluid through the pump and more important will be the depth of water pumping reached. More, an engine pump system can sometimes begin its pumping earlier and more rapidly for such configuration to the other following the required current of beginning. The technique of configuration change is simple, its price moderated.

REFERENCES

Appelbaum, J. (1989). The operation of loads powered by separate sources or by a common source of photovoltaic cells. *IEEE Transactions on Energy Conversion*, *4*(3), 351–357. doi:10.1109/60.43235.

Benmessaoud, M. T., Zerhouni, F. Z., Zegrar, M., Boudghene Stambouli, A., & Tioursi, M. (2010). New approach modeling and a maximum power point tracker method for solar cells. *Computers & Mathematics with Applications (Oxford, England)*, *60*(44), 1124–1134. doi:10.1016/j.camwa.2010.04.001.

Boudghéne Stambouli, A. (2003). *Electricity from photovoltaics and fuel cells: The expectations for an enviromental-friendly and sustainable source of energy*. Paper presented at the Environment 2010 International Symposium. Porto, Portugal.

Boudghéne Stambouli, A. (2011a). Fuel cells: The expectations for an environmental-friendly and sustainable source of energy. *Renewable & Sustainable Energy Reviews*, *15*(9), 4507–4520. doi:10.1016/j.rser.2011.07.100.

Boudghéne Stambouli, A. (2011b). Promotion of renewable energies in Algeria: Strategies and perspectives. *Renewable & Sustainable Energy Reviews*, *15*(2), 1169–1181. doi:10.1016/j.rser.2010.11.017.

Boudghéne Stambouli, A. (2011c). Algerian renewable energy assessment: The challenge of sustainability. *Energy Policy*, *39*(8), 4507–4519. doi:10.1016/j.enpol.2010.10.005.

Boudghéne Stambouli, A., & Traversa, E. (2002a). Solide oxide fuel cells (SOFCs): A review of an environmentally clean and efficient source of energy. *Renewable & Sustainable Energy Reviews*, *6*, 433–455. doi:10.1016/S1364-0321(02)00014-X.

Boudghéne Stambouli, A., & Traversa, E. (2002b). Fuel cells, an alternative to standard souces of energy. *Renewable & Sustainable Energy Reviews*, *6*(3), 295–304. doi:10.1016/S1364-0321(01)00015-6.

Busquet, S. (2003). *Etude d'un système autonome de production d'énergie couplant un champ photovoltaïque, un électrolyseur et une pile à combustible: Réalisation d'un banc d'essai et modélisation*. Ph. D Thesis. Paris, France: Ecole des Mines de Paris.

Busquet, S., Hubert, C. E., Labbé, J., Mayer, D., & Metkemeijer, R. (2004). A new approach to empirical electrical modelling of a fuel cell, an electrolyser or a regenerative fuel cell. *Journal of Power Sources*, *134*(1), 41–48. doi:10.1016/j.jpowsour.2004.02.018.

Enrique, M., Dura'n, E., Sidrach-de-Cardona, M., & Andu'jar, J. M. (2007). Theoretical assessment of the maximum power point tracking efficiency of photovoltaic facilities with different converter topologies. *Photovoltaic Energy*, *81*(3), 1–38.

Kwon, J.-M., Nam, K. H., & Kwon, B. H. (2006). Photovoltaic power conditioning system with line connection. *Transactions on Industrial Electronic*, *53*(4), 1048–1054. doi:10.1109/TIE.2006.878329.

Mutoh, N., & Inoue, T. (2007). A control method to charge series-connected ultraelectric double-layer capacitors suitable for photovoltaic generation systems combining MT control method. *Transactions on Industrial Electronics*, *54*(1), 374–383. doi:10.1109/TIE.2006.885149.

Salameh, Z. M., & Chaozi, L. (1990a). Optimum switching points for array reconfiguration controller. In *Proceedings of the 21st IEEE Photovoltaic Conference*, (vol 2), (pp. 971-976). IEEE Press.

Salameh, Z. M., Mulpur, A. K., & Dagher, F. (1990b). Two stage electrical array reconfiguration controller. *Photovoltaic Energy*, *44*, 51–55.

Weidong, X., Nathan, O., & William, D. G. (2007). Topology study of photovoltaic interface for maximum power point tracking. *Transactions on Industrial Electronics*, *54*(3), 1696–1704. doi:10.1109/TIE.2007.894732.

Zerhouni, F. Z. (1996). *Adaptation optimale d'une charge à un générateur photovoltaïque*. Master Thesis. Oran, Algeria: University of Sciences and Technology Mohamed Boudiaf.

Zerhouni, F. Z. (2005). *Contribution dans l'amélioration du fonctionnement d'un système photovoltaïque*. Paper presented at the International Conference, ICEL05. Oran, Algeria.

Zerhouni, F. Z. (2006a). Etude des l'influence de différents agents sur les caractéristiques d'un générateur photovoltaïque. *Technological and Scientific Review*, *4*, 39–44.

Zerhouni, F. Z. (2009). *Développement et optimisation d'un générateur énergétique hybride propre à base de PV-PAC*. Doctoral Thesis. Oran, Algeria: University of Sciences and Technology Mohamed Boudiaf Oran.

Zerhouni, F. Z., Benmessaoud, M. T., & Boudghene Stambouli, A. (2006b). Développement de modèle de système à énergie renouvelable pour son utilisation en électronique de puissance et optimization. In *Proceedings of JNEA 2006*. Oran, Algeria: JNEA.

Zerhouni, F. Z., Zegrar, M., Benmessaoud, M. T., Boudghene Stambouli, A., & Midoun, A. (2008). A novel method to optimize photovoltaic generator operation. *International Journal of Energy Research*, *32*(15), 1444–1453. doi:10.1002/er.1461.

Zerhouni, F. Z., Zegrar, M., Benmessaoud, M. T., Boudghene Stambouli, A., & Midoun, A. (2010). Optimum load matching by an array reconfiguration in photovoltaic generators. *Energy & Fuels*, *24*(11), 6126–6130. doi:10.1021/ef1006275.

ADDITIONAL READING

Antchev, M. H. (2010a). Energy and energy efficiency. In *Technologies for Electrical Power Conversion, Efficiency, and Distribution: Methods and Processes* (pp. 1–9). Hershey, PA: IGI Global..

Antchev, M. H. (2010b). *Storage and usage of energy. Technologies for Electrical Power Conversion, Efficiency, and Distribution: Methods and Processes* (pp. 10–31). Hershey, PA: IGI Global..

Antchev, M. H. (2010c). DC/DC conversion. In *Technologies for Electrical Power Conversion, Efficiency, and Distribution: Methods and Processes* (pp. 134–166). Hershey, PA: IGI Global..

Antchev, M. H. (2010d). Conversion of electrical power from renewable energy sources. In *Technologies for Electrical Power Conversion, Efficiency, and Distribution: Methods and Processes* (pp. 231–247). Hershey, PA: IGI Global..

Aquila, J. O., Owolabi, S. W., & Olugbemiga, I. M. (2010). Emerging technology penetration: The case of solar electricity in Nigeria. In *Nanotechnology and Microelectronics: Global Diffusion, Economics and Policy* (pp. 430–449). Hershey, PA: IGI Global. doi:10.4018/978-1-61692-006-7.ch025.

Cadman, T., & Hume, M. (2012). Developing sustainable governance systems for regional sustainability programmes and 'green' business practices: The case of 'green' timber. In *Sustainable Systems and Energy Management at the Regional Level: Comparative Approaches* (pp. 365–382). Hershey, PA: IGI Global..

Calò, C., Lay-Ekuakille, A., Vergallo, P., Chiffi, C., Trotta, A., Fasanella, A., & Fasanella, A. M. (2011). measurements and characterization of photovoltaic modules for tolerance verification. *International Journal of Measurement Technologies and Instrumentation Engineering*, *1*(2), 73–83. doi:10.4018/ijmtie.2011040106.

Canlar, T. Y. (2010). *Sustainable urban and regional infrastructure development: Technologies, applications and management*. Hershey, PA: IGI Global..

Cebi, S., Kahraman, C., & Kaya, I. (2012). Soft computing and computational intelligent techniques in the evaluation of emerging energy technologies. In *Innovation in Power, Control, and Optimization: Emerging Energy Technologies* (pp. 164–197). Hershey, PA: IGI Global. doi:10.4018/978-1-61350-138-2.ch005.

Doukas, H., Makarouni, I., Karakosta, C., Marinakis, V., & Psarras, J. (2012). EU—GCC clean energy cooperation: From concepts to action. In *Sustainable Systems and Energy Management at the Regional Level: Comparative Approaches* (pp. 288–308). Hershey, PA: IGI Global..

Fanetti, S. (2012). Renewable energy in Italy: Incentives, bureaucratic obstacles and Nimby syndrome. In *Sustainable Systems and Energy Management at the Regional Level: Comparative Approaches* (pp. 68–86). Hershey, PA: IGI Global..

Flamos, A., Karakosta, C., Doukas, H., & Psarras, J. (2010). Setting technology transfer priorities with CDM-SET: Development of sustainable energy technology transfer tool. In *Intelligent Information Systems and Knowledge Management for Energy: Applications for Decision Support, Usage, and Environmental Protection* (pp. 205–222). Hershey, PA: IGI Global..

Gelil, I. A. (2012). Globalization of the environmental issues: Response of the arab region. In *Sustainable Systems and Energy Management at the Regional Level: Comparative Approaches* (pp. 147–165). Hershey, PA: IGI Global..

Hilty, L. M. (2011). Information and communication technologies for a more sustainable world. In *Green Technologies: Concepts, Methodologies, Tools and Applications* (pp. 36–45). Hershey, PA: IGI Global. doi:10.4018/978-1-60960-472-1.ch104.

Hin, L. T. W., & Subramaniam, R. (2012). Use of policy instruments to promote sustainable energy practices and implications for the environment: Experiences from Singapore. In *Sustainable Systems and Energy Management at the Regional Level: Comparative Approaches* (pp. 219–235). Hershey, PA: IGI Global..

Hocaoglu, F. P., Gerek, O. N., & Kurban, M. (2009). Solar radiation forecasting model. In *Encyclopedia of Artificial Intelligence* (pp. 1433–1438). Hershey, PA: IGI Global..

Kim, E.-H. (2012). Government policies to promote production and consumption of renewable electricity in the US business. In *Sustainable Systems and Energy Management at the Regional Level: Comparative Approaches* (pp. 1–18). Hershey, PA: IGI Global..

Miller, W., & Birkeland, J. (2011). Green energy: Sustainable energy sources and alternative technologies. In *Green Technologies: Concepts, Methodologies, Tools and Applications* (pp. 1–16). Hershey, PA: IGI Global. doi:10.4018/978-1-60960-472-1.ch101.

Obara, S. (2010). Fuel reduction effect of the solar cell and diesel engine hybrid system with a prediction algorithm of solar power generation. In *Intelligent Information Systems and Knowledge Management for Energy: Applications for Decision Support, Usage, and Environmental Protection* (pp. 242–271). Hershey, PA: IGI Global. doi:10.1299/jpes.2.1166.

Olla, P. (2012). *Global sustainable development and renewable energy systems*. Hershey, PA: IGI Global. doi:10.4018/978-1-4666-1625-7.

Omer, A. M. (2012). Renewable energy and sustainable development. In *Innovation in Power, Control, and Optimization: Emerging Energy Technologies* (pp. 95–136). Hershey, PA: IGI Global..

Pérez-Plaza, M., & Linares, P. (2009). Strategic decisions for green electricity marketing: Learning from past experience. In *Web-Based Green Products Life Cycle Management Systems: Reverse Supply Chain Utilization* (pp. 2589–2605). Hershey, PA: IGI Global..

Sanseverino, E. R., Zizzo, G., & Scimemi, G. F. (2011). A new meta-heuristic multi-objective approach for optimal dispatch of dispersed and renewable generating units in power distribution systems. In *Knowledge-Based Intelligent System Advancements: Systemic and Cybernetic Approaches* (pp. 162–181). Hershey, PA: IGI Global. doi:10.4018/978-1-61692-811-7.ch009.

Visconti, P., Romanello, D., Zizzari, G., Ventura, V., & Cavalera, G. (2011). Design, Measurements And Characterization Of Smart Electronic Board For Pv Streetlight Based On Led And High Intensity Discharge Lamps. *International Journal of Measurement Technologies and Instrumentation Engineering*, *1*(3), 1–13. doi:10.4018/ijmtie.2011070101.

Wang, H.-L., & Ghose, A. K. (2011). Green strategic alignment. In *Green Technologies: Concepts, Methodologies, Tools and Applications* (pp. 1435–1447). Hershey, PA: IGI Global. doi:10.4018/978-1-60960-472-1.ch606.

KEY TERMS AND DEFINITIONS

Electrolysis: Is the process of decomposition of a substance into its chemical components by an electric current.

Fuel Cell: Is a device that converts the chemical energy into electricity.

Mean Square Error: EQM measures the average of the squares of the errors.

Photovoltaic Modules: A system which uses photovoltaic cells to convert sunlight into electricity.

Power's Management and Control Device: Is an equipment used to control the power between each component.

Proton exchange membrane (PEM): Fuel cells take hydrogen and air and produce power, heath and water.

Renewable Energy: Is energy which comes from natural resources which are renewable naturally.

Solar Cells Array: The modules are linked together to form an array.

This chapter was previously published in Sustainable ICTs and Management Systems for Green Computing edited by Wen-Chen Hu and Naima Kaabouch, pages 399-422, copyright 2012 by Information Science Reference (an imprint of IGI Global).

Chapter 77
Are Biofuels a Factor of Sustainable Development in a Food Insecurity Context in Africa?
Case Study of Burkina Faso

Marie-Hélène Dabat
Centre de Coopération Internationale en Recherche Agronomique pour le Développement, Burkina Faso

Joël Blin
Centre de Coopération Internationale en Recherche Agronomique pour le Développement, Burkina Faso & 2iE, Institut International d'Ingénierie de l'Eau et de l'Environnement, Burkina Faso

Elodie Hanff
2iE, Institut International d'Ingénierie de l'Eau et de l'Environnement, Burkina Faso

ABSTRACT

Bearing in mind the strong link between energy and development, and given the country's heavy reliance on imported fossil fuels, this chapter discusses the opportunity for substituting fossil fuels with biofuels in a Sahelian country, Burkina Faso. Biofuel opportunities are discussed taking into account technical, agronomic, and land potentials in this country. Diversification of energy resources with biofuels would reduce the growth of fuel imports in the short term, improve overall public finances, provide a chance to develop agriculture, and provide benefits for the locals. However, if they are to generate sustainable socio-economic development, biofuel projects need to be mindful of food security and economic incentives, and should be part of national agricultural strategies. The chapter shows that a number of conditions must be met to ensure the advantages of biofuels outweigh the disadvantages: prioritising domestic use over exports; supporting the emergence of decentralised systems; localising dedicated crops in order to avoid competition with food crops; regulating the edible oil market; removing technical obstacles to production and processing; and prioritising projects implying family-farming rather than agri-business.

DOI: 10.4018/978-1-4666-4852-4.ch077

INTRODUCTION

The World Commission on Environment and Development suggests that development is sustainable where it *meets the needs of the present without compromising the ability of future generations to meet their own needs* (UND, 1986). In addition, energy has been defined by the United Nations Development Programme (UNDP) (McDade, Lallement, & Saghir, 2006) as playing a key role in sustainable development and poverty alleviation efforts. As specified in the energy objectives (NEPAD, 2001) of the New Partnership for Africa's Development (NEPAD), ensuring the provision of adequate, affordable, efficient and reliable high-quality energy services with minimum adverse effects on the environment for a sustained period is crucial for African countries. Although there are no specific Millennium Development Goals (MDG) (UN, 2005) relating to energy, it will be impossible to achieve MDGs, and among them food security, without improving the quality and quantity of energy services in the developing world (Clare, 2002; ECOWAS & WAEMU, 2006).

Many studies (ECOWAS & WAEMU, 2006; Karekezi, 2002; Martinez & Ebenhack, 2008; Sebitosi & Pillay, 2005) have shown the net positive link between energy consumption and development. For example, the African Energy Policy Research Network has demonstrated (see Figure 1) the correlation between Gross National Product (GNP) and per capita energy use in Africa (Afrepren/FWD, 2002).

While energy is not the sole factor for sustainable development, Africa needs to improve reliability and to search for more abundant, cheap energy in order to enable economic growth (IEA, 2008) and ensure the well-being of its populations. It also needs to reverse environmental degradation and health impacts that are associated with the use of traditional fuels in rural areas (Amigun, Sigamoney, & von Blottnitz, 2008; Toonen, 2009).

Energy generates electricity for a variety of applications, including domestic purposes, off-grid rural electrification, small and medium enterprises and industrial needs. Roughly 1.6 billion people, mostly in developing countries, are reported as lacking access to basic electricity services. The lack of electricity deprives people of basic necessities such as refrigeration, lighting, and communications, and undermines national competitiveness (WB, 1996). Furthermore, most African countries are highly dependent on fuel imports (ECOWAS & WAEMU, 2006). World oil reserves are being depleted at an unprecedented rate, placing considerable pressure on the

Figure 1. Modern energy use per capita (kgoe) vs GNP per capita ($) (MCPEA, 2008)

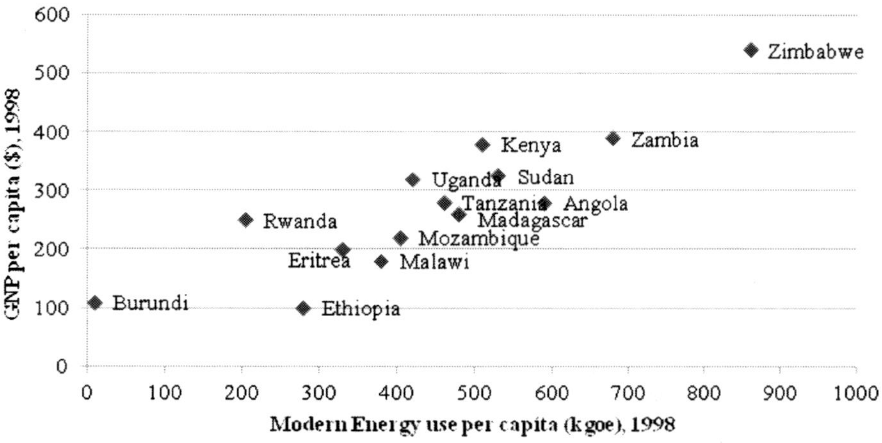

economies of oil-importing African countries in particular, and threatening them. Since the 1973 oil price shock, the prices of crude oil-based fuels have increased sharply, sometimes requiring as much as 20 percent of national income in Sahelian states (Bugaje, 2006). Recently, fuel prices have become even more unstable and have increased very sharply: up to US$ 70 in late 2006, almost reaching US$ 150 in mid-2008 (Carlsson, 2009), falling at US$ 35 at the beginning of 2009, reaching again US$ 115 in April 2011. According to the International Monetary Fund (IMF), Heavily Indebted Poor Countries (HIPC) are seriously affected by higher oil prices (Mussa, 2000). These countries are already running large current account deficits and will encounter a significant deterioration in their foreign trade balance. Since oil is a finite resource, petroleum prices will inevitably escalate when this resource becomes limiting (Carlsson, 2009; Leder & Shapiro, 2008; Schneider, 2006). This situation weighs heavily on the assessment of public finances, at the expense of public services and oil-consuming businesses. Moreover, a lack of energy availability, or excessive energy prices, means a lack of basic social services, namely healthcare, education, drinking water supplies and nutrition, as well as difficulties in developing productive activities, particularly those related to the promotion and processing of agro-pastoral products (IEA, 2000; WB, 1996).

In order to decrease reliance and pressure on fossil fuels, a number of countries, like Burkina Faso, have started to look closely at biofuels (Nonyarma & Laude, 2009) for creating a more secure supply, in a less volatile market that can be accessed domestically (Mulugetta, 2009). Biofuels are increasingly seen as one of the best fuel resources that almost any country with surplus land and labour resources can exploit through relatively low capital inputs and medium levels of management.

Thus, this creates an attractive opportunity for biofuels to boost the development of national industry and to increase incomes in rural areas, by generating new energy sources and new channels in the agricultural sector. But would the Southern countries really benefit from biofuels? This is the subject of a deep controversy and heated debate. For some, biofuels are a factor of development; for others, they compete with food products. Given the challenges, developing crops for biofuel production seems worth exploring for these countries, while taking into account the country's agricultural and land situation and understanding the socio-economic and environmental impacts (deforestation, competing land uses, use of chemicals and fertilizers, etc.) (Mangoyana, 2009).

In this chapter, the case study first will focus on the consequences of fuel energy prices on a national level in Burkina Faso, one of the poorest countries in the world, heavily dependent on fossil fuel imports. Then we take a closer look at the technical and agronomic potentials of biofuels for the country considering the effects of large-scale programmes on land use, food security and socio-economic development. Finally, it sets the context for further discussions on biofuel feasibility and sustainability in Burkina Faso in order to support a national policy framework.

The chapter makes broad loans to the publication of the same authors in Renewable and Sustainable Energy Reviews: Are biofuels an efficient technology for generating sustainable development in oil-dependent Africans nations? A macroeconomic assessment of the opportunities and impacts in Burkina Faso (Hanff, Dabat, & Blin, 2011).

ENERGY AND DEVELOPMENT IN BURKINA FASO

A Poor Country Relying on Extensive Agriculture

Burkina Faso, landlocked in the middle of West Africa, is one of the poorest countries in the world with a Human Development Index (HDI) ranking it

161st out of 169 countries in 2010 (PNUD, 2011). People living below the poverty line (established at € 111/year) amounted to 46.4% in 2003 (INSD, 2003b), and that percentage was increasing despite all the national measures and international help (1994: 44.5%, 1998: 45.3% [INSD, 2003a]), with a greater incidence of poverty in rural and peri-urban areas. It still reaches 43.9% in 2010.

The country covers 274,000 km^2 in the Sudano-Sahelian zone with an estimated population of more than 14 million people (MEF, 2009), largely farming and pastoral rural communities, while only 20% live in urban areas (10% in the capital city Ouagadougou). The country is one of the most populated in the West African region, with a strong increase in urban inhabitants, more than three times that of the rural increase (Blin, Ficini, Faugère, & Dabat, 2007). Taking a population growth rate of 2.4%, the overall population will double in 29 years (Blin, et al., 2007). Agricultural activities account for more than one third of GDP, and provide jobs and income for about 80% of the population (INSD, 2003a), with cotton being the main cash crop (Blin, et al., 2007). As Africa's leading cotton producer Burkina Faso mostly relies on that crop: 6 out of 14 million Burkinese are exclusively dependent on cotton, which accounts for 60% of national export earnings (BafD/OECD, 2008; Blin, et al., 2007). Cotton is followed by peanut, shea nut, sesame, cow pea, sorghum, millet, corn, and rice farming.

The overall economic performances of the country are linked to agricultural activities (Blin, et al., 2007), which are highly unpredictable and dependent on climate fluctuations. The country is in a transitional area, characterized by large rainfall variations and high susceptibility to drought. The land, cellulose resources and water are irregularly distributed and managed with low sustainability.

Moreover, the Intergovernmental Panel on Climate Change (IPCC) described Africa as the region most vulnerable to the impacts of projected climate changes (IPCC, 2001). Because poverty limits adaptation capabilities, agricultural systems are particularly vulnerable (Challinor, Wheeler, Garforth, Craufurd, & Kassam, 2007). Only a few reliable facts exist on West Africa concerning the extent and the consequences of climate change, but the IPCC forecasts a greater increase in temperature compared with other regions, and consequently probably more extreme climatic events (drought and flooding) (Blein, Goura Soulé, Faivre Dupaigre, & Yérima, 2008).

Despite all of these constraints, Burkina Faso's GDP (5.697 million US$ in 2005) grew by an annual average of 5% from 1994 to 2004 and by 7.5% in 2005 (inflation rate at 6.4%) (Blin, et al., 2007). However, despite many government efforts and international help to improve socio-economic conditions, the economy has remained singularly vulnerable. The overall economic performances also face a difficult domestic and international environment. Indeed, in 2007, the economy had to cope with high petroleum prices, and international decrease in exported raw material prices, mainly cotton, the depreciation of the dollar and poor agro-climatic conditions linked to climate change (Zongo, 2008). This situation impacted on economic growth, which stood at 3.6% in 2007, 5,2% in 2008 and 3,2% in 2009 as opposed to 6.1% in 2006 (Zongo, 2008). Last years, the country faced a food crisis, high imported cereals prices and persistently high oil prices, demonstrating the country's vulnerability to internal and external factors.

Thus, national sustainable development has to involve the agricultural sector substantially, in order to empower the large majority of the workforce (80%). One of the big challenges facing the country is truly related to agriculture. Currently, this agriculture is suffering from a lack of mechanization and does not ensure food security for the country each year (INSD, 2003a). The objective of the government is to increase farmer production by 5 to 10% per year, in order to limit food imports, and increase income levels by 3% per person per year (Zongo, 2008). The two current priorities of the Government are therefore agricultural intensification and diversification (DCPM, 2009). Agricultural intensification particularly calls for mechanization and better access to energy. The country also requires better access to energy in order to develop agro-processing, storage and transportation of food stuffs, in order to improve food security and income generation. Consequently, the development of biofuels in Burkina Faso is consistent with the objectives of agricultural diversification and intensification, as well as post harvest activities.

Access to Energy

The main characteristics of Burkina Faso's energy sector are

- Low consumption levels,
- Predominance of traditional energies, which exerts pressure on forests,
- Poor access to electricity,
- Modern energy prices that are prohibitive for the population, and
- Heavy dependence on imported fossil fuels, which exerts pressure on the commercial balance.

Burkina Faso's energy poverty is illustrated by the low level of fossil fuel consumption, which amounts to the equivalent of 275 litres of petrol per inhabitant per year (Laude, 2008b), equivalent to 2 or 3 steres of wood, and the ultimate energy consumption amounts to 234 kgoe/inhab. This figure is drastically low compared with other final energy consumptions: 1,145 kgoe/inhab worldwide, 5,418 in the United States and 454 on average in the CEDEAO countries (ECOWAS & WAEMU, 2006).

Burkina Faso has few natural energy resources: no petrol, no coal, limited hydraulic potentials, and over-degraded woody resources. There is a direct link between poverty and the use of traditional biomass, where a large proportion of people living on less than $2 a day use traditional biomass as their energy source (Amigun, et al., 2008). Indeed, biomass accounts for over 89% of the country's energy consumption. This biomass is distributed between fuel wood (91%), crop residues (5%), bagasse (3%), and charcoal (1%) (Ouédraogo, 2004). This has led to over-exploitation of forests by rural communities to cover their energy needs.

Oil prices are high and increasing. Prices for consumers already increased by 60% from 1995 to 2007 (OMC, 2004) and the average cost of electricity production rose by 40% over the last 4 years. As a landlocked country, almost all imported goods arrive at the ports of Cotonou, Abidjan, or Lome and have to cross by land through Benin, Ivory Coast and Togo respectively. So, Burkina Faso is doubly impacted by rising oil prices: the purchase of the barrel and coastal storage, with transportation costs that depend on fuel prices are responsible for an extra cost of 30% compared to CAF prices (Blin, Dabat, Faugère, Hanff, & Weisman, 2008).

In addition, only 5% of households in 2008 had access to electricity, compared to an average of 20% in the West African region (ECOWAS & WAEMU, 2006), with an access rate below 1% in rural areas (Blin, et al., 2007). In 2008, electricity imports from Ivory Coast and Ghana amounted to 18% of national consumption, while the national company supplied 64% of electricity consumption from thermal power stations using imported fuel, and 18% from hydroelectricity centres (Sonabel, 2008). In the near future, the strategy of interconnection (Laude, 2008b) with foreign countries would be extended to the target of 80% of national consumption, while national thermal production would only amount to 20%. This will greatly increase dependence on neighbouring countries for energy delivery, especially from Ghana and Ivory Coast.

Fossil Fuel Consumption and Importation

Imported fossil fuel is the second largest source of energy, behind wood resources, with 16% of consumption. From 1993 to 2007, overall fossil fuel consumption was almost multiplied by 3, in line with the development of electricity demand and the increase in fuel demand from the vehicle pool. Oil is mainly used for transport (63%) and electricity production in national thermal power stations (23%) (Blin, et al., 2008).

Fossil fuel imports amounted to 522,000 TOE in 2007, weighing more than 50% (around € 332 million in 2007) of the national trade balance (Laude, 2008b). Given the increasing price of fuel and growing energy needs, this dependence exerts strong pressures on the country's economy. However, according to the regional electricity integration policy, interconnected grids would provide solutions for cheaper power supplies for the medium and long terms. However, to ensure the security of supplies, a thermal production supply using at least 20,000 t of oil will remain (Nonyarma & Laude, 2009). According to the Ministry of Mines, Carriers and Energy, the electrification (and mechanization) of rural areas would create increasing demand for hydrocarbons, rising to 50,000 t in 2020 (Nonyarma & Laude, 2009). For transport energy, the current consumption of 350,000 t in 2010 would reach 680,000 t in 2020 (Nonyarma & Laude, 2009).

The increase in fuel prices, essentially consumed for transport and electricity production, leads to

- An increase in energy costs for households, the State and companies,
- Limited development of electrification, particularly in rural areas, and
- Overconsumption of woody resources in the country.

BIOFUEL POTENTIALS AND RISKS FOR THE POPULATION

The major alternative energy resources abundant throughout the African continent are solar, biomass energy and a very little wind energy potential (Bugaje, 2006). Given the national agriculture strategy (DCPM, 2009), the large rural workforce in the agricultural sector, and the availability of land, the development of new agricultural activities for diversification of energy sources and expansion of locally produced energy are goals that can be satisfied by biofuels (Amigun, et al., 2008), also called "agrofuels."

Technical Potentials

Biomass is the only renewable resource able to produce liquid fuel for engine applications. In order to meet the energy needs mentioned above, various biofuels could be produced in Burkina Faso with different characteristics, as described in Table 1. Among first generation biofuels, which are mature and applicable technologies, two main feed stocks can be utilized: sugar or starch based plants and oil-bearing plants (Blin, et al., 2008). Through the ethanolic fermentation of sugar-bearing plants, bioethanol can be produced to substitute petrol in petrol engines. However, based on the Brazilian experience, this technology is profitable with a minimum unified area of 20,000 hectares sown with a sugar cane crop and the industrial process alone needs a lot of water ($1m^3/t$) (Abamovay, 2009). Moreover, petrol is only used for transport application in Burkina Faso (Blin, et al., 2008), so bioethanol is not considered as a sustainable biofuel potential in the short term.

Vegetable oils can be obtained by simple cold pressing of oil-bearing seeds and fruits. Raw Vegetable Oil (RVO) can be used in adapted diesel engines (Sidibé, Blin, Vaitilingom, & Azoumah, 2010) or can be modified through a transesterification reaction with methanol to

Table 1. Characteristics of biofuels applicable in Burkina Faso (Blin, et al., 2008)

	Raw Vegetable Oil (RVO)	Biodiesel	Bioethanol
Production	Oil-bearing plants	Oil transesterification with methanol	Sugar or starch-based plants
Technology	Simple	Industrial process, mature for methanol etherification	Industrial process
Availability	From the village to industrial scale	Industrial scale	Industrial scale
Utilization	Mainly for static diesel engines (generators, mills, motor-pumps, etc) The use of a bicarburation kit is recommended (cheap, simple and safe).	Any type of diesel engine (transport) It can replace 100%	With petrol engines in a mixture up to 10% for classic engines or up to 100% for specific engines (flex fuel). Only for transport application
Advantages	Easy and cheap	No need to modify engines	High octane index
Limits	- Not suitable for transport - Low cetane index and high viscosity - Adaptation of the engine with dual fuelling system (0 to 100% RVO) or a mixture RVO/fuel under 30%.	- No methanol in west Africa. - Processes to produce biodiesel from ethanol not mature	- Requires energy for distillation - Only sugar cane use produces enough byproduct to generate the heat for the process - Sugar cane plantations require a lot of water

obtain biodiesel. Biodiesel is technically defined as a fuel composed of mono-alkyl esters of long chain fatty acids derived from vegetable oils or animal fats (Escobar, et al., 2009). Biodiesel is biodegradable, non-toxic and environment-friendly as compared to petrodiesel (Khan & el Dessouky, 2009). As shown in Table 1, as biodiesel has quite the same physico-chemical properties as diesel (same viscosity and cetane index) it can be used directly in all diesel engines, with an engine performance comparable to conventional diesel fuel (Haas, McAloon, Yee, & Foglia, 2006), but it has to be produced by a costly industrial process. In adapted engines, RVO can be used as fuel without any problem: either through a dual fuelling system (preheating of the fuel before injection) or blending with diesel up to 30% (Murugesan, Umarani, Subramanian, & Nedunchezhian, 2009). Diesel engines are widely used in transport, electricity generation and shaft power. The strategy suggested in this paper is to produce simple and accessible local RVO for national basic needs, such as electricity and shaft power.

The advantages of RVOs are substantial because they are easy to produce with local assets and the vegetable oil could replace petroleum oil in two main applications:

1. In static diesel engines used in villages and
2. In the national thermal power stations for electricity (Blin & Girard, 2011).

This second option could make it possible to replace 23% of the imported fossil fuels and thereby have a strong impact on the energy bill (Blin, et al., 2008). Application of this scenario may have substantial margins for expansion, because substitution could be up to 90% in the power stations, depending on the potentials and the supply of local oil. It could be flexible over time because it can be adjusted with the increasing level of RVO production in the country. Moreover, the national electricity company already has a bi-fuel kit in each power station, so no additional investment would be required in this development scenario. However, some quality standards need to be developed.

More and more developing countries are examining possibilities of substituting fossil fuels in the transport sector with locally produced biofuels (Peters & Thielmann, 2008). The objectives are to create added value locally with RVO at an acceptable price compared to biodiesel, because no costly transesterification processes with imported methanol are required. It is, moreover, easier to produce RVO, and at an acceptable price, when compared to the current price of oil, which is usually the first barrier to biofuel production and competitiveness (Peters & Thielmann, 2008).

In fact, biodiesel requires an industrial process with methanol, a component derived from petroleum activities, which is unavailable in the country and in West Africa, and uncompetitive with the current price of a barrel of petroleum oil. Given the large amount of raw materials needed for industrial biodiesel and bioethanol production, we will focus on RVO.

Agronomic Potentials

Many edible and non-edible oil-bearing plants are available and are already being cultivated in Burkina Faso, and could be cultivated to produce RVO. With the prevailing weather in Burkina Faso, the main plants potentially of interest for RVO production are *Jatropha curcas*, cotton, sunflower, peanut and soybean. In Burkina Faso, as edible oil crops are currently under-produced and food competition is a major concern, all biofuel project holders focus on non-food oil, although large-scale production of non-food oil crops requires land and labour, so may also create food competition.

Jatropha curcas has attracted a lot of attention as a promising non-food oil producing plant that is highly adapted to marginal soils and arid to semi-arid conditions in tropical and sub-tropical countries (Kumar & Sharma, 2008). Considering the large number of projects investing in *Jatropha curcas* inside the country (more than 70,000 ha of plantations claimed in 2009 [Nonyarma & Laude, 2009]), this plant could be a major raw material for the development of this scenario in the coming years. Plantations have been grown since 2007, in reaction to the European Union biofuels directive for promoting the use of biofuels for EU transport. The directive stipulated that national measures must be taken by countries across the EU aiming at replacing 5.75% of all transport fossil fuels with biofuels by 2010. On January 14th 2008, the EU cancelled its biofuels programme due to environmental and social concerns (Harrabin, 2008) (impact of biofuels on rising food prices, rainforest destruction, concern for rich firms driving poor people off their land to convert it to fuel crops, etc.) (Blin, Weisman, Hanff, & Dabat, 2011). As a consequence, project holders in Burkina Faso have focused on local and regional potentials to sell their production.

As shown in Table 2, *Jatropha curcas* is a plant with considerable potential. Plantations can prevent erosion and reclaim land. It can be grown as a live fence and under a wide range of rainfall regimes from 200 to over 1,500 mm per annum (Achten, et al., 2008; Openshaw, 2000). But *Jatropha curcas* seed yield is still a difficult issue. Actually, the mature seed yield per ha per year is not known, since systematic yield monitoring only started recently. Earlier reported figures show a very wide range (0.4–12t/ha/yr) but are not coherent (Achten, et al., 2008). The large variations in yield come from the different types of cultivation conditions, from dry wasteland to good sites with high annual rainfall (Achten, et al., 2008). In Burkina Faso, yields in the fields are around 1 to 1.5 kg/tree with family-farming. Seeds contain between 28% and 38% oil (Kaushik, Kumar, Kumar, & Roy, 2007), corresponding to 300-500 l/oil/ha (Girard & Blin, 2009).

Although *Jatropha curcas* grows naturally in Africa, this plant is still in a very early stage of agronomic development on an industrial scale, and presents a number of risks, so research and experience have to be combined to improve a number of characteristics (Achten, et al., 2008; Carlsson, 2009; Openshaw, 2000; Sujatha, Reddy,

Table 2. Jatropha curcas opportunities and risks as a raw material in RVO production (Gubitz, Mittelbach, & Trabi, 1999; Kumar & Sharma, 2008; Openshaw, 2000)

Opportunities	Risks
- No competition with edible plants - Protection against soil erosion (Openshaw, 2000) - May grow with little rainfall (Achten, et al., 2008) and on marginal soils (associated with lower seed yields) - Tolerant to drought (Gubitz, et al., 1999) - Can live up to 50 y. - Can produce biodiesel, soap and fertilizer (Gubitz, et al., 1999; Kumar & Sharma, 2008)	- Very few reliable scientific data exist either for management or environmental assessment - Perennial, needs three years to grow before production - Oil cake toxic and can only be used as combustible solid - Needs fertile lands to grow better and competes de facto with food production - Impossible to predict the level of profitability in two or three years

& Mahasi, 2008) (yield improvement and stability, reduction of toxicity, utilization of oil cake, etc.) and thus improve its economic competitiveness with hydrocarbons.

Are Biofuels a Threat to Food Security? Land Potentials and Competition

Although biofuels provide an opportunity for ending energy insecurity, is their development a threat to food security? Biofuels may threaten food security through several different mechanisms: competition between outlets (for example, whether oil is sold on the food market or on the energy market); the transfer of international prices to national prices (the effect of high prices of imported cereals such as rice on the prices of local food products, such as sorghum, millet, and maize); competition between several products for access to land and water (*Jatropha curcas* in monoculture or intercropping; sugarcane, which is a water intensive crop) or for the allocation of labour and capital within households (cash crops). What effect do these mechanisms have in Burkina Faso?

Firstly, competition between oil outlets is not an issue from now. The only edible oil produced in the country comes from cotton. At the height of cotton production, this oil only covers half of all requirements. The country imports low-cost palm oil from Ivory Coast and especially from Asia to meet demand that is increasing due to population growth. The recent international price instability for cotton fibre makes the future of this oilseed crop uncertain on the edible oil market. The promotion of new oilseed crops (sunflower, soybean, etc.) could help to meet demand for edible oil. Stimulating these sectors could be beneficial to both the food market and the energy market. Commitment by national electricity companies and fuel distribution companies to secure outlets by buying plant oils at a floor price set by the national authorities would provide incentives for oilseed crop producers. They would then be free to choose the food market if its prices were higher and to keep surplus or lower quality oils for the energy market.

Secondly, biofuels may also threaten food security if international price variations for agricultural products are transferred to national prices. Biofuels played a role in the rise in international prices in 2007-2008, although this responsibility is still debated. However, studying monthly prices of local and imported cereals from 1997 to 2009 shows that the cereal market in Burkina Faso is relatively isolated from the international market. An analysis over several decades also makes it possible to put the rise in agricultural prices into perspective. The hunger riots of 2007 were not just about the prices of cereals and oil.

Thirdly, as regarding competition for land, there is high estimated biomass potential in sub-Saharan Africa, given the surplus production capacity linked to the existence of large rangelands that could be better managed (Fallot, Girard, Dameron,

& Griffon, 2006). More than 2/3 of West Africa's cultivable lands are not farmed, including 9 million hectares of irrigated lands (OCDE, 2008). However, local potential is very variable from one place to another.

In Burkina Faso, the agricultural land sown with cereal crops every year represents about 45% of potential cropland (Nonyarma & Laude, 2009), leaving land available for new production. On this country basis, the agricultural lands sown yearly accounted for approximately 45% of the cultivable potential in 2007, estimated at 9 million hectares (Nonyarma & Laude, 2009), leaving large areas available for new crops. Irrigable lands are estimated at 165,000 ha, but only 12% are exploited (Zongo, 2008). In any case, the government wants to reserve them for food production only (Nonyarma & Laude, 2009). However, approximately 11% of lands in Burkina, mainly in the North of the country, were considered as seriously degraded in 2002.

It remains to be determined whether the land in question is actually available, in other words that it is neither cropped, nor grazed nor appropriated. Much of the land supposedly available is actually used by local populations for their survival: gathering, transhumance, collecting fuelwood, etc. Moreover, some of these lands may soon become unavailable as a result of population growth in a country that has not yet began its demographic transition (Dabat, 2011).

Thus, the question of competition between land for food or energy has to be taken into account regardless of the small territories necessary to substitute fossil fuels for national electricity demand. The land required to produce biofuels depends on the type of crop used (agronomic yield and oil content), the planned production level and the model developed. Around 15 hectares of *Jatropha Curcas* must be grown to meet the energy requirements of a rural community of 2,500 people with a multifunctional platform (Darlu, 2009). As shown in Table 3, given the productivity range for the possible biomass, no more than 2.4% of the arable land, would be necessary up to 2025 to replace 100% of fossil fuels in the thermal power stations of the national electricity company (Blin, et al., 2008).

In addition, agricultural production is

1. Family based: traditional small-scale farming with
2. Poor yield, average areas cultivated per unit of farm varying from 1.5 to 12 ha depending on the regions, and
3. Poorly equipped farms (in 2005, only 50% of farms benefit from harnessed culture) (Blin, et al., 2007).

Given farming sector organization, and regardless of agronomic options, it is fundamental to prioritize family farming rather than industrial

Table 3. The land required to replace fossil fuels for electricity production

	Years	2008	2015	2025
Annual fossil fuel (Fuel Oil) consumption to produce electricity (t) (Laude, 2008a)		71,148	26,723	16,834
Fuel Oil Net Calorific Value (NCV) (MJ/t) (Treanton, 16-17 November 2004)		40,000		
Jatropha's NCV (MJ/t) (Vaitilingom, 2006)		33,000		
Jatropha's RVO yield (t/ha) in Burkina Faso (Achten et al., 2008)		0.4		
Ha necessary for the substitution of fossil fuel to produce electricity	30% substitution	64,680	24,294	15,304
	100% substitution	215,600	80,979	51,012
% of arable land in the country considering 100% substitution (Nonyarma & Laude, 2009)		2.395	0.899	0.566

plantations. Although family farming requires more workers than industrial plantations (factor/ratio 1 to 10) (Alpha & Castellanet, 2007), this is to minimize competition for land and even water resources and to promote and/or launch other cash crops. This strategy would contribute not only to reinforce the rural world by organizing producer groups and/or processing cooperatives, but also to help people procure energy services available from the village for shaft power, and up to industrial scale for electricity production. However, biofuels may be risky for family farm organizations, because the development of new energy crops may increase the demand for land (Demirbas, 2007; Demirbas & Balat, 2006; Ilkilic & Yucesu, 2008; Unal & Alibas, 2007). In a country facing food insecurity, it is essential to be careful to the potential effects of competition between food-producing agriculture and new agriculture with a non-food purpose before developing biofuel crops (Hermelin & Lagandré, 2009). So, attention must be given on a case-by-case basis to the choice of crops, techniques and farming methods, in order to limit competition with food crops for land, water, labour and capital. However, food security is not just a matter of production capacities or levels. The causes of insecurity are found as much in badly functioning markets, inefficient agricultural policies and certain social factors as in physical availabilities and the effects of substituting different uses or factors of production. Food insecurity comes back to the more global problem of poverty.

On the other hand, synergies are possible between energy and food: intensifying production by means of mechanization and increasing yield per hectare, thereby limiting the land problem; developing food crops by processing and preserving products, thereby improving their availability over time; and facilitating access to food products through transportation, thereby improving their spatial distribution. Efforts must be made to exploit these synergies.

POSITIVE IMPACTS AND LIMITATIONS

Socio-Economic and Environmental Impacts

As discussed above, food security is a major concern when considering biofuels. The most vulnerable population is also the one mainly depending on local agriculture. Food insecurity is generally caused by poverty, in terms of income, access to education, agricultural resources, technology and credit lines for food production. Thus, rural development is an important path towards a reduction in poverty and food insecurity. Countries with land potential for biofuels have significant possibilities for developing their agricultural sector, increasing their income and substantially improving the population's living conditions (Escobar, et al., 2009). Farmers could raise their standard of living by producing oil crops in their fields (Khan & el Dessouky, 2009).

A recent Life Cycle Assessment (LCA) study focusing on West African conditions and *Jatropha curcas* biodiesel production announced a 72% reduction in Greenhouse Gas emissions (GHG) compared with conventional diesel fuel (from agricultural production to biodiesel storage) (Ndong, et al., 2009). RVO from *Jatropha curcas* sold directly on local markets would reduce GHG emissions by 45% (Ndong, et al., 2009). In fact, direct use of the oil does not change GHG emissions from *Jatropha* cultivation, but it does reduce to a larger extent the energy consumption of the process, because neither transesterification nor transportation for export is necessary. These results illustrate robust opportunities and high potential for *Jatropha* to attract carbon credits from the Clean Development Mechanism (CDM) or the voluntary market (Achten, et al., 2008), especially in sub-Saharan Africa. These results in West Africa can mostly be explained by the perennial nature of the crop and by a decentralized, non-motorized and low-input production system.

Biofuel activities would directly and keenly affect communities in which they are located (De Castro, 2007). Best-case scenarios may envisage biofuels as the major source of quality employment, providing energy services at an acceptable price, while giving rise to environmental benefits such as carbon reductions, land restoration, and watershed protection. On the other hand, one can also imagine worst-case scenarios. Crop production may lead to further consolidation of land holdings, land-use conflicts in food production, displacement of existing livelihoods and negative environmental effects. Thus, developed countries implementing biofuel programmes must be careful in order to avoid welfare losses (Peters & Thielmann, 2008).

Profitability of Biofuels

In the literature, the economic competitiveness of fossil fuels is a very common argument against renewable energies and biofuels in particular, even though the costs of producing biofuels in developing countries are very near to the world market price of petroleum fuel (Demirbas, 2009). Biofuel production costs can vary widely depending on the feedstock, conversion process, scale of production and region, and use of presscake (livestock feed, manuring, etc.), which is why financial burdens are difficult to determine globally. At the moment, the cost of producing biofuels is often significantly higher than the cost of producing fossil fuels (OCDE & FAO, 2008; Peters & Thielmann, 2008; Tatsidjodoung & Martin, 2010). However, the gap is expected to narrow with the current hike in the price of oil and given the extra cost for oil in Burkina Faso due to storage and transportation. Imported fuels in Burkina Faso cost 30% more than the CAF price due to storage in coastal warehouses and road or rail transport (which are directly indexed to the price of a barrel) (Blin, et al., 2008). This snowball effect is increasing the overall prohibitive access to energy. The costs of biofuels could also probably decline in the future, especially if it is produced by family farming even on an industrial scale, and with easy processes like RVO technologies, because no transesterification with costly processes and components is required and only few investments are necessary to develop this way. In the electricity or shaft power way, many conditions are combined in Burkina Faso to produce RVO at the same price as or cheaper than fossil fuels. In any event, if we consider the potential added value for agricultural products and the national economy, this RVO strategy appears to be profitable in the short term. Nevertheless, promotion measures are essential for ensuring substantial domestic biofuel production and demand. These promotion measures are justified by their proponents as a source of environmental benefits, fostering the security of energy supplies, and leading to job creation in the agricultural sector (Peters & Thielmann, 2008).

The "economic mechanisms" in operation here are:

- Linking commercial demand for traditional and modern fuels with the existing potential local capacity (land, labour, water, knowledge, etc.) to produce alternative modern fuels; and
- Re-orienting the financial resource flows away from "commodity purchase" (energy imports) towards "capital investments" (biofuel production facilities) and "welfare" build-up (employment and income generation).

Fiscal Aspects and Impacts of Biofuels on Macroeconomics

According to a ministerial by-law (Blin, et al., 2008), fossil fuels prices are administered according to the port of arrival, the haulage fees, and rental for storage. All products are subject to customs levies and taxes, with taxes on petroleum products and VAT for diesel fuel and petrol only. As shown in Figure 2, subsidies are only given to oil products for electricity production (€ 0.145/l for DDO and € 0.27/l for fuel oil) (MCPEA, 2008). They aim to give access to electricity at an affordable price for consumers.

In view of Figure 2 and Table 4, diversifying oil with biofuels for electricity production would automatically reduce the subsidies granted to the national electricity production and distribution company for oil products. However, concerning the transport sector, the substitution of fossil fuels with biofuels would be a major source of financial losses for the State, as taxes are quite high, and petrol, fuel oil and DDO used for transportation represent more than 60% of national oil imports. The State raises € 27 million annually from imports of oil products (based on year 2004, current price), amounting to 41% of total import taxes, all sectors combined (INSD, 2000). Subsidies only concern the fuels for Sonabel, but they reached a high amount of € 29 million in 2007, despite a State decision in 2006 setting the maximum amount of subsidies at € 23 million per year (Blin, et al., 2008). All in all, the total annual in taxes is nearly equivalent to the subsidies, so interventions nearly cancel each other out. Consequently, the substitution of fossil fuels would not impact so much on the State Budget.

A diversification of energy sources with biofuels could save on the expenditure going to oil companies and moderate fuel imports, while increasing the benefits for households, firms and administrations. Moreover, this could change unproductive products, because imported, into products that generate national income and employment.

According to the Burkinese Government, in 2008, overall fossil fuel consumption (575,409 m^3) cost 251 billion CFAF, with only 36 billion considered as national added value creation (14.3%), whereas 215 billion CFAF were needed as foreign currency to pay the fossil fuel bill (Nonyarma & Laude, 2009). This energy consumption generated 87 billion CFAF of taxes (customs levies, taxes on oil and VAT) but 16 billion CFAF were

Figure 2. Oil product price structure in euros (€) (January, 2008) (MCPEA, 2008)

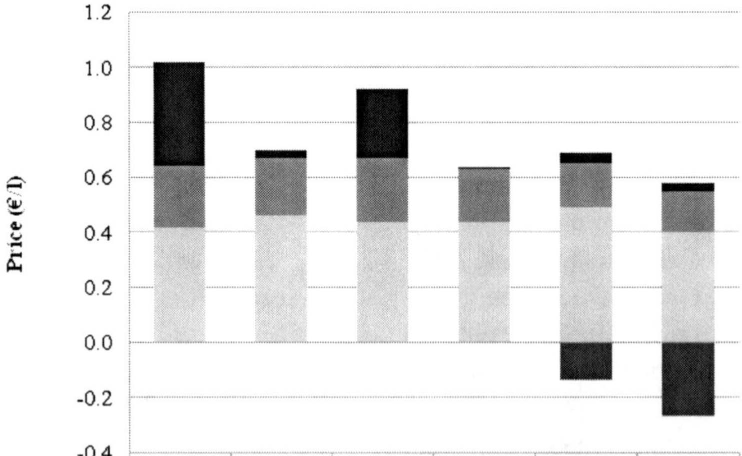

	Super 91	Paraffin	Gasoil	DDO	DDO Sonabel	Fuel Oil Sonabel
Taxes	0.380	0.030	0.250	0.003	0.040	0.030
Channels	0.220	0.210	0.230	0.190	0.160	0.150
CAF Price	0.420	0.460	0.440	0.440	0.490	0.400
Subsidy					-0.140	-0.270

Table 4. Oil product use and proportions in Burkina Faso (Blin, et al., 2008)

Oil product	Petrol	Paraffin	Gasoil	DDO	DDO Sonabel	Fuel Oil Sonabel
Ratio (%) in oil product imports*	30	5	31	11		16
Use	100% for transport	100% for rural lighting	94% for transport 5% for electricity	20% for transport (train)	80% for electricity	100% for electricity in thermal power stations

*NB: calor gas and kerosene are not taken into account considering their low contribution to the oil import mix

reused to fund electricity production (Nonyarma & Laude, 2009).

The biofuel economic structure is diametrically opposed to the imported fuels profile. Indeed, raw material production is an activity mainly generating added value (if it is family farmed), given the very limited energy costs and equipment (from 8 to 10%) (MMCE, 2009). The processing of raw materials to make RVO requires equipment (press, filter, settling equipment, etc.) and an energy consumption that will include some import costs, from 40 to 50% of production costs (MMCE, 2009). However, that percentage could be reduced with centralized processing (oil mill, cooperative, small and medium enterprise, etc.), considering the scale effect and better yields. Thus, processing will generate added value of around 50 to 60% of production costs. For rural processing, with the use of local small screw-presses powered by local RVO, the added value of the production process could amount to around 80% of production (MMCE, 2009).

Depending on a rural or an industrial scale option for production and processing, biofuel production will require around 16 to 40% of investment in foreign currency and will generate between 84 and 60% of direct added value in terms of rural incomes, wages, financial expenses, even taxes for the State, and earnings for economic operators (MMCE, 2009).

Compared to imported fuels, national biofuel production could create more significant direct added value, helping to increase national GDP (Mulugetta, 2009). This added value will contribute to rural populations by improving their standard of living and health, thereby increasing their production capacity. Incomes would also be injected into production activities, such as buying equipment, transport and services. These activities contribute indirectly to GDP and generate State earnings with an exponential effect on the economy by boosting subcontracting and finance activities. Economic simulations are roughly developed in Table 5 considering expenses of € 1 of imported fuels or € 1 of biofuels. For instance, the energy expenditure for biofuels generates 3 to 4 times more national added value, and the taxes on it would be 2 to 3 times greater than the same expenditure on imported fuels.

National Policy Framework for the Development of Biofuels

National data on fossil fuel consumption have been forecast up to 2025, taking into account parameters such as demography and the increase in access and

Table 5. Incomes provided from fossil fuel or biofuel spending (MMCE, 2009)

	Fossil fuel	Biofuel
Expenses (€)	1	1
Direct and indirect added value (€)	0.4	1.4 to 1.6
Taxes on added value (€)	0.07	0.16 to 0.22
Foreign currency need (€)	0.86	0.26 to 0.46

demand, but also the substantial interconnection strategy (Laude, 2008a). This strategy allows a very substantial reduction in oil imports (divided by 3) as early as 2010, when the first connection will start. Despite the importing of energy, fossil fuel consumption will then gradually evolve until at least 2025, returning to the same level of fossil fuel consumption as in 2009. Thus, the strategy to replace the oil used for electricity production by RVO is possible since 2010, with a relatively gentle increase in the lands required, given that 70,000 ha of *Jatropha curcas* are already planted. Moreover, in view of the available arable land in the country (Blein et al., 2008), the strategy is realistic from an agronomic and technical viewpoint (in 2005, only 36.7% of the available lands were being cultivated). Research and experience may help to improve agronomic data and choices, to achieve better yields with adequate plants and technical knowledge to allow better productivity and profitability, but also to gradually improve the level of substitution for electricity production.

In view of the projected oil demand and the volatility of oil prices, Burkina Faso has decided to commit itself 'resolutely but reasonably' to a policy of promoting biofuels (Nonyarma & Laude, 2009). Thus, the government will apply a maximum ceiling for lands reserved for energy production, which has been fixed at 500,000 ha corresponding to 5% of the total arable land (already 45% sown). The policy will promote traditional peasants' involvement in raw material supplies for national biofuel production earmarked for the domestic market. The Policy seeks to follow up and supervise the enthusiasm of developers and traditional farmers for *Jatropha curcas* (Blin, et al., 2011).

This policy will be accompanied by regulatory and tax measures to enhance the economic benefits of biofuel/hydrocarbon substitution, namely savings in foreign currency disbursements and the creation of added value on local and national levels, promoting the economic development of rural areas.

CONCLUSION

In fact, producing and using biofuels in Burkina Faso would enable the diversification of energy sources; a reduction in the dependency of the State on oil companies and of households on fossil fuel products; diversification of crops and incomes for farmers; local rural investments and better access to energy that could enable agricultural intensification by mechanization and the promotion of crops through processing, storage, and transportation of food products. In that way, biofuels could help to reduce food insecurity and be a factor of sustainable development.

The government of Burkina Faso has designed a Global Strategy for the Reduction of Poverty, with the fundamental objective of accruing human capital and triggering growth dynamics. It is also meant to enhance the competitiveness of the economy, ensure its smooth integration in the global economy and improve the living conditions of the entire nation in the long term. Energy plays a critical role in reaching these objectives by providing a vital service necessary for improving the quality of life and boosting economic progress (Singh & Sooch, 2004). Woody resources alone may not enable substantially improved growth. Shaft power and electricity are needed to develop industry and agriculture and generate incomes on a local scale. Ensuring a sustainable energy supply is a key challenge to achieve the MDGs. However, with fossil fuel dependency persisting and without an adequate energy strategy, no sustainable development would be possible. There are many reasons for biofuels to be considered as relevant technologies for sustainable development: energy security reasons, environmental concerns, foreign currency savings and socio-economic issues related to the rural sector. Biofuel production has significant macroeconomic advantages in terms of reducing the balance of trade deficit, creating direct and indirect added value and creating additional taxes on added value.

The State could benefit from this scenario, also by reducing or slowing down subsidies for electricity production and the running of its administration's operations. Local production and consumption of RVO is a favourable option to boost local potentials, promote investments and create development with accessible technologies and practices in the short term.

However, despite many potential benefits (Demirbas, 2009), biofuels will not automatically contribute to sustainable development. A sustainable process of local rural economic growth and diversification could gradually be developed with broader poverty alleviation impacts, but only within adequate and socially responsible policy frameworks.

REFERENCES

Abamovay, R. (2009). *Incertitues socioenvironnmentales dans la géopolitique de l'éthanol brésilien.* Ouagadougou, Burkina Faso: Université de Sao Paulo..

Achten, W. M. J., Verchot, L., Franken, Y. J., Mathijs, E., Singh, V. P., & Aerts, R. et al. (2008). Jatropha bio-diesel production and use. *Biomass and Bioenergy, 32*(12), 1063–1084. doi:10.1016/j.biombioe.2008.03.003.

Afrepren/FWD. (2002). *Energy database.* Nairobi, Kenya: Energy, Environment and Development Network for Africa Website..

Alpha, A., & Castellanet, C. (2007). *Défendre les agricultures familiales: Lesquelles, pourquoi?* Retrieved from http://www.alimenterre.org/ressource/defendre-agricultures-familiales-lesquelles-pourquoi.

Amigun, B., Sigamoney, R., & von Blottnitz, H. (2008). Cornmercialisation of biofuel industry in Africa: A review. *Renewable & Sustainable Energy Reviews, 12*(3), 690–711. doi:10.1016/j.rser.2006.10.019.

BafD/OECD. (2008). *Perspectives économiques en Afrique.* Nairobi, Kenya: African Development Bank..

Blein, R., Goura Soulé, B., Faivre Dupaigre, B., & Yérima, B. (2008). *Les potentialités agricoles de l'Afrique de l'ouest (CEDEAO).* Nairobi, Kenya: Fondation pour l'agriculture et la ruralité dans le monde.

Blin, J., Dabat, M., Faugère, G., Hanff, E., & Weisman, N. (2008). *Opportunités de développement des biocarburants au Burkina Faso.* Ouagadougou, Burkina Faso: Government..

Blin, J., Ficini, C., Faugère, G., & Dabat, M. H. (2007). Etude prospective sur le potentiel pour la production de biocarburants au Burkina Faso. *Cirad, 21*, 50.

Blin, J., & Girard, P. (Eds.). (2011). Guide technique pour une utilisation énergétique des huiles végétales dans les pays de la CEDEAO. Paris, France: L'Harmattan..

Blin, J., Weisman, N., Hanff, E., & Dabat, M. (2011). Vers une stratégie nationale de développement des filières biocarburant: Le cas du Burkina Faso. *Liaison Energie-Francophone, 87*, 104.

Bugaje, I. M. (2006). Renewable energy for sustainable development in Africa: A review. *Renewable & Sustainable Energy Reviews, 10*(6), 603–612. doi:10.1016/j.rser.2004.11.002.

Carlsson, A. S. (2009). Plant oils as feedstock alternatives to petroleum - A short survey of potential oil crop platforms. *Biochimie, 91*(6), 665–670. doi:10.1016/j.biochi.2009.03.021 PMID:19375482.

Challinor, A., Wheeler, T., Garforth, C., Craufurd, P., & Kassam, A. (2007). Assessing the vulnerability of food crop systems in Africa to climate change. *Climatic Change, 83*(3), 381–399. doi:10.1007/s10584-007-9249-0.

Clare, S. (2002). Energy for the poor: Underpinning the millennium development goals. *Future Energy Solutions, SW1E 5HE*, 32.

Dabat, M. H. (2011). Les nouveaux investissements dans les agrocarburants: Quels enjeux pour les agricultures africaines? *Afrique Contemporaine*, *237*, 97–109. doi:10.3917/afco.237.0097.

Darlu, J. (2009). *Analyse de faisabilité technique et économique des projets biocarburants en circuit court au Burkina Faso, étude du cas de Kinfangué, village au sud de Ouagadougou.* 2iE.

DCPM. (2009). *Campagne agricole 2009-2010: Mettre l'accent sur les filières porteuses*. Ouagadougou, Burkina Faso: Direction de la communication et de la Presse Ministérielle du Burkina Faso.

De Castro, J. (2007). *Biofuels – An overview, Final report*. DGIS/DMW/IB..

Demirbas, A. (2007). Modernization of biomass energy conversion facilities. *Energy Sources Part B-Economics Planning and Policy*, *2*(3), 227–235. doi:10.1080/15567240500402784.

Demirbas, A. (2009). Biofuels securing the planet's future energy needs. *Energy Conversion and Management*, *50*(9), 2239–2249. doi:10.1016/j.enconman.2009.05.010.

Demirbas, M. F., & Balat, M. (2006). Recent advances on the production and utilization trends of bio-fuels: A global perspective. *Energy Conversion and Management*, *47*(15-16), 2371–2381. doi:10.1016/j.enconman.2005.11.014.

ECOWAS. & WAEMU. (2006). *White paper for a regional policy geared towards access to energy services for rural and periurban populations in order to achieve the millenium development goals*. Economic Community of West African States.

Escobar, J. C., Lora, E. S., Venturini, O. J., Yanez, E. E., Castillo, E. F., & Almazan, O. (2009). Biofuels: Environment, technology and food security. *Renewable & Sustainable Energy Reviews*, *13*(6-7), 1275–1287. doi:10.1016/j.rser.2008.08.014.

Fallot, A., Girard, P., Dameron, V., & Griffon, M. (2006). The assessment of biofuel potentials on global and regional scales in the tropical world. *Energy for Sustainable Development*, *10*(2), 80–91. doi:10.1016/S0973-0826(08)60534-7.

Girard, P., & Blin, J. (2009). *Synthesis and conclusions*. Paper presented at the Biofuels: Enhancing Development or Increasing Food Insecurity. Ouagadougou, Burkina Faso.

Gubitz, G. M., Mittelbach, M., & Trabi, M. (1999). Exploitation of the tropical oil seed plant Jatropha curcas L. *Bioresource Technology*, *67*(1), 73–82. doi:10.1016/S0960-8524(99)00069-3.

Haas, M. J., McAloon, A. J., Yee, W. C., & Foglia, T. A. (2006). A process model to estimate biodiesel production costs. *Bioresource Technology*, *97*(4), 671–678. doi:10.1016/j.biortech.2005.03.039 PMID:15935657.

Hanff, E., Dabat, M., & Blin, J. (2011). Are biofuels an efficient technology for generating sustainable development in oil-dependent African nations? A macroeconomic assessment of the opportunities and impacts in Burkina Faso. *Renewable & Sustainable Energy Reviews*, *15*(5). doi:10.1016/j.rser.2011.01.014.

Harrabin, R. (2008). *EU rethinks biofuels guidelines*. London, UK: BBC News..

Hermelin, B., & Lagandré, D. (2009). Les agrocarburants: Menaces ou opportunités pour les agricultures familiales? *Ecologie et Politique*, *38*, 69–77. doi:10.3917/ecopo.038.0069.

IEA. (2000). *World energy outlook 2000*. Paris, France: International Energy Agency..

IEA. (2008). *World energy outlook, executive summary*. Paris, France: International Energy Agency..

Ilkilic, C., & Yucesu, H. S. (2008). The use of cottonseed oil methyl ester on a diesel engine. *Energy Sources Part A-Recovery Utilization and Environmental Effects*, *30*(8), 742–753. doi:10.1080/15567030701436214.

INSD. (2000). *Burkina Faso: Tableau entrées-sorties 2000*. Ouagadougou, Burkina Faso: Institut National de la Statistique et du Développement du Burkina Faso..

INSD. (2003a). *Burkina Faso: Cadre stratégique de lutte contre la pauvreté*. Ouagadougou, Burkina Faso: Institut National de la Statistique et du Développement du Burkina Faso..

INSD. (2003b). *Burkina Faso: La pauvreté en 2003*. Ouagadougou, Burkina Faso: Institut National de la Statistique et du Développement du Burkina Faso..

IPCC. (2001). Climate change 2001: Impacts, adaptation, and vulnerability. Contribution of working group II to the third assessment report of the intergovernmental panel on climate change. Cambridge, UK: Intergovernmental Panel on Climate Change.

Karekezi, S. (2002). Poverty and energy in Africa - A brief review. *Energy Policy*, *30*(11-12), 915–919. doi:10.1016/S0301-4215(02)00047-2.

Kaushik, N., Kumar, K., Kumar, S., & Roy, S. (2007). Genetic variability and divergence studies in seed traits and oil content of Jatropha (Jatropha curcas L.) accessions. *Biomass and Bioenergy*, *31*(7), 497–502. doi:10.1016/j.biombioe.2007.01.021.

Khan, N. A., & el Dessouky, H. (2009). Prospect of biodiesel in Pakistan. *Renewable & Sustainable Energy Reviews*, *13*(6-7), 1576–1583. doi:10.1016/j.rser.2008.09.016.

Kumar, A., & Sharma, S. (2008). An evaluation of multipurpose oil seed crop for industrial uses (Jatropha curcas L.): A review. *Industrial Crops and Products*, *28*(1), 1–10. doi:10.1016/j.indcrop.2008.01.001.

Laude, J. (2008a). *Evolution des consommations énergétiques au Burkina Faso*. Ouagadougou, Burkina Faso: Ministère des mines, des carrières et de l'énergie MMCE.

Laude, J. (2008b). *Le secteur de l'énergie*. Burkina Faso: Ministère des Mines, des Carrières et de l'Energie MMCE.

Leder, F., & Shapiro, J. N. (2008). This time it's different - An inevitable decline in world petroleum production will keep oil product prices high, causing military conflicts and shifting wealth and power from democracies to authoritarian regimes. *Energy Policy*, *36*(8), 2850–2852. doi:10.1016/j.enpol.2008.04.015.

Mangoyana, R. B. (2009). Bioenergy for sustainable development: An African context. *Physics and Chemistry of the Earth*, *34*(1-2), 59–64. doi:10.1016/j.pce.2008.01.002.

Martinez, D. M., & Ebenhack, B. W. (2008). Understanding the role of energy consumption in human development through the use of saturation phenomena. *Energy Policy*, *36*(4), 1430–1435. doi:10.1016/j.enpol.2007.12.016.

McDade, S., Lallement, D., & Saghir, J. (2006). *Energy services for the millennium development goals*. New York, NY: United Nations Development Programme..

MCPEA. (2008). *Annexe I à l'arrêté n° 08.002/MCPEA/SG/DGC du 10 janvier 2008 portant composition des structures de prix des hydrocarbures*. Ouagadougou, Burkina Faso: Ministère du commerce, de la Promoion de l'entreprise et de l'artisanat du Burkina Faso.

MEF. (2009). *Atelier de validation des données de population de 1997 à 2006 et des projections démographiques sous nationales de 2007 à 2020*. Ouagadougou, Burkina Faso: Ministère de l'Economie et des Finances du Burkina Faso..

MMCE. (2009). *Document de politique de développement des biocarburants au Burkina Faso*. Ouagadougou, Burkina Faso: Ministère des Mines, des Carrières et de l'Energie du Burkina Faso.

Mulugetta, Y. (2009). Evaluating the economics of biodiesel in Africa. *Renewable & Sustainable Energy Reviews, 13*(6-7), 1592–1598. doi:10.1016/j.rser.2008.09.011.

Murugesan, A., Umarani, C., Subramanian, R., & Nedunchezhian, N. (2009). Bio-diesel as an alternative fuel for diesel engines-A review. *Renewable & Sustainable Energy Reviews, 13*(3), 653–662. doi:10.1016/j.rser.2007.10.007.

Mussa, M. (2000). *The impact of higher oil prices on the global economy*. Geneva, Switzerland: International Monetary Fund..

Ndong, R., Montrejaud-Vignoles, M., Saint Girons, O., Gabrielle, B., Pirot, R., & Domergue, M. et al. (2009). Life cycle assessment of biofuels from Jatropha curcas in West Africa: A field study. *GCB Bioenergy, 1*(3), 197–210. doi:10.1111/j.1757-1707.2009.01014.x.

NEPAD. (2001). *The new partnership for Africa's development*. Abuja, Nigeria: The New Partnership for Africa's Development..

Nonyarma, E., & Laude, J. (2009). *Cadrage de la politique de développement des biocarburants au Burkina-Faso*. Ouagadougou, Burkina Faso: Ministère des Mines, des Carrières et de l'Energie MMCE.

OCDE. (2008). *Note n°2 du club du sahel et de l'Afrique de l'ouest, extrait de la publication "L'Afrique de l'ouest peut-elle se nourrir? Carburants verts, carburants du développement? Pour une meilleure cohérence des politiques en Afrique de l'ouest."* Paris, France: Organisation for Economic Co-operation and Development..

OCDE, & FAO. (2008). *Production costs of major biofuel chains*. Paris, France: Organisation for Economic Co-operation and Development, Food and Agriculture Organization.

OMC. (2004). *Examen des politiques commerciales au Burkina Faso*. Burkina Faso: Organisation Mondiale du Commerce..

Openshaw, K. (2000). A review of Jatropha curcas: An oil plant of unfulfilled promise. *Biomass and Bioenergy, 19*(1), 1–15. doi:10.1016/S0961-9534(00)00019-2.

Ouédraogo, G. (2004). *Etat des lieux de la carbonisation au Burkina Faso*. Bamako..

Peters, J., & Thielmann, S. (2008). Promoting biofuels: Implications for developing countries. *Energy Policy, 36*(4), 1538–1544. doi:10.1016/j.enpol.2008.01.013.

PNUD. (2011). *Bulletin bi-mensuel du Pnud au Burkina Faso*. PNUD..

Schneider, M. P. (2006). Plant-oil-based lubricants and hydraulic fluids. *Journal of the Science of Food and Agriculture, 86*(12), 1769–1780. doi:10.1002/jsfa.2559.

Sebitosi, A. B., & Pillay, P. (2005). Energy services in sub-Saharan Africa: How conducive is the environment? *Energy Policy, 33*(16), 2044–2051. doi:10.1016/j.enpol.2004.03.025.

Sidibé, S. S., Blin, J., Vaitilingom, G., & Azoumah, Y. (2010). Use of crude filtered vegetable oil as a fuel in diesel engines state of the art: Literature review. *Renewable & Sustainable Energy Reviews*.

Singh, K. J., & Sooch, S. S. (2004). Comparative study of economics of different models of family size biogas plants for state of Punjab, India. *Energy Conversion and Management, 45*(9-10), 1329–1341. doi:10.1016/j.enconman.2003.09.018.

Sonabel. (2008). *Rapport d'activités 2008*. Burkina Faso: Société Nationale d'électricité du Burkina.

Sujatha, M., Reddy, T. P., & Mahasi, M. J. (2008). Role of biotechnological interventions in the improvement of castor (Ricinus communis L.) and Jatropha curcas L. *Biotechnology Advances, 26*(5), 424–435. doi:10.1016/j.biotechadv.2008.05.004 PMID:18579331.

Tatsidjodoung, P., & Martin, E. (2010). *Evaluation de l'utilisation d'huiles végétales brutes à partir d'oléagineurx pour la production d'électricité au Burkina Faso*. Cirad/2iE/Agropol.

Toonen, H. M. (2009). Adapting to an innovation: Solar cooking in the urban households of Ouagadougou (Burkina Faso). *Physics and Chemistry of the Earth*, *34*(1-2), 65–71. doi:10.1016/j.pce.2008.03.006.

Treanton, K. (2004). *Energy statistics working group meeting*. Paris, France: International Energy Agency IAE..

UN. (2005). *About the millennium development goals*. New York, NY: United Nations..

Unal, H., & Alibas, K. (2007). Agricultural residues as biomass energy. *Energy Sources Part B-Economics Planning and Policy*, *2*(2), 123–140. doi:10.1080/15567240600629401.

UND. (1986). Our common future, chapter 2: Towards sustainable development. Report. New York, NY: World Commission on Environment and Development.

Vaitilingom, G. (2006). Utilisation énergétique de l'huile de coton carburant des moteurs diesels. *Cahiers Agricultures, 15*(1).

WB. (1996). *Rural energy and development: Improving energy supplies for two billion people*. Washington, DC: The World Bank..

Zongo, T. (2008). *Investir au Burkina Faso, la porte ouverte sur le grand marché de l'Afrique de l'Ouest: Note de présentation de la situation et du potentiel économique du Burkina Faso*. Ouagadougou, Burkina Faso: Premier Ministere du Burkina Faso..

This chapter was previously published in Global Sustainable Development and Renewable Energy Systems edited by Phillip Olla, pages 152-171, copyright 2012 by Information Science Reference (an imprint of IGI Global).

Chapter 78
Indicators of Land Degradation Vulnerability due to Anthropic Factors:
Tools for an Efficient Planning

V. Imbrenda
University of Basilicata, Italy & National Research Council of Italy, Institute of Methodologies for Environmental Analysis, Italy

M. D'Emilio
Institute of Methodologies for Environmental Analysis, Italy

M. Lanfredi
Institute of Methodologies for Environmental Analysis, Italy

M. Ragosta
University of Basilicata, Italy

T. Simoniello
Institute of Methodologies for Environmental Analysis, Italy

ABSTRACT

Land degradation is one of the most impacting phenomena on natural resource availability, both in quantitative and qualitative terms. In order to provide efficient tools for territorial sustainable management in areas affected by land degradation, it is important to define suitable models and indicators able to identify exposed areas and their vulnerability level, so as to provide an effective support for decision makers in identifying intervention priorities and planning mitigation/adaptation strategies. This work is focused on the evaluation at high spatial detail of land degradation vulnerability due to anthropic factors, which is a crucial issue in areas devoted to farming practices. Vulnerability is evaluated by integrating a new indicator of the mechanization level the authors recently developed, with a set of census based indicators of land management. The new indicator is independent of census data being based on land cover data; thus, it can provide a better spatial characterization and a more frequent updating compared to commonly adopted indices that are evaluated at municipal scale. By analyzing data for the whole Southern Italy, such an indicator was integrated for the first time at full spatial resolution to obtain a final vulnerability index of land management. This comprehensive index enabled a more accurate estimation of the land degradation vulnerability due to anthropic factors allowing the discrimination of priority areas within the municipal areas.

DOI: 10.4018/978-1-4666-4852-4.ch078

INTRODUCTION

A substantial part of Southern Italy (~25%) was found to be affected by high or medium-high vulnerability levels due to management factors. In particular, the heavy use of agricultural machineries and modifications of cultivated areas, mainly linked to land abandonment (~40%), represent the principal factors inducing vulnerability to land degradation in Southern Italy; thus management policies devoted to reduce intensive agriculture and promoting the requalification of abandoned areas are suggested to mitigate the territorial vulnerability of the investigated regions.

Land use planning is a fundamental discipline for the setting up of sustainable development strategies, accounting for the territorial development needs as well as for impacts induced, directly or indirectly, on the different environmental matrices: soil, water, vegetation and atmosphere. In recent years, the considerable growth of technologies, such as remote sensing and GIS software, enabled to noticeably increase the understanding of the environmental components, highlighting their peculiarities and criticalities. Geographically referred information on environmental and socio-economic components represent a heritage of knowledge which can be favorably exploited to identify vulnerable areas, distinguishing also different vulnerability levels. Such a possibility reaches even more relevance if we consider the increasingly significant impact of climate change and land transformation processes, consisting in rapid and frequent modifications of land use patterns (Hill et al. 2008; Simoniello et al. 2008; Lanfredi et al. 2004).

Among the phenomena able to deeply reduce the availability and quality of natural resources, a dominant role is accomplished by land degradation, defined as the "loss of actual or potential biological and economic productivity" (UN/FAO, 2003). It is the result of a non-sustainable resource exploitation associated with territorial structural weaknesses related to soil, vegetation, and climate (EEA, 2005; Reynolds et al. 2007).

In the context of land degradation processes, climate or climatic change act as boundary conditions and human actions are fundamental to lead to degradation (Baartman et al. 2007). This is especially true for the developed countries where anthropogenic factors (e.g., population pressing, tourism growth, agriculture intensification, industrial development) represent the main cause of land degradation phenomena (EEA, 2005). Specifically, in the past 50 years the Mediterranean area has seen an increase in population which has had a direct effect in soil sealing due to urban development into productive lands (Brandt et al. 2003). Urbanization has had serious consequences in terms of loss of areas previously devoted to forestry or agricultural practices, decline in soil quality, increase in fire risk and in salinization/alkalinization phenomena (Garcia Latorre et al. 2001; Tanrivermis, 2003; Salvati and Zitti, 2007). In addition, Mediterranean agricultural techniques have been transformed to cope with the increasing food demand so favoring intensive agriculture especially in lowland areas (Trisorio, 2005). Intensive agriculture can be positively regarded in terms of crop yield increase and technical skill improvements, but it also produces significant unfavorable effects on soil and habitat preservation (Brouwer et al. 1991). Inappropriate practices, such as deep water drainage, large-scale irrigation, heavy pesticide/fertilizer use and heavy mechanization are directly involved in the degradation of agricultural and semi-natural areas across huge territories (European Commission, 2000). Moreover, six of the soil degradation processes recognized by the European Commission (water, wind and tillage erosion, soil organic carbon decline, compaction, salinization and alkalinization, contamination and soil biodiversity reduction) are identified as threats induced by agriculture (Gay et al. 2009). Also overgrazing is considered as a driving force in generating land degradation, particularly for many areas of Southern Europe it is considered as the main threatening factor (e.g., Yassoglou, 1999).

In order to identify the most suitable strategies for minimizing/reversing land degradation phenomena, it is vital to assess their extent and magnitude. Mapping degraded areas is important to identify priority areas and the involved driving forces for supporting development plans and policy decisions devoted to a sustainable use of land resources (Hill et al. 1995). At this aim, vulnerability models are used to provide support for efficient monitoring and intervention activities in areas threatened by land degradation.

Among the different models estimating land degradation vulnerability, the ESA (Environmentally Sensitive Areas, derived from the MEDALUS project, Kosmas et al. 1999) is the most widely used approach in the Mediterranean basin thanks to its flexible structure, which permits the inclusion and exclusion of variables as a function of the investigated scale and framework (Basso et al. 2000, Contador et al. 2009). Moreover, it can be considered as an efficient early warning system for land degradation because it does not focus on a single process (erosion, compaction, salinization, crusting, etc.) but computes the synergic impact of various factors expected to cause land degradation (Salvati and Bajocco, 2011). This model consists of the combination of indices associated with soil, climate, vegetation and land management, each of which is composed by different indicators representing a specific degradation driver.

In general, the variables related to the physical-environmental system (climate, soil and vegetation) are considered as the most important for evaluating the vulnerability to land degradation (Thornes and Brandt, 1995); whereas the contribution of the anthropogenic factors (including parameters connected with agriculture and grazing), even if mentioned, is often underestimated (Juntti and Wilson, 2005; Salvati et al. 2008). This can be particularly misleading in overexploited areas, such as the Mediterranean basin. Moreover, land use planning actions mainly focus just on the anthropogenic component for mitigating risk in vulnerable areas and for safeguarding areas having a high landscape and natural value, as highlighted by UNCCD-STC (United Nations Convention to Combat Desertification - Science and Technology Committee).

Operatively, the effectiveness of the land degradation indicators is connected not only to the choice of key variables, but also to the spatial resolution and the updating rate of the available data. The use of dynamic information with an appropriate spatial detail at the investigated scales represents the core of flexible and efficient tools for sustainable management both to face early warning problems and to rationalize long term development.

Nevertheless, the most part of land management indicators developed in the last twenty years is based on census data generally provided by regional, national or international statistics agencies (in Italy the ISTAT - Italian National Institute of Statistics) and often the highest spatial detail is represented by the municipal scale (see e.g. SAR, DesertNet projects). Unfortunately, this does not permit investigations within the municipality boundaries to directly act on the threatened areas and the census data strongly bind the temporal update (once every ten years for Italian censuses).

To overcome such a difficulty, in this work we estimated land degradation vulnerability due to anthropic factors for five regions of Southern Italy by introducing a new mechanization level indicator that exploits land cover maps for evaluating soil vulnerability to compaction induced by agricultural machineries at a finer resolution than the municipal scale. This facilitates the evaluation of the mechanization levels on a realistic base assessing the specific requirements of the different agricultural areas within the investigated territory.

The final land management index, computed by integrating additional indicators, has the same spatial resolution of this new indicator and is more suitable to depict the actual local variability of the anthropic stress than indicators exclusively based on census data.

In particular, the estimated land vulnerability due to agricultural and grazing activities and the subsequent mapping of the most vulnerable areas aims at promoting restoration and/or mitigation plan and supporting a more rational set up of the environmental and territorial systems.

MATERIALS AND METHODS

Study Area

In this work, we focused the attention on five regions of the Southern Italy: Apulia, Basilicata, Campania, Calabria and Sicily (Figure 1). These regions are recognized as areas at potential risk of land degradation (Costantini, 2009). Their vulnerability is mostly due to the presence of specific climatic conditions (pronounced bi-seasonality with recurring and severe seasonal droughts followed by very intense rainfalls) as well as to the millenary human activities (e.g., agriculture, logging and livestock grazing). Moreover, in the last decades, modern intensive agriculture has exposed soils to an overexploitation in terms of heavy mechanization, large-scale use of nutrients and water requirements (Istat, 2000). Nowadays, land degradation causes a considerable economic damage to the agricultural sector: the estimated costs in Italy amounts to about 12€/ha of cultivated surface (Salvati and Carlucci, 2010).

Data and Indices

In order to evaluate the incidence of anthropic activities in generating/accelerating land degradation phenomena in Southern Italy, we adopted the ESA approach (Kosmas et al. 1999, Ferrara, 2005). Among the available indicators to assess the impact of human activities on land degradation processes, we selected four land management indicators: mechanization level, percentage of permanent grass and pasture, variation of cultivated surfaces, and grazing intensity. Our choice took into account the process representativeness and the data availability for the investigated area and scale.

The first indicator estimates vulnerability to soil compaction linked to mechanized farming practices and it is based on land cover data. The other three indicators are derived from census data and concern agricultural and grazing practices. They evaluate: changes in cultivated lands connected to intensification/abandonment processes, presence of low-impacting covers (permanent grass and pasture), and grazing impact that induces alteration on soil and vegetation by trampling and feeding of livestock.

Figure 1. Studied regions of Southern Italy

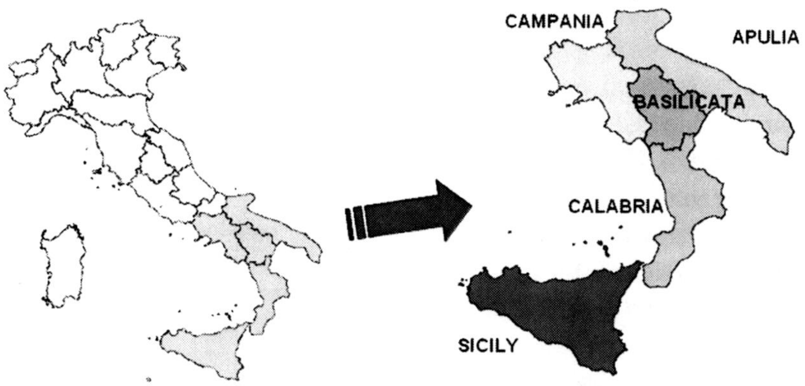

In order to compute the mechanization level indicator, we used the Corine Land Cover map 2000 acquired from SINAnet - APAT (Italian National Environmental Information System - Italian National Agency for Environmental Protection); information on machinery passes per cultivations provided by ENAMA (Italian National Agency of Agricultural Mechanization); and a 90m resolution Digital Elevation Model (source NASA, reprocessed by CSI-CGIAR, Consortium for Spatial Information Consultative Group on International Agricultural Research). The elaboration of census-based indicators was derived from 1990 and 2000 Agricultural Censuses (last available updates) at municipal scale (best available detail) obtained from ISTAT database.

By following the ESA approach, in order to combine the selected indicators, the values of each indicator were classified in a common range of vulnerability starting from 1 (very low vulnerability) up to 2 (very high vulnerability).

All the spatial analyses and basic operations were carried out in a GIS environment. The geographic information was projected to a common UTM zone 33 under the WGS84 datum.

Land Cover Based Index

The mechanization level index was elaborated by adopting a new formulation we recently developed in the framework of the Mildmap-Media project (Imbrenda et al. 2009). Such an index is a proxy for soil compaction, which is the result of persistent passes of farm machinery in agricultural areas. Generally, indicators adopted in land degradation models are calculated from census data on farm machinery ownership at municipal level. The most common indicator is the APAT one (2000) estimating the number of farm-owned machinery on agricultural surfaces (MLI_{APAT} = N/10ha of UAA). A more structured version of this indicator was proposed by Bazzoffi (2003), which integrates machinery weight per unit of area and the average number of passes per year on agricultural areas. One of the problems arising from these formulations is related to the mere evaluation of farm-owned machinery in areas where third party property management or joint ownership of farm machinery are very frequent. Moreover, they do not take into account any information on cultivation type (different cultivations require a different number of passes for various agronomic practices) and on the tractive equipment of agricultural machineries (there is a substantial difference between tyres and tracks in terms of induced soil compaction).

To overcome such problems, the adopted indicator estimates compaction vulnerability due to heavy vehicle traffic by considering both the different number of passes for each cultivation (calculated from the land cover map) and the different impact on soil produced by tyres or tracks (evaluated by using a slope map). Starting from the Corine Land Cover map (CLC2000), we associated an average number of passes, obtained aggregating data provided by ENAMA, on the single type of cultivation to each agricultural class (Table 1).

Table 1. Number of average passes for CLC2000 class, obtained aggregating ENAMA data for cultivation type

Cultivation type and corresponding CLC2000 level3 code	Number of average passes
Arable land (cereals, legumes, crops, vegetables, etc.) - 2.1.1/2.1.2	7,5
Permanent crops (vineyards, fruit trees, olive groves) - 2.2.1/2.2.2/2.2.3	7
Pastures - 2.3.1	3
Annual crops associated with permanent crops - 2.4.1	5
Complex cultivation patterns - 2.4.2	4
Land principally occupied by agriculture, with natural areas - 2.4.3	3
Agroforestry areas - 2.4.4	1
Other classes	0

In order to differentiate areas generally managed by tired vehicles from those in which tracked vehicles are used, we applied a threshold (20%) on the slope map derived from the DEM since, generally, only tracked vehicles can be employed on soil with steep slope, whereas tyres are adopted in plain areas. Tracks tend to mainly compact topsoil, that can be more easily restored, whereas tires mainly damage subsoil layers that are very difficult to restore (Keller et al. 2002, Pagliai et al. 2003). Therefore, we introduced a correction factor (*f*) attributing a lower vulnerability to areas managed with tracked vehicles ($f=1$) compared with that managed with tired vehicles ($f=1.5$).

The proposed indicator (MLI_p) has the following expression:

$$MLI_p = N_p \cdot f_p$$

where *p* indicates that the computation is carried out at the pixel resolution, N_p is the number of average passes for each Corine class, and f_p is the abovementioned correction factor accounting for track or tire use. Such an indicator has been subsequently classified in the common range 1-2 to make it comparable with the other ones, as shown in Table 2.

In order to highlight the improvements of this new formulation of the mechanization level index, we also computed the census based index of the national environmental agency (MLI_{APAT}) following the formula previously shown. The corresponding classification is reported in Table 2.

Census Based Indices

The first indicator calculates the percentage variations of cultivated surfaces (UAA_VAR) by the following formula:

$$UAA\_VAR = \frac{UAA_{t_2} - UAA_{t_1}}{UAA_{t_1}} \cdot 100$$

where $UAAt_1$ and $UAAt_2$ are respectively the starting and final value of the Utilized Agricultural Area during the analyzed period (in this case $t_1=1990$ $t_2=2000$). The indicator is classified on absolute values since both high increases and high decreases of cultivated areas can be considered equally as harmful processes (Table 3). The first one implies a reduction of natural lands, whereas the second one is generally linked to the abandonment of marginal lands, which can generate land degradation processes especially in areas characterized by structural weakness (Genesio et al. 2004; Brouwer et al. 1991).

The second indicator estimates the percentage of permanent grass and pasture surfaces (Sur_PP) within the total utilized agricultural area (UAA) through the following formula:

Table 2. Distribution of vulnerability classes for mechanization level indicators at pixel detail (MLI_p) and municipal scale (MLI_{APAT})

Vulnerability class	MLI_p Values	MLI_{APAT} Values
high	>9	>5
medium - high	7: 9	3: 5
medium	5: 7	2: 3
medium - low	3: 5	1: 2
low	<3	0: 1

Table 3. Distribution of vulnerability classes for the index of agricultural area variations (UAA_VAR)

Vulnerability class	UAA_VAR Values	
	Decreases	Increases
high	< - 50	> 50
medium - high	-50: -20	20: 50
medium	-20: -10	10: 20
medium - low	-10: -5	5: 10
low	-5: 5	-5: 5

$$PP\_UAA = \frac{Sur\_PP}{UAA} \cdot 100$$

For this indicator, data are referred to the year 2000. The highest indicator values are associated to the lowest vulnerability levels (Table 4) because permanent grass and pasture are considered covers on which the anthropic pressure has less degradation effects than on cultivated lands. Moreover, they have a protection function against erosion processes (Genesio et al. 2004).

The third indicator estimates the grazing intensity (GI), by evaluating the number of Adult Bovine Unit (ABU) on permanent grass and pasture areas (expressed in hectares), as follows:

$$GI = \frac{ABU}{Sur\_PP}$$

where ABU is calculated taking into account the unit number for the different livestock typology, homogenizing them to the size of adult bovine (Motroni et al. 2004):

$$\begin{aligned} ABU = {} & n.bovines \\ & + n.\ buffalos \\ & + n.\ equines \\ & + (n.\ goats)\ /\ 10 \\ & + (n.\ sheep)\ /\ 10 \end{aligned}$$

Table 4. Distribution of vulnerability classes for the percentage of permanent grass and pasture (PP_UAA) and grazing intensity (GI)

Vulnerability class	PP_UAA Values	GI Values
high	< 5	> 100
medium - high	5: 10	30: 100
medium	10: 30	10: 30
medium - low	30: 50	3: 10
low	50: 100	0: 3

Data used for the index computation are related to the year 2000. Overgrazing, still present in many areas of Southern Italy, causes significant damages to vegetation and soil. Specifically, the repeated trampling of livestock induces a reduction in herbaceous vegetation and an increase in soil compaction, with consequent water infiltration decrease and runoff increase. These effects altogether induce an intensification of water erosion processes (Zucca et al. 2006). Therefore, the highest vulnerability scores are attributed to the highest indicator values (Table 4).

Integration of the Indicators: Total Land Management Index

The final land management index (LMI) was calculated by integrating the four indicators previously described through the geometrical mean as follows:

$$LMI = (MLI \times UAA\_VAR \times PP\_UAA \times GI)^{1/4}$$

Table 5 reports the classification ranges and the related vulnerability classes.

Such a formula was applied considering both MLI_{APAT} and MLI_p. In the first case, we obtained a vulnerability map at municipal scale (LMI_m), where the subscript m indicates just the municipal level. In the second case, the elaborated vulnerability map has the spatial resolution (90m) derived from the MLI_p indicator and corresponding to the pixel size (LMI_p).

Table 5. Distribution of vulnerability classes for the land management index

Vulnerability class	LMI Values
high	1.8: 2
medium - high	1.6: 1.8
medium	1.4: 1.6
medium - low	1.2: 1.4
low	1: 1.2

RESULTS AND DISCUSSION

The maps of the elaborated mechanization level indicators and the land cover map are shown in Figure 2. It is possible to note that MLI_p (Figure 2(a)) provides a more representative picture of the vulnerability values that follows the actual arrangement of agricultural areas (Figure 2(c)). In particular, we found a low vulnerability level in mountain areas, widely recognized as areas less devoted to heavy agricultural practices and a medium or high vulnerability in lowlands areas (Tavoliere of Apulia, Vesuvian areas and a large part of Sicily, Calabria and Basilicata coastal areas) as well as in some hilly areas (e.g., medium and low hills surrounding Matera, largely affected by badland formations). On the contrary, the APAT indicator (Figure 2(b)) highlighted only few municipalities (especially in Campania region) having a high vulnerability value. Most of the areas devoted to heavy agricultural practices (the low Tavoliere, the Ofanto valley, the Adriatic coast in Apulia, see INEA, 2008; the Crotone's and Sibari's Plain in Calabria region and some internal areas near Catania in Sicily, see NSP of Rural Development, 2007-2013) show low or medium-low vulnerability values similarly to areas less devoted to heavy agricultural activities.

The distribution of the areas corresponding to different vulnerability classes for MLI_p and MLI_{APAT} are shown in Table 6. It is evident that the vulnerability levels of the APAT indicator are almost everywhere underestimated. Furthermore, there is no correspondence among these levels and the agricultural inclination of the investigated area. This is mainly due to the inclusion of

Figure 2. (a) Maps of the mechanization level indicators: new index at pixel detail - MLI_p; (b) Old index - MLI_{APAT}; (c) Corine land cover map 2000 with the legend showing only agricultural classes

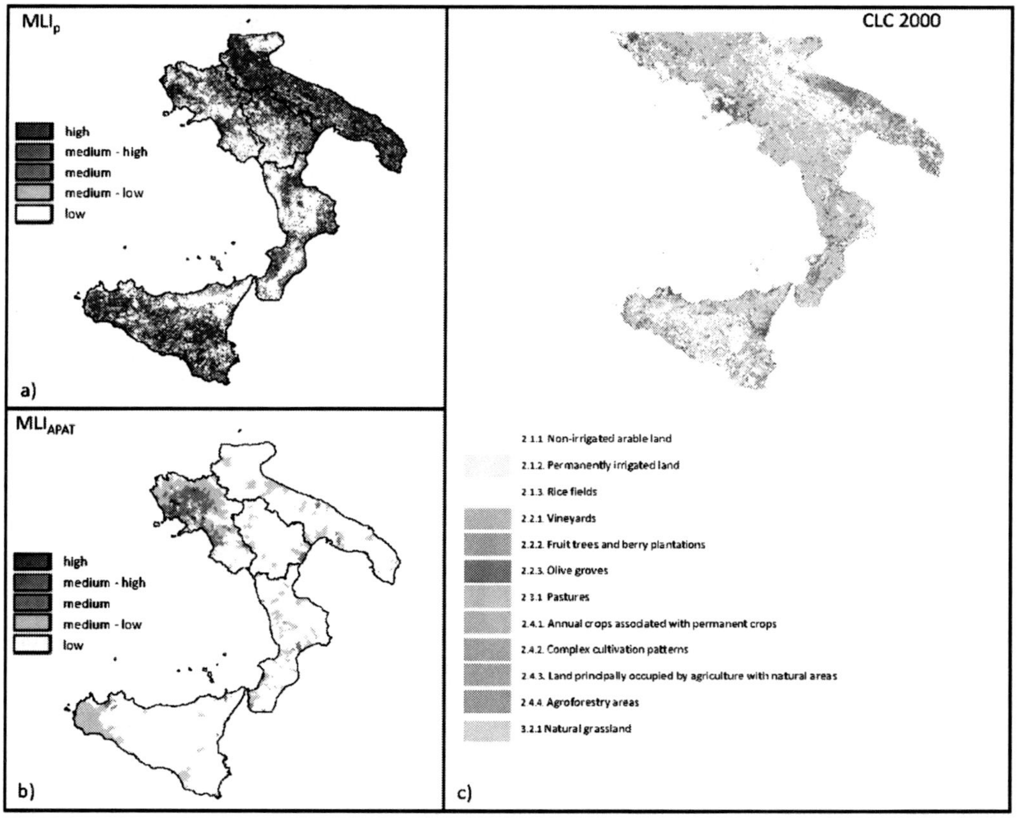

only farm-owned machinery. Unfortunately, although it is possible to use more updated and detailed databases (e.g., ACI - Automobile Club Italia) concerning the distribution of agricultural machineries, there is no possibility to correct the erroneous attribution of machineries to a single municipality (being linked to property) for considering the undifferentiated use in neighboring municipalities (especially for expensive vehicles, such as harvesters).

The maps of the other land management indicators are shown in Figure 3. We can note that UAA_VAR shows the higher vulnerability values, equally arranged in all the study area (Figure 3(a)). On the contrary, PP_UAA represents an important vulnerability factor in Apulia region, in Campania coastal areas and in south-western Sicily (Figure 3(b)). In such zones, UAA is prevalently devoted to intensive farming, rather than low-impact activities. The third indicator GI shows high vulnerability values in many municipalities of Campania region and in a strip area located among Taranto, Bari and Brindisi provinces of Apulia region (Figure 3(c)).

Table 6. Comparison between MLI_{APAT} and MLI_p in terms of spatial extent of vulnerability classes (percentage of the total area)

Vulnerability class	MLI_{APAT}	MLI_p
high	0.3	40.6
medium - high	1.5	6.8
medium	3.4	9.2
medium - low	15.2	4.7
low	79.5	38.7

The comparison of the comprehensive indices (LMI_p and LMI_m), obtained from the integration of the new mechanization index and the old one with the other census based indices, highlights a general higher vulnerability level for the first map provided at the pixel resolution (maps are shown in Figure 4).

It is evident that LMI_p allows the discrimination of different vulnerability values within any given municipality (Figure 4(a)) with obvious advantages in terms of land use planning, since it represents a helpful improvement respect to previous studies at the municipal level (Imbrenda et al. 2008). As regards to the spatial distribution of the more vulnerable areas, we observe a pattern similar to the one found for MLI_p (Figure 4(b)), underlining the importance of land management on the land degradation vulnerability. Moreover, we notice a significant increase of areas at medium-high or high vulnerability (Figure 5(a) and 5(b)).

Figure 3. Maps of the anthropic pressure indicators linked to agricultural and grazing activities: (a), UAA variation between years 1990 and 2000; (b) percentage of permanent grass and pasture on UAA; (c) grazing intensity

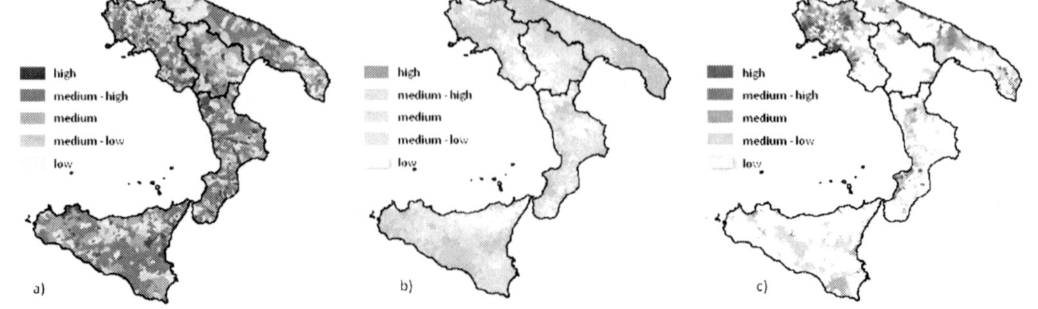

Figure 4. Maps of the two land management indices LMI_p and LMI_m obtained by integrating the land cover based (MLI_p) and the census based (MLI_{APAT}) mechanization index, respectively

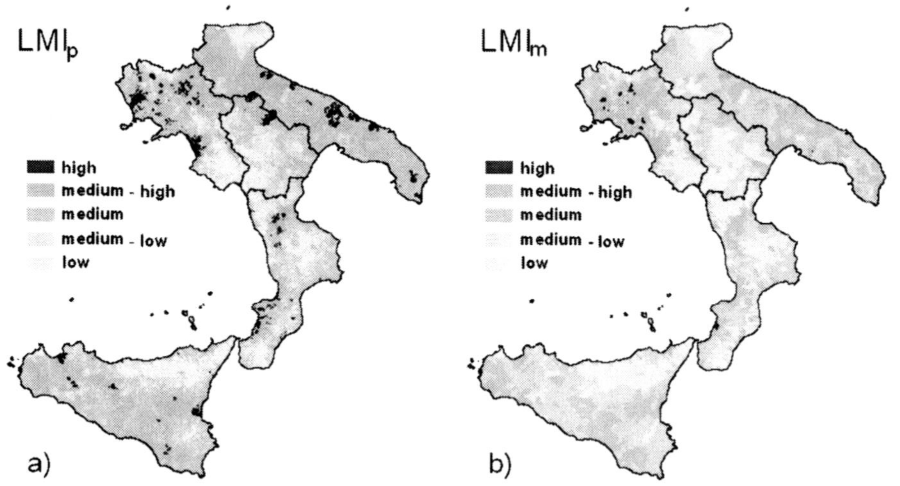

Figure 5. Distribution of the vulnerability classes for LMI_p and LMI_m (Southern Italy)

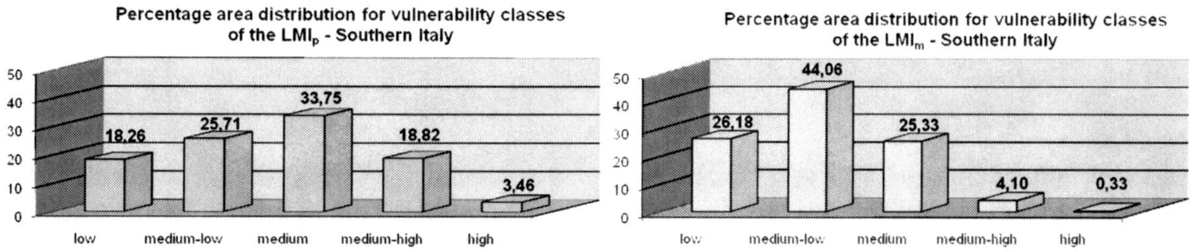

In particular, some of the more critical areas are located along the coast of Campania and Apulia regions. On the basis of the LMI_p distribution at regional level, Apulia shows the highest percentage of territory having a medium-high or high vulnerability levels (Figure 6(d)), followed by Campania and Sicily regions (Figure 6(a) and 6(b)). On the contrary, Calabria and Basilicata regions show the lowest percentage of highly vulnerable areas due to agricultural management practices. In particular, a large part of the territory of these regions (Calabria ~60%, Basilicata ~65%) is characterized by low or medium-low vulnerability levels due to the significant presence of wooded areas (Figure 6(c) and 6(e)).

CONCLUSION

In order to estimate vulnerability to land degradation due to land management factors in Southern Italy, we elaborated and integrated indicators concerning agriculture and grazing pressure. The adopted indicators were percentage variation of cultivated surfaces, percentage of permanent grass and pasture, grazing intensity and mechanization level. In particular, the latter indicator was improved according to a new approach that evaluates the vulnerability to soil compaction linked to agricultural vehicles by using a land cover map instead of data derived from censuses. Such an approach enables to estimate the number of vehicle passes as a function of the different current cultivations and the diverse

Figure 6. Frequency distribution of vulnerability levels for LMI_p in the regions of Southern Italy

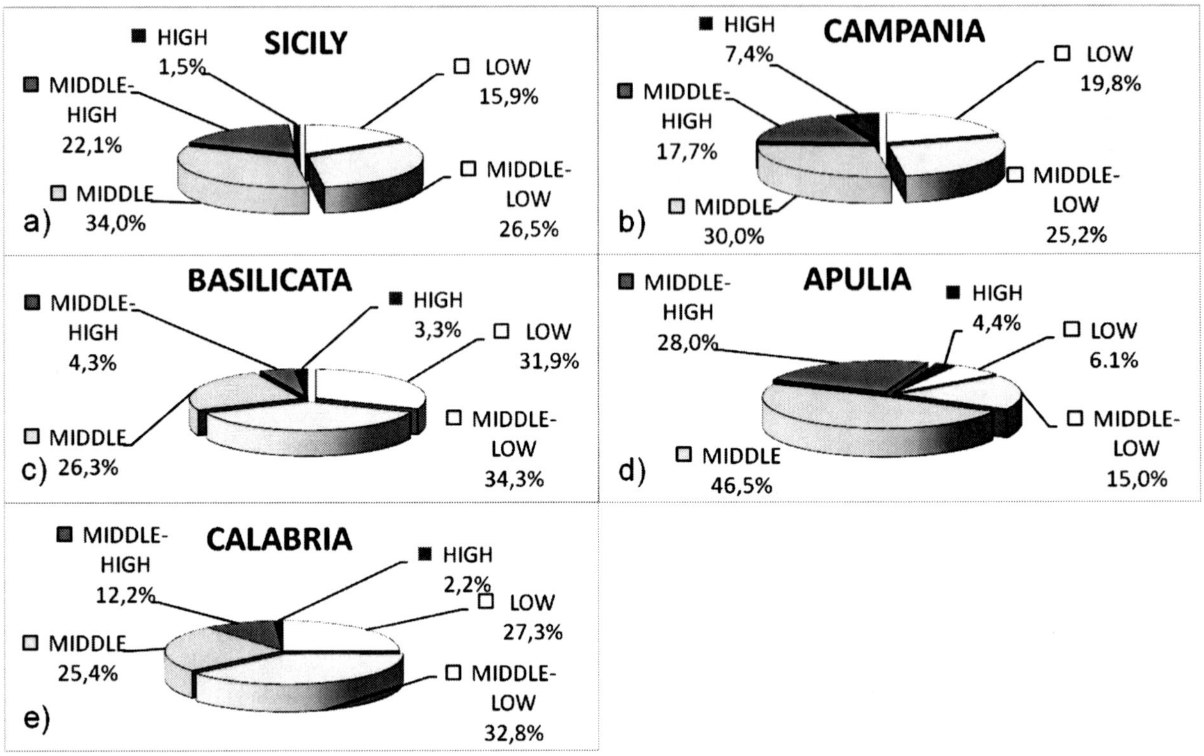

compaction effects induced by use of tracks or tires. The obtained estimations take into account the actual spatial heterogeneity of the agricultural areas that, requiring different agronomic inputs, are also subjected to different impact levels. As a consequence, the extracted vulnerability map is more faithful to the territorial complexity compared with maps derived from census-based indicators. Moreover, the high resolution mechanization level indicator offers a twofold advantage consisting in a more frequent updating (in comparison with the censuses) and in a flexible structure enabling an efficient monitoring at different observation scales (e.g., sub-regional, local) by adopting land cover maps derived from satellite data classifications at higher resolution both in number of classes and in spatial resolution (e.g., Landsat TM and Ikonos with a pixel size of 30m and 4m, respectively).

The final land management index at high spatial detail (LMI_p), obtained by integrating the three census based indicators and the mechanization index at the pixel scale (MLI_p), can better support the definition of appropriate policy instruments to contrast land degradation (restoration, reclamation, safeguard) since it allows to discriminate the various vulnerability levels within any given municipality.

The implementation of the procedure over Southern Italy showed that land management agricultural practices induce a high or medium-high vulnerability to land degradation in a substantial part of the area. In particular, land management is principally relevant for the Apulia region (>30% of the areas) followed by Campania and Sicily regions (~25%). In this context the main driving factors are linked to the heavy use of agricultural mechanization and to the changes observed in the extent of agricultural areas (intensification 6,5%, land abandonment 40,6%, Istat, 2002) suggesting that policies devoted to reduce the intensive agriculture and to promote the requalification of abandoned areas can mitigate the territorial vulnerability of Southern Italy.

ACKNOWLEDGMENT

We thank Prof. Paola D'Antonio and Dr. Francesco Genovese at University of Basilicata (DITEC) for their valuable discussions on agricultural mechanization in Southern Italy.

REFERENCES

APAT (formally ANPA). (2000). *Sviluppo di indicatori per il suolo e i siti contaminati*. RTI CTN_SSC 1/2000, 85.

Baartman, J., Van Lynden, G., Reed, M., Ritsema, C., & Hessel, R. (2007). *Desertification and land degradation: Origins, processes and solutions- A literature review*. Desire Scientific Report No. 4.

Basso, F., Bove, E., Dumontet, S., Ferrara, A., Pisante, M., Quaranta, G., & Taberner, M. (2000). Evaluating environmental sensitivity at the basin scale through the use of geographic information systems and remotely sensed data: An example covering the Agri basin (Southern Italy). *Catena*, *40*(1), 19–35. doi:10.1016/S0341-8162(99)00062-4.

Bazzoffi, P. (2003). *Annuario dei Dati Ambientali - GEOSFERA* (pp. 278–279). APAT..

Brandt, J., Geeson, N., & Imeson, A. (2003). *A desertification indicator system for Mediterranean Europe*. Bruxelles, Belgium: DESERTLINKS Project..

Brouwer, F. B., Thomas, A. J., & Chadwick, M. J. (1991). *Land use changes in Europe. Processes of change, environmental transformations and future patterns*. Dordrecht, The Netherlands: Kluwer Academic Publishers..

Contador, J. F. L., Schnabel, S., Gutiérrez, A. G., & Fernăndez, M. P. (2009). Mapping sensitivity to land degradation in Extremadura, SW Spain. *Land Degradation & Development*, *20*(2), 129–144. doi:10.1002/ldr.884.

Costantini, E. A. C., Urbano, F., Aramini, G., Barbetti, R., Bellino, F., & Bocci, M. etal. (2009). Rationale and methods for compiling an atlas of desertification in Italy. *Land Degradation & Development*, *20*(3), 261–276. doi:10.1002/ldr.908.

EEA. (2005). *The European environment, state and outlook*. Copenhagen, Denmark..

European Commission. (2000). *From land cover to landscape diversity*. Bruxelles: Belgic..

Ferrara, A. (2005). *Export system for evaluating the Environmental Sensitivity Index (ESI) of a local area: Methodology, DIS4ME, Desertification Indicator System for Mediterranean Europe*. Retrieved from www.unibas.it/desertnet/dis4me/using_dis4me/ferrara.htm

Gay, S. H., Louwagie, G., Sammeth, F., Ratinger, T., Marechal, B., & Prosperi, P. ... Schuler, J. (2009). Final report on the project ' Sustainable Agriculture and Soil Conservation (SoCo)'. Luxembourg, European Commission..

Genesio, L., Magno, R., Capecchi, V., Crisci, A., Bottai, L., Ferrari, R., Angeli, L., Gardin, L. (2004). *Final report of Desert-Net project*. Pilot Action WP10.

Hill, J., Sommer, S., Mehl, W., & Megier, J. (1995). Use of earth observation satellite data for land degradation mapping and monitoring in Mediterranean ecosystems: Towards a satellite-observatory. *Environmental Monitoring and Assessment*, *37*(1), 143–158. doi:10.1007/BF00546886.

Hill, J., Stellmes, M., Udelhoven, T., Roder, A., & Sommer, S. (2008). Mediterranean desertification and land degradation: Mapping related land use change syndromes based on satellite observations. *Global and Planetary Change*, *64*(3-4), 146–157. doi:10.1016/j.gloplacha.2008.10.005.

Imbrenda, V., Simoniello, T., Carone, T., Coppola, R., D'Emilio, M., Lanfredi, M., et al. (2008, September). *Estimation of sensitivity to land degradation due to land management: Improvements from the use of a new index of mechanization level.* Paper presented at the 5th International Conference on Land Degradation, Bari, Italy.

Imbrenda, V., Simoniello, T., Carone, T., Coppola, R., D'Emilio, M., & Lanfredi, M. etal. (2009). A new index for the evaluation of land management in the framework of land degradation assessment. In D. Maktav (Ed.), *Remote sensing for a changing Europe* (pp. 535–542)..

INEA. (2008). *L'agricoltura in Puglia. Caratteristiche e interventi per lo sviluppo rurale.* Regione Puglia..

ISTAT. (2000). *5th agriculture census.* Italian National Institute of Statistics. Retrieved from http://www.census.istat.it/wibdsi/default.asp

ISTAT. (2002). *Presentazione dei dati definitivi. Basilicata, V Censimento generale dell'agricoltura.* Roma, Italy: ISTAT..

Juntti, M., & Wilson, G. A. (2005). *Unravelling desertification: Policies and actor networks in Southern Europe.* Wageningen, The Netherlands: Wageningen Academic Publishers..

Keller, T., Trautner, A., & Arvidsson, J. (2002). Stress distribution and soil displacement under a rubber-tracked and a wheeled tractor during ploughing, both on-land and within furrows. *Soil & Tillage Research, 68*(1), 39–47. doi:10.1016/S0167-1987(02)00082-X.

Kosmas, C., Kirkby, M., & Geeson, N. (1999). *The MEDALUS project: Mediterranean desertification and land use. Manual on key indicators of desertification and mapping environmentally sensitive areas to desertification.* Brussels, Belgium: European Commission..

Lanfredi, M., Simoniello, T., & Macchiato, M. (2004). Temporal persistence in vegetation cover changes observed from satellite: Development of an estimation procedure in the test site of the Mediterranean Italy. *Remote Sensing of Environment, 93*(4), 565–576. doi:10.1016/j.rse.2004.08.012.

Latorre, J. G., Garcia-Latorre, J., & Sanchez-Picon, A. (2001). Dealing with aridity: Socio-economic structures and environmental changes in an arid Mediterranean region. *Land Use Policy, 18*(1), 53–64. doi:10.1016/S0264-8377(00)00045-4.

Motroni, A., Canu, S., Bianco, G., & Loj, G. (2004). *Environmentally sensitive areas to desertification in Sardinia.* Technical Report, 46.

Pagliai, M., Marsili, A., Servadio, P., Vignozzi, N., & Pellegrini, S. (2003). Changes in some physical properties of a clay soil in Central Italy following the passage of rubber tracked and wheeled tractors of medium power. *Soil & Tillage Research, 73*(1-2), 119–129. doi:10.1016/S0167-1987(03)00105-3.

Reynolds, J. F., Smith, D. M. S., Lambin, E. F., Turner, B. L., Mortimore, M., & Batterbury, S. P. J. etal. (2007). Global desertification: Building a science for dryland development. *Science, 316*(5826), 847–851. doi:10.1126/science.1131634 PMID:17495163.

Salvati, L., & Bajocco, S. (2011). Land sensitivity to desertification across Italy: Past, present, and future. *Applied Geography (Sevenoaks, England), 31*(1), 223–231. doi:10.1016/j.apgeog.2010.04.006.

Salvati, L., & Carlucci, M. (2010). Estimating land degradation risk for agriculture in Italy using an indirect approach. *Ecological Economics, 69*(3), 511–518. doi:10.1016/j.ecolecon.2009.08.025.

Salvati, L., & Zitti, M. (2007). Long term demographic dynamics along an urban–rural gradient: Implications for land degradation. *Biota, 8*, 61–69.

Salvati, L., Zitti, M., & Ceccarelli, T. (2008). Integrating economic and environmental indicators in the assessment of desertification risk: a case study. *Applied Ecology and Environmental Research*, *6*, 129–138.

Simoniello, T., Lanfredi, M., Liberti, M., Coppola, R., & Macchiato, M. (2008). Estimation of vegetation cover resilience from satellite time series. *Hydrology and Earth System Sciences*, *12*, 1053–1064. doi:10.5194/hess-12-1053-2008.

Tanrivermis, H. (2003). Agricultural land use change and sustainable use of land resources in the Mediterranean region of Turkey. *Journal of Arid Environments*, *54*(3), 553–564. doi:10.1006/jare.2002.1078.

Thornes, J. B., & Brandt, J. (1995). *Mediterranean desertification and land use*. Chichester, UK: John Wiley & Sons..

Trisorio, A. (2005). *Measuring sustainability. Indicators for Italian agriculture*. Rome, Italy: National Institute of Agricultural Economics..

UN/FAO. (2003). *Report on Land Degradation Assessment in Drylands - LADA*, from ftp://ftp.fao.org/agl/agll/docs/lada.pdf.

Yassoglou, N. J. (1999). Land, desertification vulnerability and management in Mediterranean landscapes. In P. Balabanis, D. Peter, A. Ghazi, & M. Tzogas (Eds.), *Mediterranean desertification: Research results and policy implications* (pp. 87–113). Luxembourg: European Commission – Directorate General Research..

Zucca, C., Canu, A., & Della Peruta, R. (2006). Effects of land use and landscape on spatial distribution and morphological features of gullies in an agropastoral area in Sardinia (Italy). *Catena*, *68*(2-3), 87–95. doi:10.1016/j.catena.2006.03.015.

KEY TERMS AND DEFINITIONS

Geographic Information System (GIS): A system devised to store, analyze and manage geographically referenced data.

Land Cover: The different physical material at the surface of the Earth including urban, agricultural and natural covers.

Land Degradation: A reduction or loss of the biological and economic productivity and complexity of lands driven by a combination of complex processes linked to both natural and anthropic factors acting at different spatial and temporal scales.

Land Management: The process of managing the use and development (in both urban and rural settings) of land resources.

Soil Compaction: A deterioration of soil structure (decrease of total pore space) as a result of persistent passes of farm machinery for agricultural purposes.

Southern Italy: Refers to a geographical area of Italian peninsula including the administrative regions of Campania, Abruzzo, Molise, Apulia, Basilicata, Calabria, Sicily and Sardinia.

Vulnerability Indicators: Synthetic indicators (numerical quantification or semiquantitative ranking) of critical ecosystem properties.

This chapter was previously published in Geographic Information Analysis for Sustainable Development and Economic Planning edited by Beniamino Murgante, Giuseppe Borruso, Stefania Bertazzon, Andrea Favretto, and Carmelo Maria Torre, pages 87-101, copyright 2013 by Information Science Reference (an imprint of IGI Global).

Chapter 79
Desert in Bengal Delta–Changes in Landscape, Changes in Livelihood:
Can Diffusion and Adoption of Sustainable Adaptation Make a Difference?

Md. Mustafa Saroar
Khulna University, Bangladesh

Jayant K. Routray
Asian Institute of Technology, Thailand

ABSTRACT

This study is aimed to assess the impacts of various climate induced events and the sustainability of adoption of coping and adaptive measures against these impacts. The study was conducted through participatory rural appraisal (PRA) and focus group discussion (FGD) in three sites of Kalapara Upazila (Sub-district), located along the Bay of Bengal in Southwest Bangladesh. Finding suggests that although flooding, cyclones, storm surges, and seasonal droughts have different attributes, they have commonality in bringing severe salinity in soil, sub-soil water, and surface water bodies that severely affects the livelihood security of natural resource-dependent coastal population. Indigenous knowledge and diffusion of vernacular technologies have helped a large majority of people to adopt various coping and adaptation strategies against the salinity problem for long. However, in a changing nature of salinity (i.e. a discrete isolated event has been turning to a perpetual event), which is resulting from climate change induced sea level rise (CC-SLR), most of their adopted coping and adaptation measures have failed to address their livelihood security on a sustained basis. To bolster their efforts toward adoption of adaptation strategies for sustainable livelihood, a range of avenues for interventions are identified.

DOI: 10.4018/978-1-4666-4852-4.ch079

ORGANIZATIONAL BACKGROUND

This research is aimed to explore the issues related to diffusion and adoption of adaptation strategies against the vulnerability of progressive intrusion of salinity which affects the sustainable livelihood of coastal inhabitants in Bangladesh. Salinity intrusion is largely governed by cyclonic and tidal surges, coastal flooding and climate change induced sea level rise (CC_SLR). The issue of adoption of sustainable adaptation strategies by the inhabitants of coast-lying and small island countries (SICs) against the long term impacts of SLR induced events drew global attention for the first time when the third assessment report of the Intergovernmental Panel on Climate Change (IPCC) warned that humanity might experience almost a one meter (88 cm) SLR by the end of this century (IPCC, 2001). Such a massive scale SLR would increase both the frequency and intensity of SLR induced events, such as cyclonic storms, storm surges, tidal surges and salinity intrusion. All of these have the potential to amplify various impacts of salinization. These networks of impacts would ultimately jeopardize the livelihood security of millions of souls who have been living in the deltaic coasts and SICs unless they successfully adopt sustainable coping and adaptation strategies. Successful adoption of sustainable coping and adaptation strategies depends on how much people are aware of the existence and pace of diffusion of such technologies. Society's readiness to adopt such techniques and technologies as part of sustainable adaptation strategies is equally important (Olsson & Jerneck, 2010). Unfortunately, countries that would suffer the most from SLR induced salinity are the ones who are due to poor adaptive capacity in lack of well-defined network of diffusion of adaptation techniques and technologies. Bangladesh with a per capita income of US$ 740 is a classic example of this duality, i.e. vary much prone to SLR induced salinity but paradoxically there exist very weak network for diffusion of techniques and technologies that relates to adoption of sustainable adaptation strategies (GOB, 2008). This chapter makes an attempt, first to present the nature of livelihood vulnerability of coastal population given the likely scenarios of climate change and SLR. Second, how they have been adopting various adaptation strategies against their impacts of climatic disasters on their livelihood security is discussed in some details. How they are planning to adopt new techniques and technologies as part of sustainable adaptation strategies is also presented. Finally, the barrier to adoption and diffusion of such sustainable adaptation techniques and technologies are identified to make policy recommendation.

Bangladesh: A Story of Desert in Delta and the Research Agenda

Bangladesh having a population over 140 million within a territory of approximately 144 thousand square kilometer (56 thousand sq. miles) is one of the most densely populated countries of the world (BBS, 2011). The economy of Bangladesh is largely agrarian; almost 75 percent of the population depend on agriculture and allied activities for their livelihood. Although per capita income is about US$ 740, yet about 40 percent population live below US$ 2/day poverty line. This South-Asian country is surrounded by India from three sides and flanked by the exposed coast of Bay of Bengal on the South. Bangladesh with a vast tract of exposed coast is one of the few countries in the Asia Pacific region most prone to progressive threat to SLR induced salinity (Figure 1). Its coastal region covers an estimated area of 36,000 sq. km (14, 000 sq. miles) (Ali Khan et al., 2000). Some 700 km (435 miles) coastline is extended from the East to West (Singh, 2001). The country has 19 coastal districts in three distinct coastal sub-regions. These three sub-regions are western, central, and eastern coastal zones. Historically the South-western part which includes the world-heritage tidal mangrove forest *Sunderbans* is most prone to multiple disastrous events due

Figure 1. Areas under oval are most prone to sea level rise; area under square: Bangladesh

to accelerated SLR. Numerous studies have projected that the coastal area in Bangladesh would experience 30 to 50 cm (12 to 20 inches) SLR by the year 2030-50 (Ahmed & Alam, 1998; World Bank, 2000). Country's National Adaptation Program of Action (NAPA) considers an SLR of 14 (5.5 inches), 32 (12.5 inches) and 88 (35 inch) cm along its coastline by the year 2030, 2050 and 2100 as the most plausible scenarios (GOB, 2005). Moreover, the country is almost flat, and one-fifth of its landmass is within one meter (3.25 feet) above mean sea level (MSL). Therefore, most parts of the low-lying non-embanked coast may be completely inundated even in the first quarter of this century.

Historically coastal regions of Bangladesh are prone to multiple climatic hazards, such as coastal flooding, cyclones, storm surges, and salinity intrusion. Yet natural resource dependent people have been living there for generations. They have developed their indigenous practices for dealing with each of these multiple hazards (Ali, 1999; Huq et al., 1998; Ali Khan et al., 2000). Their traditional knowledge of coping and adaptation has been working well against many of these climatic hazards as they were familiar with the nature of disastrous events. For instance, coastal people are familiar with normal flooding; for years through trial and error they have developed a fair understanding about the behavior of flooding, i.e. start-pick-recede. They can sense when flood water would inundate their croplands, fish farms, and other farmlands. Similarly often they can forecast the timing of tidal waves, cyclonic events and storm surges. As they are familiar with the behavior of disastrous events, they adjust their cropping cycle, fishing calendar and lifestyle to cope with and adapt to these known events. However, SLR would change the nature of these known disastrous events. Scientific advances have warned that the behaviors of most climatic events in terms of timing of onset, duration, and devastating power would be unpredictable (IPCC, 2007). Furthermore, under new circumstances the inundation would become perpetual; it would increase over time and never recede. Therefore, SLR induced salinity and flood would gradually encroach on lowlands, croplands, settlements and infrastructures, and even major parts of mangrove forest *Sunderbans (*Agrawala et al., 2003). Such intrusion of salinity by making soil, sub-surface water, and surface water completely saline would seriously impede the coastal agriculture, forestry, fisheries and other natural resource based livelihood activities (Huq et al., 1998, World Bank, 2000; Choudhury et al., 2005; Faruque & Ali, 2005). Progressive buildup of soil salinity and severe crisis of fresh water for consumption and irrigation would affect the whole coastal landscape (Ahsan et al., 2011). Delta might turn into a desert of salt as already more than 1000 sq km (about 4000 sq miles) of cultivated land and sea product culturing area have turned into salt-marshes. It means without adoption of sustainable adaptation strategies, the livelihood of 35 million natural resource-dependent coastal populations in Bangladesh would be under severe threat due to SLR induced salinity (GOB, 2008).

Despite having a long history of coping and adaptation, a substantial body of literatures indicates that coastal people's current coping and adaptation may not work as the known events would appear very differently due to climate change and SLR induced progressive salinity (World Bank, 2000; Cannon, 2002; CARE, 2003; Schaerer & Ahmed, 2004). People's current coping and adaptation strategies are largely centered on reactive responses for protection of lives, property and livelihood security. First, they face the disastrous events and then devise the strategies to cope and adapt. For future disastrous events there is need for adoption of anticipatory sustainable adaptation strategies, i.e. adaptation in advance. As SLR induced impacts of salinity are quite new compared to normal coastal flooding, people might overestimate their adaptive capacity. If so really happen, in actual situation they would find them helpless. The best way to avoid such unexpected outcome is, first to evaluate their current coping and adaptation and then to identify the possible avenues of diffusion of new techniques and technologies so that they feel confident to adopt sustainable adaptation strategies.

Although significant body of literature indicates the need for updating of current coping and adaptation strategies, hardly anyone specifies-which strategies need to be upgraded or what adaptive measures could be adopted to strengthen their current coping and adaptation against future vulnerabilities. This calls for a research study like this one which has addressed diverse sets of strategies that are employed by various occupational groups in the coastal areas of Bangladesh to secure their livelihood. Therefore, this study makes specific attempt to identify adoption of which specific adaptation strategies could offer a secure livelihood for the coastal population on a sustained basis. The policy implication of this study is, it would help policy makers and planners to design sustainable adaptation strategies by combining local knowledge with adaptation sciences and again it would help devising mechanisms for diffusion of such sustainable adaptation strategies. In fact, if peoples' accumulated experiences could be well defused and adopted after blending with modern knowledge, this would help them to cope and adapt on a sustained basis against future vulnerability because the transfer of traditional knowledge is inter-generational. It is highly likely that if present generation is equipped with upgraded version of their traditional knowledge of coping and adaptation the ultimate benefit would be enjoyed by future generations as well who are more prone to vulnerability of CC-SLR.

Theories of Diffusion and Adoption of Sustainable Adaptation Strategies for Livelihood Security

In the climate change context vulnerability refers to the degree to which a system is susceptible to and is unable to cope with adverse effects. The key parameters of vulnerability are the climatic stress to which a system is exposed, its sensitivity, and its adaptive capacity (McCarthy et al., 2001; Turner et al., 2003; Adger, 2006). In regards to assessment of vulnerability, Fussel (2007) proposed that assessment should take account of the vulnerable system, the stressors and their roots, effects on the system, and the critical time; in fact, Fussel (2007) conceptualizes vulnerability as process. However, Downing and Patwardhan (2004) and Cannon and Muller-Mahn (2010) stress the importance of identification of nature of threat, the affected areas, the population group and their resiliency, and the time period while making an assessment of vulnerability; they viewed vulnerability as outcome. Our research heavily draws on the latter conceptualization of vulnerability.

In the human system, livelihood vulnerability refers to the susceptibility to circumstances of not being able to sustain a livelihood (Alwang et al., 2001; Adger & Winkels, 2006). A sustainable livelihood comprises the capabilities, assets and activities that offer well-being for a person or household (Chambers & Conway, 1992; Ellis, 2000; Allison & Ellis, 2001). Sustainable livelihood can ensure livelihood security without degrading the resource base and the environment (Olsson & Jerneck, 2010; Below et al., 2012). The notion of sustainable livelihood has been emerged from the populous idea of sustainable development which means meeting the needs of present generation without compromising the ability of future generation to meet their own need (WCED, 1987). From a very practical sense sustainable development is the development, which sustains over time and can ensure social justice and environmental protection (Parry, 2009; Barbier, 2011; Jackson, 2011). A person's livelihood is considered stainable when the person can cope with stresses and recover from shocks and maintain (or enhance) capabilities and assets both now and in the future, while not undermining the natural resources base (Scoones, 1998; Wisner et al., 2004). Therefore, sustainable livelihood which equally means livelihood insecurity in the climate change context largely depends on the affected community's ability to adopt sustainable adaptation strategies (de Haan & Zoomers, 2005).

Adaptation is the process by which the affected individuals reduce the adverse effects of climate change on their livelihood security (Burton, 1997; Smit et al., 2000). Such processes often involve passive, reactive, or even anticipatory adjustment in behavior, and social and economic structures (Smith, 1997; Fankhauser et al., 1999; Smit & Skinner, 2002; Adger et al., 2005; Smith & Wandel, 2006). However, not all adaptive responses are sustainable (Parry, 2009; Cannon & Muller-Mahn, 2010); for instance, in a post-flood situation engaging elementary school going children in paid job to compensate the family's income loss from flood. Therefore, sustainable adaptation strategy is more important than conventional adaptation for livelihood security (Below et al., 2012). O'Brien and Leichenko (2007) and Eriksen and O'Brien (2007) conceptualize sustainable adaptation by combining aspects of both sustainability and adaptation. For instance, Eriksen and O'Brien (2007) present sustainable adaptation as a set of responses which form the overlap between poverty reduction measures on the one hand, and vulnerability reduction measures on the other. They present a thorough analysis of the links between poverty and vulnerability to climate change, arguing that sustainable adaptation interventions must focus on

- Reducing risks to current ways of securing wellbeing,
- Strengthening adaptive capacity of the poor and
- Addressing the causes of vulnerability among the poor.

A sustainable adaptation measure may focus on any one of these linkages, as long as it does not negatively affect the other two (Eriksen & O'Brien, 2007).

These conceptualizations imply that sustainable adaptation brings two extra dimensions – or two additional elements or requirements – to conventional adaptation approaches: first it should actively reduce poverty and/or vulnerability and second, it must ensure that actions are sustainable in the long term and do not undermine resilience, that is, it includes measures which do not compromise longer-term sustainable development or ability to respond to climate change and other stressors (Brown, 2011). In short, their notion of sustainable adaptation calls for adoption of technology or measures that reduce vulnerability and promote long-term resilience in a changing climate. Failure to adopt sustainable (adaptation) strategy is one of the three causes of poor coping against climatic impacts (Adger, 2010; Hinkel, 2011).

Adoption is the individual-level decision to use a new technique. Diffusion is the aggregation of a number of adoption decisions. Therefore, diffusion is the process by which an innovation (new technique) is communicated through certain channels over time among the members of a social system (Rogers, 1995). Diffusion of new technology is like that of an infectious disease. Non-adopters adopt a new technology when they come into contact with adopters and learn about the new technology. As more people adopt, the number of non-adopters declines, which increases the rate of diffusion (Geroski 2000; Comin & Hobijn, 2004). This pattern of diffusion leads to the common S-shaped curve on the rate of technology diffusion with respect to time. Five key factors govern an individual's decision to adopt: relative advantage, complexity, compatibility, trialability, and observability. These factors are only relevant after informative contact between adopter and the change agent. The change agent brings an innovation into a new social network and can effectively communicate its benefits (Rogers, 1995). In a rural setting where societies are exposed to lower level of technological innovation, it is the experienced people (not always educated) and the extension workers who play pivotal roles as change agents for adoption and diffusion of sustainable adaptation technology against the impacts of climate change.

In practice however, adoption of sustainable adaptation strategy involves adoption of both soft and hard measures and techniques which include knowhow, procedures, goods and services and equipment, as well as organizational and managerial procedures that can indefinitely and effectively work against the possible (climatic) impacts in any spatial setting while not depend on exogenous factors (Brooks & Loevinsohn, 2011). Furthermore, adoption of such sustainable adaptation strategy does not often take place instantaneously or in some cases automatically or in isolation with the total social system where a person lives (Brown, 2011; Hinkel, 2011). Therefore, adoption of sustainable adaptation strategies is a complex undertaking which is influenced by interplay of numerous factors including level of awareness and access to information, economic and financial resources at disposal, social capital, and risk taking behavior (Reddy &Painuly, 2004; Below et al., 2012). In general, the coastal people in Bangladesh are vulnerable to SLR and its associated events, such as floods, cyclones, storm surges, and saltwater intrusion. As many of these events are likely to become more frequent and intense due to CC-SLR, there is need for adoption of sustainable adaptation strategies. Sustainable adaptation is aimed to reduce the vulnerability in a sustained basis by systematically minimizing exposure and sensitivity and maximizing adaptive capacity (Smith & Wandel, 2006).

STUDY AREA AND METHODS

The study was conducted in Kalapara Upazila (a sub-district with an estimated area of 450 sq km (175 sq miles) of Patuakhali District which is located in the South-west coastal region of Bangladesh and is flanked by the Bay of Bengal (Figure 2). Many earlier studies identified *Patuakhli* district as one of the districts which is exposed to multiple hazardous events including SLR induces disasters because of vast exposed coastline. Within the Patuakhali district, "Kalapara" upazila (sub-district) was purposively selected for the study as it covers most parts of exposed coast and is susceptible to SLR and its associated events. To represent the whole "Kalapara" sub-district, three "Union Parishads (UPs)" namely "Dulasar," "Mithaganj" and "Nilganj" were randomly selected. Usually one UP includes 5-10 villages. Finally one village from each of these three UPs was randomly selected for empirical survey. The survey instrument includes three focus group discussions and one participatory rapid appraisal (PRA) session. Participatory rapid appraisal (PRA) tools included among other the community risk assessment (CRA) and vulnerability and capacity assessment (VCA) components. For comprehensive understanding of CRA and VCA see van Aalst et al. (2008). The field survey was done during January to April 2009.

The study area is predominantly rural. Almost one third of the families do not have their own cultivable land. Nonetheless, more than half of the families are primarily engaged in crop agriculture. Therefore, share-cropping is very common among the families who do not own land or who are functionally landless. Functionally landless families are those who have less than 0.2 hector (0.5 acre) of cultivable land. However, unlike most parts of Bangladesh where two to three crops are cultivated in a cropping year, here most land parcels have been used for single crop, i.e. wet-season *Aman* rice cultivation until recently when cultivation of winter vegetables, beans, legumes, nuts, and water melon were started. Apart from crop agriculture, significant proportion of families earn their livelihood from fish culture, open-

Figure 2. Study sites in Kalapara Upazila (Sub-district) in relation to map of Bangladesh

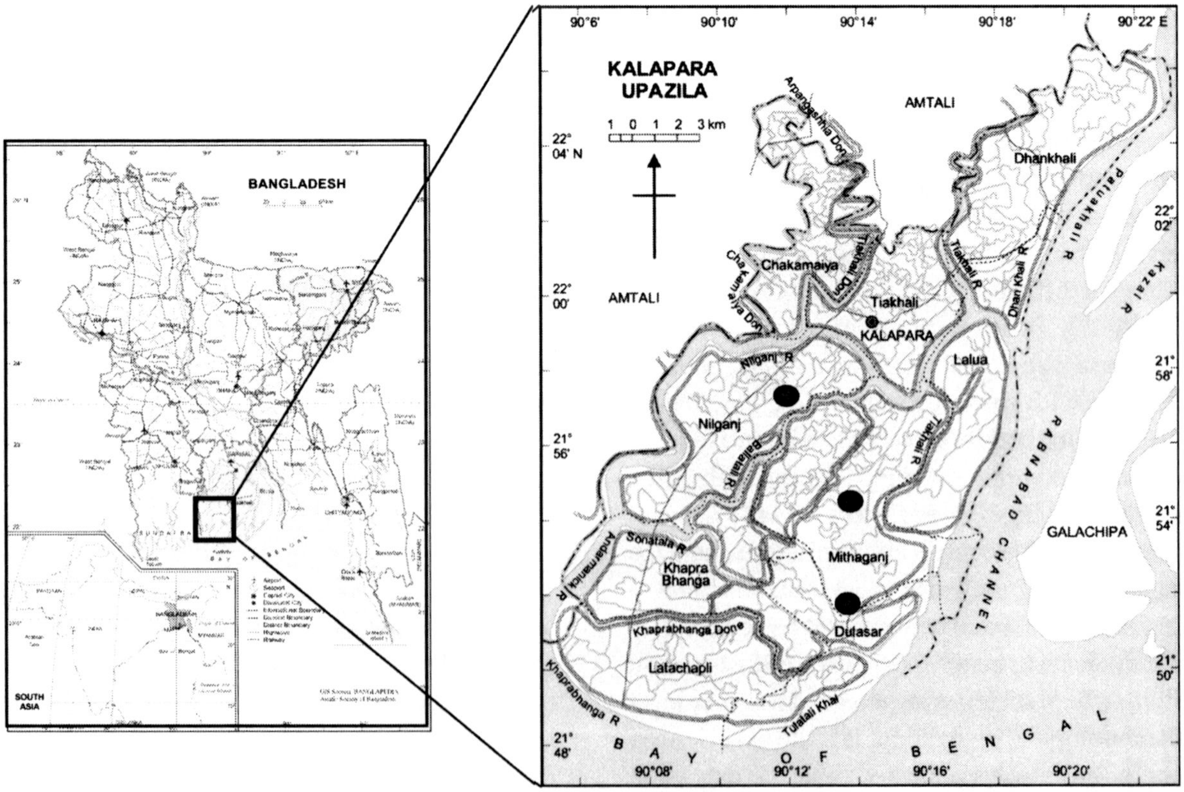

water fishing, selling human- labor, petty trading, working in rural transport sector and various seasonal on-farm and off-farm activities.

Despite diversity of occupations, the unique attribute of their livelihood is their overwhelming reliance on abundance of natural resources. Therefore, their livelihoods have been shaped by recurrent exposure to various natural hazards including intrusion of salinity. The entire area severely suffers from scarcity of fresh water throughout the winter season as the salinity buildup. Winter season starts in October and ends in April with the onset of early monsoon. Salinity starts to decrease during monsoon. People with their experience know that harvest loss is obvious due to lack of irrigation in dry winter season. This time soil salinity often exceeds 10 ds/m and the freshwater layer goes below 300-400 meter (1000-1300 feet) from the surface. Therefore, they left their thousands of hectors of arable land fallow/idle during winter which makes a significant portion of families jobless. However, throughout the cropping cycle of wet-season rice, i.e. May to December, rural unemployment rate goes down. Moreover, as this time significant numbers of people go for both freshwater and marine fishing, this often creates localized shortages of male farm laborers. This brought opportunity for female laborers to get engaged in paid farm activities. Usually the day laborers try to earn their maximum during this period to offset the loss of income in unemployed period (winter season).

RESULT AND DISCUSSION

Disasters and Changing Landscape

The entire study area is historically prone to multiple natural hazards, such as coastal flooding, cyclones, storm surges, salinity intrusion, and prolong/torrential rainfall. Apart from these, the localized nature of scarcity of fresh water for irrigation and drinking is also the major causes of changing landscape and livelihood. Although there are differences among these hazards in terms of their process of formation, timing of onset, frequency and magnitude of occurrence, yet each one has the potential to cause unprecedented damage to agricultural fields, fisheries, family assets, human health and other properties. These ultimately heavily impacts on the livelihood security of most of the marginalized population. Table 1 reports some of the key features of major disastrous events that ravaged the entire study area time and again.

Although a very diverse portfolio of occupational engagements is observed in the study area, and most people heavily depend on natural resource-based livelihood, yet some occupational groups are more affected by these natural disasters than others. From PRA and informal group discussions, it is evident that impacts of cyclones and surges are felt by the respondents irrespective of their occupational engagement. The impacts of cyclones and surges are felt severely during high tide. In fact, synchronization of cyclone with high tide often generates higher number of walls of sea-waves each having higher depth (height) which dismantle the protection walls/embankments and washout everything within a short span of time. Furthermore, stagnant water causes damage to cropland, pastures, orchards, fisheries and other production units. With the increase of surge height both the human causality and other damages increase. Similarly, the respondents mentioned that they experience higher damage to property and other assets if cyclones strike/hit during the late night. Unfortunately, it is often the cases that cyclones strike/hit during late-night and subsides at day time.

Unlike cyclones and storm surges, the impacts of other disastrous events are often not felt by various occupational groups in the same way. Impacts of coastal flooding, for instance, are more felt by people engaged in agriculture and other non agricultural activities such as small business, petty trade, rural transportation. Impacts of fresh water crisis due to high salinity in water are more felt by

Table 1. Characteristic features of natural disastrous events in the study area

Disastrous Events	Characteristics	Impacts/Affects
Coastal flooding	Occurrence: • During July-September; • When high tide synchronizes with torrential rainfall; • When spring/king tide synchronize with drainage congestion; • Breach of embankments/polders/levees • Duration: normally up to one week; in severe case up to three weeks.	• Destroy the rice seed-beds; • Damage the rice field during growing/flowering periods; • Damage the pasture land; • Destroy the culture fisheries; • Cause damage to various infrastructures; • Cause damage to settlements and dwelling units; • Interrupt physical mobility.
Cyclonic storm and surge	Occurrence: • During April- May (pre-monsoon); • This pre-monsoon cyclones are rather less frequent and less destructive • During October-November (post-monsoon); • This post-monsoon cyclones are rather more frequent and more destructive.	• Destroy the ready-to-harvest paddy; • Destroy culture fisheries; • Cause damage of settlements and dwelling units; • Cause damage of various infrastructures; • Interrupt physical mobility; • Interrupt in offshore fishing.
Intrusion of Salinity	Occurrence: • During December-April (winter dry period); • During high tide saline water travel further inland; • High amount of saline water penetrate to the soil and sub soil water tables during winter; • Salinity remains high until next monsoon which tarts in May-June.	• Destroy salt-sensitive rice seed-beds; • Interrupt transplantation/planting of winter rice; • Interrupt the cultivation/growth of salt--sensitive winter vegetables and legumes; • Damage the pasture land and cause shortage of fodder; • Localized crisis of fresh water for both irrigation and drinking.
Torrential rainfall	Occurrence: • During June-August; • Often associated with onset of Indian ocean monsoons; • When synchronize with high tide causes localized flooding; • Duration varies: few hours to couple of days	• Destroy the rice seed-beds; • Damage the rice field during growing/flowering periods in low lying parts; • Interrupt the post harvest processing of rice; • Interrupt fish-catch in the off shore areas; • Interrupt post harvest processing of fish (sun-dry processing). • Interrupt physical mobility.

Source: Author's compilation from PRA session

people engaged in agricultural and other on-farm activities. By contrast, the impacts of torrential rainfall and deep low pressure(in the nearby Bay of Bengal) are more felt by respondents who are engaged in small business and petty trade, and fish catch and fish trading activities respectively. Because deep low pressure system impedes the (fish) catches and causes problems in post harvest processing and transportation of catches. Day/casual laborers suffer the most from any kind of disastrous events. It is partly because they have only limited entitlement, i.e. physical labor which cannot be capitalized during disaster periods.

The PRA and other informal discussions with the respondents reveal that people who have been living in the low-lying parts, or close to the natural canals/creeks or naturally depressed areas often encounter higher exposure to coastal flooding. People living in these areas are also exposed to intrusion of salinity. These peoples are the most marginalized and often the landless category. They occupy the "*khash*" land (land belongs to government; for instance, land next to river banks, natural canals, on the marsh etc) for building their shelters. The marginalized people have to cope with and adapt to coastal flooding throughout the monsoon. It is partly because during monsoon and post monsoon periods, the volume of water that the rivers and canals carry often exceeds the discharge capacity of the river systems which cause severe coastal flooding. Furthermore, during pre-monsoon period often fresh water is drawn inside the embanked areas to prepare the rice seed-beds which again causes submersion condition for low-lying parts inside the polders (settlements encircled by embankment along rives/canals). In fact, often uncontrolled volume of fresh water is allowed to penetrate inside the polders to lower down the salinity level in water so that water can be used for wet-season rice cultivation. Part of that water is conserved in the cannels/natural depressions for use in dry season-cropping which also cause such inundation during monsoon.

Exposure to saline water is very common throughout the winter season especially when high tide creates backflow of salty sea-water from the sea (Bay of Bengal) to further inland. Because this time influential people take the control of sluice gates to bring saline water (from the rivers) inside the polders for their brackish-water shrimp farming. This often creates localized social conflict among the rice/crop farming community and the influential "*gheer*" (shrimp firm) owners. Whether is it coastal inundation or salinity intrusion either of the cases the ultimate victim is the mass people who experience crop failure due to submersion condition or due to acute salinity. They also experience flooding problems due to prolong rainfall accompanied by high tide. Apart from these, peasant farmers, the day laborers and the fisher groups are the worst victim of prolong rainfall. Especially if the torrential rainfall is accompanied by gusty wind, the fisher groups experience the worst while they are in the deep-sea or even in the shallow off-shore areas. This agriculture and allied occupation groups constitute nearly three-fourth of the total population.

Table 2 reports that many disastrous events often synchronize. This creates more trouble for the people especially for landless and functionally landless categories who possess tiny pieces of land. Sometime they can cope and adapt when disasters appear discretely. However, occurrence of multiple disasters at a time makes it harder for them to cope with. As in the study area most people go for single cropping, i.e. wet-season rice, they face severe difficulties during seed-bed preparation, transplantation, and in growing period. These activities need to carry out during April-June depending on the land type and configuration (Table 3). However, May-June is the period when they are hard hit by early floods, torrential rainfall, tidal surges, cyclones and storm surges (Table 2). Therefore, they often face the challenges to protect their croplands. Almost similar challenges are faced by other occupation groups, such as fish folks, fish-farming groups, and the day laborers. As this time the mobility of people is less and the boat/ferry landing stations go under elevated/flood water, the earning opportunity of other occupation groups, such as boatmen, transport workers, petty traders dwindle as well.

Horrible period they used to pass during September-October, when most of their croplands, fish-farms, fishing grounds and places of trading and working get exposed to multiple disastrous

Table 2. Multiplicity of disastrous events and threats on livelihood

Disaster and periods of occurrences	Jan-Feb	Mar-April	May-Jun	July-Aug	Sept-Oct	Nov-Dec	Most affected groups
Flooding			X	X	X		Peasant farmers
Rainfall			X	X	X		Day laborers
Tidal surge			X	X	X		Fish-farmers
Cyclone/storm surge			X		X	X	Fishermen
Salinity intrusion	X	X				X	Peasant farmers

Source: Author's compilation from PRA session

Table 3. Vulnerability of productive assets and activities to climatic hazards (excerpt from PRA)

Occurrence[a] 1 = low frequency 5 = high frequency Intensity[b] 1 = least severe 5 = most severe (- sign indicates beneficial)	**Fresh water shortage and Salinity in soil and water (5) [a]**	**Rainfall/Water logging (5) [a]**	**Cyclone/tidal surge (3) [a]**	**intensity * occurrence**	**Rank: 1 = most vulnerable 10 = least vulnerable**
***Aman* (wet-season) rice**	High salinity affects growing to flowering stage (5) [b]	Prolong water logging frequently damage seedbed/transplanted field (3) [b]	Cyclone/tidal surge damage mature to ready to harvest field (3) [b]	**49**	**2**
***Boro* (winter) rice (now rarely cultivated)**					
Summer season's vegetables	High salinity affects planting of most summer vegetables during Apr-May. (4)	Even occasional and light rainfall helps growth of most summer vegetables during Jun-July. (-5)	Cyclone and tidal surge inundate newly planted vegetable field during May-Jun & completely damage. (5)	**10**	**9**
Winter season's vegetables	High salinity affects planting and growth of most winter vegetables during Nov-Feb. (3)	Even occasional and light rainfall help planting and growth of most winter vegetables during Nov-Dec. (-3)	Cyclone and tidal surge inundate newly planted vegetable field during Oct-Nov & completely damage. (5)	**15**	**8**
Culture fisheries/ Shrimp *gheer*	0	Heavy rainfall in Jul-Sept leads to lower salinity in *gheers* which result in poor harvest (3)	Cyclone and tidal surge with abnormal height severely damage productivity (5)	**30**	**6**
Fish catch/fishing	0	0	Fisher groups loss many income earning days due to very bad weather during May-June & Oct-Nov. (3)	**9**	**10**
Collection of fuel wood	0	0	Fuel wood is collected whenever they enter into Sundarbans, which is minimum in May-June & Oct-Nov. due to very bad weather. (3)	**9**	**10**
Collection of honey	0	0	Honey is collected during Apr-Jun which is affected by Strong wind in May-June. (3)	**9**	**10**
Cattle/Goat rearing	High salinity degrade the pasture land during dry season & creates shortage of fodder. (3)	Heavy rainfall induced water logging during Aug-Sept damage pasture & restricts mobility of animals. (3)	Cyclone and tidal surge with abnormal height wash away livestock during May-Jun & Oct-Nov. (5)	**45**	**3**
Poultry/Duck (Swan) production	0	Heavy rainfall induced water logging during Aug-Sept makes poultry rearing difficult. (3)	Cyclone and tidal surge with abnormal height wash away poultry during May-Jun & Oct-Nov. (5)	**30**	**6**

continued on following page

Table 3. Continued

Occurrence[a] **1 = low frequency** **5 = high frequency** **Intensity[b]** **1 = least severe** **5 = most severe (- sign indicates beneficial)**	**Fresh water shortage and Salinity in soil and water (5) [a]**	**Rainfall/Water logging (5) [a]**	**Cyclone/tidal surge (3) [a]**	**intensity * occurrence**	**Rank:** **1 = most vulnerable** **10 = least vulnerable**
Wage labor	Due to higher salinity demand of labor in crop agriculture is the worst now. (5)	0	Following the cyclonic events localized demand of labor increases for reconstruction and rebuilding activities. (-3)	**16**	**7**
Water for drinking & domestic uses	Fresh water crisis is acute round the year due to salinity. (5)	Only wet season (monsoon & post-monsoon) fresh water are relatively easy to harvest or fetch from community managed pond. (-2)	Fresh water crisis is acute during & aftermath of cyclone. (5)	**30**	**6**
Social and economic mobility (Market, School/health facilities)	0	5	5	**40**	**4**
Housing/Settlements	Salinity severely affects building materials used in plinth, floor and roof. (3)	2	Mud house and similar weaker houses collapse due to wind & water action during cyclone & surge. (4)	**37**	**5**
Public Health & Sanitation	Water-born disease, i.e. skin infection due to use of high saline water. (3)	Water-born disease, i.e. diarrhea, skin infection due to use of contaminated water & unhygienic sanitation practice. (5)	Water-born disease, i.e. diarrhea, skin infection due to use of contaminated water & unhygienic sanitation practice. (5)	**55**	**1**
Total	170	45	183		
Rank	**2**	**3**	**1**		

Note: [a] Occurrence of a disaster is constant; [b] Intensity of a disaster vary depending on the affected resource/sector/crops (- sign indicate positive/beneficial effect)

events. This time although the problem of late flood is not so severe for the rice growers but the cyclonic episodes with strong surges often cause damage to nearly mature paddy and severely restrict the fish catch both in the off-shore and deep-sea areas. Then again, during November-December deadliest cyclones with powerful surges often severely damage the ready-to-harvest paddy in the field (Table 3). Fisher groups completely stop going deep-sea or even off-shore areas for fishing. Their incomes drop sharply for few weeks of the cyclonic period. Probably it is the January-April quarter when most people in the study areas do not face substantial threats from disastrous events. However, this time they face severe and prolong challenge in their livelihood security due to extreme salinity in soil, sub-soil water table and salinity in surface water bodies. Actually, salinity starts to build up from late October- November and reaches its pick in April (Table 2).

Salinity: The Single Most Important Threat to Sustainable Livelihood

The coastal region as a whole and the study area in particular are in shortage of fresh water for drinking and similar uses due to higher presence of salinity. While in other parts of the country hand tube well/shallow tube well with a maximum depth of 50-100 meter (150-300 feet) is enough to ensure year round sufficient supply of fresh (non-saline) water, in the study area, however, this depth range from 300-400 meter (1000-1300 feet) which entails higher cost. The reasons of the presence of saline water throughout the subsurface water column could be attributed to lateral movement of sea-water in addition to infiltration of saline water from surface sources. During high tide, saline water accumulates in the surface water bodies, such as natural depression, canals. This saline water is stored and conserved by influential peoples to use in their saline-water shrimp farms. Furthermore, few syndicates of influential shrimp farm owners are responsible for on purpose ingression of salt water in the locality throughout the year which severely affects the crop agriculture, horticulture, vegetable production, freshwater fish culture and even production of animal fodders (Table 3). This practice, in fact, has been altering the whole landscape of the coast- making delta a desert.

Most parcels of farmlands are kept fallow throughout the winters due either to lack of fresh water irrigation or presence of extreme salinity in soil and water table. It is not that there is no water in the lowland or in the network of channels or even in the river system. There is plenty of water indeed. As the influential people- the "*gheer*" (shrimp farm) owners control over the sluice gates, they bring saline water inside the embanked areas. This water cannot be used for agricultural crop production as the salinity level often exceed 10 parts per thousand (PPT) or soil salinity exceed 10 ds/m. There were sporadic attempts to store fresh water in the networks of channels and lowlands during monsoon or post monsoon periods to use in agriculture and fresh-water fish production. However, as the influential owners of shrimp farms close (lock) the sluice gates during monsoon or post monsoon to stop penetration of fresh (non-saline) water from the river system to the embanked areas those attempts were unsuccessful. Rather these shrimp farm owners by taking the control of sluice gates preserve saline water inside the embanked areas.

With the increased pace of salinity intrusion, people in the study area have witnessed massive changes in land use, agricultural practices, vegetable cultivation, food habit and even lifestyle. Their key observations include- rapid disappearance of most local varieties of rice, disappearance of small- indigenous fish species, disappearance of grazing fields, orchards, vegetable gardens. They have seen the introduction of shrimp, exotic fish species. People's habits of use of cooking fuels and drinking water also have been changed overtime. Earlier they used agricultural residuals as cooking fuel; now they completely rely on timber/wood from the *Sundarbans* mangrove forest (Table 3). They can recall- there was no family who does not have its own vegetable garden for meeting family consumption throughout the year. Now they cannot do any more due to soil salinity and flooding. As a consequence the whole locality has been turning to a desert in a saline affected delta.

Adoption and Diffusion of Adaptation Strategies for Sustainable Livelihood

Since people have been coping and adapting to many of these disastrous events time and again they have developed various sets of coping and adaptive measures to secure their lives and livelihood. Some measures are very generic, i.e. are used in multiple situations, while others are used in specific circumstances. People engaged in various agricultural activities usually employ multiple measures. They have their own cropping calendar, which includes the suitable time of land preparation, seed-bed preparation, transplantation, watering/irrigation, weeding, applying chemical inputs, growing and flowering period's care, harvesting, post harvest processing and storing. They usually frame all these stages of cropping cycle based on natural season. They often adjust timing of some of these activities/stages of cropping cycle taking into account of the variation in day to day weather and climatic events. At a glance one may find it a very simple undertaking; however, closer insight reveals that here involves well defined adoption and diffusion of soft technologies and techniques which are presented in brief in the sections below. However, the processes of such adoptions and diffusions are very slow, incremental, and path-dependent- often pass-over from one generation to the next.

Local people know from their constant interaction with the local climate that the dry-season crops including rice (dry-season rice) could not be produced in harsh environmental conditions (i.e. dry spells and extreme salinity in soil and water), therefore, most small and peasant farmers used to keep their land parcels idle during the dry season. As a livelihood strategy, they used to employ their talent (local technical knowledge), time, effort and resources to have a good harvest from wet-season *Aman* rice so that they can compensate their dry-season's loss of income. However, a good harvest of rain-fed wet-season *Aman* rice is not risk free as well. A good harvest is often ruined by multiple climatic events. For instance, seed-bed preparation is delayed or abandoned due to delay in rainfall; seed-bed and transplanted plants go under flood water if monsoon induced rainfall starts earlier than normal, full-grown paddy is infested (by pests) if black-cloud persist for long (2-3 days) without rainfall, ready-to-harvest paddy is damaged by hailstorms or cyclonic surges if delayed in harvesting; all these virtually render harvest loss. Therefore, they adopt a host of adaptation strategies for sustainable production which ranges from selection of appropriate seed-varieties to land type to irrigation method to appropriate timing for various stages of *Aman* rice production (cropping cycle). Some of these, they have adopted for long while for many others, they are still in the early stage of adopting, and the processes of diffusion of these techniques are rather slow. For example, for relatively highland they have adopted moisture stress (i.e. drought tolerant) variety of traditional *Aman* rice and for relatively low land they have adopted flood tolerant variety of traditional *Aman* rice. Now of course, within a short span of time farmers are adopting BRRI dhan 57 (drought tolerant rice) and BRRI dhan 52 (flood tolerant rice) for relatively highland and low land respectively. Similarly they have adopted the strategy of using their relatively highlands for seed-bed preparation if the monsoon induced rainfall starts earlier than normal to avoid damage or failure of seed-beds that are often resulted from water logging. On the contrary when they suspect late monsoon they use their lowlands for seed-bed preparation. In fact, when they understand that monsoon/rainfall would be delayed and they cannot wait further for rain, they bring freshwater to their lowlands from natural canals although often difficult to prepare seed-bed (Table 3).

However, their total cropping cycle is heavily affected if late monsoon is synchronized with flood-water which usually comes from upper catchments. In such a case, most of the peasant farmers experience either partial or complete damage to transplanted field crops in their growing stage. To cope with this situation they often go for replanting. Although in most cases replanting works but there are cases when replanted (as delayed in production cycle) field crops encounter one more deadly event i.e. storm surges which wash away/wipeout mature/full grown paddy. This usually happens in the month of October-November. To cope with this devastating situation, they have adopted the strategy of early harvesting of paddy; sometime even harvesting partially mature fields. As the paddies are not in full-grown/ready-to-harvest stage, the shelf-life of such paddy is often shorter and the farmers encounter problems while storing and preserving them. Subsistence farmers do not want to take the risk of further loss, they prefer selling immediately, often at a lower price.

Entire study area experiences severe crisis of fresh (non-saline) water throughout the winter season. In fact, after the month of October the salinity in both surface water bodies and subsurface water table increase tremendously. Higher level of salinity persists for 4 to 5 months. Only after the onset of rainy season/monsoon, the level of salinity starts to decline. Acute salinity in subsurface water table and in soil did not permit cultivation of most winter crops including winter-season rice. Even the shortage of animal/cattle fodders was acute throughout the winter due to higher salinity in soil. During this period to earn a living a significant portion of the peasant farmers used to go for fishing in nearby rivers and in the Bay of Bengal. Furthermore, a significant proportion of people used to enter *Sundarbans* mangrove forest for collection of both timber and non-timber forest products, such as honey, wax etc. Seasonal diversification of livelihood earning avenues was getting popular. However, recently, many families most of who are beneficiaries of NGOs have adopted new adaptation strategies against moisture stress condition and soil salinity to use their land for winter crops (rice, beans, vegetables) cultivation. With the extension supports of agricultural department and local NGOs, many peasant farmers now adopted high-value non-cereal crop agriculture. Accordingly, they have started to produce winter beans, legumes, nuts, sweet potatoes, and watermelons that demand less water. In fact, to cope with and adapt to the salinity problem, they have adopted some other beans and vegetables that demand very less water and can tolerate certain level of salinity. Furthermore, they have adopted aeroponics vegetable production in handing bases. Although in most cases productivity is less than normal, yet the peoples are happy that this production meets their own consumption need. In a few cases, beneficiaries of few large NGOs, such as BRAC and SPEED Trust took group-loan to buy mechanical plough and deep irrigation equipment. With this support they have started to produce dry-season *Boro* rice in some pockets in the study area. Interestingly, they have adopted the drought tolerant rice- BRRI dhan 28 and 29, which can survive against 2 weeks of moisture stress condition. The adoption and diffusion of all these adaption strategies are influenced by the change agents, such as extension workers of agriculture department and the local NGOs.

After agriculture and related farm activities, fishing is one of the dominant sources of livelihood. In fact, those who have been living close to the shoreline for generations heavily depend on both inland and marine fishing for their livelihood. Although small fish-folks catch fish year round, the commercial/large fishermen catch fish few months in two distinct seasons. During winter, the commercial fishermen employ manpower to catch marine fish in the offshore and in the deep-sea areas. On the other hand during rainy season, they employ more number of manpower to catch mostly fresh-water fish in the rivers and estuaries, and also in the shallow offshore areas of the Bay of Bengal. This activity allows many unemployed youth to earn their living throughout the rainy season when they cannot sell their labors in the crop agriculture, which is heavily impacted by the monsoon induced coastal flooding. Similarly many unemployed agriculture laborers also get earning opportunities during winter when they cannot sell their labor in agricultural activities as winter-season cropping is very limited in the study area.

Although adoption of modern technology is rather rare in general among the fishermen, yet some progresses have been observed among the commercial fishermen. The commercial fishermen send their fishing teams in the offshore areas or in the deep-sea for one to two weeks in one go. Their fishing trawlers are equipped with fishnets, fish storage and preservation facilities, navigation tools, food and other logistic supports. The commercial fishermen get money in advance from traders having fish-depots in the city. Based on specific terms of contact, they share the revenues. During rainy season enormous quantity of *hilsha* fish is caught as this fish becomes abundant in the river-estuaries and the bay during this time. Since the shelf-life of this fish is very short, it requires careful preservation before it reaches in the hand of consumers. Knowing the difficulty of long-term preservation, they adopt their indigenous knowledge to process the fish differently. Usually during full-moon (including the days before and after) and new-moon (including the days before and after) when their catches exceed their storage and preservation capacity, they employ more number of female workers to semi-process and to prepare salted fish. At one phase of such processing, they use the sunlight/sunshine to completely dry the semi-processed fish before it reaches to the hand of consumers. This is how the fishing-based livelihood operates in the study area. In fact, salted-fish preparation activities have been immerged as new and sustained sources of livelihood for a significant proportion of destitute women.

Although fishing activities provide both primary and supplementary source of livelihood for a significant proportion of population, these activities are equally threatened by complex interplay of various natural disastrous events. During the rainy season, especially in May-June, their catches are interrupted by prolong low-pressure system, torrential rainfall, gusty winds and cyclonic storms. Similarly in the month of October-November and even in December when the Bay of Bengal becomes very rough and hostile due to formation of numerous cyclonic events, they encounter difficulties while catching fish. For one episode of such low-pressure system or cyclonic storms, the fisher groups suspend/postpone their fishing activities for a week or so (Table 3). As a result number of fish catch-days and amount of catch as well decline during these two critical periods of a calendar year.

Over the years some improvements have been observed in fish-based livelihood. The subsistence fish-folks cope with the adversities of fish-catch during bad weather by adopting some adaptation strategies. For instance, they used to intensifying their effort to catch during non-disturbance periods, avoid off-shore fishing and limit catch in the inland water bodies during bad weather. Over the years they have developed their local knowledge; they are familiar with the timing when the Bay of Bengal might get hostile. They know what fish species would be abundant in a particular time/month of both dry season and wet season; they know the shelf-life of most fish species that they catch. Depending on the circumstances, they organize their fishing efforts. For instance, they have adopted the strategy of maximum catch from the bay no matter what kind of fish species they can catch during the winter (dry) season. This time they take with them various types of fishing gears and nets that suit usually for different kinds of marine fish. As this time fish are not that abundant in the shallow offshore, they have to travel further in the deep-sea; therefore, only the commercial and relatively big merchants who can afford trawlers with better navigation system send their teams for winter-season catch. In most cases, the fishing teams stay in the sea for 14 days or even more in one go. Part of the catches they sell afresh and remaining quantity sell as dry fish. Although the adaptation strategies that they have adopted can ensure livelihood and living for both the workers and the fish-merchants, however, their catching behavior is unsustainable. Fortunately, only handful of big-merchants has adopted this unsustainable strategy of catch during the winter-season.

On the other end, throughout the wet-rainy season, which is the prime time of fish catch (both inland and marine), both the fish-folks, and commercial fishermen adopt a strategy of selective catch. This time they adopt the strategy of *Hilsha* catch as it is abundant and has price advantage over other fish. Therefore, they do not need to carry various types of fishing gears and nets; they mostly concentrate their effort to catch in the off-shore shallow bay areas in addition to the nearby rivers and estuaries. However, as the shelf-life of this *Hilsha* fish is relatively shorter if not well preserved, they pay maximum attention to preservation techniques. As preservation unit, they use 16 cubic feet volume insulating ice-boxes. The strategy they have adopted to prepare this 16 cubic feet icebox made of locally available hard-wood and insulating materials seems suit with the local condition and indigenous knowledge. Although the adoption and diffusion of this technique seem relatively faster and well accepted yet there is room for improvement. Apart from these coping and adaptation measures that relate to fish catch, and preservation and storage, they have adopted other techniques of post-harvest processing. During rainy days or during low pressure system or cyclonic storms, they cannot dry their semi-processed fish by using sunlight. To cope with this situation they have adopted the use of special types of preservation and drying devices (oven) made of locally available materials and techniques. Although, these techniques are cheap but question remains about their food-safety and public health related aspects.

Adoption of Strategies to Secure Lives and Properties

As the entire study area is subject to multiple hazards, the respondents adopt a range of portfolio of adaptation strategies to protect their homesteads, dwelling units, and other family assets from the damaging impacts of those disastrous events (Table 4). Some of the adaptation strategies that they have adopted work against multiple situations; many others, however, are only effective against a particular event of specific nature. Then again, although most adaptation strategies are anticipatory in nature some strategies are essentially reactive in nature. As regards severe shortage of salt free fresh water, the people have been adopting two types of measures. First one is related to use of pond-sand filer (PSF) for removing basically suspended impurities. The PSF is a package type slow sand filter unit developed to treat surface water, usually pond water for domestic water supply. Treated water from a PSF is usually bacteriologically safe or within tolerable limits (Kawahara, et al. 2004). With the direct supports of local NGOs, community people have installed PSF unit near or on the bank of pond where they preserve rainwater throughout the rainy season. During the winter when fresh water is very scarce due to presence of excessive amount of salt in tube well water, the water from this PSF is collected on a community managed rationing system. The other method that they have adopted lately is the rainwater harvesting through installation of simple harvesting apparatus on the roofs of their dwelling units. Throughout the rainy season they collect rainwater and preserve in plastic containers for use in dry season. Adoption and diffusion of rainwater harvesting technology is rather slow as compare to the FSP technology. It is probably related to the cost that an individual family needs to bear if they want to have rainwater technology installed.

They consider their housing as not merely a place of sleeping or having rest; to them it is the central unit of decision making for production and earning for livelihood security; it is the store house of both productive and non-productive assets having market and non-market (intrinsic) values. Therefore, they adopt a series of or incremental adaptation strategies for the protection of housing/settlement. For instance, as a first line defense against flood and torrential/prolong rainfall, they adopt the strategy of building homestead on relatively high-ground. As the second line defense, they adopt the strategy of further raising the already high-ground by piling up column of soil on the home-stead site. Therefore, at the beginning while they erect their dwelling units on the homestead, their homesteads are always 1.5 to 2 meters (4.5 to 6.5 feet) above the normal ground level. This height is enough to keep their dwelling units free from normal floods. Then again to secure from one-in-five year moderate floods, in many instances dwelling units are

Table 4. Adoption and diffusion of adaptation strategies against natural disasters (Excerpt from PRA)

Occurrence 1 = low frequency 5 = high frequency	**Salinity**	**Rainfall/Flooding**	**Cyclone/Tidal surge**
Adaptation strategies	Land that are heavily affected by salinity is used for shrimp farming; Other land parcels are used for *Aman* rice although that yields very poor; Homestead land and in some cases agricultural fields are used for winter vegetables, beans, nuts, watermelon production keeping provision for manual irrigation; Harvest rain water from individual house and fetch water from community managed fresh-water pond; Enter into *Sunderbans* located almost 100 km (60 miles) away to extract non-timber forest products and go for offshore and deep-sea fishing; also seasonally migrate to other areas.	Try to maximize the use of rain water in homestead or infield vegetable garden; Delay in planting of *Aman* rice in lowland to avoid immediate short-duration inundation; Construct housing unit on elevated base; Try to build terrace in homestead vegetable garden to regulate rainfall induced water; Store/preserve agricultural residuals & similar fodders to feed cattle and goats during wet season; Construct latrine on elevated ground to avoid discomfort and public health hazards.	Take shelter/refuge on Water Development Board's embankments or cyclone shelters during emergency; Adjust resource extraction from *Sunderbans* and its adjoining offshore/bay areas depending on weather/climatic conditions; Women and children maintain their own calendar to take the advantage of low-tide catch of wild fish (shrimp) fry; Try to repair and rebuild housing units during dry season to ensure maximum protection against normal cyclonic/windy condition.
Specific needs	Maintenance of sluice gates and canals to stop intrusion of saline water inside embanked settlements from the nearby river-systems; Excavation and regular maintenance of natural canals to store and conserve rainwater for use in various purposes.	Maintenance of sluice gates and canals to ensure smooth flush/drain out of excessive water from the embanked settlements to the tidal river system.	Maintenance of Water Development Board's embankments along *Andhar Manik* river-system to protect settlements from higher wave-surges and cyclonic-surges; Introduction of special social safety nets and livelihood programs for coastal areas.

Source: Author's compilation from PRA session

erected with an additional plinth level of 0.5 to1 meter (1.5 to 3.5 feet) high above the already raised platform of the homestead. Of course, the height of the plinth level varies for each of the components of dwelling unit depending on their importance. For example, for living room or room used as store house of household goods, food stuffs, and fuel, the plinth level is kept at least 1 meter (3.25 feet) above the raised platform of the homestead. But for other components, such as cattle-shed, the height of the plinth level is rather low. While such earthen platforms are good to protect the dwelling units from the normal flood water, those are however not effective against massive force of flood water that usually result from breach of embankments or over toppling of embankment. Either of the cases, the raised earthen platform is often washed away by the erosive force of flood or surge-water. Therefore, the sustainability of adaptation strategy is in big question mark. Here the diffusion and adoption of indigenous adaptation strategies are rather path-dependent- inter generational. There is room for innovation that can offer flood resistant housing unit at an affordable price.

Similarly, as the people are quite familiar with the timing of cyclonic storms, directions of wind flow, and the nature of damage it can cause to their property and wealth, they adopt their coping and adaptation strategies accordingly (Table 4). As a first-line defense, they plant wind-breaker trees around their homesteads. They have specifically adopted the practice of planting of deep-rooted plants' species to create obstruction against the directions of wind flow to minimize direct exposure to cyclonic force. They extensively plant few palm varieties, especially the walls of *Talmyra* palm, and coconut trees in this regards since they work as strong windbreakers against normal cyclonic force. Along with the windbreaker trees, they plant fast-growing trees, and trees that have commercial value. Fast-growing trees are harvested at a short circle to meet the need of fuel and other domestic uses. The commercial trees are harvested at longer interval and get hard cash, which they utilize for special purposes; even for meeting the part of expenses of marriage of daughters and sons. Although they prefer local species while planting fast-growing trees, some of the exotic species have already been adopted in the study area as because of their faster growth and price advantage.

As most dwelling units are made of locally available traditional materials, these can hardly resist against the force of wind during cyclonic event. As a second-line defense, they have adopted the techniques of orientation of their main dwelling unit in such a way that it does not directly obstruct the free flow of wind. In fact, due to the good defense of windbreaker trees and proper orientation, the walls of their dwelling units can avoid direct friction of wind flow. However, their roofs are the most fragile part of their dwelling units. Most roofs are exposed to wind flow. To avoid higher friction with the wind flow, they adopt the techniques of special treatment of their roofs; they bend the roofs from all sides which look like the exterior part of the bottom of country boat. Such smoother shape of roof helps them avoiding/minimizing the friction with wind flow. While the rich people build their dwelling units using brick and concrete materials, relatively less well-off people construct wood and corrugated iron-sheet made double storied dwelling units. The upper floor is relatively low height. The ground floor/basement is used as living unit and the upper floor is for storing and preservation of food grains, seeds, emergency stock of fuel, and other commodities alongside use as living unit. During normal flooding or in some cases, for instance, in one-in-five year magnitude of flooding, the family members take shelter in the upper floor. In the event of mild cyclone, they also take shelter in the upper floor. Although the people get some degrees of protection against normal cyclonic force by adopting the adaptation techniques mentioned above, however, their sustainability is not out of question. Hence, we observe many events of human causality, which could probably be avoided through adoption and diffusion of sustainable adaptation strategies. An important point of start could be blending of local technical knowledge with modern technology.

Non-Adoption of Measures and Resulting Livelihood Vulnerability

This study has mapped out the potential sources of livelihood vulnerability. Respondents believe that any failure to address the vulnerability as cited below would surely trigger huge out migration from the locality.

- Permanent loss of income from current main sources;
- Permanent impairment in access to various social services, such as local health care, markets, schools etc;
- Loss of free/low-cost access to saline free water for drinking and other uses;
- Threatened food security (production and availability); and
- Permanent outmigration of close relatives.

Although people have been coping with and adapting to many natural disasters by adopting various techniques, however, they think that the above causes would trigger huge out migration. It is primarily because in the changing climatic condition marginalized section of the coastal community, such as functionally landless families, share croppers, peasant farmers, day laborers and even fish folks would fail to ensure sustainable livelihood due to repetitive exposure to multiple disastrous events. For example, the adaptation strategies the share croppers and peasant farmers have been adopting could work at best against a discrete event of flood of certain duration and depth. However, in the context of changing climate same flood or salinity intrusion, cyclonic surges may turn into a perpetual event of frequent nature. This kind of repetitive exposure surely would exceed their coping threshold if they failed to adopt new strategies of sustainable adaptation. Similarly, another dominant occupational group-the fish-folks could not ensure good catches in the off-shore or deep-sea during heavy rain that accompanied with gusty wind and cyclonic storm. In fact, for a single episode of such disturbance in weather they loss 7-10 days of earning opportunity. Similarly they face serious problem with post harvest fish processing in absence of proper sunshine/light. Therefore, their income avenues as well as livelihoods are seriously impacted by harsh climatic condition. In the context of changing climate as those disturbances would be even more frequent and intense, these would eventually turn their livelihood strategy as an unsustainable one in the absence of adoption of new innovation. In the changing climate other occupational groups, such as petty traders, small businessmen, transport workers, personal service providers as well would experience livelihood crisis due to severe impairment in the mobility of peoples, goods and services.

People in the coastal area in general can cope with idiosyncratic shock, however in the event of climate change such shock could be repetitive which the respondents could not manage to cope and adapt unless they adopt new strategies. However, adoption of sustainable adaptation strategy is not a single shot exercise; it requires times for innovation (that suits with local circumstance), development of awareness (among people about new innovation), dissemination of early adopters' testimonies (about effectiveness of innovation) and diffusion. Keeping this in mind, various stakeholders including government, NGOs, research institutions, and the communities should come forward with a clear vision to face the future challenges; otherwise, adoption of sustainable adaptation for development and environmental protection would remain as an elusive idea only as we have seen in some parts of developing Asia (Keskinen & Varis 2012; To et al. 2012).

Making Adoption of Sustainable Adaptation Strategies Happen

Although most respondents are reluctant to receive any training to upgrade their current coping and adaptation, yet they believe substantive visible initiatives from concerned bodies to protect their livelihood from the current as well as future climatic events may encourage the community to adopt sustainable adaptation strategies (Table 5). Community perceived that both hard and soft measures would be required to adopt by the public authority to make people believe that they would not need to evacuate the locality in the years to come. They further added that if hard measures are adopted by the government, the community would adopt various measures and techniques related to their social, economic and behavioral aspects of life to make their private adaptation strategies sustainable. It seems like a prisoner dilemma- public agencies and the development partners want to see: community move first; by contrast, community people want to see other's (e.g. public) initiatives first.

In fact, close consultation reveals that community believes,

- Construction of sea wall would surely protect their farmland, freshwater fish farms, vegetable gardens, and open water fishing grounds/canals from the intrusion of salinity;
- Judicial regulation of sluice gates could allow regulated penetration of fresh/sweet water during monsoon and post monsoon

Table 5. Measures that can encourage people to adopt sustainable adaptation strategies

Measures	Target beneficiary	Characteristic of the beneficiary group(s)	Implications	Stakeholders and Implementation
Structural measures:				
Construction of safe shelters	Coastal people in general	People with very low socio economic standing, such as landless families, peasant farmers, wage laborers, fish folks particularly who do not have land.	Many of the existing cyclone shelters are located close to the shoreline and would be non-functional. Construction of new shelters in strategic locations would encourage individual family to adopt adaptation *in situ* at private level as those safe-shelters would provide them maximum protection against future exposures to climatic events.	Government agencies, development partners and the local community need to act in coordination to implement this program under long term planning. Community people should have more roles in maintenance.
Construction of new sea-walls/ embankments/ levees	People living along the most exposed parts of the coastal area.	People who have been living in the marginalized area and whose livelihood as well as food security would be severely affected due to higher exposure to multitude of new form of hazards, including coastal inundation, cyclonic surges and intrusion of salinity etc.	Seawalls/embankments/levees those were constructed during 1960s would no more be capable of protecting people from the future exposures. Moreover, over the last couple of decades many new settlements have been developed in the open area of the coast. New defence walls of these types would minimize the future exposure which would encourage the local people to adopt their own adaptation strategies to secure their livelihood on a sustained basis	Government agencies, development partners and the local community need to act in coordination to implement under long term planning. Local government, local NGOs and the community should ensure proper maintenance.
Re-excavation of illegally occupied canals and construction of new sluice gates	People living along the most exposed parts of the coastal area.	Small peasant farmers and other occupational groups whose food security and access to fresh water would be severely affected due to higher exposure to salinity intrusion.	After recovery of currently (illegally) occupied canals could be re-excavated which would encourage the small and peasant farmers to adopt cooperative strategy of adaptation for conserving and rationing of increased amount of fresh water for use in dry-season crop agriculture. Similarly improved system of sluice gates would help regulating the penetration of saline water, which would ensure better access to fresh water and reduce people's sensitivity to fresh water crisis.	Local government, local NGOs and the community could implement and monitor the condition under a medium term plan.
Non-structural measures:				
Special social safety nets for coastal area	People in need of special supports to secure livelihood in the coastal environment.	Landless, small peasant farmers and others, such as day laborers, petty traders, fish folks who would be more susceptible to income shock due to future catastrophes would be benefited.	Current social safety nets are for the people in special needs, such as elderly/aged people, vulnerable women irrespective of their place of living. Although there is special provision of relief and rehabilitation for disaster victims, yet a comprehensive special social safety net for the targeted occupational groups in the coastal area would enhance their adaptive capacity against future impacts of SLR and its associated events.	Both central and local government, local NGOs and local the philanthropic foundations could implement and monitor the progress under short to medium term plan.

continued on following page

Table 5. Continued

Measures	Target beneficiary	Characteristic of the beneficiary group(s)	Implications	Stakeholders and Implementation
Community food security program for coastal area	People in need of special support to secure livelihood in the coastal environment.	Landless, small and peasant farmers, and others low income groups who often encounter harvest failure or loss of income due to extreme climatic events would be benefited.	Food security is a major concern in most parts of the coast as single cropping (wet season crop) is practiced due to acute salinity in dry season. Intrusion of salinity due to future SLR would even more diminish the productivity of soil, which would aggravate the food security of even more people re in a particular time of a year. A well functioning community food security program would help people to adopt measures to better manage the crisis on a sustained basis.	Government agencies, development partners and the local community need to act in coordination to implement under short to medium term plan. Proper implementation would help community to adopt other measure at private level to enhance adaptive capacity.

Source: Author's compilation from PRA session

period which could be conserved in lowlands, natural depressions and surface water bodies to use in the dry season for irrigation and drinking;

- Re-excavation of illegally occupied canals could be used to trap fresh water during monsoon and post monsoon for use in the dry season agriculture;
- Current localized practices of rainwater harvesting for domestic use could be updated for increased water harvesting efficiency at a minimal cost (Table 5).

Furthermore, they added that they believe implementation of these measures would encourage adoption of sustainable adaptation strategies by private individuals who otherwise would migrate out from the locality given the future impacts of climate change. Other measures that can assist people to adopt sustainable adaptation strategies against various impacts including the increased salinity are presented in the Table 5.

While there are obvious benefits from exposure minimization measures, such as construction of new sea- walls, operation of new sluice gates and re-excavation of illegally occupied (dead) canals, however, costs might be substantial. Implementation of exposure minimization measures of these kinds are beyond the private adaptive measures. It requires planned-public initiatives to ensure distributive justice and burden sharing among all. This would help people to adopt various sustainable adaptive measures at private level to manage the situation.

ADOPTION AND DIFFUSION OF SUSTAINABLE ADAPTATION: CHALLENGES AND WAY FORWARD

Scientific advances already have established a clear link among climate change induced accelerated SLR and various hydro-metrological disastrous events. The accelerated SLR would probably increase the frequency and magnitude of coastal flooding, salinity intrusion, cyclonic storms and surges, and coastal erosion in coastal areas of Bangladesh. Although respondents in the study area are familiar with many of these disastrous events, in the changing scenarios of future climate, their known patters of such disastrous events may appear very differently than what were in the past. Coastal flooding and salinity intrusion, for instance, are projected to encroach over much larger areas than now. The more horrific thing is-such flooding would be perpetual. Once an area goes under accelerated SLR, it would never return to its earlier state. It means newer areas would go under saline water as the time elapses. Peoples are familiar with intermittent/seasonal nature of inundation of their farmlands, lowlands, water-bodies which return to their original states after some time of inundation. However, peoples are not familiar with the perpetual nature of inundation that is more likely in the event of accelerated SLR. This is a big challenge for future livelihood security.

Most respondents are familiar with seasonal nature of intrusion of saline water in the soil, sub-soil water tables and in surface water bodies. Although such salinity reaches its pick during the dry season and makes the land unsuitable for cultivation of most winter-crops, the salinity level starts to recede as soon as the rainy season starts. However, in the changing scenario of SLR, such salinity would be acute throughout the year which may not only jeopardize year round productivity of crops most importantly this may create havoc due to acute scarcity of fresh water for domestic and other uses. This means the entire study area might be inhabitable. Similarly, as the frequency, intensity and magnitude of cyclones and storm surges are believed to increase, these would amplify the devastating power of the cyclones and surges. How people might cope with and adapt to this new circumstance is really a challenge for future.

As people have been adopting various coping and adaptation strategies against various disastrous events often successfully by and large relying on their indigenous knowledge, local cultures and traditions, they seem hardy care about the future vulnerabilities. As regards adoption of sustainable adaptation strategies against future livelihood vulnerability, a significant proportion of the respondents think that the coping and adaptation strategies they have been adopting would work fairly against the future catastrophes as well. In response to question- whether they feel the need for updating/upgrading the coping and adaptation strategies that they have been adopting, a large majority of the respondents indicated that they really do not need any external training for most adjustment related to agricultural activities. Almost similar responses were observed in regards to their felt need for upgrading of their current coping and adapting strategies that relate to non-agriculture based livelihoods. However, they think that adoption of few structural and nonstructural measures would encourage them to adopt new adaptation strategies at private level. They think that if right exposure minimization measures are in place, they could adopt their own measures to minimize their sensitivity to the future impacts by enhancing their adaptive capacity and resiliency.

However, it is really a very big question mark, why the people do not want any training to upgrade the coping and adaptation strategies that they have been adopting for long. Is it due to simple inertia to change something with which they are used to or for other reasons? Whether climate awareness or poor knowledge about the likely impacts of changing climate of the people are as well the cause of people's such behavior toward adoption of new strategies (through formal training)? We were trying to get the answer of these questions to evaluate if anything could be done to encourage adoption of sustainable adaptation strategies through education and other public campaigns. Close consultation with the respondents revealed that they are not well aware about the differential nature of threats of future events. It is partly because as the SLR induced events are new and very distant phenomenon, they have very limited or no understanding about it. Therefore, they consider the future threats as the same (threat) they have been encountering for long. Furthermore, it also appeared that some of them were probably overestimating their adaptation efficacy and effectiveness of the measures they have been adopting. It happened probably for two reasons. First, some of them probably wanted to show their courage. If it is the case, they are probably affected by illusion of control syndrome as warned by Winstin (1980).

In real situation when new threats would appear probably this category of respondents would try to acquire new (local) technical knowledge to adopt strategies that most suit with the changing circumstances. Second, cause of overestimation of adaptation efficacy by some respondents is probably due to their inflexibility to receive new ideas and innovations. If the former one is the cause, there is ample room for intervention with a coordinated program for, first to build climate awareness, second to make them aware- why the coping and adaptation strategies they have been adopting would not work against the future threats unless upgraded or replaced with new strategies, third to give them opportunity to have access (both physical and financial) to new innovations so that they can learn and eventually adopt to make their future adaptation strategies a sustainable one. In this connection all concerned, such as government agencies, research bodies, development partners and the communities must come forward and bear the shared responsibility to prepare the coastal people to adopt sustainable adaptation strategies against the future vulnerabilities.

ACKNOWLEDGMENT

Authors acknowledge the research grant provided by the Norwegian Ministry of Foreign Affairs as part of Scholarship for doing Doctor of Philosophy (PhD) of the first author at the Asian Institute of Technology (AIT) in Thailand, supervised by the second author. However, the views expressed here are the authors' alone.

REFERENCES

Adger, W. N. (2006). Vulnerability. *Global Environmental Change*, *16*(3), 268–281. doi:10.1016/j.gloenvcha.2006.02.006.

Adger, W. N. (2010). Addressing barriers and social challenges of climate change adaptation. In P. C. Stern, & R. E. Kasperson (Eds.), *Facilitating climate change responses: Report on knowledge from the social and behavioural sciences* (pp. 79–84). Washington, DC: National Academies of Science..

Adger, W. N., Arnell, N. W., & Tompkins, E. L. (2005). Successful adaptation to climate change across scales. *Global Environmental Change*, *15*(2), 77–86. doi:10.1016/j.gloenvcha.2004.12.005.

Agrawala, S., Ota, T., Ahmed, A. U., Smith, J., & van Aalst, M. (2003). *Development and climate change in Bangladesh: Focus on coastal flooding and Sundarbans*. Paris, France: OECD..

Ahmed, A. U., & Alam, M. (1998). Development of climate change scenarios with general circulation models. In S. Huq, Z. Karim, M. Asaduzzaman, & F. Mahtab (Eds.), *Vulnerability and adaptation to climate change for Bangladesh* (pp. 13–20). Dordrecht, The Netherlands: Kluwer Academic Publishers..

Ahsan, R., Karuppannan, S., & Kellett, J. (2011). Climate migration and urban planning system: A study of Bangladesh. *Environmental Justice*, *4*(3), 163–170. doi:10.1089/env.2011.0005.

Ali, A. (1999). Climate change impacts and adaptation assessment in Bangladesh. *Climate Research*, *12*(2/3), 109–116. doi:10.3354/cr012109.

Ali Khan, T. M., Singh, O. P., & Rahman, M. S. (2000). Recent sea level and sea surface temperature trends along the Bangladesh coast in relation to the frequency of intense cyclones. *Marine Geodesy*, *23*, 1–14.

Allison, E. H., & Ellis, F. (2001). The livelihoods approach and management of small-scale fisheries. *Marine Policy*, *25*(5), 377–388. doi:10.1016/S0308-597X(01)00023-9.

Alwang, J., Ersado, L., & Taruvinga, N. (2001). Changes in poverty in Zimbabwe between 1990 and 1996: Worsening outcomes under adverse conditions. *Development Southern Africa*, *18*(5), 553–580. doi:10.1080/03768350120097432.

Assan, J. K., Caminade, C., & Obeng, F. (2009). Environmental variability and vulnerable livelihoods: Minimising risks and optimising opportunities for poverty alleviation. *Journal of International Development*, *21*(3), 403–418. doi:10.1002/jid.1563.

Bangladesh Bureau of Statistics (BBS). (2011). *Population and housing census 2011: Preliminary results*. Bangladesh: Statistics Division, Ministry of Planning, Government of the People's Republic of Bangladesh..

Barbier, E. (2011). The policy challenges for green economy and sustainable economic development. *Natural Resources Forum*, *35*(3), 233–245. doi:10.1111/j.1477-8947.2011.01397.x.

Below, T. B., Mutabazi, K. D., Kirschke, D., Franke, C., Sieber, S., Siebert, R., & Tscherning, K. (2012). Can farmers' adaptation to climate change be explained by socio-economic household-level variables? *Global Environmental Change*, *22*(1), 223–235. doi:10.1016/j.gloenvcha.2011.11.012.

Brooks, S., & Loevinsohn, M. (2011). Shaping agricultural innovation systems responsive to food insecurity and climate change. *Natural Resources Forum*, *35*(3), 185–200. doi:10.1111/j.1477-8947.2011.01396.x.

Brown, K. (2011). Sustainable adaptation: An oxymoron? *Climate and Development*, *3*(1), 21–31. doi:10.3763/cdev.2010.0062.

Burton, I. (1997). Vulnerability and adaptive response in the context of climate and climate change. *Climatic Change*, *36*, 185–196. doi:10.1023/A:1005334926618.

Cannon, T. (2002). Gender and climate hazards in Bangladesh. *Gender and Development*, *10*(2), 45–50. doi:10.1080/13552070215906.

Cannon, T., & Muller-Mahn, D. (2010). Vulnerability, resilience and development discourses in context of climate change. *Natural Hazards*, *55*, 621–635. doi:10.1007/s11069-010-9499-4.

CARE. (2003). *Report of a community level vulnerability assessment conducted in South-west Bangladesh. Reducing Vulnerability to Climate Change (RVCC) Project*. Dhaka, Bangladesh: CARE..

Chambers, R., & Conway, G. (1992). *Sustainable rural livelihoods: Practical concepts for the 21st century*. IDS Discussion paper 296, Institute of Development Studies, Brighton, UK: University of Sussex.

Choudhury, A. M., Neelormi, S., Quadir, D. A., Mallick, S., & Ahmed, A. U. (2005). Socio-economic and physical perspectives of water related vulnerability to climate change: Results of field study in Bangladesh. *Science and Culture*, *71*(7-8), 225–238.

Comin, D., & Hobijn, B. (2004). Cross-country technology adoption: Making the theories face the facts. *Journal of Monetary Economics*, *51*(1), 39–83. doi:10.1016/j.jmoneco.2003.07.003.

de Haan, L. J., & Zoomers, A. (2005). Exploring the frontier of livelihood research. *Development and Change*, *36*(1), 27–47. doi:10.1111/j.0012-155X.2005.00401.x.

Downing, T. E., & Patwardhan, A. (2004). Assessing vulnerability for climate adaptation. In B. Lim, & E. Spanger-Siegfried (Eds.), *Adaptation policy frameworks for climate change: Developing strategies, policies, and measures*. Cambridge, UK: Cambridge University Press..

Ellis, F. (2000). *Rural livelihood diversity in developing countries*. Oxford, UK: Oxford University Press..

Eriksen, S., & O'Brien, K. (2007). Vulnerability, poverty and the need for sustainable adaptation measures. *Climate Policy*, *7*, 337–352. doi:10.3763/cpol.2007.0717.

Fankhauser, S., Smith, J. B., & Tol, R. S. J. (1999). Weathering climate change: some simple rules to guide adaptation decisions. *Ecological Economics*, *30*, 67–78. doi:10.1016/S0921-8009(98)00117-7.

Faruque, H. S. M., & Ali, M. L. (2005). Climate change and water resources management in Bangladesh. In M. M. Q. Mirza, & Q. K. Ahmad (Eds.), *Climate change and water resources in South Asia* (pp. 231–254). Leiden, UK: Balkema Press..

Fussel, H.-M. (2007). Vulnerability: A generally applicable conceptual framework for climate change research. *Global Environmental Change*, *17*(1), 155–167. doi:10.1016/j.gloenvcha.2006.05.002.

Geroski, P. A. (2000). Models of technology diffusion. *Research Policy*, *29*, 603–625. doi:10.1016/S0048-7333(99)00092-X.

Government of Bangladesh (GOB). (2005). *National adaptation programme of action (NAPA). Final report*. Dhaka, Bangladesh: Ministry of Environment and Forests..

Government of Bangladesh (GOB). (2008). *Bangladesh climate change strategy and action plan 2008*. Dhaka, Bangladesh: Ministry of Environment and Forests..

Hinkel, J. (2011). Indicators of vulnerability and adaptive capacity: Towards a clarification of the science-policy interface. *Global Environmental Change*, *21*(1), 198–208. doi:10.1016/j.gloenvcha.2010.08.002.

Huq, S., Karim, Z., Asaduzzaman, M., & Mahtab, F. (Eds.). (1998). *Vulnerability and adaptation to climate change for Bangladesh*. Dordrecht, The Netherlands: Kluwer Academic Publishers..

Huq, S., Rahman, A., Konate, M., Sokona, Y., & Reid, H. (2003). *Mainstreaming adaptation to climate change in least developed countries (LDCs)*. London, UK: International Institute for Environment and Development..

Intergovernmental Panel on Climate Change (IPCC). (2001). *Climate change 2001: The scientific basis. Contribution of Working Group I to the Third Assessment Report of the Intergovernmental Panel on Climate Change*. Cambridge, UK: Cambridge University Press..

Intergovernmental Panel on Climate Change (IPCC). (2007). The physical science basis, summary for policy makers. Geneva, Switzerland: WMO-UNEP.

Jackson, T. (2011). Societal transformations for a sustainable economy. *Natural Resources Forum*, *35*(3), 155–164. doi:10.1111/j.1477-8947.2011.01395.x.

Kawahara, K., et al. 2004. Pond sand filter. JICA/AAN Arsenic Mitigation Project. Report 3. Dhaka, Bangladesh: Asia Arsenic Network. ISBN: 984 32 1387 2

Keskinen, M., & Varis, O. (2012). Institutional cooperation at a basin level: For what, by whom? Lessons learned from Cambodia's Tonle Sap Lake. *Natural Resources Forum*, *36*(1), 50–60. doi:10.1111/j.1477-8947.2012.01445.x.

McCarthy, J. J., Canziani, O. F., Leary, N. A., Dokken, D. J., & White, K. S. (Eds.). (2001). *Climate change 2001: Impacts, adaptation and vulnerability. Work Group II input to the Third Assessment Report*. Cambridge, UK: Cambridge University Press..

O'Brien, K., & Leichenko, R. (2007). *Human security, vulnerability and sustainable adaptation*. Human Development Report Office, Occasional Paper 2007/9. New York, NY: UNDP.

Olsson, L., & Jerneck, A. (2010). Farmers fighting climate change—From victims to agents in subsistence livelihoods. *Wiley Interdisciplinary Reviews: Climate Change*, *1*(3), 363–373. doi:10.1002/wcc.44.

Parry, M. (2009). Climate change is a development issue, and only sustainable development can confront the challenge. *Climate and Development*, *1*, 5–9. doi:10.3763/cdev.2009.0012.

Reddy, S., & Painuly, J. P. (2004). Diffusion of renewable energy technologies – Barriers and stakeholders' perspectives. *Renewable Energy*, *29*, 1431–1447. doi:10.1016/j.renene.2003.12.003.

Rogers, E. (1995). *The diffusion of innovations* (4th ed.). New York, NY: Free Press..

Schaerer, C., & Ahmed, A. U. (2004). Adaptation to climate change in vulnerable communities: Lessons from practice in South-western Bangladesh. In A. U. Ahmed & N. Haque (Eds.), *Adaptation to climate change: Knowledge sharing for capacity building. Proceedings of Workshop held on 10 December 2003 at COP 9 Milan.* Dhaka, Bangladesh: Climate Action Network South Asia (CANSA) and RVCC.

Scoones, I. (1998). *Sustainable rural livelihoods: A framework for analysis*. Brighton, UK: IDS Publications..

Singh, O. P., Ali Khan, T. M., Murty, T. S., & Rahman, M. S. (2001). Sea level changes along Bangladesh coast in relation to southern oscillation phenomenon. *Marine Geodesy*, *24*, 65–72.

Smit, B., Burton, I., Klein, R., & Wandel, J. (2000). An anatomy of adaptation to climate change and variability. *Climatic Change*, *45*, 223–251. doi:10.1023/A:1005661622966.

Smit, B., & Skinner, M. W. (2002). Adaptation options in agriculture to climate change: A typology. *Mitigation and Adaptation Strategies for Global Change*, *7*(1), 85–114. doi:10.1023/A:1015862228270.

Smit, B., & Wandel, J. (2006). Adaptation, adaptive capacity, and vulnerability. *Global Environmental Change*, *16*(3), 282–292. doi:10.1016/j.gloenvcha.2006.03.008.

Smith, J. B. (1997). Setting priorities for adapting to climate change. *Global Environmental Change*, *7*(3), 251–264. doi:10.1016/S0959-3780(97)00001-0.

To, P. X., Dressler, W. H., Mahanty, S., Pham, T. T., & Zingerli, C. (2012). The Prospects for payment for ecosystem services (PES) in Vietnam: A look at three payment schemes. *Human Ecology*, *40*(2), 237–249. doi:10.1007/s10745-012-9480-9 PMID:22523444.

Turner, B. L., Kasperson, R. E., Matson, P. A., McCarthy, J. J., Corell, R. W., & Christensen, L. … Schiller, A. (2003). A framework for vulnerability analysis in sustainability science. *Proceedings of the National Academy of Sciences, US 100*, 8074–8079.

van Aalst, M. K., Cannon, T., & Burton, I. (2008). Community level adaptation to climate change: The potential role of participatory community risk assessment. *Global Environmental Change*, *18*(1), 165–179. doi:10.1016/j.gloenvcha.2007.06.002.

Weinstein, N. D. (1980). Unrealistic optimism about future life events. *Journal of Personality and Social Psychology*, *39*(5), 806–820. doi:10.1037/0022-3514.39.5.806.

Wisner, B., Blaikie, P., Cannon, T., & Davis, I. (2004). *At risk: Natural hazards, people's vulnerability and disasters* (2nd ed.). London, UK: Routledge..

World Bank. (2000). *Bangladesh: Climate change and sustainable development*. Report No. 21104-BD, Rural Development Unit, South Asia Region. Dhaka, Bangladesh: The World Bank.

World Commission on Environment and Development. (1987). *Our common future*. Oxford, UK: Oxford University Press..

This chapter was previously published in Cases on the Diffusion and Adoption of Sustainable Development Practices edited by Helen E. Muga and Ken D. Thomas, pages 83-117, copyright 2013 by Information Science Reference (an imprint of IGI Global).

Section 6
Emerging Trends

This section examines both future difficulties and innovative solutions for climate change and Sustainable Practices. Though the economic, ecological, and social troubles in the world today seem almost insurmountable, human ingenuity promises to offer ever more unique solutions to the problems faced by modern societies. Some notable advances explored in this section include environmental education, waste management and renewable energy, conservation of natural resources, and ecologically-friendly urban environments. The final 17 chapters of this extensive three-volume reference conclude with a detailed look at emerging trends in the field of Sustainable Practices.

Chapter 80
The University of Hawai'i at Manoa Sustainability Courtyard as a Center for Campus Engagement

John Cusick
University of Hawai'i at Manoa Environmental Center, USA

ABSTRACT

The University of Hawai'i at Manoa Sustainability Courtyard provides a venue for campus engagement to educate and increase awareness of developing solutions and/or adaptations to geopolitical and environmental challenges, particularly energy, water, and food security. Few institutions are immune to coping and addressing triple bottom line issues of energy (economy), water, food and waste management (environment), and workplace comfort and safety concerns (equity), so the limited window of time students have on university campuses is an opportunity to engage and prepare them for an uncertain future (+ education).

INTRODUCTION

What evidence will there be that society has been transformed as an outcome of all the rhetoric on sustainability of the past two decades, and how will anyone notice a difference at the University of Hawai'i at Manoa? (Cusick, 2008, p. 246)

DOI: 10.4018/978-1-4666-4852-4.ch080

This chapter describes a space on the campus of the University of Hawai'i at Manoa (UHM), the State of Hawai'i's flagship institution of higher education, which has demonstrated the potential to effectively contribute to the university's strategic planning objectives related to sustainability. The UHM Sustainability Courtyard (SC) is an example of what institutions of higher education can do, and are doing, to promote education and awareness in support of civic engagement. The cumulative efforts of the campus community, particularly over the past decade, are relevant in the development of sustainability literacy, and the SC provides a case study of how other institutions may consider the use of campus space for similar purposes.

As is often noted, the terms *sustainable development* and *sustainability* generate multiple definitions and perspectives (Barlett & Chase, 2004; Edwards, 2005). The challenges have less to do with defining terms and more with cultivating the sense of individual and collective effort required to adapt to the scale of challenges likely to confront contemporary societies. While the key concepts of all sustainability definitions focus on economic development, environmental management, and social equity, also referred to as people, planet, and prosperity (or profit), which are collectively referred to as the triple bottom line and three pillars of sustainability, the problems, as noted by David Orr (1992), are that the challenges "feed upon each other." Of relevance to the transformative use of campus spaces as described in this collection of case studies is the role of education itself. Orr suggests that "all education is environmental education" and that the practice of civic engagement to educate people to live sustainably requires that education occur "in part as a dialogue with a place" (Orr, 1992, p. 90). One such place to facilitate the paradigm shift discussed by Andres Edwards in *The Sustainability Revolution* are the campuses of institutions of higher education. "Through education, sustainability can become firmly established within the existing value structure of societies while simultaneously helping that value structure evolve toward a more viable long-term approach to systemic global problems" (Edwards, 2005, p. 23).

Consensus on the application of sustainability at an institution of higher education, let alone a working definition, can be difficult to reach for several reasons. First, many hard choices have to be made by university administrations and faculty for near-, medium-, and long-term goals, whether involving campus planning and capital improvements or research and instructional programs, while students tend to focus on near-term personal goals, such as graduating in a timely fashion and managing the costs and logistics of independent living. Second, complex issues, such as adoption of alternative sources of energy, purchasing of local and green products, and reduction and disposal of solid wastes, can require years of planning to organize and execute across a university campus. Nonetheless, any progress at UHM responding to vulnerabilities to energy, water, food, and waste management systems moves the State of Hawai'i closer to the sustainable futures so enticingly articulated by many proponents.

FRAMING THE STUDY OF PLACE

Places are records of change in space and time that allow us to investigate the active expression of human agency (Jackson, 1989); they constitute the physical setting where social relations are played out (Agnew, 1993). The concept of place is interpreted as a "center of meaning" and the "focus of human emotional attachment," understood as involving a variety of attributes, and a geography of the lived world (Relph, 1976). Places manifest themselves under conditions that emerge as distinct from one another, as reflections of contemporary concerns (Lefebvre, 1991), and can be viewed as signifiers, elements set within a cultural system. Cultural geographers focus on representations of place in their natural and social contexts to understand and explain human behaviors.

Institutions of higher education have responded to sustainability challenges with a variety of activities and initiatives ranging from commitments to lower greenhouse gas emissions and increase green purchasing to offering new courses and degree programs. As for the UHM campus, a former parking area for university vehicles was reassigned in 2006 to give the topic of sustainability a "place" on campus. The place is a commons area that occupies a central location on campus with the potential to effectively contribute through civic engagement, the recognition by individuals that they are members of a larger society and have responsibilities to promote and take actions to make differences in their communities. The reproduction of the space from wide expanses of concrete to one with a variety of services and infrastructure in an aesthetically appealing setting constitutes a new destination on campus for social interaction and student activism (Figure 1).

The interdisciplinary nature of sustainability requires broad participation of schools and colleges, departments and centers, and administration and staff to be considered a truly "engaged" campus community. The responsibility of addressing the many identified challenges cannot be the domain of a single campus unit or organization. To do so is to factionalize the effectiveness of collective action that can result from diverging agendas of individuals' short-term self-promotion objectives at the expense of achieving long-term institutional goals.

The initial momentum of the SC slightly faded in 2009-2010 due to construction in the vicinity and other campus sustainability projects gaining traction. Despite the hiatus and limited available funds, current facilities include two food kiosks with a dozen picnic tables, a stage, and a student-managed herb and vegetable garden, while student-faculty organized events in the SC over the years have included Earth Days, environmental film series, demonstration projects, and live music events. Taken as a whole, these initiatives have redefined a concrete expanse into a place to engage the campus community in the dialogue suggested by David Orr.

Figure 1. Fall 2011 sustainability courtyard event

The desired conversation "is the opposite of the specialization and narrowness characteristic of most education. The ecologically literate person has the knowledge necessary to comprehend interrelatedness, and an attitude of care or stewardship ... Knowing, caring, and practical competence constitute the basis of ecological literacy" (Orr, 1992, p. 92). However, the goal of developing a sustainability ethic as stated in the current UHM Strategic Plan requires more than a single courtyard with food kiosks serving vegetarian meals. The following sections discuss learning outcomes of the past two years and outline several visions for the space as presented by students as part of an assignment in an Environmental Practicum course taught by the author.

THE STRATEGIC CONTRIBUTIONS OF THE SUSTAINABILITY COURTYARD

Community buy-in to the goals of sustainability are made evident in the Hawai'i 2050 Sustainability Plan that recommends "putting a stake in the ground" and "creatively educating the community about the importance of living sustainably and launching programs in the schools to teach sustainability strategies" (Social Sciences Public Policy Center, 2010, p. 33). The relevance of environmental education is that "an environmentally literate person in Hawai'i is an informed, lifelong learner who values Hawai'i's uniqueness, practices environmental stewardship, and lives sustainably" (Sato & Staab, 2012, p. 2). The Hawai'i Environmental Education Alliance reports that the results of a 2000 survey demonstrated 93% of survey respondents were in favor of environmental education in the public schools. "If anything, with more attention focused on being 'green' today, public support is potentially even stronger ... the key (to success) will be to link" education with alleviating pressing issues "such as unemployment, homelessness, traffic congestion, crime and violence, illegal drugs, and the high cost of living" (Sato & Staab, 2012, p. 51).

The Hawaiian Islands are often referred to as a place of inspiration unequaled in the world, which is evident by the over seven million annual visitors who are hosted by a resident population of one million. Representations of the Hawaiian Islands as a "paradise on earth," particularly fueled by the mass tourism industry, reflect the marketability of sustainability. "We need to value our distinctiveness today. The entire world recognizes this, even if only as a tropical fantasy" (Howes & Osorio, 2010, p. 7). This sense of exceptionalism, also captured in the words of an often-sighted bumper sticker "This ain't da mainland," could be lost if proper care is not afforded natural and cultural environments. While UHM recognizes a special responsibility and obligation to do more to meet the needs of the people of Hawai'i, such responsibility is as much local as it is global in today's technologically connected societies. Assuming responsibility implies taking leadership, and institutions of higher education exemplify a commitment to leadership and innovation by embracing the processes toward developing a sustainable future.

The UHM 2011-2015 Strategic Plan identifies sustainability and technology as two of five values that guide university decision-making: "Our unique location has required Manoa to prioritize environmental, cultural and economic sustainability so we can become an international leader in this area" (University of Hawai'i at Manoa, 2011, p. 5). Among the unique challenges created by the geographic isolation of being thousands of miles from any continent are that the State of Hawai'i imports 90% of its energy needs, urbanization reduces the amount of agricultural lands available to offset food imports, and habitat loss results in the islands status as "extinction capital of the planet" with nearly three quarters of native and endemic species of plants and animals on the Endangered Species List. The concepts associated with resource management based on topographic features and natural processes associated with tropical and sub-tropical island ecosystems on volcanic islands, as was done by Native Hawaiians in ahupua'a—land divisions that generally run from mountain to sea that provided most of the needs of a community—provide organizing principles as to how contemporary societies can restore balance and avoid environmental degradation.

Smart technologies, as identified in the Strategic Plan (2011), "allow us to emerge as a stronger and more organizationally sustainable campus" and expand the reach of UHM locally and globally. For instance, innovative resource management strategies developed and tested in Hawai'i are applied to other places worldwide that are faced with similar competing demands of environmental conservation and economic development. Thus, at a time when the need is evident worldwide, universities provide education and training that societies and employers increasingly indicate are in demand to analyze and address environmental management issues. Therefore, sustainability is an investment not only in the campus community but also in the environmental quality of the places where we live, work, and recreate.

The required space and infrastructure to engage the campus community is already established within the SC. The branding of the place as the SC has been understated in terms of signage, although it is identified by name on the official campus map. The perspective of one student early on in the evolution of the space was that the SC is where the "hippies" hang out. Tibetan prayer flags, occasional evening drum and chanting circles, and weekend skateboarders aside, there may be more than meets the eye to the casual observer who passes through the SC on their way from one corner of campus to another. Cultivating a sense of place and campus stewardship to emphasize interdisciplinary approaches to knowledge, principles, and practices of sustainability is the greater challenge that will require the input of resources and labor to maximize its potential for civic engagement. Such is the importance of the SC as a place for "habitation" as referenced in a 2012 draft of the Manoa campus Landscape Master Plan.

Over the years, the SC has primarily functioned as a social space with occasional events held during the academic year. Sea Grant College administrators and staff along with the campus grounds manager were the primary force in establishing the SC in 2006. Those early efforts led to their redesigning an adjacent courtyard with picnic tables and landscaping and creating a quiet refuge in contrast to the busy flow of daily traffic through the SC. As no one single campus unit or student organization has sole responsibility for the SC, the situation is somewhat a tragedy of the commons except that less is made of its potential as a resource rather than the wholesale degradation of the place by a variety of users. The campus wireless Internet network includes the SC and allows students to socialize and/or study online. The neighboring Hawai'i Institute of Geophysics that houses Sea Grant College on one side of the SC and the Environmental Center on another have served for several years as bases for faculty-student collaborations to plan, organize, and host events and activities. These partnerships were initiated by members of the Council on Sustainability and maintained by the Manoa Sustainability Corps, both of which are faculty-student groups supported by the past two Chancellor administrations respectively.

The combined assets of open space, food kiosks, outdoor seating, stage, recycling bins, edible and ornamental landscaping, as well as two adjacent lecture auditoriums, provide communal space to host day and evening events. Previous events include film screenings and live music, open forums and vendor exhibits, and student club activities and fundraisers (Figure 2). An ongoing coalition of student clubs with academic and administrative units, particularly members of the Ecology Club, Surfrider Foundation campus chapter, Sustainable UH, Sustainable Organic Farm Training program, Cycle Manoa, and Sustainable East-West Center, with administrative support from the Manoa Sustainability Corps managed by the Vice Chancellor's Office of Physical, Environmental, and Long Range Planning, as well as students enrolled in Environmental Studies, Interdisciplinary Studies, and Honors Program courses, has helped cultivate collaboration among students as stewards of their campus. In the past academic year, students investigated and critically analyzed current UHM initiatives in support of sustainability, as well as surveyed the campus community for perspectives on how sustainability is articulated and prioritized. These efforts have been supported with funds from undergraduate and graduate student government organizations as part of campus greening initiatives.

Whether activities take place there or not, the SC serves as an inspirational space to make considerable progress on education and awareness as far as students are concerned. Headlines and articles published in the campus newspaper over the course of 2011-2012 demonstrate partial success in the campaign to increase campus engagement. Many of the articles are written by or have content provided by participating students, thus allowing them to develop a portfolio of work that enhances their employment prospects while contributing to the university administration's interest to cultivate a sustainability ethic among the campus community. The following newspaper article headlines provide a sample of the initiatives:

First Green Friday discourages dependency on plastic; Solar feasibility plan aims to take UHM off the grid; Students compete to reduce energy use in dorms; Ecology Club promotes change; UHM works toward sustainability; and Energy audit aims to decrease energy usage connected to student tuition. (see http://www.kaleo.org/)

Figure 2. Fall 2011 sustainability courtyard events flyer

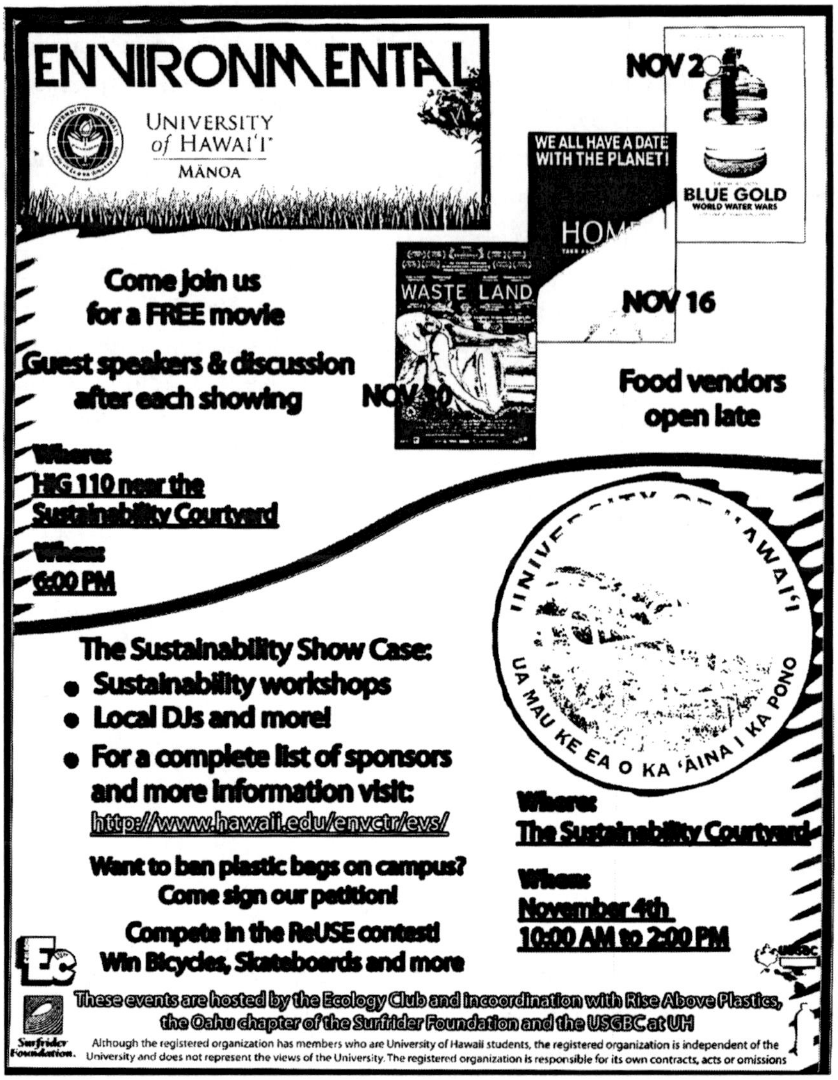

The following excerpt from one article suggests a rationale for university leaders to provide such collaborative spaces on campus:

These events are targeted at students and community members to promote understanding of environmental issues and possible progressive action. By showing these films, the club hopes to broaden knowledge of environmental problems and provide incentives for attendees to consider environmentally friendly practices ... we are in charge of our future. To irresponsibly disregard the real threats and realities is to live against what our values are, especially in Hawai'i, said Ecology Club member Keri Namimoto. (Avery, 2011)

STUDENT VISIONS FOR THE SUSTAINABILITY COURTYARD

Students active in SC-related projects are representative of the campus community as they pursue majors, minors, certificates, and/or are associated with the following departments and programs: Anthropology, Architecture, Biology, Botany, Creative Media, East-West Center, Economics, Engineering, Environmental Studies, Geography, Hawaiian Studies, Honors Program, Interdisciplinary Studies, Latin American Studies, Marine Option Program, Pacific Island Studies, Peace Studies, Philosophy, Psychology, Public Administration, Sustainable/eco Tourism, Tropical Agriculture, Women's Studies, and Zoology. In addition to student recruits from degree programs, the Environmental Studies program provides a base for the Ecology Club, one of 67 university chapters supported by the Ecological Society of America's Strategies for Ecology Education, Diversity and Sustainability (SEEDS) program. Recent activities include:

- Mentoring local high school and visiting international students,
- Securing funding for on and off-campus activities,
- Receiving scholarships to attend conferences and workshops,
- Conducting community service through paid and unpaid internships,
- Contributing volunteer hours on campus greening initiatives, and
- Providing input to campus administration on matters of mutual interest.

The collective pool of students in Environmental Studies courses and other students who are recruited through social and academic networks are actively reshaping how the SC is designed and utilized. While a wholesale radical transformation of campus is not evident at present, the trajectories for such changes are being set in motion. Diverse groups of students have enrolled in courses taught by the author in the 2010-2012 academic years and participated in projects that contribute to the research, instruction, and service missions of the Environmental Center, Environmental Studies program, and Ecology Club, as well as to support the Chancellor's advisory group, the Manoa Sustainability Corps, and the Adopt-a-Landscape program administered by the campus grounds manager.

The courses emphasized experiential approaches to learning and leadership, and highlighted principles and practices of sustainability through active engagement with campus and off-campus partners. For example, students working independently and in small groups hosted semester-long environmental film series, contributed articles to the campus newspaper and public service announcements to the campus radio station, expanded landscaping at several campus sites, and organized Earth Week events, including energy, water, and food forums in the days leading up to a celebratory Earth Day 2012 event. Students also participated in quantitative and qualitative data collection and analysis and reported research results to campus and community partners through reports and multimedia presentations, including video, posters, and Powerpoint slides that are then posted on the Environmental Studies website (see www.hawaii.edu/envctr/evs/ecologyclub_projects.html).

A final class assignment asked student groups to rethink the SC as an optimal space for learning and campus engagement. The variety of responses, in part, provides answers to the question at the top of the chapter: What will the transformation toward a sustainable future look like at UHM? Below are several excerpts from the assignment that shed light on potential transformation of spaces on the campuses of higher education institutions in the future.

One group started with the question of how to recruit visitors to the SC and listed a number of activities and demonstration projects that highlight innovative practices and technologies, thus drawing the attention of a wider audience who would be interested in unique design and application features. These include (1) a water catchment system to irrigate landscaping, (2) a demonstration aquaponics system, and (3) composting and vermiculture systems to reduce green and food wastes generated by the campus community while providing valuable nutrient-rich compost and soil for campus landscaping.

Another group focused on event planning for the SC as a strategy to recruit visitors to the space. Their suggestions included films, poetry, art, fashion and dance exhibits, and complimentary massage therapists to raise awareness of the health and wellness aspects associated with sustainable lifestyles. Group members recognized that their involvement and cooperation is critical so that the administration pays attention and makes the kinds of changes to the built campus environment advocated by and for faculty and students.

While many of the proposed changes seek to address the "low-hanging fruits" associated with reduce, reuse, and recycle, members of one group proposed a smart phone application and associated website. They noted that despite the good intentions of sustainability supporters, one strategy to address the "degree of ignorance" among the campus community would be to effectively integrate information into the lifestyle of contemporary students who want information "on the go."

Students recognized that miscommunication during the event planning phases result in missed opportunities to involve everyone in the process equitably. Leaders among students emerge, but the process of development can delay progress at vital junctures during the semester. This can impact effective promotion and decrease participation and attendance, let alone lose public relations opportunities. While some students consider attending meetings a distraction from their education, substantive meetings are important to stage successful events in order for organizers to stay on the same page and ensure that individual tasks are completed for the benefit of collective efforts.

Comments in course evaluations confirm that among the important learning outcomes for students is realization that learning outcomes are applicable to future projects and/or can be considered on the job training. Indeed, students routinely segue internships directly into jobs and entry into graduate schools. Among other learning outcomes, students identified peer communication and the equitable division of labor as critical skills developed through their roles as project managers.

Overall, students develop skills and experiences in events planning, fund raising, and professional and social networking and improve written and oral communication skills. With each passing semester, sustainability approaches the critical mass to create a positive tipping point of widespread acceptance and application on campus and in the State of Hawai'i.

LEARNING OUTCOMES AND NEXT STEPS

A primary goal for the SC is to connect students to the campus and cultivate their stewardship in the context of its location in Manoa Valley and by extension to situate campus in relation to drainage divides, mountain slopes, and coastal and near-shore environments. Within view from campus is the urban core of Honolulu that stretches from the extinct volcanic crater known as Diamond Head through Waikiki to downtown skyscrapers towering over Chinatown and continuing into the industrial waterfront that extends to the airport and Pearl Harbor. In the opposite direction are views of the valley, the upper montane forests of the Ko'olau watersheds and a residential community of 30,000 people. All of these features are within less than ten miles from campus, and yet too many students experience their education within the four walls of a classroom.

A learning community, developed over the past several years among faculty, staff, administration, and graduate and undergraduate students, and community partners provides opportunities for mentoring and networking on projects on and off campus. An experiential learning approach allows a process that, at times, admittedly resembles "organized chaos." The trade off of not conforming to a pre-arranged schedule is allowing students and instructors to respond to opportunities and events that present themselves over the course of the semester and academic year. For example, the start of semesters are generally spent preparing grant proposals and meeting submission deadlines to secure funds for the academic year, while simultaneously revising and updating résumés and cover letters to send to potential internship sponsors. Next may be the review of sustainability initiatives at institutions of higher education in the United States to provide examples of sustainability projects, then nominating movies for the film series, arranging permits and reservations for equipment, and signing up and scheduling volunteers for future events.

Students in the courses meet with and learn from key informants on campus to identify priorities for the campus community. Course activities and campus events expand on themes related to environmental philosophy, green politics, environmental economics, sustainable technologies, and environmental media as they relate to the Hawaiian Islands and elsewhere. Through collaboration, students connect to their campus and cultivate relationships that would have not been possible in traditional classroom settings. In addition to serving as a basis to understand campus sustainability, student-initiated projects enable them to articulate learning outcomes to peers and mentors alike that go beyond just having read a textbook.

Among recent outcomes of student efforts are:

- 30,000 lbs. of electronic waste collected at UHM Earth Day 2011;
- Sustainability Assessment Questionnaire survey conducted on campus each of the past four semesters to assess and monitor sustainability awareness;
- CFL lightbulb exchange fundraising events;
- Successful grant proposals to fund sustainability initiatives;
- Graduate-undergraduate student collaborations on data collection, input, and analysis;
- Restoration of landscaping impacted by recent construction projects with donations of native and non-native plants from local nurseries as well as installation of water catchment, vermiculture worm bins, and green waste management systems; and
- Film series and educational forums to highlight energy, water, food, and waste themes.

A comment from UHM Chancellor's Update May 2011 brings to light the public relations windfall that can result from student efforts. "Congratulations to everyone who helped make last month's 10th annual Earth Day Festival the largest ever on our campus and the biggest in Hawai'i. We're leading the way on providing information on how to maintain more sustainable lifestyles and ways of operating."

Civic engagement for sustainability on many campuses tends to focus on the reduce, reuse, recycle mantra; reduce energy consumption and greenhouse gas emissions, reuse water for irrigation and convert green waste into composting, and recycle as much as possible. The SC as an outdoor educational space is an opportunity to further elaborate on a Hawaiian sense of place and the university as a living laboratory to highlight the more broadly developing research, instruction, and service initiatives associated with sustainability on a 21st century campus. The fact that courtyards emphasize "habitation" rather than "movement" offers a rationale for such use of campus space so as to reflect contemporary concerns (Lefebvre, 1991). Sustainability is likely a permanent item on the agendas of generations to come, and providing places on campuses for people to gather, review, discuss, and debate challenges and solutions is not only an appropriate response by institutions of higher education but one that is increasingly expected by campus communities. In summary, the SC helps cultivate a sense of belonging to and stewardship of the UHM campus.

Sustainability is readily, and rhetorically, promoted on campuses through signage and narratives, by means of projects that may start with a flash and flare out over time, and intermittent events with unpredictable turnouts that spotlight the fact that "everyday is earth day." No matter the scale of successes and failures, Lester Brown reminds us in *Plan B 4.0 Mobilizing to Save Civilization* (Brown, 2009, p. 266) that "saving civilization is not a spectator sport."

REFERENCES

Agnew, J. (1993). Representing space: Space, scale and culture in social science. In J. Duncan, & D. Ley (Eds.), *Place/culture/representation* (pp. 251–271). London, UK: Routledge.

Avery, E. (2011). Ecology Club promotes change. *Ka Leo The Voice*. Retrieved November 5, 2011 from http://www.kaleo.org/features/

Barlett, P., & Chase, G. (Eds.). (2004). *Sustainability on campus: Stories and strategies for change*. Cambridge, MA: The MIT Press.

Brown, L. (2009). *Plan B: Mobilizing to save civilization*. New York, NY: W.W. Norton & Company.

Cusick, J. (2008). Operationalizing sustainability education at the University of Hawai'i at Manoa. *International Journal of Sustainability in Higher Education*, *9*(3), 246–256. doi:10.1108/14676370810885871

Edwards, A. (2005). *The sustainability revolution: Portrait of a paradigm shift*. Gabriola Island, Canada: New Society.

Howes, C., & Osorio, J. (Eds.). (2010). *The value of Hawai'i: Knowing the past, shaping the future*. Honolulu, HI: University of Hawai'i Press.

Jackson, P. (1989). *Maps of meaning*. London, UK: Unwin Hyman. doi:10.4324/9780203421239

Lefebvre, H. (1991). *The production of space*. Oxford, UK: Blackwell Publishers Ltd.

Orr, D. (1992). *Ecological literacy: Education and the transition to a postmodern world*. Albany, NY: State University of New York Press.

Relph, E. (1976). *Place and placelessness*. London, UK: Pion Limited.

Sato, P., & Staab, J. (2012). *Hawai'i environmental literacy plan: HELP for Hawai'i*. Honolulu, HI: Hawai'i Environmental Education Alliance.

Social Sciences Public Policy Center. (2010). *Hawai'i 2050 update: Report in response to Act 225, 2008 session laws of Hawai'i relating to sustainability*. Honolulu, HI: University of Hawai'i at Manoa.

University of Hawai'i at Manoa. (2011). *Achieving our destiny: The University of Hawai'i at Manoa 2011-2015 strategic plan*. Honolulu, HI: University of Hawai'i at Manoa.

KEY TERMS AND DEFINITIONS

Civic Engagement: The recognition by individuals that they are members of a larger society and have responsibilities to promote and take actions to make differences in their communities.

Ecological Literacy: A way of thinking that requires the "demanding capacity to observe nature with insight, a merger of landscape and mindscape" (Orr, 1992, p. 86).

Education for Sustainability: This implies learning about links between economic development, environmental management, and sociocultural equity that contributes to healthier lifestyles for communities and the planet.

Place: Cultural geographers focus on representations of place in their natural and social contexts to understand and explain human behaviors.

Sustainable Development: Human actions that do not undermine the ability for future generations to meet their resource needs.

This chapter was previously published in Cases on Higher Education Spaces edited by Russell G. Carpenter, pages 107-120, copyright 2013 by Information Science Reference (an imprint of IGI Global).

Chapter 81
Supporting Sustainability Education and Leadership:
Strategies for Students, Faculty, and the Planet

Alice Cassidy
In View Education and Professional Development, Canada

Yona Sipos
University of British Columbia, Canada

Sarah Nyrose
Boucher Institute of Naturopathic Medicine, Canada

ABSTRACT

The authors describe a workshop, called the Sustainability Education Intensive (SEI), and related programs at a major research university in Canada that served to support educators to introduce or enhance aspects of sustainability into their courses, programs, and initiatives. They present their pilot work in the context of sustainability education and leadership, both in terms of degree programs and programs for educators worldwide. The authors present the key components, steps, and associated timelines of their planning and delivery. They give examples of participant feedback and how the workshop has "stayed alive," thus representing the sustainability of this form of educational development. They provide templates that can be used or adapted at any post-secondary institution. The authors conclude with a checklist to help practitioners get started in supporting educators and leaders in the important area of sustainability education.

INTRODUCTION

In this chapter, we describe a workshop, called the Sustainability Education Intensive (SEI), and related programs at a major research university in Canada that served to raise awareness and provide skills and resources in developing sustainability-related courses, programs and initiatives at the post-secondary level across all disciplines. The target audience for the workshop and related programs included professors, instructors, graduate students, and other leaders interested in sustainability education. We present our workshop and related program in the context of sustainability

DOI: 10.4018/978-1-4666-4852-4.ch081

education and leadership programs and initiatives at other Post-Secondary Institutions (PSIs).

As you read this chapter, you may wish to ask yourself:

- What makes my PSI stand out?
- Could a new or enhanced aspect of sustainability education make the difference?
- What support for faculty, administrators, and student leaders is needed?
- What are the steps to do this, and how do I start?
- With whom do I need to collaborate?

This chapter highlights an important strategy and policy of developing, promoting and facilitating sustainability-oriented educational programs and student services in higher education. We focus on ways to successfully meet the global challenge to involve learners in sustainability-connected curricula while at the same time being a change agent at your PSI.

We provide an overview of the major components of the SEI, as well as the steps we took in its planning, design, promotion and delivery of this workshop. We describe how we got started and what we feel makes our program unique. We present some examples by which the workshop has 'stayed alive', thus representing the sustainability of this form of educational development.

A key objective of this chapter is to provide practitioners with tools to implement sustainability education and leadership at any PSI, outlining the value in doing so. While recognizing that the particulars of sustainability education must be adapted for each place of implementation, we see value in accessing the details of how a specific program was developed and implemented. Thus, we provide templates that you can adapt for your own use, as well as checklists and timelines, to provide you with step-by-step plans for a sustainability education workshop and related programs to support teachers of any discipline or experience level at your PSI.

BACKGROUND

Defining Sustainability

The Brundtland Commission[1] of 1983 defined sustainability[2] as "meeting the needs of the present without compromising the ability of future generations to meet their own needs." This definition uses the triple bottom line of social, economic and ecological. Three decades later, UBC Sustainability[3] (at the University of British Columbia, Canada) has evolved from a similar definition to one where sustainability is seen as "a matter of inter-generational justice" with respect to human use of the natural environment (http://www.sustain.ubc.ca/research/articles/matter-inter-generational-justice).

The Earth Charter[4] is an international declaration of fundamental values and principles to guide and help build a just, sustainable and peaceful global society in the 21st Century. Launched in 2000, it views a sustainable global society as one "founded on respect for nature, universal human rights, economic justice, and a culture of peace" (Earth Charter, 2000).

There is often vivid discussion over "the definition of sustainability," and though it can be helpful to explain what one means, we also acknowledge that there is no one definition that will appeal to everyone. In the SEI, we provided participants with a range of definitions and had them work with the assortment of sustainability interpretations in meaningful ways, thereby exploring and developing sustainability literacy (Cassidy et al., 2012).

The Importance of Supporting Sustainability Education and Leadership in Post-Secondary Education

Over the past thirty years, there has been an enormous increase in awareness, interest and knowledge about sustainability and its importance and value in education. As the UN Decade of Education for Sustainable Development (2005-2014) nears its conclusion, there is arguably a global consensus that post-secondary institutions have a role and responsibility to promote sustainability learning (See Sipos et al., 2008; Reid and Petocz, 2006; UNESCO, 2005; Furman and Gruenewald, 2004). The curricular reality for many university students, however, is largely unchanged (Armstrong, 2011). Why this disconnect?

To actualize sustainability learning, institutions must support professors and others in teaching and leadership roles as educators and in particular, as sustainability facilitators (Brinkhurst et al., 2011). For example, Armstrong (2011) highlights that to implement education for sustainable development, educators must be committed to this new paradigm, as well as, be skilled as "facilitator, collaborator, and fellow learner on the journey toward sustainability" (p. 5). This role is significantly different than the traditional one-way lecture mode that still characterizes much post-secondary education.

Sustainability education and leadership in post-secondary education (PSE) is an area that is growing in interest, but less so in concrete examples of how to actualize it. So, what can be done to support sustainability educators?

United Nations Educational, Scientific and Cultural Organization (UNESCO), the lead agency of the United Nations (UN) to promote education for sustainable development (the term they use interchangeably with sustainability education), has as one of its four main objectives,

Rethinking and revising education from nursery school through university to include a clear focus on the development of the knowledge, skills, perspectives and values related to sustainability is important to current and future societies. This implies a review of existing curricula in terms of their objectives and content to develop transdisciplinary understandings of social, economic and environmental sustainability. It also requires a review of recommended and mandated approaches to teaching, learning and assessment so that lifelong learning skills are fostered. These include skills for creative and critical thinking, oral and written communication, collaboration and cooperation, conflict management, decision-making, problem-solving and planning, and practical citizenship (UNESCO in Sipos et al., 2008).

Empowering faculty, staff and others to make such changes to teaching, learning, and assessment requires specific training and support in sustainability and leadership (Brinkhurst et al., 2011), as well as developing a "sustainability ethic" (Middlebrooks et al., 2009). Communities of learning and practice have been shown to be effective domains for such engagement (Furco and Moely, 2012).

In our view, the Sustainability Education Intensive and related programs described in this chapter meets these goals.

Examples of Sustainability Education Worldwide

In this section, we position our programs in the context of what has been done to date. We also provide these examples as a resource for those wishing to start or enhance similar programs at their own PSI.

The Brundtland Commission and the Earth Charter led to a proliferation of programs of study focused on the broad terrain of sustainability, including environmental education, sustainable development, and sustainability education. We have organized our examples into degree programs and programs for post-secondary educators.

Degree Programs

More and more educational institutions offer degree granting sustainability programs, with 255 programs offered over 157 different institutions in 13 different countries listed at Sustainability: Science, Policy and Practice, an online resource to which you can add your own examples (http://

sspp.proquest.com/sspp_institutions/display/universityprograms).

See Table 1 for a sample of degree programs worldwide that speaks to growing interest in sustainability-related education.

Programs on Sustainability Education for Post-Secondary Educators

Compared to the number and range of degree programs, we have found only a handful of programs that train faculty members and others to introduce or enhance sustainability into the courses and programs they teach. Most of these are found in North America.

In the United States of America, the Ponderosa Project at Northern Arizona University (NAU; http://www2.nau.edu/~ponder-p/), first offered in 1994 was designed by Geoff Chase and Paul Rowland. The main goal of this interdisciplinary faculty group effort is to incorporate environmental sustainability issues into university courses with the ultimate goal of providing future citizens the education and skills necessary to achieve sustainable communities and societies.

In 2001, the Piedmont Project was launched by Peggy Bartlett and Arri Eisen at Emory University in Atlanta, Georgia (http://sustainability.emory.edu/page/1021/Piedmont-Project). Modeled after the Ponderosa Project, with NAU leaders coming to Emory to facilitate the opening workshop, it was subsequently led by rotating teams of Emory faculty facilitators. Aiming to infuse sustainability and environmental issues across the curriculum,

Table 1. A snapshot of the variety of degree programs related to sustainability education in a selection of countries worldwide

Country	PS Institution	Program	Detail
Australia	Bond University	Certificate, Master's	Urban development and sustainability
Australia	Charles Stuart University	Master's, Ph.D.	Ecotourism and sustainable agriculture
Brazil	Federal Rural University of Rio de Janeiro	Master's	Science in Sustainable Development Practice
Canada	Dalhousie University	Bachelor's, Ph.D.	Environment, Sustainability and Society
England	University of Bath	Master's	Responsibility and Business Practice - action research based practice of leadership for sustainability
Germany	Helmholtz Interdisciplinary Graduate School for Environmental Research	Ph.D.	Society and Environment
India	Indira Gandhi National Open University	Post-graduate diploma	Sustainability Science
Japan	United Nations University - ISP	Master's	Sustainability, Development and Peace
Netherlands	Leiden University/Delft Technical University	Master's	Industrial Ecology
South Africa	Rhodes University	Master's	Environmental Education – within the Environmental Education & Sustainability Unit
South Africa	University of Cape Town	Master's	Climate Change & Development
Sweden	Blekinge Institute of Technology	Master's	Sustainable product-service system innovation
Uruguay	South American Institute for Resilience and Sustainability Studies	Inter-institution collaborations	Resilience and Sustainability Studies

it later expanded to include a program for graduate students.

The Penstemon Project at Santa Clara University (http://www.scu.edu/sustainability/education/penstemon.cfm) is designed for faculty outside of the traditional environmentally focused disciplines to find ways to incorporate sustainability into their curriculum, either as class content or in the way their class functions.

Wake Forest University offered a free conference, in collaboration with the Association for the Advancement of Sustainability in Higher Education (AASHE), to explore strategies for adaptation across the sustainability curriculum (http://sustainability.wfu.edu/conference/conference-overview/).

Some programs provide support for particular groups or disciplines. For example, the Sustainable Leadership Institute (http://www.sustainabilityleadershipinstitute.org) offers workshops for community leaders to work on sustainability concerns and possible solutions.

In Canada, the British Columbia (BC) Working Group and Network on Sustainability Education, which started in 2006, connects organizations, educational institutions, government and others for the purpose of furthering sustainability education across the province (http://www.walkingthetalk.bc.ca/). Since evolved into a Regional Centre for Excellence, they provide resources, programs and news of interest and value to a large number of stakeholders. As an example, the ten Sustainability Education Principles for British Columbia, created in 2007, have triggered much discussion and figured prominently in our own SEI, where it was of great interest to colleagues from both within and outside of the province (For the list of ten principles, see: http://www.walkingthetalk.bc.ca/node/427).

The British Columbia Institute of Technology developed a version of the Piedmont/Ponderosa workshop, called the Pacific Spirit Project that started in 2009 and runs annually for faculty members within the School of Construction and Environment (http://www.bcit.ca/construction/sustainability/pacificspirit.shtml).

Turning to the United Kingdom, the Sustainability Hub at Kingston University (http://www.kingston.ac.uk/sustainability/aboutus.html) fosters a culture of sustainable practice throughout the University through engagement campaigns, research projects and curriculum development.

It is our hope that this chapter, with its step-by-step guide to setting up a training workshop, will contribute to an increase in the number of sustainability education training programs available to PS instructors.

The Need for Sustainability Education Training

PSIs have expressed their commitment to sustainability education through signing declarations such the Talloires Declaration of University Presidents for a Sustainable Future (1990), the Halifax Declaration (International Association of Universities, 1991) and the more recent Earth Charter (2000).

From our collective experiences as teachers, students and educational developers, we have noted that while there is much interest in and growing evidence of sustainability education in terms of degree programs at the post-secondary level, there is a paucity of training programs for the faculty and instructors who need and want to implement change in their classrooms and learning environments to support the curricular reorientations.

The Sustainability Education Intensive (SEI)

In this section, we describe the main components of our workshop, with a focus on how we got started and the steps we took in its planning, design, promotion and delivery.

We designed and facilitated a multi-day workshop on sustainability education, coined the Sustainability Education Intensive (SEI), at the University of British Columbia-Vancouver (UBC),

Canada in 2009 and 2010 as a pilot. The goal of the SEI was to provide tools, resources, and structured time and support for participants to develop a Sustainability Education Plan (SEP) to integrate or enhance sustainability in their course, program or initiative. A total of 36 people participated in the two years of the SEI, representing a diversity of disciplines, and including five people from other PSIs in British Columbia and Ontario, Canada. Participants taught in a range of disciplines as diverse as agriculture, architecture, arts, botany, business, education, engineering, forestry, health, history and languages.

Participants were further supported in the development of their SEPs through studio sessions (and financial support for professional development for our UBC colleagues) in the 6-8 months following the workshop. Within one year of their participation in the SEI, participants were invited to share their results in a celebratory showcase event.

The Origins of Our Program Development

In early 2007, two of the authors (Alice Cassidy and Yona Sipos) took part in the Sustainability Across the Curriculum Leadership Workshop at Emory University in Atlanta, Georgia (see http://www.aashe.org/). This 2-day workshop, facilitated by Peggy Bartlett of Emory University and Geoff Chase of Northern Arizona University (NAU), and attended by 35 people, focused on the successes of the Ponderosa and Piedmont Projects (described above) that have helped over 200 faculty members in the United States adapt their courses to include aspects of sustainability.

Upon our return, and with funding from the UBC Sustainability Office, we set about to create and deliver a short course that was relevant and unique to UBC (Cassidy, 2011a; Cassidy, 2012). We used part of the funds to hire an undergraduate student (Sarah Nyrose).

UBC has gained a reputation as a sustainability leader in North America. AASHE recently presented it as a spotlight campus (http://www.aashe.org/node/53010) describing: the University Sustainability Initiative as uniting the university's academic and operational efforts to advance sustainability; the SEEDS (Social Ecological Economic Development Studies) academic program, in which students, faculty and staff collaborate on real-life sustainability solutions, heading toward the 500-project mark; being named Canada's first Fair Trade Campus; and the opening of its recent Centre for Interactive Research on Sustainability.

Building on this strong sustainability reputation, our goal was to further develop leadership both within and beyond our own institution.

Our intended outcomes of this work were to:

1. See an increase in number of courses, programs and initiatives that include aspects of sustainability - including environmental, social and economic – in their curricula. Responses from two years of workshop participants indicate that this happened;
2. Be change agents, raising awareness at our institution and elsewhere about the value of sustainability education and leadership workshops and related programming. The successes of workshop designers and participants indicate we met this goal.

What Made the UBC SEI Unique?

We designed our program to mirror a unique aspect of the University of British Columbia – that people in a variety of roles are active leaders in sustainability education. To our knowledge, the SEI is unique and the first of its kind for a variety of reasons:

- It was designed and facilitated by people in three different roles at post-secondary institutions: a faculty member, a graduate student, and an undergraduate student.
- We involved additional representatives in these roles, as well as staff members, post-

doctoral fellows, and community members in the initial planning and input throughout the program as members of an Advisory Board.

- People in all of these roles were invited to apply; faculty, staff, graduate and undergraduate students from a wide variety of academic disciplines all applied and participated.
- The SEP that each participant worked on, during and following the SEI, was geared to create change in his or her own post-secondary educational project, be it a credit course, a credit or non-credit program, or another kind of initiative in a formal or non-formal setting.

Undergraduate Students

UBC is noted for its undergraduate student leaders. For example, the student-initiated Sustainability Pledge (http://sustain.ubc.ca/get-involved/students/pledge) increases awareness of sustainability and inspires action towards more sustainable living. Sustainability Ambassadors (http://peerprograms.ubc.ca/the-crew/sustainability-ambassadors/), a peer program, organize, promote and deliver sustainability education programming for students. UBC students were active in the youth and student-led goBEYOND Campus Climate Network (http://www.go-beyond.ca/about-us), engaging students, faculty, staff and community partners at PSIs.

We involved undergraduate Sarah Nyrose in the initial design and delivery of the first year's SEI. After Sarah graduated, we hired Angela Willock, an alumnus from the 2009 SEI, to help us facilitate the 2010 program. Other undergraduates helped during the two years of the SEI including: as representatives in our Advisory Group; filming a promotional video; taking digital photographs; and co-leading a guest presentation related to a business course with a sustainability focus.

Other Key Stakeholders

We sought input and help from colleagues in a variety of roles and from units across and outside the University. We involved these people for initial input at networking events, through an Advisory Group, and as guest presenters to the SEI and related programs.

Design and Delivery of the SEI

We first started thinking about the potential to lead a training workshop in January, 2007. Key planning work began in early Fall, 2008, with many steps involved in the creation of the SEI that we launched in late May, 2009. Here we present the key steps involved in each stage of planning, design, delivery, and follow-up, summarized in Table 2.

Planning

We started planning soon after returning from the Sustainability Across the Curriculum Leadership Workshop in Atlanta. We had conversations with people from the Sustainability Office as well as colleagues in many Faculties and service units. During a sustainability education networking lunch, which was an event of our community of practice, in October 2008, we shared input and suggestions on our first draft of the SEI. As we wished to have involvement at all levels and to include every Faculty, Associate Deans and Directors were invited to attend or send representatives. In total, colleagues from 23 units took part, including faculty, staff and students and members of the UBC Senate and the newly formed President's Advisory Committee on Sustainability. The energy and diversity at this event led to very helpful input and new creative ideas.

In late 2008, we formed an Advisory Group, comprised of faculty, staff, post-doctoral fellows, undergraduate and graduate students, and community members to help us finalize SEI process

Table 2. Key steps and timeline for planning the SEI

Timing	Process	Detail
	Planning:	
October	Networking lunch event	Receive input and suggestions on first draft of first year's SEI
November	Budget planning	Determine all set and optional costs and plan accordingly
December	Form Advisory Group	With representation of all roles at the University plus a community representative
January	Advisory Group meeting	Fine-tune advertising and promotion plan, workshop learning goals and objectives
January-March	Advertising and Promotion:	Create Website, print and distribute small functional bookmark advertising. Promote via list-serves and networks specific to each of our target audiences
	Application and selection process:	
April	Application form Promote SEI to heads of units to nominate applicants	Via a form downloaded from Website Via emails, phone calls, word of mouth
April	Advisory Group meeting	Fine-tune lesson plan, activities, guests, resources
May	Selection process	Inform successful applicants and let them know where to meet
May	Final preparation	Collect/create material; confirm food and room and equipment bookings; confirm invited guests; confirm recycling and composting facilities
May-June	Workshop delivery:	Collect formative and summative feedback; fine-tune next day if necessary; confirm lesson plan and activities
June	Follow-up activities	Thank guest speakers; ensure bills are paid and update budget; thank nominators and all others involved
August-November	Support studio	Help SEI participants to meet their SEP goals
End of November	Collect summaries, provide funding	Collect summaries of SEP outcomes from participants and prepare them for Showcase Event. Reimburse UBC participants upon receipt of their bills and justification
	Celebrating the completion of the workshop:	
November-December	Final planning for Showcase event	Promote event online and word of mouth. Confirm guests, room and food. Take registration for event. Work with participants for final planning
January	Showcase event	University Vice-President, Academic and Provost as special guest; stations around a large room for each SEI participant to show their work; networking, lunch

and materials. Meetings of this group in January and April 2009 resulted in fine-tuning of our advertising and promotion, learning goals and objectives, application process, workshop lesson plan, and associated activities and resources.

Advertising and Promotion

Our primary goal for advertising was to reach a diverse audience of post-secondary educators that varied in discipline and in roles related to teaching, facilitation and/or leadership. Faculty, post-doctoral fellows, staff, and graduate and undergraduate students were all encouraged to apply, provided they had an interest in creating a Sustainability Education Plan (SEP) for their course, program or initiative. As funding for the SEI was from our own institution, most spaces were reserved for UBC participants (for whom the workshop was free of charge), though we saved a few spaces for people from other PSIs, charging a nominal fee to cover expenses.

Small functional bookmarks providing succinct information about the SEI (including a Website address for more detail) were widely distributed. We promoted the workshop via list-serves and networks specific to each of our target audiences.

Application and Selection Process

The SEI application form (see Appendix 1), which was also the Pre-Inquiry for research purposes, was downloaded from our Website and submitted electronically to reduce paper waste. We asked applicants to tell us a bit about themselves and their involvement in teaching, facilitation and/or leadership activities, as well as their idea about how to affect change in their course, program, or initiative.

Additionally, we invited heads of units to nominate people to take the SEI, with nominated applicants receiving special attention as we selected people for the limited spaces. The purpose of the nomination system was to help spread the word and build momentum about the program. For reasons of funding as well as ideal pedagogy, we could accept only a limited number of participants each year. We gave preference to participants who were nominated, also using the information that prospective participants provided in their applications. Satisfactory completion of the application also ensured that participants had done appropriate preparation, showing commitment that they could take part fully in the intensive 2.5 day workshop, including building on their idea through a Sustainability Education Plan (SEP) to work on throughout the 2.5 days (see Appendix 2).

We were extremely pleased with the rate of response, both in terms of numbers of applications and nominations. Each year, more people than we could accept into the program expressed interest in taking the SEI.

Workshop Goals and Objectives

We designed the workshop in accordance with the TREK 2010 planning document to further assist members of the UBC teaching community "to prepare students to become exceptional global citizens [and] promote the values of a civil and sustainable society" (http://www.vision.ubc.ca/index.html).

Learning Goals

- Foster sustainability leaders within and beyond our own institution.
- Strengthen the interdisciplinary network of sustainability educators and leaders at UBC and contribute to the development of a university-wide sustainability learning community.
- Develop relationships that enable the integration of sustainability across the curriculum.
- Provide participants with the resources and support to implement a project of their choice that is designed to affect change in their course/program.
- Create opportunities, incentives and support for integrating sustainability into courses and programs at UBC.

Learning Objectives

By the end of the short-course, you will be able to:

- Describe several sustainability education initiatives at UBC.
- Choose from the initiatives at UBC that are directly relevant to your course/program and identify how they may be incorporated.
- Identify a sustainability education project of your choosing that is designed to affect change in their course/program.
- Create an action plan of implementation to meet your goals for your project.

After the course, you will be able to:

- Apply sustainability into the discipline you teach and how you can further contribute to sustainability education at UBC.

Using these learning objectives and goals as the foundation for the development of the SEI lesson plan, we designed a workshop that modeled sustainability in all dimensions. The careful consideration of venues, activities, materials, modeling of teaching and facilitation techniques and choices for refreshments were all crucial aspects in the overall planning.

Modeling Sustainable Practice

A key concern throughout planning and delivery was to reduce waste. We did this by advertising and promoting the workshop mainly through our Website and other online sites; by communicating with applicants via email; providing as many resources as possible before and during the SEI via our blog and other electronic means. Some materials, such as worksheets that participants filled out during the SEI and copyrighted documents that we did not have permission to post online, were printed on paper with a high level of post-consumer recycled content. We provided participants with a paper folder made from 100% recycled content made by Earth in Mind from Oxford.

Recycling and composting facilities were available and promoted throughout the SEI, and lunches and snacks were vegetarian, using many local foods, and ordered from a student-based caterer. Participants brought their own mugs for hot drinks and water, and we used washable dishes and cutlery for all meals. Workshop participants commented that they noticed and appreciated the attention paid to responsible resource use.

Blog

We created the SEI Blog (http://blogs.ubc.ca/tagsustainability/) to provide participants a central hub for locating resources, exploring information about the SEI and associated programs and events offered by ourselves and others. Workshop participants gained access to the resources, including some pre-readings and material for use during the workshop, through the online Refworks program, a reference manager that hosts a variety of resources, such as journal articles, books, podcasts, and Websites.

Place-Based Learning

The first half-day of the workshop took place at the Centre for Sustainable Food Systems at UBC Farm, allowing participants to engage in a setting that is at the heart of sustainability at our institution. Later in the SEI, a field trip around campus emphasized observation of the natural world, as well as interconnections to the built environment and humans' roles in both of these. Participants expressed appreciation for being at the Farm and specifically, the opportunity to be outdoors, to learn about this important campus and community resource, and to engage in place-based experiential learning.

Participant Feedback

Spending time outdoors – at the farm and on a field trip was an opportunity to connect with (and spend time in the) natural world.

Active, Participatory, and Practical Learning

Participants were involved in individual, pair, small group and whole group activities that included collaborative learning, self-reflection, brainstorming and creativity. Participants were encouraged and supported throughout the 2.5 days to engage with the group, reflect and develop their personal Sustainability Education Plan (SEP), building from the goals they set out in their applications. Participant feedback:

Interactive and experiential. Different perspectives stimulated critical reflection.

The methods of learning and teaching were modeled - the importance of interactive learning, including cross-disciplinary and community engaged.

I loved the diversity of activities in the SEI as they helped me access my tacit knowledge as well as create new knowledge/think differently about things related to sustainability.

Video and Photos

Participants were informed of these forms of documentation during the workshop for our reports and future promotions. Digital photos were taken throughout the SEI. Sean Devlin of the student-led goBEYOND project (http://www.go-beyond.ca/about-us) created a video involving both the UBC and the BCIT workshops.

Participant Feedback

We collected formative feedback from participants at the ends of Days 1 and 2 (see Appendix 3 for sample questions and prompts), presenting summaries on subsequent days and noting where we adjusted the workshop in response. We collected summative feedback at the end of Day 3 (see Appendix 4 for the form), and also approximately one year after workshop participation (see Appendix 5 for the form). The application form presented earlier was the Pre-Inquiry. These data, collected using an ethics protocol for human subjects approved by the Behavioural Research Ethics Board at UBC application H09-00824, are the basis for papers in preparation.

An example of adjusting the workshop in response to formative feedback was allowing more time for participants to talk with one another informally, especially at lunch and during short breaks. We omitted some activities in order to spend more time and in-depth exploration of others.

An example of a change to the 2010 workshop based on feedback from 2009 participants was to re-organize an activity whereby participants explained their SEPs to one or two others, then clarified and presented to the whole group. Participant feedback from 2010:

Actually putting the SEP into a poster presentation that had to be clean and stand on its own was rather challenging. It made me focus on what was key to moving forward.

Celebrating the Completion of the Workshop

Participants received an electronic Certificate of Completion at the end of the workshop. The Certificate was intended to be both a congratulatory document, as well as a reminder to participants about their involvement and the commitments they had made through their SEPs.

We emailed colleagues who nominated participants, as well as all guest speakers, members of the Advisory Group and others who helped support the SEI, to thank them for their time and dedication to sustainability education and leadership.

Ongoing Professional Development

The ongoing professional development took the form of financial support ($500) for UBC colleagues and support studios for everyone. UBC participants used their professional development funds towards attending a training event or conference, purchasing teaching materials or hiring a student, in order to meet the goals of their SEP.

Two support studios held approximately two and five months following each SEI provided further support and networking opportunities for the participants, as they continued work on their SEP. These studios, up to two hours long, were optional to attend and included brief updates from each participant, as well as time for group discussion and paired or individual work. Participants who could not attend in person were invited to take part through the use of technology. The contributions, requests for assistance and discussion ranged from sharing a detailed mind map of all the planning and actions on a SEP to date, to 'thinking out loud' about how best to design an exercise, to setting up meetings for the fall to explore next steps with an initiative. The last studio was timed to take place just before participants needed to complete and send in their final SEP summary and, for UBC participants, their receipts and explanation of how they used the funds in order to be reimbursed (Appendix 2).

Showcase Event

We hosted the SEI Project Showcase Event seven months after each SEI to highlight participants' SEPs. During this 3-hour event, we had each participant set up a station, where they could display posters, photomontages, and electronic presentations and resources. Those who could not take part in person were invited to send us updates based on the summary they had sent us earlier.

The Showcase really became a celebratory event, with invitations going out to all who sponsored participants, guest speakers, members of the Advisory Group, colleagues, and interested members of the UBC community, including prospective applicants for a future SEI. The Showcase event following the 2009 SEI included opening remarks by the Vice-President Academic and Provost, David Farrar.

Related Programs

Sustainability across the Curriculum Community of Practice

We made full use of a network that was in place before the SEI. The Sustainability Across the Curriculum Community of Practice (CoP) was first developed in 2006 to offer events and resources and to keep in touch with over 150 members through electronic newsletters. One of 16 CoPs in place at UBC, it had clearly demonstrated success (Cassidy, 2011b).

The kinds of activities and events offered through this CoP include a regular electronic newsletter of resources and upcoming events, tied to the Website and blog. We extended invitations to members to suggest topics and help facilitate sessions. For example, during our institution's Celebrate Research Week, three CoP members, Naoko Ellis, Chemical and Biological Engineering; Santokh Singh, Botany; and Laurie Ricou, English presented mini-lessons of their actual classes, then tied it into their research and scholarly practices.

Conference Sessions and Invited Sessions

Members of the SEI design team led invited sessions and peer-reviewed sessions at conferences. Examples include Douglas College Professional Development Day, the annual conference of the Society for Teaching and Learning in Higher Education, the Gaining Ground Resilient Cities conference, and the UBC Learning Conference.

Sustaining the Success of the SEI

In addition to sending participants a survey one year after workshop completion (Appendix 5), we also contacted them to ask how they "made use" of the skills, knowledge and attitudes developed or honed during the SEI. We wanted to know if they had published, presented or further developed any of their SEI-related work. As we noted in an email letter to them,

It could be a local workshop, or conference anywhere, or something you wrote for a peer or non-peer reviewed journal, magazine or other. If so, will you give us the reference(s) and/or details? I will of course summarize it all and get it back to you.

The resulting responses show many and varied examples, showing that the SEI "stayed alive," thus illustrating the sustainability and transformational aspects and successes of this form of educational development. The complete set of results is the focus of another paper in preparation. We present a snapshot of examples here:

Changes in Curriculum

As a result of the SEI, sustainability has been introduced or enhanced in about 40 courses, programs and initiatives. One faculty participant adapted activities from the SEI were for use in a corporate social responsibility course, while another proposed the development of a virtual design space in an engineering course. One graduate student adapted her work with students at the Native Education College, while another integrated work from their SEP into a first year land and food systems course. A forestry field school for 3rd year students now includes a program developed in a SEP, a theoretical model describing how teachers could teach high school biology through an environmental.

Continuing Education

Some alumni continued their studies: an undergraduate continued her explorations in a directed studies course; a faculty member completed an international distance credit course on sustainable engineering.

Scholarly Contributions by SEI Alumni

SEI alumni have presented on their sustainability education work at conferences across Canada as well as in Kenya, Africa.

Peer-reviewed journal publications by SEI alumni based on their sustainability education work include work on a culturally relevant assessment model (Verwoord, 2011), community service-learning in the integrated medical and dental school program (Dharamsi et al., 2010) and bringing community-based research initiatives into a communication course for second-year engineering students (Berndt, 2010).

Program Design and Delivery

Shafik Dharamsi and Matthew Whiteman, alumni of the 2009 and 2010 SEIs respectively, further developed their SEPs as a program funded for two years by UBC's Teaching and Learning Enhancement Fund (TLEF; http://tlef.ubc.ca/) called the Ethics of International Engagement and Service-Learning Project (EIESL) Online Toolkit (http://blogs.ubc.ca/ethicsofisl/).

Two alumni of the 2009 SEI, Angela Willock and Maggie Baynham, further developed their SEP as a province-wide Sustainability Education across the Province (SEAP), offered in 2010 and 2011. Co-developed and hosted by a different British Columbia post-secondary institution each year in collaboration with goBeyond (http://www.gobeyond.ca/), SEAP was geared mainly for faculty members and charged participants to take part.

CONCLUSION AND FUTURE RESEARCH DIRECTIONS

We have described a workshop and related programming we developed and delivered as a pilot over two years at a large research university to provide practitioners with tools to implement sustainability education and leadership. We place our programming in the context of other initiatives to date, explaining the value of support for sustainability education and leadership through the lens of the published work in the area.

We provide templates that can be used or adapted at any institution. The key steps we followed, with associated timelines, can help others wishing to implement a program of sustainability education and leadership to get started, keep on track, and be successful agents of change.

We explored some details and the value of educational development for sustainability educators. Hocking et al. (2006), speak to the importance of "bringing together experienced sustainability facilitators, to build a shared view of what facilitation for sustainability looks like, and to consider how the frameworks, processes and skills for facilitation can be fostered in others" (p. 1). We agree completely and feel that there is much potential for future work in this area. The literature describing efficacy of support for sustainability educators is small. We encourage those working in this area of educational development to share their successes through conference sessions, publications, and workshops. It is our hope that these initiatives can lead to an increase in the number of training opportunities offered at PSIs worldwide.

Administrators and senior officials at PSIs want to make positive change in their institutions and are proud to share when they have (e.g. Talloires Declaration). They want to hear of successes elsewhere, and want to know of clear ideas on how they might adapt or implement these at their PSI. If you are a faculty member, graduate student, undergraduate student, post-doctoral fellow or staff member, know that your voice is important and make sure it is heard. Share your views on the challenges and potential solutions to increase the amount and quality of sustainability-related education at your PSI, and support for the educators who design and deliver such programs.

Timpson et al. (2006) dedicate their book, *147 Practical Tips for Teaching Sustainability: Connecting the Environment, Economy, and Society*, to the planet because it sustains and inspires us, as well as to the teachers and learners who carry on the tradition of sustainability education. It is clear that many people care about this planet we live on. Know that students at your PSI care deeply and want to learn more so they can help make positive change. Their teachers need and want support on how best to do this. And, it is not an understatement to say that the planet depends on these initiatives.

We conclude with a list of practical ideas, based on our experience. We suggest the following tips on how you can get started to implement one or more programs to support sustainability educators for at your own institution:

Checklist for Success

- Identify a couple or few people at your institution with whom you can collaborate. Start by articulating your shared understanding and vision of supporting sustainability education on your campus.
- Brainstorm to create a list of individuals, departments and other units, and networks with some connection to sustainability-related education. Ensure that you have included people in all roles and at various levels, including senior administrators.
- Talk with these people, ideally face to face, individually and in small groups. Go to them. Send emails, make phone calls, advertise on your Website that you would like to talk with as many people as are interested in the topic. (This tip and the preceding one helped us identify "champions" in a variety of roles, units and disciplines on our large campus.)
- Listen to people, in informal networking sessions, in structured meetings, via email or blog forums, through whatever timing, means or media works best for them.
- Spread the word through advertisement and promotion both electronically, in sustainably responsible hard copy, and via word of mouth; you want everyone who has an interest to know they are welcome.
- Seek input and ask for help on all stages of your plans. Collaboration helps build momentum and community. Everyone involved is made to feel part of what is being created.
- Research related programs that you can use as modelsand ways you can capture the experiences of your process and outcomes to

share it in scholarly ways. Investigate how ethics approval works at your institution. Successful application to your institution's ethics review board to undertake your study is not only the right thing to do, but can help you design surveys for maximum efficacy.

- Design well; build on what is unique about your PSI; invite the champions you have identified to come in as guests; revise and revise again. The planning, action, and reflection is all part of an iterative approach.
- Check in during your programs to see if expectations and needs of your participants are being met. Ask for formative feedback and be prepared to make change right away to address any concerns.
- Celebrate! You have accomplished a lot. Be proud of it. Finds ways to celebrate amongst facilitators, organizers, participants and all who were involved to make it the success it was.
- Reflect, after your program is over, on what facilitators and participants felt went well and what can be improved. Keep notes. Keep all feedback. Every workshop or meeting can always be improved, a little or a lot. Be ready to do that.
- Share your experiences through reports, publications and presentations at your own institution, nationally and internationally. Making your work public is a form of scholarship of teaching and learning that is not only important for you as a member of an academic institution, but to the continued development and success of your program, as well as to the growth of comparable programs elsewhere.

ACKNOWLEDGMENT

We are grateful for financial support through two key units at the University of British Columbia: the Sustainability Office (SO), the Centre for Teaching and Academic Growth (TAG)[5]. Funding for the 2010 SEI came from the Teaching and Learning Enhancement Fund (TLEF; http://tlef.ubc.ca/), made possible by UBC students. In particular, we thank Freda Pagani, Charlene Easton (as directors, SO), Gary Poole (as director, TAG), and Lorna Seppala, as Associate Director, SO, as their representative to our Advisory Group.

We are grateful to members of the Advisory Group for their valuable role in honing the SEI and for their assistance in promoting it at UBC and beyond: Jean Marcus, President's Advisory Council/Committee; Rob Van Wynsberghe, Assistant Professor, Human Kinetics, Faculty of Education; Kristin Orians, Associate Professor, Earth and Ocean Sciences, Faculty of Science; Susan Nesbit, Senior Instructor, Civil Engineering, Faculty of Applied Science; Rosie Pidcock, Undergraduate student, Sauder School of Business; Nora Timmerman, Ph.D.candidate, Faculty of Education; Ray Hsu, Post-doctoral fellow, Creative Writing, Faculty of Arts; Tirtha Dhar, Assistant Professor, Marketing Division, Sauder School of Business; John McDonald, Community member British Columbia Institute of Technology (BCIT).

Some of the references in this chapter were found by undergraduate student Braydon Walker, Carleton University, as part of a related resource repository project funded by a 2012 Educational Developers Caucus (EDC) Grant of the Society for Teaching and Learning in Higher Education (STLHE).

We are grateful to the guests who led a variety of sessions during the SEI. We thank all the participants of the SEI, each of whom brought interest and in many cases, passion for sustainability across the curriculum.

This chapter is based on a workshop and related programs designed by the authors when employed at the Centre for Teaching and Academic Growth (TAG) at the University of British Columbia, Vancouver, BC, Canada as Associate Director (Alice Cassidy), Graduate Academic Assistant (Yona Sipos) and Undergraduate Assistant (Sarah Nyrose).

REFERENCES

Armstrong, C. M. (2011). Implementing education for sustainable development: The potential use of time-honored pedagogical practice from the progressive era of education. *Journal of Sustainability Education, 2*. Retrieved from http://www.jsedimensions.org/wordpress/wp-content/uploads/2011/03/Armstrong2011.pdf

Berndt, A. (2010). Integrating community-based research initiatives into a standardized technical communication course for second-year engineering students. In *Proceedings from the Canadian Engineering Education Association*. Retrieved from http://blogs.ubc.ca/tagsustainability/files/2010/09/3081-5100-1-PB.pdf

Brinkhurst, M., Rose, P., Maurice, G., & Ackerman, J. D. (2011). Achieving campus sustainability: Top-down, bottom-up, or neither? *International Journal of Sustainability in Higher Education, 12*(4), 338–354. doi:10.1108/14676371111168269

Cassidy, A. (2011a). Sustainability education: Leading by example. *Bridges, 9*(2), 15-16. Retrieved from http://www.usask.ca/gmcte/drupal/?q=journal_index

Cassidy, A. (2011b). Communities of practice: A checklist for success. *Transformative Dialogues: Teaching and Learning eJournal, 4*(3). Retrieved from http://kwantlen.ca/TD/TD.4.3/TD.4.3.6_Cassidy_Communities_of_Practice.pdf

Cassidy, A. (2012). *In view education and professional development*. Retrieved from http://cassidyinview.wordpress.com/

Cassidy, A., Sipos, Y., & Nyrose, S. (2012). *Why did the sustainability educator cross the boundary? To walk the talk with learners*. Manuscript submitted for publication.

Dharami, S., Espinoza, Cramer, C., Amin, M., Bainbridge, L., & Poole, G. (2010). Nurturing social responsibility through community service-learning: Lessons learned from a pilot project. *Medical Teacher, 32*(11), 805–911.

Earth Charter Initiative. (2000). *The earth charter*. Retrieved from http://www.earthcharterinaction.org/content/pages/Read-the-Charter.html

Ethics of International Engagement and Service-Learning Project. (2011). *University of British Columbia*. Retrieved from http://blogs.ubc.ca/ethicsofisl/

Furco, A., & Moely, B. E. (2012). Using learning communities to build faculty support for pedagogical innovation: A multi-campus study. *The Journal of Higher Education, 83*(1). Retrieved from http://muse.jhu.edu/journals/journal_of_higher_education/v083/83.1.furco.html doi:10.1353/jhe.2012.0006

Furman, G. C., & Gruenewald, D. A. (2004). Expanding the landscape of social justice: A critical ecological analysis. *Educational Administration Quarterly, 40*(1), 47–76. doi:10.1177/0013161X03259142

Hocking, C., Ray, S., & Day, T. (2006). What does professional development and learning for environmental sustainability facilitators look like? In *Proceedings of APEN International Conference*. Beechworth, Australia: APEN. Retrieved from http://www.regional.org.au/au/apen/2006/refereed/6/2936_hockingcg.htm

International Association of Universities. (1991). Retrieved from http://www.iau-aiu.net/

Middlebrooks, A., Miltenberger, L., Tweedy, J., Newman, G., & Follman, J. (2009). Developing a sustainability ethic in leaders. *The Journal of Leadership Studies, 3*(2), 31–43. doi:10.1002/jls.20106

Reid, A., & Petocz, P. (2006). University lecturers' understanding of sustainability. *Higher Education, 51*, 105–123. doi:10.1007/s10734-004-6379-4

Sipos, Y., Battisti, B., & Grimm, K. (2008). Achieving transformative sustainability learning: Engaging head, hands, and heart. *International Journal of Sustainability in Higher Education, 9*(1), 68–86. doi:10.1108/14676370810842193

Talloires Declaration of University Presidents for a Sustainable Future. (1991). Retrieved from http://www.ulsf.org/programs_talloires.html

Timpson, W. M., Dunbar, B., Kimmel, G., Bruyere, B., Newman, P., & Mizia, H. (2006). *147 practical tips for teaching sustainability: Connecting the environment, economy and society*. Madison, WI: Atwood Publishing.

UNESCO. (2005). *United Nations decade of education for sustainable development: Re-orienting programmes*. Retrieved from http://portal.unesco.org/education/en/ev.php-URL_ID=46887&URL_DO=DO_TOPIC&URL_SECTION=201.html

Verwoord, R., Machado, J., & Mitchell, A. (2011). Supporting indigenous students through a culturally relevant assessment model based on the medicine wheel. *Canadian Journal of Native Education, 34*(1), 49–66.

ADDITIONAL READING

Cassidy, A. (2008). *The value of hands-on learning: Techniques from instructor, results from students*. Paper presented at the 4th Annual Community and Education Symposium. Vancouver, Canada.

Cassidy, A., & Sipos, Y. (2007). *Students as co-researchers and co-presenters*. Paper presented at the 27th Annual Conference of the Society for Teaching and Learning in Higher Education (STLHE) – Evolving Scholarship. Edmonton, Canada.

Marshall, J., Coleman, G., & Reason, P. (2011). *Leadership for sustainability: An action research approach*. London: Greenleaf Publishing.

Mayer-Smith, J., & Peterat, L. (Eds.). (2010). *Get growing! Activities for food and gardening learning: A teacher resource elementary and middle grades. Vernon*. Really Small Vernon Press.

Moore, J. (2005). Is higher education ready for transformative learning? A question explored in the study of sustainability. *Journal of Transformative Education, 3*, 76–91. Retrieved from http://jtd.sagepub.com/cgi/content/abstract/3/1/76 doi:10.1177/1541344604270862

Moore, J., Pagani, F., Quayle, M., Robinson, J., Sawada, B., Spiegelman, G., & VanWynsberghe, R. (2005). Recreating the university from within: Collaborative reflections on the University of British Columbia's engagement with sustainability. *International Journal of Sustainability in Higher Education, 6*(1), 65–80. doi:10.1108/14676370510573140

Podger, D. M., Mustakova-Possardt, E., & Reid, A. (2010). A whole-person approach to educating for sustainability. *International Journal of Sustainability in Higher Education, 11*(4), 339–352. doi:10.1108/14676371011077568

Sustainability Education Institute. (n.d.). Retrieved from http://www.sustainabilityleadershipinstitute.org/references_change.php

KEY TERMS AND DEFINITIONS

Educational Development: The practice of helping those who teach be better at what they do, through resources, workshops, one-on-one assistance, and other means.

Post-Secondary Institution: Any institution of learning that is attended after K-12 or primary and secondary education.

Support Educators: The workshop and related programs we describe in this chapter are to help people who teach, facilitate or lead courses, programs or initiatives become better at what they do and meet their expressed goals.

Sustainability Education and Leadership: A teaching and learning process, content or format specifically related to sustainability. Teachers and administrative leaders apply it to their own areas of expertise and roles.

Sustainability Education Intensive: The name of the 2.5-day workshop we designed and led over two years at the University of British Columbia. Shortened to SEI, this workshop supported people who teach, facilitate, or lead courses, programs or initiatives to introduce or enhance aspects of sustainability into their curriculum.

ENDNOTES

1. Brundtland Commission was formally known as the World Commission on Environment and Development.
2. Sustainability often used interchangeably with the term sustainable development, coined in the paper "Our Common Future" and released by the Brundtland Commission.
3. UBC Sustainability Office is now called UBC Sustainability http://www.sustain.ubc.ca/.
4. Originally conceived of in 1987 as part of the United Nations World Commission on Environment and Development however, launched in 2000, it views a sustainable global society as one "founded on respect for nature, universal human rights, economic justice, and a culture of peace" (Earth Charter, 2000).
5. The University of British Columbia: the Sustainability Office (SO) and the Centre for Teaching and Academic Growth (TAG) are the two units which have each since been renamed and/or restructured; see http://www.sustain.ubc.ca/and http://ctlt.ubc.ca/.
6. A note about Appendices: Please use or adapt any of these sets of questions/prompts for use in your own program. We ask for attribution, using the title of this chapter and its authors.

APPENDIX 1: PRE-INQUIRY (APPLICATION FORM)

Sustainability Education Intensive (SEI)[6]: Dates of workshop

Note: before applying, please make sure you have read the SEI Overview, including commitments if accepted, carefully. By applying, it is understood that you agree with and will follow them.

- **Application Form:** Please fill out all sections; rename the file your family name_SEI.doc (e.g. Cassidy_SEI.doc) and email it to <*email contact*> by deadline.A bit about you:
- Name:
- Institution:
- Title/role:
- Department/unit:
- Phone:
- Email address:
- Full mailing address:
- How long have you been in this role?

How long have you been involved in related roles?Please indicate the type(s) of teaching and/or leadership activities with which you are you involved. Check all that apply:

- Undergraduate coursePlease specify course code and name (e.g. BIOL 345, Human Ecology):
- Graduate coursePlease specify:
- Non-credit course, program or initiativePlease specify:
- OtherPlease specify:

Number of participants in your course, program or initiative per year or cohort:

- <20
- 20-50
- 51-100
- 101-250
- >250

Are you being nominated by a dean, head or equivalent?

- Yes (give their name)
- No

A key aspect of the 2.5 day Intensive is to create a Sustainability Education Plan (SEP) of your choosing that is designed to affect change in your course, program or initiative. How do you envision doing this? Participants from (*host institution*): you must connect your SEP to (*host institution*) activities.

Check all that apply:

- As a fundamental theme, integrated across all components of the course, program or initiative
- Into specific activities/materials/assignments
- Other; please specify: Tell us a bit more about your intended plan (in 50-100 words).

(For (*host institution*) participants only:) Tell us how you would make use of $500 in professional development funding to help make your intended plan a reality (in 25-50 words). Please answer each of the following questions (with bullets or a couple sentences):

1. Do you incorporate sustainability into your own course, program or initiative? If so, how?
2. What opportunities exist in your course, program or initiative for incorporating or enhancing sustainability?
3. What is one teaching/facilitating technique (*e.g.* active engagement, community service-learning) that you are interested to learn more about for your intended plan?
4. What do you see as the biggest challenge to incorporating sustainability into your course, program or initiative?
5. I would also like to ask or say:Please submit your application by (deadline date and time). You will be notified of your acceptance into the Sustainability Education Intensive by (date).

APPENDIX 2: SUSTAINABILITY EDUCATION PLAN

Participants were given this document at the start of the SEI with time at various points over the 2.5 days to work on it individually, in pairs and in small groups. Note that this document is presented "as is" for handing out at the workshop so you can see the spacing and font we used.

Near the end of the last day, participants used information on this document to present a "mini-poster" to share their work and ideas to date. See related materials at the end of the SEP.

- This Sustainability Education Plan (SEP) belongs to:
- Working title (as descriptive as possible):
- Note if it is for a Course/program/initiative (circle one):
  Focusing on:
- Mission or vision statement:

and/or

- Main goals, intended outcomes or learning objectives:
- People/units with whom I will be working:
- Ways I will know my project has contributed to the move to sustainability (indicators of success):
- Ways I will document my journey:

The following materials were given to participants along with the SEP to use during the SEI:

- Open-ended space for participants to jot down notes during the SEI such as Ideas, inspiration, opportunities, challenges
- Tables with empty cells to make notes on "Areas to explore more" and "Other things to think about" based on the SEI guests, resources, teaching techniques and strategies, and other parts of the SEI
- A timeline chart with dates that SEI participants had to submit receipts for reimbursement, when they had to send a final report and when a celebration event 6 months into the future was being held and a place to note "The first steps I will take on my SEP when the SEI is over."

The following material, required by all participants 5 months after the SEI was over, was sent as an email:

- **Final SEP Summary:** Required by all SEI participants, and by your institution participants in order to be reimbursed for your professional development funds.

Please complete all parts of this summary (approximately 2 pages long) and email it to contact email by due date at the latest.

- Name
- Role
- Department or unit
- Mailing address
- Email address
- Phone number
- Final project title:
- Course/program/initiative (identify which it is)
  Focusing on:
- Mission or vision statement or mail goals, intended outcomes or learning objectives (identify which is most relevant and complete to show what you set out to do):
- People/units with whom I worked:
- Ways that I feel my project has contributed to the move to sustainability (indicator of success):
- Other highlights of the SEP (including, if appropriate, unexpected outcomes)
    - A question or follow-up to this SEP that others may want to consider:
    - (If you have them) Web sites you may have created as part of your SEP, Web resources or references you used (so we can add them to our blog resources for others to use):
    - (For your institution participants only): How I spent the professional development funds and how that helped move the project forward:

APPENDIX 3: EXAMPLES OF FORMATIVE FEEDBACK

Example 1

The following words and phrases were arranged on a half-piece of paper around an oval. Here is the instruction to participants: Circle one or more words or phrases that appeal to you, and jot down a few related notes:

- Reason to return
- Peal of wisdom
- Unanswered question
- Most memorable moment
- Something new I learned
- Things we should know
- My opinion
- Something I'll use
- My recommendation
- Something else

Example 2

On a half sheet of paper, we posed the following sentences for participants to complete, spacing it out so that they had room to write their responses:

1. The best parts of today for me was…because…
2. If I could change one thing from today it would be…
3. My wish for tomorrow is…
4. I also wanted to say/observe/suggest…

The day following each of these, we started the workshop by reviewing responses and asking if there were any further comments. We tried to do this in dynamic ways, for example, to the responses in Example 1, each word or phrase plus its responses was printed on a large cut-out of a footprint and taped to the floor. Participants walked the "path of feedback" to see what others said. For the responses to Example 2 questions, one year we summarized each set on flipchart paper and showed them one by one, and in another year, we pooled all the responses to each question into a word cloud and displayed it electronically.

APPENDIX 4: SUMMATIVE FEEDBACK (DAY 3 OF WORKSHOP)

Thank you for your participation in the Sustainability Education Intensive (SEI). Reflecting on the 'short-course' that you have just taken, please answer the following questions. Please return the questionnaire within two weeks. You may choose whether to submit this post-program inquiry anonymously or not (if not, please include your name on this sheet).

- What did we do that helped you learn?
- Please check off as many as you would like:
    - Learning about sustainability education across campus.
    - Becoming part of a face-to-face network of educators at UBC interested in sustainability education.
    - Having structured space and time to work on enhancing your course, program or initiative.
    - Learning more about teaching for sustainability.
    - Learning about various pedagogies that contribute to sustainability education.
    - Accessing resources related to sustainability education.
    - Other (please specify).
- Why was this aspect/were these aspects valuable?
- What aspect(s) of the program was least helpful or most challenging to your learning? Why?
- How did you learn as a sustainability educator?
- Looking back, what aspects of your participation in the program stand out as most valuable to your learning as a person interested and engaged in sustainability education?
- Do you think you have changed as a sustainability educator through your participation in this program? If so, how?
- Do you think you have changed as an educator through your participation in this program? If so, how?
- Supporting implementation of your action plans.
- Following the short-course, the SEI offers two optional group support sessions to further advance your Sustainability Education Plan. Do you think you will attend one or both of these sessions? Why or why not?
- What additional skill development and support do you need in order to successfully implement the Sustainability Education Plan you made in the program?
- Evaluating the SEI for the future.
- What would you suggest that we change in the program (e.g., program content, group process, facilitation, etc) for future groups?
- Other comments and concerns.
- If there is anything else you would like us to know, please write on the back of this one or attach an additional sheet.

Thank you for your time and input!

APPENDIX 5: POST-INQUIRY (ONE YEAR LATER SURVEY)

Note: This was the same as the Day 3 (post-inquiry) survey with the following exceptions:
Questions not included:

- Following the short-course, the SEI offers two optional group support sessions to further advance your Sustainability Education Plan. Do you think you will attend one or both of these sessions? Why or why not?

- What additional skill development and support do you need in order to successfully implement the Sustainability Education Plan you made in the program?

New questions added in their place:

- Through your participation in the SEI, have you experienced shifts or changes in your knowledge, attitudes or skills related to sustainability?
- What changes have you experienced in your knowledge, attitudes or skills related to sustainability? Please be as specific as possible.
- Was adequate support provided to you through the SEI (including the Intensive and follow-up) to implement your Sustainability Education Project?
- If yes, please comment on the specific aspects of support that contributed to your success. If no, please indicate what would have been more useful for you.
- Please suggest improvements to the SEI (or SEI-like program) for the future (e.g., application process, program content, group process, activities, facilitation, follow-up support studios, etc.).
- What do you recommend stays the same for future SEI-like programs?
- Would you recommend the SEI to your colleagues? Why or why not?
- Any additional comments?

This chapter was previously published in Handbook of Research on Transnational Higher Education edited by Purnendu Tripathi and Siran Mukerji, pages 207-231, copyright 2014 by Information Science Reference (an imprint of IGI Global).

Chapter 82
Education Collaboration Development:
A Blended Shore Education Approach to Sustainable Development

Gabriele Strohschen
DePaul University, USA

ABSTRACT

We are living in a state of great flux. Needless to say, political, social, economic, and technological structures are changing faster than we can name and define them. As educators, we are called upon to ready adults for the challenges brought on by global changes. Educators in the 21st century are no longer knowledge producers and disseminators. Educators are involved in managing the educational process: their own and that of the adult students. Educational leadership in the knowledge society is evidenced with a curiously mixed set of skills; it is defined by emotional intelligence and spirituality; it is defined by the finely honed ability of facilitating learning in cross-cultural, multi-lingual, and inter-disciplinary settings; it is defined by a willingness to move away from the guru-stance of teaching and toward a praxis of partnering for change. Today's educator ought to be a strategic partner in the lifelong and life-wide process of learning. This chapter explores the multi-dimensional role of educational leadership, which is characterized by interdependence and calls for research on collaborative and contextual paradigms in higher education development, delivery, and management.

DOI: 10.4018/978-1-4666-4852-4.ch082

INTRODUCTION

At the root of this chapter is the premise that education is a human right and that institutions of education ought to be accountable to the learners, and serve learners within the highest, professional standards toward achieving sustainable social, civic/political, and economic access to the better life we claim higher education can provide. Within our so-termed international knowledge society, teaching within a lifelong learning context has been viewed from so many angles that clarity of approaches to education programming is often lost in the Tower-of-Babel discourse on education management. Although education for all is seen as pivotal for sustainable economic development and peaceful social co-existence of the many peoples in our world, transnational designs at providing access to sound education, training, and development have not had the anticipated impact. Universities continue to be tasked with providing the evidence for sound higher education management through research on best practices of program development and delivery. The preparation of administrators and instructors located within academia all too often does not include experience in real-life settings to appropriately familiarize prospective educators with the issues and needs of disparate stakeholders in learning settings around the world. A dominance of thought and perspective reproduces education structures globally and has led to the silencing of alternate approaches that are based on non-traditional perspectives, which might add richly to the discourse and theory building. A praxis is needed that sustains the very sustainable community development we seek through education and training endeavors.

In the historical sense, universities are defined as the places of knowledge production and dissemination, teaching and learning, and theory building that informs the practical in society[1,2]. This needs to be revisited on two levels to determine the crucial roles that such entities ought to play in our changing and varied reality. One level is the assumption that knowledge is produced through the established protocols, which are gate-kept in academia; and the other level is what kind of knowledge is to be researched: that of content, process, or context of educating. Clarifying and defining need to be done by, with, and for all stakeholders. Higher education institutions are notorious for their limited willingness and capacity to change and to change in a timely manner; and with that, universities as creators and disseminators of knowledge are vulnerable to being out-voiced by vendors of educational products, particularly by those hawking technology-based one-size-fits-all solutions.

This chapter offers a critical perspective about current approaches in education program development and implementation within the context of the reality of higher education practices as I have experienced and evaluated in Afghanistan, China, Thailand, and Kenya. By example of work on the ground in these countries, needs, challenges, and strategies for successful collaboration across countries and cultures emerged. In these countries as much as the inner city, disfranchised communities in the USA where I work, I saw how keenly education workers seek to engage ways to pose and solve problems that plague their everyday existence as much as that of their learners. When workers in the communities collaborate to develop transnational higher education programming much is learned by all involved. It is often through cross-national and intercultural exchanges and experiences that one's own reality becomes more sharply visible.

This chapter is based on my participatory action research in these countries since 2006 and my work on the blended shore education concept (Strohschen, 2009). I hope to offer real-life, current, no-nonsense strategies for consideration. The blended shore education concept provides the theoretical framework as it teased out principles and key approaches to successful transnational education program development and delivery that can sustain community development and lead to empowerment of individuals and community.

Fundamental to this blended shore education concept are the principles of popular education and critical theory as espoused, for example, by Freire (1970), Nyerere (1977, 1979, 1978), Brookfield (1995), Habermas (1970, 1971, 1971a, 1973), and Cervero (2000).

MAIN FOCUS OF THE CHAPTER

The Context for Problem-Posing within a Blended Shore Education Framework

Perspectives

As the cool mornings of the coming autumn season in Chicago entices one to get up early, I walk the dog in the morning. Past the school building we march, void of the sounds of the children. It is the autumn of 2012, and for the first time in 25 years, Chicago's teachers are on strike as I pen this chapter. I pass the red t-shirt clad teachers, who sit on the side walk, readying their protest sign, clutching their Starbuck's coffee cups, and erupting into cheers every time a passerby nods in support or a car drives past, honking in approval of the strike. A key issue prompting this action was that teachers had not been at the table for decision-making on teaching processes and school conditions. I urge the dog to move faster, past the cheering, past the giggling, striking mass.

I think back to a bustling Shanghai. Parents walk their children to school, assertively cross streets or snake their bikes through streets that drown in chaotic traffic. One child each, outfitted in sharply ironed uniform, a satchel on her back, hair neatly combed. School buildings exude an expectation of order and discipline. Expectations are sky high; children must learn, will learn. Parents support education. At a teacher training seminar at the university I meet with eager students. "Please, give us more knowledge, tell us more theories." My intended brief seminar expands to three hours and lunch thereafter for more informal exchanges. The hunger for learning to teach is strong; the opportunity to exchange ideas and learn with this visitor is seized vehemently.

I think back to the small village in northern Afghanistan. Stacks of dried dung form a wall around the school building. Students sit on dirt floors. Some lucky ones have pencils. The chalk barely stands out from the background on the board; the chalkboard nearly blind from too much use. Students chant rhymes; shout out answers in chorus; untiringly. Teachers make due. Learning takes place. Children encircle us and ask many questions; some practice English; smiles all around. We leave all the pencils, pens, and office supplies we have on us.

I think back to the mornings in Kenya, when I visit a pre-school atop a mountain in the Kisumu area. There are no chalkboards; there is but a dirt-walled, dirt-floored edifice where the children sit for their studies. Another, more affluent school sports a small, two-room building with a few posters and hand-made signs. The children's drawings hang on strings; supplies are cherished and carefully stored. There are thin straw mats, which are placed on the concrete floor for naps. The children practice their songs for graduation today. "How happy we are to learn – we thank you teacher" – Excited, they sing these words over and over, dance, smiling with anticipation. Soon, they will be able to go to school. A real school. One with benches and chalkboards and books. They will get to wear uniforms, and shoes with socks. And they will learn new songs and rhymes and do some math, first in Kiswahili and then also in English. And they will get a meal and later rush home, a tamu keki clutched in sweaty hands, saved from the meal, and kept to share with their younger siblings. But not all of them. Only a handful. Only the ones, whose families can raise the school fees, will be able to attend the school. For now all of them sing, "How happy we are to learn – we thank you teacher."

On my walk in Chicago, I think about the riches we have as teachers in the so-termed developed countries. The riches in materials and opportunity; in free public schooling; in educational opportunities for university study. And I think about the incredible waste so prevalent in developed countries, which our education systems reproduce. Without any judgment or intent to blame or find fault with ideologies or perspectives, actions or agency of the striking teachers, I tug the leash to urge the dog to move by quickly so I can rush home and pen the words to this chapter.

The Blended Shore Education Framework

The Result of Analysis and Reflection on our Praxes

Beginning with the first decade of the 21st Century, my university-supported action research and service to my profession led me to focus on education program development, implementation, and evaluation in transnational contexts. The "Handbook of Blended Shore Education: Adult program development and delivery" (Strohschen, 2009) was primarily grounded in multi-year action research and consulting work that commenced in 2004 and was conducted in Thailand, Kenya, and China. It was combined with the research and experiences of a liberatory praxis in adult education in urban environments of the USA, where I worked for over two decades in community organizing in Chicago's inner city communities of Latina/o immigrants and so-termed African-Americans, or the USA's Black population. During the grounded-theory informed phases of the research for the handbook from 2005 to 2008, 32 educators from 18 countries in Africa, Asia, the Americas, and Europe came together to describe their program innovations in education, training, and development within international/intercultural settings. Since 2010, I have added the experience of working with program evaluators of the Ministry of Education in Afghanistan and return visits to the other countries to grasp the intersection of problems and solutions for community development by means of education. I coded and analyzed data, and then synthesized the insights of these authors and mine, which are described in the handbook. Put together the research findings and reflections grounded my critical analysis of our practices in program development and led me to the creation of the blended shore education concept. These insights surfaced a matrix of essential and recurring elements of practice, which in turn became the focus of further analysis. The lens through which I examined essential and recurring elements identified in the data patterns was one of critical reflection (Brookfield, 1995) informed by constructionist ideas (Gergen, 2001), and the value of interdependence and phenomenological perspective on education research (Stanage, 1987; 1989) grounded the analysis. This approach and the conceptual frameworks used for analysis itself are at the heart of blended shore education as a way of learning and teaching with one another within a mindful (Mezirow & Associates, 2000), inter-and cross cultural praxis of adult education in international and transnational context.

This blended shore education concept, our precursor for current research in what we have termed "Metagogy," became the Blended Shore Education Framework (BSEF), not to call it a theory yet. This BSEF is scaffolded by two dimensions and four pillars. It intends to guide education practitioners in developing collaborative, interdependent program development and delivery practices that succeed across cultures and nations. These practices are deeply connected within local knowledge, local assets, and mutually examined values and assumptions, intended to mitigate the prevailing either-or approach of so-termed "best practices" in our field. This framework also validates the premises about education programs for disfranchised populations as delineated in the pillars of the Belém Framework for Action, namely, that formal education programs promote the "learning to know, learning to do, learning to be, and learning to live together" (UNESCO, 2009, p. 1; UNESCO, 1997).

Sustainable development of communities by means of education programs, which have been developed within a BSEF considers how we can benefit from thorough problem-posing that negates othering. It seeks to avoid the dichotomizing of perspectives, values, and approaches, and it embraces a both-and vista that supports collaborative endeavors in providing educational offerings to achieve what we claim to be the goals of education. The BSEF offers educationists as much as participants and administrators a loupe through which to examine the prevailing dominance of thought and perspective about education program development and delivery. One premise of the blended shore education concept is that such dominance has led to the silencing of alternate approaches and non-traditional perspectives (Strohschen, 1989). In turn, such silencing has all too often prevented the identification of relevant and meaningful education and training solutions to local and global problems, needs, and issues. Magnifying values and assumptions can aid in grasping the meaning of them for a shared understanding, and such magnification often happens when we are faced with experiences that differ vastly from what we have "always known." It was, for example, pivotal for me to examine my knowledge in the context of completely different countries' cultures, systems, and structures in order to even arrive at appropriate questions to ask when working cross-culturally and cross-nationally on program development, implementation, and evaluation. In spite of my solid academic preparation and of being an immigrant to the USA with plenty of experience in working in cultures different from my own, the experiences in many different countries in rapid succession during that first decade of the 21st Century enabled me to identify, analyze, concretize, and validate the particulars of this blended shore education approach, and to craft the BSEF subsequently. Whether such could be said to have come about due to disorienting dilemmas (Mezirow, 2000) while traveling or due to my affinity to examine experiences within a phenomenological lens (Stanage, 1987), what emerged guided my analysis in a spiraling trajectory. This continues to provide insights and problem-posing strategies. For now, the BSEF provides benchmarks that allow us to critically examine our current approaches in program development, implementation, and evaluation to determine if these are sufficiently contextualized. It offers a framework that helps analyze if the adult education program design benefits adequately from crucial input by community stakeholders to be effective for addressing the identified and expressed needs and goals of those, such programs are to serve.

Two Dimensions and Four Pillars of the Learning-Teaching Universe in the BSEF

The blended shore education concept is scaffolded by *two dimensions* and *four pillars*. The two dimensions are: one, the spectrum of relationships between the educator and student and the approaches in the learning-teaching processes in the delivery of training, education, and development programs (Strohschen & Elazier, 2005); and, two, the decision processes for the design of training, education, and development programs. The depicted scope of roles and practices in Table 1 is a framework within which to review, examine, and clarify your education practices. The question mark represents the unknown. A

Table 1. Two dimensions

Instructional Relationship X	Instructional Approach Y
• Roles of Learner and Teacher • Direction of the Learning-Teaching Metagogy	• Educational Philosophy • Purpose of the Educational Offering

key assumption underlying this scaffold to the blended shore education concept is the belief to jointly explore frontiers of education and seek to meet the needs of today's and tomorrow's learner as we move toward lived interdependence in the learning-teaching processes (see Figure 1).

Two Dimensions Depicted in the Learner-Teacher Spectrum

Four Pillars

Within the research for the handbook, the recurrent themes that emerged have traditionally and internationally been those, which have been controversial in the education discourse, particularly within our environment of globalization and commodification of education (Jarvis, 2001; Chen, 2003), by any definition for these two concepts. These themes were most succinctly exemplified in the contributions of four authors in the handbook and constitute the four pillars of the BSFE. These four pillars are representative of what contemporary adult education literature examines with respect to development, lifelong education/learning, standards, and spirituality. The educator is implored to examine her paradigmatic assumptions in those categories (Brookfield, 1995), in a suggested process of critically re-considering research-based and philosophical takes on this cluster of disparate concepts and definitions found therein.

In the BSFE, the suggested critical reflection is guided by tenets of critical theory (Habermas, 1970, 1971, 1973) and a phenomenological investigation practice as espoused by Stanage (1987). Understanding one's values and those of others different from one's own comes within constituted awareness, or what Stanage (1987) termed person (i.e., read this as "person underlined"). Person has gained clarity of self. Person is what one has become, self-aware, after having arrived at the capacity to understand self and others at the intersection of universal mutuality. According to Stanage, such common shared intersection or mutuality is intuitively ascertained (in a phenomenological sense of intuiting) and after having been engaged in the deliberate "eductive" process that is teaching and learning. This "eduction" is an interdependently conducted deconstructing and subsequent co-constructing of values and assumptions one holds, and about consciousness of the self one believes to be. Within this "eduction"

Figure 1. The learning-teaching spectrum (Strohschen & Elazier, 2005; Strohschen, 2009. Graphic rev. 2012)

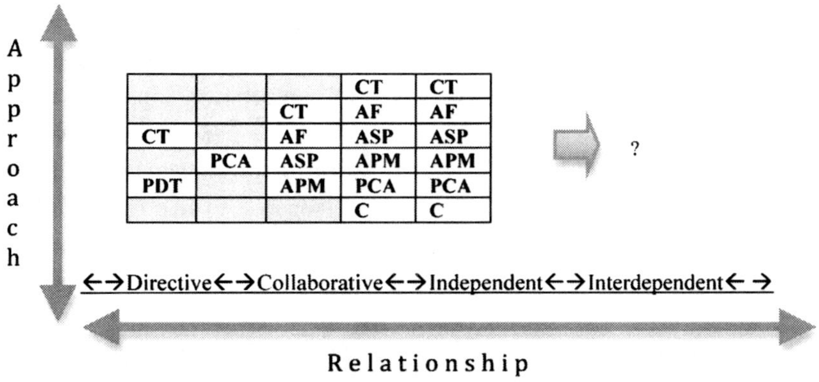

Pedagogue:	Directive Teaching, Critical Analyzing (PDT, PCA)
Andragogue:	Facilitating, Strategic Partnering, and Process Managing (AF, ASP, APM)
Cybergogue:	Using technology as catalyst and mediator in instruction (CT)
Coach:	Guiding learners to reflect on experiences
Adaptive Leader:	Strengthening decision-making and emotional intelligence (AL)

process, learning is not only mind-centered but intuition and the grasping of ego-driven actions are analyzed to bring about clarity of person (Strohschen, 2012, pp. 207-221).

The four pillar themes (see Table 2) were selected based on the exploration of these themes by Brookfield (2009, pp. 27-43), Daun (2009, pp. 45-61), Lynch (2009, pp. 63-70), and Sambuli-Mosha (2009, pp. 71-83). They provide the authors' perspectives, which bear considering and re-considering with the intent of reaching clarification on the varied definitions and meanings because these are grounded in culturally-based, personal narratives and histories, and experiences of individuals and groups. Because education philosophies and cultural values and assumptions vary, disparate concepts and definitions will emerge. Educators are implored to bracket their prevailing perspectives in order to examine their paradigmatic assumptions (Brookfield, 1995) and to embrace realities and viewpoints espoused by these authors for the duration of and purpose of thorough reflection and analysis. Again, the hoped-for outcome of such deep inquiry and reflection on these concepts is first-and foremost a solid grasp of and verification of one's own values and assumptions.

The main and hoped-for outcome of such inquiry into and critical reflection about these concepts and definitions is a deeper grasping of self-awareness. It is my contention that only within such constituted awareness, or what Stanage (1987) termed person that emerges in the "eductive" process of teaching, can an individual then interdependently deconstruct and subsequently co-construct program development and delivery approaches with others. Hence, the four pillars support clusters of concepts, models, principles, and theories within which value clarification and consideration of one's culturally-based and biased assumptions could be considered and reconsidered. Such examination and inquiry, then, is tantamount with what Freire (1970, 1998) insisted is a precursor for any educational activity, namely "problem-posing."

Table 2. Four pillars

Themes	Concepts for (re)Consideration of Assumptions and Values
Development	Hegemony/Neutrality/Intentions
Standards	Professionalism/Collaboration
Lifelong Learning/ Education	History/Constructs/Reality/Research
Spirituality	Interdependence/Indigenous Wisdoms

While considering, examining, analyzing, and then re-considering one's values and the values of those *other than self*, one's own values and assumptions are seen more clearly, and person can emerge. Collaborative program development and delivery can now be created, wherein the appropriate approach for the tasks at hand to reach identified goals are transparently selected. Teacher and learner now have a more common grasp of goals, being able to view them from identified multiple angles. We can now consider self and others – in some ways outside of a preconceived self or image of self. This type of collaboration is one gained after "educing" one's self and being in relationship with other such "selves," and it is tethered solidly to the action needed by those in whose benefit and for whose reality action is to be taken. Most importantly, this approach at least offers the opportunity for leveling skewed power relationships within the context of instructional/learning relationships. Of course, such opportunities are possible only when the premise of "learning to know, learning to do, learning to be, and learning to live together" (UNESCO, 2009, p. 1) is a shared value regarding the purpose of education as a means for community development.

The BSFE Applied

The blended shore education approach with its dimensions of learner-teacher interactions/relationship and instructional approaches and after examination of values and assumptions offers a guide for discerning the contextually most appropriate approaches to program development and delivery. Once the purpose of education or training is both based on identified needs for such educational offerings and transparency on the varying approaches, learner and teacher engage in a reciprocal dance of teaching and learning. The BSFE guides decisions on the selection of dynamic, organic, and adaptive approaches of program development and delivery, which provide fungible, proactive, and immediate responses to the changing demands on and of today's adult student and educator, irrespective of culture, location, or context. These are universals for discerning the process and content for what is to be taught and learned and how. These themes, categories, and descriptors, identified and fleshed out by the findings of the action research for the handbook, are not hugely surprising or innovative. When education is seen as a human right that has as its purpose the emancipation of the disfranchised to gain access to the knowledge and skills needed to participate in developing and implementing sustainable social, economic, and civic/political communities for the ultimate betterment of their lives, then education program development and delivery work:

- Will be grounded in values of social interdependence.
- Espouses interdependence as the guiding value in any collaboration.
- Respects indigenous wisdoms and see "indigenous" as any one particular groups' ways of knowing and ways of doing that are contextually relevant and meaningful.
- Blends such indigenous wisdoms and local knowledge with established global standards and good practices, if and when confirmed as appropriate by stakeholders, to synthesize varied approaches.
- Co-constructs educational programs based on critical reflection to make socially responsible choices.
- Embraces universal standards and competencies in the professional practice of educating adults as deemed appropriate when vetted with stakeholders.
- Acknowledges the spiritual domain of learning and incorporates delivery strategies into those geared to address traditional cognitive, conative, affective, and psychomotor domains.

FUTURE RESEARCH DIRECTIONS

Whose Reality, You Ask?

Karoly and Panis (2004) forecast of the "dramatic" consequences for learning due to the changes occurring in contemporary life meets with Jarvis' (2001) earlier confirmation that the role of the educator has been redefined as facilitator of learning. Giroux and Aronowitz (1993) passionately had made clear that the adult education industry finds itself in the midst of dynamic changes and that adult educators must navigate the turbulence by adapting their roles. Inspired by the shiny currency provided by a recurrent education market, education institutions set out to ready adults for challenges and possibilities ushered in by such change. Globally, institutions of formal teaching or higher education lead the charge, often claiming we are working toward social change by means of education.

Needless to say, we are living in a state of global flux. Political, social, ecological, cultural, economic, and technological structures are changing faster than we can name and define them. Cornel West once wrote, "The existential quest for meaning and the political struggle for freedom sit at the center of my thought" (1999). As researchers and educators, we often claim our role to be that of identifying the ways of teaching and facilitating the kind of learning adults need for their life's quests and for their emancipation. As part of the effort to play our role well, we ought to do the same. Educators have developed an abundance of theories and written scores of books on teaching and learning. Correspondingly, tomes of instructions on how to facilitate such learning exist in our field. We marvel at the latest technologies in our trade, and then apply the tools marvelously. Collectively, we reproduce the perspective that our role is to produce and disseminate knowledge, or at least disseminate ways of meaning-making and support adults in their lifelong learning. Unfortunately, goals of our institutions of higher education and its systems and structures, which are, of course, based in socio-cultural values, do not necessarily align with teachers' or learners' goals. In incongruence, all of us get trapped. We claim expertise. We position ourselves as experts. This is sanctioned by society and the academy. However, no matter how well meaning and well intentioned we do such producing and leading, without the foundation of clearly defined values, examined assumptions, and contextually appropriate practices across person-to-person and person-to-institution relationships we cannot succeed.

I posit that adult educators' leadership in today's knowledge society ought to be evidenced with a curious mix of skills; by emotional intelligence and spirituality; it ought to be defined by critically examined experience; by mindful reflection on assumptions and values; and it ought to be gauged by how finely honed is our ability of facilitating learning in cross-cultural, multilingual, and inter-disciplinary settings. Worldwide, practitioners seem to agree that educators must be willing to move away from the guru-stance of teaching subjects toward a praxis of partnering with others for change. And, still, we basically reproduce the very structures and approaches of academic preparation in ivory towers—on land and in cyberspace—for the educators and administrators of education programs, tasked with leading forth those in need of knowing what we deem, in those halls of academe, as crucial skills and knowledge. After all, our expertise is research-driven. But it is evidenced primarily by the work academicians conduct within their daytime realities. Upon whose reality and needs, however, ought we base our education and training solutions for improved life and sustainable healthy communities in transnational contexts?

The Subject and Object of Education: The Learner

The field of adult education has claimed to have a closer connection to and a deeper understanding of adults and their lifelong learning endeavors than have traditional education structures, like colleges and universities. "...our concern has been with the difference, a major one, between how faculty and students view and understand the learning/teaching process," claimed Dai Hounsell already back in 1987 (in Smith et al., p. 113). By the 1980s, adult education had been firmly established as a field of study and practice in the USA, albeit one without a unified voice as to its purpose and goals. And yet, the common denominator of education endeavors, whether formal or informal, has always been the learner.

Irrespective of the various aims and objectives claimed by the workers in this field since the early 1980s (e.g., Darkenwald and Merriam, 1982), adult educators still seem to agree with what Knowles (1964) had predicted; namely, that this movement called *adult education* might be the educational frontier of the 20th Century and beyond. Andragogues, to use the terminology for adult educators introduced in the USA by Lindeman (1926) and popularized by Knowles, distinguished themselves from pedagogues in their approaches to instructing, or what we seem to prefer to call facilitating of learning. Sets of principles and beliefs about how adults learn originated in both the emergent humanist and critical stance toward conventional education models since the Knowlesean heydays of adult education. Interest in defining and, consequently, developing techniques for facilitating adults' learning grew profusely from that era on. What should distinguish adult educators' ways of *educating* is our commitment to facilitating the learning, or, as Stanage (1998) put it, the leading forth of adults according to their expressed goals. Instead, our research adds to an instruction grab bag, expanding the repertoires of techniques and strategies, and adding tools and toys to the education program design without paying much mind to fundamental, definitional questions and relationship building to bring about needed transformation in our education systems.

We continue to teach subjects instead of adults. Such subjects are predetermined by those, more often than not, whose daytime realities do not match the daytime or nighttime realities of the learners. We measure this learning in standardized ways, developed too often too far removed from the experiences, capacities, and modes of the learners. And we award certificates, diplomas and degrees; symbols that our institution has just educated another person. The symbol, in most contexts, is but a coupon that cannot be cashed in for the sort of goods and value most learners need and desire. At the university level, we reiterate with these ways that a dichotomizing notion of a superior or universal knowledge and truth and reality exists, no matter how vigorously we may "discourse" about the content and purpose of education. With this, it is the researcher, the institution of higher education, and the reification we sanction with the worth we ascribe to such coupons that continue to constrain learning within static structures and systems of higher education. I admit to a sweeping stereotyping here; and, yes, there are exceptions to such approaches. Within the field of adult education, there are the same constraints; and there are also exceptions. But within the field of adult education, within its exceptions and exceptional practices, are options for a broader acceptance and selection of instructional relationships and instructional approaches, which rest on assumptions about the realities of learners and teachers. In some of its traditions and education philosophies, adult education as a praxis is poised to shift the paradigm in higher education and to lead learners toward learning appropriate knowledge and skills during this great, extended moment of flux of this Century.

CONCLUSION

As a world community, we seek education and training intervention to improve the quality of life for disfranchised populations; to sustain healthy, peaceful, and productive communities and economies; and to effectively educate our respective citizenry to provide for their country and be provided for by it. For education management models and systems that address the needs of our increasingly transnational development and implementation of higher education institutions to achieve such goals, we need to shift our research agenda. For one, we do need to agree on a purpose of education, e.g., is it what the Belém Framework for Action declared? For another, we need to accept a plurality of values, beliefs, and approaches in education that, while disparate and grounded in different educational philosophies, may be appropriate to particular contexts. Toward that end, we need not look for "best practices" and export or import these within an assumed superiority of excellence because they are valuable in one corner of our globe. We do not need to look for *the* model for the kind of education or management. What we do need are universals for the process of selecting and creating contextually relevant ways of teaching, learning, and managing across cultures and countries or nations in an interdependent global context that values respect and difference and still acknowledges that there are basic human rights we ought to uphold.

The praxes in the broad field of adult education provide us with a roadmap, wherein it has been accepted that many paths can lead to a desired destination (Strohschen, 2009a).This roadmap is one of depicting the process of decision-making much more than that of describing a static "best practice" content or model. This roadmap provides a sort of lay of the land, a panorama of possibilities from which to select one's preferred path for the journey. As with any journey, one has to plan, pack appropriate supplies, and be prepared to adapt one's direction at times as conditions of the roads and the climate may dictate. But the destination is clear. Always. And as long as one knows what the identified arrival point will be, one can traverse securely if one possesses the needed skills and tools to navigate the terrain or is willing to receive those from the locals. What we may want to research, therefore, focuses on determining those universals. How do we get to agreement of how to get to the shared goals of meeting human needs and supporting human rights by means of education in our world community, and how do we investigate possibilities and approaches within the context of what educationists can do? What has to change?

The BSEF and the forthcoming elaboration on the concept (i.e., Metagogy) offer the education profession a research roadmap for discerning what practices to select and how to go about the selection process. Such research may also simply validate what many learners and educators already know. More research and experimentation is needed to fine tune the universals put forth in the BSFE. They are but a starting point gleaned from the experiences provided by educators in formal, nonformal, and informal learning settings worldwide and described in the initial research findings (Strohschen, 2009). Such continued research needs to focus on how educators and administrators will need to foster cognitive, affective, physical, spiritual, and intuitive ways of teaching and learning. It should investigate how educators discern how to select and apply the tools provided us by current and emerging technologies and global changes. It should rigorously, transparently, and honestly examine what instructional relationships and approaches ought to be in training, education, and development program offerings. It should inquire what might give a global community the impetus to align its expectations with respect to education systems and leadership that meet the social, political, ecological, and economic challenges before us. It ought to find ways of solid, interdependent, and appropriate problem-posing of needs and issues as seen from the vantage point of many realities.

Knowles predicted adult education to be the movement to watch in the 20th century; internationally, we are clamoring for interdependence and a learning orientation in our educational programs (UNESCO, 2005). At this time, we have a window of opportunity as adult educators to inform the process of inquiry. The sights of young ones going to school in different countries and the exchanges with educators and administrators of programs worldwide all had me spiral back to the fundamental questions that have guided my research and action throughout the years in my field of adult education: How do we understand our disparate realities, arrive at clarity of mutual needs, and understand one another's values and assumptions so that we can have the desired impact with our education and training programs in creating and sustaining a healthy, peaceful, and productive world community? It may be time to emphasize such questions in our research about transnational higher education to find the path we will have to take together (Horton and Freire, 1990).

REFERENCES

Brookfield, S. (1985). *Becoming a critically reflective teacher*. San Francisco, CA: Jossey-Bass Inc.

Brookfield, S. (2009). Development. In G. Strohschen (Ed.), *The handbook of blended shore education: Adult program development and delivery* (pp. 27–43). New York: Springer International. doi:10.1007/978-0-387-09443-4_2

Cevero, R. M. (2000). *Power in practice: Adult education and the struggle for knowledge and power in society*. San Francisco, CA: Jossey-Bass.

Chen, Y. C. (2003). *New principalship for globalization, localization and individualization: Paradigm shift*. Paper presented at the International Conference on Principalship and School Management Practices in the Era of Globalization: Issues and Challenge. Kuala Lumpur, Malaysia.

Darkenwald, G. G., & Merriam, S. B. (1982). *Adult education: Foundations of practice*. New York: Harper & Row.

Daun, H. (2009). Lifelong learning/lifelong education. In G. Strohschen (Ed.), *The handbook of blended shore education: Adult program development and delivery* (pp. 45–61). New York: Springer International. doi:10.1007/978-0-387-09443-4_3

Freire, P. (1970, 1993). Pedagogy of the oppressed. New York: The Continuum International Publishing Group.

Freire, P. (1998). *Pedagogy of freedom: Ethics, democracy, and civic courage* (P. Clarke, Trans.). Baltimore, MD: Rowman & Littlefield.

Gergen, K. (2001). *Social construction in context*. London: Sage.

Giroux, H. A., & Aronowitz, S. (1993). *Education still under siege* (2nd ed.). Westport, CT: Bergin and Garvey.

Habermas, J. (1970). *Toward a rational society*. Boston: Beacon.

Habermas, J. (1971). *Theorie und praxis*. Frankfurt am Main: Suhrkamp.

Habermas, J. (1971). *Knowledge and human interests* (J. J. Shapiro, Trans.). Boston: Beacon.

Habermas, J. (1973). Wahrheitstheorien. In H. Fahrenbach (Ed.), Wirklichkeit und reflexion. Pfüllingen: Neske. 211–265. Reprint: 1984b, chap. 2.

Horton, M., & Freire, P. (1990). *We make the road by walking: Conversations on education and social change*. Philadelphia: Temple University Press.

Karoly, L. A., & Panis, C. W. A. (2004). *The 21st century at work: Forces shaping the future workforce and workplace in the United States*. Santa Monica, CA: Rand Corporation. doi:10.1037/e527212012-001

Knowles, M. S. (1964). The field of operations in adult education. In G. Jenson, Liveright, & Hallenbeck (Eds.), Adult Education: Outlines of an emerging field of university study. Washington, DC: Adult Education Association of the U.S.A.

Knowles, M. S. (1990). *The adult learner: A neglected species* (Revised Ed.). Houston, TX: Gulf Publishing Company.

Knowles, M. S., Holton, E. F., III, & Swanson, R. A. (1973, 2005). The adult learner (6th ed.). San Diego, CA: Elsevier.

Lynch, D. (2009). Standards. In G. Strohschen (Ed.), *The handbook of blended shore education: Adult program development and delivery* (pp. 63–70). New York: Springer International. doi:10.1007/978-0-387-09443-4_4

Mezirow, J. etal. (2000). *Learning as transformation: Critical perspectives on a theory in progress*. San Francisco, CA: Jossey-Bass Publishers.

Nyerere, J. (1974). *Freedom & development: Uhuru na maendeleo*. Dar es Salaam: Oxford University Press.

Nyerere, J. (1977). *Ujamaa-Essays on socialism*. London: Oxford University Press.

Nyerere, J. (1978). Development is for man, by man, and of man: The declaration of Dar es Salaam. In B. Hall, & J. R. Kidd (Eds.), *Adult learning: A design for action*. Oxford: Pergamon Press.

Nyerere, J. (1979). *Crusade for liberation*. Dar es Salaam: Oxford University Press.

Sambuli-Mosha, R. (2009). Spatiality. In G. Strohschen (Ed.), *The handbook of blended shore education: Adult program development and delivery* (pp. 71–83). New York: Springer International.

Smith, R. (Ed.). (1987). *Learning how to learn*. DeKalb, IL: Northern Illinois University.

Stanage, S. M. (1987). *Adult education and phenomenological research: New directions for theory, practice, and research*. Malabar, FL: Krieger.

Stanage, S. M. (1989). *Lifelong learning: A phenomenology of meaning and value transformation in postmodern adult education*. Malabar, FL: Robert E. Krieger Publishing Co., Inc.

Strohschen, G. (1989). *National literacy campaign: The Nicaraguan experience*. Chicago: Chicago Public Library.

Strohschen, G. (2009). *Handbook of blended shore education: Ault program development and delivery*. New York, NY: Springer International Verlag. doi:10.1007/978-0-387-09443-4

Strohschen, G. (2009a). Simplemente mujer: A literacy project by, with, and for women. In M. Miller, & K. P. King (Eds.), *Empowering women through literacy: Views from experience*. Charlotte, NC: Information Age Publishing, Inc.

Strohschen, G. (2012). In black and white: Transformation through examined selves. In C.J. BodenMcGill, & S. M. Kippers (Eds.), Pathways to transformation: Learning in relationship (pp. 207-221). Charlotte, NC: Information Age Publishing.

Strohschen, G., & Elazier, K. (2005). 21st century educator. Paper presented at the Quality Educational Leadership: A Partnership of East and West. Bang Na, Thailand.

UNESCO. (1997). *The Hamburg declaration on adult learning and the agenda for the future*. Hamburg, Germany: UIE.

UNESCO. (2009). *Confintea VI: Proceedings of the sixth international conference on adult education*. Belem, Brazil: Confintea.

West, C. (1999). *The Cornel West reader*. New York, NY: Basic Civitas Books.

KEY TERMS AND DEFINITIONS

Adult Education: It is part of extension education which is aimed at educating adults including educating at work place.

Andragogy: Learning strategies developed for educating adults.

Blended Shore Education: It is a framework that guides educators in developing successful collaborative, interdependent program development and delivery practices across cultures and nations.

Instructional Design: The process of identifying learning needs and development learning material in order to improve instructions to the learners.

International/Intercultural Education: Education across borders and cultures.

Metagogy: It is a process of collaborative learning amongst adults that works on interdependence of learning for the advantage of the individual as well as the community.

ENDNOTES

1 This statement rests on the premise that we agree that *adult education* has a particular knowledge base, with an identifiable history and a distinguishing philosophical framework, all of which has traditionally differed from institutions and structures that also educate adults (e.g. colleges, trade schools, corporate training departments).

2 The scope of this chapter is too narrow to do even a smidgen of justice to the rich histories – both written and unwritten- of adult education in the USA. The reader is encouraged to consider the point illustrated with this brief statement; i.e., that adult education, in its myriad manifestations, established sets of principles particular to this field. At the time, these principles of adult learning and corresponding techniques of facilitating learning were cutting edge and unique. Today, much of this knowledge base has become mainstream education/training fare, even when its originators are not credited.

This chapter was previously published in Handbook of Research on Transnational Higher Education edited by Purnendu Tripathi and Siran Mukerji, pages 512-525, copyright 2014 by Information Science Reference (an imprint of IGI Global).

Chapter 83
Environmental Education through Envkids Didactical Framework and ICT Tools

Hariklia Tsalapata
UTH, Greece

Rene Alimsi
CERETETH, Greece

Olivier Heidmann
CERETETH, Greece

ABSTRACT

Global awareness on the need to change everyday behavior towards some more environmentally friendly practices has been on the rise over the last few years in the face of emerging phenomena such as global temperature rising, desertification, extreme weather, sea water level rising, and the potential resulting disruption to everyday life for large populations. Even if in most European countries a certain amount of environmental objectives for primary education are defined, teachers in the field feel a substantial lack of supporting implementation guidelines, especially for digital deployment in the classroom. Addressing this need, the work presented in this paper aims at the development of blended learning activities that deploy an explorative and collaborative didactical framework towards environmental sustainability training for primary education through a combination of in-class instruction, virtual experimentation, storytelling practices, and on-line collaboration. The validation activities carried out so far demonstrate a positive teacher and learner reaction and an easy integration of the methodologies and demonstrators into existing school practices.

INTRODUCTION

Mark Prensky (2001, p. 1) highlights the fact that 'todays' students are no longer the people our educational system was designed to teach'. According to Prensky (2001, p. 1), students are used to interact with a ubiquitous environment where 'computer games, email, the Internet, cell phones, and instant messaging' play a critical role. Such interaction has an impact on the way students think and interpret data. Beyond all dispute students' *thinking patterns* have changed and this is something that should not be ignored during

DOI: 10.4018/978-1-4666-4852-4.ch083

the process of designing learning experiences (Tsalapatas, Heidmann, Alimisi, & Stav, 2010).

This consideration is well in line with opinions expressed by Resnick (2007, p. 18). Resnick (2007a) introduces the conception of a 'Creative Society' where human creativity is the vital turning point of any development and therefore it is important for students 'to learn to think creatively, plan systematically, analyse critically, work collaboratively, communicate clearly, design iteratively, explore and learn continuously' (Resnick, 2007, p. 22). The engagement of students in a learning experience that provides them with opportunities to learn through designing, creating, collaborating, exploring, and reflecting constitutes a challenge for new technologies and pedagogies (Alimisi &Winters, 2010; Resnick, 2007).

Crucially, there is need for changes in the schooling curriculum to be encouraged in order to make school activities more meaningful for young children (Tsalapatas, Mogli, Karagiannis, Tallvid, Protivova, Heidmann, s& Larsson, 2010; Tsalapatas, Heidmann, Alimisi, & Stav, 2010). This need has led to the emergence over the past years of learning approaches that update the existing instructional ones followed by the schools (Tsalapatas, Heidmann, Alimisi, & Stav, 2010). This need is obvious in the area of primary environmental education as well. Nevertheless in most European countries a certain amount of environmental objectives for primary education are defined, teachers in the field feel a substantial lack of supporting implementation guidelines and learning tools, especially for digital deployment in the classroom (Tsalapatas, Mogli et al., 2010). The Envkids project tries to fulfill such a need by exploiting emerging didactical models towards the introduction of value-adding learning tools for environmental education that will be easy to integrate into wider blended learning activities. With respect on the already used methods EnvKids aims at extending them and reshaping them so that meaningful learning to occur.

This paper is structured as following: First, the two core aspects of the Envkids didactical framework collaborative and explorative learning are presented. Addressing single definitions for these approaches is beyond the aims of the paper. However, issues and debates concerning the educational benefits of these learning approaches are brought into focus. Second, we briefly present the way in which the EnvKids didactical framework was further enriched with other pedagogical approaches and strategies. The implementation stage and the demonstration of the EnvKids learning tools come next. The paper concludes with some preliminary results retrieved from early stage evaluation.

COLLABORATIVE LIES AT THE ROOT OF ENVKIDS DIDACTICAL FRAMEWORK

The EnvKids project aims at the development of explorative and collaborative learning frameworks towards environmental education for primary school children (The EnvKids Project Proposal, 2009). Exploration and collaboration are seen to play a significant role in Resnick's creative spiral (2007) and it was decided to be placed in the root of the EnvKids design.

According to Smith and MacGregor (1992, p. 1) 'collaborative learning is an umbrella term for a variety of educational approaches involving joint intellectual effort by students, or students and teachers together'. The term *collaborative learning* is used in wide learning during which researchers and instructors may focus on different aspects of the approach (Daniel, 1995; Verdejo, 1996). Though collaborative learning takes on a range of forms and interpretations, in each form there is a shift away from the typical roles held by the teachers and the learners. In fact, collaborative learning constitutes 'a significant shift away from the typical teacher-centred or lecture centred milicu' in classrooms (Smith & MacGregor, 1992, p. 1). In the framework of a collaborative learning approach, teachers can find themselves to act as scafolders, as designers of intellectual experiences for students, as 'mid- wives of a more emergent process' (Smith & MacGregor, 1992, p. 1), or as Dillenbourg (1999, 2002) states as 'the mediums to make the class work in a productive direction.' The traditional lecturing process may still exist, but in the light of other processes which lay emphasis on students' discussion and active work.

Collaborative learning is connected to approaches that present learning as an active, constructive, and social process where an individual actively constructs knowledge facilitated by peer interaction, evaluation, and cooperation. Collaborative learning describes situations in which subjects are becoming mutually aware of their shared goal and are working in group interactively towards this goal, triggering learning mechanisms (Dillenbourg, 1999). There is a wide range of collaborative learning activities both in classrooms and online; collaborative learning activities can include collaborative writing, group projects, joint problem solving, tasks that utilize web- tools and software, debates, or even study teams (Dillenbourg, 1999; Dillenbourg et al., 2002).

In educational practices collaborative learning is thought to be useful for reasons connected to pedagogical perspectives such as the formation of characters (Tsalapatas, Heidmann, Alimisi, & Stav, 2010). Collaborative learning is believed to contribute in the cultivation of habits of engagement to a community because it encourages students to acquire an 'active voice' and a 'sensitive ear' (Tsalapatas, Heidmann, Alimisi, & Stav, 2010). Being able to recognize and re-solve differences that students may encounter when collaborating with others is another valuable asset for the student as a future citizen (Tsalapatas, Heidmann, Alimisi, & Stav, 2010).

However, it is worth bringing into focus the fact that there is not any guarantee that the benefits from a collaborative practice will occur in every attempt for collaborative learning. As Bennett (1996) states the fact that students have been structured in groups does not necessarily guarantee the development of collaboration. Groups can easily turn out to form a situation in which the members are working individually and not necessarily collaboratively.

EXPLORATIVE LEARNING LIES IN THE ROOT OF ENVKIDS DIDACTICAL FRAMEWORK

Explorative learning is a strategy about student exploration of knowledge, seeking for truth and information (Njoo & Jong, 1993, p. 1; Conole et al., 2009). Though researchers have proposed different definitions of inquiry and exploration (Looi, 1998; White & Frederiksen, 1998) they generally share the same opinion according to which there are at least four critical steps when con-ducting learning based on exploration/inquiry (Njoo & Jong, 1993, p. 1; Edelson et al., 1999): Generating hypothesis, collecting data, interpreting evidence, drawing conclusions.

Previous research found that learning through exploration enhances students' learning achievement, especially in the aspects of problem solving skills, ability to provide explanation for different forms of data, critical thinking, and understanding of different concepts (Chiappetta & Russel, 1982; Saunders & Shepardson, 1987; Haury, 1993). In general, learning experiences based on exploration can provide students with valuable opportunities to develop their understanding of content and practices which draw upon different subject areas (Chang et al., 2003).

Despite its merits, the implementation of explorative learning is not always a straightforward process (Chang et al., 2003). The most emergent problem is the 'limited appropriate software for explorative learning in schools' (Chang et al., 2003). A well designed and piloted exploratory learning environment may benefit both teachers' teaching and students' learning. Additionally, for inquiry learning to be more effective, it is of great significance to 'strengthen the explorative learning motives of students, to inform them with exactitude what scientific exploration skills are required, and to give them some understanding of the background knowledge' (Edelson et al., 1999, as cited in Chang et al., 2003, p. 57). Last, it is also important to provide students with the appropriate guides and scaffolding as there is considerable possibility only the brighter few students to manage to engage in the explorative learning activities (Chang et al., 2003). These points should be taken into account while designing a methodology that includes an explorative learning approach.

ENRICHING THE ENVKIDS DIDACTICAL FRAMEWORK

Through this small scale theoretical discussion collaborative learning and explorative learning were seen potentially capable of establishing a base whereupon a challenging learning experience, that speaks the language of the new generation of students, can be formed and environmental messages can be communicated. Envkids aims at incorporating meaningfully these two approaches in the context of the development of learning tools and scenarios for environmental education. However, these two approaches do not stand alone. An attempt has been made to further enrich the Envkids didactical framework with a game- based approach to learning project based learning methods and digital story telling practices (Tsalapatas, Mogli et al., 2010; Tsalapatas, Heidmann, Alimisi, & Stav, 2010).

It is worth mentioning that the game- based approach (JICS, 2007; de Freitas et al., 2006) was implemented through the development of half- baked games which appear 'as semi-finished environments, targeting at engaging students with changing their rules' (Alexopoulou & Kynigos, 2008, p. 71). The idea of the half- baked games is considered to be ideally suited to educational projects, where there is a need to guarantee a frame in which students could construct and explore specific concepts with educational value and hold an active role (Alimisi & Winters, 2010; Tsalapatas, Heidmann, Alimisi, & Stav, 2010; Tsalapatas, Heidman, Alimisi, Tsalapatas, & Houstis, 2010) The semi- finished nature of this approach could also guarantee that a level of freedom could be given to the students to develop their own thinking and to reflect upon it (Alimisi & Winters, 2010). This semi-finished nature of the games can be seen as a good way to restrain students' cognitive load and to raise students' curiosity (Alimisi & Winters, 2010; Tsalapatas, Heidmann, Alimisi, & Stav, 2010).

Project based learning methods has been also exploited as they were seen potentially capable of engaging students in sustained, cooperative explorations (BIE, 2003; Bransford & Stein, 1993) that encompass collaborative practices, authentic content, authentic assessment and reflection (Han & Bhattacharya, 2001). Last, storytelling practices have been encouraged through EnvKids didactical framework. Students are offered the opportunity to engage in 'the multiple steps of designing, creating and presenting their own digital stories,' aiming at increasing a full complement of literacy skills, such as research, presentation, writing, organizational, interviewing, interpersonal and problem-solving skills (Robin, n.d.). All these methodological approaches were considered important in the context of teaching environmental concepts in a meaningful way and raising environmental awareness.

CONTRIBUTION AND RELEVANT WORK

This project was elaborated taking into consideration comments by many teachers (all our partners from Greece, Sweden, Czech Republic and France agreed) about lack of open source digital tools and free educational resources for environmental education. It doesn't mean that no work has been done in the field of environmental educational technology, but rather than tools which support both students' environmental education and teachers in taking an active role in designing and carrying out learning activities concerning environmental issues, needs to be developed. Envkids project revolves around this objective. The main contribution of the Envkids project is four-fold:

1. To provide students with tools to foster the learning of environmental concepts
2. To support teachers in carrying out learning activities for environmental education using new technologies
3. To provide the participant community with tools for collaboration and communication and
4. To develop a methodological framework that connects learning theories, technology and environmental education.

Before presenting the Envkids toolkit, it is worth bringing into focus considerable works and studies in the field of environmental education. This section does of course not constitute a review of all the existent work done in the field. It rather aims at highlighting significant trends in the area of environmental educational technology. Tools for collaboration and communication, environments for information access and manipulation, immersive games and technologies for modeling, visual representations and simulation are identified as significant trends.

The advances in the field of information communication technology (i.e., Web 2.0 tools) have been taken into consideration in addressing environmental education a multicultural dimension. For example, the World School Network - called Eco plus- provides students with a platform where discussion about environmental issues can take place in a multinational and multicultural level (Nomura, 2004). In a similar way, the Youth Can Med, a networking youth project, which aims at introducing sustainable development in Lebanese schools, uses ICT tools and provides active youths with a platform where conversations about environmental issues are taking place ("Using ICT for environmental education," 2007). The iEarn programme –run in collaboration with young minds and Global virtual school for Sustainable Development- also centers on online communication for environmental purposes. The two participant countries in this program uses ICT tools to build a computer mediated ecological community for sharing information and knowledge about environmental, scientific and health issues (Kudrajavtsev et al., 2006).

IBM recently developed and released the serious game called CityOne that offers players the opportunity to 'optimize banking, retail, energy and water solutions via an online, sim-style game in which the player is tasked with guiding industries within a city through a series of missions' (information retrieved from the official site, http://www-01.ibm.com/software/solutions/soa/innov8/cityone/index.html). Enercities is an under- development project co-funded by the European Commission. In contrast to the previously mentioned interventions, this project does not aim at exploiting ICT tools for communication and collaboration. 'Enercities' is a 3D educational game about energy that aims to problematize children upon the balance between people, planet and profit (for more details see official site, http://www.enercities.eu/). Another useful exploitation of technology is the development of interactive environments where students can access and manipulate information. This is the case of 'Monterey Bay' where students are provided with the opportunity to virtually explore the Monterey Bay's kelp forest and to see 17 native species (more details on the official site, http://www.nationalgeographic.com/monterey/ax/primary_fs.html).

Last, allowing students to explore relationships and knowledge through media, simulations and models is also of great educational value. 'Play with Time' project falls into this category as it visually demonstrates how the world around us changes over different time periods. A web site and a traveling museum exhibit comprise this project. Tools for collaboration are also included (more details on the official site, http://www.playingwithtime.org/). Works like this can successfully bring together scientific knowledge, theory, visual representation and exploration.

FROM THEORY TO DESIGN

Drawn upon these methodological decisions three learning tools followed by learning activities and good practices scenarios were developed aiming at raising public awareness on current environmental issues (Tsalapatas, Heidman, Alimisi, Tsalapatas, & Houstis, 2010). Under the umbrella title 'my home, my town, my planet', the tool suite covers environmental issues in a learner-centered manner that starts from concepts close to home and expands to wider subjects (Tsalapatas, Heidman, Alimisi, Tsalapatas, & Houstis, 2010). The three virtual demonstrations- learning tools are briefly described.

My Home: Environmentally Responsible Practices in Everyday Life

The pilot application uses virtual experimentation and game-based learning approaches based on pre-defined alternatives to help children discover mitigation solutions that can be applied in everyday life towards environmentally friendly residential activities (Tsalapatas, Heidman, Alimisi, Tsalapatas, & Houstis, 2010). Students are given the opportunity to design a green house and to manage it in an environmentally friendly way. During this process they come across useful environmental concepts and environmentally friendly actions concerning energy consumption and everyday's life residential activities. The virtual learning application provides quick feedback on the impact of choices through a green meter/indicator in order to encourage exploration and experimentation with a variety of environmentally friendly solutions (Tsalapatas, Heidman, Alimisi, Tsalapatas, & Houstis, 2010). The aim of the game is students to understand in an explorative way the impact that different design decisions have on the green meter. A less green house is also educationally valuable as it can initiate relevant dialogue in the class. Learning activities and designed resources scaffold students during this process. The pilot application (Figure 1) takes into account recommendations introduced by Ministries of the Environment in European countries.

My Town: The Benefits of Renewable Energy Resources and Wider Policies for Greener Cities

This pilot application focuses on residential solutions and behavior at the town level related to energy production methods aiming at pollution control and better quality of life. A game-based, explorative educational approach is used in this pilot as well (Tsalapatas, Heidman, Alimisi, Tsalapatas, & Houstis, 2010). Students are invited to design and build from scratch a town that offers a good enough quality of life for its inhabitants, as little pollution as possible and smart energy management.

Three indicators reveal how successful the created town is: the inhabitants' mood, the pollution level and the energy production and management. The students are motivated by creating the most efficient town possible with the many (around 80) buildings choices available. In this way, an attempt is made to connect environmentally friendly practices to better quality of life. However, once again, a 'less green city' does not constitute a problem as it is considered as an educationally valuable result which can initiate relevant discussion in the classroom.

Figure 1. Good practices in everyday life pilot: demonstration of house improvements for energy savings; demonstration of environmentally friendly behavior at home

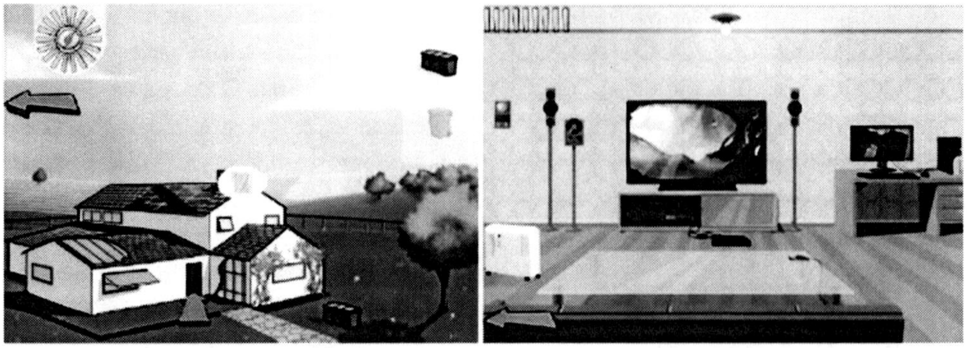

The pilot uses "sandbox" explorative approaches, in which "anything goes." In other words, learners are given the opportunity to develop from scratch their own virtual world and then introduce environmentally friendly improvements. Issues concerning clean energy production through renewable resources such as the sun, the wind, and waves, traditional energy production based on oil and gas, effective public transportation, energy saving planning are addressed (Tsalapatas, Heidman, Alimisi, Tsalapatas, & Houstis, 2010). Three feedback indicators sensitive in users' design decisions encourage reflection upon specific decisions that affect energy production levels, energy consumption, and quality of life (Figure 2) (Tsalapatas, Heidmann, Alimisi, & Stav, 2010).

My Planet: Responsible Natural Resources Management

The application pilot raises awareness on sensible management of resources such as forests, biodiver-

Figure 2. Energy production through renewable resources demonstrator with real-time feedback on energy production, energy consumption, and quality of life

sity, and water resources through project-based and story-telling practices that draw upon exploration (Tsalapatas et al., 200b). Historical, current, and projected depletion trends and consequences on climate change, quality of life, animal habitat, and the welfare of endangered species are presented to learners through age-appropriate graphical depictions in a map-based interface (Figure 3). This learning application follows a two-step learning approach which differs from the previous two (Tsalapatas, Heidman, Alimisi, Tsalapatas, & Houstis, 2010).

As a first step, the learning application presents actual scientific information on the current status quo on natural resource consumption and the related impacts on the environment. For instance, this information includes deforestation, forest cover, carbon emissions, water resources per capita for ANNEX I countries (for more information see Tsalapatas, Heidmann, Alimisi, & Stav, 2010). Comparisons are made with expected positive effects of more responsible management of resources, such as through reducing deforestation from land-use change (Tsalapatas, Heidmann, Alimisi, & Stav, 2010; Tsalapatas, Heidman, Alimisi, Tsalapatas, & Houstis, 2010)

As a second step, the application calls students to contribute to map-based educational content by collecting themselves information on more familiar environmental issues and subjects (i.e., species under threat of extinction in their areas) and creating their own stories and material. These stories can be published online by the children themselves or through teacher mediation resulting in 'a commonly-owned by all participating schools public outcome' (Tsalapatas, Heidman, Alimisi, Tsalapatas, & Houstis, 2010)

The three learning tools are designed to be used by teachers in the classroom who are willing to introduce explorative and collaborative learning practices. To facilitate the learning process, teaching material in the form of 'working sheets,' learning scenarios and good practices videos guide teachers through the exploitation of the

Figure 3. The map-based interface of the Natural Resources demonstrator, depicting graphically scientific data

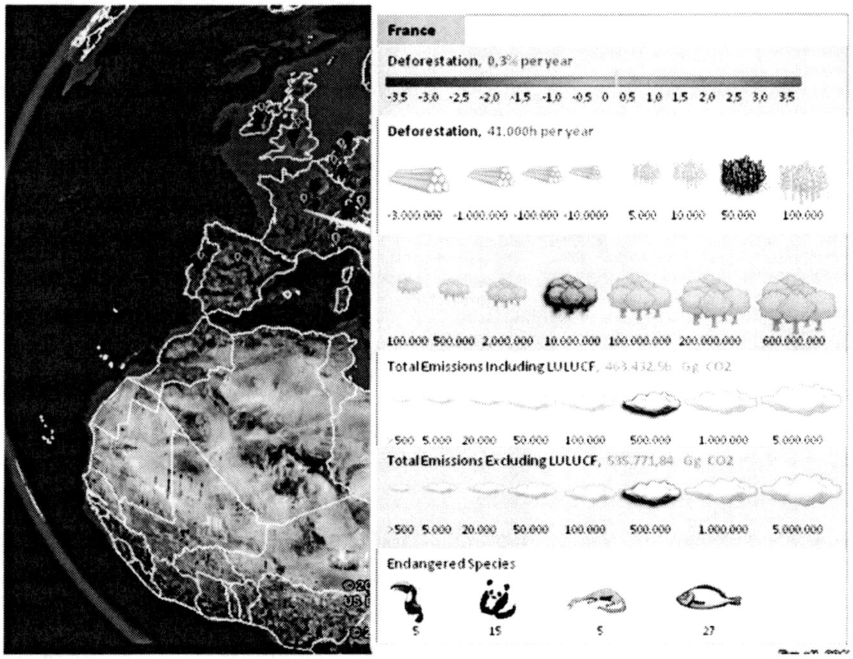

tools in the classroom setting. The interfaces of the tools are available in English, Greek, Swedish, French, and Czech. The tools have minimal technical requirements on behalf of the user: they are lightweight and deployable through the Internet; off-line versions are also available (Tsalapatas, Heidman, Alimisi, Tsalapatas, & Houstis, 2010).

Envkids Methodological Framework for Evaluation

Evaluation of the three pilot applications is taking place in participant schools in four European countries: Greece, Sweden, France, and the Czech Republic. The evaluation is a process in progress. Evaluation is carried out through teaching experiments. Cobb and Steffe (1983, p. 83) defined a 'teaching experiment' as 'a set of teaching episodes and interviews that covers an extended period of time.' A teaching experiment consists of three components: modelling, teaching episodes, and individual or group interviews. The most important aspect of the teaching experiment is 'the modelling of the students' responses into a coherent picture of the students' progress over an extended period' (Steffe & Thompson, 2000, as cited in Engelhardt, 2003). Teaching episodes involve the teacher/interviewer, an observer, and the students under investigation.

The teaching experiment is related to the Socratic teaching. In fact, 'the teaching experiment resembles a Socratic dialog' though which the elicitation of data is aimed; probing questions are formed and set moving students to verbalise their thoughts, feelings and explanations about the undertaken activities and educational processes (Engelhardt, 2003). The teaching experiment embraces also a learning cycle (Zollman, 1990). 'A typical learning cycle consists of three stages: an exploration phase, concept introduction phase, and concept application phase. In the exploration phase, students explore the concept under investigation through hands-on activities. In the concept introduction phase, an explanation of the observations that were performed in the exploratory phase is given a name and further refined. In the concept application phase, students apply the concept that they explored and later named to new situations' (Engelhardt, 2003).

The EnvKids validation activities have been inspired from the 'teaching experiment' practices. However, the idea underpinning the 'teaching experiments' is not being explicitly followed. It was opted to inform the participant external evaluators about this technique and to call them to document their perceptions after engaging students in a Socratic dialogue and a learning cycle as it is described above. However, teachers were free to implement the teaching episodes in accordance to the class needs and culture. Socratic dialog took place in combination with teacher's observations. The preliminary stage findings resulted in a documentation produced by the participant external evaluation (Alimisi et al., 2010).

The Design Enters the Classroom

Early results indicate that EnvKids didactical framework, proposed learning tools and learning activities are well accepted by the target primary education sector. Both learners and teachers so far provide positive feedback on the value-adding character of the tools for enrich the learning experience in the context of wider explorative and collaborative blended learning activities (Alimisi et al., 2010). So far positive feedback perceived on the application user interfaces, which were engaging for children, easy to use, and met expectations of a digitally-literate young generation (Alimisi et al., 2010).

A significant strength of the tools was the easy integration into the learning process, to existing explorative and collaborative practices such as excursions or school projects on the environment (Alimisi et al., 2010).The easy integration is a result of addressing environmental issues that are already the subject of on-going school activities such as biodiversity, environmentally friendly house improvements, and renewable energy sources; related learning activities are enhanced through the EnvKids digital tools that serve as complementary educational content. It was opted that teachers recognize Envkids applications as educationally beneficial tools capable of triggering mechanisms for learning and reflective practices.

The pilot helped the children understand some of the environmentally-friendly actions such as insulation, recycling, and reduction of energy[...] They [the students] could effectively see that the color of the house has indeed an impact on energy consumption (the class teacher on 'my home' pilot application).

The active participation of students on the collaborative development of part of the educational tools through digitally narrated stories (Figure 4 through 7) published in the context of a jointly owned Google Earth information layer was particularly empowering for children who saw their work publicly available.

The natural resources [my planet] pilot opened the children's eyes to a more global and worldwide issue (the class teacher on 'my planet' pilot application).

Figure 4. Students' working on the development of their stories: data collection stage

Figure 5. Students' working on the development of their stories: design stage

Figure 6. Students' collaboration on the "my home" pilot

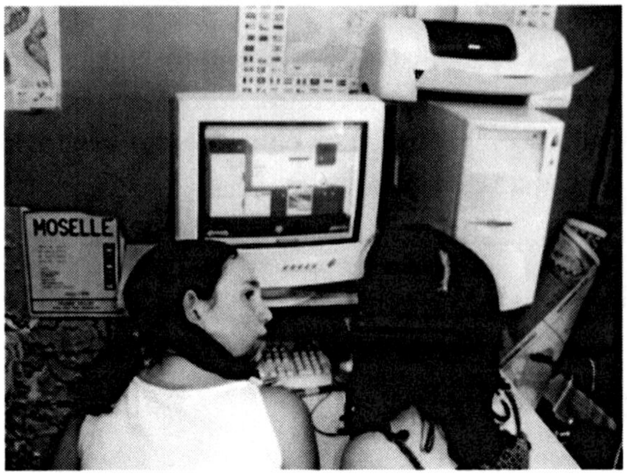

Figure 7. Students working on "my home" pilot

CONCLUSION

The students' reactions and teachers' feedback that are brought into focus briefly in this section cannot be considered as constituting a comprehensive set of final results capable of addressing clear and exact results. In fact, these preliminary data can only be seen as an initial feedback and reactions that portray work in progress. External validation activities will continue at all sites throughout the 2nd year of the project implementation period. All the three pilot applications will be tested with primary school students in a real time classroom setting. The close communication with the participant teachers will allow us to gain an insight in the learning process with the use of Envkids applications and to gather data concerning Envkids didactical framework and technological tools. Thus to this very point it is not safe to come up with brave conclusions concerning the Envkids tools and didactical framework. The feedback gained so far is encouraging and can set a basis whereupon further work could be planned (in case needed) on the reshaping of the didactical framework and the learning tools.

REFERENCES

Alexopoulou, E., & Kynigos, C. (2008). *Half-baked games as a context for understanding conditional constructs.* Retrieved from http://www.etpe.gr/files/proceedings/21/1223368001_DIDINFO08_71_80.pdf

Alimisi, R., Tsalapatas, H., Tallvid, K., Larsson, S. O., Protivova, Z., & Karagiannis, H. . . . Heidmann, E. (2010). *Evaluation activities mid-term progress report.* Retrieved from http://ohmpro.org/envkids/documents/VR_502390-2009-COMENIUS.pdf

Alimisi, R., & Winters, N. (2010). Programming playfully for a real- life problem: conditional statements on the stage of Scratch. In *Proceedings of the International Conference on Constructionism*, Paris, France.

Bennett, N. (1991). The Emanuel Miller Memorial Lecture 1990: Cooperative learning in classrooms: Processes and outcomes. *Journal of Child Psychology and Psychiatry, and Allied Disciplines, 32*(4), 581–594. doi:10.1111/j.1469-7610.1991.tb00336.x PMID:1864890

Bransford, J. D., & Stein, B. S. (1993). *The ideal problem solver* (2nd ed.). New York, NY: Freeman.

Buck Institute of Education (BIE). (2003). *Project-based learning hand-book. A guide to standards-focused project-based learning for middle and high school teachers* (2nd ed.). Novato, CA: Author.

Chang, K.-E., Sung, Y.-T., & Lee, C.-T. (2003). Web-based collaborative inquiry learning. *Journal of Computer Assisted Learning*, *19*, 56–69. doi:10.1046/j.0266-4909.2003.00006.x

Chiappetta, E. L., & Russell, J. M. (1982). The relationship among logical thinking, problem solving instruction, and knowledge and application of earth science subject matter. *Science Education*, *66*(1), 85–93. doi:10.1002/sce.3730660111

Cobb, P., & Steffe, L. (1983). The constructivist researcher as teacher and model builder. *Journal for Research in Mathematics Education*, *14*(2), 83–94. doi:10.2307/748576

Daniel, J. S. (1995). *The mega-universities and the knowledge media: implications of new technologies for large distance teaching universities* (Unpublished master's thesis). Concordia University, Montreal, QC, Canada.

de Freitas, S., Savill-Smith, C., & Attewell, J. (2006). *Computer games and simulations for adult learning: case studies from practice*. London, UK: Learning and Skills Research Centre.

Dillenbourg, P. (1999). What do you mean by collaborative learning? In P. Dillenbourg (Ed.), *Collaborative-learning: Cognitive and computational approaches* (pp. 1–19). Oxford, UK: Elsevier.

Dillenbourg, P., Schneider, D., & Synteta, V. (2002). Virtual learning environments. In *Proceedings of the 3rd Congress on Information and Communication Technologies in Education*, Rhodes, Greece (pp. 3-18).

Edelson, D. C., Gordin, D. N., & Pea, R. D. (1999). Addressing the challenges of inquiry-based learning through technology and curriculum design. *Journal of the Learning Sciences*, *8*(3-4), 391–450. doi:10.1080/10508406.1999.9672075

Engelhardt, P. V., Rebello, N. S., Corpuz, E., & Ozimek, D. (2003). The teaching experiment: What it is and what it isn't. In *Proceedings of the Physics Education Research Conference* (pp. 157-160).

ENVKIDS. (2009). *Environmental sustainability training for children through on-line simulation, exploration, and collaboration*. Retrieved from http://envkids.net

Han, S., & Bhattacharya, K. (2001). Constructionism, learning by design, and project-based learning. In M. Orey (Ed.), *Emerging perspectives on learning, teaching, and technology*. Bloomington, IN: Association for Educational Communications and Technology. Retrieved from http://www.coe.uga.edu/epltt/LearningbyDesign.htm

Haury, D. L. (1993). *Teaching science through inquiry*. Retrieved from ERIC database. (ED359048)

Johnson, D. W., & Johnson, R. T. (1996). Cooperation and the use of technology. In D. H. Jonassen (Ed.), *Handbook of research for education-al communications and technology* (pp. 1017–1044). New York, NY: Simon and Schuster Macmillan.

Kudryavtsev, A., Krasny, M., Jahi, S., Doroshenko, M., & Usova, N. (2006). Communication across continents: Integrating local and global understanding in environment and sustainable development. *Connect, UNESCO Newsletter*, 1-5.

Looi, C. K. (1998). Interactive learning environments for promoting inquiry learning. *Journal of Educational Technology Systems*, *27*(1), 3–22. doi:10.2190/L4Q4-8QMM-QPJ3-B5LJ

Njoo, M. K. H., & De Jong, T. (1993). Exploratory learning with a com-puter simulation for control theory: Learning processes and instructional support. *Journal of Research in Science Teaching*, *30*(8), 821–844. doi:10.1002/tea.3660300803

Nomura, K. (2004). Utilization of IT/ICT for environmental education. *World School Network*. Retrieved from http://www.wschool.net/

Prensky, M. (2001). *Digital game-based learning*. New York, NY: McGraw-Hill.

Resnick, M. (2007a). Sowing the seeds for a more creative society. International Society for Technology in Education, 18-22.

Resnick, M. (2007b). All I really need to know (about creative thinking) I learned (by studying how children learn) in kindergarten. In *Proceedings of the Conference on Creativity and Cognition*, Washington, DC (pp. 1-6).

Robin, M. (n.d.). *Using storytelling to enhance student learning*. Retrieved from http://www.heacademy.ac.uk/assets/York/documents/resources/resourcedatabase/id471_using_storytelling_to_enhance_learning.pdf

Saunders, W. L., & Shepardson, D. (1987). A comparison of concrete and formal science instruction upon science achievement and reasoning ability of sixth grade students. *Journal of Research in Science Teaching*, *24*(1), 39–51. doi:10.1002/tea.3660240105

Smith, B. L., & MacGregor, J. T. (1992). *What is collaborative learning? State College*. PA: National Center on Postsecondary Teaching, Learning, and Assessment at Pennsylvania State University.

Tsalapatas, H., Heidman, O., Alimisi, R., Tsalapatas, S., & Houstis, E. (2010). Virtual experiments towards explorative and collaborative environmental primary education. In *Proceedings of the Europrix Academic Network Conference*, Graz, Austria.

Tsalapatas, H., Heidmann, O., Alimisi, R., & Stav, J. B. (2010). *The EnvKids explorative and collaborative didactical framework*. Retrieved from http://ohmpro.org/envkids/index.php?option=com_content&view=article&id=68&Itemid=71

Tsalapatas, H., Mogli, M., Karagiannis, H., Tallvid, K., Protivova, Z., Heidmann, E., & Larsson, S. O. (2010). *The EnvKids explorative and collaborative learning requirements*. Retrieved from http://ohmpro.org/envkids/index.php?option=com_content&view=article&id=68&Itemid=71

Using ICT for environmental education and youth empowerment in Lebanon. (2007). *DOT-Comments, 5*.

Verdejo, M. F. (1996). Interaction and collaboration in distance learning through computer mediated technologies. In T. T. Liao (Ed.), *Advanced educational technology: Research issues and future technologies* (pp. 77–88). Berlin, Germany: Springer-Verlag. doi:10.1007/978-3-642-60968-8_4

White, B. Y., & Frederiksen, J. R. (1998). Inquiry, modeling, and metacognition: Making science accessible to all students. *Cognition and Instruction*, *16*(1), 3–118. doi:10.1207/s1532690xci1601_2

Zollman, D. (1990). Learning cycles for a large enrollment class. *The Physics Teacher*, *28*(1), 20–25. doi:10.1119/1.2342921

This chapter was previously published in Learning with Mobile Technologies, Handheld Devices, and Smart Phones edited by Zhongyu (Joan) Lu, pages 147-161, copyright 2012 by Information Science Reference (an imprint of IGI Global).

Chapter 84
Reverse Logistics for Sustainable Waste Management Processes

Fraser McLeod
University of Southampton, UK

Tom Cherrett
University of Southampton, UK

ABSTRACT

This chapter highlights some of the innovative approaches that have been taken by businesses involved in reverse logistics for the removal of waste from urban areas. The chapter reviews some of these approaches and suggests which could be used more widely, recognising the specific limitations which may restrict their applicability. These innovative approaches include: the use of delivery vehicles to take-back waste/recyclate to out-of-town facilities such as a freight consolidation or recycling centre; combining commercial and household waste collections; deploying public transport vehicles to carry specialist recyclate; using multi-modal transport; 'smart' bin technology and pipelines for the removal of waste from buildings.

INTRODUCTION

This chapter is concerned with how waste management processes may be improved through the use of innovative ideas in reverse logistics. The main focus here is on the physical transportation of waste, rather than on the management and computer systems that may also be required to support these activities. The chapter begins by describing some of the basic concepts and typical methods found in waste management practice, highlighting the main practical, organisational, financial and legal issues. The chapter then describes a number of different innovative methods that have been introduced in recent years and assesses their advantages and disadvantages. The objective of the chapter is to provide a critical review of these innovative methods and to examine their scope for more widespread application.

DOI: 10.4018/978-1-4666-4852-4.ch084

BACKGROUND

Landfilling waste is generally considered to be unsustainable due to concerns about leachate contaminating ground water, the length of time needed to stabilize materials and the potential for methane to contribute to the global greenhouse effect (Das *et al.*, 2002). Recognising this, and driven by a series of European directives, the UK government produced a 'Waste Strategy for England 2007' (Defra, 2007), which promoted more sustainable waste management processes through reduction, re-use and recycling of waste, recovery of energy and reduced landfilling of waste.

Reverse logistics refers to the "planning, implementation and control of the movement of goods, raw materials and related items from the point of consumption to the point of origin, for the purpose of recapturing value or proper disposal." (Rogers and Tibben-Lembke, 1999). Typically, the starting points for waste (or return goods) are commercial premises in urban areas (e.g. High Street shops). The end points for waste depend upon the recycling or waste treatment options being used. Waste treatment options include incineration, pyrolysis, gasification, composting, anaerobic digestion and landfilling (Williams, 2005). Materials that are most commonly recycled include cardboard, paper, metals, plastics, textiles, wood and glass, with cardboard and paper typically comprising around two-thirds of the commercial waste from High Street retailers (Maynard and Cherrett, 2010). There are often intermediate points in the reverse supply chain for the consolidation of wastes which could be at regional distribution centres (RDCs), transfer stations, or other locations, before onward transportation to final disposal. The reverse logistics channels for return goods are likely to be different from those for waste, adding complexity to the combined problem.

The alternative disposition routes used for waste and for return goods tend to add to the overall transport burden associated with waste logistics, so innovative methods to reduce this are welcome. Improvements in terms of reduced transport costs or improved recycling performance may come from improved co-ordination of reverse logistics processes and greater co-operation between the parties involved (e.g. retailers). A lack of co-ordination, sub-optimal use of vehicles and low levels of waste/recyclate backloading (use of delivery vehicles to collect waste), can result in unnecessary handling and associated increases in vehicle traffic (Shakantu *et al*, 2002).

EXISTING PRACTICE

In the UK, commercial waste is collected either by a national or local contractor, by the local authority, where they offer such a service, or in-house, using the business' own vehicles; in some cases, commercial waste may be taken, often illegally, to household waste recycling centres (Maynard and Cherrett, 2010). The frequency of collection depends on the size of the business, with the largest typically receiving daily collections and the smallest often requiring one collection each week; an average of 2.4 collections per week was estimated in a study of 83 businesses in the city of Winchester, UK (Maynard and Cherrett, 2010). The typical collection arrangements for commercial recyclable waste and residual waste from a survey of businesses on Winchester High Street are shown (Figure 1).

For returned goods, the main reverse logistics methods seen in practice (Mukhopadhyay *et al.*, 2006) are:

1. Use of delivery vehicles to take returned goods from stores to the RDC;
2. Use of delivery vehicles to take returned goods from stores back to the suppliers, in exchange for credit;
3. Use of a third party logistics (3PL) provider to take goods to the RDC; and

Figure 1. Summary of waste management systems used by businesses on Winchester High Street (Source: Maynard and Cherrett, 2010)

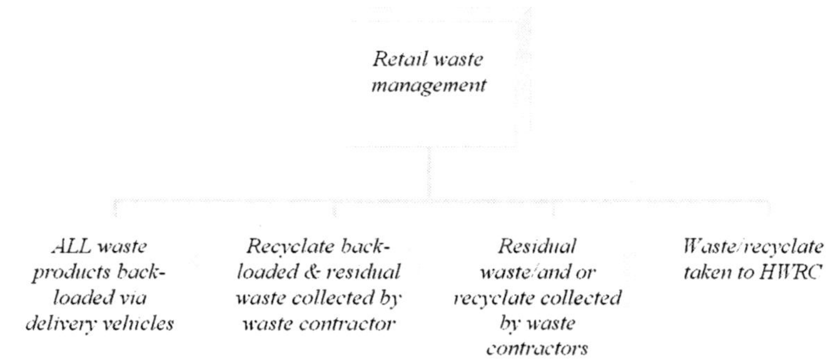

4. Outsourcing the entire returns management process to a 3PL, including refurbishment of the goods.

Where the retailer undertakes its own refurbishment of returned goods, the role of checking the goods to assess their initial status, suitability for resale and potential movement into the waste/material recovery stream may take place either at the individual store or at the RDC, depending on the level of expertise available, with the latter being most appropriate where the volume of return goods is high.

A perceived problem with existing practice is the fragmented nature of waste/recyclate collections, with a large number of waste contractors offering varying and competing waste collection services, resulting in many, part-loaded vehicles being used. In a study of businesses in Winchester (UK), 17 different waste contractors were observed operating within the city (Maynard and Cherrett, 2010). This is unlikely to be efficient from a transportation perspective and contributes to High Street congestion. The problem may be ameliorated by removing waste and/or returned goods using vehicles that already serve the city centre, such as delivery or service vehicles, e.g. refuse collection fleets used for serving domestic premises.

Levels of backloading, either for waste or return goods tend to be rather low, particularly for smaller businesses: a survey of 51 retailers across the county of West Sussex (UK) suggested that 31% of businesses 'always' had their packaging waste collected by delivery vehicles, 16% 'sometimes' and 53% 'never'; and, for return goods, 39% of businesses always had their return goods collected by delivery vehicles, 57% 'sometimes' and only 4% 'never' (Cherrett and Hickford, 2005); while across 83 businesses in Winchester (UK), 24% indicated that they used delivery vehicles to take back some element of waste and/or recyclate. Larger businesses, and, in particular, the major supermarket chains, are generally active in backloading. In most cases delivery vehicles are used to return packaging and, in some cases, other waste materials, to their depots for subsequent recycling or they can directly feed into recycling facilities either managed by supermarkets or by a third party.

There are a number of factors that determine whether backloading waste and recyclate using the delivery fleet is viable, including the size of the vehicle, vehicle utilisation, space and facilities at the location waste is to be taken to, and the frequency of vehicle deliveries in relation to the waste collection requirements. While it may be desirable to see levels of backloading increase, it should be recognised that there are a number of reasons why more backloading does not take place in practice, including: prioritising the outbound delivery service and not risking the possibility that backloading activities will cause delays; tight scheduling of return journeys; lack of knowledge of the backloading opportunities that may exist; and the use of specialised vehicles (e.g. temperature controlled compartments) or handling equipment that may be unsuitable for general backloading work (McKinnon and Ge, 2006). An analysis of grocery supply chains suggested that suitable backloads could be found for only 2.4% of empty journey legs, representing 2% of overall mileage (McKinnon and Ge, 2006). Multi-drop operations are often not well suited to returns management due to the numbers of time-critical deliveries made, the pressure on vehicle scheduling and possible cross-contamination (Fernie and Hart, 2001). Vehicle scheduling is a complex task, particularly where there are constraints such as delivery time windows. Incorporating backloading makes vehicle scheduling yet more complex and may act as a barrier to wider take-up but algorithms to solve variants of the combined 'backhauling and linehauling' problem have been proposed by Mitra (2005), Reimann and Ulrich (2006), Goetschalckx and Jacobs-Blecha (1989).

Another perceived problem is poor recycling performance, particularly for smaller businesses: the Winchester surveys indicated that 64% of businesses did not receive any dedicated recyclable waste collections. This may be due to financial reasons (cost of collection; cost of separating the waste), lack of space or difficulty in finding a suitable recycling service. Space issues may be influenced by the types and sizes of containers used to store waste. In Winchester, the most common receptacle used was an 1100L bin, while some businesses were required by their waste contractor to use sacks. It was observed that where waste contractors offered a recycling service, this was limited to a single day of the week, which may not suit all businesses. From the waste contractor's perspective, insufficient demand is often a core issue impacting on service frequency.

LEGAL CONSTRAINTS

The waste management activities that retailers and waste contractors can undertake are governed by various national and European directives, laws and regulations. The main issues are summarised in this section.

Duty of Care

Retailers and carriers are under a 'duty of care' to ensure that waste is stored, treated and disposed of safely and that incompatible wastes are not mixed. All carriers of waste in the UK must be registered with the Environment Agency, with some exemptions (e.g. charities, local authorities). Where waste is transferred from one company (e.g. a retailer) to another (e.g. a waste contractor) then waste transfer notes describing the waste type and origin are completed, signed and required to be kept for a period of at least two years by both parties.

Pre-Treatment Requirements

As part of the Landfill Directive (1999/31/EC), all waste destined for landfill must first undergo treatment in the form of either a physical, thermal or biological process. Commercial waste producers can meet these requirements by either separating different waste streams for recycling, or by making arrangements for their waste contractors to treat the waste before disposing of it. In addition, co-disposal of hazardous waste with non-hazardous waste at landfill sites is banned, as is the disposal of liquid wastes, used tyres, healthcare and infectious clinical wastes, animal-by-products, explosive, corrosive, flammable, oxidising and chemical wastes (Triantafyllou and Cherrett, 2010a).

Hazardous Wastes

Producers of hazardous wastes must comply with stringent regulations ensuring safety and proper treatment and disposal. Hazardous wastes are defined on the basis of the 'European Waste Catalogue (EWC) 2002' and include fluorescent tubes (containing mercury), non-edible oil and fat, certain medicines and clinical waste, batteries and electrical and electronic equipment containing hazardous wastes. Hazardous waste often requires specialist collections and this can lead to a proliferation of vehicles used: in a study of Southampton's West Quay shopping mall, it was estimated that hazardous waste collections were made by 65 different waste contractors, although the vast majority (90%) of these were electrical/maintenance contractors replacing fluorescent tubes during routine maintenance visits a few times a year (Triantafyllou and Cherrett, 2010b).

Waste Electrical and Electronic Equipment

Under waste electrical and electronic equipment (WEEE) regulations, retailers and distributors are required to inform customers of how they can dispose of WEEE free of charge. The retailer/distributor can either do this by taking goods from customers directly (in-store, through the post, or collection from the customer) or by joining the Distributor Takeback Scheme (DTS), managed by Valpak Ltd on behalf of the UK government. This scheme allows customers to return WEEE to a designated collection facility (DCF), which comprise civic amenity sites (household waste recycling centres, waste transfer stations), not-for-profit/charity organisations and commercial sites. Although retailers/distributors who join the scheme must make financial contributions towards the provision of the WEEE collection facilities, this may be a more cost-effective option than organising their own reverse logistics. The treatment, recovery, recycling and disposal of WEEE is the responsibility of the producers of the goods, who are required to join a producer compliance scheme (PCS), approved by the Environment Agency. The PCSs are required to collect WEEE from all DCFs free of charge (Ongondo and Williams, 2011). The distances that WEEE has to travel are highly dependent on the goods type and location of the treatment facilities: as an example, mobile phones purchased from a shopping centre in Southampton were typically returned to a processing hub in Essex some 113 miles away (Triantafyllou and Cherrett, 2010b).

Packaging and Packaging Wastes

Packaging regulations set design standards with the aim of increasing recovery, reuse and recycling levels of commercial packaging and minimising use of hazardous materials. Packaging waste regulations require UK businesses with an annual turnover of more than £2 million and handling more than 50 tonnes of packaging annually to reach specified recycling and recovery targets and to set up systems to allow customers to return packaging. The targets for 2011 and 2012, set by Defra (2010), for various materials, are shown alongside the achieved recycling figures for 2009 (Figure 2). These latest available statistics show that recycling targets for wood, aluminium and paper are already being achieved, whereas improvements are demanded in the recycling of plastics, steel and glass.

CONCEPTS FOR IMPROVING THE EFFICIENCY OF REVERSE LOGISTICS PROCESSES

Backloading to an Out-of-Town Facility

The UK Department for Transport has promoted backloading through its Freight Best Practice programme (Freight Best Practice, 2005), high-

Figure 2. UK business recycling targets (for 2011 and 2012) and achieved figures for 2009 (Source of data: Defra, 2010)

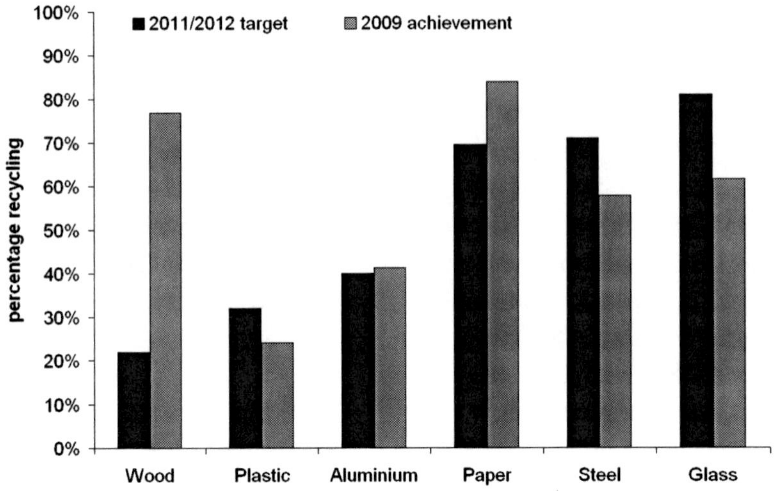

lighting the potential efficiency gains to be made. Opportunities for backloading waste appear to exist, given the 2009 road freight statistics which showed that empty running across rigid lorries was 28.9% with an average 50% lading factor (ratio of actual goods moved to the maximum tonne-kms achievable if the vehicles, whenever loaded, were loaded to their maximum carrying capacity) (Department for Transport, 2010). Several online load matching companies have been created to facilitate backloading (e.g. www.truckspace.co.uk, www.haulageexchange.co.uk and www.courierexchange.co.uk). These allow hauliers and logistics providers to find and bid for backloading work by providing visibility of compatible loads across different supply chains.

Existing backloads most commonly involve a retailer returning goods or waste packaging from their own stores back to an RDC for subsequent re-use, recycling or disposal. However, an alternative option, which may reduce the transport footprint associated with waste management in urban centres, is to use delivery vehicles to take waste to an out-of-town facility, such as a consolidation centre, a recycling facility or a waste transfer station. There are already examples of this and in the UK, in 2008, it was reported that the Broadmead consolidation centre had received some 14.5 tonnes of consolidated cardboard and plastic from 64 retailers in Bristol (Hapgood, 2008).

The advantages and disadvantages of using consolidation centres have been described in detail by Browne *et al.*, (2007), by the EU-funded BESTUFS project (2007) and by TTR (2010), among others, mainly from the perspective of the forward distribution channel, although similar considerations may also apply to reverse logistics. The main anticipated benefit is a reduction in the number of freight vehicles, particularly large and/or polluting ones, entering busy central areas. Reported statistics for the Bristol Broadmead consolidation centre, in 2007, included a 75% reduction in vehicle movements, equating to an annual savings of 178,000 vehicle kilometres, 20.3 tonnes of CO_2, 660 kg of NOx and 19.7 kg of PM_{10} (Hapgood, 2008). A potential drawback is that the location of the consolidation centre may require some vehicles to divert considerably from their normal routes, incurring additional cost and time.

To provide an example of the requirements for a consolidated take-back scheme, an analysis of the cardboard collection requirements of Winchester High Street businesses was made (Maynard and Cherrett, 2010). This indicated that each business generated around 1300 litres of cardboard each week, on average, equating to a total for the whole High Street of around 139,000 litres of cardboard, or 131 roll-cage equivalents. From this, it was then estimated that 15 collections per week would be needed to remove waste from the High Street, based on the use of rigid-bodied delivery vehicles with a gross vehicle weight of 18 tonnes and an average spare capacity of 26% of the deck area ($5.6m^2$), taken from Department for Transport (2006) statistics. The storage requirements, prior to collection, would depend on the frequency of collections: for example, if collections were only once a week then a minimum storage area of $83m^2$ (131 roll cages x $0.63m^2$ per roll cage) would be required. However, in practice, some businesses may choose not to participate in such a co-operative take-back scheme, as they may already backload their own waste or may be satisfied with their existing arrangements, so the collection requirements may be lower than this.

Consolidation of return goods, particularly small items, may also be facilitated through the emerging networks of attended and unattended collection-delivery points (McLeod and Cherrett, 2009). The use of attended premises, typically convenience stores, petrol stations and post offices (e.g. Kiala, Royal Mail, Parcel Force, Redpack Network) or unattended locker banks or boxes (e.g. ByBox, Bearbox) for returning products and recyclate back up the supply chain are already well established in the service industry. These systems are being used successfully by the service sector whereby used parts destined for reconditioning or cannibalisation can be returned to service centres via an attended or unattended collection point in the vicinity of the customer, reducing the amount of time the service engineer has to spend collecting and returning parts.

Coordinated Waste Collection from a High Street

The existing fragmented nature of waste collections from retailers in urban areas could be ameliorated with better co-ordination and centralisation of activities, similar in concept to the co-ordination that normally takes place at shopping malls, where all waste is centrally managed. If High Street retailers worked together, with support from the local authority, they could form common waste collection arrangements. This co-ordination would result in fewer waste collection vehicles being used and greater utilisation of those vehicles. Recycling performance may also improve as was seen for centralised co-ordination of returned goods at a shopping centre in Southampton which increased the amount of waste diverted from landfill from a mean of 26 tonnes per month to 143 tonnes per month between the periods January 2006 to April 2007 and May 2007 to August 2008, respectively, using a range of equipment to store and consolidate waste (pallets, roll cages, bins, balers and compactors) (Triantafyllou and Cherrett, 2010a).

A survey of 83 Winchester High Street business managers suggested that 45% of them would consider collaborating with their neighbours to improve material recovery (Maynard and Cherrett, 2010). Whether they would decide to do so would likely depend on:

- Perception of the anticipated benefits in terms of cost savings or environmental performance.
- Availability of storage space and vehicle access.
- Organisational issues associated with how the scheme would be operated.
- How the scheme meets their legal obligations.
- Their willingness to collaborate with other retailers who may be potential competitors.
- Financial values of recyclable materials (currently in decline).

- Whether local or central government would provide incentives (e.g. tax breaks, priority access for delivery vehicles) to join such as scheme.

Collection of Commercial Waste alongside Household Waste

A potential way to improve waste logistics in an urban area is to combine waste collections from commercial premises, particularly small businesses, with household waste collections. Since household collections are the responsibility of waste collection authorities (WCAs) in the UK, this could only be undertaken by WCAs, or their appointed waste contractors. Such combined collections could be targeted towards recyclate or could also include residual waste. The main aims of combining collections in this way are to increase the amount of recyclable materials that are recovered, reduce the number of collection vehicles travelling in urban areas and to reduce the vehicle mileage associated with these collections. In a study of waste collections from 577 commercial premises and 25,600 households in Winchester, involving detailed modelling of existing rounds and development of new round designs, a vehicle mileage savings of 4.7% was estimated for combined collections compared to the existing separate vehicle rounds (McLeod and Cherrett, 2010). The results also suggested that the new round designs could accommodate collections from an additional 288 commercial customers per week, indicating scope for expansion of the existing business.

The number of WCAs offering commercial waste services (either collection and/or bring site provision) has increased in recent years with research suggesting that 65% and 43% of WCAs in England offered residual trade waste and recyclable trade waste services, respectively, in 2010 (WRAP, 2010). Perceived disincentives to collect trade waste are that while residual trade waste counts towards landfill targets, recyclable trade waste may not be directly counted towards recycling targets. This may change as the UK government has announced its intention to review definitions of commercial and municipal waste. Fully or near-fully integrated commercial and household collections in the UK are only undertaken by a few WCAs, including New Forest, Newham, Caerphilly and Ceredigion. The vast majority of WCAs undertake these collections separately from one another or only collect a small amount of commercial waste on the household rounds where it is expedient to do so (e.g. in remote areas) (Sanders, 2010). Although combining commercial and household collection rounds may have the potential to create more efficient vehicle rounds, this may not be suitable for all WCAs. For example, such a concept is not appropriate when a 'kerbside sort' system, using compartmentalised vehicles, is employed for household waste collections. Also, combined collections can be difficult to implement when an alternate weekly collection system (recyclable waste and residual waste collected from households in alternate weeks) is in use, since the commercial collections normally have to be made weekly, or more frequently. Other potential difficulties may arise if there are fundamental differences in how existing household and commercial rounds operate (e.g. start times, shift lengths, types of vehicle and lifting equipment).

Innovative Systems for Waste Logistics

Other innovative ideas for removing waste and recyclate from urban areas include the following.

Use of Public Transport Vehicles

There may be scope for using some public transport vehicles (trains, trams, buses) to carry certain elements of waste, thereby reducing the use of lorries in urban areas and conferring benefits in terms of fuel use, emissions and road congestion. Perhaps the best-known practical example is the Cargotram system in Zurich, introduced in 2003 (BESTUFS, 2005). The Cargotram system serves nine stations on the existing tram network, with a four-weekly service at each station, and carries bulky waste and WEEE, placed in standard refuse containers and loaded onto flat wagons, to a waste transfer station. The service is provided, free of charge, to public transport users, local residents, cyclists and pedestrians. The Cargotrams operate between 3pm and 7pm each day and are timetabled to avoid causing any delays to passenger trams. The main driving force behind this scheme was the municipal waste disposal and recycling authority, with the support of the public transport operators. Although this scheme does not take commercial waste, one can envisage the expansion of such a concept to incorporate carriage of commercial waste at non-peak times. Similar concepts can be envisaged for light rail systems. Trains are used to transport larger container-loads of waste and can support multi-modal operations. The concept of using buses for the dual purpose of transferring waste materials naturally raises legitimate concerns with regard to public safety. Such a system might only be suitable for the movement of specialist recyclate returns (e.g. small-sized engineering parts, as part of a service company returns network) but there will always be issues of segregating passengers and material on the same vehicle, available capacity and timetable compatibility. Dual use of rural buses to carry passengers and freight goods was reported in some regions of Germany (Schaefer, 2003), while in the UK the Royal Mail has a 'Postbus' service (www.postbus.royalmail.com) which allows passengers to travel on some of its postal vehicles; however, both of these cases are rather small scale (few routes, limited days of the year, restricted capacity, rural areas only) and it is difficult to assess the scalability of this concept to large urban areas.

Use of Inland Waterways or Short Sea Shipping

Existing examples of waste being moved by barges on inland waterways include use of canals and the River Thames in London (STRAW, 2006), with further plans to use inland waterways to remove waste from the Olympic Park during the 2012 Games (London2012.com, 2009), and on the River Meuse in Liege, Belgium (BESTUFS, 2005); however, there is considerable potential for waterborne transport, including short sea shipping, to be used more extensively for freight and this may become increasingly important in the future with the distances that waste has to be transported likely to increase as the number of waste management facilities grows (STRAW, 2006). Use of waterborne transport for carrying waste requires suitable loading/unloading areas, vehicles and lifting equipment (Figure 3). As with the use of public transport vehicles, the main perceived benefits are the reduction in truck use in city centres with associated reductions in emissions and energy use.

Figure 3. Roll-on-off waste container being loaded onto a barge by a standard hook-lift vehicle (Source: Wood, Hall, Heward Ltd., London)

Use of Multi-Modal Vehicles

Where waste is to be transferred from a road vehicle to another mode of transport (e.g. train, barge) then an innovative and efficient way to do this is the use of a so-called multi-modal refuse collection vehicle (MMRCV), which has a demountable container and loads either from the side, front or rear. This type of vehicle was introduced in waste collection practice in 2003, most notably in the Netherlands and Germany, and, in 2008, was reported to be collecting around 40% of Germany's municipal waste (Transport for London, 2008). Use of such vehicles in the UK is not yet common, with fully-implemented examples only in Aberdeenshire and Bromsgrove and trials in London and Horsham (Transport for London, 2008). It was reported, by Transport for London (2008), that Aberdeenshire Council had been able to replace its fleet of 37 standard refuse collection vehicles with only 19 side-loading MMRCVs due to the efficiencies gained from being able to replace full containers with empty ones. In trials undertaken by Transport for London, a rear-loading MMRCV was preferred, as side-loading would be difficult in narrow streets full of parked cars. Their trials identified benefits in terms of reduced mileage, faster loading speeds and elimination of the need for waste transfer stations and 'double' handling of waste.

Use of Pipelines

A system introduced in Helsinki in 2001 by Puzair Ltd, a Finnish company specialising in environmental and building maintenance technology, used specially installed pipelines and pneumatic equipment to transport waste bags from apartment blocks and a neighbouring shopping centre to communal bins, with the installation taking place at the same time as the construction of the flats and shopping centre (BESTUFS, 2005). This rationalisation of waste collection was reported to involve fewer truck trips, improve working conditions for drivers and save residents the nuisance associated with trucks visiting their courtyards. Subsequently, a 'biowaste' version of this system was developed, designed to remove food and other biological waste from buildings such as restaurants, canteens, hospitals, shopping centres, airport terminals and residential apartment blocks (www.ecosir.com). Issues that would affect the suitability of such a system include cost (not reported) and how readily it can be installed in existing buildings.

Use of Smart Bins

Where waste collections are made from communal bins with varying fill rates, it would be useful to know the point at which bins are reaching capacity (>75% full) in order to schedule vehicle collections more efficiently. The aims are to avoid scheduling visits to bins that are under a threshold fill level (e.g. 50%) and to ensure that bins are visited before they overflow, as this can lead to waste being left outside the container, illegally dumped elsewhere or in the case of valuable recyclate, losing out on a potential donation to another recycler. This understanding has led to the innovative idea of equipping bins with some form of detection sensor and communications device to report this information to a central office. The advantages of dynamic waste collection schedules over fixed schedules were demonstrated, both analytically and using computer simulation, in Malmö, Sweden, where 3300 recycling containers were fitted with level sensors and wireless communication equipment (Johansson, 2006). For dense collection areas it was concluded that the best policy was to schedule bins to be collected when full, whereas for less dense areas it became more expedient to include collections from 'almost' full bins as well. The authors noted that, in practice, despite having sensor information available to them, many waste collection managers had not used the information to adapt their schedules, suggesting, perhaps, a need for training in being able to use dynamic information effectively. In the UK, similar technology was reported (www.smartbin.com) to have reduced vehicle trips by over 37% for a company making collections from textile banks. The data collected from such remote monitoring systems may also be useful for reporting purposes and in identifying occasions of theft, where materials have some value.

FUTURE RESEARCH DIRECTIONS

The 'Internet of Things' (Valhouli, 2010) means that, through ubiquitous forms of computing, all devices, items, products can be perpetually 'on' and connected to the internet. The developments in this area have significant implications for reverse logistics practices in the future. The ability to fit objects with smart tags and transmitters means that individual items (such as a bin, or a specific part in a photocopier) can have an IP (Internet Protocol) address, enabling them, with minimal processing power, to communicate with other 'things' in their vicinity (thing-to-thing), to be interrogated by users (person-to-thing) or inform of their situation/condition (thing-to-person). This means that the analogue bar code, used uniformly across the retail sector will enable 'things' to be tracked from cradle to grave using open platforms. Individual 'end-of-life' items will be able to report their status, identify the nearest refurbishment point and potentially find links with other similar items in the vicinity which could constitute a full load for collection.

As new treatment options for waste and return goods become available, there is an increased need for identifying and monitoring individual items within the reverse supply chain. The enhanced 'visibility' of 'things' within the reverse supply chain afforded by this technology could vastly improve when and how reverse movements occur, and further research into the impacts of this is warranted. This also creates a need for research to evaluate the most promising reverse logistics options from the point of view of feasibility, practicality, economic sustainability, material recovery, transport efficiency and environmental impact. In this regard, there will always be trade-offs between conflicting objectives. Increased recycling activity often results in increased transport, with negative impacts on fuel use and emissions and such negative impacts need to be accounted for in any scheme appraisal. Areas for future research should include assessment of collaborative transport mechanisms such as joint domestic and commercial waste collection schemes, co-ordinated take-back across centralised distribution systems, use of non-road transport and smart bin technology to improve waste collection scheduling.

CONCLUSION

This chapter has described innovative approaches for the removal of waste and/or returned goods from urban centres. Some of these have been based on technological innovations, such as smart bins, pneumatic pipes and multi-modal vehicles with demountable containers, while others have been based on improved collaboration and methods of working between retailers, waste contractors and local authorities, with central government also having a key role to play in enabling and/or encouraging changes to be made. Redefining how commercial waste contributes to recycling and LATS targets could provide a greater incentive for local authorities to collect trade waste which may improve the currently fragmented nature of commercial waste collection. Similarly, this fragmentation may be ameliorated if there was improved co-ordination and co-operation between High Street retailers in organising how their combined wastes are dealt with.

For this to happen would most likely need the involvement of the local authority, or some other third party 'controller', to co-ordinate take-back of the consolidated waste streams through specific channels, so reducing the amount of separate contractor activity.

REFERENCES

Allen, J., & Browne, M. (2010). *Road freight transport and sustainability in Britain, 1984-2007.* Retrieved January 6, 2011, from www.greenlogistics.org/themesandoutputs/wm9/downloads/Sustainablefreighttransport-nationalfinal.pdf

BESTUFS. (2005). *Deliverable D2.1. Best Practice Handbook.* Retrieved November 19, 2010, from www.bestufs.net/download/BESTUFS_II/key_issuesII/BESTUFS_BPH.pdf

BESTUFS. (2007). *Good Practice Guide on Urban Freight Transport.* Retrieved November, 2010, from www.bestufs.net/download/BESTUFS_II/good_practice/English_BESTUFS_Guide.pdf

Bridgewater, E., & Anderson, C. (2003). *CA Site WEEE Capacity in the UK. Network Recycling.* Retrieved January 6, 2011, from www.dti.gov.uk/files/file30304.pdf

Browne, M., Woodburn, A. & Allen, J. (2007). Evaluating the potential for urban consolidation centres. *European Transport [Trasporti Europei], 35*, 46-63.

Cherrett, T. J., & Hickford, A. J. (2005). Freight and the economy. The effects of freight movements associated with retailers common to Chichester, Horsham, Worthing and Crawley. Transportation Research Group, University of Southampton. Final report for West Sussex County Council.

Das, K. C., Smith, M. C., Gattie, D. K., & Hale Boothe, D. D. (2002). Stability and quality of municipal solid waste compost from a landfill aerobic bioreduction process. *Advances in Environmental Research, 6*(4), 401–409. doi:10.1016/S1093-0191(01)00066-1

Defra (2007). *Waste strategy for England 2007.* Retrieved January 6, 2011, from www.defra.gov.uk/environment/waste/strategy/strategy07/documents/waste07-strategy.pdf

Defra (2010). *Packaging*. Retrieved January 12, 2011, from www.defra.gov.uk/environment/waste/producer/packaging/

Department for Transport. (2006). *Key Performance Indicators for Non-Food Retail Distribution*. Freight Best Practice Benchmarking Guide.

Department for Transport. (2010). *Road freight statistics, 2009.* Retrieved January 12, 2011, from www.dft.gov.uk/pgr/statistics/datatablespublications/freight/goodsbyroad/roadfreightstatistics2009

Fernie, J., & Hart, C. (2001). UK Packaging Waste Legislation; Implications for Food Retailers. *British Food Journal, 108*(3), 187–197. doi:10.1108/00070700110386737

Freight Best Practice. (2005). *Make backloading work for you.* Retrieved January 6, 2011, from www.freightbestpractice.org.uk/make-backloading-work-for-you

Goetschalckx, M., & Jacobs-Blecha, C. (1989). The vehicle routing problem with backhauls. *European Journal of Operational Research, 42*(1), 39–51. doi:10.1016/0377-2217(89)90057-X

Hapgood, T. (2008). Short term actions to reorganize transport of goods. Presentation made at ITS Freight Interest Group meeting, University of Southampton, 27 May 2008.

Johansson, O. M. (2006). The effect of dynamic scheduling and routing in a solid waste management system. *Waste Management (New York, N.Y.), 26*, 875–885. doi:10.1016/j.wasman.2005.09.004

London2012.com (2009). *Recycling waste boosts sustainable transport target.* Press release 25 June 2009. Retrieved January 10, 2011, from www.london2012.com/press/media-releases/2009/06/recycling-waste-boosts-sustainable-transport-target.php

Maynard, S., & Cherrett, T. J. (2010). *Characteristics of Retail Waste Logistics on Winchester High Street*. Retrieved January 10, 2011, from www.greenlogistics.org/news/retailwastelogistics.htm

McKinnon, A. C., & Ge, Y. (2006). The potential for reducing empty running by trucks: a retrospective analysis. *International Journal of Physical Distribution & Logistics Management*, *36*(5), 391–410. doi:10.1108/09600030610676268

McLeod, F. N., & Cherrett, T. J. (2009). Quantifying the environmental benefits of collection/delivery points. *OR Insight*, *22*(3), 127–139. doi:10.1057/ori.2009.2

McLeod, F. N., & Cherrett, T. J. (2010). Encouraging sustainable 'take-back': The scope for joint domestic/commercial waste collections. *Proceedings of the Logistics Research Network conference (CD)*, Harrogate, 8-10 September, 2010.

Mitra, S. (2005). An algorithm for the generalized vehicle routing problem with backhauling. *Asia-Pacific Journal of Operational Research*, *22*(2), 153–169. doi:10.1142/S0217595905000522

Mukhopadhyay, S. K., & Setaputra, R. (2006). The role of 4PL as the reverse logistics integrator. *International Journal of Physical Distribution & Logistics Management.*, *36*(9), 716–729. doi:10.1108/09600030610710872

Ongondo, F., & Williams, I. D. (2011). *A critical review of the UK household WEEE collection network. Accepted for publication in*. Waste and Resource Management.

Reimann, M., & Ulrich, H. (2006). Comparing backhauling strategies in vehicle routing using ant colony optimization. *Central European Journal of Operations Research*, *14*(2), 105–123. doi:10.1007/s10100-006-0163-8

Rogers, D. S., & Tibben-Lembke, R. S. (1999). *Going backwards: Reverse logistics trends and practices*. Pittsburg, USA: Reverse Logistics Executive Council.

Sanders, R. (2010). *Local Authority Commercial Waste Collections: A more sustainable approach to waste management?* Unpublished M.Sc. dissertation, University of Southampton, UK.

Schaefer, C. (2003). *MULTIBUS – A new and innovative approach for bus systems in rural areas. Association for European Transport*. Retrieved January 10, 2011, from http://etcproceedings.org/paper/download/888

Shakantu, W., Tookey, J. E., & Bowen, P. A. (2002). *Defining the role of reverse logistics in attaining sustainable integration of materials delivery with construction and demolition waste management.* In Proceedings of Creating a Sustainable Construction Industry. Conference of CIB W10: Creating a Sustainable Construction Industry in Developing Countries, South Africa. Retrieved January 10, 2011, from http://buildnet.csir.co.za/cdcproc/docs/3rd/shakantu.pdf

STRAW. (2006). *Sustainable Transport Resources and Waste*. Retrieved November 19, 2010, from www.massbalance.org/downloads/projectfiles/2150-00494.pdf

Transport for London. (2008). *Developing a multi-modal refuse collection system for London. Project report by the Freight Unit.* Retrieved January 10, 2011, from www.tfl.gov.uk/microsites/freight/documents/publications/multi-modal-refuse-system-london-nov-2008.pdf

Triantafyllou, M., & Cherrett, T. J. (2010a). *Waste Management and 'Take-Back' Processes in a Dedicated Shopping Centre - Results from the West Quay shopping centre Managers Survey*. Retrieved January 10, 2011, from www.greenlogistics.org/SiteResources/MT-Waste Report-2010.pdf

Triantafyllou, M., & Cherrett, T. J. (2010b). The logistics of managing hazardous waste – A case study analysis in the retail sector. *Proceedings of the Logistics Research Network conference (CD)*, Harrogate, 8-10 September, 2010.

TTR. (2010). *Freight consolidation centre study*. Report for the Department for Transport by Transport and Travel Research Limited. Retrieved January 10, 2011, from www.dft.gov.uk/pgr/freight/research/freightreport/

Valhouli, C. A. (2010) *The Internet of Things: Networked Objects and Smart Devices*. The Hammersmith Group Research Report, February 2010. Retrieved February 23, 2011 from http://thehammersmithgroup.com/images/reports/networked_objects.pdf

Williams, P. T. (2005). *Waste Treatment and Disposal*. Chichester, UK: John Wiley & Sons, Ltd. Retrieved January 10, 2011, from http://onlinelibrary.wiley.com/doi/10.1002/0470012668.fmatter/pdf

WRAP. (2010). *Local authority trade waste and recycling survey 2010*. Retrieved January 10, 2011, from www.wrap.org.uk/downloads/TW_survey_report_v1.9.a428e3e4.9521.pdf

This chapter was previously published in Supply Chain Innovation for Competing in Highly Dynamic Markets edited by Pietro Evangelista, Alan McKinnon, Edward Sweeney, and Emilio Esposito, pages 224-237, copyright 2012 by Business Science Reference (an imprint of IGI Global).

Chapter 85
Energy from Waste: Present Scenario, Challenges, and Future Prospects towards Sustainable Development

Kalpana Arora
Indian Institute of Technology, India

Ashwani Kumar
Indian Institute of Technology, India

Satyawati Sharma
Indian Institute of Technology, India

ABSTRACT

Considering the confrontation of waste disposal and minimizing Green House Gas (GHG) emission, technologies of Waste To Energy (WTE) production seem appealing. It provides one key solution for two major concerns regarding energy crisis and waste management. Energy from biomass can be seen as a promising alternative for fossil fuels, which are getting scarce and more costly day by day. Since a significant amount of organic waste from agriculture, industries, and community sources is collected annually, it can be convertible to useful energy forms like biohydrogen, biogas, bioalcohols, etc., through various Waste-To-Energy Routes (WTERs) for sustainable development. The adoption of this WTE technology will help the world not only in saving the traditional energy resources, but also in reducing GHG emission, and lowering environmental impact. With all these advantages, WTE industry is expected to experience a noticeable growth in the coming years and make greater contribution in supplying renewable energy. The review presents the technical, economical, and environmental aspects of various WTE techniques and focus on the benefit that this thermochemical conversion is a step forward towards sustainable development.

DOI: 10.4018/978-1-4666-4852-4.ch085

INTRODUCTION

Lowering the severity of climate change by appeasing Green House Gas (GHG) emission and reducing fossil fuel based power production in a sustainable way are the prime concerns for the environmentalists. Enormous increases in energy demand due to developing social, economic, and industrial status leads to the degradation of the environment at threatening levels (Caton, et al., 2010). Synchronically, the waste generation is also aggrandizing continuously. Therefore, explorations of alternative energy strategies have recently become important, particularly for future world stability.

Considering above all apprehensions and a willingness to expand the domestic renewable energy base have driven the need for buttress energy and environmental strategies across the world. Renewable waste materials from agriculture (Mahamat, et al., 1989; Jain, et al., 1992; Abbasi, et al., 1991; Bardiya, et al., 1996), industries (Deivanai & Kasturi Bai, 1995; Lay, 2001; Shin, et al., 2004; Yu, et al., 2002; Yokoi, et al., 2001), and domestic (Van Ginkel, et al., 2005; Okamoto, et al., 2000; Lay, et al., 1999; Lay, 2001; Kalia, et al., 1994; Grady, et al., 1999) sources are convertible to useful energy forms like biohydrogen, biogas, bioalcohols, etc., through Waste-To Energy Routes (WTERs) a viable waste treatment option capable of delivering clean energy as compared to conventional fossil fuels (Kothari, et al., 2010).

Waste-To-Energy (WTE) processes recover the energy from the waste through either direct combustion (e.g., incineration, pyrolysis, and gasification) or production of combustible fuels in the forms of methane, hydrogen, and other synthetic fuels (e.g., anaerobic digestion, mechanical biological treatment, and refuse-derived fuel). Incineration and gasification are the two primary WTE technologies that have been used successfully throughout the world. It is estimated that about 130 million tonnes of MSW are combusted annually in over 600 WTE facilities worldwide, producing electricity and steam for district heating and recovered metals for recycling (Themelis, 2003). WTE incineration has long been accepted as a solid waste management option, complementing landfilling and composting (American Society of Mechanical Engineers, 2008; Denison, 1996; Themelis, 2003; United Nations Environment Programme, 1996).

WTE is gaining increasing popularity in throughout the world primarily for its ability to reduce the volume of solid waste that requires landfilling, it also lessens the dependency on fossil fuel and Greenhouse Gas (GHG) emissions. This article provides an overview of the situations and the development of WTE routes and its future potentials. The chapter is structured as follows. The next section Background provides an overview of the waste generation, its composition, and utilization techniques until now. Further, it gives an outlook about the issues, controversies, and problems related to the waste to energy conversion. Future research directions were discussed later proving that different WTE routes having potentials towards the energy crisis and sustainable environment.

BACKGROUND

Waste Generation and Energy Crisis: Global Problem

A clear appreciation of the quantities and characteristics of the waste being generated is a key component in the development of robust and cost-effective solid waste management strategies. Although amongst some of the more developed countries within the region the quantification and characterization of waste forms the basis for management and intervention, elsewhere little priority is given to the systematic surveying of waste arising and the quantities, distinctiveness, seasonal variations and future trends of waste generation are poorly understood.

Rates and composition of solid waste generation vary from country to country depending on the economic situation, industrial structure, waste management regulations, and life style. The accessibility and quality of data on solid waste generation as well as subsequent treatment also vary significantly from country to country. Solid waste is generated from households, offices, shops, markets, restaurants, public institutions, industrial installations, water works and sewage facilities, construction and demolition sites, and agricultural activities (Pipatti, et al., 2006). The data pertaining to solid waste generation is shown in Figure 1. In general, the developed countries generate much higher quantities of waste per capita compared to the developing countries of the region. However, in certain circumstances the management of even small quantities of waste is a significant challenge depending upon small land areas and resultant lack of disposal options. A breakdown of solid waste types and sources is provided in Table 1.

The advantages and limitations of the major waste disposal technology options, landfilling, composting, and incineration, are compared in Table 2. Incineration of solid waste in WTE fa-

Figure 1. Economics, environmental, and social statistics (source: OECD Factbook, 2008)

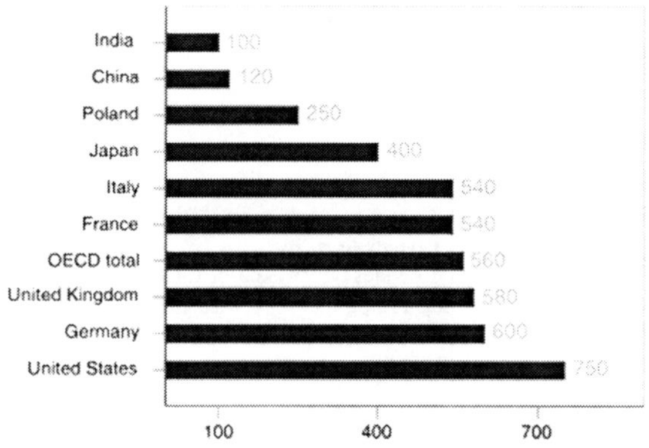

Table 1. Breakdown of solid wastes and its sources

Source	Typical waste generators	Types of solid wastes
Residential	Single and multifamily dwellings	Food wastes, paper, cardboard, plastics, textiles, leather, yard wastes, wood, glass, metals, ashes, special wastes (e.g. bulky items, consumer electronics, white goods, batteries, oil, tires), and household hazardous wastes
Industrial	Light and heavy manufacturing, fabrication, construction sites, power and chemical plants	Housekeeping wastes, packaging, food wastes, construction and demolition materials, hazardous wastes, ashes, special wastes
Commercial	Stores, hotels, restaurants, markets, office buildings, etc.	Paper, cardboard, plastics, wood, food wastes, glass, metals, special wastes, hazardous wastes
Institutional	Schools, hospitals, prisons, government Centers	Same as commercial
Construction and demolition	New construction sites, road repair, renovation sites, demolition of buildings	Wood, steel, concrete, dirt, etc.
Municipal Services	Street cleaning, landscaping, parks, beaches, other recreational areas, water and wastewater treatment plants	Street sweepings, landscape and tree trimmings, general wastes from parks, beaches, and other recreational area, sludge
Process	Heavy and light manufacturing, refineries, chemical plants, power plants, mineral extraction and processing	Industrial process wastes, scrap materials, off specification products, slag, tailings
All of the above should be included as "municipal solid waste."		
Agriculture	Crops, orchards, vineyards, dairies, feedlots, farms	Spoiled food wastes, agricultural wastes, hazardous wastes (e.g. pesticides)

Table 2. Comparison of the major MSW management technology options: landfilling, composting, and incineration

Technology	Advantages	Disadvantages
Landfilling	An universal solution that provides ultimate waste disposal; Relatively low cost and easy to implement; Complements with other technology options for handling the residual waste; Can derive landfill gas as a byproduct for household and industrial uses; Costs incurred incrementally as landfill expands.	Cost increases significantly with liner, leachate collection and removal system, and stricter regulations; Requires large area of land; Does not achieve the objectives of reducing volume of MSW and converting MSW into reusable resources; May result in secondary pollution problems, including groundwater pollution, air pollution and soil contamination; May serve as breeding ground for pests and diseases; Long post closure care obligations and unknowns exist, and sets long-term restrictions on site land use; Site location may be limited by the local geology and natural stability of the underground soil; Due to public acceptance and space limitation, landfills are often far away from the places where waste is generated, necessitating long distance transport of the waste.
Composting	Converts decomposable organic materials into an organic fertilizer; Reduces the amount of waste to be landfilled and integrates well with landfilling and materials recovery/recycling.	Takes up more space than some other waste management technologies; Can be costly to implement and maintain, and has no environmental or economic advantages compared to incineration; Requires waste size reduction and some degree of waste separation/processing; There are issues with public perception, such as odor and bioaerosol emissions during the composting process, and the control of disease producing organisms, weeds and insects; Quality of the fertilizer produced is low and volume is disproportionately large, resulting in poor market demand; Compost product may cause soil pollution by heavy metals and pathogens.
Incineration	Incineration Provides substantial reduction (by 90%) in the total volume of waste requiring disposal in landfill; Requires minimal pre-processing of waste; The bottom ash from incineration is biologically clean and stable, and can be used in road building and the construction industry; A very stable process, and virtually all wastes can be burned and the burning process can be adequately controlled; Heat from combustion can be used as energy source for generation of steam and/or electricity; Incineration facilities can be located near residential areas, thereby reducing costs of transporting MSW to locations of waste disposal; Air emissions can be well controlled; More optimal land use and more efficient integration of resources than landfilling.	High capital and operational and maintenance costs, compared to other, non incineration options; Significant operator expertise is required; Air pollution control equipment is required to treat the flue gas, and the fly ash needs to be disposed in hazardous waste landfills; More raw material have to be used to replace those that have been incinerated, and it does not save energy in the long run as resources are not recycled; May some time discourage recycling and waste reduction; Public perception is sometimes negative, primarily with dioxins emission.

Source: Cheng and Hu (2010)

cilities prevents the possible aqueous and gaseous pollution associated with landfilling and provides a source of reliable renewable energy. As a proven, environmentally sound technology, WTE has been used extensively in Europe and developed countries in Asia such as Japan and Singapore (Cheng & Hu, 2010).

Thermal treatment with heat recovery (Waste-to-Energy, WTE) is preferred due to a number of advantages. Naturally, the public acceptance of increased use of incineration necessitates effective elimination of the environmental impacts. Thus, the regulations concerning the quality of side products from the incineration processes (namely flue gas, wastewater, and solid residues) are becoming increasingly stringent. Originally, simple incineration facilities were developed into complex processes. To be able to meet strict emission limits, each new generation of incineration plants involved newer and more effective solutions. The discharge rate of pollutants (including extremely hazardous e.g. dioxins (McKay, 2002; Grosso, et al., 2007; Parizek, et al., 2008) is very low in, 'up-to-date' facilities. Comparing emission limits valid for waste incinerators and other large combustion plants (i.e. those combusting gaseous, liquid, as well as fossil or biomass-based fuel) we can state that new WTE systems are the cleanest and most reliable sources of energy in the form of heat as well as electrical power (Stehlik, 2009).

Waste as a Potential Alternative

The problem is having a solution in itself. Thermal processing of waste does not only mean its disposal; it is also material and energy utilization (waste-to-energy). Various types of waste can be classified as a renewable source of energy (e.g. sewage sludge, wood pulp). It is necessary to find effective ways to utilize other types of waste whenever possible. Utilization of treated municipal waste and sewage sludge as alternative fuel for the cement and lime production industries are an example of material and energy utilization of waste (wasteless technology). During clinker production, most of the pollutants (e.g. heavy metals, S, Cl, or alkali) are fixed in the clinker lattice (Bebar, et al., 2003). Sludge incineration can be a complementary solution mainly for wastewater treatment plants with large processing capacity. This solution of sludge management has significant environmental benefits, because sludge is not landfilled as is common in many countries. Incineration of suitably prepared sludge (mainly sufficient dewatering) can also be economical, when the off-gas energy is used for heat and/or electric power production. Ash can be used as a secondary building material (e.g. for road construction) as well, but its utilization depends on its quality. Replacing 'normal' fuels with mining gas that has no other use is an example of waste gas utilization. It is a specific application that was realized in one pulp plant situated in a former mining area (Oral, et al., 2005). In this case mining gas substitutes for natural gas in the thermal treatment unit for sludge incineration. For this purpose, a special "dual" burner for combustion of both fuels was designed and installed (Stehlik, 2009). Different processing routes for solid waste are shown in Figure 2.

A variety of waste sources like urban, agriculture, industrial sectors, vegetable markets, etc., generate huge quantities of solid waste containing a sizeable proportion of biodegradable-organic matter with Municipal Solid Waste (MSW) having largest proportion. This material, if processed anaerobically, will not only generate significant quantity of biogas, i.e. about 250–350 m^3/tonne of waste (NEERI Report, 1996) and manure but will also reduce the load on landfilling and will in turn prevent the degradation of environmental quality due to uncontrolled decomposition of organic matter in the landfills. Bouallagui et al. (2003) studied both fruit and vegetable wastes together for biogas production (Bardiya, et al., 1996; Bouallagui, et al., 2003). Similar type of studies were done by Ranade *et al.* (1987), Math-Alvarez *et al.* (1993), and Pavan *et al.* (2000)

Figure 2. Alternative processing routes for waste (source: Manders, 2009)

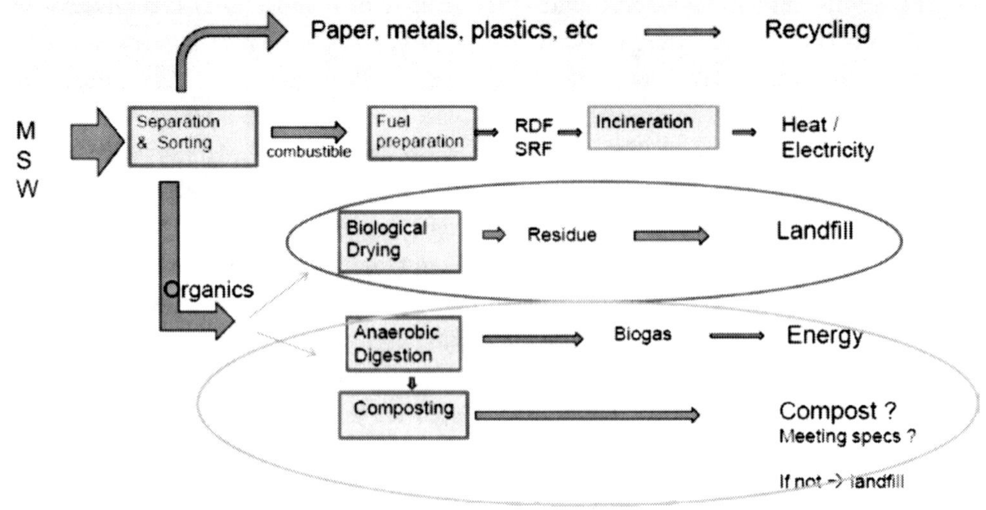

taken market waste (rotten vegetables, fruit skins, potatoes, onion, etc.) and household solid waste respectively intensively used for methane production. Demirel and Scherer (2009) did the laboratory-scale study to investigate the long-term anaerobic fermentation of an extremely sour substrate, an energy crop, for continuous production of methane (CH_4) as a source of renewable energy. The sugar beet silage was used as the mono-substrate, which had a low pH of around 3.3–3.4, without the addition of manure. The mesophilic biogas digester was operated in a Hydraulic Retention Time (HRT) range between 15 and 9.5 days, and an Organic Loading Rate (OLR) range of between 6.33 and 10 g VS^{-1} l^{-1} d^{-1}.

Anaerobic digestion process takes place in a warmed, sealed airless container (the digester), which creates the ideal conditions for the bacteria to ferment the organic material in oxygen-free conditions. The digestion tank needs to be warmed and mixed thoroughly to create the ideal conditions for the bacteria to convert organic matter into biogas (a mixture of carbon dioxide, methane, and small amounts of other gases). The proper design and scale-up of anaerobic reactors requires knowledge regarding correlations between reactor configuration and process efficiency to set the limits for the biochemical operating parameters. The production of biogas through anaerobic digestion offers significant advantages over other forms of waste treatment, including:

- Less biomass sludge is produced in comparison to aerobic treatment technologies.
- Successful in treating wet wastes of less than 40% dry matter (Mata-Alvarez, 2002).
- More effective pathogen removal (Bendixen, 1994; Lund, et al., 1996; Sahlstrom, 2003). This is especially true for multi-stage digesters (Kunte, et al., 2004) or if a pasteurization step is included in the process.
- Minimal odour emissions as 99% of volatile compounds are oxidatively decomposed upon combustion, e.g. H_2S forms SO_2 (Smet, et al., 1999).
- High degree of compliance with many national waste strategies implemented to reduce the amount of biodegradable waste entering landfill.
- The slurry produced (digestate) is an improved fertilizer in terms of both its availability to plants (Tafdrup, 1995).

- A source of carbon neutral energy is produced in the form of biogas.

Hence, the production of biogas through waste-to-energy route technologies and sources is gaining importance for supplementing the fuel/energy requirements.

Technology Routes Followed

Current Waste Management Practices and Its Shortcomings

As indicated in Table 3, the current practices employed in the management of solid waste vary considerably between the low, middle, and high-

Table 3. Comparison of typical solid waste management practices

Activity	Low income	Middle income	High income
Source reduction	No organized programmes, but reuse and low per capita waste generation rates are common.	Some discussion of source reduction, but rarely incorporated in to any organized programme.	Organized education programmes are beginning to emphasize source reduction and reuse of materials.
Collection	Sporadic and inefficient. Service is limited to high visibility areas, the wealthy, and businesses willing to pay.	Improved service and increased collection from residential areas. Larger vehicle fleet and more mechanization.	Collection rate greater than 90 per cent. Compactor trucks and highly machined vehicles are common.
Recycling	Most recycling is through the informal sector and waste picking. Mainly localized markets and imports of materials for recycling.	Informal sector still involved, some high technology sorting and processing facilities. Materials are often imported for recycling.	Recyclable material collection services and high technology sorting and processing facilities. Increasing attention towards long-term markets.
Composting	Rarely undertaken formally even though the waste stream has a high percentage of organic material.	Large composting plants are generally unsuccessful, some small-scale composting projects are more sustainable.	Becoming more popular at both backyard and large-scale facilities. Waste stream has a smaller portion of compostables than low and middle-income countries.
Incineration	Not common or successful because of high capital and operation costs, high moisture content in the waste, and high percentage of inerts.	Some incinerators are used, but experiencing financial and operational difficulties; not as common as high-income countries.	Prevalent in areas with high land costs. Most incinerators have some form of environmental controls and some type of energy recovery system.
Landfilling	Low-technology sites, usually open dumping of wastes.	Some controlled and sanitary landfills with some environmental controls. Open dumping is still common.	Sanitary landfills with a combination of liners, leak detection, leachate collection system, and gas collection and treatment systems.
Costs	Collection costs represent 80 to 90 per cent of the municipal solid waste management budget. Waste fees are regulated by some local governments, but the fee collection system is very inefficient.	Collection costs represent 50 to 80 per cent of the municipal solid waste management budget. Waste fees are regulated by some local and national governments, more innovation in fee collection.	Collection costs can represent less than 10 per cent of the budget. Large budget allocations to intermediate waste treatment facilities. Upfront community participation reduces costs and increases options available to waste planners (e.g. recycling and composting).

Source: World Bank (1999)

income countries. The extent of application and the effectiveness of these practices are reviewed in the subsections that follow.

Many countries face critical problems with regard to waste management. A range of common shortcomings has been identified, including insufficient government priority and political support for action; lack of finance; inadequate long-term planning, indiscriminate disposal of waste; poor handling and disposal of hazardous and biomedical wastes; insufficient recycling and reuse; ineffective legislation and institutions; lack of skilled personnel; and poor monitoring and enforcement.

The prevailing view in many countries, particularly in respect of industrial waste, is that it is not possible to constrain the growth of economy by forcing industry and municipalities into introducing sophisticated and expensive waste treatment and disposal technologies. The short-sighted nature of such a policy, with implications for the long-term problems and costs of waste treatment has yet to be realized or understood in many countries in the region.

Private sector incentives and initiatives in waste management in many countries are still rare and the responsible authorities are seldom willing to see the provision of waste management services given into private hands. There is a general lack of funding which may be used to establish a waste management system operated by private contractors. In addition, there is also a very low level of public awareness and participation regarding waste management. This is because people are not sufficiently informed about the health and economic benefits of proper waste collection, treatment and disposal.

Much of the existing infrastructure and facilities for the waste collection, treatment, and disposal have not kept pace with the economic development in recent years. It is not the lack of knowledge but the lack of finance and administration that is the main reason for the growing inefficiency of waste management practices in the region. Particularly in smaller cities and rural areas (where the patterns of consumption have also changed), the existing standards of waste collection, treatment and disposal remain very low. Adoption of inappropriate technologies creates many problems in the region. There are countless examples of plant failure. Many problems exist especially for municipal solid waste collection, processing and disposal in the cities of Asia and the Pacific. Waste collection services are often sporadic as they rely upon insufficient numbers of vehicles, which are often old, under maintained and unreliable. Collection is irregular: again open dumping at the roadside, along and in water channels, rivers, and along railways is quite normal. Disposal is often to uncontrolled open dumping sites and in many countries, industrial hazardous waste and biomedical waste are brought to the same dumpsites. Waste management practices are most effective where they form part of a robust and integrated approach to the collection and disposal of all generated wastes. At present, however, waste management is given relatively low priority in many countries of the region despite increasing loads that stretch the already limited resources of waste collection and disposal agencies.

Waste to Energy

New technologies make it possible to utilize organic waste in a new way to achieve higher power efficiency, to store energy or to produce fuels for transportation. Technologies of interest include second-generation biofuel production, gasification/pyrolysis, anaerobic digestion and improved incineration. In a system perspective, the new technologies have potential benefits, such as the possibilities of regulating the production of electricity, heat, and transport fuels and thereby increasing the flexibility of the system. Various types of wastes from agricultural (plant and animal wastes), industrial (sugar refinery, dairy wastes, confectionary waste, pulp and paper, tanneries, and slaughter houses) and residential (kitchen waste and garden waste) sectors are the potential renewable energy sources to attain sustainability and for Switch over to Waste-To-Energy Routes (WTERs).

Biogas Technology (BT)

In 1776, for the first time, the Italian Physicist, Volta, demonstrated methane in the marsh gas, generated from organic matter in bottom sediments of ponds and stream. Under anaerobic conditions, the organic materials are converted through microbiological reactions into gases (biogas) and organic fertilizer (manure). Biogas and manure are the end products obtained from BT whereas conventional composting process produces only manure as the product after decomposition of solid organic waste. Thus, comparatively BT could be considered as better option for its compactness, cleaner operation and better product range (i.e. both gas as energy source and processed solid waste as manure). Methane is the main constituent of biogas. About 90% of energy of substrate is retained in methane. It is used mainly for cooking, lighting and in internal combustion engines to power water pumps and electric generators. The residual product from digestion can often replace manure or artificial fertilizer. The most economical benefits are minimizing environmental pollution and meeting the demand of energy for various purposes. In India, the Ministry of New and Renewable Energy (MNRE) (Government of India) has declared a National Master Plan in 1994, which incorporates BT as one of the major waste-to-energy options to be developed and adopted in the country (MNES India; Kothari, et al., 2010).

Tricase and Lombardi (2009) estimated that the total available animal waste is about 132 millions m^3/year, which is capable to produce approximately 2.2 Gm^3/year of biogas. Providing adequate and clean energy is considered to be one of the main challenges for sustainable development in the world. Biomass may be a feasible energy source that can reduce the country's dependency on imported oil. Abu-Abu- Ashour et al. (2010) studied that the main sources of biomass in Jordan are the animal waste and the olive waste known locally as "Jift." Preliminary studies carried out by the National Energy Research Centre of Jordan show that biogas from animal and domestic waste can save up to 4% of imported oil which is equivalent to 130,000 tonne per annum (see Figure 3).

When biodegradable municipal waste is digested, 19 kg of nitrogenous fertiliser/tonne of Dry Substance (DS) and 4 kg of phosphorous fertilizer/tonne DS can be replaced by by-products from digestion (the dry substance is around 30%) (Berglund & Börjesson, 2003). This is equivalent to an energy saving of about 291 MJ/tonne organic waste, due to the substitution of artificial fertilizers (The assumption is made that the energy use when manufacturing fertiliser is 45 MJ/kg nitrogen and 25 MJ/kg phosphorous.). The biogas produced in the digestion plant is equivalent to 3.7 GJ/tonne organic waste. The energy needed to operate the biogas plant is for electricity 0.4 GJ/tonne and heat 0.5 GJ/tonne (Holmgren & Henning, 2004).

Hydrogen Energy

As a sustainable energy supply with minimal or zero use of hydrocarbons, hydrogen is a promising alternative to fossil fuels. It is a clean and environ-

Figure 3. Different sources of waste feedstock's for biogas energy

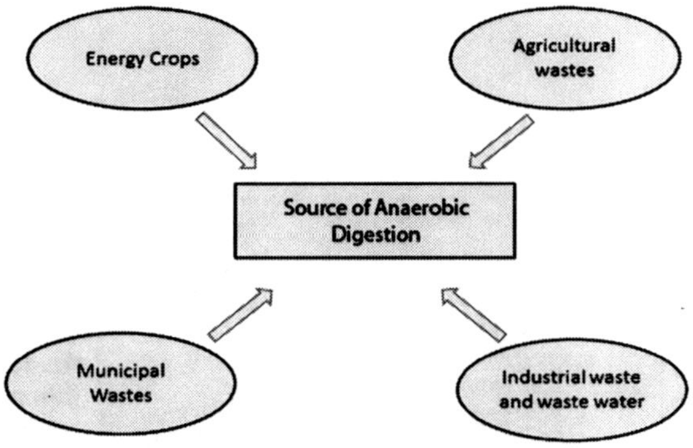

mentally friendly fuel that produces water instead of greenhouse gases when combusted. It seems to be logical conclusion for numerous environmental problems like acid rain and greenhouse effect.

There are many feedstocks and techniques available to harness hydrogen from fossil fuel, water, and biomass as shown in Figure 4. Among these, electrolysis of water, steams reforming of hydrocarbons and autothermal processes are well-known methods for hydrogen gas production, but not cost-effective due to high energy requirements. But, carbohydrates rich, nitrogen deficient solid wastes such as cellulose (Kalia, et al., 1994; Liu, et al., 2003; Islam, et al., 2009; Magnusson, et al., 2009) and starch (Yokoi, et al., 1998, 2001; Kanai, et al., 2005; Lay, 2000; Ren, et al., 2006; Yu & Mu, 2006; Lin, et al., 2006) containing agricultural and food industry wastes and some food industry wastewaters such as cheese whey (Veziroglu & Barbir 1998), olive mill, and baker's yeast (Veziroglu & Barbir, 1998) industry wastewaters can be used for hydrogen production by using suitable bioprocess technologies or other WTERs.

This bioprocess for hydrogen production referred as biological hydrogen. Pure and mixed cultures (Chen, et al., 2009; Yang, et al., 2007) known to produce hydrogen from carbohydrates include species of *Enterobacter* (Rachman, et al., 1997; Fabiano & Perego, 2002), *Bacillus* (Kalia, et al., 1994), and *Clostridium* (Kumar & Das, 2000; Hawkes, et al., 2002; Skonieczny & Yargeau, 2009) in different reactor configurations like continuous, membrane bioreactor and UASB reactor has been reported in the literature (Chen, et al., 2009; Ren, et al., 2006; Lin, et al., 2006).

Gasification and Pyrolysis

One of the criteria that are of primary interest when it is to evaluate a method for agricultural waste treatment is the raw material moisture content and the stoichiometric relationship C/N. For moisture contents lower than 50%w/w and ratios C/N > 30, thermochemical treatment of biomass is applicable in the form of direct combustion, pyrolysis, or gasification, otherwise it is better to be lead to biochemical treatment (aerobic/

Figure 4. Selected feedstock and process alternatives

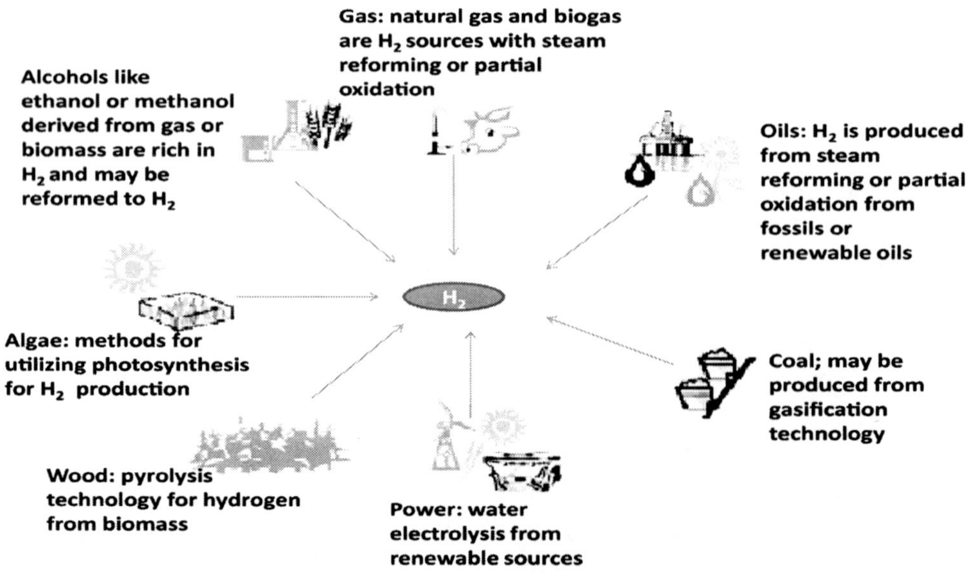

anaerobic digestion, composting, etc.) (Skoulou & Zabaniotou, 2007).

Biomass combustion (Vamvouka & Zografos, 2004; Scala & Chirone, 2004) is the most ancient method for energy production that mankind ever used. Even nowadays, combustion contributes to almost 85% in energy production methods. Combustion is characterized by a chemical reaction between a fuel and usually air in excess levels and produces a significant amount of energy in the form of heat, but it is known that it also contributes unfavorably to environmental pollution through ash generation, gaseous emissions, etc. Co-combustion of woody wastes with some amounts of fossil fuels promotes renewable source utilization, while it reduces the consumption of fossil fuels, ensures annual feeding of biomass–fossil fuel blend and may lead to money and energy saving.

On the other hand, pyrolysis (thermal cracking) (Encinar, et al., 1998; Ozbay, et al., 2001) takes place under high temperature (up to 500 °C), with slow heating rates and under environments with no air, in order to prevent biomass combustion. Imposing high temperatures, under oxygen-limited atmosphere, biomass bonds degrade and as a result, a gaseous mixture composed mainly of CO, CH_4, and H_2 is produced. Also, owing to the temperature that is not high enough to break long bonds in carbon chain, tars (bio-oils) are produced in significant quantities. Finally, and after biomass devolatilization, char (black solid residue) is produced. Many basic and applied scientific studies on the conversion of lignocellulosic biomass (such as wood and agricultural residues) into oil, gas, and char have been conducted. Many types of organic waste materials have been applied as feedstocks in the pyrolysis process, such as waste vegetable oil, waste plastic, municipal waste, waste tires, and WLO (Waste Lubricating Oil). Recent scientific research activities are focused on flash pyrolysis used in bio-oil production, with the advantages of less gas and char production (Singhabhandhu & Tezuka, 2010; Skoulou & Zabaniotou, 2007).

Biomass gasification (Bridgwater, 2003; Kim, 2003; Demirbas, 2005) can be considered as a form of pyrolysis, which takes place in higher temperatures and produces a mixture of gases with H_2 content ranging 6–6.5%. Its scope is to optimize and maximize gas production. Firstly, drying and pyrolysis take place, while in second stage, synthesis gas is produced. The reason is that CO and H_2O from combustion are converted to CO_2 and H_2. When imposing high temperatures, biomass bonds break down and reform a mixture of permanent gases consisting mainly of H_2, CO, CO_2, and CH_4. The percentages of those gases in the mixture depend on several factors like the type of gasification agent (air, oxygen, steam, or mixtures of them), biomass physicochemical nature (moisture, particle size, heating value, etc.). Comparing with air, when steam is introduced to reform CH_4 to H_2 and CO, higher percentage of hydrogen is achieved, and the produced gas has higher heating value. Additionally, gasification in combination with fuel cells, internal combustion engines, or gas turbine systems could be extremely attractive for electricity generation and is expected to be an important part of biomass energy production with environmental friendly way. Generally, as it concerns the by-products from different thermochemical treatments, e.g. gasification against combustion gives better limits in CO_2 emissions and it is widely known that CO_2 is considered ''neutral'' with respect to air pollution problems. It does not increase CO_2 concentration in atmosphere, as the carbon dioxide released from gasification is already the inherent amount that biomass gained from atmosphere with photosynthesis.

Gasification under certain practice (Integrated Gasification combined Cycle) gives higher efficiencies (45–50%) than that is usually achieved via combustion (25–35%) (Maniatis, 2004). On the other hand, dioxin contents that results from animal manure combustion can be avoided under certain and strict control of good combustion conditions. Finally, pyrolysis can lead to biofuel utilization, with the advantages that a liquid fuel is able to offer (easy storage for short time and easy transport) or char material exploiting as active carbon (Figure 5).

Bioethanol Production

Interest in ethanol was renewed in the 1970s, when oil supply disruptions in the Middle East became a national security issue and America began to phase out lead (an octane booster) from gasoline. The American Oil Company and several other major oil companies began to market ethanol as a gasoline volume extender and as an octane booster (US Department of Energy). Ethanol was blended directly into gasoline in a mix of 10 percent ethanol and 90 percent gasoline, called gasohol.

Emission of green house gases is increasing everyday with the fast depleting oil resources. Utilization of fossil fuels in the form oil, natural gas, and coal, which modern society relies on for energy contributes significantly to global warming. Alternative fuels made from renewable resources, such as fuel ethanol, provide numerous benefits in terms of environmental protection, economic development, and national energy security (Yang & Yanpin, 2007). Much research is been poured in finding an alternative fuel through biological ways because of the positive environmental benefits of biofuels. Potential feedstocks for biofuel

Figure 5. Agricultural/animal waste thermochemical treatment overview (source: Skoulou & Zabaniotou, 2007)

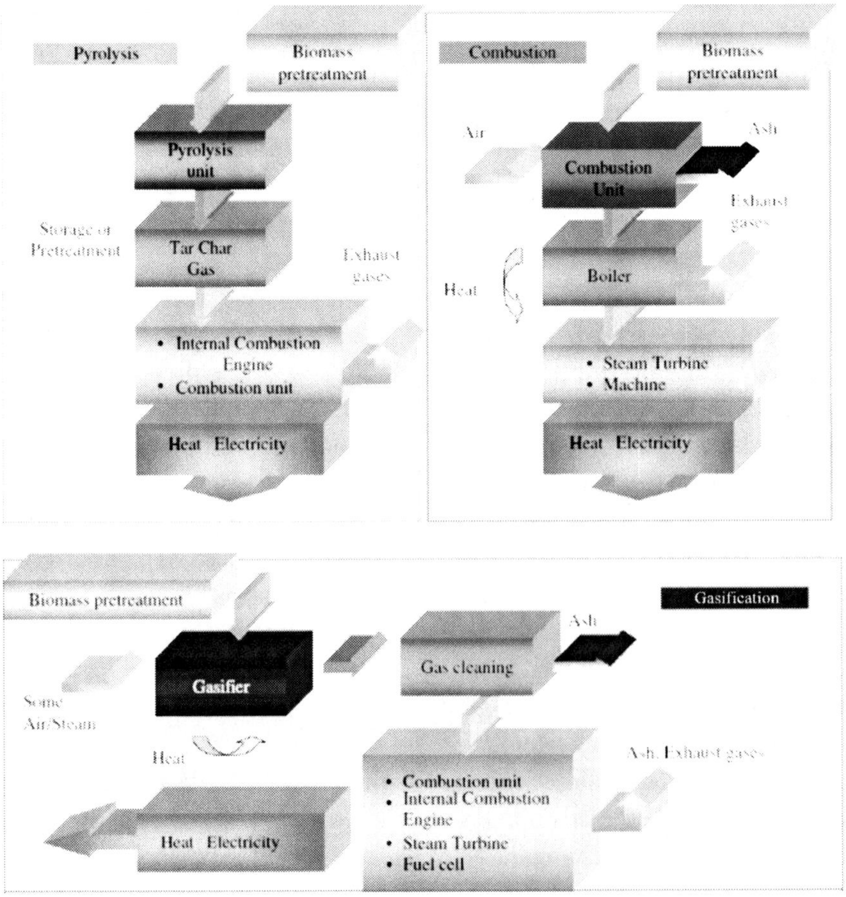

production include the cellulosic biomass as well as waste materials, occurring in abundance outside human food chain, which can be obtained throughout the year and are relatively inexpensive. Cellulosic materials are renewable natural biological resources and generation of biobased products and bioenergy from such substances is important for the development of humans (Zhang, et al., 2006). Various industries across the world generate huge volumes of cellulosic waste which have an immense potential to be utilized for the production of several bio-products (Das & Singh, 2004).

These dedicated substances provide a low-cost and uniquely sustainable resource for production of many organic fuels and chemicals that can reduce greenhouse gas emissions, enhance energy security, improve the economy, dispose of problematic solid wastes, and improve air quality (Wyman, 2003). There are mainly two processes involved in the conversion: hydrolysis of cellulose in the lignocellulosic biomass to produce reducing sugars, and fermentation of the sugars to ethanol (Sun & Cheng, 2005). Cellulose can be effectively hydrolyzed and depolymerized into fermentable sugars by the enzyme cellulase. Cellulase based strategies make the process of biorefinery more economical by means of utilizing cheaper substrates for enzyme synthesis (Arja & Suominen, 2002; Zhang, et al., 2006). Currently most commercial cellulases are produced from *Trichoderma* sp. and *Aspergillus* sp. (Kirk, et al., 2002). Cellulase is a term usually used to describe a mixture of cellulolytic enzymes whose synergistic action is required for effective breakdown of substrate to its monomeric units. The action of cellulases involves the concerted action of

- **Endoglucanase(s):** Randomly attacks the internal, β1,4-linkages,
- **Cellobiohydrolase:** Cleaves off cellobiose units from the nonreducing ends of the glucan, and
- **β 3-glucosidase:** Hydrolyzes cellobiose to glucose (DiPardo, 2007).

Most of the β-glucosidic bonds in naturally occurring lignocellulosic materials are inaccessible to cellulase enzymes and also cellulose in naturally occurring materials is closely associated with hemicellulose and other structural polysaccharides, and carbohydrate-rich microfibrils are surrounded by a lignin seal. The substrate usually requires a pretreatment process before being subjected to enzymatic breakdown which is aimed at increasing the susceptibility of cellulose to enzyme. Current leading pretreatment technologies include the use of dilute acid, hot water, flow through, Ammonia Fiber Explosion (AFEX), ammonia recycle percolation, lime, microwave, steam explosion and "organsolv" which are intensively investigated (Kitchaiya, et al., 2003; Lee, et al., 2002). The overall performance of cellulase mixture is highly dependent on the residual lignin present along with cellulose. Cellulosic residues such as wheat straw, corn stover, rice straw, corn cobs and other agricultural wastes are the most available renewable resources which store derived energy in their chemical bonds (Silverstein, et al., 2007; Arthe, et al., 2008). Lignocellulosic biorefinery feedstock's are summarized in Figure 6.

Incineration

In general, incineration with energy recovery is the most common method for recovering the energy stored in waste. The majority of the MSW incinerators use moving grate technologies and they are designed to handle large volumes of MSW with no pre-treatment. This type of incineration (as shown in Figure 7), which is commonly defined as Mass Burn Incineration (MBI), includes large-scale combustion of MSW in a single stage chamber unit where complete combustion or oxidation occurs (Williams, 2005, p. 251). In the Combined Heat and Power (CHP) incinerators, the residual heat is recovered and exported to

Figure 6. Lignocellulosic biorefinery feedstock (source: ECOREFINE Biorefinery, 2007)

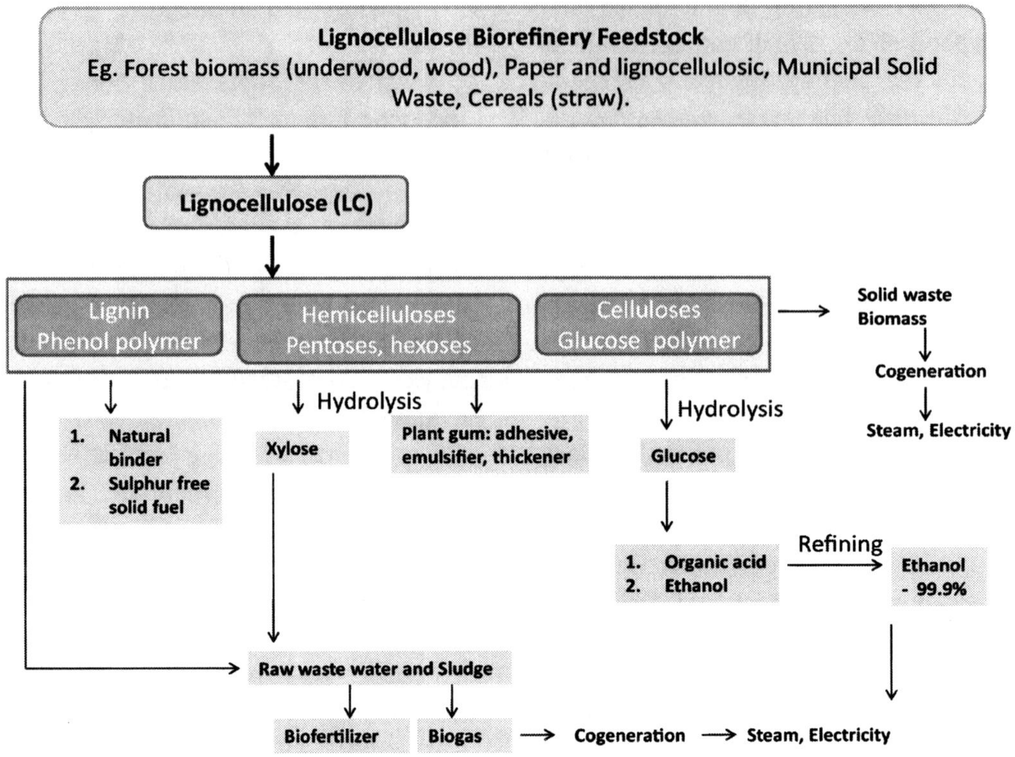

adjacent industrial premises or districts for space heating, hot water supply and other duties. Another technology used for energy recovery from MSW is MBT that partially processes mixed MSW by mechanically removing some parts of the waste and by biologically treating others. There are numerous possible permutations of MBT with different outputs. Apart from metals, potential products could be compost, stabilized waste for landfill or SRF (Solid Recovered Fuel) (Juniper Consultancy Services Ltd., 2005a). In the case of SRF production, the most common configuration incorporates bio-drying prior to mechanical treatment. Bio-drying drives-off moisture from the waste using the biological activity in an aerobic in-vessel system but does not fully bio-stabilize the waste. The reduction of moisture and the degradation of a part of the more volatile biodegradable fraction of the waste, increase the calorific value of the produced SRF rendering it like this a very attractive option for thermal treatment with energy recovery or co-incineration in industrial processes (Juniper Consultancy Services Ltd., 2005b). A relative new technology for the treatment of MSW is MHT. This technology uses a thermal or steam based pre-treatment process prior to mechanical separation of MSW into more useable fractions. A commonly used steam based technique to treat the waste is via autoclave technology that has been used to sterilize medical waste for many years (DEFRA, 2004).

In this plant, saturated steam at 160 °C is first injected into the rotating vessel containing the waste over a period of 15 min, which may vary depending on the specific heat capacities of the waste materials being processed and the amount of waste in the autoclave. The vessel will be maintained at high pressure for a period of up to 45 min to allow the process to fully ‘cook’ the waste, thereby breaking waste down into its or-

Figure 7. MSW management systems via incineration

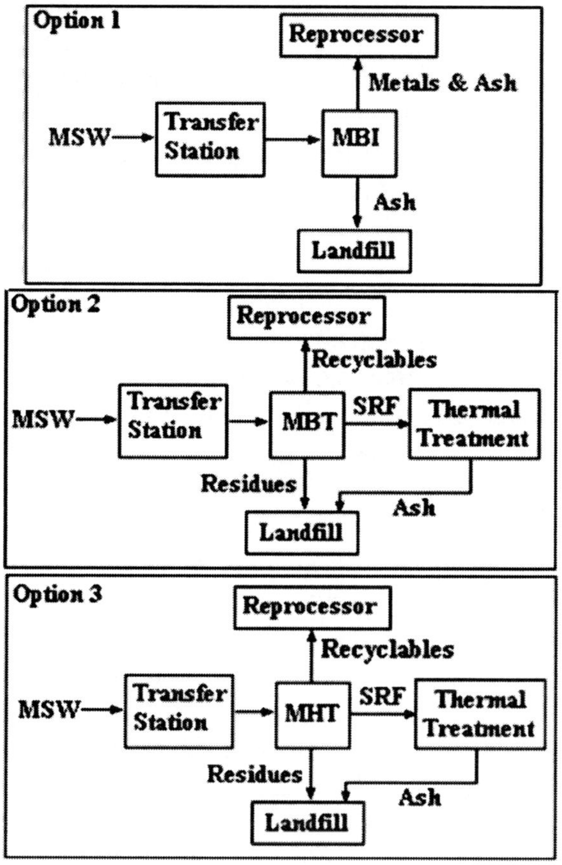

ganic and inorganic constituents. The full cycle of loading, treatment and sorting lasts up to 90 min. Processed waste is then treated by a series of screens and recovery systems (e.g. debris roll screen, over-band magnet, eddy current separator and air classifier) to achieve secondary recycling and separation of the predominantly cellulose fibre, with residue left for landfilling. The resulting fibre comprises the putrescible, cellulose and lignin elements of the waste stream and fibre could be used as SRF or after further treatment as compost (WTDC, 2006).

Many studies assessed the GHG emission impacts of MBI and found that not only it can offer significant reduction of CH_4 emissions compared to land filling, but at the same time it can provide additional net GHG emission savings due to the recovery of energy (Porteous, 2005). Despite these benefits of this technology in terms of climate change, there is a strong opposition against MBI, as it is claimed to undermine recycling of valuable resources and is considered hazardous for the human health, even though various studies assessed the health risks associated with MBI and found that health risks from MBI are considered generally acceptable and in particular cancer risks are 5 times less than land filling (Moy, et al., 2008). MBI with energy recovery or combustion of sorted waste (the so-called energy fraction of waste) could be a potential treatment for the huge amount of waste, as well as MBT and MHT configured for SRF production. However, MHT is a relatively new technology and its climate change impact has not been assessed yet. On the other hand, there are studies that have assessed the GHG impact of MBT plants that use aerobic biological treatment or composting for the treatment of waste. These processes are used mainly for the stabilization of waste or production of compost. However, the most commonly used biological treatment method in MBT plants configured for production of SRF is bio-drying and as this configuration has not been examined in detail in previous studies, it should be assessed as well (Papageorgiou, et al., 2009).

ISSUES, CONTROVERSIES, PROBLEMS

Economic Aspects

Agricultural and animal waste management practices are very important for generating income and employment and could turn to be an economic way of energy production. Low-cost by-products of agriculture and animal breeding activity could enhance rural economies and keep them alive. This possibility seems to be more attractive when biomass is exploited under careful planning and in combination with a well-established energy production technology. With the rising fossil fuel prices, biomass conversion in energy will cost less in future and its utilization will contribute to money saving and environmental protection (Demirbas, 2005). Decentralized or modular form of bioenergy production would probably help those enterprises' economic viability. Disposal of the waste is the greatest challenge as it contributes to the 23% of the total cost of WTE plan. Also, it could set free from labour cost (approximate 5.8%), which is one of the main factors that make energy production from biomass quite expensive. The lack of appropriate mechanization for biomass residue collection, transporting problem for big raw biomass volumes, the lack of knowledge and inexistence in applicability of waste management are some of the factors that at the moment dominate in biomass negative exploitation for energy production reasons. Currently there is no organized practice in agricultural and animal wastes management and as a result, their exploitation is still set aside (Skoulou & Zabaniotou, 2007).

Keeping all the points in mind, something that should not set aside the fact that biomass remains one of the cheapest fuel sources, but during the last decade, and under the pressures of environmental protection and commitments in reduction of greenhouse gases and biomass management instructions, there is a considerable interest in thermochemical technology (gasification and pyrolysis) application for energy and related material production from biomass resources. Biomass wastes fix annual availability, reduce its cost as a fuel, and give better efficiencies. Sometimes biomass itself is cost-free and clearly does not depend on fuel price pressures. CH_4 and H_2 gases that formed could be used for heat and electricity production or burned in boiler.

Problems and Prospects of WTE

Along with the rapid economic growth, world having urgent needs for environmentally sound waste management technology and clean energy. WTE is playing, and will continue to play, an increasingly important role in MSW management in the near future. Nonetheless, development of WTE industry in World faces several major challenges. Energy production of agricultural and animal wastes has a relatively low impact in local environment, including not only humans but also animals. First of all, as it concerns the human health, there is a potential of lung-breathing illnesses, which are caused due to the smoke and gaseous emissions inhalation. The risk of uncontrolled fires is always on question, threatening the ecosystem, especially during the summer season and when temperatures are high and agricultural activities are in their peak. Additionally, burning those wastes always hide the danger or releasing some possibly dangerous chemicals that were absorbed from vegetable tissues, or consumed by animals. It introduces some risk in releasing heavy metals to the environment. Also, sound pollution can occur in some extent due to the machinery that is used for waste collection and management (tractors, trailers, etc.). Large amounts of wastes accumulated on field, mainly from animal wastes, create odour and aesthetic problems. Many pathogen bacteria are generated in animal manures and are carried in underground waters through manure leachates, attract vectors, and impose a serious odour emitting problem.

Capital and Operating Costs

Compared to other MSW treatment technologies and power generation from other renewable resources, WTE requires high capital investments and operational costs. Incineration equipments are expensive and have high operating and maintenance costs; equipments based on domestic technologies cost much less but are generally limited to relatively low capacities. Most of the WTE facilities in the developed countries have been built in the economically more developed cities, funded by the municipal governments and some time with assistance from foreign loans, although participation of the private sector is increasing.

They are predominantly supported and operated by municipal governments. It seems difficult for developing countries and under developed countries to invest such a huge amount on incineration technology. Countries should invest in the research and development of domestic technologies and incineration equipment, especially the large capacity incinerators, to help make WTE incineration more affordable by municipalities across the country.

Equipment Corrosion

Due to the lack of waste sorting and material separation, MSW contains relatively high levels of chlorine and sulfur (Zhang, et al., 2008), which form acid gases (HCl and SO_2) during combustion and can cause serious corrosion of the boilers. Even though higher efficiencies can be achieved by operating WTE plant heat recovery boilers at higher temperatures, the rate of high temperature corrosion will also increase. This problem is significant for gasification and incineration process. To reduce the acid gas corrosion attack, the boilers are typically operate at rather low steam conditions (400 °C, 4 MPa), resulting in relatively low overall thermal cycle efficiencies (~20%). On the other hand, although the capital cost for boiler tubes made of highly resistant materials is higher, the superheater can be operated at higher temperature and their lifespan can be significantly extended. The increased energy recovery efficiency and reduced superheater maintenance may lead to greater long-term economic benefits than the system built with less corrosion resistance. To combat the corrosion problem and to improve the power generation efficiency of WTE plants, more efforts are to be spent on improving the process conditions in the boiler and on developing less expensive corrosion-resistant alloys.

Air Pollutant Emissions

A range of air pollutants, particularly dioxins, are produced in MSW incineration and can be released into the atmosphere in significant quantities if the incinerator and the flue gas cleaning system are not properly designed and operated. Nonetheless, it is necessary for WTE facilities to continue improve the incinerators and flue gas treatment systems to further reduce their air pollutant emissions. For better protection of public health, there should be tighter limits on emissions of dioxins and other air pollutants from incineration facilities (towards the European and U.S. standards), which is expected to lead to a wave of development and implementation of new air pollution control technologies.

Fly Ash Management

Management of fly ash from MSW incineration has not received adequate attention, while improper disposal can potentially cause secondary environmental pollution from the fly ash. Despite the significant waste volume reduction, considerable amount of solid residues (i.e., bottom ash, fly ash, and air pollution control residue) are generated at different points in the process of MSW incineration. After appropriate stabilization treatment, the bottom ash is often used as a construction material. In contrast, fly ash is a hazardous waste due to the enrichment of dioxins and heavy metals, and must be disposed accordingly (Li, et al., 2004; Yan, et al., 2006).

In China itself, MSW incineration currently produces approximately 200,000 tonnes of fly ash each year, most of which is dumped in open pits or sold to private parties (Nie, 2008). Meanwhile, similar to bottom ash, fly ash can also be reused in construction applications after decontamination/stabilization treatment (Shi & Kan, 2009; Haiying, et al., 2007). Significant efforts should be spent on developing appropriate dioxin destruction and heavy metal stabilization technologies (e.g., vitrification, sintering, thermocatalyic, and hydrothermal treatments) and strengthening the regulations on fly ash management and reuse.

FUTURE RESEARCH DIRECTIONS

There are several limitations to the development of waste substrates as an immediate energy resource. Foremost is the fact that current anaerobic digestion technologies are not sufficiently efficient to recover usable energy at a cost that is currently competitive with fossil fuel technologies. If we have to utilize renewable waste as a bio-energy resource, substantial improvements in the efficiency of energy recovery will have to be made, or additional value-added benefits related to waste diversion or GHG mitigation will have to develop to make these processes more commercially viable. With fossil energy sources eventually dwindling and becoming increasingly more expensive, waste-to-energy routes are likely to have future attraction. The improvements in fermentation system yields obtained from better preprocessing and recovery techniques would also help improve their viability in general. Additionally, valuing the collateral avoidance of renewable waste residue related environmental problems, such as municipal waste landfill, GHG emissions, as achieved through utilizing MSW, can also influence sustainability of WTE approaches. With these various points in mind, renewable waste substrate can currently cover the greatest practical potential for energy production in world. Industrial waste rich in organic contents are produced in large amounts in urban areas with high-energy needs, and thus, could provide usable energy on a sustainable basis to some communities. Large livestock feedstock from agriculture sector close to urban and rural centers also offer a potential for co-digestion of manure with Organic Municipal Solid Waste (OMSW) from domestic sector, producing usable energy and value-added fertilizer, as well as reducing their environmental problems.

For incineration and gasification, research and development must be initiated in this field, as these technologies are affordable by developed countries and some developing countries only due to its high cost. Moreover, there should be a check over the release of harmful chemicals and heavy metals to the environment. Lots of efforts and research is required in this area as these thermo-conversions are efficient ones. Energy required for the substrate pre-treatment, purification of the biogas and hydrogen, reactor maintenance, manpower, etc., also limited the WTER technologies. Although low-yield and the rates of energy production may be overcome by selecting and using more effective organisms or mixed cultures, developing more efficient processing schemes, optimizing the environmental conditions, improving the light utilization efficiency and developing more efficient reactors in WTERs. The significant potential of biogas and hydrogen energy yields shown here through utilizing microbial technologies for existing and otherwise unused renewable waste resources supports the establishment of a focus on this area, including additional research and development on conversion technologies and the creation of commercial incentives to encourage the industrial development of this potential. Hence, with scientific and engineering advancements, limitations for WTER technologies can be solved and viewed as a key and economically viable component to a renewable based economy.

CONCLUSION

The clearest conclusion of the review is that waste should be used for energy purposes. This review finds that WTER products in place of fossil fuels have an excellent potential for high energy content, such as by anaerobic digestion, biological hydrogen production, gasification/pyrolysis and incineration. With the significant benefits of environmental quality and reduction of GHG emissions, MSW is increasingly accepted as a clean source of energy WTE solves the problem of waste disposal while recovering the energy from the waste materials, and the pollutant emissions can be controlled to low levels. With the significant benefits of environmental quality and reduction of GHG emissions, waste is increasingly accepted as a clean source of energy. Research and technology development focusing on corrosion phenomena, flue gas control, fly ash management, and beneficial reuse of residues will further drive the growth of WTE industry.

This paper highlights the potential of WTE as an effective waste management option and as an energy source. WTE can minimize the amount of waste sent to landfill and by virtue of its biomass content can contribute to achieving the world's renewable energy and electricity targets. In the future, the cost of landfilling waste is, due to land scarcity and disamenity, likely to increase further thus making energy recovery from waste even more cost-effective. On the other hand, although recycling remains the most climate-friendly waste management option, the inability to recover energy value can be a major concern of this option. Hence, by attempting to approach the waste treatment levels in best practice countries, WTE can form an important part of waste management strategy as well as energy and environmental policies. However, considerable work and studies are required to be done to increase the efficiency of the system, to make it more economic and reduce the harmful pollutant leakage to the environment.

REFERENCES

Abbasi, S. A., Nipaney, P. C., & Panholzer, M. B. (1991). Biogas production from the aquatic weed pistia (pistia stratiotes). *Bioresource Technology*, *37*, 211–214. doi:10.1016/0960-8524(91)90185-M

Abu-Ashour, J., Qdais, H. A., & Al-Widyan, M. (2010). Estimation of animal and olive solid wastes in Jordan and their potential as a supplementary energy source: An overview. *Renewable & Sustainable Energy Reviews*, *14*, 2227–2231. doi:10.1016/j.rser.2010.03.001

American Society of Mechanical Engineers. (2008). *Waste-to-energy: A renewable energy source from municipal solid waste*. New York, NY: American Society of Mechanical Engineers.

Arja, M. O., & Suominen, P. (2002). Enhanced production of trichoderma reesei endoglucanases and use of the new cellulase preparations in producing the stonewashed effect on denim fabric. *Applied and Environmental Microbiology*, *8*, 3956–3964.

Arthe, R., Rajesh, R., Rajesh, E. M., Rajendran, R., & Jeyachandran, S. (2008). Production of bio-ethanol from cellulosic cotton waste through microbial extracellular enzymatic hydrolysis and fermentation. *Electronic Journal of Environment. Agriculture and Food Chemistry*, *7*(6), 2984–2992.

Bardiya, N., Somayaji, D., & Khanna, S. (1996). Biomethanation of banana peel and pineapple waste. *Bioresource Technology*, *58*, 73–76. doi:10.1016/S0960-8524(96)00107-1

Bebar, L., Kermes, V., Puchyr, R., Stasta, P., Stehlik, P., & Oral, J. (2003). *Waste as alternative fuel for cement and lime production industry*. Paper presented at Fourth International Symposium on Waste Treatment Technologies (Thermal, Non-Thermal, and Gas Clean-Up). Sheffield, UK.

Bendixen, H. J. (1994). Safeguards against pathogens in Danish biogas plants. *Water Science and Technology*, *30*, 171–180.

Berglund, M., & Börjesson, P. (2003). *Energy analysis biogas system (energy analysis of biogas systems)*. Unpublished Report 44. Lund, Sweden: Lund University.

Bouallagui, H., Ben, C., Marounani, R., & Hamdi, M. (2003). Mesophilic biogas production from fruit and vegetable waste in a tubular digester. *Bioresource Technology*, *86*, 85–89. doi:10.1016/S0960-8524(02)00097-4 PMID:12421014

Bridgwater, A. V. (2003). Renewable fuels and chemicals by thermal processing of biomass. *Chemical Engineering Journal, 91*, 87–102. doi:10.1016/S1385-8947(02)00142-0

Caton, P. A., Carr, M. A., Kim, S. S., & Beautyman, M. J. (2010). Energy recovery from waste food by combustion or gasification with the potential for regenerative dehydration: A case study. *Energy Conversion and Management, 51*, 1157–1169. doi:10.1016/j.enconman.2009.12.025

Chen, W. H., Sung, S., & Chen, S. Y. (2009). Biological hydrogen production in an anaerobic sequencing batch reactor: pH and cyclic duration effects. *International Journal of Hydrogen Energy, 34*, 227–234. doi:10.1016/j.ijhydene.2008.09.061

Cheng, C. H., & Hu, Y. (2010). Municipal solid waste (MSW) as a renewable source of energy: Current and future practices. *Bioresource Technology, 101*, 3816–3824. doi:10.1016/j.biortech.2010.01.040 PMID:20137912

Das, H., & Singh, S. K. (2004). Useful by-products from cellulosic wastes of agriculture and food industry-A critical appraisal. *Critical Reviews in Food Science and Nutrition, 44*(2), 77–89. doi:10.1080/10408690490424630 PMID:15116755

DEFRA. (2004). *Introductory guide: Options for the diversion of biodegradable municipal waste from landfill*. London, UK: Enviros Consulting Limited.

Deivanai, K., & Kasturi Bai, R. (1995). Batch biomethanation of banana trash and coir pith, short communication. *Bioresource Technology, 52*, 93–94. doi:10.1016/0960-8524(95)00016-8

Demirbas, A. (2005). Potential applications of renewable energy sources, biomass combustion problems in boiler power systems and combustion related environmental issues. *Progress in Energy and Combustion Science, 31*, 171–192. doi:10.1016/j.pecs.2005.02.002

Demirel, B., & Scherer, P. (2009). Bio-methanization of energy crops through monodigestion for continuous production of renewable biogas. *Renewable Energy, 34*, 2940–2945. doi:10.1016/j.renene.2009.05.013

Denison, R. A. (1996). Environmental life-cycle comparison of recycling, landfilling, and incineration: A review of recent studies. *Annual Review of Energy and the Environment, 21*, 191–237. doi:10.1146/annurev.energy.21.1.191

DiPardo, J. (2007). Outlook for biomass ethanol production and demand. In *Energy Information Administration/Outlook for Biomass Ethanol Production and Demand,* (pp. 1-14). Retrieved from ftp://tonto.eia.doe.gov/features/biomass.pdf.

Encinar, J. M., Beltran, F. J., Ramiro, A., & Gonzalez, J. F. (1998). Pyrolysis/gasification of agricultural residues by carbon dioxide in the presence of different additives: Influence of variables. *Fuel Processing Technology, 55*, 219–233. doi:10.1016/S0378-3820(98)00052-6

Fabiano, B., & Perego, P. (2002). Thermodynamic study and optimization of hydrogen production by Enterobacter aerogenes. *International Journal of Hydrogen Energy, 27*, 149–156. doi:10.1016/S0360-3199(01)00102-1

Grady, C. P. L., Daigger, G. T., & Lim, H. C. (1999). Biological wastewater treatment. In *Biological Nutrient Removal* (pp. 487–497). New York, NY: Marcel Dekker.

Grosso, M., Cernuschi, S., Giugliano, M., Lonati, G., & Rigamonti, L. (2007). Environmental release and mass flux partitioning of PCDD/Fs during normal and transient operation of full scale waste to energy plants. *Chemosphere, 67*, 118–124. doi:10.1016/j.chemosphere.2006.05.114 PMID:17222439

Haiying, Z., Youcai, Z., & Jingyu, Q. (2007). Study on use of MSWI fly ash in ceramic tile. *Journal of Hazardous Materials*, *141*, 106–114. doi:10.1016/j.jhazmat.2006.06.100 PMID:16889889

Hawkes, F. R., Dinsdale, R., Hawkes, D. L., & Hussy, I. (2002). Sustainable fermentative hydrogen production: Challenges for process optimization. *International Journal of Hydrogen Energy*, *27*, 1339–1347. doi:10.1016/S0360-3199(02)00090-3

Holmgren, K., & Henning, D. (2004). Comparison between material and energy recovery of municipal waste from an energy perspective: A study of two Swedish municipalities. *Resources, Conservation and Recycling*, *43*, 51–73. doi:10.1016/j.resconrec.2004.05.001

Islam, R., Cicek, N., Sparling, R., & Levin, D. B. (2009). Influence of initial cellulose concentration on the carbon flow distribution during batch fermentation by Clostridium thermocellum ATCC 27405. *Applied Microbiology and Biotechnology*, *82*, 141–148. doi:10.1007/s00253-008-1763-0 PMID:18998122

Jain, S. K., Gujral, G. S., Jha, N. K., & Vasudevan, P. (1992). Production of biogas from Azolla pinnata R. Br. and Lemna minor L.: Effect of heavy metal contamination. *Bioresource Technology*, *41*, 273–277. doi:10.1016/0960-8524(92)90013-N

Juniper Consultancy Services Ltd. (2005a). *Mechanical-biological-treatment: A guide for decision makers – Processes, policies and markets*. London, UK: Juniper.

Juniper Consultancy Services Ltd. (2005b). *Mechanical-biological-treatment: A guide for decision makers-Processes, policies and markets: Annex A*. London, UK: Juniper.

Kalia, V. C., Jain, S. R., Kumar, A., & Joshi, A. P. (1994). Fermentation of bio-waste to H2 by Bacillus licheniformis. *World Journal of Microbiology & Biotechnology*, *10*, 224–227. doi:10.1007/BF00360893

Kanai, T., Imanaka, H., Nakajima, A., Uwamori, K., Omori, Y., & Fukui, T. (2005). Continuous hydrogen production by the hyperthermophilic archaeon, Thermococcus kodakaraensis KODI. *Journal of Biotechnology*, *116*, 271–282. doi:10.1016/j.jbiotec.2004.11.002 PMID:15707688

Kim, H. Y. (2003). A low cost production of hydrogen from carbonaceous wastes. *International Journal of Hydrogen Energy*, *28*, 1179–1186. doi:10.1016/S0360-3199(02)00290-2

Kirk, O., Borchert, T. V., & Fuglsang, C. C. (2002). Industrial enzyme applications. *Current Opinion in Biotechnology*, *13*, 345–351. doi:10.1016/S0958-1669(02)00328-2 PMID:12323357

Kitchaiya, P., Intanakul, P., & Krairiksh, M. (2003). Enhancement of enzymatic hydrolysis of lignocellulosic wastes by microwave pretreatment under atmospheric pressure. *Journal of Wood Chemistry and Technology*, *23*(2), 217–225. doi:10.1081/WCT-120021926

Kothari, R., Tyagi, V. V., & Pathak, A. (2010). Waste-to-energy: A way from renewable energy sources to sustainable development. *Renewable & Sustainable Energy Reviews*, *14*, 3164–3170. doi:10.1016/j.rser.2010.05.005

Kumar, N., & Das, D. (2000). Enhancement of hydrogen production by Enterobacter cloacae IIT-BT08. *Process Biochemistry*, *35*, 589–593. doi:10.1016/S0032-9592(99)00109-0

Kunte, D. P., Yeole, T. Y., & Ranade, D. R. (2004). Two-stage anaerobic digestion process for complete inactivation of enteric bacterial pathogens in human night soil. *Water Science and Technology*, *50*, 103–108. PMID:15536996

Lay, J. J. (2000). Modeling and optimization of anaerobic digested sludge converting starch to hydrogen. *Biotechnology and Bioengineering*, *68*, 269–278. doi:10.1002/(SICI)1097-0290(20000505)68:3<269::AID-BIT5>3.0.CO;2-T PMID:10745195

Lay, J. J. (2001). Biohydrogen generation by mesophilic anaerobic fermentation of microcrystalline cellulose. *Biotechnology and Bioengineering*, *74*, 281–287. doi:10.1002/bit.1118 PMID:11410852

Lay, J. J., Lee, Y. J., & Noike, T. (1999). Feasibility of biological hydrogen production from organic fraction of municipal solid waste. *Water Research*, *33*, 2579–2586. doi:10.1016/S0043-1354(98)00483-7

Li, M., Xiang, J., Hu, S., Sun, L., Su, S., Li, P., & Sun, X. (2004). Characterization of solid residues from municipal solid waste incinerator. *Fuel*, *83*, 1397–1405. doi:10.1016/j.fuel.2004.01.005

Lin, C. N., Wu, S. Y., & Chang, J. S. (2006). Fermentative hydrogen production with a draft tube fluidized bed reactor containing silicone-gel-immobilized anaerobic sludge. *International Journal of Hydrogen Energy*, *31*, 2200–2210. doi:10.1016/j.ijhydene.2006.05.012

Liu, H., Zhang, T., & Fang, H. P. P. (2003). Thermophilic H_2 production from cellulose containing wastewater. *Biotechnology Letters*, *25*, 365–369. doi:10.1023/A:1022341113774 PMID:12882554

Lund, B., Jensen, V. F., Have, P., & Ahring, B. (1996). Inactivation of virus during anaerobic digestion of manure in laboratory scale biogas reactors. *Antonie Van Leeuwenhoek International Journal of General and Molecular Microbiology*, *69*, 25–31. doi:10.1007/BF00641608 PMID:8678476

Magnusson, L., Cicek, N., Sparling, R., & Levin, D. B. (2009). Continuous hydrogen production during fermentation of a-cellulose by the thermophilic bacterium Clostridium thermocellum. *Biotechnology and Bioengineering*, *102*, 759–766. doi:10.1002/bit.22092 PMID:18828175

Mahamat, A. Y., Gourdon, R., Leger, P., & Vermande, P. (1989). Methane recovery by anaerobic digestion of cellulosic materials available in Sahel. *Biological Wastes*, *30*, 181–197. doi:10.1016/0269-7483(89)90120-1

Manders, J. (2009). *Waste-to-energy as the dominant waste processing route across Europe*. Athens, Greece: Confederation of European Waste-to-Energy Plants (CEWEP).

Mata-Alvarez, J. (2002). *Biomethanization of the organic fraction of municipal solid wastes*. New York, NY: IWA Publishing.

Math-Alvarez, J., Mtzviturtia, A., Llabresluengo, P., & Gecchi, F. (1993). Kinetic and performance study of a batch 2-phase anaerobic digestion of fruit and vegetable wastes. *Biomass and Bioenergy*, *5*(6), 481–488. doi:10.1016/0961-9534(93)90043-4

McKay, G. (2002). Dioxin characterisation, formation, and minimisation during municipal solid waste (MSW) incineration Review. *Chemical Engineering Journal*, *86*(3), 343–368. doi:10.1016/S1385-8947(01)00228-5

Ministry of New and Renewable Energy. (2011). *Website*. Retrieved from http://mnes.nic.in/.

Moy, P., Krishnan, N., Ulloa, P., Cohen, S., & Brandt-Rauf, P. W. (2008). Options for management of municipal solid waste in New York City: A preliminary comparison of health risks and policy implications. *Journal of Environmental Management*, *87*, 73–79. doi:10.1016/j.jenvman.2007.01.032 PMID:17379391

Nie, Y. (2008). Development and prospects of municipal solid waste (MSW) incineration in China. *Environmental Science and Engineering*, *2*, 1–7.

OECD. (2008). *Environment directorate: OECD key environmental indicators*. Retrieved from http://www.oecd.org/dataoecd/20/40/37551205.pdf.

Okamoto, M., Miyahara, T., Mizuno, O., & Noike, T. (2000). Biological hydrogen potential of materials characteristic of the organic fraction of municipal solid wastes. *Water Science and Technology*, *41*, 25–32. PMID:11381999

Oral, J., Sikula, J., Puchyr, R., Hajny, Z., Stehlik, P., & Bbar, L. (2005). Processing of wastes from pulp and paper plant. *Journal of Cleaner Production*, *13*, 509–515. doi:10.1016/j.jclepro.2003.09.005

Ozbay, N., Putun, A. E., Uzun, B. B., & Putun, E. (2001). Biocrude from biomass: pyrolysis of cottonseed cake. *Renewable Energy*, *24*, 615–625. doi:10.1016/S0960-1481(01)00048-9

Papageorgiou, A., Barton, J. R., & Karagiannidis, A. (2009). Assessment of the greenhouse effect impact of technologies used for energy recovery from municipal waste: A case for England. *Journal of Environmental Management*, *90*, 2999–3012. doi:10.1016/j.jenvman.2009.04.012 PMID:19482412

Parizek, T., Bebar, L., & Stehlik, P. (2008). An persistent pollutants emission abatement in waste-to-energy systems. *Clean Technology and Environmental Policy*, *10*(2), 147–153. doi:10.1007/s10098-007-0135-2

Pavan, P., Battitoni, P., Gecchi, F., & Mataalvarez, J. (2000). Two phase anaerobic digestion of source sorted OFMSW (organic fraction of municipal solid waste): Performance and kinetic study. *Water Science and Technology*, *41*(3), 111–118. PMID:11381981

Pipatti, R., Sharma, C., & Yamada, M. (2006). Waste generation, composition and management data. In Waste, IPCC Guidelines for National Greenhouse Gas Inventories, (Vol 5), (pp. 2.1-2.23). IPCC.

Porteous, A. (2005). Why energy from waste incineration is an essential component of environmentally responsible waste management. *Waste Management (New York, N.Y.)*, *25*, 451–459. doi:10.1016/j.wasman.2005.02.008 PMID:15869989

Rachman, M. A., Furutani, Y., Nakashimada, Y., Kakizono, T., & Nishio, N. (1997). Enhanced hydrogen production in altered mixed acid fermentation of glucose by Enterobacter aerogenes. *Journal of Fermentation and Bioengineering*, *83*(4), 358–363. doi:10.1016/S0922-338X(97)80142-0

Ranade, D. R., Yeole, T. Y., & Godbole, S. H. (1987). Production of biogas from market wastes. *Biomass*, *13*, 147–153. doi:10.1016/0144-4565(87)90024-2

Ren, N., Li, J., Li, B., Wang, Y., & Liu, S. (2006). Biohydrogen production from molasses by anaerobic biohydrogen production from soluble starch. *International Journal of Hydrogen Energy*, *31*, 2147–2157. doi:10.1016/j.ijhydene.2006.02.011

Renewable Fuels Association. (1999). *Ethanol industry outlook 1999 and beyond*. Washington, DC: Renewable Fuels Association.

Sahlstrom, L. (2003). A review of survival of pathogenic bacteria in organic waste used in biogas plants. *Bioresource Technology*, *87*, 161–166. doi:10.1016/S0960-8524(02)00168-2 PMID:12765355

Scala, F., & Chirone, R. (2004). Fluidized bed combustion of alternative solid fuels. *Experimental Thermal and Fluid Science*, *28*, 691–699. doi:10.1016/j.expthermflusci.2003.12.005

Shi, H., & Kan, L. (2009). Leaching behavior of heavy metals from municipal solid wastes incineration (MSWI) fly ash used in concrete. *Journal of Hazardous Materials*, *164*, 750–754. doi:10.1016/j.jhazmat.2008.08.077 PMID:18838222

Shin, H. S., Youn, J. H., & Kim, S. H. (2004). Hydrogen production from food waste in anaerobic mesophilic and thermophilic acidogenesis. *International Journal of Hydrogen Energy*, *29*, 1355–1363. doi:10.1016/j.ijhydene.2003.09.011

Silverstein, R.A., Chen, Y., Sharma-Shivappa, R. R., Boyette, M. D., & Osborn, J. (2007). A comparison of chemical pretreatment methods for improving saccharification of cotton stalks. *Bioresource Technology*, *98*(16), 3000–3011. PMID:17158046

Singhabhandhu, A., & Tezuka, T. (2010). Prospective framework for collection and exploitation of waste cooking oil as feedstock for energy conversion. *Energy*, *35*, 1839–1847. doi:10.1016/j.energy.2010.01.004

Skonieczny, M. T., & Yargeau, V. (2009). Biohydrogen production from wastewater by Clostridium beijerinckii: Effect of pH and substrate concentration. *International Journal of Hydrogen Energy*, *34*, 3288–3294. doi:10.1016/j.ijhydene.2009.01.044

Skoulou, V., & Zabaniotou, A. (2007). Investigation of agricultural and animal wastes in Greece and their allocation to potential application for energy production. *Renewable & Sustainable Energy Reviews*, *11*, 1698–1719. doi:10.1016/j.rser.2005.12.011

Smet, E., Van Langenhove, H., & De Bo, I. (1999). The emission of volatile compounds during the aerobic and the combined anaerobic/aerobic composting of biowaste. *Atmospheric Environment*, *33*, 1295–1303. doi:10.1016/S1352-2310(98)00260-X

Stehlik, P. (2009). Contribution to advances in waste-to-energy technologies. *Journal of Cleaner Production*, *17*, 919–931. doi:10.1016/j.jclepro.2009.02.011

Sun, Y., & Cheng, J. J. (2005). Dilute acid pretreatment of rye straw and bermudagrass for ethanol production. *Bioresource Technology*, *96*(14), 1599–1606. doi:10.1016/j.biortech.2004.12.022 PMID:15978993

Tafdrup, S. (1995). Viable energy production and waste recycling from anaerobic digestion of manure and other biomass materials. *Biomass and Bioenergy*, *9*, 303–314. doi:10.1016/0961-9534(95)00098-4

Themelis, N. J. (2003). An overview of the global waste-to-energy industry. *Waste Management World*, *3*, 40–47.

Tricase, C., & Lombardi, M. (2009). State of the art and prospects of Italian biogas production from animal sewage: Technical-economic considerations. *Renewable Energy*, *34*, 477–485. doi:10.1016/j.renene.2008.06.013

United Nations Environment Programme. (1996). *International source book on environmentally sound technologies for municipal solid waste management*. Osaka, Japan: UNEP.

U.S. Department of Energy. (2011). *History of biofuels*. Retrieved from http://www.ott.doe.gov/biofuels/history.html.

Vamvouka, D., & Zografos, D. (2004). Predicting the behavior of ash from agricultural wastes using combustion. *Fuel*, *83*(14–15), 295–304.

Van Ginkel, S. W., Oh, S. E., & Logan, B. E. (2005). Biohydrogen production from food processing and domestic wastewaters. *International Journal of Hydrogen Energy*, *30*, 1535–1542. doi:10.1016/j.ijhydene.2004.09.017

Veziroglu, T. N., & Barbir, F. (1998). *Hydrogen energy technologies: Emerging technology series*. Vienna, Austria: UNIDO.

Williams, P. (2005). *Waste treatment and disposal* (2nd ed.). West Sussex, UK: John Wiley & Sons Ltd. doi:10.1002/0470012668

WTDC. (2006). *Website*. Retrieved from http://www.environment-agency.gov.uk/wtd/.

Wyman, C. E. (2003). Potential synergies and challenges in refining cellulosic biomass to fuels, chemicals, and power. *Biotechnology Progress*, *19*(2), 254–262. doi:10.1021/bp0256541 PMID:12675557

Yan, J., Chen, T., Li, X., Zhang, J., Lu, S., Ni, M., & Cen, K. (2006). Evaluation of PCDD/Fs emission from fluidized bed incinerators co-firing MSW with coal in China. *Journal of Hazardous Materials*, *135*, 47–51. doi:10.1016/j.jhazmat.2005.12.007 PMID:16442229

Yang, B., & Yanpin, L. (2007). Perspective: The promise of cellulosic ethanol production in China. *Journal of Chemical Technology and Biotechnology (Oxford, Oxfordshire)*, *82*, 6–10. doi:10.1002/jctb.1637

Yang, P., Zhang, R., McGarvey, J. A., & Benemann, J. R. (2007). Biohydrogen production from cheese processing wastewater by anaerobic fermentation using mixed microbial communities. *International Journal of Hydrogen Energy*, *32*, 4761–4771. doi:10.1016/j.ijhydene.2007.07.038

Yokoi, H., Saitsu, A., Uchida, H., Hirose, J., Hayashi, S., & Takasaki, Y. (2001). Microbial hydrogen production sweet potato starch residue. *Journal of Bioscience and Bioengineering*, *91*, 58–63. PMID:16232947

Yokoi, H., Tokushige, T., Hirose, J., Hayashi, S., & Takasaki, Y. (1998). Hydrogen production from starch by mixed culture of Clostridium butyricum and Enterobacter aerogenes. *Biotechnology Letters*, *20*, 143–147. doi:10.1023/A:1005372323248

Yu, H., Zhu, Z., Hu, W., & Zhang, H. (2002). Hydrogen production from rice winery wastewater in an upflow anaerobic reactor by mixed anaerobic cultures. *International Journal of Hydrogen Energy*, *27*, 1359–1365. doi:10.1016/S0360-3199(02)00073-3

Yu, H.-Q., & Mu, Y. (2006). Biological hydrogen production in a UASB reactor with granules: II: Reactor performance in 3-year operation. *Biotechnology and Bioengineering*, *94*, 988–995. doi:10.1002/bit.20923 PMID:16615161

Zhang, P., Michael, Y. H., Himmel, E., & Mielenz, J. R. (2006). Outlook for cellulose improvement: Screening and selection strategies. *Biotechnology Advances*, *24*, 452–481. doi:10.1016/j.biotechadv.2006.03.003 PMID:16690241

Zhang, Y., Chen, Y., Meng, A., Li, Q., & Cheng, H. (2008). Experimental and thermodynamic investigation on transfer of cadmium influenced by sulfur and chlorine during municipal solid waste (MSW) incineration. *Journal of Hazardous Materials*, *152*, 309–319. doi:10.1016/j.jhazmat.2007.08.054 PMID:17697749

This chapter was previously published in Global Sustainable Development and Renewable Energy Systems edited by Phillip Olla, pages 271-296, copyright 2012 by Information Science Reference (an imprint of IGI Global).

Chapter 86
Encouraging the Development of Renewable Energy:
The Role of Cooperatives

Eric Viardot
EADA Business School, Spain

ABSTRACT

The development of Renewable Energy (RE) has become a major societal challenge. Even though RE technology is improving fast, the general public has been slow to adopt it. Meanwhile many RE cooperatives have been created; they act as social entrepreneurs in utilizing the social capital of a community to engage the public. Thus, the purpose of this chapter is to consider how they can contribute to the adoption of RE. The author uses an exploratory research design with a multiple case approach drawing on nine RE cooperatives. The chapter discusses a number of actions implemented by cooperatives in order to develop the adoption of RE. These are related to the diminishing of the RE costs, the use of educational campaigns, and the setting of RE projects at local level; the specific status of cooperatives is another facilitating factor. The author concludes by drawing future directions for research and stressing the cultural challenges for the RE cooperative model.

INTRODUCTION

Energy consumption is accused of being the largest contributor to environmental degradation. The biophysical impacts implicit in our consumption decisions are well illustrated by the combustion of fossil fuels (Parker, et al., 2005). Within the past decade, environmental degradation and climate change have emerged as important issues on the global political agenda. So much interest has been generated that recent initiatives to reduce the amount of greenhouse gases, CO_2 in particular as the suspected key contributor to climate change, have been pursued on both national and international levels (Meyer & Koefold, 2003). The increased concern has lead to several international

DOI: 10.4018/978-1-4666-4852-4.ch086

conventions, such as the UN Framework Convention on Climate Change, the Kyoto Accord, and the Copenhagen Summit on Climate Change, designed to create legally enforceable greenhouse gas reduction targets among participating nations. In spite of the fact that the agreements have been met with unsatisfactory amounts of commitment and success, the political stimulus created by these events has reinvigorated the drive towards the adoption of RE as a method to reduce harmful emissions. The European Union has committed itself to its 20-20-20 targets; a 20% reduction in greenhouse gas emissions from 1990 levels with 20% supply of energy from renewable sources by the year 2020 (Danish Energy Agency Factsheet, 2010a). Within the context of geopolitical agreements and target setting, the level of RE development varies at the national level. Some nations are pursuing RE more aggressively and have achieved greater success.

In addition to the positive impact RE could have on the environment, is their relevance to energy security. One of the desired goals of sound energy policy is the provision of secure, diverse, sustainable and competitive energy supplies (Strachan & Lal, 2004). Energy security can be improved through local sourcing, reducing energy imports and the associated conflicts over fossil fuels (Federal Ministry for Environment, 2011). Another key component of energy security is the decentralization of electricity generation. Given that the distributed power generated from renewable sources satisfies these security concerns, one would be surprised at the unenthusiastic reaction to it.

However, the adoption of these sources is being resisted. The problem of convincing government, business, and consumers of the importance of RE sources is being manifest in a lackluster rate of uptake. While politicians and the business world tend to focus on the ramifications of reduced energy security, a potentially more acute consequence of inaction will be the environmental impact.

Economic inertia may justify the slow acceptance of RE as modern economies have been designed for cheap, concentrated, and centrally distributed fuel sources. The infrastructure to support this economic model has developed over decades, and it will take decades to transition towards distributed energy. From a purely economic perspective, RE is less cost effective than traditional fuel sources, and cannot compete in a free-market economy. Ignoring the economic cost of renewable energy, the issue of reliability is often referred to as a disadvantage.

RE sources are diffuse and intermittent. While they can be harnessed using rudimentary technologies, they are less easy to predict and control. The technologies which have been innovated to extract the power have increased in complexity over time to the point where they are considered to be high tech. The inherent consumer uneasiness for high tech products combined with inconsistent power generation has made trust in RE difficult to foster. These criticisms could be surmounted if one was dealing with a more educated and conscientious consumer. The lack of informed decisions being made stems from a lack of education on the long-term impact of our energy consumption choices, and a general apathetic attitude towards environmental issues. Clearly, there are hurdles to overcome if RE is to achieve widespread acceptance in a timely fashion.

In addition to public policy and private initiatives, cooperatives have recently appeared with the goal to promote the use of RE. One noteworthy example is the Samso Project in Denmark. Samso was developed by a cooperative with local residents and the municipality as its members, and exists as an island off the coast of Jutland with electricity supplied entirely from alternative sources.

In Germany, one of the most well known RE cooperatives is Greenpeace Energy. Founded in 1999, it boasts 18,000 members, 95,000 customers, and an annual turnover in 2009 of 86€ million. This cooperative has achieved a sufficient purview to plan, construct, and operate its own power plants, and, to date, nine have been commissioned producing a combined capacity of 33.7 MW (DGRV, 2010).

A last illustration is the case of the German village of Juhnde; residents went as far as to source all of their heat and electricity from biomass, thus rendering the village autonomous in terms of energy. The biomass is supplied by local agriculture and forestry, and the government grant supplied the capital necessary to purchase the generation equipment. This project was facilitated through the formation of a village cooperative with more than 70% of the inhabitants becoming members, each contributing 1,500€ to receive voting rights and have their homes connected to the district heating. An evaluation of the project reveals how the active participation of the village population, combined with a well functioning social network, lead to the creation of a self-sustaining community (IEA Bioenergy, 2010).

Overall, Danish cooperatives own about 15% of wind turbines today (Danish Wind Turbine Owners Association, 2010), while as of 2010, there were 95 energy and water cooperatives in operation in Germany. Thus this chapter will address the question of how cooperative can be an appropriate, business model to encourage the adoption of RE sources.

Our chapter is structured as follows: we start with a presentation of public policy on RE as it delineates the context of intervention of RE cooperatives; then we introduce the main characteristics of RE cooperatives in terms of organization and status; next we build on the literature analysis about the barriers to development of RE in order to introduce our research in term of methodology and scope; we then discuss the key findings about the role of cooperatives in the development of RE before drawing future directions for research and concluding about the cultural challenge that the RE cooperative model is facing.

PUBLIC POLICY ON RENEWABLE ENERGY

In addition to concern for the environmental effects of traditional electricity generation methods, a second force behind the growing interest in alternative energy sources is the trend towards the restructuring of many public power-generating institutions (Rowlands, et al., 2002). The privatization of the electricity market has allowed consumers to now choose where they would like to have their electricity sourced from. Privatization, which began in the 1980's, was expected to encourage the implementation of RE sources as market forces would stimulate the industry.

However, if left to market forces, RE technologies are not as cost competitive as fossil fuels, such as coal and natural gas. As a way of quantifying this statement, one can examine the cost of electricity in Germany. The cost of procurement of base-load power, as measured on traded markets was set at 5.7 cents/kWh. This figure can be compared with the remuneration rates for new wind or solar power plants at 30 cents/kWh and 8 cents/kWh respectively (Federal Ministry for Environment, 2011). Premium pricing programs were introduced to allow customers who wished to buy green power the opportunity to do so for a marginal fee. The extra cost would be used to finance the construction of RE projects. These programs received some success, though not enough to have a substantial impact on the market.

The public's perception of these programs was that they were of a token nature and were being used to green the image of utilities, rather than support a young industry. The shares of residential customers who have switched their electricity utility since the dissolution of the regional monopolies remains low; in 2002 these shares were 3% in the U.S., 15% in Norway, 14% in Sweden, 5% in Finland, and 4% in Germany (IEPM, 2002).

In order to assist the market uptake of RE and create a sustaining demand for the technology, markets have relied on governments to subsidize the industry. Without government intervention the short-termism, or boom/bust disjointed direction of industry development policy would leave little chance for the alternative energy industry to establish itself and drive down the procurement prices as forecasts are expecting (Jones, 2009). Short-term concerns over how a nation's global economic competitiveness will be affected by the shift towards RE sources are balanced by the proponents of a long-term energy policy. Business groups from the latter school of thought believe that the risk of fossil fuels becoming increasingly scarce will put nations that delay the adoption of greenhouse gas reduction measures at a disadvantage in the long-term through abatement measures and energy costs (ABRCC, 2006). Two nations in particular, Germany and Denmark, are generally considered to be the early movers in sustainably sourced electricity.

Over the past 30 years, Denmark has developed an expansive RE portfolio. Wind energy comprises the vast majority of the power generated from renewable sources and currently represents roughly 19% of the total Danish electricity supply. This translates to more than 5000 wind turbines with a total capacity of 3,200 MW. The Danish government has stated they would like to increase the share of RE in power generation to 30% by 2020 (Danish Energy Agency Factsheet, 2010b). The Danish government has been supporting its RE industry since 1976. Between 1976 and 1989, the state granted investment support of up to 40% of the cost of a wind turbine installation. Between 1991 and 2009, a price premium was offered on green power and the deployment of wind turbines increased dramatically during this period. Upon the privatization of the electricity supply market in 1999, until present day, a FIT program has been in place to support the industry (Danish Energy Agency, 2010). Public support for RE in Denmark has resulted in the creation of what are known as eco-cities. These are communities with ambitious sustainability targets apart from national and regional environmental regulations.

Germany is also considered to be a good model of how public policy and community support can intensify the movement towards sustainable energy supply. Although the German RE industry did not take off until the early 1990's, it has grown into a significant economic stimulant since then with a total turnover in 2009 of 33.4€ billion, and an industry work force of 300,500. In 2009, the installed RE capacity was measured at 45.3 GW. This capacity supplied 16.1% of electricity consumption, and 10.1% of total final energy consumption, which takes into account the enormous amount of energy required for transportation. Germany has set ambitious targets, far exceeding the EU directives, for the growth of sustainably sourced energy. The nation is aiming for a 30% share of RE sources in gross electricity consumption by 2020, and an 18% share in total final energy consumption.

Historically, the nations with the most robust RE industries have achieved their advancements through Feed-In-Tariff (FIT) programs coupled with Power Purchase Agreements. A power purchase agreement dictates that the Local Distribution Company (LDC) must purchase sustainably sourced electricity from a private producer at an above market price. This agreement is supported by a FIT contract guaranteeing a fixed price paid to the producer for every kilowatt of electricity produced over the length of the contract. Given that it is public money financing the FIT program, the government mandates LDC's to marginally increase the energy costs to all electricity consumers. This differential cost, also known as a public service obligation, provides the government revenue necessary to maintain the FIT program (Danish Energy Agency, 2010). The differential cost is relatively small in relation to the affluence and disposable income characteristic of consumers in developed nations. In Germany, the added procurement cost for utilities under an obligation to purchase privately produced power totaled 4.6€ billion. This sum is passed on to all consumers based on their level of consumption, with the exception of those businesses granted exemption, resulting in a 1.1 cent/kWh price increase. Given that the average annual household energy consumption in Germany is 3500 kWh, the annual differential cost to consumers is a marginal 39€, or roughly 5% of their electricity cost (Federal Ministry for Environment, 2011).

Since FIT programs provide long term, government backed, contracts, a secure investment environment is created. These programs have proved to be very successful in Denmark, Germany, and Spain. Recently, the Province of Ontario, Canada, has introduced a FIT program modeled after the German initiative.

COOPERATIVES AND RENEWABLE ENERGY

The Cooperative Business Model

Cooperative organizations have been participants in global economics since the middle of the 19th century. They are a well established business form that has persisted throughout the changing political atmospheres of the past 200 years. The unique functionality of a cooperative has made them well suited for various economic activities, one of which is the recent development of renewable energy. In order to set the context of a RE cooperative, it is first necessary to describe the general business form and its attributes.

The general paradigm of cooperatives can be simply described as an autonomous association of persons untied voluntarily to address their common economic, social, and cultural needs through a jointly-owned and democratically-controlled business form (Danish Wind Turbine Owners Association, 2010). Cooperatives are generally classified within the nonprofit economic sector, or the so-called third sector.

The guidelines by which cooperatives operate can be summarized by a set of 7 principles (International Cooperative Alliance, 2010):

1. **Voluntary and Open Membership:** any individual, without discrimination, willing to assume the responsibilities of membership.
2. **Democratic Member Control:** The principle of one member-one vote is enforced in primary cooperatives. Members actively participate in policy setting and decision making in a democratic process. The elected management representing the organization is accountable to the members.
3. **Member Economic Participation:** Members democratically control the capital of their cooperative. Potential surpluses can be used for dividends, internal investment, supporting external activities and setting up reserves.
4. **Autonomy and Independence:** If contracts are entered into with private enterprises or public institutions, the cooperative ensures that democratic control by its members is not jeopardized.
5. **Education, Training and Information:** Cooperatives assume the responsibility of educating the general public as to the benefits provided by their organization.
6. **Cooperation among Cooperatives:** Cooperatives are charged with strengthening the cooperative movement through exchange and collaboration with other cooperatives.
7. **Concern for the Community:** The economic motives of a cooperative are primarily designed to encourage sustainable development of their communities.

RE Cooperatives as For-Profit Social Ventures

The majority of RE Cooperatives are For-profit social ventures, i.e. organizations concerned with the double bottom line of the creation of economic value, and the creation of social value. RE cooperative are for-profit social venture as they are producing and selling electricity derived from renewable sources. They are maintaining a reasonable economic return, while simultaneously reducing pollution, the dependency on oil as well as the damage wrought from drilling in sensitive ecosystems.

There are two principle benefits of the for-profit structure for a cooperative. First, being incorporated as for-profit enables access to the private capital market for investment funds, thus reducing the demand for scarce public and philanthropic funds that could be used to subsidize more needy projects. The second benefit is that for-profit ventures have been proven to respond quicker to changing market conditions. This market responsiveness can be an advantage for spreading innovations at a faster rate (Hansmann, 1996).

There is certainly complexity in combining two seemingly incompatible objectives, as the social value being created is more difficult to measure, and may be overshadowed by more quantifiable financial figures. There is also the pressure to dismiss inefficient social objectives if they infringe on the competitiveness of the firm, or vice versa (Dees & Anderson, 2003). But Cooperatives are well suited to deal with these issues thanks to their democratic control mechanisms.

LITERATURE ANALYSIS: THE BARRIERS TO DEVELOPMENT OF RENEWABLE ENERGY

Given the relative infancy of the RE industry, the literature accumulated on the promotion and marketing of this technology is recent, and for the most part exploratory. In researching for this chapter several recurring themes were uncovered describing the state of knowledge on the barriers to RE uptake. These themes ranged from the more tangible and easily analyzed topics of cost and reliability, to the less conclusive topics of demographics, education, environmental awareness, consumer attitudes, purchasing behaviour, immediacy and personal relevance, partnerships, and community-based marketing. The significance of each of these topics to the development of RE will now be addressed.

Cost

Cost is of paramount importance to the consumer considering supporting renewable energy. In concert with cost as a consideration of consumers is the reliability of renewable sourced power to meet their electricity needs.

Consumer studies have returned cost as a particularly important priority when dealing with an electricity supplier, trumping environmental considerations (MORI, 2001; Losch, 1998). With respect to price premiums offered by electric utilities for renewably sourced power, price was listed among the top three motivating factors for why a customer would potentially switch electricity supplier (Rowlands, et al., 2003; Rowlands, et al., 2004).

Even within a FIT market, individual consumers are being asked to make large initial expenditures in return for lower energy costs over time, and a 10 year wait period, on average, before the principle is recovered (Guardiola, Gabay & Moskowitz, 2009).

Reliability

Reliability concerns are obvious when one questions the ability of an energy source that is not always present to be a continuous and dependable supplier of electricity. In the United Kingdom, for example, the fact that the wind is not completely reliable as a source of energy, coupled with the lack of effective and efficient power storage systems raises issues for wind energy. Back-up, or base-load, power would need to be maintained, and while the cost of supporting technologies, such as batteries, will fall in the future, it remains today a significant barrier (Strachan, et al., 2006). Two of the published articles already mentioned citing cost (Losch, 1998; Rowlands, et al., 2004) also name reliability as a priority for consumers when dealing with their electricity supplier. While reliability is certainly a legitimate factual consideration, many of the barriers as described in the literature are founded on a fundamental lack of knowledge. Awareness building subsequently emerges as a powerful promoter, especially if delivered appropriately.

Education, Awareness, and Demographics

Education levels have been linked to greater concern for the environment, and greater likelihood of participation in environmental protection activities (Granzin & Olsen, 1991). In attempting to profile a potential green energy purchaser, Rowlands et al. (2003) used the demographic characteristics of income, education, age, gender, and greater knowledge about energy issues in their community. Interestingly enough, only gender and age were disproved as positively correlating characteristics, although the author cautioned that the sample was not completely representative of the Canadian population. Wealthier, more formally educated and more energy aware consumers were more likely to purchase green energy.

Similarly, Parker et al. (2005) have shown that an awareness of the environmental effects of one's choices leads to changes in energy consumption patterns, including from which sources they receive their electricity). A more recent report published by the Danish Wind Turbine Owners' Association in 2009, stressed that wide support within that nation's population for wind energy (90% of those polled) was not translating into accelerated project development due to uncertainty and skepticism at the local level as a result of a lack of information in the planning phase of a project.

Consumer Attitudes and Behaviors

Renewable energy's environmental and risk diversification benefits are public goods and tend to be undervalued by electricity producers, utility regulators, legislators, and consumers (Guardiola, et al., 2009). Consumers in particular have been rather lackluster in adopting renewable energy. The intangibility of the product, relatively new technology, and non-conformity within the marketing practices of RE firms have led to skepticism by consumers. Consumers claim that implementing energy saving practices or equipment takes too long (Barr, et al., 2003; Parker, et al., 2005) or believe that their individual contribution will have no effect (Wiser, 1998).

Using reasoned action theory to analyze consumer attitudes towards RE. Bang et al. (2000) showed was that those who were categorized as being more concerned for the environment, exhibited more willingness to pay a premium price for renewable energy; and consumers with a strong belief in the positive consequences will be more likely to pay a premium price.

A second theory, the contingent valuation method, exposed a behavioral trait among some energy consumers known as free-riding behavior (Byrnes, et al., 1999; Wiser & Pickle, 1997). Contingent valuation method measures the value a group of people place on a resource, or service, by asking them the maximum amount they would be willing to pay for this item. Free-riding behavior is the belief that a program will be implemented with or without one's support, thus providing one with all the public benefits without any personal cost, was named as the reason for this discrepancy. Free-riding was also cited as a concern by Blank et al. (1999). The research report discovered that some groups viewed voluntary payment programs as inappropriate since RE investments are a public good and should receive ubiquitous support. This sentiment has been decisively addressed through the trend towards FIT stimulus programs.

Finally, according to Rowlands et al. (2002), consumers appear most willing to pay a premium price for wind and solar power sources as wind and solar are almost unanimously considered green sources of power. But there are disparate opinions regarding the level of environmental impact for hydro, natural gas, biomass, and incineration.

METHODOLOGY

The research design used is an exploratory research design adopted in order to discover how RE cooperatives are using marketing to escape the barriers to the uptake of RE. A multiple case approach was deemed most appropriate and more robust than a single case study as the potential benefits of data richness, depth and quality, compensate for the associated shortcomings of limited representativeness and generalization (Eisenhardt, 1989; Strauss & Corbin, 1994; Yin, 2003; Collis & Hussey, 2003).

The sample selected to be interviewed was assembled purposefully. Nine RE cooperatives were selected as the source of primary data for this research because it was necessary to learn about the operations of these organizations within their communities. External sources to the organization could be used for triangulation; however, their input could be supplementary at best.

RE cooperatives is still a relatively new phenomenon within the cooperative realm, they are confined to markets where government incentives are subsidizing the use of RE. So, at first, we contacted 10 cooperatives in Canada and 10 in Europe. Based on preliminary attempts to contact cooperatives in Europe and North America, and secure a willing participant for the interview process, it was discovered that the European firms were less willing to cooperate. The language barrier certainly inhibited the development of a rapport. It is also speculated that the cultural barrier, or conversely, the cultural familiarity and loyalty of Canadian cooperatives, caused European firms to be more reluctant to provide assistance.

Ultimately, seven Canadian, one Danish, and one British cooperatives were selected to be within the sample, as the other ones were not available for answering our questions. Each of the cooperatives in the sample was pursuing, or had already installed, either solar panels or wind turbines. Table 1 provides the nominative list of the selected coops.

The question of who was going to be asked to procure information from each cooperative in the sample was approached as follows. If someone was listed on the website as being the person to contact for information or inquiries, they were contacted. When no one was specified, the decision was left to the organization as to who would provide the interview, or complete the questionnaire if an interview was not possible. This was deemed an appropriate method to ensure that an authority on the topic with direct experience in the functioning of the cooperative would be reached.

Table 1. Alphabetical list of the 9 cooperatives

Name	Location	Date of Creation
Barrie Windcatchers	Ontario, Canada	2009
Countryside Energy Coop Inc	Ontario, Canada	2005
Energy4All	Cumbria, England	2002
LIFE Cooperative	Ontario, Canada	2008
Middelgrunden Wind Turbine Cooperative	Copenhagen, Denmark	1996
Power Up R.E. Coop (PURE)	Ontario, Canada	2004
Superior R.E. Coop (SREC)	Ontario, Canada	2002
Sustainable Ottawa Community Energy Cooperative	Ontario, Canada	2007
Toronto Renewable Energy Cooperative (TREC)	Ontario, Canada	1999

In the questionnaire, which was posed to each key informant in the sample cooperatives, the interview questions focused on the barriers to development of RE, or hindrances. The key informants were asked how their cooperative was addressing each of the hindrances outlined in the questionnaire. This questionnaire can be seen in the Appendix.

Given the broad geographic region being sampled, the interviews were conducted remotely over the telephone. The hindrances, and their associated questions, were divided into six themes: physical, financial, educational, ontological, social, and legal. The reason for this was to group similar questions, providing the interviews with a logical flow. Within this framework, a semi-structured interview style was adopted to allow some flexibility in the order of questions, the adherence to each individual question and to allow the informant to digress if it was deemed that the conversation was proceeding along a relevant path. If additional questions arose that had not been predicted prior to the interview, they were asked at the end. Each interview lasted between 45 and 60 minutes. The interviews were recorded, with the consent of the informant, to facilitate transcription and coding of the responses. Due to the semi-structured nature of the interviews, transcription was not performed on the entire conversation, but, rather, as each question posed was directly addressed, the response was noted. Furthermore in order to increase the rigor of the information being analyzed, we performed a content analysis (Rokeach, 1979) of the interview conversations to cross-check the potentially unreliable information of the informants with other written sources: publicly available records such as websites, annual reports, cooperative by-laws, and independent assessments.

MAIN RESULTS

Insights from the explorative interviews help identify three main categories of actions which are implemented effectively by the RE cooperatives in order to lower the barriers to the adoption of RE. They are related with the diminishing of the RE costs, the use of educational campaigns, and the setting of RE projects at local level. Additionally, the specific status of cooperatives contributes in reducing the hurdles to the uptake of RE.

Diminishing the Costs

The cooperative business model allows lessening the impact of the cost of RE systems. Considering the minimum investment into a RE project was \$100, only the most destitute of consumers are excluded. The low investment requirement essentially eliminates the need to access credit in order to make an investment. The risk of investing in RE is limited by the legal restrictions on borrowing placed on a cooperative, the government backed FIT contracts, and feasibility studies to determine the expected electrical output of a system. The return on investment one could expect through participation in a cooperative was more attractive than conservative investment vehicles; yielding a tax protected annual return averaging 5-6%. One area where RE cooperatives did not appear to be capable of surmounting a known barrier was with respect to the recovery time of an investment. Based on the average return offered, an investor can expect to wait at least ten years before his principle would be recovered. For an investor who no longer can afford to maintain a share in a project, the internal share exchange that is organized by each of the cooperatives alleviates any worry of stranded capital. The third group of questions posed to the sample dealt with their activities to educate the public.

As a way of assisting the low income consumer group, several cooperatives would bundle systems for multiple projects to take advantage of bulk purchase discounts, and pass those savings onto the investors. Those with difficulty accessing credit could rely on the support of the cooperative to negotiate bank loans. The community developers and mature stage cooperatives were able to leverage their partnerships with local vendors, in the form of preferential rates and service exchanges, to the benefit of their investors. The most interesting way a cooperative was able to reduce the risk of a large project was organized by TREC. The cooperative had partnered with a large private developer who agreed to front all the money for the installation of a large rooftop solar project. TREC had secured a guarantee from the developer stating that TREC could then buy out an agreed upon portion of the electricity being produced as soon as it was able to raise the share capital from the community.

Education Campaigns

Cooperatives are extremely good with education campaigns. The educational methods employed are diverse. They range from traditional media (print, radio, television, and web) to more directed forms, such as public meetings, expositions, community engagement through efficiency programs and sponsoring of related initiatives. Numerous innovative educational initiatives can be mentioned across all cooperative categories. Several of which incorporated educational material in the form of academic and technical studies as a way of lending credibility to the information being presented to the public. Those cooperatives that had already installed a system were able to provide the public with live tours, giving substance to these projects in the minds of the public. The systems that were already in operation were producing data on their output and were being used to show the reliability of the power supply.

Furthermore, by marketing through public presentations, expositions, and interpersonal connections (word of mouth, personal references) the marketing costs are kept low without sacrificing the effectiveness of the campaign.

One should also noticed that the community cooperatives are also well suited to deal with the unique issues that will motivate their residents as the coops are organized and managed by members of the community they serve.

One of the outcomes of educating a community is the utilization of social capital in order to increase the capability of a community to produce meaningful and sustainable development. Education is both a determinant of and determined by the society in which it is located, acting as an agent of change, and is in turn changed by society (Fagerlind & Saha, 1985). By educating the public, RE cooperatives are enhancing the receptiveness of that population for the deployment of a system in their neighborhood.

Local Projects

Cooperatives contribute to the emergence and implementation of local projects, which reduce a major barrier to the adoption of RE. It is the impression amongst consumers that their personal choices have no impact on the fight against environmental degradation. The local projects serve to enlighten a community, and make them cognizant of where their energy comes from. When the projects are clearly visible and producing power, the impact that an individual's energy choices have becomes apparent. Intuitively, cooperatives remove the worry that installing RE in a neighborhood will ostracize an individual; community power is organized by, and developed for, the local population. The issue of distaste for the appearance of the technology is difficult to overcome since it is derived from a personal preference; however, the democratic decision making process used to decide where a system will be installed manages this concern in the most equitable way possible. A cooperative is also ideally designed to remove the perturbance that an investor would be faced with if they were to undertake the installation process privately. All project management and administrative work is the responsibility of the management team.

Finally, the legal preoccupations that coincide with all the administrative work necessary to develop a RE project are managed entirely by the cooperatives' executive team. Any personal liability that may deter a potential investor is minimized through comprehensive insurance and warranty coverage.

Specific Status of Cooperatives

Finally, one should note that the specific status of cooperatives also contributes to diminish some obstacle to the adoption of RE. For instance, cooperatives are ideally designed to remove the perturbation that investors would be faced with if they were to undertake the installation process privately because all project management and administrative work is the responsibility of the management team.

Secondly, as members' personal properties are not necessarily involved in a coop, because it is a non-profit organization, it lessens the worry about a potential diminishing resale value of a property equipped with an RE system.

The RE coop distinctive status also eases another possible issue, the distaste for the appearance of the technology, which is difficult to overcome since it is derived from a personal preference; however, the democratic decision making process used to decide where a system will be installed manages this concern in the most equitable way possible.

FUTURE RESEARCH DIRECTIONS

The results of this chapter were made possible thanks to the willingness of the sample cooperatives to participate. Whether or not cultural familiarity was in fact a determinant, the sample that emerged was weighted heavily with Canadian organization. It would be beneficial if a similar study were to be performed with a focus on German and Danish cooperatives, as these nations have a longer history of RE use, and their citizens show a greater affinity for the cooperative business model pertaining to energy production.

A second limitation was created by the minimal range of technologies being analyzed. RE cooperatives are also prevalent in biogas and biomass sources. Wind and solar were approached in this chapter for their simplicity and notoriety. Biogas and biomass are also important to a comprehensive RE portfolio, and should be given attention. Technologies such as hydropower and geothermal heating and cooling are also worth discussing. The remaining RE (wave, tidal and osmotic pressure) could certainly be evaluated once applications of these technologies are successfully deployed and a body of data has been established.

CONCLUSION

In this chapter, we have demonstrated that RE cooperatives create the promotional factors necessary to convince the public of the value in engaging in a community based project and are effective at reducing the barriers to personal involvement in a project.

Yet, the cooperative business model is not without its limitations as the majority of consumers are accustomed to their needs being met through public institution, or private interests. In some countries, cooperatives are even viewed with a pseudo-communist stigma, which has been demonized by private enterprise. Reconditioning the public to accept the idea that a cooperative organization can serve them better than the private—or the public—model will take time and precedent.

REFERENCES

ABRCC. (2006). *The business case for early action*. Retrieved from http://www.businessroundtable.com.au/pdf/F078-RT-WS.pdf.

Bang, H., Ellinger, A. E., Hadjimarcou, J., & Traichal, P. A. (2000). Consumer concern, knowledge, belief, and attitude toward renewable energy: An application of the reasoned action theory. *Psychology and Marketing*, *17*(6), 449. doi:10.1002/(SICI)1520-6793(200006)17:6<449::AID-MAR2>3.0.CO;2-8

Barr, S., Ford, N. J., & Gilg, A. W. (2003). Attitudes towards recycling household waste in Exeter, Devon: Quantitative and qualitative approaches. *Local Environment*, *8*(4), 407–421. doi:10.1080/13549830306667

Bioenergy, I. E. A. (2010). *Biogas in the society*. Retrieved from http://www.iea-biogas.net/Dokumente/casestudies/biogas_village.pdf.

Blank, E., Mayer, R., & Swezey, B. (1999). The grassroots are greener: A community-based approach to marketing green power. *Renewable Energy Policy Project*. Retrieved from http://www.repp.org.

Byrnes, B., Jones, C., & Goodman, S. (1999). Contingent valuation and real economic commitments: Evidence from electric utility green pricing programmes. *Journal of Environmental Planning and Management*, *42*(2), 149–166. doi:10.1080/09640569911190

Collis, J., & Hussey, R. (2003). *Business research: A practical guide for undergraduate and postgraduate students*. London, UK: Palgrave Macmillan.

Danish Energy Agency. (2010). *The history of Danish support for wind power.* Retrieved from http://www.ens.dk/enUS/supply/Electricity/Conditions_for_production_plants/Subsidies_for_generation_of_electricity/Sider/Forside.aspx.

Danish Energy Agency Factsheet. (2010a). *Denmark - A success story about wind power*. Retrieved from http://www.ens.dk/Documents/Faktaark/Engelske%20faktaark/vindmøller%20engelsk%20240709.pdf.

Danish Energy Agency Factsheet. (2010b). *Energy makes its mark on the European agenda*. Retrieved from http://www.ens.dk/en-US/Info/news/Factsheet/Documents/EU.pdf%20engelsk.pdf.

Danish Wind Turbine Owners Association. (2010). *Cooperatives- A local and democratic ownership to wind turbines*. Retrieved from http://www.dkvind.dk/eng/faq/cooperatives.pdf.

Dees, J. G., & Anderson, B. B. (2003). *For profit social ventures: Social entrepreneurship*. Dublin, Ireland: Senate Hall Academic Publishing.

DGRV. (2010). *Facts and figures annual report*. Retrieved from http://www.greenpeace-energy.de/index.html.

Eisenhardt, K. M. (1989). Building theory from case study research. *Academy of Management Review*, *14*(4), 532–550.

Fagerlind, I., & Saha, L. J. (1985). *Education and national development: A comparative perspective*. Oxford, UK: Pergamon.

Federal Ministry for Environment. (2011). *Development of renewable energy sources in Germany*. Retrieved from http://www.erneuerbare-energien.de/inhalt/5996/42456/.

Federal Ministry for Environment. (2011). *Electricity from renewable sources - What does it cost?* Retrived from http://www.esru.strath.ac.uk/EandE/Web_sites/01-02/RE_info/hec.htm.

Granzin, K. L., & Olsen, J. E. (1991). Characterizing participants in activities protecting the environment: A focus on donating, recycling, and conservation behaviors. *Journal of Public Policy & Marketing*, *10*(2), 1–27.

Guardiola, S., Gabay, G., & Moskowitz, H. (2009). Renewable energy: Tapping and typing the citizen's mind. *Humanomics*, *25*(4), 254–267. doi:10.1108/08288660910997629

Hansmann, H. (1996). The changing roles of public, private, and nonprofit enterprises in education, health care, ad other human service. In V. R. Fuchs (Ed.), *Individual and Social Responsibility: Child Care, Education, Medical Care, and Long Term Care in America*. Chicago, IL: University of Chicago Press.

IEPM. (2002). *IEPM news*. Augusta, ME.

International Cooperative Alliance. (2010). *What is a co-op?* Retrieved from http://www.ica.coop/al-ica/.

Jones, S. (2009). The future of renewable energy in Australia: A test for cooperative federalism. *The Australian Journal of Public Administration*, *68*(1), 1–20. doi:10.1111/j.1467-8500.2008.00606.x

Losch, M. E. (1998). *Iowa electric utility customer survey: Perceptions of electric utility restructuring*. Iowa City, IA: The University of Iowa Social Science Institute.

Meyer, N. I., & Koefold, A. L. (2003). Danish energy reform: Policy implications for renewables. *Energy Policy*, *31*(7), 597–607. doi:10.1016/S0301-4215(02)00145-3

MORI. (2001). *Experience of the competitive market: The domestic electricity and gas markets*. London, UK: Market and Opinion Research International.

Parker, P., Rowlands, I. H., & Scott, D. (2005). Who changes consumption following residential energy evaluations? Local programs need all income groups to achieve Kyoto targets. *Local Environment*, *10*(2), 173–187. doi:10.1080/1354983052000330761

Rokeach, M. (1979). From individual to institutional values: With special reference to the values of science. In M. Rokeach (Ed.), *Understanding Human Values* (pp. 47–70). New York, NY: Free Press.

Rowlands, I. H., Parker, I. H., & Scott, D. (2002). Consumer perceptions of "green power." *Journal of Consumer Marketing*, *19*(2/3), 112. doi:10.1108/07363760210420540

Rowlands, I. H., Parker, P., & Scott, D. (2004). Consumer behaviour in restructured electricity markets. *Journal of Consumer Behaviour*, *3*(3), 272–283. doi:10.1002/cb.140

Rowlands, I. H., Scott, D., & Parker, P. (2003). Consumers and green electricity: Profiling potential purchasers. *Business Strategy and the Environment*, *12*(1), 36. doi:10.1002/bse.346

Strachan, P. A., & Lal, D. (2004). Wind energy policy, planning and management practice in the UK: Hot air or a gathering storm? *Regional Studies*, *38*(5), 551–571. doi:10.1080/0143116042000229311

Strachan, P. A., Lal, D., & Von Malmborg, F. (2006). The evolving UK wind energy industry: Critical policy and management aspects of the emerging research agenda. *European Environment*, *16*, 1–18. doi:10.1002/eet.400

Strauss, A. S., & Corbin, J. (1994). Grounded theory methodology. In N. K. Denzin, & Y. S. Lincoln (Eds.), *Handbook of Qualitative Research* (pp. 273–285). Thousand Oaks, CA: Sage.

Wiser, R., & Pickle, S. (1997). *Green marketing, renewables, and free riders: Increasing customer demand for a public good*. Berkeley, CA: Berkeley National Laboratory. doi:10.2172/645498

Wiser, R. H. (1998). Green power marketing: Increasing demand for renewable energy. *Utilities Policy*, *7*(2), 107–119. doi:10.1016/S0957-1787(98)00005-8

Yin, R. K. (2003). *Case study research: Design and methods*. Thousand Oaks, CA: Sage.

KEY TERMS AND DEFINITIONS

Barriers to Development: In business strategy, the barriers to development are the various forces which are preventing the growth of a product or a technology or a company. In the case of RE the barriers to adoption are mostly related to costs, reliability of the technology, education, demographic and attitudinal characteristics of the consumers.

Cooperatives: Cooperatives are autonomous association of persons untied voluntarily to address their common economic, social, and cultural needs through a jointly-owned and democratically-controlled business form.

Feed-in-Tariff (FIT) Programs Power Purchase Agreements (PPA): Feed-in-Tariff (FIT) programs coupled with Power Purchase Agreements (PPA) are public programs design to encourage the adoption of sustainably sourced electricity. A PPA dictates that the local distribution company (LDC) must purchase sustainably sourced electricity from a private producer at an above market price. This agreement is supported by a FIT contract guaranteeing a fixed price paid to the producer for every kilowatt of electricity produced over the length of the contract.

RE Cooperatives: RE cooperatives are cooperative whose major goal is to promote the use of RE. Some famous Re cooperatives are the Samso Project in Denmark and Greenpeace Energy in Germany.

Renewable Energy (RE): Comes from naturally replenished resources such as sunlight, wind, tides, rain, and geothermal heat.

Technology Adoption (or Uptake): The rate of technology adoption by users or customers is a key driver of the success of any technology driven innovation such as RE. The concept of technology adoption is a central concept in the Technology Acceptance Model (TAM) provides the theoretical background to analyze the development of new technologies.

APPENDIX: QUESTIONNAIRE

Technological Hindrances

1. **Unfamiliarity with the technology**- How does a R.E. Coop familiarize the public with the technology involved in harnessing renewable energy sources, and thereby reduce the fear associated with unfamiliar technology?
2. **Lack of awareness of the need for RE**- How does a R.E. Coop enlighten the public of the justification for using R.E.?
3. **Lack of awareness of the benefits, for the community and the environment**- How does a R.E. Coop explain the environmental and quality of life benefits of R.E.? In real terms...in terms that will be relevant to the public being addressed.
4. **The opinion that the technology is unreliable**- How does a R.E. Coop prove the reliability and dependability of the technology they are championing?
5. **Believe the technology has harmful side effects (ie: bats, birds killed by wind turbines)**- How does a R.E. Coop dispel the myths, or ease the concerns of the detrimental side-effects of R.E.?
6. **Skepticism with respect to the credibility of the industry**- How does a R.E. Coop address the skepticism that prevails over the R.E. industry regarding its merit and importance?

Ontological and Social Hindrances

1. **Indifference towards protecting the environment**- How does a R.E. Coop help to overcome the indifference, or apathy, towards personal stewardship of the environment?
2. **View the implementation of RE systems as an unnecessary hassle**- How does a R.E. Coop improve the ease with which R.E. systems are deployed?
3. **View the technology as unattractive**- How does a R.E. Coop counter the opinion that R.E. systems are unattractive and will reduce the aesthetic appeal and value of a community?
4. **Consumers do not feel that their personal energy consumption choices will have a real impact.** How does a R.E. Coop help to empower consumers, and show them that their personal choices have a meaningful impact?
5. **Detachment from the source of electricity**- How does a R.E. Coop forge the connection between the consumer and the source of his/her electricity?
6. **Worry about fitting in with the community**- How does a R.E. Coop alleviate the worry that owning R.E. systems alienates one from his community?

Financial and Legal Hindrances

1. **Cost**- How does the inscription into a R.E. Coop address the concerns that R.E. systems are too costly; to the point where it is not affordable? How does a R.E. Coop make systems more affordable?
2. **Investment is viewed as too risky**- How does participating in a R.E. Coop reduce the risk involved with a R.E. system purchase?

3. **Not sufficiently satisfied with the returns offered**- How do the returns offered by a R.E. Coop compare to the returns of traditional investment vehicles (Savings Account, RRSP, GIC's, Equity, the returns of a R.E. investment independent of a Coop,..)?
4. **The investment is viewed as requiring too much time before the capital invested is recuperated and begins yielding returns**- How does a R.E. Coop reduce the time between the initial investment, and time at which surpluses are realized?
5. **Inability to access sufficient credit**- How does a R.E. Coop remove to need to acquire large amounts of credit in order to take part in R.E. projects?
6. **The capital that could be (has been) invested in a RE project is needed elsewhere**- How does a R.E. Coop remove the risk of stranded capital in an investment?
7. **Worry about the resale potential of a house/property if a RE system is installed**- How does a R.E. Coop remove the worry that the resale value of a property is diminished if a R.E. system is present?
8. **Consumers may be frightened away by the studies, assurances and permits required before a system can be installed**- How does a R.E. Coop assist with the administrative work involved in R.E. systems?
9. **Liability in case there is a problem**- How does a R.E. Coop minimize the liability any one person faces through their participation in a R.E. project?

Physical Hindrances

1. **Unsuitable location for the installation of a RE system**- How does a R.E. Coop accommodate for the constraint of not owning a feasible site on which to deploy a R.E. system?

This chapter was previously published in Global Sustainable Development and Renewable Energy Systems edited by Phillip Olla, pages 1-15, copyright 2012 by Information Science Reference (an imprint of IGI Global).

Chapter 87
Value Creation with Wood-Based Energy Business Models

Wim Westerman
University of Groningen, The Netherlands

Jeffrey Paays
University of Groningen, The Netherlands

Satu Pätäri
Lapeenranta University of Technology, Finland

ABSTRACT

Renewable energy sources are increasingly coming into use. This notably counts for non-food based biomass sources, including those producing wood-based energy. A firm that (re-)directs activities to this area can potentially create value. The external and internal environment of the firm sets the outline for the value creation potential. Important in the value creation process are dynamic capabilities to alter resource bases. The authors show this with an example in the mature Finnish pulp and paper industry. A Delphi study helps to identify the main industry- and company-level factors that influence the forest-based energy sector, its value-creation potential, and the future roles of forest and energy companies in it. The results of a study on a power plant in The Netherlands shows how the economic value creation for a specific firm can be assessed. The plant is designed to "vaporize" biomass input, in our case so-called "B-wood," to a bio-oil that can be used to generate Green Power. The economic value analysis draws upon a risk-oriented production unit positioning analysis that is modeled with multiple scenarios. The results show that innovative business models can fruitfully be identified with multiple experts in sophisticated Delphi rounds and dynamic resource-based approaches appear promising. The authors also find that bio-refinery production chains are economically feasible with relatively small production units with multiple applications. The chapter gives way to a management view that integrates resource-based and economic value perspectives on wood-based energy sources for firms. In that sense, it adds to the literature on value creation with innovative business models.

DOI: 10.4018/978-1-4666-4852-4.ch087

1. INTRODUCTION

A combination of fear and hope, and not in the least rising energy prices, has given birth to a renewable energy sector that is readily becoming able to rival its fossil fuel counterpart. New business models are being explored, but not all of them seem to create value. Take for example wood-based energy. Traditionally, it was viewed as a by-product of production processes. The question has become, however: how can value be created with innovative business models in this area? This chapter offers a view that may help firms with their actual decision-making and it also brings academics a view on a timely topic.

The structure of this chapter is as follows. In section 1, the background of our study is provided. Literature on value creation with business models is treated in section 2. The research design is outlined in section 3. In section 4, a business identification perspective on wood-based energy is illustrated with a study on the Finnish Pulp and Paper Industry (PPI). In section 5, an economic value view is exemplified with a feasibility study on a wood-fed Dutch bio-oil plant. In section 6, our conclusions on value creation with wood-based energy are presented.

2. BACKGROUND

Many industries are confronted with structural changes. An example of this includes the forest industry, which is not mainly anymore about cutting wood, producing pulp and paper, reusing or doing away with the waste and planting trees. Another example is the waste industry, which deals with a lot more these days than separating waste, recycling the valuable parts and burning the rest. Lastly, today's energy sector does not just convert fossil fuels into a bit of usable power that is partly dissipated on its own turn. The forest industry, and especially the PPI, finds out about ways to utilize its resource stock more economically and to use by-products more profitably, the waste industry diversifies and multiplies its inputs and the energy sector shows up with increasingly efficient production processes and with an intensified focus on renewable energy. However, exploitation of the emergent business opportunities requires the knowledge and resources of multiple and diverse actors.

Adding to both Paays and Westerman (2008) and Pätäri and Westerman (2011), we focus on value creation in the bioenergy sector, which emerges at the interface between the forest, waste and energy industries. *Bioenergy* is renewable energy derived from biomass, here in particular from wood-based biomass. *Bioenergy products* include both bioenergy (e.g., heat and power) and biofuels (e.g., ethanol). This study explores innovative business models related to biomass-for-energy in terms of *what* they are, *how* forest, waste, and energy companies can exploit them and *how* their value creation can be assessed. The changes relating to bioenergy within competitive environments carry the potential for a radical transformation of industries that create opportunities to both new entrants and competitors (see Lei & Slocum, 2002). The emerging business challenges and opportunities can be identified through scanning the competitive landscape and the structural industry changes. This scan may serve as an input to create and renew the resource base of the firms involved in the bioenergy sector.

The investments identified to develop resource bases can be evaluated regarding their placement in the production chain and the business process, as well as in the internal and external environment. In the end, there is value creation if the present value of the expected stream of future cash flows that an investment provides to its owners and stakeholders is positive (Dorsman, et al., 2009). Bioenergy investment decisions are costly and subject to a high level of uncertainty with respect to future developments and economic value creation, since in the early stages of an industry there is a lack of information about the course of the cash flows. That is why in the bioenergy sector business risk and financial risk is a central factor in determining the economic value.

3. LITERATURE REVIEW

We begin our review of the literature on value creation with resource-based approaches that enable to identify value-creation opportunities. Next, we discuss the economic perspective on value creation with concrete investments.

3.1. Dynamic Resource-Based View

The resource-based view of the firm connects resources and sustainable competitive advantage, and identifies the characteristics needed for company resources. According to Wernerfelt (1984, p. 60), a resource is *anything which could be thought of as a strength or weakness of a given firm*. According to Barney (1991) and Eisenhardt and Martin (2000), resources refer to assets, abilities, and competences (i.e., skills, knowledge, and information) that enable the firm to conceive of and implement unique value-creating strategies in order to create competitive advantage. The resources can be either tangible, intangible or human in nature (Lockett, et al., 2009). Barney (1991) identifies four ("VRIN") attributes that a resource must have simultaneously in order for it to become the basis of competitive advantage: Valuable (V), Rare (R), Inimitable (I) and Non-Substitutable (N). Additionally, superior resources and capabilities need to be developed, protected, and deployed in order to create and preserve sustainable competitive advantage (Amit & Schoemaker, 1993).

Whereas the standard resource-based approach stresses the *possession* of resources, increasing emphasis is being given to their *exploitation*. Teece et al. (1997, p. 516) coin dynamic capabilities as *the firm's ability to integrate, build, and reconfigure internal and external competences to address rapidly changing environments*. Within the dynamic-capabilities framework, the competitive advantage of a firm is thought to rest on its managerial and organizational processes (routines, patterns, and learning), which are shaped by the specific asset *positions* (intellectual property, complementary assets, customer base) and the available evolution *paths* (business policy alternatives).

In an innovative business identification setting, a 'dynamized' extension of the resource-based view is in place. While departing from the classic focus on intra-firm resources, it goes beyond traditional industry and firm boundaries with a broader understanding of critical resources. Accelerating competition, the opening up of new markets and ongoing globalization urge firms to look for new business models that enable to create value with sustained competitive advantage. As to the forest, waste, and energy sectors, diverse combinations of market parties are active in producing bioenergy and biofuels from wood-based biomass in particular.

3.2. Economic Value Perspective

The application of pyrolysis in the bioenergy sector is a significant innovation. Its feasibility primarily depends upon its economic value. The economic value of an investment is positive if long-term revenues are realised that exceed the capital costs (Koller, et al., 2010). Innovative investments are subject to much uncertainty with respect to future developments. That is why risk becomes the central factor in determining the economic value. Rappaport (1986) distinguishes between business risk and financial risk. The latter is related to the mix of financing instruments, the equity capital, and the loan capital. Business risk is related to the uncertainties in the internal and external environment.

By analyzing the *production chain and the internal organization*, the internal risks of the selected investment can be identified. Technology-driven cost and investment sum projections form the basis of the analysis. Risk aspects of the internal organization can be determined with so-called key performance indicators. An analysis of the *external organization* shows the external risks. A technological analysis highlights the competitive advantage potential. Threats and opportunities can be analysed with Porter's five forces model (1980). The five competitive forces are the substitute products, the buyers' power of negotiation, the suppliers' power of negotiation, the rivalry among the current competitors, and the threat of new entrants. Macro-factors can be defined and analyzed from the sector's perspective.

With the above, one can develop an understanding of the relevant value drivers: performance variables that influence the economic value. The investment sum indicated by the plant's projection is the amount that has to be financed. The optimal financing mix indicates the *capital structure* that minimizes the (weighted average) cost of capital, including the return requirements of the financial market parties: the shareholders and the loan capital providers. The expected future free cash flow (Koller, et al., 2010) is the operational cash flow (after taxes). The evaluation of the investment in both a static sector (*base case scenario*) and in a dynamic sector ("*scenario planning*") modeling environment completes the economic value perspective (Schlosser, 2002).

4. RESEARCH DESIGN

In this chapter, we will develop and exemplify a view on value creation with wood-based energy business models. We apply multiple research methods in our research strategy on the emerging industry that uses these sources. After having devised the methodological strategy, we will show how the data for this study were collected.

4.1. Research Strategy

The key research methods and data-generation techniques used in this study include structured expert inquiries and themed interviews on an industry level (Delphi study), as well as a firm-level study of business and financial risks along with a dynamic modeling analysis (feasibility study). This kind of approach gives way to multiple triangulations (Yin, 2008) with research methodologies, quantitative and qualitative data sources, investigators with various core backgrounds, as well as theories used. This way, more complete and confirmed results on the same phenomenon can be produced. The amalgamation of two studies provides insight in a complete value creation process.

The Delphi study conducted in Finland was aimed at understanding bioenergy business models that are enabled at the interface of the forest and energy industries. Emphasis was laid on identifying the value creation potential of business models as well as challenges and threats that influence them. Key factors that affect the bioenergy business were mainly investigated at an industry level. The Delphi procedure as a whole produced a rich set of data, including both closed numerical responses and written open comments, next to many data from interviews.

The follow-up of a business identification procedure is an economic analysis. A specific investment on the interface of the waste and energy industries was examined on its business risks related to uncertainties regarding a proposed plant's production chain and its internal production organization, as well as to uncertainties in its business sector and the macro environment. The economic value creation was studied in relation to the risks identified. To obtain a general indication of the feasibility, first, the value was roughly assessed by means of a "base case scenario." Next, a link was made with the risks by developing three scenarios and examining these on the expected value creation.

Overall, this study is of an explorative kind. A setting that combines qualitative and quantitative research methods and, thus, studies the research topic from different angles, with careful personal involvement of the researchers, is suitable for studying a phenomenon that is just emerging and that has a scarcity of comprehensive historical and financial data. A further typical characteristic of the research strategy is that it combines industry- and firm-level analysis: it starts with a wide examination of changes within external operational environments to identify business opportunities and moves on to a very detailed analysis of the risks of an actor that flows into economic scenarios.

4.2. Data Collection

Business Identification

Specialist opinions can give guidance with the identification of the value creation potential of business models. When applying a Delphi method, the opinions of many experts are sampled, typically using IT techniques. After the first survey round, the participating experts are provided with the results of previous round. However, participant and response anonymity is warranted. The iteration may go over several rounds, typically two to three rounds, thereby deepening the understanding of the phenomenon studied. The Delphi process at hand was conducted in the second half of 2006. It was coordinated at Lappeenranta University of Technology in Finland (Pätäri, 2009).

The Delphi study was held in three rounds of online panel inquiries, added by themed expert interviews. The time scale was 10 years, and the perspective was primarily that of Finland. The experts were challenged to share their opinions on bioenergy business models that emerge at the interface between the pulp and paper and energy industries. It was aimed to find out what would have the strongest influence on the future development of a certain business opportunity. Therefore, differences of opinion were analyzed, thereby especially focusing on identifying the most focal issues. However, the mean values of the panelists' comments were calculated as well.

The first questionnaire of the Delphi study was open-ended. Following this, four expert interviews were held. The second and third rounds included more specific questions on business models regarding bioenergy production with forest-based biomass. The questionnaires were pre-tested, and the experts received feedback after each round. The panelists consisted of three groups: business managers and executives of forest and energy companies, representatives of the joint organizations of the two industries and university employees. Thirty-six panelists were included in the first Delphi inquiry. There were 11 experts in the second round and 10 in the third round. The responses of the panelists were always anonymous.

Economic Analysis

A concrete innovative investment deals with suppliers, a production chain and a business process, as well as with risks. Market structures are relevant at a micro- (firm), meso- (industry), and macro- (country) level. The investment amount and the financing structure are examined as well. Next, an economic model enables the positioning of a now data-rich base case. Here, virtual contracts provide the conditions for the purchase of the inputs and the sale of the outputs. As soft future scenarios are translated into hard prognoses, learning effects can be realised as well (Fahey, et al., 1998).

The investment project studied was initiated by a five party Dutch consortium. It included the prospective investor, a consultancy agency, a specific foundation, a Dutch province, and a regional stimulation body. The consortium lacked sufficient knowledge of the possibilities of setting up a plant. That is why it initiated a preliminary study of the availability of biomass feeds (Paays, 2004). Whereas this study provided clarity on concrete project options with pyrolysis (an evaporation technique), there was still no insight into the degree to which they were economically attractive and feasible.

Therefore, semi-structured interviews were held with the consortium members and experts in mainly the fields of energy and waste. Specific databases in these fields were also repeatedly accessed and a complex spreadsheet model was developed. The focus being on the assessment of business and financial risks, the model had to carry multiple sensitivity analyses and four scenario analyses. The objectified information flew into a feasibility study on the economic value of a real plant (Paays, 2005). Follow-up data analysis and stakeholder discussions lasted for about three years.

5. IDENTIFYING BIOENERGY BUSINESS MODELS: FINNISH PERSPECTIVE

This section first sheds light on the changes taking place in the forest and renewable energy industries. Next, we explore the value creation possibilities arising at the interface of these two sectors. Whereas the focus is primarily that of Finland, the general picture is applicable on a much wider scale, however.

5.1. The Forest and Renewable Energy Sectors

The pace of change within the forest industry has been overwhelming. One of the key drivers has been increased competition from low-cost global producers and the globalization of the industry. Other primary drivers include changing customer preferences, downward price pressures on paper products, significant overcapacity due to low demand, changing environmental standards, price increases of the most important input factors, growing shareholder expectations, as well as a lack of capital to counter these challenges. Consequently, many firms perform financially poor. Due to the adverse economic environment with increased competitive pressure, the PPI industry has witnessed many mergers and acquisitions and other, mainly cost cutting, efforts to combat the weak financial performance (Clement, 2001; Ghosal & Nair-Reichert, 2009; Pätäri, 2009). Yet, in order to gain and sustain competitive advantage in the longer run, the PPI firms need to innovate with value-creating business models.

The negative effects of fossil fuel inflammation on the environment and depletion of the fossil fuel stock have increased the global interest in renewable energy sources (Saidur, et al., 2010). As a result, the share of renewable energy should rise. In this respect, the 1997 Kyoto Protocol urged the developed countries to reduce their overall greenhouse gas emissions below their 1990 levels by the years 2008–2012 (UNFCC, 1998), but it seems that rather an increase will appear. Ten years later, the '20-20-20' target was set by the European Council. It aims to reduce the emissions by at least 20 per cent and to attain a 20-percent share of renewable energy and to make a 20% improvement in energy efficiency by 2020 (European Commission, 2008). According to the International Energy Agency (IEA) (2010), renewable energy sources accounted for approximately 13% of the world total primary energy supply in 2008. The main sources include combustible renewables and waste (this is mainly biomass) followed by hydro (see Figure 1). However, as it can be seen from the figure, oil and gas together dominate the world's energy supplies.

Biomass-based energy is expected to play a big role in reaching the European Union's targets on renewable energy, as a three-fold increase in the bioenergy production is needed. In Finland, biomass covers some 20% of the primary energy supply, thus making it one of the leading bioenergy-using countries in the world along with Sweden (Ericsson, et al., 2004). The forest sector (consisting of pulp and paper industries, forestry, and wood-products industries), which contributed slightly over 4% to the national GDP in 2009 (Finnish Forest Research Institute, 2010), is both the main provider and the main user of forest-based

Figure 1. Fuel shares of total primary energy supply in 2008 (IEA, 2010)

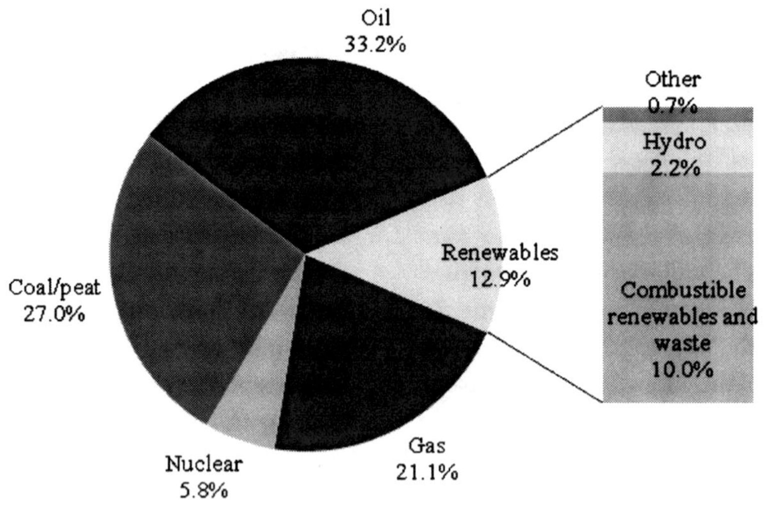

energy in Finland. Traditionally biomass has been regarded in the sector as just a by-product of the pulping process, but increasingly the forest-based biomass is collected systematically and used in pilot plants for heat and energy generation. Moreover, the Finnish geo-political and techno-industrial structures, together with key persons in the forest and energy sectors, have created favorable conditions for bioenergy (Ericsson, et al., 2004; Hetemäki, et al., 2006). This being said, however, the rapid development of renewable energy technologies and the Pan-European liberalization of the national energy sectors are boosting the dynamics of the energy sector and open up opportunities for new actors (Elliott, 2000; Jørgensen, 2005). In the next section, we illustrate the evolving bioenergy business models.

5.2. Value Creation through Joint Innovation

One of the most promising opportunities to improve the business models of the PPI companies is to transform their traditional pulp and paper mills into integrated forest biorefineries (Mansoornejad, et al., 2010). Demirbas (2009) and Cherubini (2010) view a biorefinery as a facility that integrates biomass conversion processes and equipment to produce a spectrum of value-added products including biofuels, power, and chemicals from biomass. PPI companies have a good starting position to engage in biorefining. They can produce a variety of bioproducts in conjunction with pulp and paper products. Forest biorefinery process technologies can in principle be integrated into existing pulp and paper mills. Although the innovations needed when implementing forest biorefineries are not spectacular, significantly reduced capital costs may be the outcome. In addition, the established Finnish infrastructure for collecting and processing biomass resources provides a strong basis that can be built on (Consonni, et al., 2009; Mansoornejad, et al., 2010). In the words of one actor representing the forest industry:

It is important to strive for biofuels production, which is induced by the pressures to reduce CO2 emissions. Another driver is to have an as affordable energy system as possible. Renewable energy originates from our raw material bases and it is favorable in other respects too.

However, many authors stress that knowledge and resources of actors outside of the forest industry are needed to succeed in the biorefining business (see e.g., Rodden, 2008). Despite of having profound expertise in the procurement and use of forest fuel and in the related technology, the forest sector lacks knowledge on processing the material into energy or fuels and supplying the products to end-users. However, this is the kind of knowledge that the energy industry possesses. The industries can thus unite their capabilities and develop joint business models on bioenergy. According to several experts in the Delphi study, it would be beneficial for the forest and energy industries to collaborate: *If forest-based biomass will be utilized in energy production, cooperation with the forest industry is sensible*. Especially cooperation potential was witnessed in the raw material procurement field: *At least the procurement of raw material is one such thing in which cost efficiency can be achieved through collaboration*.

There are many benefits of integrated forest biorefinery business models. Yet, their implementation will not be straightforward and much innovative efforts are needed. Firstly, there are sizable investments in new infrastructure required to produce, store and deliver the bioenergy and biofuels to customers (Demirbas, 2009). Secondly, development of technology and skills is needed. Thirdly, the scale of biorefining facilities has to be considered. Processing biomass in relatively small-scale wood power facilities might be efficient and practical as a first step in the bioenergy business. Nevertheless, the economic value perspective seemed to be larger in forest biorefinery concepts. As Cherubini (2010, p. 1414) notes: *a whole range of different-sized installations* may occur in the bioenergy business. Fourthly, new management practices and manufacturing culture are needed as the core businesses of companies will change when they engage in biorefining business models (Mansoornejad, et al., 2010).

6. THE ECONOMIC VALUE OF A DUTCH BIOENERGY PLANT

This section describes a study of the economic value creation of a pyrolysis plant planned in Groningen, the Netherlands. This particular plant is designed to pyrolyse biomass input (vaporise) to bio-oil that can be used to generate Green Power (as opposed to non-replenishable grey power). The proposed plant has a small scale and, therefore, its radius is limited regionally.

6.1. Production Chain and Business Process

Pyrolysis was developed in the 1950s to extract liquid fractions from coal and waste. The technique was refined since by using modern materials, control systems, and energy circulation techniques. In the 1980s, pyrolysis became in use to separate liquid from biomass. Essential is here, that drying techniques are to be used. If wood is to be pyrolised, the output can consist of bio-oil, as well as of coal and combustible gasses.

A preliminary study by Paays (2004) to identify biomasses in the province of Groningen showed four concrete investment options. A project option involving a supplier of secondary ("B-") wood and a purchaser of bio-oil was selected for further elaboration. In order to obtain an insight into the relation between the required inputs and the manufactured outputs and to make projections of the costs and proceeds, a simulation was made of the mass and energy streams.

The assessment of the production chain links based on economic criteria indicated that the drying process could be executed internally, but that shredding the feed as well as transporting both the B-wood and the bio-oil had to be outsourced. With respect to the plant's business process the investor projected, in addition to the pyrolysis process, a drying machine and a combined heat and power installation. Besides bio-oil, the plant would be engaged in the production of Green Power.

The projection was mainly based on interviews with the consortium members and experts mainly in the field. It contained a plant with an output of 1.5 tons of bio-oil per hour and investment expenditures of about 2.7 million Euros. The expected economic life span was 15 years. The plant would run round-the-clock, 24 hours a day for 365 days a year. The production staff was estimated at 7 workers, of whom 6 were operators. The plant processed shredded B-wood, acquired at a tariff of €0/ton (on wet pulp basis). The production capacity was 2 ton/hour (on organic pulp basis).

The projection included two buffer stocks, to guarantee a continuous production, which in initial value analyses were actually considered negligible. This approach could be risky if they appeared to be of essential importance in a later stage. If the stocks were deemed significant, another hidden risk factor would be the possibility of bio-oil becoming unsalable. This would upwardly distort the value to be attained.

Critical factors undermining the future exploitation of the plant include the degree of capacity utilisation, which runs the risk of being estimated too optimistically, the quality of the biomass, which in actual practice may be more inconstant, as well as the efficiency of the reactor and the efficiency of the combined heat and power installation that transfers the biomass, coals, and gases into bio-oil and Green Power. Factors such as these have a sizeable influence on the economic value to be attained.

6.2. Threats, Opportunities, and Competitive Advantage

The bioenergy sector has as its core activities the production and trading of commodities, attaining a competitive advantage on costs, the utilisation of capacities, and the competition on market share (Siemons, 2002). The market for biomass feeds is geographically restricted to the (surrounding) countries in Europe.

Pyrolysis competes with four other technologies in processing wood-based biomasses. These are (small and large-scale) gasifier installations, Waste Incineration Installations (WIIs), co-combustion and co-firing options for coal and gas plants, and Hydrothermal Conversion Installations (HCIs). The conventional technologies are expected to rise in costs (especially WIIs). With respect to innovative technologies, it is expected that the operational costs will significantly decrease.

Major value-decreasing uncertainties arise in the sector's direct environment and the macro-environment. In the long run, this sector may grow slowly and develop overcapacity, especially when the entrants are not organised well. High exit thresholds may prevent actors from swiftly exiting the sector. Yet, if the biomass market becomes subject to scarcity, the costs of the biomasses, which are still priced negatively, could rise proportionately and become highly positive as the sector capacity increases. If demand is larger than supply, suppliers may not deliver at negative prices anymore. Moreover, they may negotiate about the cost division of the activities within the chain.

Political uncertainties may also pose a threat. Without funding, or if the sector as a whole is freed from governmental interference, the purchaser could deny buying oil on the existing terms. Moreover, the purchaser may decide to integrate backwards, taking over the control of the entire biomass chain. Investment opportunities especially seem to be present if investments are made in an early stage. As long as the competitive advantages are not yet fully exploited, biomass contracts can still be closed at relatively favourable rates prices and government funding is ample available. As a result, the entry barriers are still relatively low.

To obtain a first indication of the competitive advantage of pyrolysis, the cost-related quantitative parameters of the available technologies can be compared mutually. Pyrolysis scores above average on the various performance indicators, indicating that the technology is feasible. Pyrolysis requires relatively low investment levels, its capacity utilisation is high (the production of fuels is generally more robust), the costs of preparatory treatment and transport are relatively low and, moreover, pyrolysis can easily be included in a biomass - Green Power conversion chain (the oil can, without significant investments, be used in existing burner installations in power plants).

6.3. Financing of the Project

In order to investigate whether the investor's financing proposal was feasible and optimal, it was considered in terms of project financing, as the assessment criteria for this type of financing were the strictest. A sensible course of action was to cover as many risks as possible by entering into partnerships and to ascribe risks to these partners and to other possible sponsors. In this way a hybrid form of financing could be realised that softened the financing terms and thereby increased the plant's feasibility.

At the time of the study, banks perceived projects with a compensation system of tradable Green Power certificates and market-based price-making processes riskier than projects that were compensated by Green Power funding guaranteed by the government. Banks mainly finance projects on the basis of cash flows, and they use the Debt Service Coverage Ratio (DSCR) in determining their financing terms. In calculating these conditions, they exclude cash flows resulting from the compensations on the delivery of Green Power. Over the first 7 years of the investment, the DSCR required of 2 on average is largely covered at an interest rate of 6%.

At the same time, also the providers of equity make assessments on the basis of the Green Power compensation system. They demand a higher cost-effectiveness of projects that receive compensation by means of tradable Green Power certificates with market-based pricing. Investors perceive the system of government-guaranteed compensation for Green Power in the Netherlands as instable and associate it with high risks. They expect that in the mid-term the system will change into a system of tradable Green Power certificates based on supply and demand. Within the sector, the equity providers regard returns of 15 to 20% as common. This study started from a return requirement of 20%.

In establishing the value of the base case scenario it was assumed that 30% of the project could be financed with a UKR-investment subsidy (UKR = Unieke Kansen Regeling – unique opportunity regulation). Credit financing, financing the investment with loan capital, initially seemed a realistic option. Moreover, this type of financing appeared to be optimal for a maximum value creation. This resulted in an average weighted capital cost rate of around 9%.

6.4. The Economic Value Base of the Case

In a static world, the base case, it was presumed that the plant could be funded by the MEP-subsidy (Regeling Milieukwaliteit Elektriciteitsproductie—environmental quality of power production regulation) during its whole life span, consisting of 97 Euros per megawatt hour of generated Green Power. Further, it was assumed that the B-wood was shredded, having a normal price of zero Euro, while the bio-oil could be sold without further costs at a price of €4,50/GJ (the opportunity costs of natural gas). Moreover, the investment subsidy would amount to 30% of the initial investment.

The shareholder value that the project could create with the base case is around 1 million Euros after tax. This means that according to the base scenario the project seems sufficiently feasible. It seems that in principle the main value drivers are the bio-oil and biomass proceeds as well as the MEP-subsidy returns. Applying the Vamil-regulation (Willekeurige Afschrijving Milieu-investeringen—random depreciation environmental investments) to the base case results in a value creation increase of about 9%. The value drivers served as starting point for a further analysis of value in a dynamic environment.

6.5. Scenario Planning

Entering and exiting moves will play a significant role in the industry. It determines the capacity in the sector and the resulting price of wood-based biomasses. The model presumes that entrants to the sector aim to create economic value in the long term. They calculate the value of the project, assuming that the future cash flows can be obtained perpetually. If the value is positive, the party enters with a predefined capacity. This line of reasoning applies to existing capacity: if the value is negative one will exit.

The principal scenario variables (value drivers) modelled are the MEP-subsidy, the UKR-investment subsidy, the price development of natural gas and grey power, the moment of exit of WIIs, the moment the innovative HCI-process starts, and the entrance and exit speed of new capacity and existing capacity. While striving to be as unbiased and objective as possible with the analysis, risks were stressed by taking a pessimistic rather than an optimistic perspective. Scenarios presented were similar to those used by the CPB (Centraal Planbureau—Dutch central planning agency) (Bollen, et al., 2004).

The scenarios roughly depend on two causal factors related to the value drivers: the economic climate and the policy of the government. The four scenarios covered an optimal risk area. Two realistic bi-polar scenarios were labeled "optimistic" and "pessimistic." An unrealistic extreme scenario served to gain an insight into a world without government subsidies. The fourth scenario was labeled "normal," being the most plausible one in view of the sector's development.

It appears that the most pessimistic scenarios still have no major consequences. Again, it seems that pyrolysis, as opposed to other competitive technologies, has a sustainable competitive advantage. Pyrolysis is much more robust, since it is less dependable on subsidy for operating the plant. The abolishment, or in the extreme case, the non-existence of another form of Green Power compensation, does not cause the project to fail to a significant degree. Pyrolysis is better capable of competing with the production chain of fossil fuels, which are all related to the price of natural gas.

Much more relevant to the success of pyrolysis is the way in which the biomass price develops. Biomass is the plant's feed and an increase of the biomass price means a relatively large increase in the operational costs, resulting in a reduction in the operational margin. Moreover, it means that, if stocks are relevant (which they are in the case of positive biomass prices), the cash outflows shift in relation to the cash inflows, leading to a decrease in value creation.

6.6. Value Creation in a Dynamic Environment

Table 1 shows the results of the shareholder value assessment under various conditions. The first column indicates the result of the case base evaluation. It assumes that during the plant's life span contracts can be established with the actors in both the production chain and in the environment of the plant: developments in the prices of the biomass feed, the bio-oil, and the Green Power are only caused by inflation. The second column presents the results of the assessment of the four different scenarios and the assignment of the probability rates to them. The row in the middle gives the estimated expected value. The effect of a deviating future scenario (more optimistic or more pessimistic) on value creation is included in the top and bottom row of the table.

Compared to the base case, which arrives at about 1 million Euros, the expected value halves when including uncertainty in the assessment. The plant's value is slightly less than 0.5 million Euros with a bandwidth of about 0.1 million

Table 1. Results of determining the shareholder value under various conditions (in €)

	Evaluation of the Results Under:		
	Static Environmental Conditions (base case)	**Dynamic environmental Conditions (stocks excluded)**	**Dynamic Environmental Conditions (stocks included)**
Estimated shareholder value in an optimistic future scenario		571,700	423,809
Estimated expected shareholder value	**1,040,054**	**475,119**	**328,107**
Estimated shareholder value in a pessimistic future scenario		362,957	216,102

Euros. The effects of keeping biomass and bio-oil stocks are still excluded here. The final column presents the results of the scenarios including the effects of keeping stock. By having stocks available (four weeks of biomass, 8 weeks of bio-oil plus an extra investment in storage capacity of € 150,000) the value again decreases by 30%, resulting in 0.3 million Euros. Despite this large decrease, sufficient return can still be made with the invested shareholders capital. In the scenario planning that includes stocks the return is still 27%.

These projections support the investor in taking the proper decision. Yet, the scenarios may reflect a future that is too pessimistic. With the MEP-subsidy meanwhile being abolished, it was presumed in the ultimate model that no compensation will exist. Otherwise, the economic value may highly increase. Furthermore, if more and better contracts can be realised, the value will shift to that in the base case. Else, it will tend to move in the direction of the other extreme. The scenarios were executed on the basis of linear depreciations. If Vamil-subsidy could be obtained, it will, moreover, increase the value.

7. CONCLUSION

In this explorative study, we have examined the emerging interface between the forest, waste and energy sectors, the opportunities for innovative business models that arise in response to growing global interest towards renewable energy sources and the economic value to be created with a concrete investment project in bioenergy. Our business identification study refers to Finland. This country has large and incompletely exploited biomass resources from forests. Moreover, its innovative and cooperative industrial and technical structures make Finland an excellent candidate for bioenergy production. Our economic valuation study deals with a pyrolysis plant in the Netherlands, a densely populated country with much wood-based waste.

As to the application of primary wood, the technology can provide both the forest and energy sectors with value creating business models. According to the Finnish Delphi study, the erection of biorefineries in small local units close to the biomass origin seems to be the most practical first step. The panel of experts regarded joint innovation with the energy industry as lucrative due to the complementary capabilities of the industries. The joint production of pulp together with bioenergy and bio-based fuels would be more beneficial than the sole production of pulp or bio-based fuels and energy. Successful adoption of biorefinery business models depends upon optimalization of the value from the resources that are economically disposable to the mills. The new, innovative products and services must also meet the customer needs.

The success of the cooperation between the forest and energy sectors in the bioenergy field depends on the development of innovative technological and marketing skills. However, the collaboration is still in its infancy. It would, thus, be fruitful to further study business models on this interface by focusing on the organization of biorefineries and the mutual make-or-buy decisions of the forest and energy companies. Moreover, research focusing on the required capabilities for managing joint innovation projects within the bioenergy field would be useful, as there is a wide range of actors that can be involved with bioenergy.

As to the application of secondary wood, it can be concluded that the production of bio-oil and the generation of Green Power by means of pyrolyzation is an attractive bioenergy project option. A particular regional Dutch chain of production has a competitive advantage and can, without high exploitation subsidies, survive economically on its own. Under certain conditions, a pyrolysis plant is feasible. Its economic value depends on the degree to which business risks and financial risks resulting from uncertainties can be covered. It is, thus, important to find as many sponsors as possible who can guarantee the coverage of these risks.

When setting up a plant, it is important to make sound estimations of the operational parameters and investment variables. In addition, price and subsidy analyses within the production chain should be economically interpreted and used in a timely manner. Further, the provision of specified information can enlarge the scope for financing coalitions, thereby lowering the capital costs. Next, as the project's realisation approaches, the study has to be extended towards softer aspects, such as environmental effects. All in all, economic value is more than simple figures, and making investments remains a matter of trust and having an eye for risks.

Whereas our examples are both on value creation with wood-based energy, neither one of them covers the whole perspective from business idea generation to full-grown implementation. These development processes, if really being with innovative business models, often take many years. Several kinds of trajectories may be explored simultaneously and successively, whereby the parties involved may change as projects evolve over time. Getting and keeping access to such projects is a wish that the present authors share with others. Yet, we believe that the perspective of our chapter is valid.

In this chapter, we adopted an integral perspective on bioenergy production with wood-based energy business models. Biorefinery production chains are economically feasible with relatively small production units with multiple applications. Innovative business models can fruitfully be identified with multiple experts in sophisticated Delphi rounds and dynamic resource-based approaches appear to look promising. The economic value of biorefineries can be assessed with a risk-oriented production unit positioning analysis, reflected in a dynamic finance-oriented modeling analysis.

REFERENCES

Amit, R., & Schoemaker, P. J. H. (1993). Strategic assets and organizational rent. *Strategic Management Journal, 14*(1), 33–46. doi:10.1002/smj.4250140105

Barney, J. (1991). Firm resources and sustained competitive advantage. *Journal of Management, 17*(1), 99–120. doi:10.1177/014920639101700108

Bollen, J., Manders, T., & Mulder, M. (2004). *Four futures for energy markets and climate change*. The Hague, The Netherlands: CPB.

Cherubini, F. (2010). The biorefinery concept: Using biomass instead of oil for producing energy and chemicals. *Energy Conversion and Management, 51*(7), 1412–1421. doi:10.1016/j.enconman.2010.01.015

Clement, D. (2001). Pulp friction. *Fedgazette, 13*(2), 8–10.

Consonni, S., Katofsky, R. E., & Larson, E. D. (2009). A gasification-based biorefinery for the pulp and paper industry. *Chemical Engineering Research & Design*, *87*(9), 1293–1317. doi:10.1016/j.cherd.2009.07.017

Demirbas, M. F. (2009). Biorefineries for biofuel upgrading: A critical review. *Applied Energy*, *86*(1), S151–S161. doi:10.1016/j.apenergy.2009.04.043

Dorsman, A., Scholtens, B., & Westerman, W. (2009). Preface to the special issue on energy and value. *Frontiers in Finance and Economics*, *6*(2), 96–100.

Eisenhardt, K. M., & Martin, J. A. (2000). Dynamic capabilities: What are they? *Strategic Management Journal*, *21*(10/11), 1105–1121. doi:10.1002/1097-0266(200010/11)21:10/11<1105::AID-SMJ133>3.0.CO;2-E

Elliott, D. (2000). Renewable energy and sustainable futures. *Futures*, *32*(3-4), 261–274. doi:10.1016/S0016-3287(99)00096-8

Ericsson, K., Huttunen, S., Nilsson, L. J., & Svenningsson, P. (2004). Bioenergy policy and market development in Finland and Sweden. *Energy Policy*, *32*(15), 1707–1721. doi:10.1016/S0301-4215(03)00161-7

European Commission. (2010). *Energy 2020 – A strategy for competitive, sustainable and secure energy*. Retrieved May 2011, from http://www.energy.eu/

Fahey, L., Robert, M., & Randall, M. (1998). *Learning from the future, competitive foresight scenarios*. New York, NY: John Wiley & Sons.

Finnish Forest Research Institute. (2010). *Finnish statistical yearbook of forestry*. Retrieved May 2011, from http://www.metla.fi

Ghosal, V., & Nair-Reichert, U. (2009). Investments in modernization, innovation and gains in productivity: Evidence from firms in the global paper industry. *Research Policy*, *38*(3), 536–547. doi:10.1016/j.respol.2008.10.010

Hetemäki, L., Harstela, P., Hynynen, J., Ilvesniemi, H., & Uusivuori, J. (Eds.). (2006). Suomen metsiin perustuva hyvinvointi 2015, Katsaus Suomen metsäalan kehitykseen ja tulevaisuuden vaihtoehtoihin. [Prosperity founded on Finnish forests, A review on the development of Finnish forestry and visions of the future]. Working Papers of the Finnish Forest Research Institute. Retrieved May 2011, from http://www.metla.fi

IEA. (2010). Key world energy statistics 2010. *International Energy Agency of Paris, France*. Retrieved May 2011, from http://www.iea.org

Jørgensen, U. (2005). Energy sector in transition – Technologies and regulatory policies in flux. *Technological Forecasting and Social Change*, *72*(6), 719–731. doi:10.1016/j.techfore.2004.12.004

Koller, T., Goedhart, M., & Wessels, D. (2010). *Valuation: Measuring and managing the value of companies* (5th ed.). New York, NY: John Wiley & Sons.

Lei, D., & Slocum, J. W. Jr. (2002). Organization designs to renew competitive advantage. *Organizational Dynamics*, *31*(1), 1–18. doi:10.1016/S0090-2616(02)00068-2

Lockett, A., Thompson, S., & Morgenstern, U. (2009). The development of the resource-based view of the firm: A critical appraisal. *International Journal of Management Reviews*, *11*(1), 9–28. doi:10.1111/j.1468-2370.2008.00252.x

Mansoornejad, B., Chambost, V., & Stuart, P. (2010). Integrating product portfolio design and supply chain design for the forest biorefinery. *Computers & Chemical Engineering, 34*(9), 1497–1506. doi:10.1016/j.compchemeng.2010.02.004

Paays, J. X. (2004). *Pyrolyse in Groningen: Onderzoek naar de haalbaarheid van een pyrolysefabriek in de provincie Groningen* [Pyrolysis in Groningen: A study on the feasibility of a pyrolysis plant in Groningen]. Groningen, The Netherlands: Rijksuniversiteit.

Paays, J. X. (2005). *Pyrolyse in Groningen: Analyse van een investeringsbeslissing in de opkomende bio-energiesector* [Pyrolysis in Groningen: Analysis of an investment decision in the emerging bioenergy sector]. Groningen, The Netherlands: Rijksuniversiteit.

Paays, J. X., & Westerman, W. (2008). Economische waarde van een fabriek van groene stroom. [Economic value of a green power factory]. *Maandblad voor Accountancy en Bedrijfseconomie, 82*(12), 507–514.

Pätäri, S. (2009). *On value creation at an industrial intersection – Bioenergy in the forest and energy sectors.* (PhD Thesis). Lappeenranta University of Technology. Lappeenranta, Finland.

Pätäri, S., & Westerman, W. (2011). Value creation with wood-based energy sources. In A. Dorsman, Ö. Arslan, M. Karan, & W. Westerman (Eds.), *Financial Aspects in Energy* (pp. 197–211). Berlin, Germany: Springer Verlag. doi:10.1007/978-3-642-19709-3_11

Porter, M. (1980). *Competitive strategy.* New York, NY: The Free Press.

Rappaport, A. (1986). *Creating shareholder value: The new standard for business performance.* New York, NY: The Free Press.

Rodden, G. (2008). Actions catching up to words. *Pulp & Paper International, 50*(6), 33–37.

Saidur, R., Islam, M. R., Rahim, N. A., & Solangi, K. H. (2010). A review on global wind energy policy. *Renewable & Sustainable Energy Reviews, 14*(7), 1744–1762. doi:10.1016/j.rser.2010.03.007

Schlosser, M. (2002). *Business finance.* Harlow, UK: Prentice Hall.

Siemons, R. V. (2002). *A development perspective for biomass-fuelled electricity generation technologies.* (PhD. Thesis). University of Amsterdam. Amsterdam, The Netherlands.

Teece, D. J., Pisano, G., & Shuen, A. (1997). Dynamic capabilities and strategic management. *Strategic Management Journal, 18*(7), 509–533. doi:10.1002/(SICI)1097-0266(199708)18:7<509::AID-SMJ882>3.0.CO;2-Z

UNFCC. (1998). *Kyoto protocol to the United Nations framework convention on climate change.* Retrieved May 2011, from http://unfccc.int

Wernerfelt, B. (1984). A resource-based view of the firm. *Strategic Management Journal, 5*(2), 171–180. doi:10.1002/smj.4250050207

Yin, R. K. (2008). *Case study research: Design and methods* (4th ed.). London, UK: Sage.

KEY TERMS AND DEFINITIONS

Business Model: Rationale of an opportunity for creating value for stakeholders.

Delphi Method: Sampling the opinions of many experts, typically using IT techniques, whereby panelists are provided with the results of previous rounds, while warranting participant and response anonymity over all of the iterations.

Economic Value: Value as regarded from the perspective of financial stakeholders.

Innovation: A course that leads to new or improved processes.

Modeling Analysis: Creating and researching data by way of a computerized analysis.

Positioning Analysis: Analyzing the location of an organization in its internal and external environment.

Resource-Based View: View of the firm that connects resources and sustainable competitive advantage, and identifies the characteristics needed for company resources.

Wood-Based Energy: Energy developed from forest or waste wood sources.

This chapter was previously published in Industrial Dynamics, Innovation Policy, and Economic Growth through Technological Advancements edited by I. Hakan Yetkiner, Erkan Erdil, and M. Teoman Pamukcu, pages 292-306, copyright 2013 by Information Science Reference (an imprint of IGI Global).

Chapter 88
Impacts of High Wind Power Penetration on the Frequency Response Considering Wind Power Reserve

Bakhtiar Badmasti
Islamic Azad University—Marivan Branch, Iran

Hassan Bevrani
University of Kurdistan, Iran

Ali Hessamy Naghshbandy
University of Kurdistan, Iran

ABSTRACT

With high wind power penetration, imbalance between generation and consumption is increased, and as a consequence the frequency deviation from the nominal value will be magnified. For frequency control purpose in conventional power systems, the synchronous generators are only responsible. However, it is expected that the renewable energy sources (RESs) are also needed to contribute in frequency control issue in near future. In this paper, impacts of high wind penetration on the frequency control are studied, and an approach is introduced for creating the secondary reserve for doubly fed induction generator (DFIG) wind turbines. This reserve can contribute to load frequency control (LFC) task. Simulation results on updated standard IEEE 39-bus system are presented and it is shown that the wind turbines can contribute in LFC service, successfully.

INTRODUCTION

Nowadays, using renewable energy sources (RESs) such as wind, solar and tidal power for generation of energy is increasing. Among all RESs, the wind power has the maximum contribution to the renewable power generation. China, USA, Germany, Spain and India at the end of 2010 had maximum installed capacities in the world with 44.7 GW, 40.18 GW, 27.21 GW, 20.67 GW and 13.06 GW, respectively; it is expected that, at the end of 2020, total capacity in the world will reach to 1500 GW (World Wind Energy Association WWEA, 2011). With growth of wind energy, the

DOI: 10.4018/978-1-4666-4852-4.ch088

power system is faced with new problems such as new challenges in frequency control (Bevrani, Ghosh, & Ledwich, 2010) and power reserve estimation. With increasing wind power in a power system, the initial frequency drop may increase following a load disturbance (Erlich, Rensch, & Shewarega, 2006).

Currently, using of Doubly Fed Induction Generators (DFIGs) is more usual than the other wind turbine technologies, because they can easily provide ancillary services and active/reactive power control. Naturally, the DFIG wind turbines cannot contribute to the inertial response because the rotational speed of them is not directly connected to the network (Lalor, Ritchie, Rourke, Flynn, & O'Malley, 2004; Ekanayake & Jenkins, 2004; Lalor, Mullane, & O'Malley, 2005). But, adding a supplementary loop control to the DFIGs, they can release their kinetic energy and contribute to inertial response and primary control (Lalor, Ritchie, Rourke, Flynn, & O'Malley, 2004; Lalor, Mullane, & O'Malley, 2005; de Almeida & Lopes, 2005; Morren, de Haan, Kling, & Ferreira, 2006; Anaya-Lara, Hughes, Jenkins, & Strbac, 2006; Conroy & Watson, 2008; Chowdhury & Ma, 2008; El Mokadem, Courtecuisse, Saudemont, Robyns, & Deuse, 2009; Tarnowski, Kjar, Sorensen, & Ostergaard, 2009; Ping-Kwan, Pei, Banaka, & Boon Teck 2009; Erlich & Wilch, 2010).

Wang, Sun, Li, and Ooi (2010) have proposed a voltage and frequency control method for DFIG. This method makes DFIG equivalent to a synchronous generator and controls active and reactive power of stator by controlling the voltage magnitude and frequency of the rotor. Kaneko, Uehara, Senjyu, Yona, and Urasaki (2011) used an integral control method for wind farms to reduce frequency deviation in a small power system. In this method, wind farm achieves the frequency regulation objective based on two control schemes: load estimation and short-term ahead wind speed prediction. The proposed methods have some drawbacks such as reduction of wind farm output power, as well as increased pitch action. Bhatt, Ghoshal, and Roy (2010) have introduced an extra frequency control support function which affects the rotational speed of rotor in the case of frequency disturbance and reduces the frequency dip by releasing short term transient active power. For optimal dynamic frequency performance, Thyristor controlled phase shifter (TCPS) and Superconducting Magnetic Energy Storage (SMES) are also used in coordination with DFIG frequency control loop. Mokadem, Courtecuisse, Saudemont, Robynsand, and Deuse (2009) have introduced a primary frequency control based on fuzzy logic for variable-speed wind generators. This method controls the torque generator and pitch angle to keep primary reserve. Senjyu, Kaneko, Uehara, Yona, Sekine, and Kim (2009) have presented an output power control of wind turbine using pitch angle control. In this approach, wind turbines achieve flexible output power control and therefore frequency deviation reduces. Xiangyu, Heming, and Yi (2010) have added a PD controller to the DFIG, and they have introduced a new control loop for participation of DFIG in the frequency control, but this loop only provides the primary frequency control.

Mauricio, Marano, Gomez-Exposito, and Martinez Ramos (2009) used a method for frequency regulation by releasing kinetic energy of rotating mass and, with a communication between the wind turbine and the conventional generator a coordination is also created. Ullah, Thiringer, and Karlsson (2008) have discussed the capability of providing a short-term active power support of a wind farm for the primary frequency control utilizing the kinetic energy of the rotated mass. Khaki, Asgari, Sirjani, and Mozdawar (2008) have shown that coefficient of power of wind turbine can be changed through two different criteria defined for tip speed ratio parameter (the ratio of blade tip speed to wind speed), and from this viewpoint, a novel structure has been proposed for primary frequency control that controls turbine speed reference. Abe, Ohba, and Iwamoto (2006) introduced the disturbance observer for decreasing the instantaneous power variations of the wind power generations and indicated that proposed method can improve the quality of the response and are effective in the control of the output variation. Ramtharan, Ekanayake, and Jenkins (2007) indicated that by adding two control loops, an enhanced inertial response can be obtained and also, the possibility of wind turbine to provide of primary and secondary responses has been discussed.

Lingling, Zhixin, and Osborn (2009) indicated that with coordination of pith angle and HVDC rectifier, wind turbine can participate in the load-frequency control (LFC) and in this method, HVDC rectifier senses frequency deviations and corresponding it, the active power flow increases or decrease at the rectifier control loop. Morren, de Haan, and Ferreira (2006) have investigated application of fuel cells to provide primary frequency control with contribution of wind turbine. The authors have specified the quantity of the installed capacity of fuel cells, as function of the percentage of wind energy, in order to provide proper primary frequency control. Morren, Pierik, and Haan (2006) have compared two strategies for the contribution of DFIGs in the frequency control and have mentioned that, the primary control loop has more advantages than the inertial response loop. Bevrani, Daneshfar, and Daneshmand (2010) used Multi-agent Bayesian Networks for Load frequency control (LFC) in the presence of wind power without any controller for DFIGs.

Since, wind as a primary energy of wind turbines cannot be exactly predicted, the estimation of secondary reserve provided by conventional units in the presence of wind power is difficult. Following the increasing of wind power penetration, it is expected that wind turbines provide secondary reserve in cooperation with synchronous generators.

Ackermann and Morthorst (2005) estimated that for maintaining frequency in nominal value in Germany and Denmark countries, with 1000 MW growth of wind power, about 6.6 MW reserve is necessary in each minute. Provision of secondary reserve by wind turbine could be important for the reliability and security of power systems. Yingcheng and Nengling (2010), and Sun, Zhang, Li, and Lin (2010) have provided a comprehensive review on contribution of a wind turbine to the frequency control and have mentioned that there may be a need to revise secondary control for compensation of frequency disturbances due to wind generation. De Almeida and Lopes (2007) introduced an idea for primary and secondary frequency control which is examined for primary control, only.

In this article, the mentioned idea is completed and is applied on a standard IEEE 39-bus test system using de-loading power curve and coordination of it with pitch angle control. A framework is introduced for the contribution of wind power to the secondary frequency control. Finally, the following four scenarios are examined on the updated IEEE 39-bus case study using Matlab/Simulink/SimPowerSystem software:

1. Without wind power
2. 10% wind power penetration
3. 20% wind power penetration
4. 40% wind power penetration

PITCH ANGLE CONTROL AND SECONDARY CONTROL OF DFIG

Generally, a DFIG unit has three parts: wind turbine, induction generator (IG) and back to back converter. Stator side of IG is directly connected to the network and rotor side of IG is connected to the network through a back to back converter. Rotor side of the convertor is used for the control of active power and stator side of the converter is used for the control of DC bus voltage. For performing secondary control in the DFIG systems, firstly wind power reserve should be provided. In this paper, pitch angle control and de-loaded power curve are used for creation of the wind power reserve in a DFIG. In Figure 1, a general structure for secondary frequency

Figure 1. General structure for frequency control with contribution of wind turbine

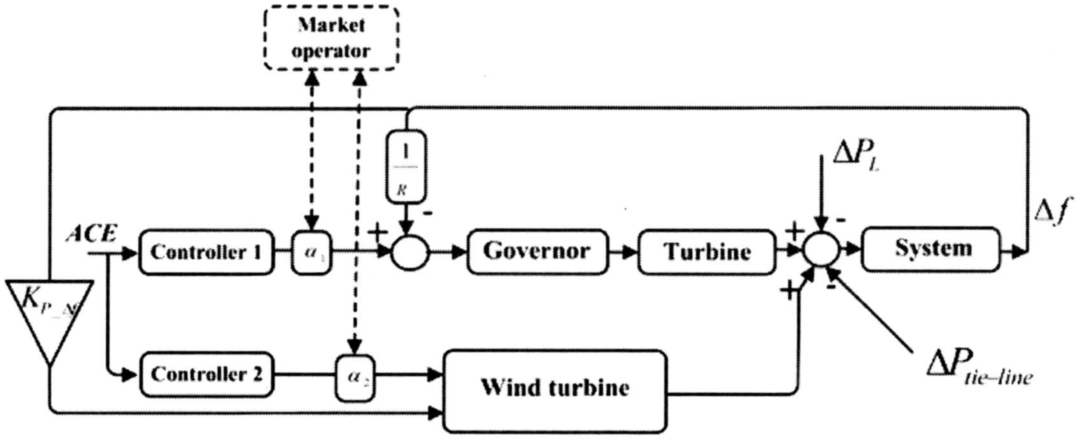

control mechanism by contributing wind turbines and conventional units is presented. Parameter of α_i in this figure is the LFC participation factor. In a given control area, the sum of participation factors is equal to 1.Here, it is assumed that the LFC system has two participators (α_1, α_2).

Determination of α_i is an economic and technical issue, because α_i dependson the costs of generation, congestion problems and ability of various units for power generation. Here, it is assumed that the participation factors are defined by LFC market operators and the area control error (ACE) is used to feed the controllers. The DFIGs use the maximum power curve (MPC) for tracking power based rotational speed. In case of using de-loaded power curve instead of the MPC, they do not use their full capacity at any time. Figure 2 shows the MPC and 90% of the MPC for

Figure 2. Maximum and de-loaded power curve

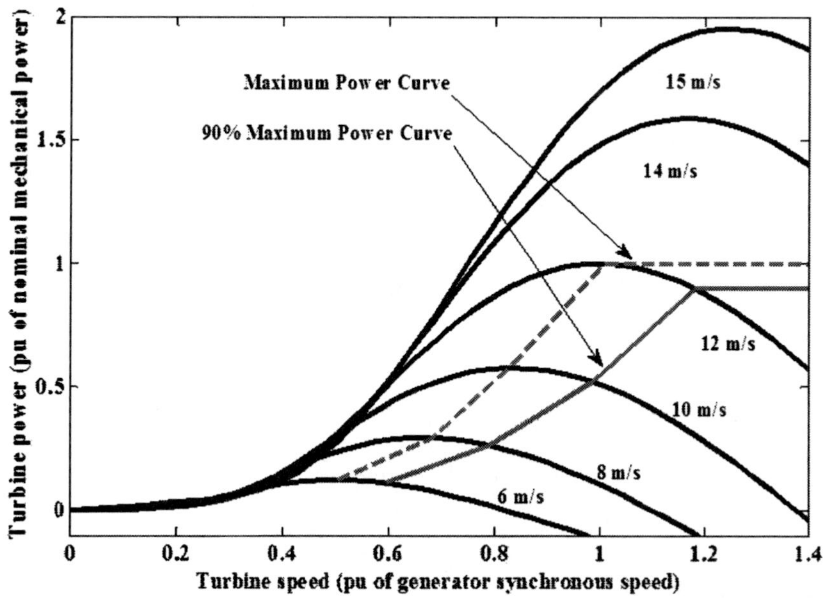

DFIG wind turbines. It can be observed when MPC reduces to 90%, rotational speed of wind turbine increases. To avoid increasing in speed, rotational speed should be limited by pitch angle regulation. When frequency control is activated, the pitch control should limit the output power of wind turbine. Therefore, set point of pitch control is fixed at 1, and the created reserve should be released when system frequency deviates from the nominal value.

In this article, similar to the real power systems, the conventional generating units use the proportional-integral (PI) controller for secondary frequency control loop. Figure 3 shows a control framework for the DFIG unit with the pitch angle control. From Figure 3, it can be observed, when DFIG uses de-loaded power curve and a disturbance occurs, the set point of power (P_{ref}) is shifted from the nominal value by blocks of primary and secondary frequency control (shown in the figure).The DFIGsdo not have controllers for the frequency regulation purpose, but in this paper primary and secondary controls are considered to the control framework.

CASE STUDY

In this paper, some simulations are performed on the updated IEEE 39 bus system. This system includes 10 synchronous generators that are equipped with excitation system, power system stabilizer (PSS) and primary frequency control system.

Data and parameters about generators, transformers and transmission lines are given in Bevrani, Daneshfar, and Daneshmand (2010). Figure 4 shows the case study with three wind farms added to the system. This system is divided to three areas that are shown with dashed lines in Figure 4. Total generated active power and load in each area are given in Table 1. Data and parameters of DFIG wind turbine are:

Table 1. Total generation and load in each area

Power / Area	Generation(MW)	Load (MW)
Area 1	199	265.25
Area 2	162.1	124.8
Area 3	235	193.28

Figure 3. Control structure for rotor side of DFIG with pitch angle control and frequency control

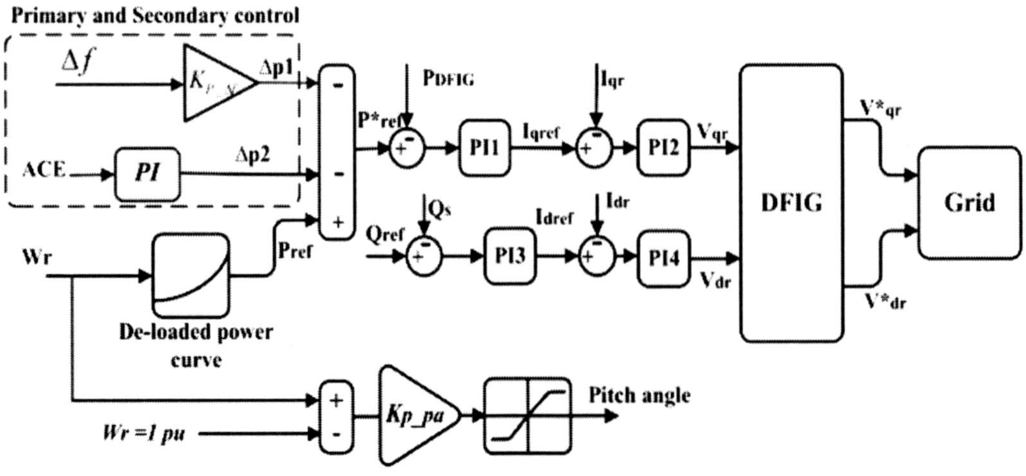

Figure 4. Case study: Updated IEEE 39-bus system with wind farms

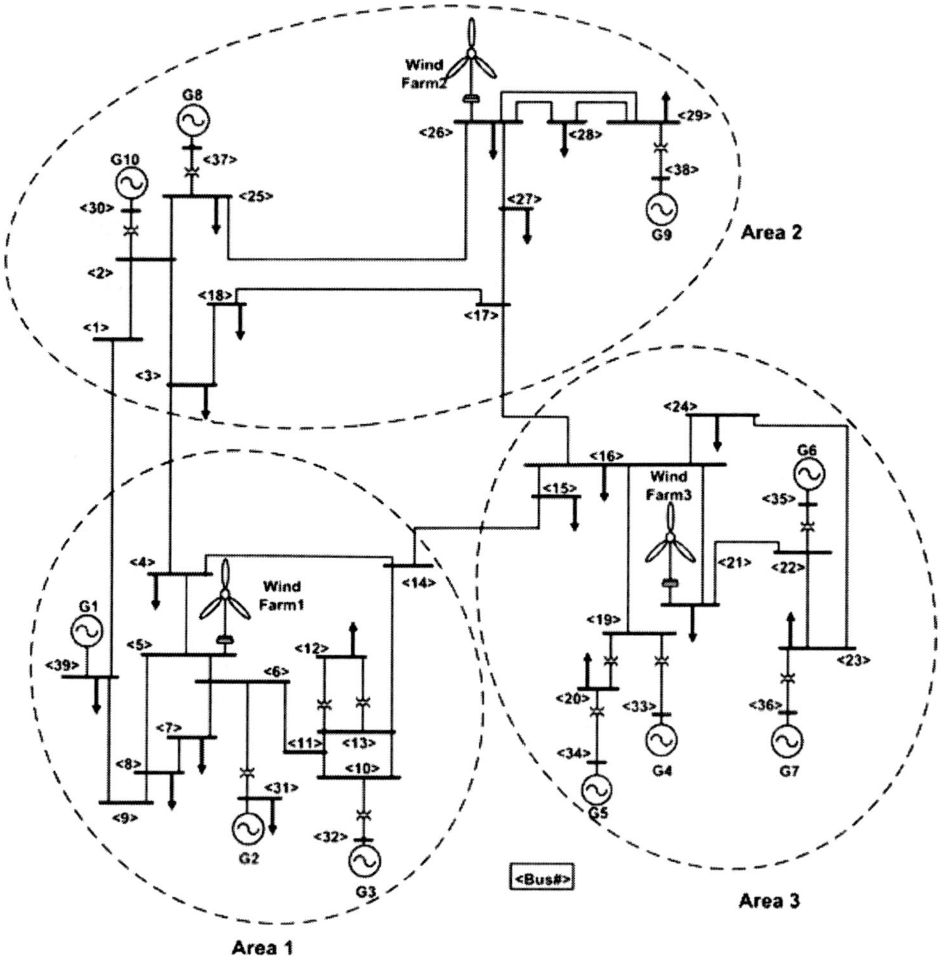

Sn=1.5/0.9 MVA,
Vn=575 V, H=5.04s,
Rs=0.00706 pu,
Ls=0.171 pu,
Rr=0.005 pu,
Lr=0.156 pu,
Lm=2.9 pu,
shunt capacitor=10000e-6.

In each area, one synchronous generator is responsible for LFC issue (area 1: G1, area 2: G9, area 3: G4).

SIMULATION RESULT

In this section, four scenarios which described in the introduction are simulated. The pattern of wind speed that is used for wind farms is given in Figure 5. Three PI controllers are used for secondary frequency control in the control area that is shown in Table 2. As a sever disturbance, an extra load(10 MW) is added in each area at t=5sec.

Scenario 1: Without Wind Power

In this scenario, no wind power is penetrated to the system. Following adding of extra load to

Figure 5. Wind speed pattern

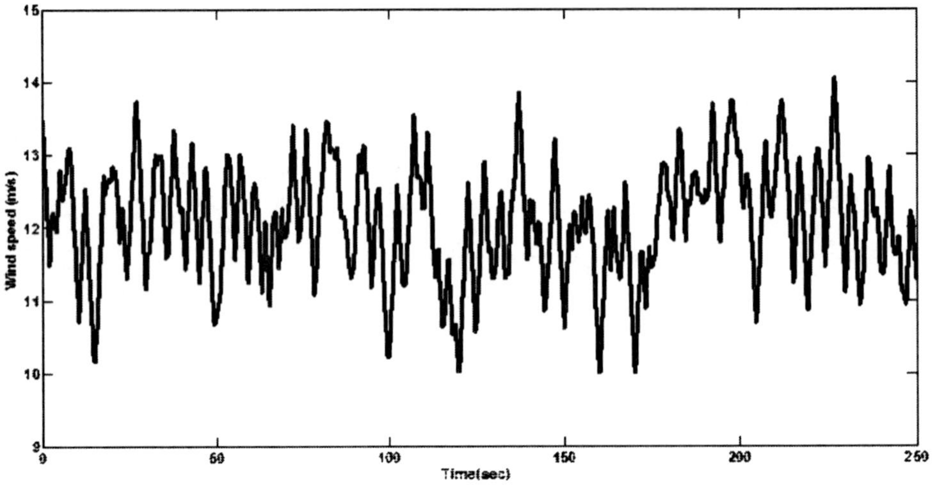

Table 2. Parameters of LFC controller in synchronous generators

Area \ Parameter	K_P	K_I
G 1	-325	-20
G 4	-330	-23
G 9	-340	-25

the system; firstly, frequency reduces from the nominal value, but this reduction is compensated by primary and secondary control of synchronous generators. Figure 6 shows the frequency of system, ACE signal, and active power of G1. It can be observed that PI controllers of the synchronous generators (G1, G4 and G9) are able to properly restore the ACE signal and frequency deviation to the nominal value and therefore, the active power of generators should be increased.

Scenario 2: 10% Wind Power Penetration

In this scenario, 10% of the generation in each area is generated by the wind farms. For example, in area 1, 199 MW power is generated and sharing of wind power is about 20 MW. In this scenario, 10% from the ten percent of wind power penetration is also considered as reserve wind power for LFC. Simulation results are shown in Figure 7 and Figure 8.

Figure 7 compares the first scenario and second scenario (without and with wind power reserve) for frequency and ACE signals. It can be observed that with penetration of wind power, deviation of wind speed leads to deviation the active power of wind farms and therefore frequency, ACE and output power of synchronous generators oscillate irregularly (Figure 8). It can be seen that with penetration of wind power, the frequency drop is reduced. But, after occurrence of the disturbance, deviation of wind speed leads to increase in deviations of the ACE signal and frequency. With contribution of wind power reserve, the system performance (deviation in frequency and ACE signals) are improved. With contribution of wind farms in the LFC, it is expected that output power of synchronous generators reduces. Figure 8(a) shows the secondary control of wind farm 1 causes an increase in its output power, as well as Figure 8(b) shows a decrease in conventional generated power.

Figure 6. Scenario 1: (a) Frequency deviations, (b) ACE signal, (c) Output power of G1-Solid (only primary control), dashed (primary and secondary control)

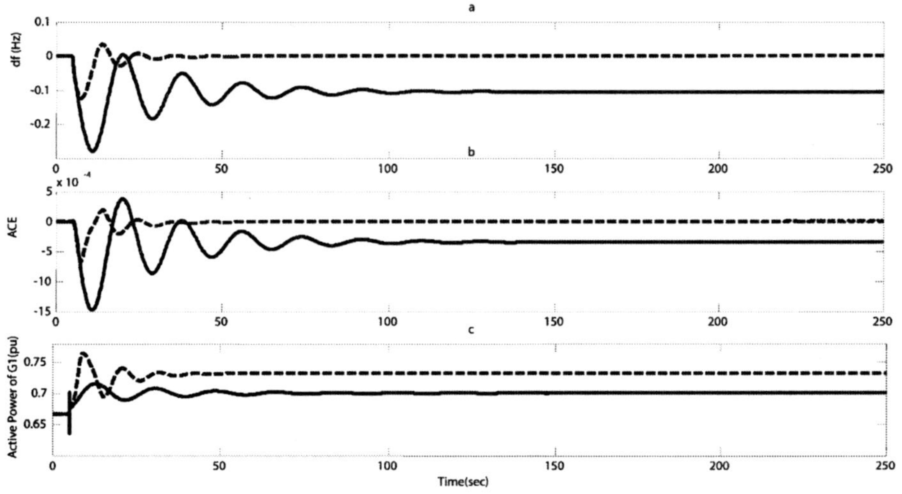

Figure 7. (a) Frequency deviations, (b) ACE signal; Solid (scenario 2: with 10% wind power reserve), dashed (scenario 2: without wind power reserve), dotted (scenario 1).K_{P_WF}=36, K_{i_WF}=20, $K_{P\_\Delta f}$=30

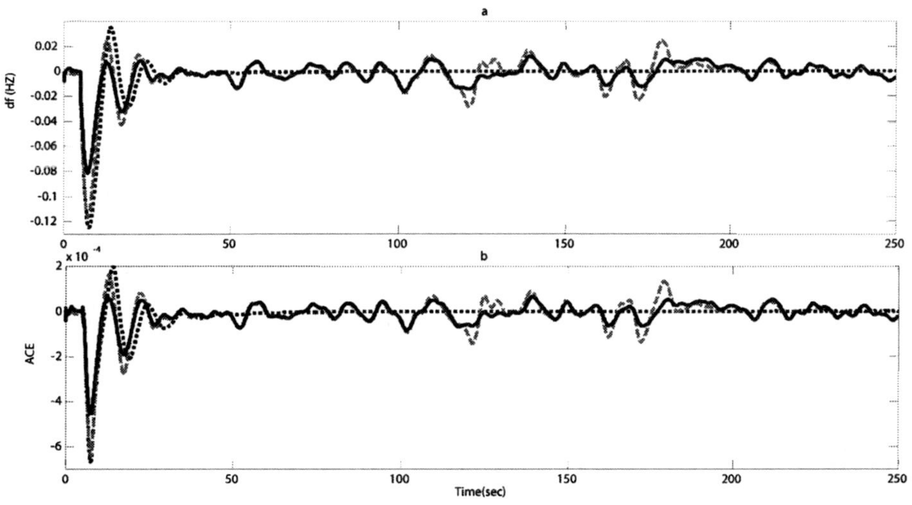

Scenario 3: 20% Wind Power Penetration

Amount of wind power penetration in this scenario is twenty percent. Figure 9 compares frequency and ACE for the first, second and third scenarios. It is seen, when penetration is increased, similar to scenario 2, the frequency drop due to disturbance is decreased. For twenty percent penetration, two rate values are used for reserve providing: 10% and 20%. From Figure 10, it can be observed that in the case of using wind power reserve, maximum drop of frequency is reduced and deviation of frequency created by wind speed is also significantly

Figure 8. Scenario 2: (a) Active power of G1(MW), (b) Active power of wind farm 1; Solid (with wind power reserve, inactive frequency control by wind farm), dashed (with 10% wind power reserve). $K_{P\_WF}=36$, $K_{i\_WF}=20$, $K_{P\_\Delta f}=30$

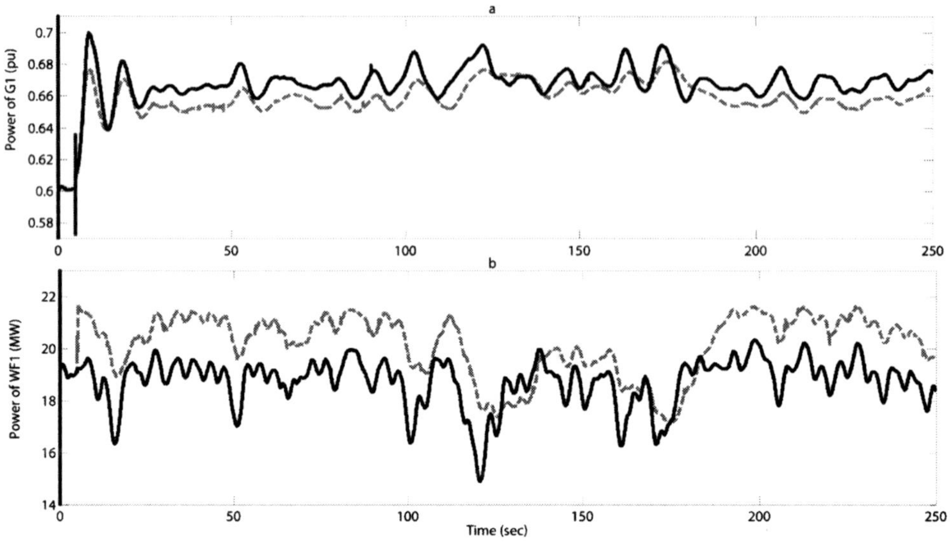

reduced. For 20% reserve, two different values for PI controllers are applied (Figure 10). It can be observed that output of wind farm, when PI parameters are 10 and 5, is smoother but in comparison with next choice (K_{P_WF}=25 and K_{i_WF}=15), the amount of contribution in secondary control is decreased. Figure 10 shows the importance of PI parameters selection. In other words, proper choosing of PI parameters can create a coordination between wind turbines and synchronous

Figure 9. (a) Frequency deviations, (b) ACE signal; Solid (scenario 3), dashed (scenario 2), dotted (scenario 1)

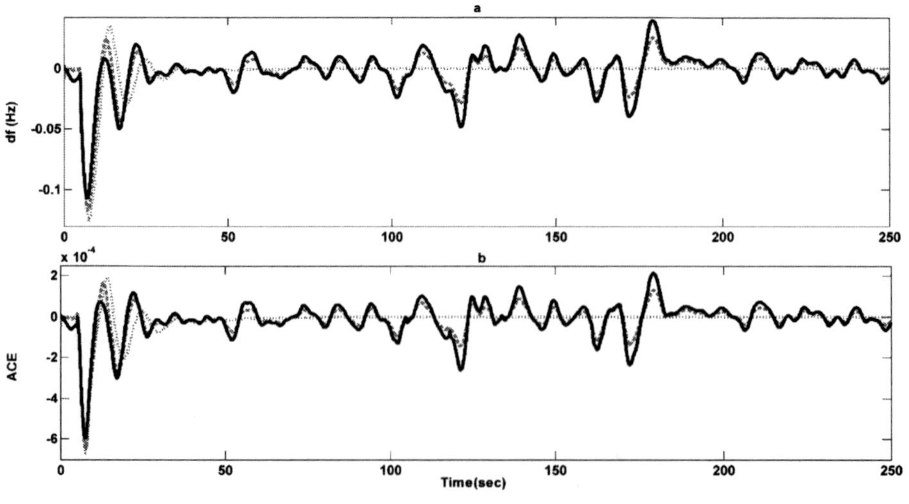

Figure 10. Scenario 3: (a) Frequency deviations; Solid (with 20% wind power reserve, K_{P_WF}=25, K_{i_WF}=15, $K_{P\_\Delta f}$=30), dashed (with 20% wind power reserve, K_{P_WF}=10, K_{i_WF}=5, $K_{P\_\Delta f}$=30), solid thick (with 10% wind power reserve), dotted (without wind power reserve)

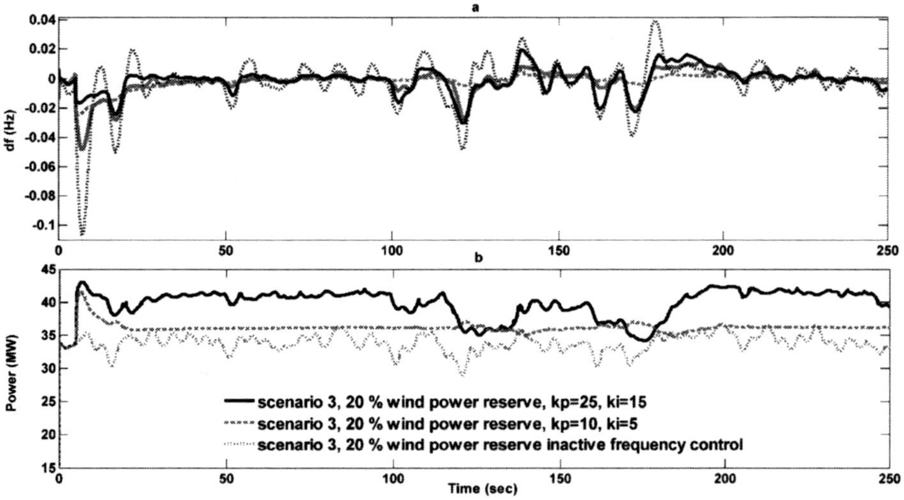

generators. By definition of participation factors (Figure 1), contribution of wind turbines in the frequency control can be adjusted. In this section it can be seen that by using 10% wind reserve, frequency drop and oscillations can be improved and it is not needed to increase the wind power reserve to 20%.

Scenario 4: 40% Wind Power Penetration

This scenario provides 40% as most important scenario in this work. For performing this scenario, G2, G5 and G8 are replaced by three wind farms. Figure 11 shows an increase in frequency devia-

Figure 11. (a) Frequency deviations, (b) ACE signal; Solid (scenario 4), dashed (scenario 3), dotted (scenario 2)

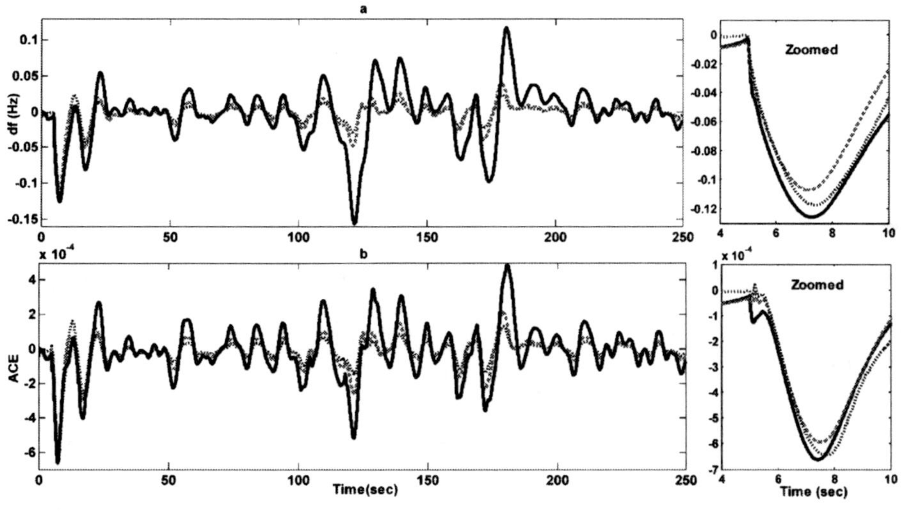

tion. The results of scenarios 2 and 3 presented a less frequency drop in comparison of scenario 4. Because, following replacement of three conventional generating units by the wind farms, the sum of kinetic energy and total system inertia to contribute to the frequency control is decreased.

From Figure 11(a), it can be observed that frequency deviations is more higher and even in some occasions, the resulted frequency drop from wind speed deviations is worse than the sudden occurrence disturbance but from Figure 12 it can be observed that with 10% wind reserve, frequency drop decreases and also using only primary control by wind farms is enough to compensate the frequency drop (in the case of a sufficient conventional power reserve).But using only primary control of wind farms in higher wind power penetration is not sufficient and the wind power contribution in secondary control is needed. Figure 12 shows that by using the proposed approach using the PI controllers, the wind farms can provide secondary reserve to contribute in LFC similar to synchronous generators. Figure 13(a) and Figure 13(b) show the output power of wind farm 1 for 20% and 10% wind power reserve, respectively; and it can be observed that the output power for 20% provides a smoother and higher contribution level.

DISCUSSION

In this paper, it is shown that frequency control can be provided by wind farms and conventional units using PI controllers, simultaneously (controller 1 and controller 2 in Figure 1). In the performed second and third scenarios, 10% and 20% are considered for penetration wind power, respectively. From Figure 7 and Figure 8 it can be seen that increasing the penetration of wind generation can increases the system robustness and reduces frequency deviation after occurrence a disturbance. In the Fourth scenario, 40% wind power is penetrated, but in contrast with previous

Figure 12. Scenario 4: Frequency deviations; Solid thick (20% wind power reserve, complete frequency control, K_{P_WF}=15, K_{i_WF}=10, $K_{P\_\Delta f}$=43.), solid (10% wind power reserve, only primary of wind farms, $K_{P\_\Delta f}$=30), dashed (10% wind power reserve, complete frequency control, K_{P_WF}=10, K_{i_WF}=5, $K_{P\_\Delta f}$=30), dotted (without wind power reserve)

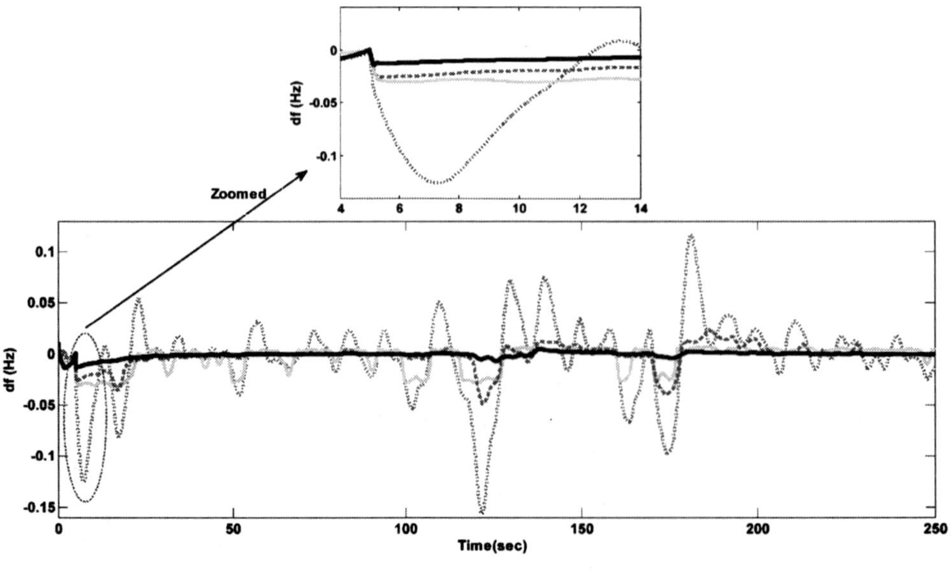

Figure 13. Scenario 4: Active power of wind farm 1; (a) 20% wind power reserve (K_{P_WF}=15, K_{i_WF}=10, $K_{P\_\Delta f}$=43, (b) 10% wind power reserve(K_{P_WF}=10, K_{i_WF}=5, $K_{P\_\Delta f}$=30)

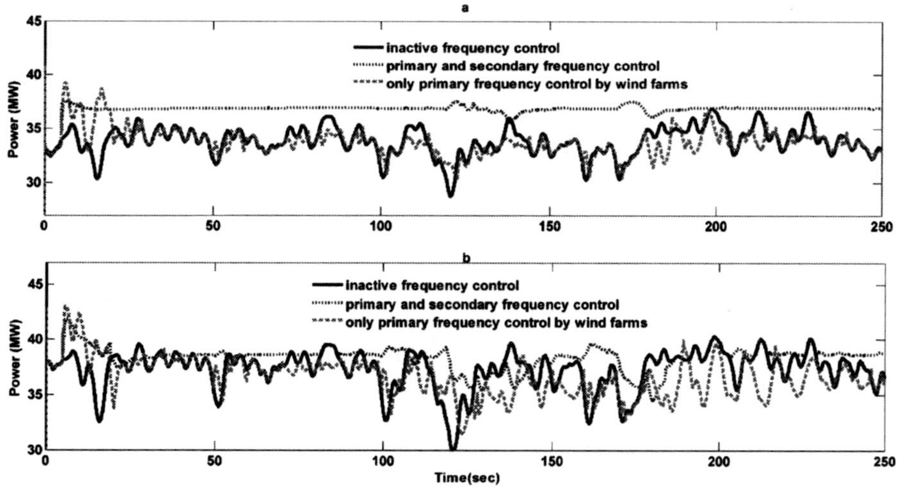

scenarios, the frequency drop due to disturbance is increased (Figure 11). In this scenario, three conventional units are replaced by wind farms and therefore total inertia of the system is reduced. Results show that the proposed method improves frequency performance of power system by proper contribution of DFIGs in the LFC system (Figure 7a, Figure 10a, and Figure 12). For further development (future works), following points are important:

- In the introduced method, de-loaded power curve is used for the DFIG and therefore, output of the wind farms is decreased from nominal installed capacity. This subject must be investigated from economic perspective.
- From economic and security point of views, determination of participation factors for units is an important issue. By proper definition of participation factors, the power system operator can limit the amount of output unit. Since the wind turbines use the power electronic devices, they have a high control capabilities and therefore, appropriate determining of participation factors realize optimum frequency control and high contribution of wind units in a given power network.
- The values K_p and K_i of the PI controllers are tuned by trial and error. It is suggested to use a more effective optimization approach for tuning of PI control parameters for coordination of the DFIGs and synchronous units to provide the secondary frequency control.
- In this paper, several different levels are considered for wind power reserve but none of them are optimum. Getting the best values for reserve in the wind units and synchronous units concerning economic and security issues is a very important issue.

CONCLUSION

One of new challenges in power system operation and control is the frequency control issue in the presence of renewable energy systems. Following the growth of wind power in the world, the necessity of creating secondary reserve by DFIG wind turbines is increased. In this paper, 4 scenarios are performed and it is investigated that wind power penetration increases the system frequency deviations are increased. If only conventional units are responsible for the LFC task, security of power system decreases and causes blackout in part of power system. In this paper, firstly a control framework for creating reserve in DFIGs is introduced and similar to the conventional units, the DFIGs are contributed in LFC by using simple PI controllers. Simulation results presented the DFIGs satisfactorily contribute in the LFC task.

ACKNOWLEDGMENT

The authors would like to thank P.R. Daneshmand from university of Kurdistan for his help to complete this work.

REFERENCES

Abe, K., Ohba, S., & Iwamoto, S. (2006). New load frequency control method suitable for large penetration of wind power generations. In *Proceedings of the IEEE International Power Engineering Society General Meeting.*

Ackermann, T., & Morthorst, P. E. (2005). *Economic aspects of wind power in power systems wind power in power systems* (pp. 383–410). New York, NY: John Wiley & Sons.

Anaya-Lara, O., Hughes, F. M., Jenkins, N., & Strbac, G. (2006). Contribution of DFIG-based wind farms to power system short-term frequency regulation. *IEEE Generation. Transmission and Distribution, 153*(2), 164–170. doi:10.1049/ip-gtd:20050264

Bevrani, H., Daneshfar, F., & Daneshmand, P. R. (2010). Intelligent automatic generation control: Multi-agent Bayesian networks approach. In *Proceedings of the IEEE International Symposium on Intelligent Control* (pp. 773-778).

Bevrani, H., Daneshfar, F., Daneshmand, R. P., Wang, L., Singh, C., & Kusiak, A. (2010). Intelligent power system frequency regulations concerning the integration of wind power units. In L. Wang, C. Singh, & A. Kusiak (Eds.), *Wind power systems: Applications of computational intelligence* (Vol. 0, pp. 407–437). Berlin, Germany: Springer-Verlag. doi:10.1007/978-3-642-13250-6_15

Bevrani, H., Ghosh, A., & Ledwich, G. (2010). Renewable energy sources and frequency regulation: survey and new perspectives. *IET Renewable Power Generation, 4*(5), 438–457. doi:10.1049/iet-rpg.2009.0049

Bhatt, P., Ghoshal, S. P., & Roy, R. (2011). Coordinated control of TCPS and SMES for frequency regulation of interconnected restructured power systems with dynamic participation from DFIG based wind farm. *Renewable Energy, 40*(1), 1–11.

Chowdhury, B. H., & Ma, H. T. (2008). Frequency regulation with wind power plants. In *Proceedings of the Power and Energy Society General Meeting on Conversion and Delivery of Electrical Energy in the 21st Century* (pp. 1-5).

Conroy, J. F., & Watson, R. (2008). Frequency response capability of full converter wind turbine generators in comparison to conventional generation. *IEEE Transactions on Power Systems, 23*(2), 649–656. doi:10.1109/TPWRS.2008.920197

de Almeida, R. G., & Lopes, J. A. P. (2005). Primary frequency control participation provided by doubly fed induction wind generators. In *Proceedings of the 15th Power Systems Computation Conference*, Liège, Belgium (pp. 1-8).

de Almeida, R. G., & Lopes, J. A. P. (2007). Participation of doubly fed induction wind generators in system frequency regulation. *IEEE Transactions on Power Systems, 22*(3), 944–950. doi:10.1109/TPWRS.2007.901096

Ekanayake, J., & Jenkins, N. (2004). Comparison of the response of doubly fed and fixed-speed induction generator wind turbines to changes in network frequency. *IEEE Transactions on Energy Conversion, 19*(4), 800–802. doi:10.1109/TEC.2004.827712

El Mokadem, M., Courtecuisse, V., Saudemont, C., Robyns, B., & Deuse, J. (2009). Experimental study of variable speed wind generator contribution to primary frequency control. *Renewable Energy, 34*(3), 833–844. doi:10.1016/j.renene.2008.04.033

El Mokadem, M., Courtecuisse, V., Saudemont, C., Robyns, B., & Deuse, J. (2009). Fuzzy logic supervisor-based primary frequency control experiments of a variable-speed wind generator. *IEEE Transactions on Power Systems, 24*(1), 407–417. doi:10.1109/TPWRS.2008.2007005

Erlich, I., Rensch, K., & Shewarega, F. (2006). Impact of large wind power generation on frequency stability. In *Proceedings of the IEEE Power Engineering Society General Meeting*.

Erlich, I., & Wilch, M. (2010). Primary frequency control by wind turbines. In *Proceedings of the IEEE Power and Energy Society General Meeting*.

Kaneko, T., Uehara, A., Senjyu, T., Yona, A., & Urasaki, N. (2011). An integrated control method for a wind farm to reduce frequency deviations in a small power system. *Applied Energy, 88*(4), 1049–1058. doi:10.1016/j.apenergy.2010.09.024

Khaki, B., Asgari, M. H., Sirjani, R., & Mozdawar, A. (2009). Contribution of DFIG wind turbines to system frequency control. In *Proceedings of the International Conference on Sustainable Power Generation and Supply*.

Lalor, G., Mullane, A., & O'Malley, M. (2005). Frequency control and wind turbine technologies. *IEEE Transactions on Power Systems, 20*(4), 1905–1913. doi:10.1109/TPWRS.2005.857393

Lalor, G., Ritchie, J., Rourke, S., Flynn, D., & O'Malley, M. J. (2004). Dynamic frequency control with increasing wind generation. In *Proceedings of the IEEE Power Engineering Society General Meeting*.

Lingling, F., Zhixin, M., & Osborn, D. (2009). Wind farms with HVDC delivery in load frequency control. *IEEE Transactions on Power Systems, 24*(4), 1894–1895. doi:10.1109/TPWRS.2009.2030256

Mauricio, J. M., Marano, A., Gomez-Exposito, A., & Martinez Ramos, J. L. (2009). Frequency regulation contribution through variable-speed wind energy conversion systems. *IEEE Transactions on Power Systems, 24*(1), 173–180. doi:10.1109/TPWRS.2008.2009398

Morren, J., de Haan, S. W. H., & Ferreira, J. A. (2006). Primary power/frequency control with wind turbines and fuel cells. In *Proceedings of the Power Engineering Society General Meeting*.

Morren, J., de Haan, S. W. H., Kling, W. L., & Ferreira, J. A. (2006). Wind turbines emulating inertia and supporting primary frequency control. *IEEE Transactions on Power Systems, 21*(1), 433–434. doi:10.1109/TPWRS.2005.861956

Morren, J., Pierik, J., & de Haan, S. W. H. (2006). Inertial response of variable speed wind turbines. *Electric Power Systems Research, 76*(11), 980–987. doi:10.1016/j.epsr.2005.12.002

Ping-Kwan, K., Pei, L., Banakar, H., & Boon Teck, O. (2009). Kinetic energy of wind-turbine generators for system frequency support. *IEEE Transactions on Power Systems, 24*(1), 279–287. doi:10.1109/TPWRS.2008.2004827

Ramtharan, G., Ekanayake, J. B., & Jenkins, N. (2007). Frequency support from doubly fed induction generator wind turbines. *IET Renewable Power Generation, 1*(1), 3–9. doi:10.1049/iet-rpg:20060019

Senjyu, T., Kaneko, T., Uehara, A., Yona, A., Sekine, H., & Kim, C.-H. (2009). Output power control for large wind power penetration in small power system. *Renewable Energy*, *34*(11), 2334–2343. doi:10.1016/j.renene.2009.04.004

Tarnowski, G. C., Kjar, P. C., Sorensen, P. E., & Ostergaard, J. (2009). Variable speed wind turbines capability for temporary over-production. In *Proceedings of the IEEE Power & Energy Society General Meeting* (pp. 1-7).

Ullah, N. R., Thiringer, T., & Karlsson, D. (2008). Temporary primary frequency control support by variable speed wind turbines- potential and applications. *IEEE Transactions on Power Systems*, *23*(2), 601–612. doi:10.1109/TPWRS.2008.920076

Wang, Z., Sun, Y., Li, G., & Ooi, B. T. (2010). Magnitude and frequency control of grid-connected doubly fed induction generator based on synchronised model for wind power generation. *IET Renewable Power Generation*, *4*(3), 232–241. doi:10.1049/iet-rpg.2009.0088

World Wind Energy Association (WWEA). (2011). *World wind energy report*. Retrieved from http://www.wwindea.org/home/images/stories/pdfs/worldwindenergyreport2010_s.pdf

Xiangyu, Z., Heming, L., & Yi, W. (2010). Control of DFIG-based wind farms for power network frequency support. In *Proceedings of the IEEE International Conference on Power System Technology* (pp. 1-6).

Yingcheng, X., & Nengling, T. (2010). Review of contribution to frequency control through variable speed wind turbine. *Renewable Energy*, *36*(6), 1671–1677. doi:10.1016/j.renene.2010.11.009

Zhang Sun, Y., Sui Zhang, Z., Jie Li, G., & Lin, J. (2010). Review on frequency control of power systems with wind power penetration. In *Proceedings of the International Conference on Power System Technology* (pp. 1-8).

This article was previously published in the International Journal of Energy Optimization and Engineering (IJEOE) 1(3) edited by Pandian Vasant, pages 32-47, copyright 2012 by IGI Publishing (an imprint of IGI Global).

Chapter 89
International Development Partnerships and Diffusion of Renewable Energy Powered Lighting Technologies in Off-Grid Communities in Developing Countries:
Exploratory Study in Talamanca, Costa Rica

Inna Platonova
University of Calgary, Canada

ABSTRACT

Worldwide, 1.4 billion people virtually live in darkness after sunset. New lighting technologies, such as light emitting diodes (LED), powered by renewable energy allow non-electrified communities to access for the first time in their lives clean, durable, affordable, and higher quality lighting service. The international NGOs play an important intermediary role in diffusion of these technologies to off-grid communities and commonly operate via development partnerships. With the goal of providing insights into the nature of these partnerships and factors that influence their effectiveness, the exploratory case study was conducted which examined and compared development partnerships promoting renewable energy powered lighting technologies in off-grid indigenous communities in Talamanca, Costa Rica. The study acknowledged the catalytic role of the international NGOs and emphasized the centrality of locally embedded organizations and their capacities in successful implementation of development interventions. A set of factors was identified that contribute to the effectiveness of the development partnership in renewable energy in off-grid communities.

DOI: 10.4018/978-1-4666-4852-4.ch089

INTRODUCTION

Sustainable energy services are central to economic development and poverty alleviation (Modi et al, 2006). Yet 1.4 billion of the world population do not have access to electricity (EIA, 2010b). Majority of people without electricity live in remote rural areas far from a national grid (EIA, 2010a). Off-grid renewable energy technologies[1] can be particularly suitable for these areas given the prohibitive costs of expanding grid into remote and rural areas, growing environmental concerns over conventional energy, the abundance of renewable energy resources in most of developing countries, reduced costs of renewable energy technologies and their improved efficiency and reliability (EIA, 2010a; World Bank, 2008).

Off-grid renewable energy appears to be most relevant for lighting, refrigeration, telecommunications, applications in agro-processing, water pumping and other productive uses with lighting being the most common application. Among the poorest of the poor, lighting is frequently the most expensive energy item they have to acquire, amounting up to 30% of disposable income (Lighting Africa, n.d.). Despite the expenditures, many households receive little value in return. Fuel-based lighting sources they rely upon, such as kerosene, are inefficient, provide limited and poor quality light, pose significant health and fire hazards. Furthermore, fuel-based lighting is responsible for about 190 million metric tons of carbon dioxide per year, equivalent to one-third the total emissions from the United Kingdom (Mills, 2005).

New lighting technologies, such as light emitting diodes (LED), powered by renewable energy allow non-electrified communities to access for the first time in their lives clean, durable, affordable and higher quality lighting service (Adkins, Eapen, Kaluwile, Nair, & Modi, 2010; Peon, Doluweera, Platonova, Irvine-Halliday, & Irvine-Halliday, 2005). The benefits of adequate and reliable lighting can be profound. Ability for children to study after dark enables education and literacy. Replacement of unhealthy and unsafe fuel-based lighting sources improves indoor air quality and health. Ability to extend work hours after sunset allows to increase scarce disposable income. Avoiding burning kerosene for lighting also can positively impact environment through reduction in carbon dioxide emissions and less deforestation (Schultz, Platonova, Doluweera, & Irvine-Halliday, 2008). These are some of the benefits new lighting technologies powered by renewable energy can offer to the off-grid communities.

So far, renewable energy powered lighting solutions have been introduced in a number of developing countries around the world. Without being exhaustive, I note few examples of these efforts. Light Up The World, an international NGO, pioneered the use of LED lighting powered by renewable energy in developing countries. For the past decade, together with partners and supported by donors the organization has introduced its technology and approach in over 50 countries in the developing world (Schultz et al., 2008). LED technology has been also successfully introduced as part of the Millennium Development Project in Malawi (Adkins et al., 2010). IFC and World Bank launched a program which supports development and deployment of affordable, clean, and efficient modern lighting and energy solutions for Sub-Saharan Africans who are currently denied access to the electricity grid (Lighting Africa, n.d.).

The international development non-governmental organizations (NGOs)[2] have been playing an increasingly important role in the diffusion of the renewable energy technologies in developing countries (Biagini & Sagar, 2004). These NGOs facilitate the provision of technologies that address basic energy needs, provide funding, build necessary capacities, apply innovative delivery mechanisms and commonly operate via partnerships with other actors. This chapter deals with the partnership aspect of this work partly because development partnerships are increasingly promoted as vehicles to deliver development goals (Caplan, Gomme, Mugabi, & Stott, 2007). In fact, working in partnerships has become a common practice for organizations in addressing various development challenges and it is a common mode for the international development NGOs working in developing countries (Horton, Prain, & Thiele, 2009).

The notion of partnership has received wide attention from the academics and practitioners in various disciplines and professional domains, such as management and organizational development, public policy and public management, international development, social policy, science, technology and innovation, to name a few. Nevertheless, a lack of empirical analysis of partnership remains a knowledge gap in this area. For instance, Horton et al. (2009) indicate knowledge gaps at the level of individual partnerships, at the organizational and domain levels. Available studies are often rooted in particular disciplines that influence their perspectives and lack cross-referencing and cross-fertilization that is needed to enrich our understanding of the phenomenon of partnerships (Horton et al., 2009). Furthermore, compared to other development issues, NGOs have received far less systematic research attention and available literature is often too normative in nature. Therefore, some argue that this topic is under-researched. The questions on effectiveness, accountability and scalability of NGO activities remain largely unanswered (Mitlin, Hikey, & Bebbington, 2007).

To address some of these gaps, an exploratory case study was conducted to examine and compare two development partnerships driven by international NGOs in off-grid indigenous communities of Talamanca, Costa Rica. The goal of the study was to shed light on the nature of the development partnerships driven by NGOs in diffusion of renewable energy technologies in off-grid communities in developing countries and identify factors that impact the effectiveness of these partnerships. The chapter first discusses existing development partnership approaches emphasizing their strengths and weaknesses in capturing the complexity of development partnerships and addressing the question of effectiveness. Then, the partnerships in Talamanca are described and compared and the effectiveness factors are identified. In conclusion, research and practice implications are discussed.

DEVELOPMENT PARTNERSHIP APPROACHES

Development partnerships are commonplace in the context of international development policy and practice (Lewis & Kanji, 2009). While the notion of partnership is widely utilized by scholars and practitioners in a variety of ways and meanings, development partnerships can be generally understood as

an agreed relationship based on a set of linkages between two or more agencies within a development project, involving a division of roles and responsibilities, a sharing of risks, and the pursuit of joint objectives [...] (Lewis, 2007, p. 96).

These partnerships are widely promoted by international donors and development agencies and are espoused by the state, private sector and civil society.

There are a number of reasons why such partnerships are desired. First and foremost, the partners are driven by the desire to improve the results of development interventions (Morse & McNamara, 2006). The nature of the development challenges is often very complex making it impossible to address them in isolation (Brinkerhoff, 2002a). From an international NGO perspective, partnerships can help leverage donor funding, increase organizational accountability, gain credibility and legitimacy, build capacity and capabilities, reduce information and transaction costs and access scarce resources. Partnerships depend on and foster trust among stakeholders which enhances the ownership of the development processes which in turn improves their effectiveness and sustainability (Fowler, 1998).

Without being exhaustive, several approaches can be distinguished in the development partnerships literature with specific reference to development NGOs. One such approach is the normative perspective on North-South partnerships (Fowler, 1998). The power dimensions, accountability, trust and mutuality are at the core of this understanding of the development partnerships (Fowler, 1998; Lister, 2000). One of key themes is the criticism of North-based NGOs for failing to form equitable or "authentic" partnerships with their counterparts in the South (Ahmad, 2006; Fowler, 1998). Some of the most common factors for failure relate to matters of power imbalance, lack of accountability and inequality (Fowler, 1998). This approach focuses on partnership as an end in itself and is criticized for being difficult to operationalize, having little practical value and being subjective and values-based (Brinkerhoff, 2002b).

The social issues cross-sector partnership literature presents a more pragmatic approach towards development partnerships and views them as means to achieving specific objectives (Brinkerhoff, 2002b). It examines partnerships between NGOs and government, government and business (public-private partnerships), NGOs and business, and tri-sector partnerships that address social issues, such as poverty alleviation, education, health, and other (Selsky & Parker, 2005). Selsky and Parker, 2005 consolidate the vast multidisciplinary literature that addresses this type of partnership through the lens of resource dependence, social issues and societal sector conceptual "platforms" which have strong instrumental focus. While these platforms help identify "functionally useful dimensions" and success factors affecting partnerships outcomes, the authors conclude that the partnerships are often approached in a linear way through the input-output systems models, which

neglect the embeddedness of social processes within the larger institutional structures studied in institutional theory, as well as processes of nonlinear emergence as studied in complexity theory (Selsky & Parker, 2005, p. 865).

Another approach is the Base of the Pyramid (BoP)[3] and subsistence markets partnerships (Rivera-Santos & Ruffin, 2010)[4]. This literature, which focuses on development partnerships driven by businesses that target pro-poor markets, suggests that serving the BoP requires cooperation between large corporations and small firms, civil society organizations, development agencies, governments and the poor themselves (Prahalad, 2005). Such partnerships are largely to be affected by the institutional environment of the pro-poor markets (Riviera-Santos &Ruffin, 2010). For example, lack of formal market institutions at these markets is likely to be compensated by greater involvement of government and civil society. We are likely to observe a greater diversity of organizations, in particular, non-market agencies (such as NGOs), involved to fill the gaps in the governance structures of pro-poor markets. The partnerships are to rely on informal trust-based governance mechanisms (Riviera-Santos & Ruffin, 2010).

The existing development partnerships approaches (Table 1) are useful in expanding our understanding of the development partnerships, their rationale, dynamics, and success factors, however they often focus on process of partnering rather explaining and evaluating how the partnership leads to specific development outcomes. Their liner trajectory also ignores the complexity of the interactions within the partnerships and their broader settings, goals and interactions. To address these concerns, the question on the nature of partnerships is addressed through the lens of innovation system approach and the question of effectiveness brings together the process and outcome views.

Table 1. Development partnerships approaches

Partnership Category	Development Partnerships Approaches			
Type	North-South partnerships (e.g. Fowler, 1998)	Base of Pyramid partnerships (e.g. Rivera-Santos & Ruffin, 2010)	Social issues cross-sector partnerships (e.g. Selsky & Parker, 2005	NGO development partnerships in diffusion of technologies (Author)
Common Initiator	North agency: either donor or operational NGO	Business firms, often multinational enterprises	Government, business, NGOs	International NGOs or local actors
Common conceptual constructs	Power and equality, authenticity	Impact of distinct BoP institutional environment, need for intermediaries	Functional dimensions, success factors	Partnership dynamics and partnership outcomes
Main disciplinary focus	Development studies	Business and management	Multidisciplinary	Interdisciplinary

Source: Author's compilation based on literature review and own research.

INNOVATION PERSPECTIVE

The partnerships that are the focus of this research are part of the technology diffusion processes that involve a myriad of actors and linkages. At first, we understood technology diffusion as a linear innovation model according to which a straightforward path from basic research and development to technology commercialization and adoption can be followed (Schumpeter, 1959). Today, technology diffusion is broadly viewed as a complex process involving a broad range of private and public actors which constitute national innovation systems (OECD, 1997).

National innovation systems (NIS) can be defined as

[...] a set of interrelated institutions; its core is made up of those institutions that produce, diffuse and adapt new technical knowledge, be they industrial firms, universities, or government agencies. The links between these institutions consist of flows: knowledge, financial, human (people being the bearers of tacit knowledge and know-how), regulatory, and commercial (Niosi, 2002, p. 291).

The relationships between the NIS agents are viewed as carriers of knowledge and through the interaction among them new knowledge is produced and learnt (Johnson & Lundvall, 2003). Regulatory, policy, international, infrastructure and market conditions have major impacts on innovation and therefore are integral part of the NIS (Gu & Lundvall, 2006). The NIS concept has expanded to include industrial, local, regional and even international systems of innovations (Gu & Lundvall, 2006; Niosi, 2002). Since its beginnings in the late 1980s, the innovation system (IS) approach has become influential in academia, international and policy arenas (Sharif, 2006).

Though the notion of NIS originated in developed countries, it has been expanding to include developing countries (Lundvall, Vang, Joseph, & Chaminade, 2009; Muchie, Gammeltoft, & Lundvall, 2003). The NIS approach challenges the limitations of the traditional approach, which views developing countries as 'technology users' and focuses on North-South technology transfer with learning being imitative as opposed to innovative (Mytelka, 2003). NIS emphasizes the importance of building capabilities in developing countries to absorb technology, adopt them to local conditions and innovate (Dolun, 2005). Building upon indigenous knowledge and local governance strengths is essential (Johnson & Lundvall, 2003).

The innovation system (IS) framework can prove useful in understanding the diffusion process involving international NGOs and their partnerships and see these partnerships in a larger system of actors and interactions and how these impact the process and outcomes of the partnerships. It is used in this study as an analytical tool which looks at a partnership in terms of its actors, the roles they play, linkages and interactions taking place, capacities and learning.

The idea of using the IS framework for the analysis of partnerships in developing countries is not new, for example, Thorsteinsdóttir et al. (2010) examined the South-South research partnerships in health biotechnology and Hall (2006) worked on public-private partnerships in agricultural innovation in developing countries. However, they look at partnerships within the innovation systems rather than apply IS as analytical tool to partnerships themselves. In general, application of IS framework in the context of developing countries remains untested and the role of NGOs in innovation systems is not well understood (Hall, Bockett, Taylor, Sivamohan, & Clark, 2001).

PARTNERSHIP EFFECTIVENESS

While there is an underlying logic that the partnerships lead to the development outcomes, these partnerships are often analysed in their own right rather than in terms of how they are actually achieving the development outcomes (Geldof, Grimshaw, Kleine, & Unwin, 2011). A recent study on the Information and Communication Technologies for Development (ICT4D) concluded that the partnership literature mostly ignored the development outcome angle of the partnerships (Geldof et al., 2011). Similarly, Dowling, Powell, and Glendinning (2004) conducted research on partnerships involved in provision of health and social services and found that research was mainly focused on the process issues or on functioning of partnership in terms of addressing shared goals, rather than outcome issues, e.g. changes in service delivery, and impact on service users. They acknowledge this discrepancy as a major knowledge gap.

In light of this research, my argument is that if we want to improve the outcomes of the partnership-based development interventions, we need to be concerned with both process issues and outcomes issues. Therefore, I propose an approach to the analysis of the effectiveness of the international NGO-driven development partnerships involved in diffusion of off-grid renewable energy technologies in developing countries which brings together the process view and the development outcome view. I introduce respectively the notions of the partnership dynamics (process issues) and partnership outcomes (outcome issue) as constructs that can explain the effectiveness of partnerships in a more comprehensive way (Chart 1). The partnership dynamics focuses on the relationship between the partners, the management of the partnership, decision making process, communication. The partnership dynamics can be measured in terms of partner satisfaction with the goal achievements and relationship (Ashman, 2001a), and benefits to partners (Brinkerhoff, 2002), e.g. financial, reputational benefits and other. The partnership outcomes, on another hand, can be understood in terms of goal achievement (Ashman, 2001a), project sustainability (Ashman, 2001b), scalability of the initial efforts (Ashman, 2001b)[5] and impacts on beneficiaries. The measures and their application will be further explained later in the chapter.

METHODOLOGY

The case study was chosen as a preferred method for the research, because:

- The research raises the "why" question (why are some partnership more effective?);
- The research focuses on contemporary phenomenon within a real-life context; and
- The researcher has little influence over events (Yin, 2009).

The research included an exploratory multiple case study in Talamanca, Costa Rica.[6] This type of case study was chosen because the research topic is not well understood and there is a need for an exploratory work to provide the base for further investigation. Furthermore, this approach helps address the complexity of the phenomenon which may not be approached by other forms of research, e.g. experimental research. At the same time, choosing multiple cases enables the researcher to investigate the differences within and between cases with the goal of replication of findings across cases (Yin, 2009). Such approach is also considered as robust and reliable (Baxter & Jack, 2008). Furthermore, case studies have been a frequent choice for the research in the pertinent fields. Recent examples include: a case study of service learning partnerships in New England (Dorado, Giles, & Welch, 2009), a study of business and NGO partnerships in Lebanon (Jamali & Keshishian, 2009), a study of Dutch development partnerships (Kolk, van Tulder, & Kostwinder, 2008), and a study on cross-sector partnerships in South Africa and Zambia (Rein & Stott, 2008).

To do my exploratory case study I travelled to Costa Rica in summer 2010. The case study goal was to explore the characteristics of international development NGO driven partnerships serving off-grid communities, when and why they are formed and what makes them effective. I examined and compared two development partnerships driven by the same international NGOs that are implementing off-grid solar energy lighting programs with the indigenous communities in the Talamanca canton, the Limón province, Costa Rica. The reason I chose to compare these partnerships is because while seemingly having comparable chances for success and having similar context and actors involved, the initiatives unfolded differently and achieved varying degrees of success.

During my field work I also participated in micro-finance and technical trainings and joined installations of solar lighting systems in the remote community called Los Cocos as part of one of the initiatives in Talamanca. These activities allowed me to gain first-hand insights into the development work and observe interactions between the partners. It also helped to build rapport with my research participants.

I started with the review of documents and archival records of the partnerships. Documentation sources included partners' agreements, studies, project proposals, and project reports. Archival records included baseline surveys, impact assessments, maps and charts. Documentation and archival records analysis aimed at building a detailed account of studied partnerships and projects, identifying key participants of each project, understanding the context for each case, determining progress of each partnership and factors responsible for their successes or failures. However, during this analysis it became clear that other methods were needed in order to obtain the detailed data for my case study.

These data were gathered during the key informant semi-structured interviews with the representatives of the respective organizations where they were asked to comment on the project goals, choices for partners, partnership goals, the progress of partners in achieving these goals, what works and what does not work, what contributes to successes, and what poses challenges. I also conducted semi-structured interviews with beneficiaries to help understand partnerships' benefits for the communities and the relationship between beneficiaries and implementing partners. I also met with other organizations and individuals in Costa Rica working in the off-grid renewable energy area. Key informant interviews were conducted with these organizations and individuals asking about their experience with off-grid energy projects and partnerships in Costa Rica. These interviews helped understand the broader context in which the partnerships in question are taking place. The interviews were transcribed and then coded. Descriptive codes were applied to organize responses. These codes included categories and subcategories based on research questions, literature review and theoretical framework (precoding) and were revised based on interviews (postcoding).

The conclusions in this case study reflect the specific context of partnerships in Talamanca and may not be generalized to other partnerships in other contexts. However, the value of this study is that it provides insights into the nature and effectiveness of these partnerships as perceived by both international development NGOs and their local partners. Existing empirical studies often focus on one side of partnerships, for example the view of North-based NGOs or South-based NGOs. These insights can be further explored through qualitative and quantitative research. It also provides preliminary guidance for the international development NGOs and their local partners in their collaborative effort to bring renewable energy technologies to off-grid communities.

EXPLORATORY CASE STUDY IN TALAMANCA, COSTA RICA

Background

Talamanca is the poorest canton in Costa Rica located in the Limón province south of the country. It scored 9.75 in the Social Development Index in the last national census with score of 10 signifying least development (APPTA, n.d.). Similarly to the rest of Costa Rica, poverty in Talamanca is concentrated among the indigenous and Afro-Caribbean populations. The municipality of Talamanca (2003) reports that Talamanca is the poorest canton in Costa Rica, with about half of the population (49.7%) living in poverty (Whelan, 2005). Total population of Talamanca is 33,569 with indigenous population (primarily Bribri and Cabécar indigenous people) accounting for 12,118 (INEC, n.d.). The district of Bratsi in the canton, where the Bribri and Cabécar indigenous territories are located, has the highest concentration of poverty within Talamanca.

The Bribri and Cabécar indigenous territories lack access to basic services, including access to health care, education, transportation and energy. Infant mortality and malnutrition in the area is much higher than the national average (Whelan, 2005). The education levels are very low with the average grade level achieved is the fourth grade. Only 14% of the population complete secondary education. Illiteracy rates are as high as 30% among the Bribri and 50% among the Cabécar (Whelan, 2005). Indigenous people are primarily involved in agriculture producing cacao, banana, plantain, and an increasing number of producers are raising animals for consumption and sales: pork, hens, bovine cattle and fish (APPTA, n.d.). The income levels are among the lowest in Costa Rica. The average monthly income of a family of four in Talamanca is approximately USD $100 making it difficult to meet their basic needs (Whelan, 2005). Until very recently the communication and transportation infrastructure was practically nonexistent in the area. Even though the infrastructure conditions have improved since then, the issues of poor socio-economic conditions remain.

While country-wide electrification rates are high (98.6% - the highest in Central America), many households in Talamanca have no access to electricity. One of the reasons is their remote location and low population density which makes grid extension difficult. However, even when the grid is being expanded, connecting to this grid is prohibitively expensive. One can spend over CAD $200 for a connection kit and CAD $10 for each meter of wiring. Somebody who lives 50 meters from the grid would have to spend about CAD $700-800 for connecting. In words of one of the beneficiaries and also a coordinator of the educational project with the indigenous communities in Talamanca: "Some people might be moving closer to the road where grid will be, but there will be still bunch of people who will not move. There will be demand for solar and less centralized technologies. I am not going to move closer to where the grid will be."

Off-grid households are relying on kerosene and candles to meet their lighting needs. In addition, people use flashlights that are running off dry-cell batteries. Recycling for these batteries is non-existent in the area and many of them are being disposed around the homes and into the rainforest which is a concern. Often a kerosene lamp will be used in the kitchen and candles - for studying. People will use light for few hours in the evening before going to bed at around 7-8 pm and in the morning after waking up at around 3 am. Lighting expenditures for a family of four can be as high as CAD $30 per month - up to one third of their annual income. Using kerosene and candles is not only expensive, but these are sources that are inefficient, have negative health, safety and environmental impacts. To address the issue of access to sustainable energy, international NGOs facilitated provision of solar based LED lighting in remote communities in Talamanca through ACEM and APPTA Solar Energy Lighting Initiatives described below.

ACEM Solar Energy Lighting Initiative

The first program is led by a local NGO called Association for Science and Moral Education (ACEM) which has been running an educational program for development called Preparation for Social Action (PSA) in Talamanca since 2001. PSA is a two year tutor-delivered program for youth to improve their knowledge in areas such as math, science and technology, language and communication, and community service. Its tutors and learners throughout the Talamanca form a natural network of people from the communities. The solar energy lighting initiative commenced in 2008. ACEM saw a natural fit between solar lighting and education. As Luis Dumani, ACEM Executive Director notes:

We know we are in indigenous community. We were estimating that if we bring better light, they have better light for studying, they could study more, the study will be more meaningful, deeper, their spaces of cultural interaction will be improved, better environment to communicate, to do things.

The organization recruited and trained technicians, most of whom were the PSA tutors. The PSA tutors/technicians became the technical backbone for the project. The initiative adopted a microfinance approach where families redirect payments previously used for candles and kerosene toward the solar lighting system for their home. The microfinance approach is an important aspect of the project sustainability. As systems are repaid in full, household disposable income increases which improves the economic situation of families. The payback period for the systems in this project has been so far less than two years. Other important outcomes include children spending more time after sunset doing their homework and reading books, and families having extended hours for productive and social uses. ACEM established a formal contract system where a beneficiary signs the contract which specifies responsibilities of ACEM and a beneficiary. There are also receipts used and issued for each purchase. A local coordinator hired by ACEM is fully responsible for administration and management of this project on ground. As of the end of September 2010 ACEM has installed 378 systems in 15 communities in Talamanca.

APPTA Solar Energy Lighting Initiative

The other program is implemented by APPTA, an organic producers' cooperative founded in 1987. Currently the association consists of 1067 producers of which 80% are indigenous Bribri and Cabécar people. Women constitute 38% of the members of the association. The cooperative has been involved with solar energy since 2004. APPTA also saw a good match between this project and its main mission. As Walter Rodriguez, Director of Marketing and Development of APPTA indicates:

APPTA has two big objectives: 1) improve the standards of our members and, at the same time 2) protect the natural environment. Through lighting project we meet both objectives.

All activities have been primarily performed on a volunteer basis and there was no paid staff dedicated to this initiative. The organization is fully consumed by the matters related to their main operations as a cooperative and complexities of this business have not allowed to prioritize the solar program so far. There is a full realization on the part of the organization that for this project to move forward it should be integrated into its strategic priorities.

Most recent development is that a group of technicians has been trained and the project is now overseen by the Executive Director of the cooperative. New installations started after I completed my field work. The visit to the homes where systems were installed showed that people are satisfied with the system and note positive educational and social benefits of having solar lighting. However there were some technical issues that have not been addressed. For example, some households had also the ACEM system and the APPTA system was running of the battery from the ACEM system. If there was no such system in place, some APPTA systems would not work as the batteries needed a replacement which has not been performed. As of August 9, 2010 the organization has installed 117 systems.

International NGOs as Partners

Important aspect of both initiatives is that the partnership is driven by two international NGOs: Light Up The World and Y Service Club. Light Up The World (LUTW) has pioneered the use of solid state lighting solutions powered by renewable energy for the world's poor who do not have access to the electric grid and rely on kerosene, candles and traditional biomass for their lighting needs. Since its inception a decade ago, LUTW has been involved in numerous projects that resulted in installations of thousands of lighting systems in 50 countries throughout the developing world. At the core of LUTW approach are the partnerships with major manufactures, NGOs, local communities and academia. LUTW as an intermediary bridges the gaps in the technological supply chain which fails to reach the poorest of the poor. On ground, LUTW is working with local grassroots organizations which understand the local environment for technological and social innovations, connected to and trusted by the community members. LUTW contributes to planning and project design, helps to identify and procure necessary equipment, facilitates shipping of the equipment from overseas and logistics of component procurement locally. It equips the local partners and stimulates local social innovations in distribution and servicing that are essential in the sustainable energy service delivery in the context of poverty.

Y Service Clubs are community service clubs working towards developing global citizens, building culture and peace internationally, promoting cross-cultural, cross-border international relationships, building a web of relationships and knowledge about other places and other cultures through connections with clubs all over the world. The club that is involved in this project is the Owen Sound Y Service Club based in Ontario, Canada. Its role is to provide training on solar system concepts and applications as well as infrastructure development for the projects. The organization is also actively involved in fundraising and conducting training and monitoring visits to Costa Rica since March 2007. In case of APPTA, there is one more partner, an individual David Wiwchar, Vice Principal of Vision Lively District Secondary School based in Ontario, Canada. Mr. Wiwchar is the founder of the Light Up Costa Rica program. He runs an educational program where he brings Canadian high school students to Costa Rica during March break to be involved in solar lighting projects. Through this work and his previous experiences and contacts in Costa Rica, Mr. Wiwchar introduced the idea of solar lighting to areas of need and facilitated linkages with the implementation partners.

NATURE OF PARTNERSHIPS

Actors and Linkages

At first glance, there is a myriad of actors involved in the presented partnerships in variety of ways, the IS framework helps to map out actors within these partnerships, and highlight their linkages, and interactions. The framework also helps position these partnerships within a larger system of actors and interactions that are part of the technological diffusion process. Figure 1 draws out a map of actors, their linkages and interactions in the studied partnerships which are explained hereafter.

The development partnership domain constitutes actors that are the focus of this research. These are the international NGOs and local organizations working with the target communities. They can be called "the boundary partners" - organizations that are directly interacting within the development project and whose behaviour the partnership can have opportunities to influence (Earl et al, 2001). The boundary partners have close linkages and frequent interactions. Stronger linkages are manifested through regular and open communication, regular exchange of knowledge about project progress and sharing of learning from the field, open discussion of challenges and the ways to overcome them.

The international NGOs in the studied partnerships play an important catalytic role, while their local partners play a central role in the solar lighting projects and are closely linked with the

Figure 1. Map of actors and linkages in development partnerships in Talamanca, Costa Rica

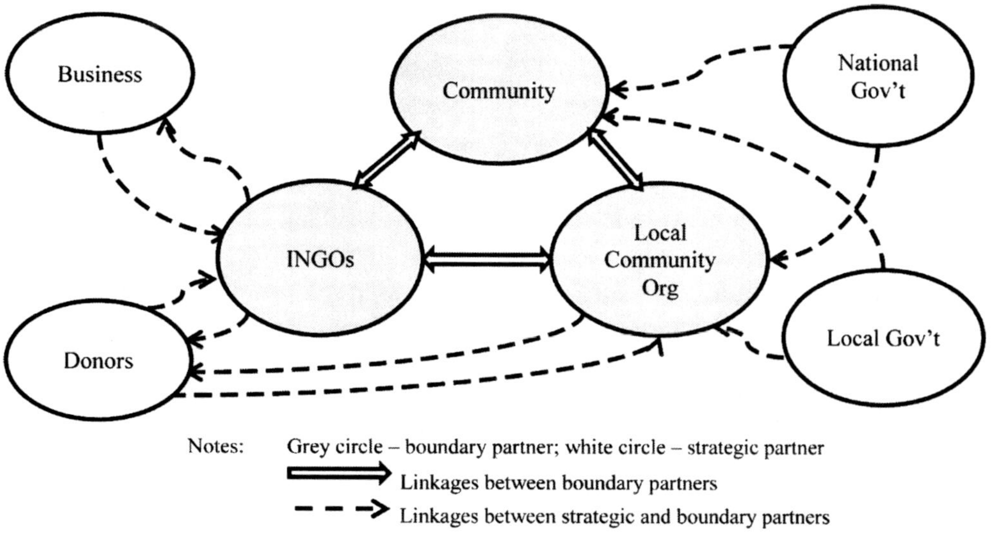

beneficiaries. Participating international NGOs facilitate access to financial, management and technical resources and help build capacities that are needed by local organizations to implement the projects and ensure their sustainability. Local implementing organizations, an NGO and an organic producers' cooperative, which have presence in the community, are central to the project since they understand local environment for technological innovation and its diffusion. They are closely connected with and are trusted by the community. They are leading the planning and implementation of the projects, and are fully responsible for monitoring and evaluation activities. Local organizations provide feedback to the international NGOs on the project progress and engage in the exchanges that help move projects ahead, deal with the challenges and nurture successes.

LUTW as a supply chain intermediary allows organizations like ACEM and APPTA to access best in class solar and LED technology at a lower cost. This is possible due to LUTW's established relationships with the manufacturers of components (indicated as business in Figure 1) and funding from donors. These are contractual relationships between an NGO that acts as a distributor and a manufacture that plays a role of a supplier. Using the language of Earl et al, (2001) these are strategic partners that provide resources to the initiatives but whose behaviour the initiatives do not aim to change. In return business partners gain corporate social responsibility benefits and opportunity of their products to be introduced to new markets. Other strategic partners include donors which provide needed funding for the international NGOs and the local NGO. The linkages between these actors include financial flows to the NGOs and reporting from the NGOs to donors.

The map also highlights both national and local government as participating actors in the larger system. The local partners have linkages with the local authorities which play a facilitating role in the projects. However, the linkages with the national government were noted as weak. The government policies did impact some aspects of the projects, for example, in case of shipping through customs regulations. However, the national authorities were largely unaware of these partnerships and initiatives. The boundary partners indicated that it would be important to coordinate with the government, in particular, ICE, Instituto Costarricense de Electricidad - a Costa Rican publicly owned energy utility responsible for electrification and be aware of the government plans for grid extension into the area. Local organizations could benefit from accessing certain funds and/or technical expertise from ICE. In turn, the government could rely on these actors to cover Talamanca communities which will not be reached by the grid and expand its own capacities in the off-grid rural electrification by learning from these projects. ICE already collaborates with an NGO in renewable energy electrification of remote communities in other areas. There are unrealized opportunities for national government involvement with the studied initiatives which can be a win-win scenario for energy regulators and providers, local NGOs and communities.

Drivers

The exploratory study showed that community service was the key driver for the partnering organizations. While responding to specific community needs, an NGO activity is often related to market or government failures. In case of the NGO-driven initiatives in Talamanca, it is a response to the government not being able to extend the grid to some communities, which is primarily due to economic reasons. It is also about market failure where the market has not acknowledged that there is a demand for certain services.

The international operational NGOs in studied partnerships play the role of "hands-on donors" where they do not only (co-)fund development projects but also help local organizations build capacities necessary to sustain these projects. Among local NGOs and community-based organizations in developing countries, there is a certain level of resource dependencies in technology, capacity and financing which necessitate involvement of the organizations that can provide such resources. In the case of the partnerships in question, international NGOs bring project management and technical expertise, access to technology at low cost, and funding - resources that are required for a local organization to be successful in implementing a project.

International NGOs, on the other hand, need to work with the local organizations in order to ensure the necessary closeness to the communities, understand the communities' needs and realities, and gain trust of the communities' members. It is costly in terms of time, human and financial resources for an international NGO without a local base or presence to build strong community relationships within a short period of time. Therefore, partnering with the locally embedded organizations is a key in this type of initiatives.

Challenges

Both international NGOs and their local implementation partners are often vulnerable due to financial constraints and reliance on external funding. The funding availability is limited in terms of its amount and duration. Most frequently one to two year grants are available. Every year the NGOs in question are searching for funding, which distracts from their main development operations. This situation is exacerbated by financial crises and economic slowdowns where governments are more likely to cut foreign aid spending that particularly affects international development NGOs. Some private donors mainly in developed countries are also turning their support towards their domestic communities. Securing funding for international development projects is becoming more difficult. The international NGOs and the local NGO I interviewed emphasized lack of financial sustainability as a key constraint to their operations.

Another set of barriers is related to logistics and infrastructure. The climate in Talamanca is challenging and devastating rains often shut down the communications and flood the communities which make community access impossible. Even basic roads were non-existent until a year ago. Currently, some communities are accessible by gravel roads, however access remains difficult in many cases. These constraints present major challenges for local organizations to reach these communities and for international NGOs to conduct monitoring and evaluation visits and as well capacity building activities on ground.

Cross-cultural communication and lack of communication capacities can be challenging as well. Phone and/or e-mail are often not available and/or are not appropriate to deal with the project communities in Talamanca which makes it difficult for international NGOs to engage remotely. You have to be face-to-face with people on a regular basis to get things going. Another challenge is the coordination among multiple actors within a project and managing expectations. Having regular and open communication as well as a formal agreement specifying the roles and responsibilities are central in mitigating this problem.

PARTNERSHIP EFFECTIVENESS

The effectiveness of studied partnerships was measured in terms of partnership dynamics and outcomes as low, moderate or high. The partnership dynamics was measured in terms of to what extent the partners were satisfied with the goal achievement and with the relationship and benefits to partners, including financial, reputational and technological. Financial benefits included ability of the partnership to draw funding for the project, reputational benefits were about whether the partnership increased the profile and status of the partnering organizations, technological benefits referred to ability to access technical expertise and technology.

The partnership outcomes were measured in terms of how sustainable the project is: if there is a viable distribution system in place to reach remote off-grid communities, and whether there are necessary human, financial and technological capacities to sustain the benefits of the initial projects to the users. Scalability was referred to the number of users and communities reached. Scalability is high when more than 1000 beneficiaries have been reached. Moderate - from 500 to 1000 beneficiaries impacted, and low - fewer than 500 beneficiaries reached.

The benefits to beneficiaries included economic, social and environmental benefits. Economic benefits were associated with savings from not using kerosene and candles – those can be observed after the systems are paid off. Social benefits included educational benefits, and ability to socialize in the evening. Environmental benefits were about improved indoor air quality, and using less fuel-based sources for lighting (such as kerosene). Table 2 compares two studied partnerships using these effectiveness measures. Clearly ACEM Solar Lighting Infinitive emerged as a more effective partnership overall, however it must be noted that the key differences were noted primarily in the partnerships outcomes. As far as partnership dynamics, both partnerships scored high. This preliminary evidence confirms the need to go beyond the process view as it did not explain in this case study why one partnership led to the desired outcomes more than the other one.

It is important to note that the analysis has focused on partnerships' dynamics and outcomes since the inception of partnerships and up to the time of the field work in August, 2010. There were several key changes made in APPTA right before my field work that had potential to change the trajectory of the outcomes. Walter Rodriguez, Director of Marketing and Development of APPTA explains:

There is a team of local inspectors who check on certification. Some of them have been trained to go out, install, maintain [systems] and we pay them for this work. [...] They [APPTA] institutionalize it [solar lighting initiative] for the first time this year.

Table 2. Comparing effectiveness of development partnerships in Talamanca

Effectiveness Measures	ACEM Solar Lighting Initiative (2008-08/2010)	APPTA Solar Lighting Initiative (2004-08/2010)
PARTNERSHIP DYNAMICS		
Partner satisfaction with the goal achievement	+++	++
Partner satisfaction with the relationship	+++	+++
Financial benefits	+++	++
Reputational benefits	++	++
Technological benefits	+++	+++
PARTNERSHIP OUTCOMES		
Goal achievement	+++	++
Project sustainability	+++	+
Scalability	+++	+
Impact on beneficiaries	+++	++

+++ High; ++ Moderate; + Low

Since these were very recent changes, for example, the technical training has been delivered to the group of inspectors 4 weeks before the interviews; it was hard to assess at that point how this new organizational system would work and what the outcomes would be. However, these changes are important developments that can have direct impact on effectiveness of the partnership involving this local partner. The following sections explain the factors that came out of this study that contributed to the partnership effectiveness in terms of its dynamics and outcomes.

Partnership Dynamics

The analysis of partnerships dynamics in Talamanca identified a set of factors that contribute to an effective partnership, such as: shared values and goals, complementary expertise and capacities, confidence and trust, clear roles and responsibilities, regular and open communication, personality match and continuity of staff. Thus, Christoph Schultz, Projects Director of LUTW talks about dynamics success factors:

[...] the partnership makes sense because we are trying to achieve the same things and they [partners] have certain skills and capacities to do things that we can't do and we have certain skills, capacities. We complement each other. [...] Functional partnerships are result of essentially communication. [...]Commonalities of business or communication cultures [are important]. [...] clear understanding of roles and responsibilities.

These views were widely shared by all partners. The organizations also emphasized the importance of personalities. For example, Walter Rodriguez from APPTA talked about partnership from the dynamics point of you: *I feel like it always functioned. We have always clicked.* Luis Dumani from ACEM points: *In my experience, institutional relationships depend a lot on individuals, personalities.* He also raises a question about what happens if either of the individuals from partnering organizations leaves. Clearly a continuation of staff is important as relationships take time to build. It takes time and effort to build trust and confidence. Furthermore, individuals involved in these partnerships also retain tacit knowledge about their work which cannot be easily transferred to a new comer.

These findings confirm much of the general understanding of what makes a partnership successful coming from the partnership management literature and also development studies. International development research also emphasizes equality, trust, shared commitment, agreed division of tasks and responsibilities, compatible objectives, mutual agreement, transparency, shared contributions, shared rewards, win-win benefits, accountability, flexibility as some of the success factors (Ashman, 2001; Clarke-Okah, 2004; Lewis, 1998). Partnership management studies highlight as success criteria: mutual trust, resource dependency, commitment symmetry, shared common goals, converging cultures, leadership, resource exchange, schedule, communication, transparency, accountability, representation, and participation (Austin, 2000; Brinkerhoff, 2002; Dorado, Giles, & Welch, 2009; Jamali & Keshishian, 2009; Kolk et al., 2008; Morse & McNamara, 2006; and Tennyson, 2003). These factors largely coincide with those identified through the case study. Though, the importance of some criteria differed possibly due to differentiation in the organizational type and context. The partnerships in question were influenced a lot by individual dynamics and the personality match and continuity of staff were identified as important factors. Human factors often receive less attention in the partnership literature. To explain the role of human factors in the dynamics of partnerships, partnership literature can benefit from insights on these aspects from the personal and organizational psychology.

Partnership Outcomes

The analysis of the partnerships showed that their ability to achieve development outcomes was dependent on three sets of factors related to partnership outcomes:

- Local champion,
- Community, and
- Technology.

It was critical to have an established and committed local partner that had experience with the community members and their trust; developed effective network in the community; was able to institutionalize learning and build on it. ACEM was such champion for the partnership. The organization was able to identify the beneficiaries, secure financial resources, hire local staff, train technicians and establish the system of payment collection and address the lighting needs of 10% of target population within relatively short period of time. This was possible primarily because of the initial capacities of ACEM developed through the successful education program that the organization has been running in the area for almost a decade, through their full commitment to the solar energy initiative and ability to prioritize it and allocate resources and staff. The network of tutors as part of the educational program of ACEM has effectively become a solar energy network and was the key asset for the solar lighting project. The tutors/technicians are from the community and enjoy a great degree of trust from the beneficiaries.

Furthermore, ACEM developed administrative procedures for this project through contract and receipt system as well as built fundraising capacities. It had a committed leadership and staff fully responsible for coordination and management of the project. Accountability and transparency were noted as key in having trust of beneficiaries, partners and donors. Most importantly, ACEM approached solar lighting systems not as the gadget, but as a knowledge which can be taught, learnt, shared, modified and improved. This conceptualization allowed integration of this project into their main educational activities in Talamanca. In words of the ACEM Executive Director:

Our mission is to develop capacity. We use education as a tool for development. I did not want just to bring the equipment to a community and that is it. I was thinking how we could combine both things.

In contrast, APPTA, while having effective network as part of its operations as a cooperative was not able to fully benefit from it. It had difficulty integrating the project into the main activities and prioritizing it. As Walter Rodriguez from APPTA explains:

It has been very difficult to sustain organization because banana price has fallen and there was a lack of market. For many years I tried to get the project through, but it has not been possible to prioritize it.

The organization enjoys a great degree of the community trust and is much respected among community members. However, there was no paid staff dedicated to engage with the community in this initiative. There are some developments highlighted before that indicate that the project is becoming institutionalized, the group of technicians is being trained and a general manager is taking on a role of a leader for this project.

Another factor is that ACEM was particularly successful in institutionalizing its learning processes. As Luis Dumani of ACEM notes:

We have a good, identified and proven methodology based on reflection, action and consultation study that we have been implementing for a couple of years already.

ACEM is able to learn from what happens and acts based on this learning. ACEM has different spaces and methodology for learning. During bi-weekly meetings the local coordinator and tutors discuss day-to-day matters and reflect upon their work. Every month or two there is a meeting among all tutors and the Executive Director to reflect and consult about more strategic issues related to advancement of the program. The tutors use "the reflection book" as a tool for learning and tutoring. Every three months they write a report on their learning and reflections.

One example of their learning methodology in practice is the pilot project that ACEM started with. They initially installed 60 solar powered lighting systems without charge controller[7] trying to keep the cost of a system down. The batteries started failing as people would drain the batteries by using extensively the lights, even when they did not need them. The technicians communicated this problem to the coordinator and the discussion started on how to improve the design of the system. A charge controller was added to the system and users were trained on how to use it. Since this improvement there has not been a single problem reported with the batteries. If there was no mechanism in place to learn from mistakes and act upon, the organization would keep installing inferior systems.

Another interesting finding is related to the number and diversity of actors. The more organizations were involved, the harder it was to coordinate the interactions. This proposition contradicts some assumptions from the BoP literature suggesting that the BoP partnerships are likely to involve a greater diversity and number of actors and sectors in order to address the complexities of the BoP and subsistence marketplaces (Rivera-Santos & Ruffin, 2010). These assumptions are developed for the private sector led partnerships and may not hold true in the case of international operational NGO driven partnerships. While partnerships will always involve transaction costs associated with the coordination of interactions among partners, for small and medium international operational NGOs which are continuously struggling due to lack of stable financing for their operations, coordination costs will have to be kept to a minimum which may necessitate much more focused project partnerships with smaller number of actors involved. Greater diversity of actors in a partnership is also associated with the higher transaction costs since organizations have to align their diverse strategies, approaches, processes, and goals.

As for community aspects, what was important is that community members embraced the projects; they were actively involved in the projects and had strong sense of ownership. Community trust in local implementing organization and the project was crucial. Understanding natural environment (climate, terrain, natural resources) of the communities, opportunities and constraints it may present was important in terms of developing appropriate project delivery strategy and choosing suitable technological solutions.

As for technology, the key characteristics that impact the adoption of technologies were as follows: solar lighting systems provided significant advantages in terms of quality of light in comparison to kerosene lamps and candles; they were easy to use; potential users could observe the benefits from using the new technology; the payback period was one-two years after which the system was nearly maintenance free. The system design has been changing as the project progressed which directly accounted for the user specific needs.

What came out strongly from this research is that technological diffusion is very much about capacity building and knowledge. As Brian Minielly, President of Owen Sound Y Service Club points:

What we are doing is building human capacity and transferring knowledge and skill and vision and raising expectations and building their [communities] capacity to take it over and run with it and helping them find that there are a lot of things they can do. That they have the resources and that they have the intelligence. All they need is a chance and maybe some technical help from somebody who has done it before to help them see that possibilities are there that they had not considered before and go with it. That means that you do not just dump technology in the community; you bring technology if you can have the resources to bring the technology, but it is more important to bring the support for the people and their growth and development as individuals and as communities. [...] Development is something that has to be defined in terms of people you are working with and you are working for.

CONCLUSION AND RESEARCH IMPLICATIONS

The exploratory multiple case study in Talamanca, Costa Rica has shed light on NGO-driven development partnerships in the diffusion of renewable energy powered technologies in off-grid communities in developing countries. The theoretical framework brought together insights from the NGO partnerships literature, development studies and innovation literature. Based on this literature existing development partnerships approaches were highlighted and an approach to address the complexity of studied partnerships within their broader settings, goals and interactions was proposed. The approach included elements of the innovation system framework to explain the nature of these partnerships and introduced the dichotomy of partnership dynamics and partnership outcomes to address the effectiveness question.

Shared values and goals; complementary expertise and capacities; confidence and trust; clear roles and responsibilities; effective communication; personality match and continuity of staff were identified as a key dynamics effectiveness criteria confirming much of the previous research on partnerships while highlighting specifics of this type of partnerships. The research emphasized the importance of human factors in the dynamics of partnerships and called for the exchange between partnership literature and personal and organizational psychology that can provide further insights into these aspects.

The research acknowledged the catalytic role of the international operational NGOs facilitating access of a local organization to technical, financial, human and management resources necessary for the diffusion of the technologies in off-grid communities. The study emphasized the centrality of local NGOs and other community-based organizations in successful implementation of development interventions. Partnering with a local organization that had existing financial, human and technological capacities as well as a close relationship with the target communities was viewed as a faster and more effective approach. The development outcomes of the partnerships improved when local organization was capable to institutionalize its learning processes.

The research findings contribute to the literature that provided theoretical underpinnings for the study. The research adds to the innovation literature by providing a better understanding into the role of innovation actors that are largely ignored, such as NGOs. Furthermore, it looks at the role of these actors in technological diffusion in the off-grid communities in developing countries – the context which is also not well explored. The research also shows the usefulness of the innovation system framework in analyzing the development partnerships. As for the partnership and development literature, the research expands our understanding of the NGO-driven development partnership in technological diffusion in the context of poverty. Further research in this area can enrich our knowledge of the development partnerships and their roles in addressing development challenges. Mixed methodology combining qualitative and quantitative research can help explore this topic in other contexts in light of energy poverty and other development challenges, for example, access to clean water, health services and other.

The insights from this research can have important practical implications. The research signals the importance of international NGOs as intermediaries in the technological supply chains and local organizations as the intermediaries with the community. For international operational NGOs it is essential to partner with reputable local organizations that are close to the community, have and are able to enhance their capacities for delivering renewable energy projects. For local partners the key is building local capacities, thinking, designing and implementing for sustainability, institutionalizing their learning. Both international NGOs and local community-based organizations should focus on securing stable sources of financing for their operations.

ACKNOWLEDGMENT

The exploratory study presented in this chapter was completed as part of the Research Internship 2010 at the International Development Research Centre (IDRC) in Ottawa, Canada. The author wants to acknowledge IDRC for their support and her mentor during the Internship, Dr. Veena Ravichandran who helped to design and implement the case study, as well as Dr. Fernando Santiago who guided further analysis of the study under the Professional Development Award (PDA) 2011. The author acknowledges the Innovation for Inclusive Development program of the IDRC for hosting her during the Internship and PDA and providing continuous support and encouragement. Special gratitude goes to all my research participants from Canada and Costa Rica for facilitating my field work and participating in the study. Dr. Larissa Muller, author's PhD supervisor is also acknowledged for her comments on the draft. The views and opinions expressed are those of the author alone.

REFERENCES

Adkins, E., Eapen, S., Kaluwile, F., Nair, G., & Modi, V. (2010). Off-grid energy services for the poor: Introducing LED lighting in the Millennium Villages Project in Malawi. *Energy Policy*, *38*, 1087–1097. doi:10.1016/j.enpol.2009.10.061

Ahmad, M. (2006). The 'partnership' between international NGOs (non-governmental organisations) and local NGOs in Bangladesh. *Journal of International Development*, *18*(5), 629–638. doi:10.1002/jid.1251

APPTA. (n.d.). Retrieved from http://www.appta.org/j15/

Ashman, D. (2001a). Civil society collaboration with business: Bringing empowerment back in. *World Development*, *29*(7), 1097–1113. doi:10.1016/S0305-750X(01)00027-4

Ashman, D. (2001b). Strengthening north-south partnerships for sustainable development. *Nonprofit and Voluntary Sector Quarterly*, *30*(1), 74. doi:10.1177/0899764001301004

Austin, J. (2000). Strategic collaboration between nonprofits and businesses. *Nonprofit and Voluntary Sector Quarterly*, *29*(1), 69–97. doi:10.1177/089976400773746346

Baxter, P., & Jack, S. (2008). Qualitative case study methodology: Study design and implementation for novice researchers. *Qualitative Report*, *13*(4), 544–559.

Biagini, B., & Sagar, A. (2004). Nongovernmental organizations (NGOs) and energy. In C. J. Cleveland (Ed.), *Encyclopedia of energy* (Vol. 4). Elsevier Inc. doi:10.1016/B0-12-176480-X/00448-4

Brinkerhoff, J. M. (2002a). Assessing and improving partnership relationships and outcomes: A proposed framework. *Evaluation and Program Planning*, *25*(3), 215–231. doi:10.1016/S0149-7189(02)00017-4

Brinkerhoff, J. M. (2002b). Government–non-profit partnership: A defining framework. *Public Administration and Development*, *22*, 19–30. doi:10.1002/pad.203

Caplan, K., Gomme, J., Mugabi, J., & Stott, L. (2007). *Assessing partnerships performance: Understanding the drivers for success. Building Partnerships for Development*. BPD.

Clarke-Okah, W. (2004). *Partnerships in sector-wide programming in education in Tanzania: Narratives of experience* (PhD Dissertation). McGill University, Canada.

Dolun, M. (2005). *Role of universities in business generation in Kenya*. Globelics Conference, Tshwane University of Technology, South Africa.

Dorado, S., Giles, D. Jr, & Welch, T. (2009). Delegation of coordination and outcomes in cross-sector partnerships: The case of service learning partnerships. *Nonprofit and Voluntary Sector Quarterly*, *38*(3), 368. doi:10.1177/0899764008316055

Dowling, B., Powell, M., & Glendinning, C. (2004). Conceptualising successful partnerships. *Health & Social Care in the Community*, *12*(4), 309–317. doi:10.1111/j.1365-2524.2004.00500.x PMID:15272886

Earl, S., Carden, F., & Smutylo, T. (2001). *Outcome mapping: Building learning and reflection into development programs*. Ottawa, Canada: IDRC.

Fowler, A. (1998). Authentic NGDO partnerships in the new policy agenda for international aid: Dead end or light ahead? *Development and Change*, *29*(1), 137–159. doi:10.1111/1467-7660.00073

Geldof, M., Grimshaw, G. M., Kleine, D. J., & Unwin, T. (2011). *What are the key lessons of ICT4D partnerships for poverty reduction?* Systematic Review Report for DFID. ICT4D Collective. Retrieved from http://www.gg.rhul.ac.uk/ict4d/workingpapers/DFID_ICT_SR_Final_Report.pdf

Gu, S., & Lundvall, B.-A. (2006). Policy learning as a key process to the transformation of the Chinese innovation systems. In B.-A. Lundvall, P. Intarakumnerd, & J. Vang-Lauridsen (Eds.), *Asia's innovation systems in transition*. Cheltenham, UK: Edward Elgar.

Hall, A. (2006). Public private sector partnerships in a system of agricultural innovation: Concepts and challenges. *International Journal of Technology Management and Sustainable Development*, *5*(1). doi:10.1386/ijtm.5.1.3/1

Hall, A., Bockett, G., Taylor, S., Sivamohan, M. V. K., & Clark, N. (2001). Why research partnerships really matter: Innovation theory, institutional arrangements and implications for developing new technology for the poor. *World Development*, *29*(5), 787. doi:10.1016/S0305-750X(01)00004-3

Horton, D., Prain, G., & Thiele, G. (2009). *Perspectives on partnership: A literature review*. (Social Sciences Working Paper No. 2009-3). International Potato Centre.

International Energy Organization (IEA). (2010a). *Comparative study on rural electrification policies in emerging economies*. Paris, France.

International Energy Organization (IEA). (2010b). World energy outlook 2010. Paris, France.

Jamali, D., & Keshishian, T. (2009). Uneasy alliances: Lessons learned from partnerships between businesses and NGOs in the context of CSR. *Journal of Business Ethics*, *84*(2), 277. doi:10.1007/s10551-008-9708-1

Johnson, B., & Lundvall, B.-A. (2003). National systems of innovation and economic development. In M. Muchie, P. Gammeltoft, & B.-A. Lundvall (Eds.), *Putting Africa first: The making of African innovation systems*. Aalborg, Denmark: Aalborg University Press.

Kolk, A., van Tulder, R., & Kostwinder, E. (2008). Business and partnerships for development. *European Management Journal*, *26*(4), 262–273. doi:10.1016/j.emj.2008.01.007

Lewis, D. (1998). Nongovernmental organizations, business, and the management of ambiguity. *Nonprofit Management & Leadership*, *9*(2), 135–151. doi:10.1002/nml.9202

Lewis, D. (2007). *The management of non-governmental development organizations* (2nd ed.). UK: Routledge.

Lewis, D., & Kanji, N. (2009). *Non-governmental organisations and development*. London, UK: Routledge.

Lighting Africa. (n.d.). Retrieved from http://www.lightingafrica.org/

Lister, S. (2000). Power in partnership? An analysis of an NGO's relationships with its partners. *Journal of International Development*, *12*(2), 227–239. doi:10.1002/(SICI)1099-1328(200003)12:2<227::AID-JID637>3.0.CO;2-U

Lundvall, B.-A., Vang, J., Joseph, K. J., & Chaminade, C. (2009). *Innovation system research and developing countries. Handbook of innovation and developing countries: Building domestic capabilities in a global setting*. Edward Elgar Publishing.

Martinot, E., Chaurey, A., Lew, D., Moreira, J. R., & Wamukonya, N. (2002). Renewable energy markets in developing countries. *Annual Review of Energy and the Environment*, *27*(1), 309. doi:10.1146/annurev.energy.27.122001.083444

Mills, E. (2005). The specter of fuel-based lighting. *Science*, *308*(5726), 1263–1264. doi:10.1126/science.1113090 PMID:15919979

Mitlin, D., Hikey, S., & Bebbington, A. (2007). Reclaiming development? NGOs and the challenge of alternatives. *World Development*, *35*(10), 1699–1720. doi:10.1016/j.worlddev.2006.11.005

Morse, S., & McNamara, N. (2006). Analysing institutional partnerships in development: A contract between equals or a loaded process? *Progress in Development Studies*, *6*(4), 321. doi:10.1191/1464993406ps146oa

Muchie, M., Gammeltoft, P., & Lundvall, B.-A. (Eds.). (2003). *Putting Africa first: The making of African innovation systems*. Aalborg, Denmark: Aalborg University Press.

Mytelka, L. K. (2003). The dynamics of catching up: The relevance of an innovation system approach in Africa. In M. Muchie, P. Gammeltoft, & B.-A. Lundvall (Eds.), *Putting Africa first: The making of African innovation systems*. Aalborg, Denmark: Aalborg University Press.

Niosi, J. (2002). National systems of innovations are "x-efficient" (and x-effective): Why some are slow learners. *Research Policy*, *31*(2), 291–302. doi:10.1016/S0048-7333(01)00142-1

Organisation for Economic Co-operation and Development (OECD). (1997). *Diffusing technology to industry: Government policy and programmes*. Paris, France: OECD, Working Group on Innovation and Technology Policy.

Peon, R., Doluweera, G., Platonova, I., Irvine-Halliday, D., & Irvine-Halliday, G. (2005). Solid-state lighting for the developing world – The only solution. *Proceedings of the Society for Photo-Instrumentation Engineers*, *5941*, 59410N.

Prahalad, C. K. (2005). *The fortune at the bottom of the pyramid: Eradicating poverty through profits*. Wharton School of Publishing.

Rein, M., & Stott, L. (2008). Working together: Critical perspectives on six cross-sector partnerships in Southern Africa. *Journal of Business Ethics*, *90*, 79–89. doi:10.1007/s10551-008-9915-9

Rivera-Santos, M., & Rufín, C. (2010). Global village vs. small town: Understanding networks at the Base of the Pyramid. *International Business Review*, *19*, 126–139. doi:10.1016/j.ibusrev.2009.07.001

Schultz, C., Platonova, I., Doluweera, G., & Irvine-Halliday, D. (2008). Why the developing world is the perfect market place for solid state lighting. *Proceedings of the Society for Photo-Instrumentation Engineers*, *7058*, 705802.

Schumpeter, J. A. (1959). *The theory of economic development: An inquiry into profits, capital, credit, interest and the business cycle*. Cambridge, MA: Harvard University Press.

Selsky, J., & Parker, B. (2005). Cross-sector partnerships to address social issues: Challenges to theory and practice. *Journal of Management*, *31*(6), 849–873. doi:10.1177/0149206305279601

Sharif, N. (2006). Emergence and development of the national innovation systems concept. *Research Policy*, *35*(5), 745–766. doi:10.1016/j.respol.2006.04.001

Stoneman, P. (1986). Technological diffusion: The viewpoint of economic theory. *Richerche Economiche*, *40*, 585–606.

Tennyson, R. (2003). *The partnering toolbook*. London, UK: The International Business Leaders Forum and the Global Alliance for Improved Nutrition.

The National Institute of Statistics and Census of Costa Rica (INEC). (2000). *Population census 2000*. Retrieved from http://www.inec.go.cr/Web/Home/pagPrincipal.aspx

Thorsteinsdóttir, H., Melon, C. C., Ray, M., Chakkalackal, S., Li, M., & Cooper, J. E. et al. (2010). South-South entrepreneurial collaboration in health biotech. *Nature Biotechnology*, *28*, 407–416. doi:10.1038/nbt0510-407 PMID:20458300

Whelan, M. P. (2005). *Reading the Talamanca landscape: Land use and livelihoods in the Bribri and Cabécar indigenous territories.* Thesis of Magíster Scientiae in Tropical Agroecology. Tropical Agriculture Research and Higher Education Center (CATIE), Turrialba, Costa Rica. Retrieved from http://orton.catie.ac.cr/repdoc/a0840i/a0840i.pdf

World Bank. (2008). *Designing sustainable off-grid rural electrification projects: Principles and practices*. Washington, DC: World Bank.

World Bank. (n.d.) *Nongovernmental organizations and civil society/overview*. Retrieved from http://library.duke.edu/research/subject/guides/ngo_guide/igo_ngo_coop/ngo_wb.html

Yin, R. K. (2009). *Case study research: Design and methods* (4th ed.). Los Angeles, CA: Sage Publications.

ENDNOTES

1 Renewable energy refers to modern technologies based on solar, wind, biomass, geothermal, and small hydropower. This definition excludes traditional biomass and large hydropower (Martinot et al, 2002).

2 International NGOs are defined as organizations, which are typically headquartered in developed countries and carry out operations in more than one developing country (World Bank, n.d.). It must be noted that we see a rise of international NGOs based in developing countries. For this research I focused on international development operational NGOs only, which are defined by the World Bank (n.d.) as organizations, whose primary purpose is the design and implementation of development-related projects.

3 C.K. Prahald (2005) introduced the notion of the bottom (base) of the (economic) pyramid (BoP) which consists of 4 billion people

living on less than $2 dollars per day. He argued that multinational enterprises can help these people out of poverty by making profits from the poor.

4 This type of partnership may be viewed by some as the variation of cross-sector partnerships. However, for the purpose of this research, it is important to distinguish it as it specifically targets partnerships in the context of poverty and discusses impact of the BoP markets' specificities on partnership functioning.

5 Ashman, 2001b (p.1101) uses the term "development reach" which refers to "the numbers of people or institutions that benefited from the project." The reach can be understood also as an ability of the project to expand/scale up beyond the pilot stage and reach users in other communities.

6 This research has been granted the ethics approval from the Conjoint Faculties Research Ethics Board for the University of Calgary. All research activities have carefully followed the approved research protocol, including informed consent process.

7 A charge controller limits the rate at which electric current is added to or drawn from electric batteries.

This chapter was previously published in Comparing High Technology Firms in Developed and Developing Countries edited by Tomas Gabriel Bas and Jingyuan Zhao, pages 145-166, copyright 2012 by Information Science Reference (an imprint of IGI Global).

Chapter 90
Demand-Side Response Smart Grid Technique for Optimized Energy Use

Fouad Kamel
University of Southern Queensland, Australia

Marwan Marwan
Queensland University of Technology, Australia

ABSTRACT

The chapter describes a dynamic smart grid concept that enables electricity end-users to be acting on controlling, shifting, or curtailing own demand to avoid peak-demand conditions according to information received about electricity market conditions over the Internet. Computer-controlled switches are used to give users the ability to control and curtail demand on a user's premises as necessary, following a preset user's preferences. The computerized switching provides the ability to accommodate local renewable energy sources as available. The concept offers further the ability to integrate charging electrical vehicles during off-peak periods, helping thus substantially improving the utilization of the whole electricity system. The approach is pursuing improved use of electrical energy associated with improved energy management, reduced electricity prices and reduced pollution caused by excessive use of combustion engine in transport. The technique is inherently restricted to take effect in frame of energy tariff regimes based on real-time price made to encourage and reward conscious users being proactively participating in holistic energy management strategies.

INTRODUCTION

The traditional user-supplier rapport in the electrical energy market has historically evolved following a strategy implying whenever a load is switched on it is expected to be fulfilled by the supplier at the expected time and quality. Growing electrical demands followed by constantly growing supply led to troubled electrical services manifested mainly by daily and seasonal excessive peak and low demands. Those chronic peaks on electrical networks are usually associated with compromised quality, risk of forced outages and

DOI: 10.4018/978-1-4666-4852-4.ch090

high-priced energy supply; while low-demands on the other side might be driving some power plants to be operating at critical economic viability. Demand-side-response techniques are helping electricity users to become proactively participating in averting detrimental conditions presently prevailing in the electricity sector (Kamel, 2009b). Coordinated strategies shall help achieving improved use of electrical power plants and electricity infrastructure, besides integrated use of different types of energy sources. (Chua-Liang Su & Kirschen, 2009) proposed a day-ahead market-clearing mechanism that allows consumers to submit complex bids. Those bids are expected to give consumers the opportunity to specify constraints on their hourly and daily consumptions in the same way as generators can specify the operating constraints on their generating units. It is a day-ahead market with complex bids and offers whose objective is to maximize the social welfare. The social welfare is described as the difference between the value that consumers attach to the electrical energy that they buy and the cost of producing this energy. Not all consumers have the ability or the motivation to adjust their demand as a function of price. Part of the demand will therefore remain perfectly inelastic. Therefore, consumers were classified into two types, price-taking and price-sensitive. Price-taking is those consumers who have, in theory, an infinite marginal value as otherwise the consumers would have placed price responsive bids with a finite marginal value attached.

Plant Capacity Factor

The plant capacity factor (PCF) or the so also called utilization factor of a power plant is, by definition, the proportion of the actual electrical energy generated yearly by the plant to the quantity of electrical energy, which would be generated if the plant was operated at rated power for full year's time (8760 hours) as reported by Brinkmann (1980). The factor has a direct influence on the energy cost as can be deduced from the following equation of the fixed charge method according to De-Meo (1978), Leonard (1977), Chobotov (1978) and Clorefeine (1980):

$$c_E = c_{tr}\ FCR/(T_o\ PCF) + c_{op} \qquad (1)$$

where c_E is the cost of energy generated, c_{tr} cost of installed power including taxes during the installation period, FCR fixed charge rate of the capital, normally 15...18% a year according to Leonard (1977) and Leonard (1978), T_o =8760 (h) the hours per year, PCF plant capacity factor and c_{op} the operation and maintenance cost of the plant. For plants operating 24 hour/day, 7 days a week, i.e. 8760 hour/year PCF is a unity, which produces the least possible energy cost and best economic conditions. For power plants operating any less than 8760 hour/year the PCF will respectively be lower (below unity) what drives the cost of the produced energy to be accordingly higher, Equation (1).

Figure 1 illustrates the impact of the plant capacity factor on the cost of the produced energy. The calculation is made on the basis of the cost of the installed power c_{tr} = \$1000/kW, capital fixed charge rate FCR = 0.17 and the operation and maintenance cost of the plant c_{op} = \$0.02/kWh. It is evident that a power plant operated at low plant capacity factor e.g. PCF = 0.1 (this is 2.4 hour/day) will be producing energy at \$150/MWh, while operated continuously for 24 hour will produce energy at a cost of \$25/MWh.

Figure 1 depicts the importance of operating power plants and electrical network infrastructure at elevated plant capacity factor, close to the unity, in order to verify best economic performance. Various efforts on avoiding peak demands on the electrical network are mainly aiming at leveling demand throughout the year in order to achieve as high plant capacity factors as possible for all electrical power components. Additionally, leveling demands is pursued to avert or delay the urgency to expand generation capacity and network

Figure 1. Impact of the plant capacity factor on the cost of the produced energy (Kamel, 2009b)

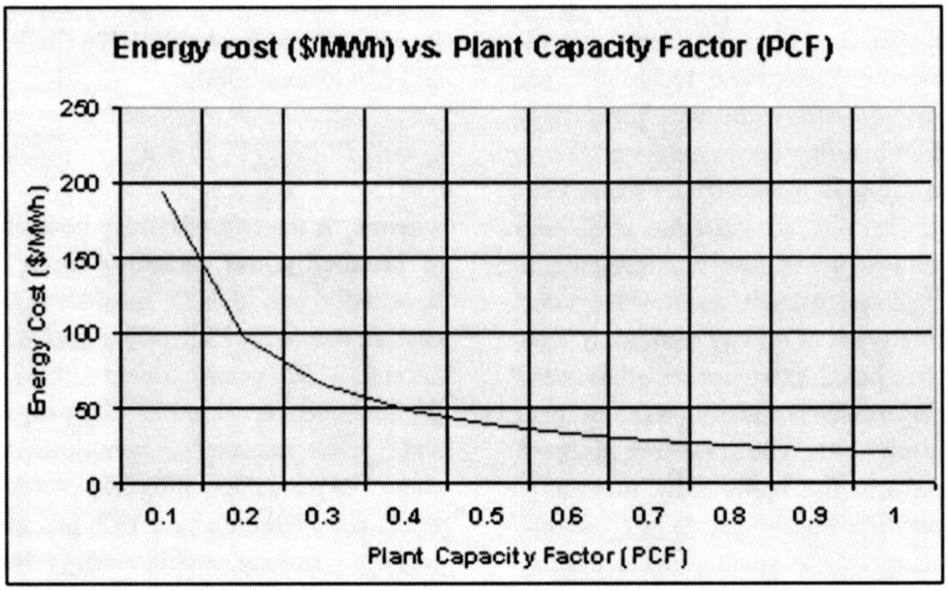

infrastructure to cope with consistently rising peak demands. Important methods to avert or delay peak demands are represented in demand-side response activities, both on the supplier and on the user's side.

Demand Side Response and Smart Grid Technologies

Demand Side Response (DSR)

Demand Side Response (DSR), as described by Albadi & El Saadany (2007) can be defined as the changes in electricity usage by end-use customers from their normal consumption patterns in response to changes in the price of electricity over time. Parvania & Fotuhi-Firuzabad (2010) describe DSR as tariff or program established to motivate change in electric consumption by end-use customers in response to change in the price of electricity over time. Further on, DSR programs provide means for utilities to reduce the power consumption and save energy, maximize utilizing the current capacity of the distribution system infrastructure, reducing or eliminating the need for building new lines and expanding the system as described by Dam, Mohagheghi & Stoupis (2008).

Vos (2009) described demand response as an integral part of the smart grid, is a cost effective, rapidly deployed resource that provides benefits to utilities and customers. Some advantages of DSR according to Greening (2010) are: increased economic efficiency of the electricity infrastructure, enhanced reliability of the system, relief of power congestions and transmission constraints, reduced energy prices and mitigated potential market power. Further on, demand response can help reduce peak demands and therefore reduce spot price volatility as illustrated by Nguyen (2010). In addition, Hyung Seon & Thomas (2008) described demand-side-response participation would help electricity power markets operate in a more efficient way. Based on a review of current utility programs, the Electric Power Research Institute (EPRI) estimated that demand response has the potential to reduce peak demand in the U.S. by 45,000 MW as reported by Walawalkar,

Blumsack, Apt & Fernands (2008). Therefore, the implementation of DSR programs is expected to improve economic efficiency in the electricity market.

In the United Kingdom, various techniques have been used to develop load electricity management. One of the methods is called responsive demand or demand side management program and was developed in the early 1960s (Hamidi, Li & Robinson, 2009). This system served to maintain the security of electricity supply and limited the facilities for electricity generation, transmission and distribution. This program has been participating in improving the economy, security and reliability of the electricity industry as well as eliminating the environmental concerns (Hamidi, Li & Robinson, 2009). However, later in 2007 the British Government initiated the building of the "Energy Demand Research Project" which focused on the actual benefits of demand response for consumers (Torriti, Hassan & Leach, 2009).

The British Government is currently considering the economic benefits of the Demand Side Response program, as this system requires a high implementation cost. Besides that, the government must first conduct a reform of the electricity industry to support this program, for instance: by restructuring the electricity price and market, transmission and distribution as well as the retail sector. According to Torriti, Hassan & Leach (2009) much of the debate around the economic potential of Demand Response focuses on the actual benefits of DR for consumers, and it provides some benefits and weaknesses for both the government and the user. Hence, there are five technology specifications that can potentially compromise such as: a minimum meter specification, smart enables meters that substitute old meters, dumb meters combined with smart boxes, retrofit devices and clip-on costumer display unit (Torriti, Hassan & Leach, 2009).

Similar to what has occurred in UK, in Finland, Interruptible Programmes as a part of demand side response model have been used as disturbance reserve for several years (Torriti, Hassan & Leach, 2009). Utilisation of the demand response program is more effective to overcome peak load, breakdown and manage electricity supply to all customers. This plan is not just applied by small-costumers but also has been used by large-scale industry. Therefore, in 2005 total Demand Response potential in Finnish large-scale industry was estimated at about 1280 MW, which represents 9% of the Finnish power demand peak (Stam, 2005). Following that, in 2008 the Finnish main electricity utility invested in an advance metering reading system to automatically read, control and manage all 60,000 of its customer metering points (Torriti, Hassan & Leach, 2009).

In Korea, the Demand Side Management program has been used for several years. In the 1970s, several programs were introduced in Load Management, for instance: night thermal-storage per rates program (1972), inverted block program (1974), the seasonal tariff (1977) and the time of use tariff (1977) (Jin-Ho, Tae-Kyung & Kwang-Seok, 2009). However, this program has not reached the maximum results to control load demand for peak demand sessions. Therefore, 2006, after the revision of the law, the government announced its 3rd National Electricity Demand Forecast and Supply Plan which addressed the government's main concerns about the Demand Side Management (Jin-Ho, Tae-Kyung & Kwang-Seok, 2009). In China, demand side management started from 1990s (Zhong, Kang & Liu 2010).

In Australia, implementation of the DSR program has been conducted several years ago. In late 2002, the Energy Users Association of Australia (EUAA) conducted a trial to demonstrate the benefits of a DSR aggregation process which would enable electricity consumers to respond to both the extreme prices and extreme peak demand (Fraser, 2005). This experiment was conducted by end-users to determine the value of an effective DSR for electricity consumers and its impact in terms of supporting an energy saving program. This trial was supported by the Victorian, New

South Wales and Commonwealth Government, as well as the CSIRO, to implement a Demand Side Response Facility Trial (Energy User Association of Australia Document, 2010).

In the experiment described above, the Australian Government through the EUAA involved customer to participate in the DSR trial. This experiment was conducted in three regions that fall under the National Electricity Market operation, New South Wales, South Australia and Victoria (Jones, 2004). These areas are regarded to represent the electricity load in Australia, and the results obtained show some significant benefits of using DSR for consumer and electricity providers. Hence, in December 2003 the Ministerial Council for Energy advised the Council of Australian Governments (COAG) on the need for further reform of the energy market to enhance active energy user participation (Jones, 2004).

The energy users association of Australia targeting a demand-side-response action, according to Fraser (2005), summarizes that, for example, South Australian electricity consumers only use the highest 10% of their maximum electrical demand on the network less than 0.5% of the time per year, i.e., for about 40 hours per year. The report is stating further: while the electricity consumers are insulated from price volatility by 'flat' electricity prices, they are also paying a significant and undisclosed (hard to evaluate) premium in their retail electricity prices to cover the retail supplier's costs of managing the risks of the extreme price volatility.

Demand Side Response Models

Many different economic models are used to represent Demand Side Response programs. In the report of the strategic plan of the International Energy Agency (2010) DSR is divided into two basic categories, namely the time based program and the incentives based program. According to Aalami H.A, Moghaddam M.P & Yousefi G.R (2008) the specific types of time based program are: time of use, real time pricing (RTP) and critical peak pricing; Federal Energy Regulatory Commission (2006) reports, while the specific types of incentive based program consist of direct load control (DLC), Interruptible/curtailable (I/C), demand bidding (DB), emergency demand response program (EDRP), capacity market (CAP) and ancillary service markets (A/S) programs. An overview of selected DSR models: I/C program, the EDRP, TOU and the proposed model, as presented in Figure 2. In the following a brief description of three popular market available models: I/C program, the EDRP and TOU.

Interruptible and Curtailable Program

The interruptible/Curtailable program has traditionally been one of the most common demand-side-management (DSM) tools used by electric-power utilities, which customers sign an interruptible-load contract with a utility to reduce their demand at a fixed time during the system's peak-load period or at any time requested by the utility (Yu, Zhang, Chung & Wong, 2005). This service provides incentives/rewords to customers participating to curtail electricity demand. The electricity provider sends directives to the user for following this program at certain times. The user must obey those directives to curtail their electricity when being notified from the utility or face penalties. For example: the customer must curtail their electricity consumption starting from 6:00 pm – 7:00 pm; those customers who are following will get a financial bonus/reword to their electricity bill from the utility. In California the incentive of I/C program was $700/MWh/month in 2001 as reported in (Aalami H.A, Moghaddam M.P & Yousefi G.R, 2009).

Emergency Demand Response Program

Emergency demand response program is energy-efficient program that provides incentives to customers who can reduce electricity usage for a

Figure 2. Models of DSR programs: a) Interruptible/Curtailable, b) Emergency Demand Response, c) Time Of Use and d) the Proposed Model (Marwan & Kamel, 2010b)

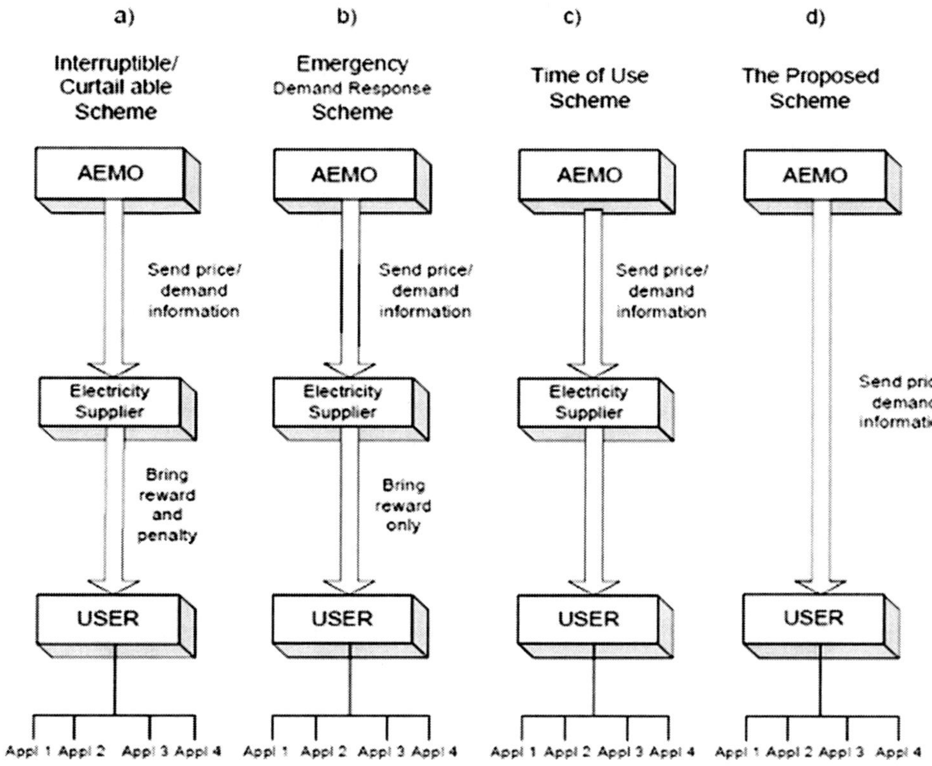

certain time; this is usually conducted at the time of limited availability of electricity. According to Covino (2003) emergency demand response program provide participant with significant intensive to reduce load. To participate on this program, all customers are expected to reduce their energy consumption during the events. Tyagi & Black (2010) described this program will determine which houses must be included in the event to minimize cost and disruption, while alleviating the overload condition. When asked to curtail, and verified to have performed, the resource is paid the higher of $500/MWh (Rahimi & Ipakchi, 2010). In New York, emergency demand response program allows participant to be paid for reducing their energy consumption upon notice from the New York Independent System Operator that an operating reserves deficiency or major emergency exists (Lawrence & Neenan, 2003).

Time of Use Program

According to Na & Ji-Lai (2006) Time of Use is one of the important demand side management methods, TOU demand side will response to the price and will change the shape of the demand curve. Further on, Time of Use rate is the most obvious strategy developed for the management of the peak demand in the world, which is designed to encourage the consumer to modify the pattern of electricity usage (Wen-Chen, Yi-Ping & Tzu-Hao, 2007). For applying this program, the utility does not provide reward or penalty to customer. To participate, all customers are required to remove their energy consumption during peak session to off-peak session as soon as their receipt information from the utility. Kirschen suggested (2003) in this type of contract, the rate is fixed for the duration of the contract but depends on the time of the day. As compared to the flat rate contract, some of the risk is shifted from the retailer to the consumer because the consumer has an incentive to consume during periods when the rates are lower.

Calculation of Energy Consumption for Maximized Benefit

Different models are used in Demand Side Response program planning as described by the Federal Energy Regulatory Commission (2006). In the report of the strategic plan of the International Energy Agency (2010), DSR is divided into two basic categories, namely, the time based program and the incentives based program. In Aalami H.A, Moghaddam M.P & Yousefi G.R (2008) the following approach is used to calculate the savings in electricity expenditure based on price, demand, incentives and costumer benefits associated with the DSR program used.

The change in energy consumption $\Delta d(t)$ at the time t when the user changes demand from $d(t)$ *to* $do(t)$ is:

$$\Delta d(t)= d(t)- do(t) \tag{2}$$

For participating in the DSR program, the total incentives $P\Delta d(t)$, when $A(t)$ is paid as incentive to the costumer at the time t for each kWh load reduction, can be calculated as the following:

$$P(\Delta d(t))=A(t).\Delta d(t) \tag{3}$$

The total penalty $PEN\cdot\Delta d(t)$ when the consumer does not commit to the obligations as a member of the DSR program, $pen(t)$ is the penalty per kWh at the time t and $D(t)$ is the DSR program contract level of consumption in kWh, will be accounted as the following:

$$PEN\cdot(\Delta d(t))=pen(t).\{D(t)-[d(t)-do(t)]\} \tag{4}$$

The total customer's benefit S a member of the DSR program can make at a certain time t is made up of income and expenditures. The contract could be mentioning a benefit from joining the program so that e.g. $B(d(t))$ is the customer income during that period from the use of $d(t)$ kWh, at the same time the customer could be receiving additional incentives $P(\Delta d(t))$ as described in Equation (2). The cost of the consumed electricity and any penalty for not following the program, if applicable, will be deducted from the income as the following:

$$S=B(d(t))+P(\Delta d(t))-d(t).r(t)-PEN(\Delta d(t)) \tag{5}$$

where, $r(t)$ is the rate the customer pays per kWh electricity at that time.

To maximize the customer benefit the slope $\partial S/\partial d(t)$ should be equal to zero, accordingly:

$$\frac{\partial S}{\partial d(t)} = \frac{\partial B(d(t))}{\partial d(t)} + \frac{\partial P(\Delta d(t))}{\partial d(t)} - r(t) - \frac{\partial PEN}{\partial d(t)} = 0 \tag{6}$$

This leads to:

$$\frac{\partial B(d(t))}{\partial d(t)} = r(t) - A(t) + pen(t) \tag{7}$$

The benefit function most often used according to Schweppe, Caramanis, Tabors & Bohn (1988) is the quadratic benefit function:

$$\partial B(d(t)) = Bo(t) + ro(t).\{d(t) - do(t)\}.\left\{1 + \frac{d(t) - do(t)}{2\beta(t)do(t)}\right\} \tag{8}$$

where $Bo(t)$ Benefit when $d(t) = do(t)$ (\$); $ro(t)$ Nominal rate for electricity consumption (\$/kWh); and $\beta(t)$ Elasticity parameter.

$$\beta(t,h) = -\frac{ro(h)}{do(t)}.\frac{\partial d(t)}{\partial r(h)} \tag{9}$$

The elasticity parameter β is a unit-less factor indicating how strong the energy demand depends on energy price, i.e. the effect the energy price on demand. The multiplier $\frac{ro(h)}{do(t)}$ helps transforming the parameter into a unit-less factor by referring to initial known conditions.

By differentiating Equation (8) and solving for $\frac{\partial B(d(t))}{\partial d(t)}$ and substituting the result in (7)

$$r(t) - A(t) + pen(t) = ro(t)\left\{1 + \frac{d(t) - do(t)}{2\beta(t)do(t)}\right\} \quad (10)$$

When the customer is participating in a DSR program, the customer consumption *d(t)* for a maximized benefit can be calculated from Equation (10) as the following:

$$d(t) = do(t)\left\{1 + \beta(t,t)\frac{[r(t) - ro(t) - A(t) + pen(t)]}{ro(t)}\right\} \quad (11)$$

Integration Renewable Energy Sources and Electricity Vehicles to DSR Smart Grid System

Integrating renewable energy with power generation is a new way to improve reliability, sustainability and cost effectiveness on the electrical network. According to Hammons (2006) the key challenges that need to be considered in the composition of future network include distributed generation and the integration of renewable energy sources, such as biomass, wind and solar. The utilization of renewable energy is expected to be leading to the harmony between humans and nature with low pollution and sustainable accessibility to resources as described in Figure 3.

Some countries have applied smart grid technology for renewable energy utilization. In Japan, using solar power generation systems in every ordinary house is the most active project in smart grid researches as described by Li & Yao (2010). The significant amount of installed wind power in the German power system in 2008 produced more than 22 GW producing between 1500 - 7700 GWh/month as reported in Hammons (2006).

On the other hand, electricity vehicles (EV) technology brings impacts to the electrical distribution grid. According to Clement-Nyns, Haesen & Driesen ((2010) the vehicle can not only charge, but also discharge and thus inject energy to the grid. In addition, there are social, environmental and economic advantages in switching to electricity vehicles as described by Anna Cain, Iain MacGill & Bruce (2010).

The User's DSR Concept

This concept helps developing the technical tools to be independently implemented and managed by end-users to enable control energy consumption accustomed to user's preferences. The proposed DSR Smart-Grid scheme is targeting flattening load profiles by averting periods of peak demands helping thus redistributing the load profile throughout the day/year. The scheme helps electrical generating plants and electrical infrastructure to be operated 24 hour/day achieving thus higher

Figure 3. Smart grid network diagram (Energy Efficiency and Conservation Authority, 2009)

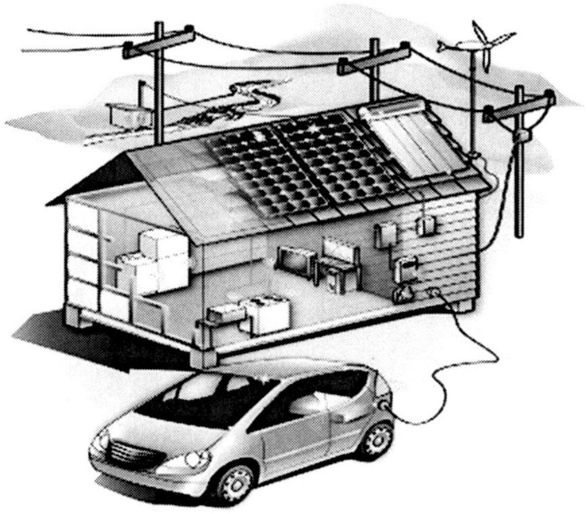

Figure 4. The proposed user's smart grid demand-side-response scheme (DSR) balancing conventional supplier's smart grid DSR programs

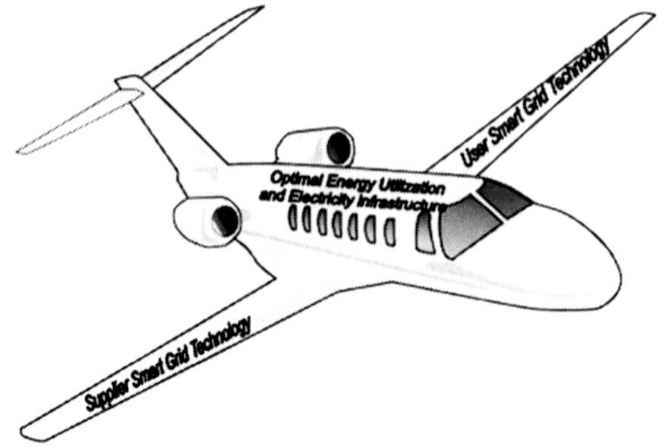

utilization factor, enhanced efficiency and considerably reduced energy price. The presented user's side DSR concept is providing the needed balance in the electrical system to complement the efforts undertaken by suppliers to mitigate peak demands and improve supply reliability and stability, Figure 4.

The concept is presenting a low-cost DSR technique implemented at user's premises, which assists electricity end-users to be shifting loads around the clock averting peak-demand periods and making use of on-site renewable energy sources as appropriate. This shall help users to be engaged in mitigating peak demands on the electricity network. The proposed concept comprises a technical set-up of a programmable internet relay, a router, solid state switches in addition to the suitable software to control electricity demand, Figure 5. The software's on appropriate multimedia tool (CD Rom) offers users optimized control of energy consumption.

The concept enables commercial and industrial customers on fluctuating energy prices to be

Figure 5. The user's DSR concept

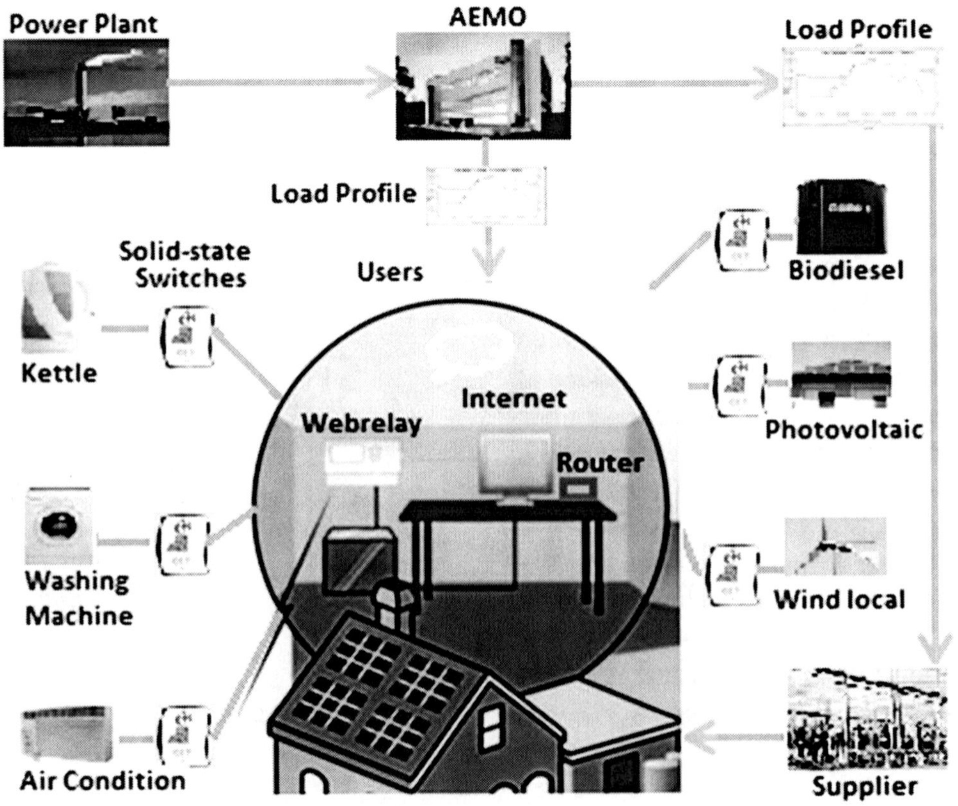

achieving immediate financial savings. For residential customers on flat-rate tariffs, in contrast, users are gaining financial benefits from reducing energy consumptions at peak-demand periods. Residential customers on different tariffs, where energy price differs with day time and network conditions (e.g. night tariffs), they are gaining financial benefits also from shifting loads from day- to night-times, when electricity is cheaper.

The scheme uses a router and a programmable internet relay and solid-state switches to control electrical demand at the user's premises. The relay is programmed to receive and act upon information received from the AEMO on the internet about demand/price conditions. Figure 6 illustrates the control regime, where three appliances are controlled by three solid-state switches receiving on/off signals from the relay.

Consumers use local computers to set-up their preferences for appliance profile usage and priorities, e.g. Table 1. The profile of appliances identifies when an appliance is run according to electricity price or network conditions (national demand). Pursuant to the order from the relay to a solid-state switch, household appliances connected to that switch can be turned on/off.

All control systems above are implemented by a shell script under a Linux operation system. Figure 7 shows the pseudo code of the controller that is executed with each interaction.

Table 1 illustrates an example of an appliance profile. All control systems above is implemented by a shell script under a Linux operation system.

ELECTRICITY INDUSTRY DEVELOPMENT IN AUSTRALIA (CASE STUDY)

In Eastern and Southern states of Australia the Australian Energy Market Operator is managing the power flow across the Australian Capital Territory, New South Wales, Queensland, South Australia, Victoria and Tasmania. The AEMO

Figure 6. Control Regime

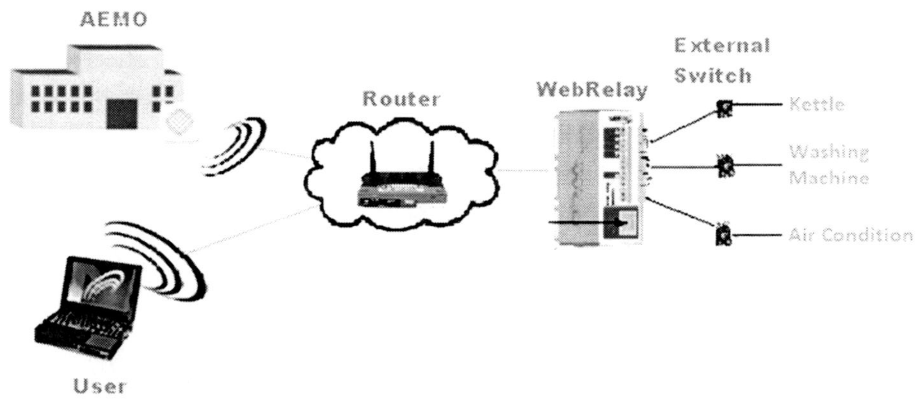

Table 1. Example of appliance profile

Appliance	Start After	Finish Before	Session Time
Kettle	08.00 AM	04.00 PM	Off-peak Session
Washing Machine	06.00 AM	10.00 AM	Off-peak Session
Air Condition	06.00 AM	04.00 PM	Off-peak Session

Figure 7. Pseudo code of the control loop

```
echo $name
echo $startAfter
echo $finishBefore

  if [ $startAfter -lt $nowSec -a $finishBefore -gt $nowSec ]; then
     case $name in
              Kettle) wget -q http://192.168.1.2/state.xml?relay1State=1;;
              WashingMachine) wget -q http://192.168.1.2/state.xml?relay3State=1;;
              Aircondition) wget -q http://192.168.1.2/state.xml?relay4State=1;;
         *) echo "Better put a usefull error message" ;;
     esac
   else

     case $name in
              Kettle) wget -q http://192.168.1.2/state.xml?relay1State=0;;
              WashingMachine) wget -q http://192.168.1.2/state.xml?relay3State=0;;
              Aircondition) wget -q http://192.168.1.2/state.xml?relay4State=0;;
         *) echo "Better put a usefull error message" ;;
       *) echo "Better put a usefull error message2 " ;;
     esac
   fi

done
```

is regularly updating energy demand and prices current situations publicly on the internet. The presented concept is using programmable internet relays and controllable electronic switches to be proactively responding to demand/price peaks and congestions conditions. The technique is globally valid in other electricity markets under similar operating conditions. Figure 8 depicts an example of an actual energy demand and prices situation. The price curve is closely following the demand curve. Electricity prices are typically at their lowest level at night during times of low demand (off-peak). Prices are rising daily according to two major peak demands in the morning and evening. It is to be noted that electricity prices in Figure 8 are wholesale regional references prices (RRP) i.e. prices traded to electricity suppliers. Electrical energy at this stage still needs to be transmitted to the different localities then distributed to end-users. End-users could be industrial, commercial or residential. Usually industrial and commercial end-users are contracting electricity suppliers on special agreements to provide the service satisfying their requirements. For most residential electricity customers, electricity pricing typically follow one set price "flat-rate." Residential users are often also offered a night tariff, where electricity prices are substantially reduced. The night tariff corresponds off-peak times.

Figure 9 illustrates the frequency of electricity demand supplied in Queensland during the year 2008 as extracted from data of the Australian Energy Market Operator (2009). The figure indicates mainly the fact that the higher the load above the base load the lesser likely the extent of their duration will be. Base load power stations are those operated twenty four hours/day throughout the year corresponding to 8760 h/year, a utilization factor of 1. Those power stations are providing the most economic operation at the least possible energy price. Any loads exceeding the base load are usually covered by other power

Figure 8. Wholesale electricity price in AUD $/MWh and demand in MW for a typical day in Queensland on 5th May 2009(Australia Energy Market Operator, 2009)

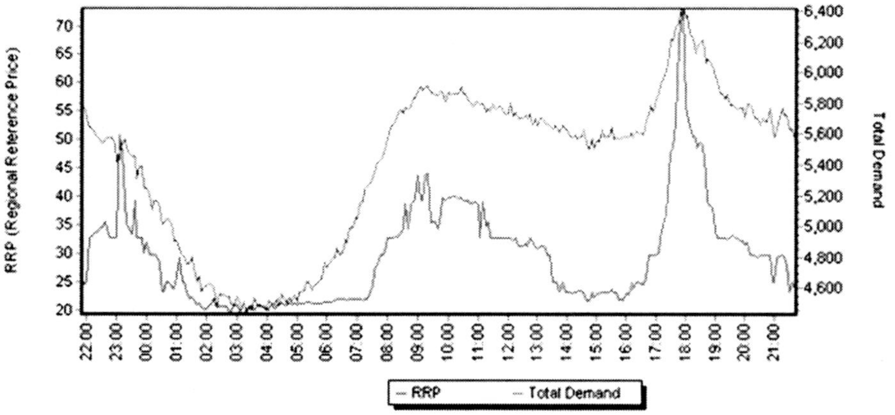

plants operated for shorter periods of time at utilization factor lesser than 1, thus generating energy at high prices. Accordingly, the intermittent operation of the expensive to-run power plants makes their operation even more expensive.

Figure 10 illustrates the occurrence of the regional reference wholesale price RRP in Queensland during the year 2008; extracted from the Australia Energy Market Operator (2009). The figure indicates mainly that low-priced supplies are taking place at very high occurrences of more than 80% a year, while high prices happen at lower occurrences. For instance, prices around AUD $20/MWh are occurring at frequencies of about 80%, while prices of over $50/MWh have occurrences of less than 10%.

Based on data of 31 December 2008, Queensland total electricity generating capacity was 12487

Figure 9. Occurrence of electrical energy demand Queensland during 2008. Peak demand 8413 MW, base-load 4100 MW and total supplied electrical energy 52.18 TWh (Kamel, 2009b)

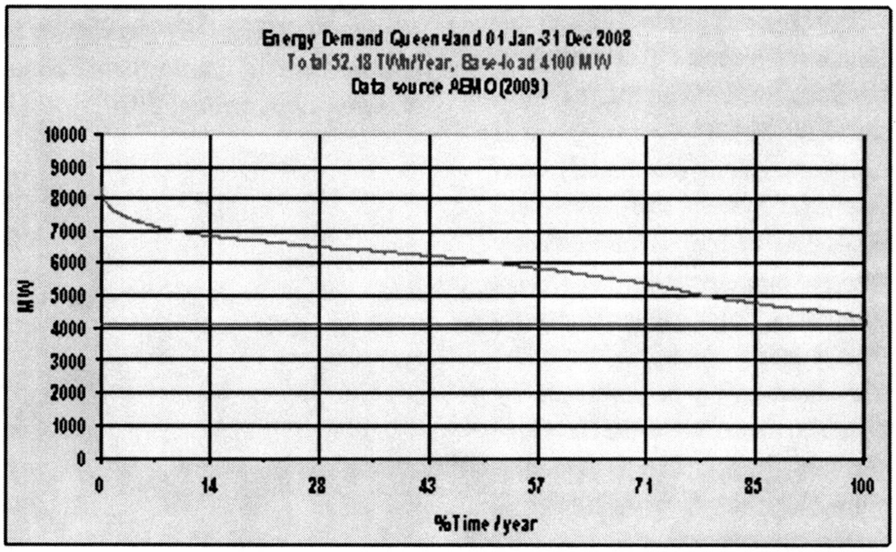

Figure 10. Electricity wholesale price RRP in Queensland in 2008 (Kamel, 2009b).

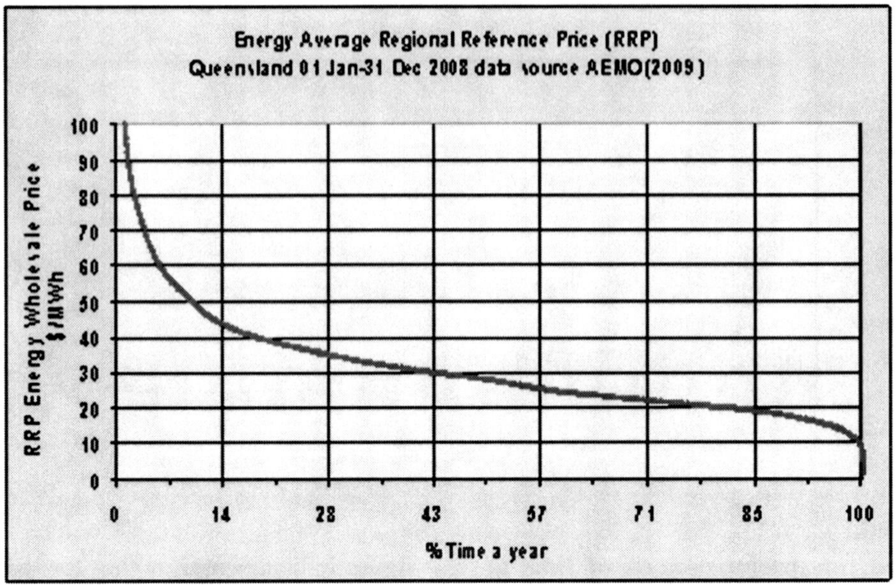

MW; coal-fired power stations provided 70% of this total capacity, while gas-fired electricity accounted for 17% and renewable energy accounted for around 5% as stated by the Department of Employment Economic Development and Innovation (2010). These power generations are used to provide electrical energy for all consumers in the Queensland area: residential, commercial and industrial. However, the amounts of energy produced from various generators depend on market demand, price and availability of sources. Figure 11 illustrates electricity generation in Queensland according to the Department of Employment Economic Development and Innovation (2010).

Most of the power stations are directly connected to the transmission system. The Queensland electricity transmission system is provided by Powerlink, licensed to operate more than 12,000 kilometres of Queensland high voltage transmission network, transporting electricity from the generators to the distribution networks as by (Department of Employment Economic Development and Innovation, 2009). The distribution network is carrying electricity from the transmission system to consumers. In Queensland, ENERGEX and ERGON energy are purchasing electrical energy from the Energy Market and distributing it to the customer. ERGON e.g. provides energy at several tariff options to end users. For example, Tariff 11 for all domestic consumption 18.84 ¢/kWh, while the night rate Tariff 31 for all consumption 7.7 ¢/kWh and the economy Tariff 33 for all consumption 11.32 ¢/kWh (Ergon Energy, 2010).

Figure 11. Electricity generation in Queensland (Department of Employment Economic Development and Innovation, 2010)

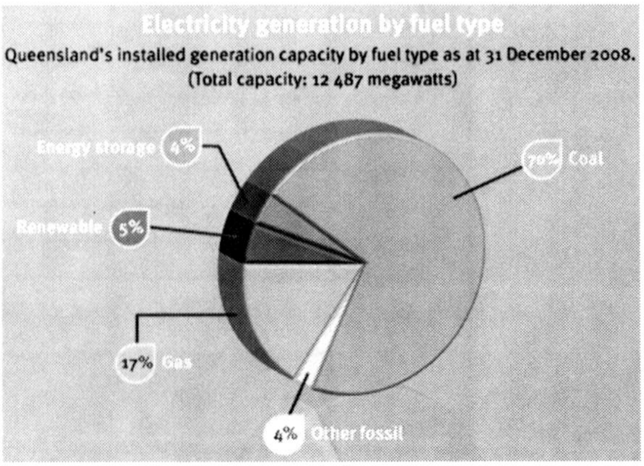

According to Queensland Competition Authority (2000), the total energy consumption in Australian grew at an annual rate of 2.6% for the 25 years to 1997/1998. In the 2007- 2008 period according to Department of Employment Economic Development and Innovation (2009a) the annual electricity consumption in Queensland has grown by over 29% or approximately 10500 GWh, making Queensland the second highest consumer of electricity in Australia. This indicated that Queensland has a significantly greater number of high energy users than any other state, most of these in regional Queensland.

Narayan & Smith (2005) describes since the beginning of the 1990s, Australia's electric power industry has undergone a series of structural reforms. In Queensland, the electricity industry was restructured on 1 July 1998 to prepare the industry for participating in the competitive National Electricity Market, which is responsible for structure, rules and regulations in the delivery of energy to customers (Department of Employment Economic Development and Innovation, 2009b). The National Electricity Market Management Company (NEMMCO) Limited was the Wholesale Market and Power System Operator for the Australian NEM. NEMMCO was established in 1996 to administer and manage the NEM, develop the market and continually improve its efficiency and as of 1 July 2009 was replaced by the Australian Energy Market Operator.

To improve governance, and enhance the reliability and sustainability of the State's electricity system, the Australian Government has created a collaborative electricity and gas industry in the form of the Australian Electricity Market Operator (Australian Energy Market Operator, 2010), which commenced operation on 1 July 2009. The AEMO is managing power flows across the Australian Capital Territory, New South Wales, Queensland, South Australia, Victoria and Tasmania. Western Australia and the Northern Territory are not currently connected to this market primarily because of their geographic distance from the rest of the market. AEMO's responsibilities include wholesale and retail energy market operation, infrastructure and long term market planning demand forecasting data and scenario analysis as described in (Australian Energy Market Operator, 2010). The electricity market comprises of a wholesale sector and a competitive retail sector. All electricity dispatched in the market must be traded through the central spot market. The Market structure of NEMMCO/AEMO can be presented as in Figure 12.

Marwan & Kamel (2010a) summarized in Figure 13 an example of classic fluctuations in electricity price in Queensland, from 22 May 2008 to 22 May 2009. This graph illustrates that the average price during that time was in the range of \$50/MWh (¢5/kWh) Regional Reference Wholesale Price (RRP), however, extreme prices occurred exceeding \$500/MWh (¢50/kWh). The graph indicates also that excessive demands are occurring regularly in all states on the interconnected power network. Fraser (2005) stated that customers, even those bound by flat-rate contracts, must bear the additional cost for managing the corresponding extreme prices.

Controlled Scenario Using the User's DSR scheme

In order to evaluate the effect of the proposed scheme on electricity energy saving the electricity price/demand in Queensland for the period 10th-12nd May 2010 has been used. In the following, ten scenarios have been formulated to demonstrate the results as presented in Figure 14 and summarized in Table 2.

Scenario 1: In this scenario users are shifting 375 MWh peak electricity usage occurring between 17:00 pm-19:00 pm towards the time period 19:00 pm-21:30 pm when energy demand and prices are low. All participants are suggested to set-up the electricity profile to stop chosen appliance from running dur-

Figure 12. The Market structure of NEMMCO/AEMO (Department of Resources Energy and Tourism, 2009)

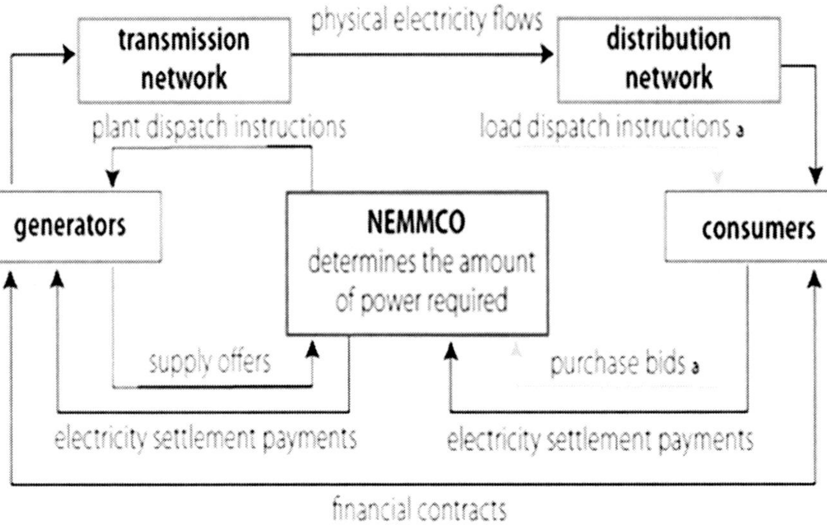

Figure 13. Fluctuation of electricity price in Queensland (Marwan & Kamel, 2010a)

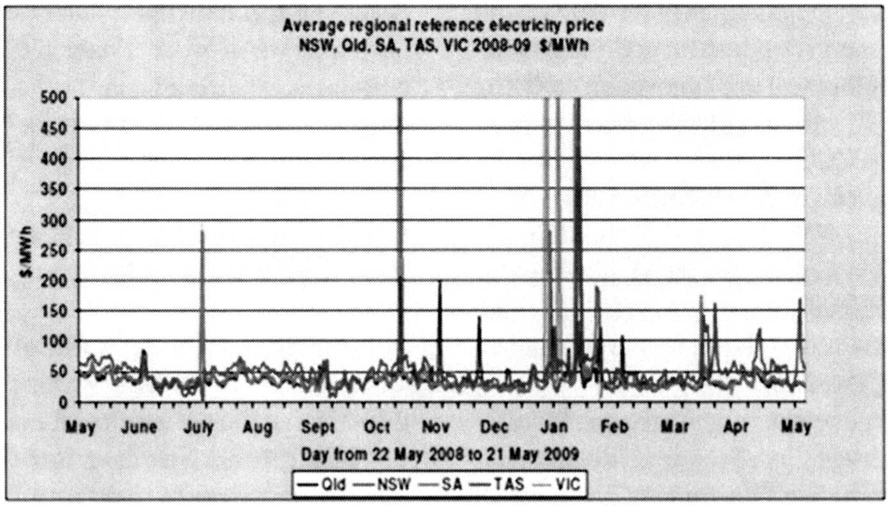

ing that time. For example, kettle, washing machine and air condition could be effectively operated at optional times of the day. No savings in energy cost due to applicable day-time tariffs. However, the scheme was still able to remove congestions out of peak demand areas.

Scenario 2: Users are shifting peak demand of 375 MWh occurring between 17:00pm-19:00pm to the period between 21:30 pm to 23:30 pm. Achievable savings $21149 per day.

Scenario 3: Users are shifting peak demand of 375 MWh occurring between 17:00pm-19:00pm to the period between 23:30 pm to 01:00 am. Achievable savings $28200 per day.

Figure 14. Controlled Scenarios

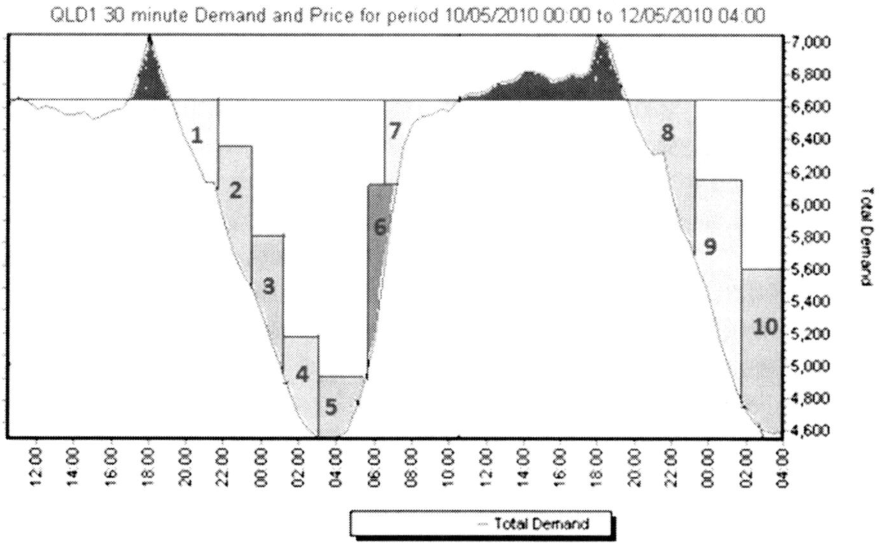

Table 2. Results of some scenarios

Scenario NR	Time to Curtail Load	Time to Reconnect Load	Load to Curtail	Day Tarrif 18.84 ¢kWh	Night Tarrif 11.32 ¢kWh	Saving ($)
1	17.00 pm to 19.00 pm	19:00 pm to 21:30 pm	375	70650	NA	NA
2		21:30 pm to 23:30 pm	375	70650	49501	21149
3		23:30 pm to 1:00 am	375	70650	42450	28200
4		1:00 am to 3:00 am	375	70650	42450	28200
5		3:00 am to 5:30 am	375	70650	42450	28200
6		5:30 am to7:00 am	375	70650	42450	28200
7		6:30 am to 10:30 am	375	70650	67125	3525
8	10.30 am to 19.30 pm	19:30 am to 23:30 am	1730	325923	277147	48785
9		23:30 pm to 01:30 am	1730	325923	195836	130096
10		1.30 am to 4:00 am	1730	325923	195836	130096

Scenario 4: Users are shifting peak demand of 375 MWh occurring between 17:00 pm-19:00 pm to the period between 01:00 am to 03:00 am. Achievable savings $28200 per day

Scenario 5: Users are shifting peak demand of 375 MWh occurring between 17:00 pm-19:00 pm to the period between 03:00 am to 05:30 am. Achievable savings $28200 per day

Scenario 6: Users are shifting peak demand of 375 MWh occurring between 17:00 pm-19:00 pm to the period between 05:30 am to 07:00 am. Achievable savings $28200 per day

Scenario 7: Users are shifting peak demand of 375 MWh occurring between 17:00 pm-19:00 pm to the period between 06:30 am to 10:30 am. Achievable savings $3525 per day

Scenario 8: Users are shifting peak demand of 1730 MWh occurring between 10:30 am-19:30 pm. All participants are suggested to set-up the electricity profile to stop some appliance to run during this time. User can run chosen appliances between 19:30 pm to 23:30 pm. Achievable savings in energy cost $48785.

Scenario 9: Users are shifting peak demand of 1730 MWh occurring between 10:30 am-19:30 pm to the period between 23:30 pm to 01:30 am. Achievable savings $130096 per day.

Scenario 10: Users are shifting peak demand of 1730 MWh occurring between 10:30 am-19:30 pm to the period between 01:30 am to 04:00 am. Achievable savings $130096 per day.

Economic Model

While the concept is designed to be targeting flattening the national electrical demand throughout the year the concept involves an economic model based on the maximization of financial benefits to electricity users. The scheme is applicable in regions managed by the Australian Energy Management Operator (AEMO) and other regions under similar conditions. Usually the electricity price is high during peak demands and low at off-peak periods. The concept allows customers controlling consumption to avoid peak demands periods. In case the user is on other DSR program with the supplier, the scheme is still allowing additional savings besides the benefits and saving already achievable through the DSR agreement.

For commercial and industrial customers on fluctuating energy prices implementing the scheme will enable achieving immediate financial savings. For residential customers on flat-rate tariffs, in contrast, users are gaining financial benefits from reducing energy consumptions at certain times a day; mainly averting peak-load periods or using off-peak night tariffs.

Multimedia Tool

The concept includes a multimedia tool (CD-Rom) run on local computers at user's premises to allow implementing and operating the DSR model continuously. The CD-Rom is containing an introductory part to address the electricity peak demand issue to the user, a programming part, where the user will be recording operation profiles for the different appliances and an executive part, where the information is transferred to the programmable relay, which on turn sending signals to electronic switches to operate the different appliances.

Significance

Figure 15 shows customers fully curtailing energy withdrawals at any energy price above $55/MWh as example. Figure 16 shows achievable energy savings in Queensland by curtailing energy demand over a certain energy prices.

Referring to Figure 16, the technique is able to remove 6.1 TWh/year from a total of 52.18 TWh/year if users are setting switches to curtail own loads at any regional reference prices above $50/MWh; this is a percentage of 11.7% of the total demand. In case users chose to curtail loads at $40/MWh, the savings will be 11.1 TWh/year; a percentage of 21.2%, and 24.8 TWh/year at a $30/MWh curtailment; a 47.5%.

Figure 17 depicts the case where coordinated strategies are able to lead customers to defer loads from times of peak-demand to times of low-demands. Such a procedure shall help flatten the total energy demand to meet a constant average of 5941 MW for Queensland, achieving considerable improvement in the system utilization and thus in the whole system economics. In such a procedure the technique enables deferring 3.26 TWh/year from peak to off-peak times.

Figure 15. Day electricity demand curtailed for wholesale regional reference price not to exceed AUD $55/MWh in Queensland in 2008 (Kamel 2009a)

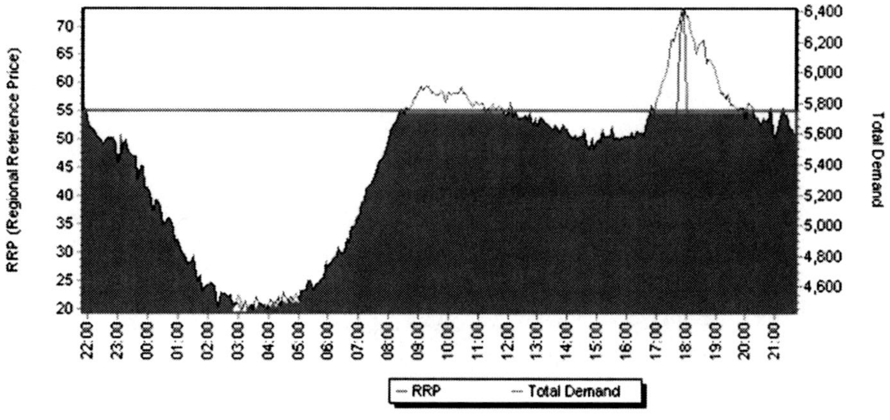

Figure 16. Achievable energy savings by curtailing energy demand over a certain energy prices in Queensland (Kamel, 2009a)

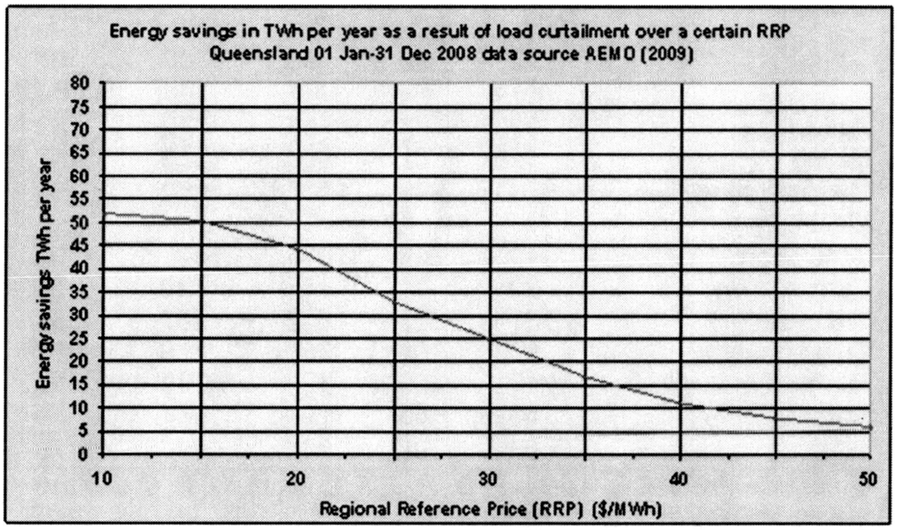

Queensland Government (2009) is describing: transport is the fourth largest source of Queensland's greenhouse gas emissions, contributing 10.4 per cent to Queensland's total emission profile and the Queensland Government will invest $1.4 million to undertake a trial of low-emission diesel-electric buses in the public transport fleet. The Department of Resources Energy and Tourism (2009) is indicating 1359 PJ (377.5 TWh) were used for transport in Australia in 2006-07 while 1695 PJ (460.83 TWh) for electricity generation (1TWh = 3.6 PJ) as illustrated in Figure 18.

The technique offers further the present electrical supply system to be supplying the transport sector for the use of charging electric vehicles at low-demand times (at night). Figure 19 depicts the possibility to utilize 21.52 TWh/year of electrical capacity, mainly of peak-load power stations,

Figure 17. Occurrence and average electrical energy demand Queensland (Kamel, 2009a)

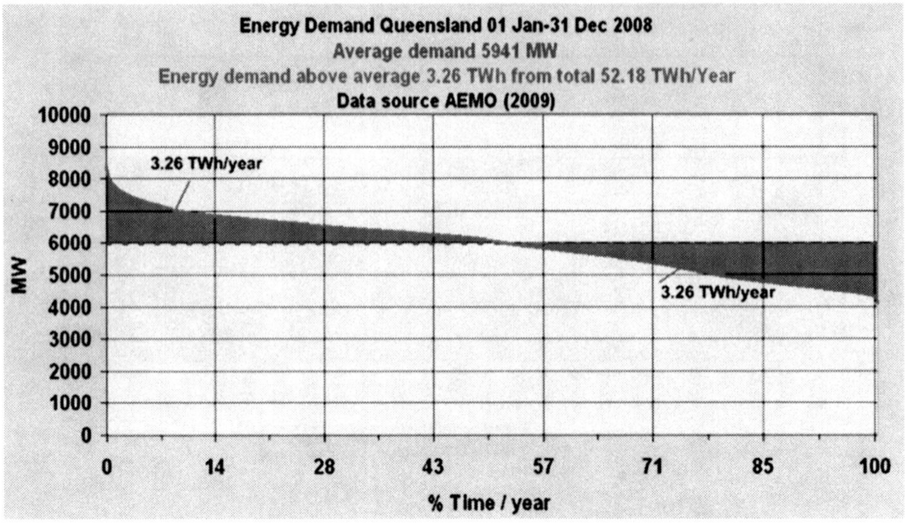

Figure 18. Energy consumption in Australia (Department of Resources Energy and Tourism, 2009)

	1974-75 PJ	1979-80 PJ	1989-90 PJ	1999-00 PJ	2006-07 PJ
Agriculture	39	47	55	72	92
Mining	65	81	160	273	457
Manufacturing	928	965	1067	1192	1369
Electricity generation	540	743	1066	1427	1695
Construction	29	38	41	29	26
Transport	701	825	1012	1267	1359
Commercial a	87	104	151	219	252
Residential	246	262	322	392	442
Other b	59	66	69	77	78
Total	2695	3131	3944	4946	5770

a includes ANZSIC Divisions F, G, H, J, K, L, M, N, O, P, Q and the water, sewage and drainage industries;
b includes consumption of lubricants and greases, bitumen and solvents, as well as energy consumption in the gas production and distribution industries.
Source ABARE, Australian energy statistics

otherwise not used. The procedure helps enhancing the utilization of present electrical power stations to approach a plant capacity factor close to the unity, achieving thus an optimal use of power plants.

FUTURE RESEARCH DIRECTION

For future research direction on DSR method, more development of modeling both on-site renewable energy sources and electricity vehicles connected

Figure 19. Electrical energy demand Queensland Vs maximum utilization of generating capacity (Kamel, 2009a)

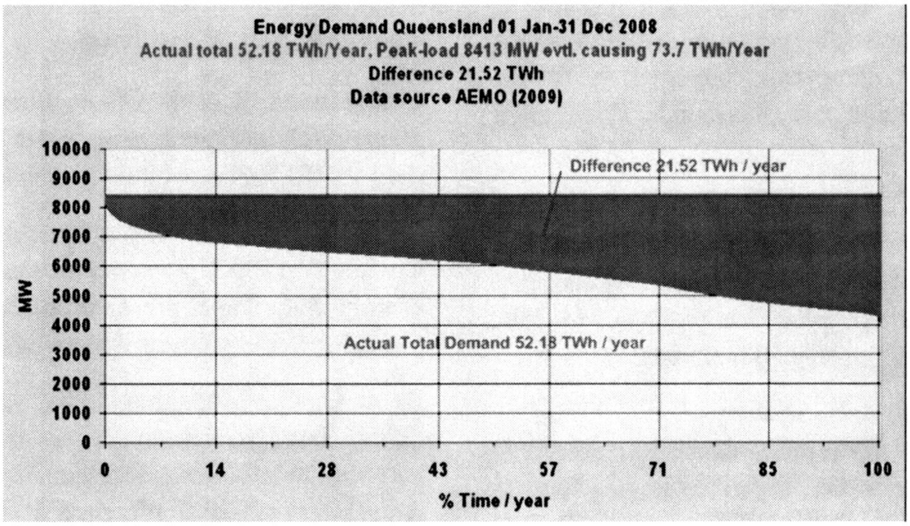

to the distribution grid on the Australian national electricity market, will be considered as follows:

- Optimization of on-site renewable energy sources to distribution grid under DSR method in Australian national electricity market.
- Deployment of plug-in electricity vehicles and their impacts to the Australian national electricity market.

CONCLUSION

The concept is aiming to achieve moderated energy demand, reduced energy prices and curbed price volatility, improved grid usability and reliability, and reduced energy consumption. The concept is making use of the internet and modern communication technologies to maximize benefit for users and suppliers. Additionally, the concept is providing additional capacity more quickly and more efficiently than new supplies. The flexibility provided lowers the likelihood and consequences of forced outages as well. By reducing significant peaks, the concept is averting the need to use the most costly-to-run power plants, driving electricity costs down for all electricity users. And most importantly, by enabling end-users to observe electricity prices and congestions on the electrical network it allows them to be positively sharing responsibility by reducing and optimizing energy consumption and realizing electricity savings. The concept can be considered a complementary effort to concurrent energy supplier's efforts to mitigate electrical peak demands and the associated technical and economic detriments. It allows electricity end-users to "smoothen out" significant peaks by curtailing or shifting demand, avoiding or delaying investments in new infrastructure. A wide deployment of the scheme will allow a quite flattened load profile representing thus an optimized use of the electricity generation and distribution infrastructure.

REFERENCES

Aalami, H. A., Moghaddam, M. P., & Yousefi, G. R. (2008). *Demand response model considering EDRP and TOU programs*. Paper presented at the The Transmission and Distribution Conference and Exposition.

Aalami, H. A., Moghaddam, M. P., & Yousefi, G. R. (2009). Demand response modeling considering interruptible/curtailable loads and capacity market programs. *Applied Energy*, *87*(1), 243–250. doi:10.1016/j.apenergy.2009.05.041

Albadi, M. H., & El-Saadany, E. F. (2007). *Demand response in electricity markets: An overview*. Paper presented at the Power Engineering Society General Meeting, 2007. IEEE.

Australia Energy Market Operator. (2009). *Current trading interval price and demand graph Queensland*. Retrieved from http://www.aemo.com.au/data/GRAPH_30QLD1.html

Australian Energy Market Operator. (2010). *About AEMO*. Australian Government. Retrieved from http://www.aemo.com.au/aboutaemo.html

Brinkmam, K. (1980). *Introduction in the economic of electrical energy*. Braunsching, Germany: Vieweg Verlag.

Cain, A., MacGill, I., & Bruce, A. (2010). *Assessing the potential impacts of electricity vehicles on the electricity distribution network*. Paper presented at the The 48th Australian Solar Energy Society (AuSES), Solar 2010, Canberra.

Chobotov, Y. (1978). *Analysis of photovoltaic total energy system*. Paper presented at the Proc IEEE Photovoltaic Specialist Conference, Wasington DC.

Chua-Liang, S., & Kirschen, D. (2009). Quantifying the effect of demand response on electricity markets. *IEEE Transactions on Power Systems*, *24*(3), 1199–1207. doi:10.1109/TPWRS.2009.2023259

Clement-Nyns, K., Haesen, E., & Driesen, J. (2010). The impact of vehicle-to-grid on the distribution grid. *Electric Power Systems Research*, *81*(1), 185–192. doi:10.1016/j.epsr.2010.08.007

Clorefeine, A. S. (1980). *Economic feasibility of photovoltaic energy system*. Paper presented at the 14th IEEE Photovoltaic Specialist Conf., San Diego, California.

Covino, S. (2003). *Demand side response 21st century style*. Paper presented at the Power Engineering Society General Meeting, 2003, IEEE.

Dam, Q. B., Mohagheghi, S., & Stoupis, J. (2008). *Intelligent demand response scheme for customer side load management*. Paper presented at the Energy 2030 Conference, 2008. ENERGY 2008. IEEE.

De-Meo, E. A. (1978). Prespectives on utility central station photovoltaic applications. *Solar Energy*, *21*, 177–192. doi:10.1016/0038-092X(78)90020-8

Department of Employment Economic Development and Innovation. (2009). *Transmission and distribution*. Brisbane, Australia: Queensland Government. Retrieved from http://www.dme.qld.gov.au/Energy/transmission_and_distribution.cfm

Department of Employment Economic Development and Innovation. (2009a). *Electricity*. Brisbane, Australia: Queensland Government. Retrieved from http://www.dme.qld.gov.au/Energy/electricity.cfm

Department of Employment Economic Development and Innovation. (2009b). *National electricity market*. Brisbane, Australia: Queensland Government. Retrieved from http://www.dme.qld.gov.au/Energy/national_electricity_market.cfm

Department of Employment Economic Development and Innovation. (2010). *Electricity generation in Queensland*. Brisbane, Australia: Queensland Government. Retrieved from http://www.energyfutures.qld.gov.au/electricity_generation_in_queensland.cfm

Department of Employment Economic Development and Innovation. (2010). *Electricity generation in Queensland*. Brisbane, Australia: Queensland Government. Retrieved from http://www.energyfutures.qld.gov.au/electricity_generation_in_queensland.cfm

Department of Resources Energy and Tourism. (2009). *Energy in Australia*. Canberra, Australia: Abare. Retrieved from http://www.ret.gov.au/energy/Documents/facts%20statistics%20publications/energy_in_aus_2009.pdf

Energy Efficiency and Conservation Authority. (2009). *Domestic-scale distributed generation guidance for local government*. Wellington.

Energy User Association of Australia Document. (2010). *Energy User Association of Australia document*. Australian Government. Retrieved from http://www.euaa.com.au/index.htm

Ergon Energy. (2010). *Domestic tariff*. Retrieved from http://www.ergon.com.au/home/electricity_for_your_home/ep_domestic.asp

Federal Energy Regulatory Commission. (2006). *Assesment of demand response and advanced metering*. Washington, DC. Retrieved from http://www.ferc.gov/legal/staff-reports/demand-response.pdf

Fraser, R. S. (2005). Demand side response in the national electricity market case studies: End use customer awareness program.

Greening, L. A. (2010). Demand response resources: Who is responsible for implementation in a deregulated market? *Energy*, *35*, 1518–1525. doi:10.1016/j.energy.2009.12.013

Hamidi, V., Li, F., & Robinson, F. (2009). Demand response in the UK's domestic sector. *Electric Power Systems Research*, *79*(12), 1722–1726. doi:10.1016/j.epsr.2009.07.013

Hammons, T. J. (2006). Integrating renewable energy sources into European Grids. *Proceedings of the 41st International Universities Power Engineering Conference*, 2006. UPEC '06.

Hyung Seon, O., & Thomas, R. J. (2008). Demand-side bidding agents: Modeling and simulation. *IEEE Transactions on Power Systems*, *23*(3), 1050–1056. doi:10.1109/TPWRS.2008.922537

International Energy Agency. (2010). *Strategic plan for the International Energy Agency demand-side management program 2004-2009*.

Jin-Ho, K., Tae-Kyung, H., & Kwang-Seok, Y. (2009). *Roadmap for demand response in the Korean electricity market*. Paper presented at the Power & Energy Society General Meeting, 2009. PES '09. IEEE.

Jones, T. E. (2004). *Australian example of demand side management actions*. Australia.

Kamel, F. (2009a). *Sharing communication network resources for a user-controlled electrical energy consumption*. Paper presented at the QUESTnet, Gold Coast Australia.

Kamel, F. (2009b). *User-controlled energy consumption in a transparent electricity system*. Paper presented at the 47th Annual Conference of the Australian and New Zealand Solar Energy Society, Townsville Queensland Australia.

Kirschen, D. S. (2003). Demand-side view of electricity markets. *IEEE Transactions on Power Systems*, *18*(2), 520–527. doi:10.1109/TPWRS.2003.810692

Lawrence, D. J., & Neenan, B. F. (2003). *The status of demand response in New York*. Paper presented at the Power Engineering Society General Meeting, 2003, IEEE.

Leonard, S. L. (1977). *Mission analysis of photovoltaic-major mission for the mid-term 1988-2000*. (Solar Energy Conversion-San 1101/PA 8 - 1/3, 3).

Leonard, S. L. (1978). *Central station power plant application for photovoltaic*. Paper presented at the Solar Energy Conversion 13th IEEE Photovoltaic Sp Conference.

Li, Z., & Yao, T. (2010). *Renewable energy basing on Smart Grid.* Paper presented at the Wireless Communications Networking and Mobile Computing (WiCOM), 2010 6th International Conference on IEEE.

Marwan, M., & Kamel, F. (2010a). *Demand-side response load management modelling encountering electrical peak demands in Eastern and Southern Australia - Smart Grid tools.* Paper presented at the Australasian Universities Power Engineering Conference AUPEC 2010, Christchurch New Zealand.

Marwan, M., & Kamel, F. (2010b). *User-controlled electrical energy consumption towards optimized usage of electricity infrastructure*. Paper presented at the Southern Region Engineering Conference, Toowoomba Australia.

Na, Y., & Ji-Lai, Y. (2006). *Optimal TOU decision considering demand response model.* Paper presented at the Power System Technology, 2006. PowerCon 2006. International Conference on IEEE.

Narayan, P. K., & Smyth, R. (2005). Electricity consumption, employment and real income in Australia evidence from multivariate Granger causality tests. *Energy Policy*, *33*(9), 1109–1116. doi:10.1016/j.enpol.2003.11.010

Nguyen, D. T. (2010). *Demand response for domestic and small business consumers: A new challenge.* Paper presented at the Transmission and Distribution Conference and Exposition, 2010 IEEE PES.

Parvania, M., & Fotuhi-Firuzabad, M. (2010). Demand response scheduling by stochastic SCUC. *IEEE Transactions on Smart Grid*, *1*(1), 89–98. doi:10.1109/TSG.2010.2046430

Queensland Competition Authority. (2000). *Electricity demand forecast*. NSW, Australia. Retrieved from http://www.qca.org.au/files/QLDElectricityDemandForecast.pdf

Queensland Government. (2009). *ClimateQ: Toward a greener Queensland -Transport moving towards a low carbon future*. Retrieved from http://www.climatechange.qld.gov.au/__data/assets/pdf_file/0013/24061/ClimateQ_Report_chapter15.pdf

Rahimi, F., & Ipakchi, A. (2010). *Overview of demand response under the smart Grid and market paradigms.* Paper presented at the Innovative Smart Grid Technologies (ISGT) 2010, IEEE.

Schweppe, F. C., Caramanis, M. C., Tabors, R. D., & Bohn, R. E. (1988). *Spot electricity price.* Boston, MA: Kluwer Academic.

Stam, E. (2005). *Demand response activities in Finlad.* Paper presented at the A Nordic Conference on Enhancing and Developing Demand Response in the Energy Market, Copenhagen.

Torriti, J., Hassan, M. G., & Leach, M. (2009). Demand response experience in Europe: Policies, programmes and implementation. *Energy*, *35*(4), 1575–1583. doi:10.1016/j.energy.2009.05.021

Tyagi, R., & Black, J. W. (2010). *Emergency demand response for distribution system contingencies.* Paper presented at the Transmission and Distribution Conference and Exposition, 2010 IEEE PES.

Vos, A. (2009). *Effective business models for demand response under the Smart Grid paradigm.* Paper presented at the Power Systems Conference and Exposition, 2009. PSCE '09. IEEE/PES.

Walawalkar, R., Blumsack, S., Apt, J., & Fernands, S. (2008). *Analyzing PJMs economic demand response program.* Paper presented at the Power and Energy Society General Meeting - Conversion and Delivery of Electrical Energy in the 21st Century, 2008 IEEE.

Wen-Chen, C., Yi-Ping, C., & Tzu-Hao, L. (2007). *The competitive model based on the demand response in the off-peak period for Taipower System.* Paper presented at the Industrial & Commercial Power Systems Technical Conference, 2007. ICPS 2007. IEEE/IAS.

Yu, C. W., Zhang, S., Chung, T. S., & Wong, K. P. (2005). Modelling and evaluation of interruptible-load programmes in electricity markets. *IEEE Proceedings Generation. Transmission and Distribution, 152*(5), 581–588. doi:10.1049/ip-gtd:20045138

Zhong, J., Kang, C., & Liu, K. (2010). *Demand side management in China.* Paper presented at the Power and Energy Society General Meeting, 2010 IEEE.

ADDITIONAL READING

Almeida, A., & Rosenfeld, A. (1988). *Demand-side management and electricity end-use efficiency*. Kluwer Academic Publishers.

America's Energy Future Panel on Electricity from Renewable Resources, & Council, N. R. (2010). *Electricity from Renewable Resources: Status, Prospects, and Impediments*: National Academies Press.

Barnett, D., & Bjornsgaard, K. (2000). *Electric power generation: a nontechnical guide*. PennWell Books.

Barsali, S., Ceraolo, M., & Possenti, A. (2002). Techniques to control the electricity generation in a series hybrid electrical vehicle. *Energy Conversion. IEEE Transactions on, 17*(2), 260–266.

Budde, P., & Whittle, R. (2007). *Smart grid: energy management and broadband*. Paul Budde Communication.

Caves, D. W., Herriges, J. A., Hanser, P., & Windle, R. J. (1988). Load impact of interruptible and curtailable rate programs: evidence from ten utilities [tariff incentives]. *Power Systems. IEEE Transactions on, 3*(4), 1757–1763.

Chen, C. S., & Leu, J. T. (1990). Interruptible load control for Taiwan Power Company. *Power Systems. IEEE Transactions on, 5*(2), 460–465.

Chua-Liang, S., & Kirschen, D. (2009). Quantifying the Effect of Demand Response on Electricity Markets. *Power Systems. IEEE Transactions on, 24*(3), 1199–1207.

Collins, M. M., & Mader, G. H. (1983). The timing of EV recharging and its effect on utilities. *Vehicular Technology. IEEE Transactions on, 32*(1), 90–97.

Daryanian, B., Bohn, R. E., & Tabors, R. D. (1989). Optimal demand-side response to electricity spot prices for storage-type customers. *Power Systems. IEEE Transactions on, 4*(3), 897–903.

Deep, U. D., Petersen, B. R., & Meng, J. (2009). A Smart Microcontroller-Based Iridium Satellite-Communication Architecture for a Remote Renewable Energy Source. *Power Delivery. IEEE Transactions on, 24*(4), 1869–1875.

Fox-Penner, P. (2010). *Smart Power: Climate Change, the Smart Grid, and the Future of Electric Utilities*. Island Press.

Garg, H. (1987). *Advances in Solar Energy Technology: Collection and storage systems*. D. Reidel Pub. Co.

Goswami, D. (2007). *Advances in Solar Energy: An Annual Review of Research And Development*: Earthscan in association with The American Solar Energy Society.

Hajimiragha, A., Caizares, C. A., Fowler, M. W., & Elkamel, A. (2010). Optimal Transition to Plug-In Hybrid Electric Vehicles in Ontario, Canada, Considering the Electricity-Grid Limitations. *Industrial Electronics. IEEE Transactions on*, *57*(2), 690–701.

Hamilton, K., & Gulhar, N. (2010). Taking Demand Response to the Next Level. *Power and Energy Magazine, IEEE*, *8*(3), 60–65. doi:10.1109/MPE.2010.936352

Hori, Y. (2004). Future vehicle driven by electricity and Control-research on four-wheel-motored "UOT electric march II." *Industrial Electronics. IEEE Transactions on*, *51*(5), 954–962.

Ibitoye, F. I. & Adenikinju, A. (2007) Future demand for electricity in Nigeria. *Applied Energy, 84*(5), 492-504. doi: DOI: 10.1016/j.apenergy.2006.09.011

Jazayeri, P., Schellenberg, A., Rosehart, W. D., Doudna, J., Widergren, S., & Lawrence, D. et al. (2005). A Survey of Load Control Programs for Price and System Stability. *Power Systems. IEEE Transactions on*, *20*(3), 1504–1509.

Kreith, F., & Goswami, D. (2007). *Handbook of energy efficiency and renewable energy*. CRC Press.

Masiello, R. (2010). Demand Response the other side of the curve [Guest Editorial]. *Power and Energy Magazine, IEEE*, *8*(3), 18–18. doi:10.1109/MPE.2010.936206

Molina-Garcia, A., Bouffard, F. & Kirschen, D. S. (2010) Decentralized Demand-Side Contribution to Primary Frequency Control. *Power Systems, IEEE Transactions on, PP*(99), 1-1.

Momoh, J. (2011). *Smart Grid: Fundamentals of Design and Analysis*. John Wiley & Sons, Limited.

Ochoa, L. F. & Harrison, G. P. (2010) Minimizing Energy Losses: Optimal Accommodation and Smart Operation of Renewable Distributed Generation. *Power Systems, IEEE Transactions on, PP*(99), 1-1.

Roos, J. G., & Lane, I. E. (1998). Industrial power demand response analysis for one-part real-time pricing. *Power Systems. IEEE Transactions on*, *13*(1), 159–164.

Rotering, N. & Ilic, M. (2010) Optimal Charge Control of Plug-In Hybrid Electric Vehicles in Deregulated Electricity Markets. *Power Systems, IEEE Transactions on, PP*(99), 1-1.

Sheen, J. N., Chen, C. S., & Yang, J. K. (1994). Time-of-use pricing for load management programs in Taiwan Power Company. *Power Systems. IEEE Transactions on*, *9*(1), 388–396.

Sovacool, B. (2008). *The dirty energy dilemma: what's blocking clean power in the United States*. Praeger Publishers.

Talukdar, S., & Gellings, C. (1987). *Load management*. IEEE Press.

Warkentin, D., & Warkentin-Glenn, D. (1998). *Electric power industry in nontechnical language*. PennWell.

KEY TERMS AND DEFINITIONS

Electrical Generation: Is the process of creating electricity from other forms of energy.

Electrical Vehicle: Is as an electric driven vehicle, uses one or more electric motors for propulsion. Electric vehicles include electric cars, electric trains, electric lorries, electric airplanes, electric boats, electric motorcycles and scooters and electric spacecrafts.

Energy Management: Is a system of computer-aided tools used by operators of electric utility grids to monitor, control, and optimize the performance of the generation and/or transmission system.

Market Conditions: Are characteristics of electricity market, such as electricity price and demand, which could be introduced to all customers.

On-Site Renewable Energy: Is an energy sources generated from natural resources, such as biodiesel, photovoltaic and wind. These energies are provided by customers.

Smart Grid: Is a form of electricity network utilizing digital and computer technology. A smart grid delivers electricity from suppliers to consumers using two-way digital communications to control appliances at consumers' homes; this saves energy, reduces costs and increases reliability and transparency.

Switch: Is an electrical component that can break an electrical circuit, interrupting the current or diverting it from one conductor to another.

This chapter was previously published in Innovation in Power, Control, and Optimization edited by Nadar Barsoum, Jeffrey Webb, and Pandian Vasant, pages 137-163, copyright 2012 by Engineering Science Reference (an imprint of IGI Global).

Chapter 91
Management and Utilisation of Natural Resources in Special Nature Reserves:
A Case Study

Božo Drašković
Institute of Economic Sciences, Serbia

Jovan Zubović
Economics Institute, Serbia

Ivana Domazet
Institute of Economic Sciences, Serbia

ABSTRACT

Special nature reserves are particularly regulated land areas designated for protection and utilization of natural resources at their disposal. In this chapter, the authors present a pioneering attempt to analyze management efficiency and utilization effectiveness of natural resources in protected nature reserves in Serbia based on an empirical research. Moreover, the authors explore the regulations that define the protection, management, and conservation of natural values and goods in Serbia, with a purpose of evaluating opportunities for development and implementation of Environmental Accounting (EA). EA is still at an early stage of development, especially from the standpoint of its incorporation into the developed standard models of company accounting. There still exist several unexplored problems to be solved at the level of national accounts, such as the expression of environmental damage, resource consumption, sustainable use of natural capital, evaluating environmental benefits, and free gifts of nature. The chapter is completed with a case study based on an empirical study in the field of environmental management in one protected area in Serbia, Zasavica.

DOI: 10.4018/978-1-4666-4852-4.ch091

INTRODUCTION

A historical context of allocating some geographical areas with all the amenities of natural resources in a special regime of protection has a long history. The first such area in the world delimited by state regulations for a special regime of protection and preservation in 1872 was The Yellowstone national park in the U.S. (US Congress, 1872, p. 392). This example later led many countries in the world to apply similar regulations. Hence, today there is almost no country that has not at least some area with natural values marked with the special protection regime. In Serbia there are several areas which are by special legislation defined and declared as special protected areas. Nature reserves have a special regime of protection and specific model for use of resources at their disposal. In this work we will analyze different aspects of conflict of interest and regulations that, in connection with the economic exploitation of resources on the one hand, and the ecological approach that limits the exploitation of resources in order to maintain the special natural values biodiversity and protected areas on the other side.

In the first part, we analyze normative aspects of natural values and natural resources, its utilization, and maintenance regimes of economically usable and unusable natural resources and values. A special aspect of the analysis in this paper is devoted to the limitations with regard to private, corporate and state property rights over the forest and land resources that are located within environmentally protected areas, since the regime of protected areas is largely diversified.

In the second part, we present protected areas in Serbia with respective management model for their utilization. Standard legal management model for specific protected areas in Serbia is based on an approach by which it is limited or even completely prohibited to extract natural resources that are allocated inside the protected area. Limiting the use of resources requires creation of the control model for the exploitation of resources and monitoring process which tracks the quality of water, air, movement of plant populations and animal species at specific time intervals. Restricting private rights requires consideration of introduction of benefit models for the owners who refrain from the use of some economically valuable goods.

In the third part of the chapter, we analyze some aspects and features of incorporating environmental and economic results of monitoring in particular environmental accounting. Finally, we present the partial results of an empirical research in a form of a case study of the population living in the area in and around special nature reserve "Zasavica" in Serbia.

NORMATIVE ASPECTS OF THE ENVIRONMENT PROTECTION AND NATURAL VALUES IN SERBIA

Legal protection of the environment and natural resources is a complex process that must be harmonized with international standards, recommendations and conventions (Drašković, 1998). Normative regulation of environmental protection is being harmonized with the recommendations and directives in European Union. Environmental law is placed in an international legal framework more than other areas of law. It has beenestablished primarily due to the fact that a large number of environmental problems are both regional and global its character. The first aspect of its content and framework in Serbia's legislation in this area is determined by the membership of Serbia in the UN and relevant regulations embroidered by the acceptance of environmental protection. Especially significant are international aspects of accepting a standard regulations introduced by: the Stockholm Declaration of the United Nations on the Human Environment (United Nations, 1972), the Rio Declaration on Environment and Development(United Nations, 1992),the Aarhus convention (UNECE, 1998), the Convention on biological diversity (UNEP, 1992), etc.

The objects of legal protection of natural resources and values in Serbia are, above all, protection of soil, water, air, forests, flora and fauna and natural diversity in general. Such protection incorporates the regulation which restricts the dynamics of using individual plants and animal species, as well as the control of utilization of certain environmentally hazardous materials.

Serbia's membership in the Council of Europe bring along the obligation to ratify all of the conventions that protect human rights and its key segment, the right to a healthy environment (Council of Europe, 1993), (Council of Europe, 1998), (Council of Europe, 1982), (European Council, 2004). Another important aspect in creating the content and principles of environmental protection in Serbia is a general political commitment to EU accession process. In the process of harmonization of relations with the EU the adoption of standards related to conservation and environmental protection, implementation of the goals of energy efficiency, waste management and reduction of harmful substances is associated.

In accordance with the above international legislation, the legislation in Serbia in the field of environmental protection has included basic principles and standards, as well as the protection of the rights to live in a healthy environment, a sustainable economic development, the rational use of resources, and the integrity of the biosphere (Pešić, 2008). Serbian regulation is based on the adopted principles which were developed on international and domestic practice. One of the key standards which areapplied in the legal legislation is related to the liability and the cost to legal entities and individuals to use natural resources and harm the environment. The legislation incorporates the principles of "polluter pays" and the "user pays."

The basic legal act in the Republic of Serbia is the Serbian Constitution. The Constitution is the fundamental legal document and is the basis for all laws, including those governing the protection of the environment. One of the key constitutional principles is the proclamation that everyone has the right to a healthy environment, as well to timely and complete information about conditions of the environment. The Constitution authorizes the Republic of Serbia to organize and implement environment for sustainable development, to create protection systems and improve environmental protection, to improve flora and fauna as well as production, trade and transport of weapons, toxic and flammable, explosive, radioactive and other hazardous substances. The Constitution imposes obligations concerning the implementation of environmental principles to decentralized autonomous province, cities, and municipalities.

Serbian national legislation provides specific legal and financial measures that are introduced with a goal of ensuring the implementation and enforcement of regulations, which require sustainable use of natural resources and prevention or reduction of the effects of pollution. When the result of an economic activity or behavior of an individual is a breach of regulations, and hence the cause is a damage to the environment, there must exist the repressive measure, which aims to prevent further degradation of the environment and provide compensation for the damage made. The mechanism of protection is implemented through financial and even criminal punishment for perpetrators for the harm they made to the environment.

In the structure of environmental legislation in Serbia are also included the following laws: Law on Environmental Protection, Law on Impact on the Environment; Law on Strategic Environmental Impact and Law on Integrated Pollution Prevention and Control of Environmental Pollution. In addition to these general legal framework for environmental protection, there are special legislations regulating specific areas, such as the Law on Air Protection, Law on Water Protection, Law on Protection against Ionizing Radiation and Nuclear Safety, the Law on Non-Ionizing Radiation Protection and a series of laws, regulations, decisions, rules, which regulate in detail the legal aspects of environmental protection. In order to illustrate the rich, multidimensional and comprehensive legislation in the field of environmental protection, we suggest further reading listed at the end of the chapter were is presented a list of a number of laws and by-laws of the Republic of Serbia.

The supervision in application of the Law on Environmental Protection and other relevant regulations is under the Ministry in competence for the protection of the environment. The inspection is performed by the environmental inspector. As the inspection body, environmental inspections primarily base their work on the Law on Public Administration and further on in the Law on Environmental Protection. In accordance with the Law on Environmental Protection there were enacted other special laws, such as: the Law on Strategic Environmental Impact, Impact Assessment Act on Environment and the Law on Integrated Prevention and Control of Environmental Pollution, which also indirectly regulate the activities of the inspection.

The system of supervision for implementation of environmental regulations is split into three levels. At the first and the highest level are Republic institutions, followed by institutions that are at the level of the Autonomous Region, and finally the institutions located at the level of municipalities and cities - local government units that carry out control of the obligations under their jurisdiction. Legislation has given the inspectors broad jurisdiction in control of quality management and sustainable use of natural resources, implementation of the proposed measures, compliance with standards, quality and technological facilities, import and export and other significant aspects that make up an integrated system of environmental protection in Serbia.

The right to manage designated protected areas within the defined rules is given by the Government of Republic of Serbia. In the selection of a manager there is de facto no discrimination from the perspective of ownership structure of the managed land or the entity to which the control is given. There are different types of ownership of the land areas that are placed under the protection regime and hence being management. They might be in private, cooperative or state ownership. The legal entity to which the management deed is given most often gains no property rights in particular protected area. From the propriety point of view, ownership which belongs to various persons (legal and natural) remains unchanged, but some restrictions are imposed for use of those assets to their owners. An example of this is restriction for the exploitation of resources, use of mineral fertilizers, the discharge of waste material, mode of use of building land etc.

PROTECTED AREAS IN SERBIA

In the Republic of Serbia under the regime of environmental protection is an area which in total covers 522,120 hectares of land, forests, pastures, ponds, bodies of water which are located in 463 areas, comprising 5.91% of the territory of Serbia. Protected areas are managed by different entities, NGOs, private companies and public companies in state ownership. It is necessary to emphasize that one public company "Srbijašume" manages 96 protected areas that cover 216,804.88 hectares comprising 41.52% of the total protected areas in Serbia.

The Government of the Republic of Serbia had in 2007 adopted a National Sustainable Development Strategy which defined the national priorities for environmental protection, especially related to economic development. Within the strategy, a specific consideration was given to the objectives of raising the level of protection, preservation and enhancement of biodiversity, increasing the area under strict protection to the level of 10% of the total land area of Serbia, extending the network of protected areas and establishment of ecological corridors. By the Regional Plan for Development of Serbia for the period 2010-2020, there has been set a goal that total area under the regime of special environmental protection is increased to 12%. Management of protected areas is defined as an activity of general interest.

Under the regime of protected areas in Serbia belong several different types which include: national parks, strict nature reserves, natural monuments, protected habitats, protected landscapes, cultural sites of exceptional quality and nature parks.

- **National Parks:** Are areas with more diverse ecosystems of national importance with special landscape features and cultural heritage of the region whose people can live in harmony with nature. National parks are intended to preserve the existing natural values and resources, geological and biological diversity. They can host different forms of scientific, educational, spiritual, aesthetic, cultural, tourist, health and recreational activities, as well as other activities in accordance with the rules and principles of conservation and environmental protection.
- **Strict Nature Reserves:** Are areas, which comprise of the unmodified natural features that include representative natural ecosystems. They are areas designated for preservation of nature with its genetic structure and maintenance of ecological balance. In them it is allowed to conduct scientific research under the condition that it does not disturb the natural features, phenomena, processes and values.
- **Special Nature Reserves:** Are areas with the unaltered or slightly altered state of nature, which is important because of its uniqueness, rarity, or representativeness and which is also the habitat of endangered species of wild fauna and flora. These areas are typically without villages or urban areas with rare population in which people live in line with nature. These areas are intended for preservation of the existing natural features, genetic structure, ecological balance, and for monitoring of natural phenomena and processes. In these areas, scientific researchers are permitted and visits of the picnickers are controlled in order to preserve the traditional, inherited natural resources. Special nature reserves by its contents may be forestall, zoological, geological, hydrogeological, hydrological, or botanical.
- **Natural Monuments:** Are smaller, unchanged, or partially changed natural areas or complexes, objects or phenomena that are clearly physically distinct, recognizable, and unique. They are representative by geomorphological, geological, hydrological, botanical, or other features. Natural monument can also be the areas where people created or formed some type of botanical values that can be used in scientific, cultural, educational, aesthetic and tourist purposes. There are different types of natural monuments: geological, geomorphological, speleological (cave, pits, etc.), hydrological (complete or parts of the rivers, lakes, ponds, waterfalls), botanical (rare and important examples of plant life, individual trees or groups of trees, alleys, parks, botanical gardens).
- **Protected Habitats:** Are areas that include one or more types of habitats which are important for the preservation of single or multiple populations of wild species and their communities.
- **Protected Landscapes:** Are the areas that have distinct and special appearance with significant natural biological -ecological, aesthetic and cultural—historical values. A key feature of these landscapes is that they are the result of the interaction of nature and natural resources on one side and traditional human activities on the other side. Protected landscapes maybe a natural areas or cultural landscapes of outstanding

features. These areas have significant biological—ecological and aesthetic values, where traditional local population life and natural ecosystems are not significantly disrupted.

- **Cultural Sites of Exceptional Quality:** Are the areas that have recognizable features, strengths, as well as cultural and historical values that have evolved over time due to the interaction of nature and natural resources of a given area and the traditional ways of life of residents who lived or live in these areas.

According to the above division of protected areas in Table 1 it is possible to find the information on their number and size as well as the information on the participation of the largest single managing company in Serbia – public company "Srbijašume."

Out of the 463 protected areas with an area of 522 thousand hectares, 5 national parks are covering an area of approximately 159 thousand hectares or 30.45% of the total protected area. Additional 12 nature parks account for 41.32% of the total surface of protected areas. In the third place in size are 17 special nature reserves occupying the territory of 16.74% of the total protected areas in Serbia. Hence, all other 429 forms of protected areas account for only 11.49% of the total surface of protected areas. As noted before, the largest individual managing company of protected areas is a public company "Srbijašume" governing 41.52% of the total protected area.

Given that the National Parks are public companies as well, in addition to managed area by "Srbijašume," public companies manage altogether 375.790 hectares, or approximately 72% of the total protected areas in Serbia.

The initial definition of regulations for the management of protected areas have been passed

Table 1. Natural protected areas in Serbia in 2011

#	Type of Protected Area	Serbia		PC "Srbijašume"	
		Quantity	Size (ha)	Quantity	Size (ha)
1.	National parks	5	158.986,36		
2.	Nature parks	12	215.760,57	3	197.261,00
3.	Regional nature parks	4	361,86	1	296,64
4.	Protected landscapes	12	33.406,80	3	4.348,96
5.	Sites of special beauty	4	12.105,63	3	11.265,00
6.	Forest	1	19,65	1	19,65
7.	Strict nature reserves	42	2.207,28	35	561,89
8.	General nature reserve	4	60,49	4	60,49
9.	Special nature reserve	17	87.410,14	2	123,50
10.	Natural monuments	327	7.681,00	36	2.222,44
11.	Memorial natural monuments	19	2.394,67	5	442,95
12.	Areas surrounding cultural sites	16	1.725,55	4	202,36
13.	Strictly protected species	1681			
14.	Protected species	821			
Total 1-12		463	522.120,00	96	216.804,88

Source: The table is prepared on the basis of data obtained from the Acts of declaration of protected areas, data from "Srbijašume" and the Institute for Nature Conservation of Serbia in 1991 by The Law on Environmental Protection, where it was defined that simultaneously with the adoption of an act to declare an area as protected area, shall be selected the managing company. Selection of the manager is followed by defining of the obligations it has, especially in the aspects of management and obligations concerning the protection of natural values of protected areas.

Management of Protected Areas

As noted above, some of specially protected areas, which have the status of national parks, are managed by the state owned public companies. Managers and caretakers of natural resources and protected areas, apart from public companies, may be private joint stock companies, limited liability companies, civic associations, such as hunting and hiking clubs, tourist organizations, fishing farms, museum, religious organizations and institutions, NGOs, and scientific institutions such as the Faculty of Biology and Forestry.

Amendments to the Law on the Natural Protection, which occurred in 2009, determined that management of protected areas and goods is well defined as an activity of general interest. The law defines precisely the rights and obligations for management. The manager according to the legislation must:

- Preserve a protected area and implements the agreed regime of protection.
- Enhance and promote the protected area.
- Deliver a management plan and an Act of internal control and safe-keeping which is determined in the Acton the Protection.
- In accordance with the special regulations of labeling, mark the protected area, set borders and regime for protection.
- Ensure smooth evolution of natural processes and sustainable use of protected areas.
- Give consent to carry out scientific research, perform surveying, filming, erection of temporary structures in a protected area and provide additional approvals in accordance with law and regulations on internal order and managing services.
- Provide oversight over the implementation of conditions and measures of nature protection.
- Monitor the movement and activities of the visitors and provides trained guides for tourist visits.
- Keep records on the state of natural resources and provides data to the Department of Nature Conservation.
- Keep records of human activities, actions and processes that represent an endangering and damaging factor to the protected area and shall submit that data to the Institute and the Ministry.
- Keep record on real estate data relevant to the management of protected areas.
- In cooperation with national and regional institutions prevent all actions and activities that are defiant with the laws on the protection and which may represent a threatening factor for devastation of the protected area.
- Deliver an Acton the cost for the utilization of natural resources.
- Perform all other duties prescribed by law and regulations on protection.

The managing activity and sustainable use of protected area can be performed by the managers-legal entities that are registered for the activities in the field of nature protection, and whose business is close to the environmental matters. The manager must own an organization and have employees educated for protecting and guarding services, services of promotion, improvement, and sustainable development of protected areas.

ECOLOGICAL ACCOUNTING

Research on the possibility of constituting a special segment of accounting called environmental accounting or green accounting had been implemented during the last two decades. The concept itself is still at an early stage of development. The motivation for the implementation of the Environmental Accounting (EA) or the Green Accounting (GA) originated primarily from the desires and needs of certain, mainly developed, countries to develop the methodology and techniques for monitoring and the implementation of the concept of sustainable development. In fact in the late twentieth and early twenty-first century in the world are growing movements and awareness of the need for economic and social development to be reconsidered from the point of its sustainability. Pioneer attempts to uncover economic aspects of exhaustible resources was made by (Hoteling, 1931)The basic subject of the concept of sustainability is to observe relationships between current consumption of resources and their survival in the future (van den Bergh, 1996). A key task of environmental accounting should be to determine how threatened is the sustainable development, especially regarding exhaustion and degradation of resources in the future, by collecting and quantifying economic parameters and development trends. The standpoint of new methodology is that in conjunction with current development and exploitation of resources should be kept in mind that non-renewable resources are limited, and finally could be exhausted.

There is a large number of definitions of sustainable development, which can be reduced to one of the most common, which is from the standpoint of substance and meaning quite acceptable, and it is formulated in the Bruntland Commission Report (World Commission on Envirnoment and Development, 1987), in which sustainable development is defined as: "development that meets current needs with no danger for future generations to be disabled to meet their own needs." That primarily promotes the sustainable use of natural resources, and the nature, which is inherited from previous generations.

The essence of the concept of sustainability is the need for constant maintenance of reserves of natural capital over time (World Commission on Environment and Development, 1987). Hence, the environmental accounting assumes that the application of national accounting principles in recording and balancing should include all elements of property (assets) at the state level, as it follows: natural capital, human capital and manufactured or created capital. In current general accounting practice and bookkeeping methods implemented in the national accounts, produced equity is taken as a basic element both at micro and macro levels. These accounts are the basis for reporting and tracking changes in GDP. Such accounting technique does not include (and trace) spending of natural capital and goods, which are often treated as a free resource or external social cost. The problem of recording the cost of any such exploitation of natural resources or natural resources that can be degraded due to economic activity and pollution is growing. Therefore, it became increasingly clear that as a result of a decrease in reserves of non-renewable resources we must take account of the "natural capital spending." The second type of natural resources, which is expressed by the term "human capital," is also not accounted for in accounting records and in effective consumption and renewal. Like in the case of natural capital, in the case of human capital there is still no harmonized approach that would give directions for their empirical accounting records, accounting reporting and its incorporation as a segment of the overall social accounting, social wealth, and the welfare state.

In environmental accounting sustainability can range between weak and strong. In the case of weak sustainability, the goal is the sustainability of the overall capital structure, which means that total assets value should remain constant, while the growth of one type of capital can go to the expense of another kind. In accordance with that principle, the reserves of natural capital may be reduced, and the environment in some parts may be degraded. At the same time at the expense of reducing natural capital and natural resources, it is necessary to have an increase in the amount of produced capital or increase of other type of natural capital through savings and regeneration. The problem of weak sustainability is that to growth of created capital can result in permanent loss or permanent degradation of certain parts of the environment or resources. In that case they cannot be restored later regardless of the availability of funds that could be invested in the revitalization of natural resources and capital. Hence that phenomenon is irreversible. Actually, there is a one-way flow in which it is possible to use natural resources and make transfer of natural capital into cash or financial capital. In that process, non-renewable natural resources are depleted, and the result of their consumption is increase in generated cash or capital. Reverse process is usually not possible, especially in the case of non-renewable natural capital and natural resources that are permanently destroyed so that there is no possibility of natural self-reparation.

Another sustainability model is called "strong sustainability." In this concept it is assumed that capital in its structure remains unchanged in the future. This approach is especially true for natural resources or resources that have the ability to regenerate with or without external stimulus. A good example of this approach is the case of controlled exploitation of forests. Harvesting can be performed only to the extent of annual natural forests growth. The same principle applies in the case of "common resources" such as catching fish, which can be pursued only to the level of annual fish reproduction level, which can ensure that fish stocks remain unchanged. The model of strong sustainability model can be applied in pollution protection as well. In that case it is required that disposal of waste must not exceed the assimilation capacity of the environment: water, soil and air, and self-renewing capacity of the environment.

Development of environmental accounting should provide to a variety of community environmental groups, businesses, and national economy policy makers:

- More comprehensive monitoring, data collection instruments, and data processing on natural resources, both those that have some market value, and those which are classified as natural common goods and services.
- Identification of the factors that have a major impact on the relationship between economy and ecology.
- Comprehensive research to create a model with a dynamic formula which should explains how economic growth affects the achievement of environmental goals.
- Develop concepts and establish a database for strategic planning of development directions, in order to achieve sustainable development policy in the future and for future generations.

Implementation of environmental accounting with accounting records of natural resources began to be developed and initially implemented in some countries in 1970. A pioneer country in application of this type of accounting is Norway. It has been almost two decades of development and experimental application, when in the early 90s, the United Nations, OECD, and later the World Bank and Eurostat, initiated and supported a group of professionals who have had a role to define the key elements of the methodology for establishing a consistent system of environmental accounting. As a result of this work it was created the initial methodology entitled "System of Economic and Environmental Accounting" (SEEA). Further on the United Nations had issued instructions for application of environmental accounting (United Nations, 1993).

As a result of the above studies and support by the institutions there have been constituted four basic components of environmental (green) accounting:

1. Assets (reserves) accounting of natural resources, which keep record on reserves of natural resources and monitors their changes in national accounting.
2. Accounting of materials and pollution flow (flow of energy and resources), which provides information about the level of industrialization and the use of energy and materials as inputs into production, finished products and services and generated pollution and solid waste in the production process. This accounting is associated with a table "Sources and use of the resources" in the "National Accounting System" and it is used to create input-output (IO) tables.
3. Environmental and resources protection, management of costs in accordance with national accounting that are generated in industry, governments, and households with a purpose of protecting the environment or management of resources.
4. Environmentally adjusted macroeconomic aggregates, which include indicators of environmental sustainability adjusted to Net Domestic Product (NDP).

The first component, the Assets Accounting of natural resources monitors exploitation of natural resources during certain time intervals and delivers yearly reports. Moreover, it shows changes from previous levels and reserves that may be utilized in the future. Reserves may also include current and potential reserves. Assets accounting of natural resources is kept in quantity and in value. Quantitative recording tracks changes that occur during the year as a result of utilization. For plants and animal species, not only changes are recorded, but natural growth and decay as well. There are basically two recording models for the value of natural capital: the Net Present Value (NPV) and the net price. Assets accounting is harmonized with the standard accounting models that are used by companies and which are coded at the international level. Recording of assets and natural capital creates a problem in the environmental sectors, which do not have pricing indicators for certain forms of natural capital. This is especially true, for example, in the case of common resources, such as clean water or clean air.

The second component, the Accounting of Materials and Pollution Flow is also in accordance with the standard company accounting concept. It requires recording and monitoring of consumption of materials and energy and the subsequent pollution created in industry and consumption of all final products. Currently the most advanced is recording of production of energy, especially fossil fuel consumption and, accordingly, the emission of greenhouse gases. Monitoring and recording of pollution caused by burning fossil fuels was developed in those countries that under Annex I of Kyoto agreement, accepted the obligation to reduce emissions for 5% by the year 2012. Other records kept under this accounting include water accounting and water pollution, origin and quantity of solid waste and other forms of environmental degradation. The problem of extensive emission of carbon dioxide as a global problem opened the debate on the rights of developed countries on the quantity of emissions, as well as on the rights of developing countries to increase their quota of participation in the global pollution, which inevitably increases with their economic development.

Next component of the environmental accounting system tracks costs associated with environmental protection. These costs are sorted in the following three groups:

- The costs of environmental protection and resource management by public or private sector.
- Activities environmental services industry.

- Taxes and subsidies for environmental protection and use of resources.

Development of environmental accounting system, although still in its initial stage, is especially focused on the macro level and the problem of correct calculation of real GDP. Current accounting system, for example, treats costs for elimination of environmental damage as an item that affects the growth of GDP. In order to eliminate the above drawback, primarily from the standpoint of correct monitoring of sustainable development at the national level, there is a need for the correction of conventional macroeconomic indicators that show changes in GDP and NDP. The records of the levels of standard of physical and monetary development indicators should be adjusted for the costs of exploitation of natural resources in a form of natural capital depreciation costs, and investments in environmental protection and the elimination of negative consequences should also be included as a correction factor in GDP growth.

Development system of the green accounting is in the initial stage both in the company's micro-level and national macro-level. It is facing one of the key issues of how to include complex environmental dimensions of natural capital and goods in the standard system of company and social accounting.

Case Study: Special Nature Reserve "Zasavica"

Special Nature Reserve Zasavica was placed under the state protection regime as a particular natural resource of the first category in 1977. The reserve is located in the Republic of Serbia in the southern region of Srem called Mačva near the town Sremska Mitrovica. SNR is a wetland area with water meadows and forests of the total area of 1825 hectares. The reserve stretches along river Zasavica, whose overall course has a total length of 33.1 km. Through the area also runs a streamBatar, canals and Jovac and Prekolac, and the entire wetland system is connected with the Sava River. In the area of tar, canals, and Jovac and Prekolac, the entire wetland system is connected with the Sava River.

Empirical research was conducted in early 2012. The study interviewed residents of villages located on the edge of a special nature reserve. There have been 103 interviews with persons of different ages and gender. The study focused on the attitudes of the population in relation to natural resources located in SNR Zasavica. Here we present some findings of the research.

When asked to define which are the key natural value in the reserve interviewees responded:

- The most valued was unspoiled nature (15.64% of respondents), followed by the rare species with 14.66%, an eco-tourism development as a business opportunity with 14.66%, and then follows the clean air that valued 11.73%, the water supply 8.47% of the respondents, 6.51% healthy foods, and the arrangement of space 3.58% of the respondents.
- Rare species are valued relatively modest as a specificity of Zasavica. Beaver leads with 10.75%, mangulica (type of pig) 9.45%, and Podolian cattle with 2.93%.
- On the question "how affected local population is by nature reserve" only 14.56% of respondents stated that they suffered some kind of damage, while 85.44% of respondents expressed that they believe that there was no harm suffered (Figure 1).

Regarding the attitude of respondents towards possible change in the status of protected area by conversion into agricultural land, 8.7% of respondents expressed the view that support change of SNR Zasavica into agricultural land. In contrast as much as 91.3% of respondents expressed the view that there should be no change of the status of this nature reserve (Figure 2).

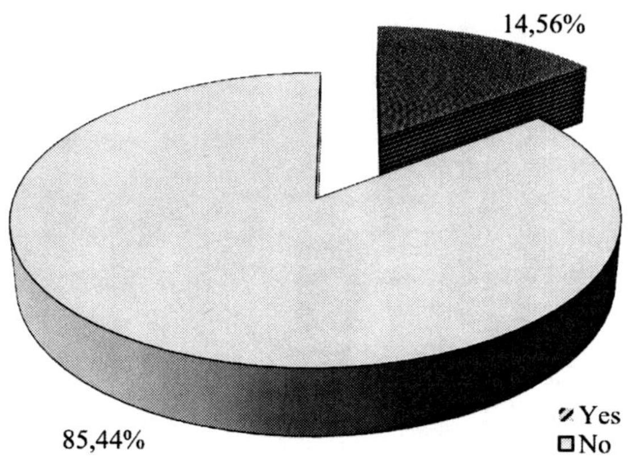

Figure 1. How is local population affected by nature reserve

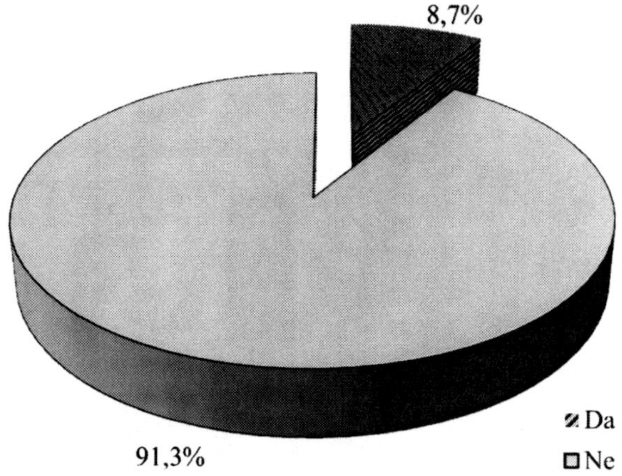

Figure 2. Changing the status of the natural reserve

The responses regarding interviewee's assessment of the value, expressed in money, of nature reserve Zasavica are very interesting? From a total of 103 persons interviewed only one respondent had no opinion about the value. Problem in evaluation of Zasavica and expressing its value in money was specifically expressed during the survey. Most respondents were not able to state the monetary value of Zasavica. Specifically, 65% of them said that the value "cannot be expressed in money." That means that total value of natural resources in the SNR is outside of the range of the monetary expression. From the remaining 35% of persons who expressed their value in money, around 11.7% valued it in the range of €2-4 million, 11.7% as more than €5 million, and in the range of € 0.5-1.5 million a total of 10.7% of respondents (Figure 3).

Unlike the approach of a neutral evaluation of nature reserves worth, the question at what price they would be willing to buy the reserve, most respondents gave the monetary value. Some 96 respondents gave numeral answer while 7 had not. The dominant response of the respondents was that they would pay over € 5 million (Figure 4).

As seen of Figure 4 to the maximum value Zasavica of over 5 million was attributed by 39.8% of the respondents. Around 20.4% of respondents

Figure 3. Estimated value of "Zasavica"

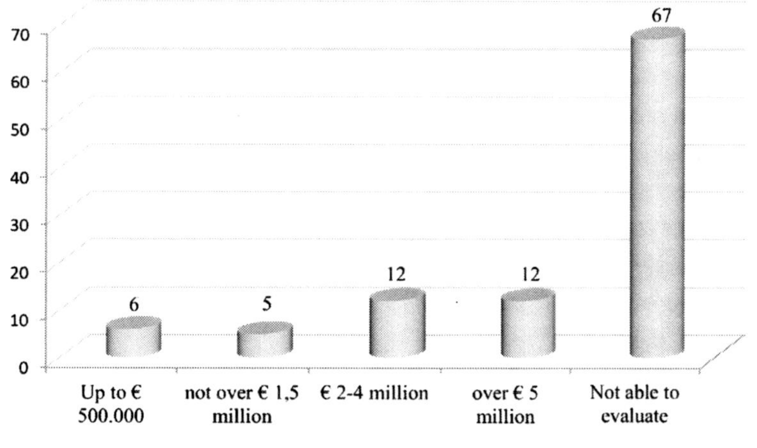

Figure 4. Price that would pay for "Zasavica"

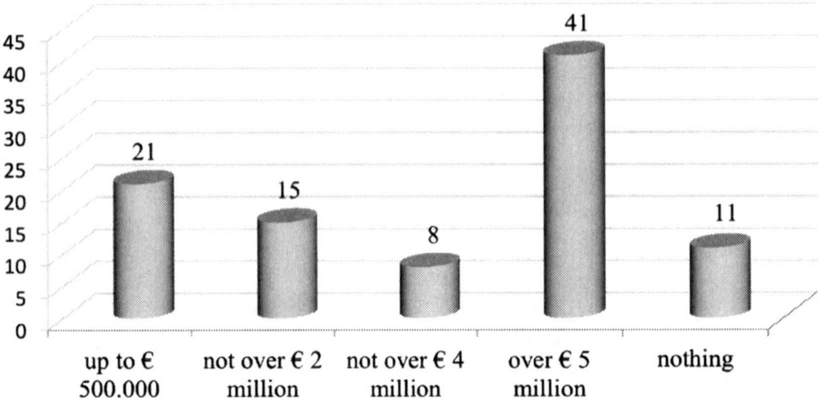

gave an answer that the value is lower than € 500.000, as long as 14.6% gave it the value in the range of € 2-4 million, while 7.8% would be willing to pay between € 4-5 million. There is a relatively high share of respondents who expressed a view that they would not be willing to pay anything for Zasavica with 10.7% of respondents.

To the hypothetical question to define their behavior in the event that they became owners of a special nature reserve, and would they allow the use of specially protected reserves as a common good, the answer was given 102 persons The views of respondents are greatly affected by introduction of the assumption of changed ownership over the natural reserve. In the event that they become owners of Zasavica, majority of 77.5% of respondents would close their private park to visitors. Only 22.5% of respondents, after transforming the nature reserve into private property would leave it open for the visitors (Figure 5).

When asked about behavior in the event that they become owners of the reserve, most respondents argued that they would have not drain the swamps and stub the forests, what would have paved the way for change of the ecosystem of the reserve. The absolute majority of the respondents said that after they became owners of nature reserves they would have not carried out modifications in the allocation of the reserve, which expressed in figures means that 97.1% of them would retain the status as a natural resource Zasavica. Only 3.9% of respondents would change the status of the nature reserve, and so it would cut down forests, drain marshes and plowed into arable land (see Figure 6).

Most respondents expressed their opinion regarding the willingness to allocate even a modest, very little funding to support the survival of nature reserves. This question tests the "willingness to pay" for the maintenance of natural resources without a change of ownership. The obtained answers can be classified into two major groups, the ones who gave their answer in terms of money, and those who believe that Zasavica should support itself from its own income. In the analysis of these attitudes should be taken into account the state of purchasing power and income of the respondents, and that fact that in the region

Figure 5. Private property and visitors

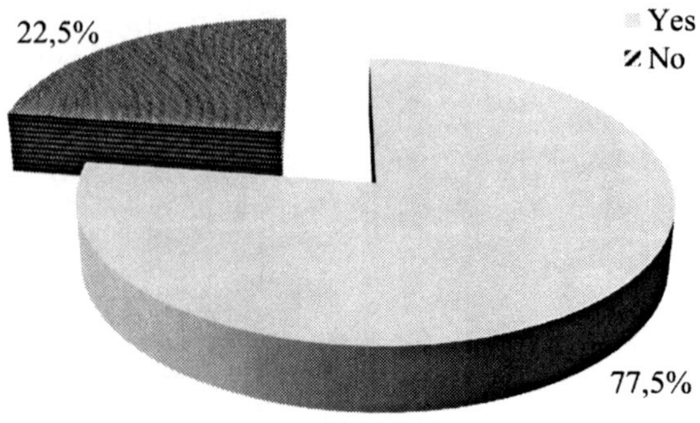

Figure 6. Willingness to financially support development of nature reserve

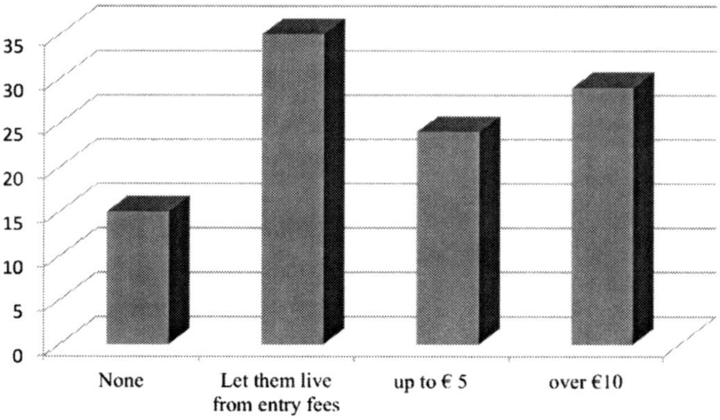

lives population with low incomes. If we set aside the respondents who would let Zasavica support itself, the attitudes that they are not willing donate any funds have 48.6% of respondents. The opinion that they are willing give some money to help Zasavica has 51.5% of respondents, out of which 23.3% would give 50€ annually and 28.2% of respondents would donate over € 100.

RECOMMENDATIONS AND CONCLUSION

Environmental legislation in Serbia is very developed, and it complies with international environmental standards and EU directives. In the period of last ten years there has been developed a system of institutional funds and the Agency for Environmental Protection, and the effects of their activities have initial positive impact on improvement of the protection and use of resources. Special protected areas are organized in the form of national parks, special nature reserves, strict nature reserves, natural monuments, and protected habitats, landscapes of exceptional characteristics and cultural landscapes of outstanding features.

Introduction of environmental accounting in operational terms, both the micro and macro level is still in the initial development phase. A particular problem which requires first to be conceptually and then empirically resolved is finding ways of integrating the new green accounting systems into general accounting, recording and reporting.

Therefore, we recommend intensive education of the general public about basic principles of ecology and also taking necessary measures to strengthen the capacity of institutions, state and local governments as well as strengthening the capacity of businesses, all in order to create conditions for greater efficiency in achieving the objectives of environmental policy and sustainable development. In this regard, we propose measures and activities related to:

- Redefining and redesigning the system of economic instruments (including broadening eco-funds at all levels).
- Quantifying the need for financial resources.
- Augmentation, increase of the capacity and more efficient operation of local environmental funds.
- Introduction of systematic monitoring of the implementation of economic instruments in the Statistics office of Serbia.
- Introduction of economic incentives (exemptions and tax incentives) to all entities that invest in pollution reduction and the introduction of cleaner technologies (investment tax credit).

- Improvement of the current practice in using eco funds with the obligation to repay them, which means to offer favorable loans, especially by providing bank guarantees for loans from domestic and foreign commercial banks.
- Tightening of the current system of fines for civil offenses, economic offenses and criminal offenses.

Finally the results of empirical research conducted in a special Zasavica confirm the willingness of people to conserve natural reserve, without changing its purpose, but the unwillingness of the personal investment to support its development. Such results reflect general public attitude on environmental issues which appeals for comprehensive and intensive training in order to change some negative attitudes which may be stated as: "I understand the problem of global destruction of natural resources, I am aware of the problems of sustainable growth, and I support any solution that does not cost me personally. In this case I have to pay for the sustainability of ecosystems and natural resources - let the next generation worry about the ecosystem, because now I live in a time of global economic crisis and ecology in this context is a secondary matter."

ACKNOWLEDGMENT

This chapter is written as a part of research projects numbered III47009 (European integrations and social and economic changes in Serbian economy on the way to the EU) and OI179015 (Challenges and prospects of structural changes in Serbia: Strategic directions for economic development and harmonization with EU requirements), financed by the Ministry of Science and Technological Development of the Republic of Serbia.

REFERENCES

Council of Europe. (1982). *Culutre, heritage, diversity.* Retrieved 07 07 2012, from http://www.coe.int/t/dg4/cultureheritage/nature/bern/default_en.asp

Council of Europe. (1993). *Council of Europe treaty office.* Retrieved 07 07, 2012, from http://conventions.coe.int/Treaty/en/Treaties/Html/150.htm

Council of Europe. (1998). *Council of Europe treaty office.* Retrieved 07 07, 2012, from http://conventions.coe.int/Treaty/Commun/QueVoulez-Vous.asp?NT=172&CM=1&CL=ENG

Drašković, B. (Ed.). (1998). *Ekonomija prirodnog kapitala, vrednovanja i zaštuta prirodnih resursa.* Belgrade: Institute of Economic Sciences.

European Council. (2004). *Culture, heritage, diversity.* Retrieved 07 07, 2012, from http://www.coe.int/t/dg4/cultureheritage/heritage/landscape/default_en.asp

Hoteling, H. (1931). The economics of exhaustible resources. *The Journal of Political Economy, 39*, 137–175. doi:10.1086/254195

Pešić, R. V. (2008). Ekonomska aktivnost i stanje životne sredine. In *Strategija održivog razvoja Srbije.* Belgrade: Serbian Government.

UNECE. (1998). *UNECE web site.* Retrieved 06 07, 2012, from http://www.unece.org/env/pp/welcome.html

UNEP. (1992). Retrieved 06 08, 2012, from http://www.cbd.int/convention/

United Nations. (1972). *United Nations environment programme.* Retrieved 05 06, 2012, from http://www.unep.org/Documents.Multilingual/Default.asp?documentid=97&articleid=1503

United Nations. (1992). *United Nations environment programme*. Retrieved 05 06, 2012, from http://www.unep.org/Documents.Multilingual/Default.asp?documentid=78&articleid=1163

United Nations. (1993). Handbook of national accounting: Integrated environmental and economic. New York: United Nations Department of Economic ans Social Information and Policy Analysis, Statistical Division.

US Congress. (1872). *An Act to set apart a certain tract of land lying near the head-waters of the Yellowstone River as a public park. U.S. Statutes at Large*. Boston: Little, Brown and Company.

van den Bergh, J. (1996). *Ecological economics and sustainable development: Theory, methods, and applications*. Lyme, CT: Edwards Elgrad.

World Commission on Envirnoment and Development. (1987). *Our common future*. New York: Oxford University Press.

ADDITIONAL READING

Burningham, D., & Davies, J. (1995). *Green economics*. Oxford: Halley Court, Jordan Hill.

Goodstein, E. S. (2011). *Economics and the environment*. Hoboken, NJ: John Wiley & Sons, Inc.

Hussen, A. (2005). *Principles of environmental economics*. New York: Routledge.

Lange, G.-M. (2003). *Policy applications of environmental accounting*. Washington, DC: The World Bank Environmental Department.

Milne, M. J., & Gray, R. (2007). Future prospects for corporate sustainability reporting. In J. Unerman, J. Bebbington, & B. O'Dwyer (Eds.), *Sustainability accounting and accountability* (pp. 184–207). London: Routledge.

Ponting, C. (1991). *A new green history of word: The environment and the collapse of great civilizations*. New York: Penguin Books.

The Law on Air Protection of Serbia. (2009). Retrieved from http://80.93.243.155/en/upload-centar/dokumenti/zakoni-i-nacrti-zakona/zakoni/english/Law%20on%20air%20protection.pdf

The Law on Environmental Protection of Serbia. (2009). Retrieved from http://80.93.243.155/en/upload-centar/dokumenti/zakoni-i-nacrti-zakona/zakoni/english/Law%20on%20EIA.pdf

The Law on Packaging and Packaging Waste of Serbia. (2010). Retrieved from http://80.93.243.155/en/download-687/upload-centar/dokumenti/zakoni-i-nacrti-zakona/zakoni/english/Law%20on%20packaging%20and%20packaging%20waste.pdf

The Law on Strategic Environmental Impact Assessment of Serbia. (2010). Retrieved from http://80.93.243.155/en/upload-centar/dokumenti/zakoni-i-nacrti-zakona/zakoni/english/Law%20on%20SEIA.pdf

The Law on the Protection of Environmental Noise of Serbia. (2010). Retrieved from http://80.93.243.155/en/upload-centar/dokumenti/zakoni-i-nacrti-zakona/zakoni/english/Law%20on%20environmental%20noise%20protection%20cons%20ver.pdf

The Law on Waste Management of Serbia. (2010). Retrieved from http://80.93.243.155/en/download-1116/upload-centar/dokumenti/zakoni-i-nacrti-zakona/zakoni/english/Law%20on%20waste%20management.pdf

Unerman, J., Bebbington, J., & O'Dwyer, B. (Eds.). (2007). *Sustainability accounting and accountability*. London: Routledge.

KEY TERMS AND DEFINITIONS

Ecological Accounting: Financial aspects of maintenance, protection and utilisation of nature reserves.

Ecological and Economical Monitoring: The components of ecological accounting.

Environment Protection Costs: Cost associated with protection and sustaining of natural resources.

Environmental Assets: Objects that provide sink, service and resource functions, as well as usable and non-usable benefits.

Natural Resources: The substances found in nature, which are considered valuable in their relatively unmodified natural form.

Protected Areas: Legally protected nature reserves.

Special Nature Reserves: Land areas with special regimes for protection and utilization of natural resources at their disposal.

Value of Nature Reserves: Differentiated factors specific for the areas designated as nature reserves.

This chapter was previously published in Sustainable Technologies, Policies, and Constraints in the Green Economy edited by Dorel Dusmanescu, Andrei Jean-Vasile, Turek Rahoveanu Adrian, and Jonel Subic, pages 106-122, copyright 2013 by Information Science Reference (an imprint of IGI Global).

Chapter 92
Promoting Rainwater Harvesting (RWH) in Small Island Developing States (SIDS):
A Case in the Grenadines

Everson J. Peters
University of the West Indies, Trinidad and Tobago

ABSTRACT

A sustainable development path is critical to most small island developing states, which are often environmentally fragile and have little natural resources. For decades, the small Grenadine islands in the Eastern Caribbean have depended on rainwater harvesting (RWH) for water supply. The RWH systems have improved from the use of household drums and communal cisterns to individual cisterns such that water per capita use at some households is at a similar level to that of developed countries. This improvement was brought about through self-financed projects and the use of local construction know-how. A small project to promote RWH in Green St. Vincent was reviewed to procure an appreciation of the diffusion process. The RWH experiences in these islands as they relate to quality, design, and cost can be relevant to the regional efforts of promoting RWH. The chapter proposes a framework for enhancing the promotion of RWH. The chapter recognizes the importance of RWH in disaster management in the islands.

ORGANIZATIONAL BACKGROUND

Freshwater, which is fundamental to life and health, is a diminishing natural resource. Currently, access to freshwater in many parts of the world is limited. The Dublin Statement on Water and Sustainable Development recognized that the scarcity and misuse of fresh water pose a serious and growing threat to sustainable development and to the protection of the environment. Sustainable development, which was first conceptualized as a solution to environmental problems, has been defined in many ways. The most notable defini-

DOI: 10.4018/978-1-4666-4852-4.ch092

tion is from *Our Common Future*, also known as the Brundtland Report (WECD, 1987). Here it is defined as *development that meets the needs of the present without compromising the ability of future generations to meet their own need.* This requires that the world be seen as a system—a system that connects space and time.

Given that most societies desire to achieve economic development to secure higher standards of living, as well as seek protection and enhancement of their environment, now and for their children. Sustainable development tries to reconcile these two objectives (HMSO, 1994). To ensure that development is sustainable, its costs should not be transferred to future generations or, at least, an attempt should be made to compensate for such costs (Pearce, 1993). Since the Brundtland report and the Rio Summit, sustainable development has transitioned from being an interesting yet at times contested idea to a widely endorsed concept and an almost universally adopted guiding principle (Drexhage and Murphy, 2010).

For many years the World Health Organization (WHO) has been monitoring water supply and sanitation worldwide. At the 1992 UN Conference on Environment and Development, 178 nations agreed to Agenda 21, a set of social and economic development goals resulting from the Rio Declaration on the Environment and Development (UNCED, 1992). In 2000, the world community adopted the Millennium Development Goals (MDGs), which provided a framework and an impetus for the world to accelerate the pace of development and improve human welfare. As water and sanitation are important issues to human welfare, the MDGs gave special consideration to these issues in Goal 7 and set targets aimed at halving the proportion of people who are unable to reach or to afford safe drinking water by 2015. Improving the development and management of freshwater resources is critical to meeting the MDGs. Since Agenda 21, there have been a number of international and regional conferences and seminars aimed at making decisions and seeking financial support for meeting the MDGs. Nonetheless, achieving the MDGs remains a challenge for many countries in the face of small state budgets and limited donor support. The Caribbean and Latin America, however, are on track to meeting the MDGs (WHO and UNICEF, 2004). In preparation for the 20th anniversary of the formulation of Agenda 21 or the Rio+20 UN Conference on Sustainable Development (20-22 June 2012), world leaders intend to secure renewed political commitment for sustainable development, assess current progress and the remaining gaps in the implementation of the outcomes of the major summits on sustainable development, and address new and emerging challenges.

In Latin America and the Caribbean (LAC), over 75 million people have no access to safe drinking water (IADB, 2005). In the Caribbean Community (CARICOM), access to freshwater is better than that for the rest of the LAC. However, population growth, rural to urban migration and environmental issues such as climate change and sea level rise are expected to negatively impact water quality in the future. Therefore, the overall sustainability of access to safe freshwater will be a future challenge for planners and policy makers in their quest to promote sustainable development. In small island developing states (SIDS), policy makers and managers in the water sector are being challenged to meet the MDGs given the increasing trends of extreme events, limited freshwater availability and demands for competitive water uses. This challenge is most evident in the very small Grenadine islands of the Eastern Caribbean where surface water is non-existent and poor quality groundwater is limited. In these islands, rainwater harvesting (RWH) is the main source of potable and non-potable water supply.

In many small islands, a total annual rainfall of 1000 mm (40 inches) is generally sufficient for RWH. However, the seasonality of the rain can raise concerns about the reliability of RWH, particularly during the dry season. In some islands, RWH has not generally been given the attention it deserves. It has often been looked upon as a technology of last resort, to be used only when no other alternatives exist. In the Eastern Caribbean, due to chronic water scarcity, some islands have developed the technology of RWH to ensure water security. Some others are now in the early stages of implementing RWH projects for water augmentation or to be used as a standalone source due to increased pressures on their traditional water resources. Peters (2006) reported that in the case of Carriacou and Petite Martinique there are sufficient rainfall amounts and roof areas on the islands to theoretically satisfy a per capita consumption of over 120 l/d/person. Despite a period of relatively little attention in other parts of the world, RWH has regained center stage as a philosophy as well as a set of technologies for water management and supply (Smet and Moriarty, 2001). This may be considered a viable long-term strategy to address the increasing pressure on freshwater resources in many parts of the world.

UNDP (2009) highlighted the global importance and the numerous positive benefits of RWH. The technology is low cost and highly decentralized thus empowering individuals and communities to manage their water supply and sanitation at the local level. RWH also has the potential to increase food production. In many countries, RWH has been identified as an important component in strategic programmes for climate resilience and only a few investment programmes that are aimed at adaptation to climate change are developed without including RWH. In some developed countries, RWH facilities have been promoted to solve the problems of extreme flooding and drought events and to achieve sustainable water management in urban areas (Mun et al., 2008).

RWH in its broadest sense can be defined as the "collection of runoff for its productive use." Notwithstanding the advantages of RWH, contemporary diffusion/adaption of the technology in relation to installation of systems has not been rapid. Instead most efforts have been limited to public awareness through conferences (like the Bi-annual International Conference on RWH Catchment Systems), local and international seminars, and the development of prototype systems for particular locations and research. In most cases where systems already exisit and new ones are being installed, there are some common issues related to the cost of development, sizing of systems and confidence in the quality of the water, that have been identified as limiting the rate of adoptation of RWH. Inspite of some of the challenges associated with RWH, a pertininent question is to what extent can RWH provide opportunities to enable sustainable development, increase human well-being and ensure environmental protection?

This chapter traces the development of RWH in the Grenadines by using household information on RWH systems from Carriacou and Petite Martinique that was collected through household surveys. For the purposes of this chapter the focus of RWH is on household or domestic supply. The experiences of RWH in the Grenadines, where islanders were virtually single-handedly responsible for the growth of RWH, can be useful in long-term planning and design of RWH programs elsewhere. The chapter further reviews initiatives on RWH promotion and highlights issues and constraints that are considered fundamental to its successful promotion. Moreover, it recommends the use of RWH as a viable tool for coping with future climate change as it relates to household water supply, saltwater intrusion of the coastal dugwells and agriculture in small islands.

BACKGROUND

A Brief History of RWH Rainwater Harvesting Technology

Rainwater harvesting is an old technology that dates back to some 4000 years. The existence of these systems in the Negev Deserts of Israel for more than 2,000 years has been well documented. Roman villas and cities were planned in such a way to take advantage of rainwater for drinking and air-conditioning. Cisterns were fairly common in rural areas in the United States America (USA) until the 1920s. A survey of American State Health Departments revealed that there are an estimated 253,815 cistern systems still in use in the United States (Lye, 1992). Ten years later, Lye (2002) estimated 100,000 residential rainwater harvesting systems are in use in the United States.

In Latin America and the Caribbean, particularly in some small islands where the annual precipitation is between 800 mm (32 inches) and 1500 mm (59 inches), water has been collected from roof tops and stored in clay containers, steel drums, wooden casts, concrete tanks and plastic barrels. In some remote and mountainous areas where rainfall is high but rivers and springs are far away, people are forced to harvest rainwater for domestic uses.

Notwithstanding the long history of rainwater harvesting from roof tops, it is only in the last two decades that concerns for the environment, decreasing availability of water through the conventional sources and increasing cost of producing conventional supplies have increased the interest in the use of RWH. For example, Thailand and Kenya have led their respective continents in demonstrating the enormous potential of this technology. In Thailand, more than 10 million 2 m^3 ferrocement roof catchment tanks for domestic water supply have been constructed since 1985, while in Kenya, the construction of large (10 to 100 m^3) ferrocement tanks have gained popularity in many regions with thousands being built at schools, clinics and private homes (Gould,1999). Other developing countries in Asia such as Bangladesh, India and the Philippines, as well as many semi-arid countries on the African continent, are showing rapid growth in the use of RHW technologies as a way to improve water security. Industrialized countries are also interested in rainwater collection as demonstrated by recent developments in Australia, Hawaii and Singapore. Japan has traditionally used rainwater catchment systems and is currently making adjustments to meet the requirements of urban environments.

For more than three centuries, rooftop catchments and cistern storage have been the basis of domestic water supply on many small islands in the Caribbean. Although the use of rooftop rainwater catchment systems has declined in some countries, it is estimated that more than 500,000 people in the Caribbean islands depend, at least, in part on such supplies. The islands that still use rainwater heavily include Antigua and Barbuda, the Bahamas, the US and British Virgin Islands (USVI and BVI respectively), the Turks and Caicos and the Grenadines (of both St. Vincent and Grenada). RWH is necessary in areas having significant rainfall but are lacking any kind of conventional centralized government supply system, and also in areas where good quality fresh surface water or groundwater is lacking. On islands like Carriacou, rainwater is the only water supply so it is used for everything, from drinking to irrigation. In other parts of the Caribbean, rainwater may be used for gardening or laundry with the public water supply reserved for drinking. In Jamaica, a few hotels use RWH as a supplementary water supply which reduces water utility costs up to 35% (Meade and Gonzalez-Morel, 1999). In the Bahamas, rainwater is also used as a supplementary supply for large apartment buildings, hotels and restaurants.

RWH technology is appropriate for all areas in the Caribbean but caution may be required as in some cases water quality can be affected by atmospheric pollutants and material deposited on the catchment surface. In addition, concerns about consistent availability can also arise very frequently in the Caribbean. As much as 80% of the rain falls during June to December (rainy season) where as little or no precipitation is observed during the remaining months (dry season). In some islands, the annual and regional distribution of rainfall can differ significantly. Serious supply problems occur when the components of the rooftop rainwater catchment systems (RRCS) are not sized appropriately. Therefore, there is a need for criteria on the proper sizing for the components of a new RRCS or upgrading existing systems. There is also a need for suggestions on how to manage these systems.

Rainwater Harvesting Technology (RWH)

RWH is a technology used for collecting and storing rainwater for multiple uses. Roofs, ground surfaces as well as intermittent or ephemeral watercourses can be utilized to collect and store water for domestic and other purposes. The basic set-up used in Grenadine (Figure 1) is composed of a water-collection system, a storage cistern and a water distribution system. It mainly needs:

- Rain, and plenty of it preferably;
- A catchment area;
- Conveyance (gutters and pipes);
- Storage (jars, tank; underground cisterns);
- Treatment (optional); and
- Pump or pressure tank for distribution (where possible a gravity system can be used).

In many parts of the world, rainwater harvesting provides the only source of water for household use and for institutions such as schools and community centers. Household systems are decentralized and every household is responsible for the management of its system. For communal systems, which serve entire communities, local government or a community management entity is responsible for the management of the system. In the Caribbean, water for the communal cisterns is collected from the rooftops of public buildings such as schools, churches and community centers.

METHODOLOGY

In order to develop a model to promote a technology like RWH, it is useful to consider the lessons learned from successful implementation. The Grenadines in the Eastern Caribbean (Figure 2) represent an organic laboratory where RWH technology has grown and islanders have taken the responsibility for its development. The material for this chapter comes from the study of RWH in the islands for the past 8 years. The history of RWH is reviewed in order to distill lessons,

Figure 1. The basic setup of rooftop RWH system used in the Grenadines (Source: CEHI, 2009)

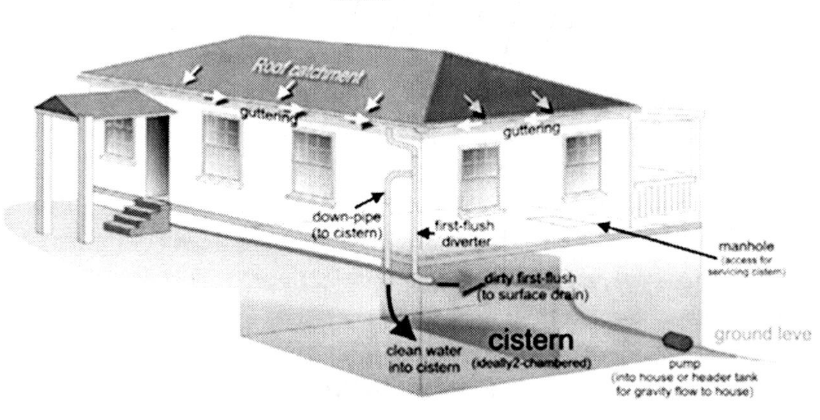

Figure 2. Map of the Grenadines in the Eastern Caribbean

which, if embraced, can facilitate the promotion of the technology in an era when modern society is preoccupied with sustainable development and green economy. Household surveys were undertaken mainly in Carriacou, Petite Martinique and Union Island complimented by national census data from Grenada and St. Vincent, to trace the rise in RWH in these islands. Survey questions were aimed at obtaining information on household composition, storage facilities type, construction material, size, age, treatment of the facilities and general household water use practices. The water quality tests of cistern water were carried out using H_2S field tests supplemented by detailed laboratory tests. At least 10% of all the households, randomly selected, were used in the surveys and water quality tests. To evaluate the potential of RWH to meet the current and future needs of islanders, simulations of the performance of cisterns, under different demands and roof area, were carried out using 30 years of rainfall data for the locations.

To better understand how the initiatives of RWH might perform under current conditions a project was undertaken at a representative site in a new low-income housing development at Greenhill, St. Vincent. In the study, level of awareness of RWH; attitudes towards features of the domestic water supply; collection and use of RWH in the community; willingness-to-invest in RWH systems; and inhibitors to the uptake of RWH were examined. This study particularly informs the later section of this chapter on the constraints to RWH promotion.

The Development of Rainwater Harvesting in the Grenadines

Background

In the Southern Caribbean, the Grenadine Islands form a chain of small islands, which represent the exposed summit of peaks that are situated on a single narrow bank of submerged volcanic mountains. These islands came into existence in the late Oligocene period, sank or eroded away during the Pliocene period and were completely submerged during the Pleistocene period. The islands are mostly volcanic in origin, of steep rugged topography with a main mountain chain running almost north to south. Carriacou is the most southerly of the Grenadines, situated 42km north of Grenada, at latitude 12.5 degrees. It has an area of 34km^2 (13m^2) and is the largest of the chain of islands between the Grenada mainland and St. Vincent. Petite Martinique, which is the smallest of the Grenada Grenadines and lies about 4km (2.5 miles) from Carriacou with an area of 2.7km^2. The islands experience relatively low rainfall of about 1000mm annually. Moreover, greater than eighty percent of the year is mostly sunny with mostly drier conditions.

Residents on the islands have always depended on RWH. In the late 1700s, the first cisterns were built for use by Europeans in the great houses. For example, the Craigston Estate, a cistern (Figure 3(a)) with capacity of 190m^3 (50,000 gallons), was built in 1785 out of fieldstone rubble with an inner yellow brick liner that would later be replaced with a plastered liner (Slade, 1984). This cistern is still in use. After the abolition of slavery, the freed slaves built a few smaller cisterns from stone and white lime mortar with capacities of 10m^3 to 15m^3. Today, there are three main types of rainwater storage systems in the Grenadines. These include drums with capacities of about 0.170m^3, plastic tanks with capacities of 0.75-3m^3 and concrete cisterns (underground and above ground) with an average capacity of 30m^3 (Figure 3(b)). Modern engineering techniques are routinely used in the construction of reinforced concrete cisterns.

Within the past 100 years, there has been a number of construction methods used for RWH cisterns in the Grenadines. The earliest types of cisterns were built from locally available stone and white lime mortar. These cisterns have a life design expectancy of over 100 years; however, there are many over150 years old still in operation. The cisterns of the 1950s and 1960s were made from concrete and boulders. The next generation

Figure 3. (a) First cistern built in 1785 at Craigston, Carriacou and (b) RWH system with pumping and gravity feed internal plumbing

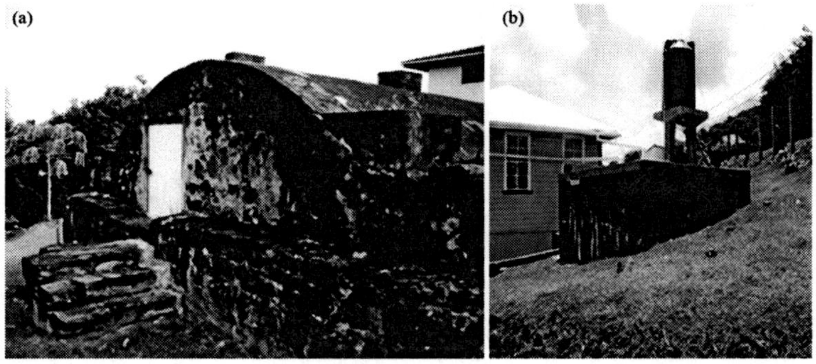

of cisterns was constructed from hand mixed poured concrete without steel bar reinforcements.

Unlike the US Virgin Islands and some other territories, installation of RWH systems is not required by law in the Grenadines. However, the increasing dependence on tourism of these islands would soon force, at least, public facilities to establish quantity and quality standards for cisterns. With annual rainfall averaging about 1000mm per year and the daily domestic per capita consumption estimated at a maximum of 136 litres (36 gallons) for local population plus livestock, the demand is close to the estimated domestic daily per capita for developed countries (130 litres) (Environmental Agency, 2008).

Growth of Cistern Use

RWH systems depend on the rainfall frequency, intensity, and the size of the catchment area. Rainfall needs to be frequent with high intensities and the catchment areas need to be large to make the system feasible and allow the cisterns to be filled. During the 1980's and 1990's, a survey of 200 households in Carriacou and Petite Martinique showed that rapid socioeconomic development in the Grenadines has provided opportunities for the construction of new cisterns. There has been a continuous growth in the cistern stock on the islands with a peak in the 1980s as shown in Figure 4(a). Over the past half a century, the average size of cisterns has also steadily increased (see Figure 4(b)). Furthermore, since the 1980s, the available cistern storage has almost tripled, primarily due to the large number of returning nationals who have retired from North America and the United Kingdom. The average construction sizes of these cisterns are 68m^3 (18,000 gallons, US). Currently, 91% of the 2300 households in Carriacou and Petite Martinique have concrete cisterns. In these households, where cisterns provide the main storage, plastic tanks are used to supplement these cisterns or are used as overhead tanks to facilitate gravity flow in the internal plumbing systems. About 6% of the residents depend on polyvinyl chloride (PVC) and high-density polyethylene (HDPE) tanks with minimum capacity of 1.5m^3 (400 gallons), with the remainder using drums and other smaller containers to harvest rain.

Development Costs

In the promotion of RWH facilities, the most critical issue is economic efficiency. The capital cost of RWH systems is highly dependent on several factors, which include:

- Volume of rainwater to be collected
- Type of surface for rainwater collection,
- Existing structure and fittings for collection and flow,

Figure 4. 40 Year trend in Grenadine (a) cistern stock and (b) average cistern size constructed in Carriacou

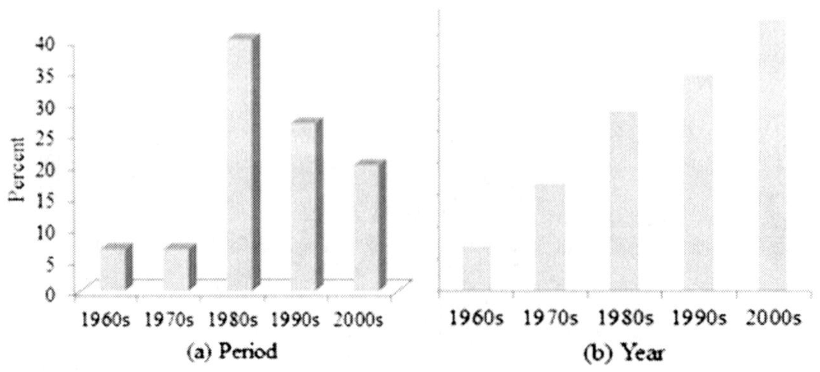

- Type and capacity of the harvested water storage device,
- Method and system of rainwater retrieval for use from storage device.

In the Grenadines, the capital costs are as high as US$67.00 per m^3 (US$304.50/1000 gallons) of water storage for reinforced concrete cisterns. However, the cost of water delivered to the households during the life-time of the cistern is only US$1.07 per m^3 (US$4.86/1000 gallons). This is only slightly more expensive than that for water produced from surface water sources on the Grenada mainland where it is approximately US$0.85 per m^3 (US$3.90/1000 gallons) (Peters, 2006). Yet, the costs can be significantly reduced in locations where RWH systems are less sophisticated and local materials are used. For example, the cost of harvested rainwater in Asia, which varies from US$0.17 to US$0.0.37 per m^3 (US$0.77 to US$1.68 per 1000 gallons) of water storage, is relatively low compared to many countries in Africa (Lee and Visshher, 1990).

Communal RWH Storages

Communal cisterns or RWH storages are served by their concrete catchments adjacent to the cistern or by catchments from public buildings such as schools. Table 1 shows some of the characteristics of communal cisterns in the Grenadines. Most of these cisterns were built before the 1960s. In Carriacou, the large network of communal cisterns was used routinely for everyday purposes. In Union Island, Canaoun and Bequia, communal storage facilities are utilized mainly during the dry seasons when periods of extreme water shortage are exhibited. In 2011, a solar diesel system was installed in Bequia. This facility may reduce the importance of the communal facilities.

Up to the early 1970s, the communal cistern was a place of village socialization. Water was distributed 3 or 4 times per week during early mornings (6:30 am to 8:00 am) or on afternoons (3:30 pm to 6:00 pm). Each household, irrespective of size, was given about 96 litres (25 gallons) per week in the rainy season and 40 to 60 litres (10 to 15 gallons) in the dry season. Based on distances from the communal cisterns, it was possible for some households to have access to two or three communal cisterns by allowing different members of the household to fetch water from different communal cisterns simultaneously. The village women and children would congregate as they waited for their buckets to be filled. Typical filling rates would be about 11 litres (3 gallons) per minute.

The Decline in the Use of Communal Cisterns

During the 1970s and 1980s, the use of communal cisterns declined as the number of residential cisterns increased. As households became more self-sufficient in their water needs the number of households requiring the use of these communal cisterns, particularly during the dry rainy season, declined. As a result, there was a fall in the num-

Table 1. Communal cisterns in the Grenadines

Island	Population	No. of public Catchments	% of households with private cisterns	Communal storage in m^3 (million gallons)	Private storage in m^3 (million gallons)
Bequia	5,000	1	80	818 *(0.150)*	NA
Union Island	1935	3	75	509 *(0.112)*	112980 *(24.9)*
Canaoun	~1100	3	80	413 *(0.091)*	NA
Carriacou	7,000	33	93	5250 *(1.155)*	205640 *(45.2)*

ber of households with a vested interest in the operations of such facilities. As a consequence, full community involvement in the operation and maintenance of these facilities was lacking. The continuing need for communal cisterns was evident during the 2009/2010 drought, when water from these communal cisterns was needed but was unavailable or unusable due to poor water quality. Notwithstanding the apparent recent success in the establishment of household rainwater supply, the need for maintaining communal cisterns is important, at least, for emergency purposes.

Water Quality Issues

In these islands when average and above average rainfall is exhibited there is ample storage and quantities available for domestic use. In terms of water quality, there is a public presumption that harvested rainwater is safe and that the risk of illness arising from its consumption is low. This presumption is possibly supported by the observation that there are few documented incidences of serious illnesses linked to the drinking of rainwater stored in cisterns. In many cases these reports are from, visitors to the islands suffering from diarrhoea as a result of consuming un-boiled rainwater that was stored in cisterns.

The difference in the level of risk for residents and visitors may be explained by an acquired immunity by locals to common pathogens from long-term use and exposure to untreated cistern water. This apparent resilience of residents in the use of rainwater stored in cisterns is supported elsewhere. In South Australia, about 42% of residents use rainwater in preference to mains water without any apparent effect on the incidences of gastrointestinal illnesses (Heyworth, 2001). Other studies in Thailand and Australia showed that among small children there was no greater risk for gastroenteritis in drinking rainwater instead of treated mains water (Heyworth, 2004; Wirojanagud, et al., 1989). Notwithstanding the confidence in the quality, concerns have been raised elsewhere about the safety of harvested rainwater for drinking (Fujioka and Chinn, 1987, Krishna, 1989) that should not be ignored.

To assess the quality of water stored in cisterns in the Grenadines, random testing of water stored in 250 cisterns (15% of all cisterns) was carried out in 2008. The assessment included the measurement of the presence of faecal coliform using the H_2S test following methods described by Manja et al. (1982) and Venkobachar et al. (1994). The H_2S strip test is based on the readily observable formation of an iron sulphide precipitate in water on a paper strip that has formed by the reaction of H_2S with iron. The intention of the test is to detect bacteria associated with faecal contamination. This is evidenced by the activity of these microorganisms in reducing organic sulphur to H_2S gas, which ultimately reacts with iron to form a black iron sulphide precipitate (Allen and Geldreich, 1975). Sobsey and Pfaender (2002) reviewed the scientific basis, validity, available data and other evidence for and against the H_2S test. While some studies showed that the method is sometimes superior and gives results comparable to the tests for traditional bacterial indicators of faecal contamination (Sobsey and Pfaender, 2002) there can be some shortcomings. For example, there is a concern that the test may detect bacteria not associated with faecal contamination and its attendant pathogens.

The results of the field sampling using H_2S showed that between 80% and 90% of the cistern water was of good quality or is absent of faecal coliforms. This high quality water is consistent with households that use basic chlorination to treat the water stored in the cisterns. More than 70% of the samples of the untreated cistern water in the islands had faecal coliform counts of less than 10 CFU/mL. This suggests that in the majority of cases that water in storage is safe while contamination may be due to poor household practices in collecting and storing rainwater. It was found that about one third of the households use chlorine tablets or liquid bleach without specified dosage while most households installed some protection against mosquitoes and wild life.

The physical quality of the water stored in the cisterns was found to be free from odour and colour. The maximum total suspended solids (TSS) was less than 3 mg/L at the end of the dry season, which may be due to the low water level and churning of the water in the tank. The low level of TSS in stored rainwater in the islands helps to reinforce the perception of good quality water. This perception may lead to reduce care taken by users to ensure safe potable water.

The concerns of water safety in the islands, particularly, as it relates to visitors and the tourism sector, requires additional scrutiny be placed on the quality of drinking water to ensure that the World Health Organization (WHO) guidelines are met. There have been recent efforts to monitor the water at selected hotels. However, there are neither regulations nor set guidelines or standards for the storage of rainwater for drinking in the islands.

To overcome the previously mentioned shortcomings, some regulations to promote the monitoring of stored rainwater and encourage timely implementation of remedial action need be established. This should include proper assessments of the risks of contamination to be followed up with actions to mitigate any unacceptable risks. Such practices already exist elsewhere such as in New Zealand (NZ Ministry of Health, 2005), where stored rainwater supplies of individuals that are serving more than 1500 person-days (e.g. more than 25 persons for 60 days) must have some form of continuous monitoring. This would capture those small bed and breakfast type establishments, which are typical in the Grenadines. Individual supplies serving households may be exempt from continuous monitoring but this can be encouraged, where possible, to maintain the system in such a way that the risks for contamination is minimised. Ultimately, the best approach would be to implement a Water Safety Plan System (WSPS) as described by the WHO guidelines (WHO, 2004). This WSPS should be a long-term strategy for the water and health authorities in small islands where RWH is the main source of domestic supply.

Sizing Issues

In the islands the main factor that seems to determine the size of cistern to be built is the wealth of the household. In the case of low income households, where the household size tends to be above the average of 3.8 persons, the size of the cistern or tank is adequate during the rainy season but dries up during the dry season. On the other hand, in the case of upper income households, cisterns tend to overflow for most of the year as consumption is much smaller than the supply, indicating that these cisterns are too large for such households. There are many ways to estimate an appropriate size of a cistern. For example, the following situations and worked example best describe the possible approaches taken to estimate the appropriate cistern size.

Situation A: RWH as a Secondary Source

In areas where RWH is not the primary source of potable water, regulations provide for construction of cisterns depending on the size of the roof, the intended use of the structure and the number of floors in the building. In Barbados, for example, new residences built with floor areas of 140-185m^2 must have a water tank of 13.6m^3 capacity and those over 185m^2 require a tank of 27m^3.

Situation B: Simple Approach

A simple approach suggested by the University of Warwick is based on the consumption rates of the inhabitants of a building without taking the rainfall amount or roof size into consideration. This method is applicable in cases where sufficient rainfall and catchment area are available. It is a rough guide to estimate the tank size as illustrated by the following example:

A Worked Example

The storage requirements (T) can be obtained by using the formula for annual consumption (litres) multiplied by the average number of dry days within a year. For example, using Equations (1) and (2) and the conditions outlined below the proper cistern can be obtained:

$$\textit{Annual consumption}\text{ (litres)} = C \times n \times 365 \qquad (1)$$

where C is the consumption per person per household (litres) and n is the number of persons per household.

$$\textit{Storage Requirement}\text{ (T)} = \textit{Annual Consumption} \times (DP/365 \text{ days}) \qquad (2)$$

where DP is the average extended period of observed days with no rainfall or dry days (days)

Consumption per person per day, C = 100 litres
Number of people per household, n = 5
Longest average dry period = 30 days

Thus;

Annual consumption = 100 x 5 x 365 = 182500 *litres*

Storage requirement (T) = 182500 litres x 30days/365days = 15000 *litres*

The storage needed for this dwelling should be able to hold 15,000 litres of water. Another simple guide to estimating the size is to use the following formula:

Roof Area x Annual Rainfall = Maximum Available for Capture (3)

Most Days Without Rain x Average Daily Water Use = Storage Vol. Required on Day One of Dry Spell

If for example, the maximum days without rain was 80 days and the average daily usage was 500L, the volume on hand on day one of 80 days would have to be 40,000L. It is very difficult to predict the required volume, because often a dry period can start when the tank is not full and there are days when the volume of water collected would be less than 500L. However, this method is a guideline that can have a safety factor added to it to increase requirement. This safety factor can also be considered when adding storage for change in water use over time.

Situation C: Inadequacy of Simple Methods

The inadequacy of the simple methods can be addressed by using a methodology where the variation in rainfall, household size, and consumption patterns during the year (including dry season conservation practices) are considered. Peters (2004) considered such an approach based on the conditions of the Grenadines by simulating the performance of a cistern under a number of conditions. In the simulation, the cistern was allowed to run dry once in 15 years. A monograph was produced for the estimation of appropriate cistern size.

Figure 5 is a monograph of the cistern design criteria without considerations for climate change. It indicates that for a known contributing roof area, selected on the x-axis for example 220m^2 and with a permanent household population of 6, the size of the cistern should be 14m^3 (3000 gallons) in capacity. If the house contributing roof area was half the size of 110m^2 the required tank volume would be about 41m^3 (9000 gallons).

Constraints to RWH Growth

Although the adoption of household rainwater harvesting is currently booming in some parts of the world, what actually motivates this adoption is not well understood (White, 2007). Greater emphasis has been placed on the promotion of

Figure 5. Cistern selection criteria (Tank size in m³ (gallons)), source Peters (2003)

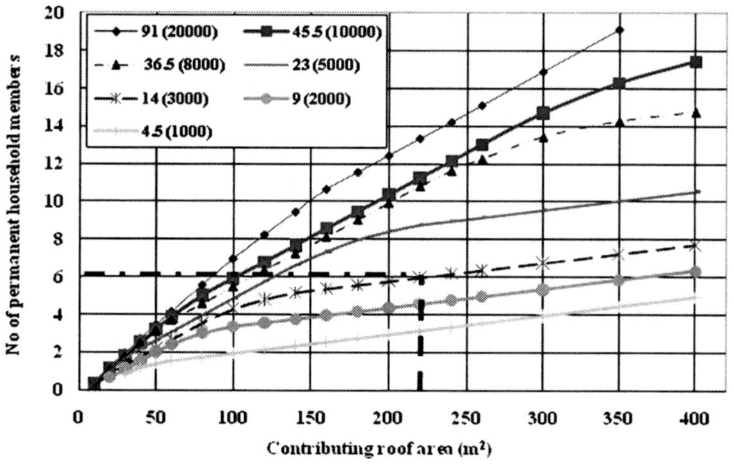

RWH for agriculture and hence there is a better understanding in this area. He et.al. (2007) found that non-consideration of household socioeconomic factors can lead to inappropriate targeting of RWH projects.

This section examines the opportunities and challenges of RWH for communities in small islands. It reviews the research findings on the promotion of RWH in small residential communities on small islands and highlights the work on the Greenhill village of St. Vincent (Peters and Mandeville, 2011).

Research on the constraints to RWH promotion in the Caribbean is limited. Peters (2006) explored the potential of RWH in the Grenadines. It was accepted that RWH was the most viable freshwater source and the main concern was related to availability during severe dry seasons. There was a preference of residents for rainwater as the first choice and hence the willingness of residents to invest in the construction of storage facilities. Hoyte (2009) found that in Barbados, there is hesitation towards the adoption RWH for domestic water supply despite the efforts of regional and international organizations. He reported that this was mainly due to improvement in the standards of living, adoption of the western culture and an improved water distribution system. He also reported that in Barbados there was a general perception that RWH was costly and therefore the willingness to adopt the technology tended to be motivated by the availability of incentives provided by the government.

The work of Cardona (2006) on assessing the feasibility of domestic RWH in Malta showed constraint patterns similar to those in the Caribbean. Cardona pointed out that some constraining factors include low price of water, low political profile of RWH projects and high initial capital costs. However, it was pointed out that proper planning, enforcement, education and regulation, as well as policy integration, seemed to be the key to unlocking the reluctance to use RWH.

Green Hill Case Study

Green Hill is a typical village developed by state agencies in St. Vincent in the Eastern Caribbean. As the village is located at a high elevation (340 m above sea level) it suffers from low pressures in the public supply system and is quite suited as a pilot for RWH. The village has 56 houses (2 and 3-bedroom units with roof areas of 64.6 m² and 82 m² respectively).

While 75% of households considered RWH a good method for providing supplementary water only 37% or about half of these households reported that they practised RWH in some form. Despite the high awareness of the technology and the frequent interruptions of public supplies, only about 42% of households made efforts to store water. In cases where rainwater was collected, the storage containers were usually small and ranged from 0.02 m^3 plastic buckets to 0.2 m^3 plastic drums. The use of larger storage tanks is uncommon and this was attributed to the low income characteristic of the village and low affordability which prevented residents from using the technology to the fullest. Some non-economic concerns, which residents identified, discouraged the use of harvested rainwater. These included the fear of contamination by air pollution, fear of the presence of harmful bacteria and the fear of mosquito breeding.

At Greenhill, the average household water bill is relatively low as compared with other utilities. The highest annual water bill was in the range of US$225.00 to US$270.00 which is about 4% of annual household income. The average household water bill is US$137.00. Over 30% of households use less than 4.5m^3 of water monthly from the public source and supplement this with stored rainwater.

Prior to the study in 2010, only about 32% of the households gave serious consideration to investing in RWH as an alternative source of water. While water quality perception may negatively affect the willingness to invest, the survey results showed that a perception of high installation and maintenance costs plays a part in the slow adoption of the technology. Nonetheless, in addition to their current monthly bill, up to 44% of households were willing to make a minimum investment of 5% of the household's income in RWH technology. While this investment is small, it shows that there is a positive factor for advancing the use of the technology. The findings for Greenhill are consistent with that from other regions. McPhail (1993) showed that in five small Moroccan cities households were willing to allocate 7% to 10% of total household expenditures for improved services despite already having reliable and free standpipe service. Generally, households are willing to invest in RWH through programs that involve some government support. At Greenhill, households identified four types of incentives that can be used such as direct subsidy in the form of cash, duty concessions on material, rebate on installation costs and reduction in property taxes to reflect the installation of RWH facilities. From an economic standpoint there is no significant difference in these preferences

In some parts of the Caribbean, there is legislation for the use of rainwater harvesting. In the US Virgin Islands and Barbados, there is mandatory law requiring private residence and businesses to construct cisterns for the capture and storage of rainwater from rooftop. Enforcing the use of the water, however, is not always easy as seen in Barbados. For islands now promoting the use of RWH, appropriate laws and enforceable regulations must be in place to ensure that initiatives in RWH result in the desired effects.

RESULTS AND DISCUSSION

Initiatives for Diffusing RWH in SIDS

Over the last two and a half decades, there has been a growth in the promotion and diffusion of RWH. Many international and regional agencies have been investing in the promotion and diffusion of RWH. Diffusion is defined here "the process by which an innovation is communicated through certain channels over time among the members of a social system" (Rogers, 2003). Different approaches have been applied. Government, Non-Government Organizations (NGOs) and community organizations play different but important roles in mainstreaming RWH technologies as a viable alternative. The International Rainwater Catchment Systems Association (IRCSA) was founded in 1989 to promote and advance rainwater catchment systems technology with respect to planning, development, management, science, technology, research and education worldwide; and to establish an international forum for scientists, engineers, educators, administrators and those concerned in this field. The association also worked towards developing international guidelines on this technology (IRCSA, 2012). In 1994, the American Rainwater Catchment Systems Association (ARCS) was founded to promote rainwater catchment systems in the United States. In 1995, People for Rainwater (PR) was formed in Japan to promote RWH culture around the world by conducting rainwater fair, developing social systems and technical standards, developing rainwater education programme (People for Rainwater, 2012), and by organizing conferences, forums and seminars. PR also serves as a driving force for the RWH Museum, Rain Library and other educational and social matters. Further, PR is involved in publications related to RWH such as Rain Encyclopedia. This voluntary group has become a main player to promote RWH in Tokyo. The experiences of PR suggest that it is easier to persuade people to participate in non-profitable activities.

In 2002, the International Rainwater Harvesting Alliance (IRHA) was created following recommendations formulated during the World Summit for Sustainable Development in Johannesburg. The mandate called for the federation and unification of the disparate rainwater harvesting (RWH) movement around the world, to promote rainwater as a valuable water resource and to build on achievements in this field for the fulfillment of the MDGs (International Rainwater Harvesting Alliance, 2010). Under an initiative of the United Nations Environmental Programme (UNEP), the Rainwater Partnership was established in 2004 to promote and mainstream rainwater harvesting activities, that is, the integration of rainwater harvesting in water policies and thereby enabling the use of rainwater in all sectors. The results of these initiatives are reflected in a new thinking and investments in rainwater harvesting and management. For example, UNDP as the lead organization was involved in promoting RWH technologies for water supply and sanitation in schools in Tanzania through projects that costs less than US$10,000 (GEP Small Grants Programme, 2008). Another example of the success can be found in South Korea where the Ministry of Land Transport and Marine Affairs announced that the new cities will be designed such that rainwater is collected instead of drained (Han, 2009).

In the Caribbean, attempts at diffusing RWH have taken different forms. For example, in Barbados all new buildings must incorporate a rainwater collection system, the size of which is dictated by the floor area. Buildings cannot receive their certificate of completion without this installation. There are however, no measures to ensure that such systems are actually utilized (Government of Barbados, 2008) and the use of the harvested water by households is low. On the other hand, in 2008, 31 RWH systems were installed in 21 households and 10 public facilities (including 7 schools and 2 health centers) under a project partly funded by the European Union in a rural area of Saint Lucia. This initiative was embraced by the community as water stress was reduced. The impact of the project was immediate as it was embraced by the community which experienced significantly less water shortages.

In 2005, The Caribbean Environmental Health Institute (CEHI) teamed up with UNEP on an initiative to promote the use of RWH in the Eastern Caribbean. UNEP had implemented similar pilot projects in Asia, Africa and the Pacific and facilitated the formation of a rainwater partnership and extended the initiative to the Caribbean. It was hoped that, through CEHI, using the lessons learnt from other areas, especially from the Pacific Small Island Developing States (PSIDS), there would be a wider use of RWH technology. In a first phase, Grenada was used as a pilot to develop a National Programme to promote RWH. The lessons learnt from the promotion of RWH in

Grenada were used to inform the Draft Strategy for the Caribbean. As a result of the Grenada initiative, a Regional RWH Programme was developed for the Caribbean and it included the production of public awareness material such as posters, public service announcements on the radio, a feature-length video and a technical brochure.

In 2009, CEHI designed and created a fully functional RWH model that could be adopted by water-scarce communities in the Caribbean. Demonstration models showing best practices for RWH in Antigua and Barbuda were developed. These models show RWH practitioners and the public how to improve the quality of harvested and stored water by incorporating certain elements into the design of the systems.

The model provides a practical example of how an actual rainwater harvesting system would operate as it is designed to allow water to be introduced into the system and persons can therefore have a live demonstration on how the system works. The model also includes interpretive information panels and pockets for toolkits that contain posters, leaflets, brochures, technical documents, handbooks and other material on rainwater harvesting. The success of the efforts of CEHI in mainstreaming RWH in the region would become evident in the future.

In the Grenadines, initiatives to promote RWH focused on increasing public awareness on quality issues associated with RWH and projects to improve current systems particularly for the low income islanders. Since 2003, there have been a number of externally funded public awareness projects (CEHI, 2006). Recent promotion initiatives in the Grenadines include the Saint Vincent de Paul of Grenada (GSVP) which was involved in a five year project (1998 to 2003) in providing assistance to low income families of 75 plastic tanks and the construction of 30 concrete cisterns (Peters, 2006). The success of the project secured a commitment to expand the project (St. Vincent de Paul Society, England and Wales, 2009).

Since 2008, the Government of Grenada has provided support to farmers in RWH in the form of PVC water tanks for small livestock and vegetable production. In 2012, a number of farmers benefited from a Japanese funded World Bank project to develop RWH systems for irrigation (Government of Grenada, 2012). The aim of this project was to diffuse the technology widely on the islands. In the St. Vincent Grenadines, there have also been proactive approaches to promoting RWH. For example, in Union Island, similar projects to improve the access to large rainwater tanks were funded by some small non-governmental agencies. Further, as the Government of St. Vincent and the Grenadines has incorporated RWH in its management plans for a sustainable solution for water needs (Government of SVG, 2011).

Perhaps, the most important initiative to promote RWH in the Grenadines and the wider region is found in the prominent place that it has achieved in the integrated water resources management (IWRM) that has been adapted as the way towards sustainable water resources. A survey of all the recent national roadmaps for water management shows that RWH is a central feature. To accelerate the diffusion and adoption of RWH as viable sustainable practices, it is suggested that factors for the success of RWH in the Grenadines should be studied and lessons learnt be incorporated in the process. Experiences show that RWH, which starts from small beginnings, can expand significantly to meet the real demand of people, where there is no or poor quality groundwater and no surface water nearby (Rwabambari, 2005).

Framework for Promoting RWH

What is Necessary for RWH Promotion and Development?

As a tool for sustainable development, promotion of RWH does not lead to a singular or prescribed outcome. It is an approach that requires many steps, trial and error, openness and learning. It is a continuous process to ensure the best available path is taken to achieve benefits for all, at any given time. The promotion of RWH, is therefore a 'journey' in which different organizations seek to mainstream RWH as a fundamental element of sustainable water resources development in all aspects of their business. In addressing what is necessary for the promotion and development of rainwater utilization, the UNDP proposed four considerations:

- **A Systematic Approach:** In which RWH is incorporated iand integrated into water resources management;
- **Implementation Policies:** Through which government takes leadership;
- **Technology Development and Training:** Which support rainwater use through efficient and affordable devices; and
- **Networking:** Which encourages regional exchanges amongst public servants, citizens and industry representatives involved in RWH (UNDP 2002).

In developing a project or programme to promote RWH there should be a good understanding of the existing water resource conditions, objectives that are realistic, broad participation driven by key targeted stakeholders and activities that are targeted to lead to the desirable outcomes.

Understanding the Situation?

There are a number of socioeconomic, political, physical and climatic conditions that can influence the success of a RWH promotion. Additionally, there are many strengths and weaknesses of RWH in the particular location which would influence the ability to promote the technology. These must be considered at the design stage of a project or programme to facilitate better planning for and implementation of RWH projects. An analysis of the situation can be undertaken by considering the strengths, weaknesses, threats and opportunities (SWOT) for developing RWH in the community or region that is being targeted. An example of the SWOT analysis for the Grenadines that was carried out in 2011 is given in Table 2. The SWOT analysis should be accompanied by a needs assessment. Communities and regions may be at different stages in the development of RWH technology and would require different promotional approaches. In some communities, public water supply may be adequate and affordable and so RWH may at first look unattractive. In such communities, however, the environmental concerns such as water conservation, at source flood control, and sustainability of the local water resource are likely to be of greater importance than enhancing water supply. In other cases where public water supply is available but is unreliable or expensive, RWH can be used for augmenting water supply. In other cases, where RWH is the main source of water, the need might be to improve the quality of the water, manage water in times of low rainfall, or target the low-income households by increasing their storage capacities.

Defining the Scope

Many RWH promotion projects have been unsuccessful because the geographic area for the project was too large, the wrong set of beneficiaries were selected or the promotion type did not fit the needs of the community. For example, wide scale public awareness on RWH in the Grenadines would be useless since RWH is already part of the cultural DNA of the community. For these islands, promotion should be aimed at increasing quantity, improving quality, reducing costs of RWH establishment and strengthening RWH where it is already practiced. In communities where RWH is not routinely practiced, low income households or households in isolated locations can be the targeted beneficiaries. The size of promotion initiatives should match the needs of the community and the amount of available funding.

Table 2. SWOT analysis of RWH in the grenadines

Strengths **RWH for household supply** • Low cost technology • Provides water at the point of consumption • It is decentralized, empowering individuals and communities to manage their water • Islanders have developed adequate skills to build their own systems. This reduces costs and encourages more participation, ownership and sustainability at the community level • Not affected by local geology or topography (households at high elevations are not disadvantage as those connected to centralized public systems) • Almost all roofing material used on the islands is acceptable for collecting water for household purposes	**Weaknesses** **RWH for household supply** • Limited by the amount of rainfall and the size of the catchment area and storage reservoir • Supply is sensitive to droughts: Occurrence of long dry spells and droughts can cause water supply problems • The cost of rainwater catchment systems is almost fully incurred during initial construction • Proper operation and regular maintenance is a very important factor that is often neglected • Rainwater quality may be affected by air pollution, animal or bird droppings, insects, dirt and organic matter • Leakage from cisterns can cause the deterioration of load bearing slopes • Vulnerable to cracking in earthquake prone areas • Rainfall harvesting systems increase construction costs and may have an adverse effect on home ownership. Systems may add 30% to 40% to the cost of a building. • Some designs may need pumping for utilizing stored water in the internal plumbing systems • It is not included in water policies for the islands
Opportunities • To use the technology in developing a green economy • To use technology in reducing urban flooding • To encourage participation in managing communal systems • To incorporate the use of the technology in disaster management (hurricanes and droughts) • To develop comprehensive policies and guidelines for the use of RWH in the islands • To use the islands to showcase the use of RWH and a model for development	**Threats** • Climate change can impact on the current storage facilities. Increased storage capacities add to construction, retrofitting and operating costs and may make the technology economically unfeasible • Increased use of motor vehicles can lead to household water pollution

Mobilizing the Stakeholders

Success of RWH promotion initiatives improves where there is a creation of synergies between the major stakeholders involved in the RWH sector and they are able to influence policy at all levels: international, regional and national. Identifying the key local stakeholders for a RWH promotion project or programme is important. The main stakeholders are likely to be:

1. Targeted beneficiaries (users of the system);
2. National or local government;
3. Community Based Organizations (CBO);
4. Local Non-Government Organizations (NGOs);
5. Local water providers;
6. Private individuals and companies that provide related services; and
7. Fund providers.

In selecting beneficiaries to be involved in the project, it is useful to identify small groups that are highly motivated, have a substantial need, and are willing to invest their time and or money in projects. Targeted beneficiaries are important in determining and setting more realistic developmental objectives which can ultimately have a greater chance of being achieved. For example, it has been observed that on main islands like Grenada and St. Vincent, where RWH is not routinely applied the majority of owners of the RWH systems are persons who lived on the smaller islands that routinely practiced RWH for extended periods. This suggests that the promotion of RWH may be enhanced be identifying such persons in the community to be project sponsors for pilot projects.

CBOs and NGOs are often already engaged with communities that are high on a priority list for RWH development. They may be working with the beneficiaries in some other areas such as health, job creation and poverty alleviation. These areas of focus are generally suited to be integrated with RWH promotion activities. In addition, community groups and local NGOs are usually in good positions to identify the best group of beneficiaries suited for these types of projects.

Greater success in RWH projects have been realized where the local and national governments are highly supportive. Government can provide logistic support as well as incentives which can help reduce costs in the establishment of RWH systems. In small island communities, small projects like RWH are not always politically attractive unless the local politician discerns that the benefits would be translated into votes. It is therefore best to identify an area in the community where RWH can have significant impact on people's lives. Success in this community can be a motivation for the sponsoring of a RWH project by the political directorate.

The local water providers, which in most small islands are government owned corporations, should be involved in RWH as a successful RWH programme can impact on their operations both positively or negatively. Often, it is in the interest of the water provider to invest in such projects because when there is a high need for RWH, particularly for domestic supply, is a good indicator that the provider is unable to meet the full needs of its customers.

Special attentions should be placed on the involvement of the youth in "Rainwater Awareness Programme for School Children" with RWH incorporated into the school curriculum. By encouraging youth participation the project is also helping to build greater awareness and support for such projects which enhances sustainability.

Setting the Objectives

A general aim of a RWH programme is "To contribute to the conservation of the water resources of the island through adoption of sustainable water management and conservation technologies." In developing a realistic RWH initiative more specific objectives are required. This can be facilitated by SWOT analysis and needs assessment to determine the objectives of a particular RWH project or programme. For example, in a case where public water supply is generally satisfactory but there is need for addressing environmental concerns, the objective would be to improve public awareness. On the other hand, in a case where RWH is already established but there are concerns with the quality of water, the main objective may focus on quality issues. At a regional workshop on promoting RWH in the Caribbean, experts in the field of water resource management (WRM) from the region as well as regional and international agencies identified the following themes that can be adapted to suitable objectives of RWH promotion (CEHI, 2006).

- Policy and incentives for RWH;
- A programme for the protection, monitoring and treatment of RWH systems;
- Support to the less fortunate households in the acquisition of RWH systems;
- Refurbishment of selected communal cisterns;
- Propagating ongoing initiatives such as RWH and food security programme;
- Putting emphasis on public awareness programmes for RWH at the national and regional levels;
- Regional policy planning for RWH;
- Involvement of NGOs and other private sector agencies; and
- Accessing RWH equipment and supplies.

Developing Appropriate Activities

The minimum elements of a good RWH Programmes would include:

- Awareness raising;
- Capacity building;
- Legislative and policy formulation; and
- Administration, monitoring and evaluation.

These programmes should be conducted with the widest involvement of the local community. This leads to increasing awareness on RWH practice among the community members. Besides, the community level involvement, especially hands-on in construction activities, creates a culture of RWH which becomes a symbol of the community.

Consideration can also be given to activities that can lead to:

1. A website for the capture and dissemination of best practices in RWH (example http://www.cehi.org.lc/);
2. The toolkit for public awareness and technical information which can include instructions on technical aspects such as design of RWH systems and selection of pumps (example http://www.cehi.org.lc/Rain/Rainwater%20Harvesting%20Toolbox/index2.htm);
3. Engagement of manufacturers of RWH supplies and equipment to be involved in the promotion;
4. Use of the media including the Government Information Systems to assist in of the airing of the Public Awareness Programmes for RWH;
5. Incorporating RWH in the school curriculum ;
6. Identifying and or developing a competent agency in RWH responsible for providing technical assistance;
7. Building capacity within agencies to develop a skills bank to provide technical assistance in RWH to territories;
8. A responsible body for enforcement and monitoring of RWH;
9. A package of tax incentives;
10. Water conservation programmes that may be linked to the provision of incentives for RWH;
11. Harmonization of RWH with the municipal system; and
12. Inclusion of water quality, engineering and legal issues in the national water plans.

Funding

Funding for RWH diffusion is extremely important. In the past, the biggest barrier to wide adoption of RWH seems to have been the scant funding it has received from international NGOs and finance organizations (Cain, 2010). This is however changing. As an adaptation to climate change, many international agencies and governments from developed countries are providing funding for RWH. For example, in 2011 Germany and Australia pledged US$625M to the Global Environmental Facility (GEP) for climate change adaptation. Such funds were made available to RWH projects (GEF, 2011). As seen in this study, funds for RWH can be from own sources, government or local and international agencies. Great success is likely where funds from own-sources are utilized over time, like seen in Carriacou and Petite Martinique. However, for low income households, financial support is required from either government or charitable organizations. The diffusion in this case may be slow, taking a number of years to reach full and optimal utilization in the community. In any case, to speed up the diffusion process a combination of funding sources should be explored.

RWH now straddles the concepts of renewable resources, sustainable development and green economy. Once a good RWH project can be developed obtaining funding is likely to be successful. Previous attempts at promoting RWH in the region have been driven by external agencies such as UNEP, UNDP and USAID of which a substantial amount is spent on administrative functions. A better balance between administrative functions and direct benefits is required. Where it is possible, potential beneficiaries should be encouraged to contribute financially to projects.

Investment in RWH at the community level should be framed against the following considerations (UNEP/OAS, 1997):

- Is there a real need for an improved water supply in terms of reliability and quality?
- Are present water supplies easily accessible or contaminated, or both?
- Are suitable roofs and/or other catchment surfaces available for capture of rainwater?
- Does the average annual rainfall exceed 400 mm? (Note: this consideration is a benchmark applied in arid countries and is not generally applicable in the Caribbean where rainfall exceeds that amount)
- Does an improved water supply feature prominently in the community's list of development priorities?

If the answers to the questions are yes, then it is a good indication that the RWH is promotable.

Other Emerging Areas of RWH in the Caribbean

The success at mainstreaming RWH for household consumption in the Grenadines can be used to broaden the use of the technology for non-household purposes. The application of RWH as an adaptation tool for climate change, particularly in relation to flood mitigation, disaster management (hurricanes and drought) and as an integral part of the new green economies that are being proposed is already recognized

RWH and Flood Control

In many Caribbean islands, local councils and physical development planners are challenged to solve drainage problems in the urban and suburban areas. They must address questions such as:

1. How to create space for flow and storage making optimal use of high value and dwindling land resource?
2. How to control ad hoc development in the absence of law enforcement?

The one dimensional view of traditional storm water management is giving way to a more integrated and comprehensive perspective. This perspective is captured by terms like integrated storm water management, rainwater management and sustainable urban drainage systems. Over the past decade, the term Integrated Storm Water Management Planning (ISMP) has gained widespread acceptance by local governments and the environmental agencies in North America which consider a comprehensive, ecosystem-based approach to rainwater management. Peters and Lewis (2007) simulated the use of RWH tanks at households in a suburban district in Trinidad and found that Rooftop Rainwater Harvesting (RTRWH) can reduce the peak flow for a 50yr storm by about 27%.

RWH and Disaster Management

An important consideration which warrants investment in RWH systems is the potential for enhancing water security following natural disasters (notably hurricanes). The experiences of some islands after hurricanes show that RWH can mitigate the hardships from water shortages. This was evident in 2004 and 2005 after experiencing Hurricanes

Ivan and Emily respectively in Grenada. While these hurricanes left the infrastructure of the public water supply in shambles, RWH systems were virtually untouched. A key lesson from the Hurricane Ivan experience is that households that practiced RWH using concrete or metal cisterns and had minimum storage capacities of around 4 m^3 were able to meet their needs during the first few weeks after the hurricane (Peters, 2010). It was also observed that the severity of the water shortages at emergency shelters was less in those shelters that had water storage facilities that could be replenished from rooftop catchments. This suggests that communal rainwater storage systems, using schools and community centers, could be established to provide a reliable system for communities during the early periods after a hurricane disaster.

FUTURE RESEARCH

While much progress has been made in mainstreaming RWH in the Grenadines, there are yet many opportunities waiting to be harnessed. To optimize such opportunities, research is required to understand the consumption partners of water in a RWH domain and how good conservation practices can be used to improve efficiency. Also, in the area of agricultural use, research on integrating RWH with renewable energy for improved irrigation efficiency can be beneficial. Finally, research into the application of low costs technologies like solar disinfection of rainwater for improving the safety of drinking water in these RWH islands, should be explored.

CONCLUSION

The benefits of promoting a new sustainable water resource system are not always readily obtained as many social and economic issues may need to be adequately addressed. The recent financial and economic crisis and the beginning of the decline of trust in liberalization and globalization can lead to some renewed receptivity for a sustainable development paradigm which embraces the efforts made by local people in solving their own problems.

RWH is a big success in the small Grenadine islands of the Eastern Caribbean. While there are obvious constraints of the technology, the way in which islanders have developed and managed water with almost total dependence of RWH represents a useful model for development elsewhere. Once the technology can satisfy a current need and people have the knowhow, it may not be problematic to obtain financing for promoting the technology. The future potential of RWH in small islands is great and promises more than the satisfaction of household supplies under normal conditions. RWH is likely to be an indispensable tool for flood reduction, hurricane disaster mitigation and to the new concept of a green economy. Successful RWH promotion initiatives would require, understanding the situation, defining the scope of the initiative, identifying the group of beneficiaries that are most likely to be fully engaged, formulating the right set of objectives and selecting the best set of activities. Ultimately, RWH needs to be considered as a part of a long term and overall environmental plan towards more sustainable development. The advantages to the population for water security and to the wider environment by preventing flooding and erosion of the watersheds are incalculable.

REFERENCES

Allen, M. J., & Geldreich, E. E. (1975). Bacteriological criteria for groundwater quality. *Ground Water, 13*(1), 5–52. doi:10.1111/j.1745-6584.1975.tb03064.x

Cain, N. L. (2010). A different path: The global water crisis and rainwater harvesting. *Consilience: The Journal of Sustainable Development, 3*(1), 187–196.

Cardona, C. (2006). *An integrated approach towards assessing the feasibility of domestic rainwater harvesting in Malta*. Unpublished MSc. Thesis, School of Geography and the Environment, University of Oxford. Oxford. United Kingdom.

Caribbean Environmental Health Institute (CEHI). (2006). Promoting rainwater harvesting in the Caribbean. *Regional Strategic Planning Workshop Proceedings*, Caribbean Environmental Health Institute (CEHI), British Virgin Islands. Retrieved December 2, 2011, from http://www.cehi.org.lc/Rain/print/Regional%20Strategic%20Planning%20Workshop%20Proceedings%20BVI.pdf

Caribbean Environmental Health Institute (CEHI). (2009). *Caribbean rainwater harvesting handbook*. Retrieved November 13, 2011, from http://www.unep.org/ecosystemmanagement/Portals/7/Documents/em-rainwater-handbook-caribbean.pdf

Drexhage, J., & Murphy, D. (2010). *Sustainable development: From Brundtland to Rio 2012*. Background paper, prepared for consideration by the High Level Panel on Global Sustainability, 19 September 2010, International Institute for Sustainable Development, United Nations Headquarters, New York. Retrieved January 2, 2012, from http://www.un.org/wcm/webdav/site/climatechange/shared/gsp/docs/GSP1-6_Background%20on%20Sustainable%20Devt.pdf

Environmental Agency (UK). (2008). *Water and the environment international comparisons of domestic per capita consumption*. Retrieved January 10, 2012, from http://publications.environment-agency.gov.uk/PDF/GEHO0809BQTD-E-E.pdf

Fujioka, R. S., & Chinn, R. D. (1987). The microbiological quality of cistern waters in Hawaii. *Proceedings of the 3rd. International Conference on Rain Water Cistern Systems*, Khonkaen University, Thailand, (pp. 1-13).

Global Environmental Facility (GEF). (2011). *Germany, Australia pledge more adaptation resources for the poor and vulnerable*. Retrieved February 3, 2012, from http://www.thegef.org/gef/press_release/Germany_Australia_pledge_LDCF_SCCF

Gould, J. (1999). *Contributions relating to rainwater harvesting*. Paper prepared for the World Commission on Dams Secretariat (WCD) [Assessment of Water Supply Options, WCD, Cape Town, South Africa.]. *Thematic Review, IV*, 3.

Government of Barbados. (2008). *Road map towards integrated water resources management planning for Barbados, Barbados*. Retrieved February 10, 2010, from http://iwcam.org/documents/iwrm-roadmaps/draft-iwrm-roadmap-barbados/at_download/file

Government of Grenada. (2012). *Farmers in Carriacou getting assistance*. Retrieved March 15, 2012, from http://www.gov.gd/egov/news/2012/apr12/20_04_12/item_3/farmers_carriacou_assistance.html

Government of St. Vincent and the Grenadines. (2011). *Strategic programme for climate resilience, proposed investment program components for PPCR funding*. Kingstown, St. Vincent: Government of Saint Vincent and the Grenadines.

Han, M. (2009). Promotion of rain cities in Korea – Policies and case studies. *Proceedings of the 14th. International Conference on Rain Water Cistern Systems*, Kuala Lumpur, Malaysia.

He, F.-X., Cao, H., & Li, F.-M. (2007). Econometric analysis of the determinants of adoption of rainwater harvesting and supplementary irrigation technology (RHSIT) in the semiarid Loess Plateau of China. *Agricultural Water Management, 89*, 243–250. doi:10.1016/j.agwat.2007.01.006

Heyworth, J. (2001). The diary study of gastroenteritis and tank rainwater consumption in young children in South Australia. *Proceedings of the 10th International Conference on Rain Water Catchment Systems* (pp. 1-9). Mannheim, Germany, September.

Heyworth, J. (2004). *An epidemiologic study of childhood gastroentritis and consumption of tank rainwater, an untreated water supply in South Australia*, Unpublished PhD Thesis, Flinders University, Australia.

HMSO. (1994). *Sustainable development: The UK strategy*. London, UK: HMSO.

Hoyte, H. A. (2009). *The social, economic and institutional barriers of rainwater harvesting in Barbados*. Unpublished MSc. Thesis, Center of Resource Management and Environmental Studies, University of the West Indies, Cave Hill, Barbados.

Inter-American Development Bank (IADB). (2005). *The millennium development goals in Latin America and the Caribbean: Progress, priorities and IDB support for their implementation*. Washington, DC: IADB. Retrieved August 4, 2009, from www.iadb.org/idbdocs.cfm?docnum

International Rainwater Harvesting Alliance. (2010). Who are we? Retrieved May 15, 2012, from http://www.irha-h2o.org/?page_id=19

Krishna, J. (1989). Cistern water systems in the U.S. Virgin Islands. *Proceedings of the 4th International Conference on Rain Water Cistern Systems*, Manila, Philippines, (pp. 1-11).

Lee, M. D., & Visscher, J. T. (1990). *Water harvesting in five African countries*. UNICEF, Occasional Paper Series, Occasional Paper 14 sponsored by IRC International Water and Sanitation Centre, The Hague, The Netherlands.

Lye, D. J. (1992). Microbiology of rainwater cistern systems: A review. *Journal of Environmental Science and Health, Part A: Environmental Science and Engineering and Toxicology, 27*(8), 2123–2166. doi:10.1080/10934529209375845

Lye, D. J. (2002). Health risks associated with consumption of untreated water from household roof catchment systems. *Journal of the American Water Resources Association, 38*(5), 1301–1306. doi:10.1111/j.1752-1688.2002.tb04349.x

Manja, K. S., Maurya, M. S., & Rao, K. M. (1982). A simple field test for the detection of faecal pollution in drinking water. *Bulletin of the World Health Organization, 60*(5), 797–801. PMID:6983930

McPhail, A. A. (1993). The "five percent rule" for improved water service: Can households afford more? *World Development, 21*(6), 963–973. doi:10.1016/0305-750X(93)90054-D

Meade, B., & Gonzalez-Morel, P. (1999). *Improving water use efficiency in Jamaican hotels and resorts through the implementation of environmental management systems*. Retrieved May 3, 2012, from http://www.ucowr.org/updates/pdf/V115_A6.pdf

Mun, J. S., Ki, J. H., & Han, M. Y. (2008). *Economical analysis on rainwater harvesting (RWH) system of buildings and residential complex in the Republic of Korea*. Retrieved January 15, 2012, from http://iwahq.org/ContentSuite/upload/iwa/Document/P16.pdf

NZ Ministry of Health. (2005). *Draft guidelines for drinking-water quality management for New Zealand 2005*, 2nd ed. Wellington, New Zealand: Ministry of Health, Retrieved August 8, 2010, from http://www.moh.govt.nz/moh.nsf/pagesmh/4387?

Open Organisation of American States. OAS. (2006). *Final report of the first Inter-American meeting of ministers and high-level authorities on sustainable development*. Retrieved December 19, 2010, from http://www.oas.org/dsd/Documents/REPORT.pdf

Pearce, D. (1993). *Blueprint #: Measuring sustainable development*. London, UK: Earthscan.

People for Rainwater. (2012). *People for rainwater*. Retrieved February 3, 2012, from http://www.skywater.jp/profile1_e.html

Peters, E. J. (2004). Sizing criteria for domestic rainwater cisterns in the Grenadines. *West Indies Journal of Engineering*, *27*(1), 68–78.

Peters, E. J. (2006). Rainwater potential for domestic water supply in Grenada, West Indies. Proceedings of the ICE - Water Management, *159*(WM3), 142-154.

Peters, E. J. (2010). Impact of Hurricane Ivan on Grenada Water Supply. Proceedings of the ICE - Water Management, *163*(2), 57-64.

Peters, E. J., & Lewis, S. (2007). *Rainwater harvesting for reducing on-site flooding: A case study*. Paper presented at the 16th Annual CWWA Conference and Exhibition: Integrating Water and Sanitation in Sustainable Development, St. Kitts October 1-5, 2007.

Peters, E. J., & Mandeville, S. (2011). *The viability of rooftop RWH for Green Hill, St. Vincent*. Paper presented at the 20th Annual CWWA Conference and Exhibition, Caribbean Cooperation: The Future of Water and Waste Management in the Region. 2-7th October, 2011, Gosier, Guadeloupe.

Rogers, E. (2003). Diffusion of Innovations (5th Ed). New York: Free Press.

Rwabambari, C. (2005). Women Groups Harvesting Rainwater and influence National Policy: Ugandan case study, In "Mainsteaming Rainwater Harvesting," *Proceedings of the 12th International Rainwater Catchment Systems Conference*. International Rainwater Catchment Systems, New Delhi, India.

Slade, H. G. (1984). Craigston and Meldrum Estates, Carriacou, 1769-1841. *Proc Soc Antiq Scot*, *114*, 481–537.

Small Grants Programme, G. E. P. (2008). *Promotion of rainwater harvesting technology for water supply and sanitation in schools- Mwang'halanga Primary School*. (TAN/CWI/07/05). Retrieved December 15, 2010, from http://sgp.undp.org/index.cfm?module=Projects&page=Advanced Search

Smet, J., & Moriarty, P. (2001). DGIS Policy Supporting Paper: Rooftop Rainwater Harvesting. IRC Delft, Retrieved July 12, 2008, from www.cepis.ops-oms.org/bvsarg/i/fulltext/dgis/dgis.pdf

Sobsey, M. D., & Pfaender, F. K. (2002). *Evaluation of the H_2S Method for Detection of Fecal Contamination of Drinking Water*. Geneva, Switzerland: World Health Organization.

St. Vincent de Paul Society (England and Wales). (2009). Celebrating our 350th Anniversary, *Vincentian Concern*, Issue 31. Retrieved March 3, 2012, from http://www.svp.org.uk/docs/vc%20 31.pdf

The International Rainwater Catchment Systems Association (IRCSA). (2012). *About us*. Retrieved May 5, 2012, from http://www.ircsa.org/index.htm

UNEP/OAS. (1997). *Source Book of Alternative Technologies for Freshwater Augmentation in Latin America and the Caribbean, United Nations Environment Programme and Unit of Sustainable Development and Environment General Secretariat*. Washington, D.C.: Organization of American States.

United Nations. (1994). *Report of the International Conference on Population and Development*, Cairo, 5-13 September 1994. United Nations, New York.

United Nations Conference on Environment and Development, UNCED (1992) Rio Declaration on the Environment and Development. Retrieved April 4, 2012, from http://www.unesco.org/education/information/nfsunesco/pdf/RIO_E.PDF.

United Nations Development Programme. UNDP (2002). Rainwater Harvesting And Utilisation: An Environmentally Sound Approach for Sustainable Urban Water Management: An Introductory Guide for Decision-Makers. Retrieved April 19, 2010, from http://www.unep.or.jp/ietc/Publications/Urban/UrbanEnv-2/index.asp

United Nations Environment Programme. UNEP (2009) Rainwater harvesting: A lifeline for human well-being. (Ed) Barron J., United Nations Environment Programme/SEI, Stockholm, Sweden. Retrieved April 19, 2010 from http://www.unwater.org/downloads/Rainwater_Harvesting_090310b.pdf

Venkobachar, C., Kumar, D., Talreja, K., Kumar, A., & Lyengar, L. (1994). Assessment of bacteriological water quality using a modified H_2S strip test. *Journal of Water Supply: Research & Technology - Aqua, 43*(6), 311–314.

White, I. (2007). A theoretical framework for understanding factors that contribute to household adoption of rainwater harvesting in south east Queensland, Retrieved August 23, 2011, from http://www.ctahr.hawaii.edu/hawaiirain/Library/papers/White_Ian.pdf

Wirojanagud, W. Mungkarndee. P. & Hovitcitr, P. (1989). Evaluation of Rainwater Quality: Heavy Metals and Pathogens, Proceedings of 4th International Conference on Rain Water Cistern Systems, Manila. Retrieved September, 4, 2008, from http://www.eng.warwick.ac.uk/ircsa/4th.html

World Commission on Environment and Development. WCED (1987). Our common future. Oxford: Oxford University Press.

World Health Organization and United Nations Children Fund (WHO & UNICEF). (2004). Meeting the World MDG Drinking Water and Sanitation: A Midterm Assessment of Progress. Retrieved May 4, 2012, from http://www.unicef.org/publications/files/who_unicef_watsan_midterm_rev.pdf

World Health Organization (WHO). (2004), Guidelines for drinking-water quality, third edition incorporating first addendum Volume 1 – Recommendations, World Health Organisation, Geneva Retrieved November 22, 2009, from http://www.who.int/water_sanitation_health/dwq/gdwq3rev/en/index.html

This chapter was previously published in Cases on the Diffusion and Adoption of Sustainable Development Practices edited by Helen E. Muga and Ken D. Thomas, pages 403-438, copyright 2013 by Information Science Reference (an imprint of IGI Global).

Chapter 93
The Sustainable Waterfront

Matthew Bradbury
Unitec Institute of Technology, New Zealand

ABSTRACT

This chapter discusses how a methodology for the development of a sustainable waterfront might be shaped through an understanding and consideration of environmental remediation technologies. The chapter begins by considering the history of waterfront development in Europe and America and how this model has become codified into a generic real estate process. The author develops a critique of this model from an environmental perspective. The use of sustainability as an ideology to provide a framework for critique but also strategies and techniques for moving towards a new model of waterfront development are explored. The development of a possible hypothesis for the design of a sustainable waterfront is developed, followed by a speculative case study of a waterfront project that explores how the design hypotheses might be tested. The chapter concludes by speculating on the ways in which a study of urban ecology, in particular urban biodiversity, could enhance the finding of the case study towards the development of a waterfront design process that could contribute to the ecological health of the 21st century city.

INTRODUCTION

The massive growth in the development of cities is the global phenomenon of the 21st century, (Brugmann, 2010; West, 1993) already half the world's population live in urban environments. At the same time a growing awareness of the importance of the environment is increasing becoming part of everyday consciousness. A growing consensus over the effects of CO2 emissions on the global environment has important implications for the shape and core drivers of the world economy. (Hodson, 2010) The effects of this colossal urbanisation and the enormous consumption of resources in both the building and maintenance of the new cities, has had a deleterious effect on the environment. The challenge of finding a way for cities to become more sustainable is a subject that has increasing

DOI: 10.4018/978-1-4666-4852-4.ch093

occupied researchers and design practitioners. Contemporary waterfront development has been the lens through which many of these urban trends have played out. The way in which 19th industrial ports and infrastructure have been transformed to new kind of urban development, the waterfront, is a critical part of 20th century urbanisation. This mode of waterfront development, first presented in the Baltimore Waterfront project, (Norris 2003) has become a globally accepted model.

A new dialogue between sustainability and urban development has resulted in a rethinking of the waterfront building programme towards one that is directed by sustainable building guidelines.

The author will argue that the tools presently advanced to transform waterfronts to a more sustainable model ignore some specific issues that are unique to waterfront developments, in particular environmental issues. The underlying environmental problems of many waterfront developments, polluted stormwater, brown field sites, and polluted dredging, are often elided. The author argues that to develop a sustainable waterfront, the complex environmental problems that are present in many waterfront developments need to be bought forward, to be privileged, and the full array of known remediation technologies explored and utilised. It is in this way that the conventional urban waterfront programme; public building, apartments, offices, and public spaces, can be re-examined and re invented to provide a more equitable and sustainable city.

The objectives of the chapter are;

- To demonstrate the importance of waterfront development in contemporary urbanism.
- To show how this development type relies heavily on a prescriptive and generic/real estate formula.
- To show the importance of the environmental/ecological investigation of waterfront sites, in particular the environmental problems of the contemporary waterfront such as; terrestrial and marine pollution, stormwater flows, and sea level rise,
- To establish a new methodology of sustainable waterfront development that draws on four intersecting areas of research, waterfront development models, urban sustainability, environmental remediation, and urban ecology.

BACKGROUND

The world wide structural transformation of waterfronts from ports to pleasure zones is still best described by Hoyle's (Hoyle, 1981) (five-phrased development model. The five phases describe how the European port and waterfront has developed over the last 700 years. The first part is the primitive city port phase, from the medieval city port; most notable, the Italian trading ports, to the beginning of the European industrial port. The second phase is the expanded city port, the result of the territorial expansion of the traditional city port utilising the technology of the industrial revolution to build a new port infrastructure. The 19th century port became a heavily industrialised site that was a crucial part of the global supply chain developed by European colonialism in the 19th century. The port infrastructure assumed a common form around the world, an infrastructure of reclamation and wharfs, backed with warehousing, and surrounded by a new residential district of dockworkers.

The third phase, the modern industrial city port, is part of the development of the large-scale industrial infrastructural processes of the mid 20th century economy; manufacturing, oil storage and refining, and the start of containerisation. The consequences of containerisation were not only a new cargo typology, but the invention of a new port infrastructure with radical implication for the connection of port to city. With the necessity of a new waterfront infrastructure design, the huge sums of money that needed for this reorganisation, and the chance of port owners to break powerful waterfront unions, many ports relocated away from the original city sites. One of the best-known examples of this splitting of city and port is the relocation of the Port of London from the Docklands next to the City, to Tilbury in the 1960's (Williams, 1993)

The bifurcation of port and city is the fourth phase of Hoyles development model. This period, lasting from the 1960's to the mid 1980's, saw the abandonment of the 19th century port infrastructure; the brick warehouses and wharfs. The new city ports were characterised by oil refineries, ro ro handling systems, and break-bulk cargo infrastructure.

The last phase in Hoyles model, is the redevelopment of the old industrial waterfront, from the mid 1980s to the present day. This was initially generated by the proximity of abandoned waterfront sites to the existing city, providing an irresistible real estate opportunity to turn old industrial land that had lost its use value into a valuable development opportunity.(Meyer, 1999)

The first project that manifests this particular impetus is the Baltimore Inner Harbour development (Norris 2003). This project initiated in 1970. The project was driven by the proximity of the harbour to the city, the harbour had silted up, resulting in operators moving away and abandoning the old port infrastructure. The new master plan was developed in the 1970s, the first buildings were completed in the late 1970's. The buildings were a mixture of public and privately financed projects. The first building constructed was the publically financed Baltimore Convention Centre finished in 1979, followed by the HarbourPlace mall developed by the Rouse company, opened in 1980, then the National Aquarium in 1981, Pier 6, an outdoor entertainment facility, in 1991, the Power Plant, a multi use building with bookshops, restaurants, and a gym, in 1997-8, and the Port Discovery, a children museum, in 1998. Along with this development programme was the construction of eleven hotels and two sports stadium. The building along the harbour edge are all linked with new public space, including promenades, squares, and parks. The success of the Inner Harbour development has been aggressively promoted with claims that the project receives more visitors a year than Disneyland, produces $786 million dollars a year in revenue and has generated 16,000 jobs.

The design impetus for the transformation of the abandoned industrial port into the waterfront of personal consumption is accomplished through the appropriation of another type of waterfront development, the promenade. The promenade is a new urban form invented in 18th century England. As part of English societies fashion/concern for bodily health, a number of therapeutic sites were founded where the wealthy could avail themselves of a variety of cures for contemporary ailments. These therapeutic regimes often include water, drinking spa water, as visitors to Bath, or swimming and breathing salt air at Brighton. The particular therapeutic regime in Brighton engendered a particular urban response, the creation of a sea promenade. Visitors walked along a new sea terrace and breathed the salty air. The promenade was constructed to ensure protection from tidal encroachment, with the establishment of this seawall/promenade came real estate opportunities, the building of hotels and houses along the new promenade. A good example in Brighton is the development of Kemptown, an important piece of early 19th century urbanism (Dale 1947). The urban building block developed in Brighton was a four-story terrace house; this type could accommodate a large range of functions; shops, hotels, residential, offices, and the terrace forms a 'wall' to the seaside promenade. This particular urban type was to be duplicated in many parts of the world from the 18th to the early 20th century.

The insight/invention of the 1980's waterfront was the conflation of the promenade type with the opportunity presented by the abandoned 19th century industrial waterfront. The result was a new urban model, a promenade and urban spaces on the water's edge framed with an architectural assemblage of apartments, hotels, bars and restaurants, and programmed with an event structure of shopping, eating, sleeping and entertainment. The formula has been so successful that it has been replicated around the world. (Gastil & ebrary Inc., 2002)

The thirty year history of this new waterfront development model has been exhaustively documented, (Breen & Rigby, 1996) however certain authors have noted a structural shift. Vallega (Vallega, 2001)) has suggested that two phases in the development of the urban waterfront can be discerned. The first phase, from the 1960s to the mid 90s is titled the Baltimore syndrome, where the transformation of port to waterfront is structurally driven by a combination of local imperatives, employment, and real estate development. Vallega suggests that the second phase of waterfront development, from the mid 90s to the present day, has been influenced by the ideas of sustainable development.

Vallega see the impetus behind sustainability development as being driven from two global events, Global Change, which includes climate change, and Globalisation. Global change, the awareness of larger environmental forces that have been ignored by conventional economy systems started to become more important during the 1990's. Driven by concerns about climate change, sustainable development has become codified as guidelines that are currently being mandated for many urban projects. In studying the development of the sustainable urbanism the history of sustainability as a critical trope deserves examination. It is generally agreed that the initiation of the generalised discourse around sustainability began with the publication of the Brundtland Commission. (WCED, 1987) This document defined sustainable development as, 'development that meets the needs of the future generation without compromising the ability of future generations to meet their own needs'. The consequences of this report on how cities might be developed in a sustainable manner had many variations, one example is the goals developed by the European Environment Agency in 1995 (UNCED, 1992). Urban development was to;

- Minimise the consumption of space and natural resources.
- Ration and efficiently managing urban flows.
- Protect the health of the urban population.
- Ensure equal access to resources and services.
- Maintain cultural and social diversity.

In the 2000's effort was directed towards the development of techniques to model and assess the ways in which urban development could be made more sustainable. In the paper, Modelling Urban Sustainability Klaus Spiekemann and Michael Wegner (Spiekerman, 2003) discuss the development of sustainable development methodologies that lead to groups of indicator evaluation and presentation systems. The authors begin the discussion by positing a number of sustainability indicators that they derive from three larger systems; environmental, social/cultural, and economic. The authors propose a methodology for both modelling and testing urban policies for their impact on sustainability.

Four indicator models are used to calculate sustainability, these are; the raster model, the economic indicator model, the justice indicator module and the sustainability indicator model. Each of these indicators models contributes a distinct evaluation to the process, for example, the raster model collects and calculates the environmental indicators, the economic indicator contributes a cost benefit analysis, and the justice indicator investigates the social equity implications. The authors go on to demonstrate how such a model would work in evaluating land use transport. The findings of the paper do not preclude the testing of this methodology on other urban policies.

Developing the notion of a assessable metric for urban development, the authors, Mohamed

El-Haram et al (El Harim 2007) discusses the development of a methodology (ISAT) in the paper, Development of an Integrated Sustainability Assessment Toolkit. ISAT brings together a number of methodologies that allow the assessment of urban policies through sustainability criteria. The authors begin by canvassing the history and development of the idea of sustainability, the importance and contribution of the notion of urban sustainability, the history of assessing urban sustainability and the development of guidelines and models. The authors summarise these evaluation models into two groups, a baseline approach and the objective lead approach. The baseline approach firstly assesses the indigenous site then assesses the impact of the intervention from the indigenous viewpoint. In the objectives lead approach the assessment is based on how much the process and result of an urban intervention is sustainable. Various critiques of these models are discussed; the conclusion is that some integration of the two models is desirable. This leads to the explication of the ISAT system. The system concentrates on the built environment. The author's emphasise the important of understanding the scale and life cycle of any element of the built environment. Once a particular project intervention is located within its context, a sustainability assessment can be carried out according to criteria in three groups; environment, social and economic. The actual techniques for these assessments are discussed, the vast ranges and scale of sustainable development assessment tools and a list of generic criteria are advanced, and some of these criteria include, spatial scale, sectors of use, project type, and interface operating platforms. While these particular procedures are developed for the built environment, they would seem to able to be adapted to other urban systems.

In the paper, Environmental Sustainability Metrics: A Provisional Set, (Shane, 2000) the authors Megan Shane and Thomas Graedel investigate the way in which environmental sustainability could be measured. The authors suggest that the three environmental components of the city are; natural resources, infrastructure, and people. These can be broken down into ten subcategories to which a quantitative metric can be applied, these are; air, water, solids, transportation, energy, resources, use, population, urban ecology, liveability, and lastly a qualitative measure, environmental management. The author outline a list of essential input into the running of a modern city and suggest a number of measurable goals for each category to achieve sustainability.

Water supply should be 150 litres per day. The aim of solid waste generation is 1 kg a day; transportation metrics are measured by the percentage of housing units within 360m of public transport. Energy is calculated as a per capita energy efficiency goal of 2-5 tons of oil a year. Resources are based on a metric based on a ratio of the generation of solid waste to the local recycling; a sustainable urban population is based on a population density of 100-170 people per ha. A functioning urban ecology become possible when the percentage of open space to built space in a city reaches 1.75 to 2.75 ha per person, (a more nuanced scale may be that of the ecological footprint, that is how much land and water are needed to support the needs and concurrent wastes of the city). Liveability is the last quantitative measure that the authors advance, they use the United Nations development programme, human development index, and a goal of.911 is adapted as a goal for an environmentally efficient city. Turning to quality measurement, the authors canvass six measures, the first is air quality management, ten-second watershed management, the development of an efficient public transport network, the adoption of renewable energy sources, the reduction of resource consumption, and the desire for densification and liveability. The author's use these criteria to evaluate a given urban project at a number of scales.

What this research demonstrates is that in order to proceed towards the development of sustainable city, a reconceptulisation of the city as a larger environment needs to take place. We could use the term 'landscape', as a phrase that is geographically located and grounded within an ecology that works at a number of scales; hydrological, horticultural, and topographical to describe this new territory.

Vallega's second driver behind the development of the sustainable waterfront is Globalisation. Bunce and Desfor (Bunce, August 2007) suggest that it is critical to frame the contemporary waterfront as part of a larger world where broader transformational practices occur at a variety of scales. The result is that the waterfront is a kind of test bed, for 'the production of nature, patterns of social entanglement and the political – economic configuration in cities'. The authors use the trope of ecology to explore two important aspects of waterfront development, political ecology and social ecology. Both these model seek to escape the idea of the waterfront as a static fixed geography territory. Rather, the political waterfront is a field of competing forces, located in shifting sands of global capitalism.

The authors use the notion of a social ecology to point to the social construction of nature. 'Nature is an integral component of the history of power relationships and economic production on urban waterfronts'. Nature and society are not part of some enlightenment dualism but instead, unavoidably connected, an assemblage of inseparable social processes and material forms of nature. It is in this light that the authors consider the new 'green' vision of sustainable waterfront development. The authors cite a number of studies on contemporary waterfront development that proudly parade there sustainable and green credentials while morbidly pointing out the pretentions both on the actual scale of the development and as part of a larger political economy.

Moving from the broader concepts of how to measure and critique the development of a sustainable city to investigate the ways in which a sustainable waterfront might develop, means confronting both the economic drivers and the environmental reality of the contemporary waterfront. The underlying economic drivers of waterfront development are primarily real estate speculation. Any tendency to consider the contemporary move towards a more sustainable waterfront as some kind of benign or altruistic shift needs to frame this particular phenomenon within the larger economic shifts of the 1990s. The triumph of globalization after the fall of communism saw an unfettered capitalism dragged the world, ready or not, into a globally connected capitalist hegemony.

There are however the indisputable environmental problems of many waterfronts, the legacy of an industrial past. Contaminated stormwater and sewage entering harbours, contaminated seabeds, the accumulation of debris from hundreds of years of shipping, and contaminated terrestrial sites, are just some of the issues. These deep and complex environmental problems are becoming more prominent as the public understanding of the importance of a sustainable future for the city becomes more advanced. For the underlying structural mechanism of waterfront development to recognise these fundamental problems, a more systematic investigation into the ways in which waterfront development can be made more sustainable has to take place.

MAIN FOCUS OF THE CHAPTER

Issues, Controversies, Problems

Contemporary waterfront development presents certain commonalities; a specific economic machinery in the planning of waterfront development and, simultaneously, specific environmental problems. Understanding the generic procedures followed in contemporary waterfront development helps to frame the environment under which the search for ways to make a sustainable waterfront must operate..

Comprehensive development studies of waterfronts are generally made up of two parts; design concept/master plan and market study. These two parts can be carried out simultaneously, or in sequence, their timing feeds into each other, the net effect is a comprehensive connection between design and economics (FCA, 2005). Many urban design master plans follow a particular urban model developed by the new urbanist movement in the 1980s (Kunstler, 1994). Based on traditional American urban settlement patterns originally developed in the 18th century and replicated in the 1920s and 30s, new urbanist city planning favours a grid street layout with public spaces determined by a classical urban typology of square, park, and promenade, all architecturally defined by buildings. Civic buildings are located in these spaces, spatial devices such as axis underscore their importance. Classical urban planning principles, symmetry and proportions help order the civic plan. The best-known examples of new urbanist planning are the towns of Seaside and Celebration in Florida (Grant, 2006). While these towns have been praised for their success in terms of residential and retail development, new urbanist designers have found the incorporation of commercial offices and industrial buildings more problematic. Urban designers has fallen back on modernist planning ideas by using a functional spatial zoning of industrial and commercial uses. Looking specifically at the effect of this ideology on waterfront development, a generic formula often emerges. The waterfront project frequently occupies an old industrial site next to a water body; lake, stream, river, or sea. The formula uses two design devices; generally there is the provision of a public space alongside the water body, a walk, a promenade, or a park. The body of the site is generally subdivided by a road grid, separating the site into a number of different sections, these blocks are further segregated by a zoning imperative. The location of residential, commercial, industrial, and retail zones are driven by the proximity of the zone to appropriate infrastructural elements, for example, residential and retail areas are generally placed next to the water edge, industrial areas are placed next to an transport arterial route.

The accompanying market analysis is usually divided into two parts, the first is an analysis of the existing market. This looks at a regions demographic, work opportunities, and economic future. The second part of the analysis looks at the markets existing real estate history, in particular, residential and commercial sales. For this analysis, past records are used to determine which real estate products are liable to do well in a certain area, for example single story houses versus apartments. Similarly, commercial real estate products can be usefully divided into groups, offices, power centre, retail and so on. The second part of any market analysis is how the development might be financed; in America there are a number of way of achieving this essential equation. One important model, particularly useful for joint ventures between developers and local authorities is for the cost of any development to be offset against the subsequent creation of taxes on land improvement. The immediate funding requirement can be met in a number of ways including bond issues. The following case study demonstrates how some of these themes are developed in a real world situation.

District of Mission Waterfront Study Project: Mission Landing

In 2006, the District of Mission, BC, Canada, commissioned MVH Urban Planning and Design (MVH, 2006) to produce an urban master plan for the development of a waterfront site adjoining the town. The site is defined by the Lougheed Highway to the north, the Fraser River to the south, the Mission raceway to the west and the CPR ROW to the east. The site is situated on a flood plain though partially protected by the Mission dyke along the Fraser River. Historic use of the site was residential in the 19th century changing to commercial and industrial, especially forestry related operations, in the 20th century.

The economic drivers behind the proposed waterfront development were; the riverfront location, the redundancy of the present industrial uses, and good accessibility from the existing downtown rail and commuter routes. Other strengths the project presents are; the opportunity to increase the commercial and residential tax base through the revitalisation of the waterfront and the way the development will act as an investment spur for other nearby development. The price of the residential component in the project will be cheaper than in neighbouring areas, the site presents a strategic location close to Abbotsford (a regional economic driver), and the project will allow the new home owners to connect to a wide range of outdoor regional activities.

The original design concept by MVH divided the site into five precincts, west, west central, north, east, and core. The north precinct is located nearest to the existing down town though separated by a railway and the Lougheed Highway. This area is proposed for residential use with the suggestion of building large residential towers along the highway boundary. The east precinct is proposed for civic functions that are to be grouped around a large internal park/commons space. Core is proposed to be a mix of commercial and residential uses, the area has a direct relationship to the river, as does the west central precinct, the proposed function here is a mix of residential and business park. The west precinct is proposed to be devoted to a business park. The proposed design conditions along the river, changes from an urban promenade with a variety of wharfs and basins in the core and east precincts to a more park like edge in west central and west precinct.

An urban design and market analysis review in 2010 (GCE., 2010) supported the broad outline of the original study. The market research found support for the residential and civic components of the original scheme, located in the core, north and east sectors, however the proposal for a business park in the west central and west precincts was found to be commercial unviable, an auto mall and power centre were proposed respectively for these sites.

The urban design review by Patrick Cotter Associates (GCE., 2010) broadly concurred with the original design *parti,* suggesting some refinements by; breaking the urban blocks up into a smaller, finer grained grid, making better connections to the existing downtown, and suggesting denser, taller, residential construction, up to 5 stories high, adjoining the river. In summary, the proposed master plan treats the site as a tabula rasa, the project is divided into functional zones each with a specific real estate use, the river is treated as a spectacle that is addressed by a pedestrian promenade.

In addition to the master plan and market analysis an environmental study of the waterfront site was also carried out by Aplin and Martin Consultants in 2009 (Martin, 2009). This study identified five specific environmental problems with the site.

- The ground of the north and east precincts is heavily contaminated, probably as a result of the previous industrial use.
- The dyke protecting the site was deemed inadequate as it is physically discontinuous and the construction methods were unknown.
- The underlying geology of the riverbank is susceptible to slipping in an earthquake.
- The groundwater table is very close to the surface.
- Contaminated stormwater is being discharged into the river.

The consultants describe a number of remediation measures for these serious environmental problems. To address the possible failure of the dyke and the risk of flooding, the authors suggest filling the flood plain; this however would require something in the order of 1. 4 million tones of fill, an expensive and time consuming solution. The author's recommendation for the contaminated sites, the proposed centre of the civic and residential precincts is not to avoid construction in these areas, as the necessary remediation of the contaminated ground would be too expensive. This study demonstrates the disconnect between the real estate/master plan conception of the waterfront development and the actual environmental particularities of the site.

While waterfront developments may be shy of connecting the necessity of large-scale environmental remediation with the unsustainable development of the design master plan, many contemporary Ports have carried out considerable research into the identification of environmental problems in new Port developments and the economic benefits of remedial measures

The Portfields project (EPA, 2004) is one example of how the economic development of Port infrastructure can be advanced with the development of particular environmentally sustainable goals. The Portfield project was initiated in 2004 as a joint venture amongst a number of federal agencies in America, including the US Environmental Protection Agency and the National Oceanic Atmospheric Administration. The Portfields project makes an explicit link between the remediation of environmental problems found in many Port sites and the improvement of the social dynamics of port communities. Three Port project sites were chosen; Bellingham in Washington, New Bedford in Massachusetts, and Tampa in Florida. They share similar environmental issues; the safe disposal of dredging and the remediation of brown field sites. The projects also have the same community aspirations; plans for community revitalization, air and water quality improvement, habitat protection and the provision of new green spaces.

The Tampa Port project (Authority, 2005) provides one of the most interesting case studies. Four specific environmental problems are detailed and remediation techniques are canvassed. The issues in Florida are; brown field sites, disposal of port dredging, the recreation of indigenous marine flora habitats, and stormwater remediation. Three brownfield sites are explored; Port Ybor, a former shipyard for the Department of Defence, has serious contamination problems with petroleum solvents and metals present in the ground. Future plans for the site, once it has been remediated, are for a cruise ship terminal, light industry, offices, and warehousing. Tampa Scrap processing is a 10-acre scrap yard, contaminated with petroleum, solvents and metals, after remediation it will be used as a break bulk terminal. Tampa Shipbuilding and Repairs is still a working site, polluted with petroleum, solvents and metals, remediation will enable the business to expand operations. The second project is the restoration of a native flora as part of a wave energy study. A problem with shoreline erosion is to be partially rectified through the creation of an artificial longshore bar that will reduce the effects of wave energy and help in the restoration of a 450-acre sea grass colony that will be planted behind the bar. Disposal of dredging from the ship channels in Tampa Bay is the third project. Traditionally the dredged sediment have been deposited in two artificial islands, 2D a 500-acre island and 3D a 400-acre island, located to the south of the port between the McGill AFB and Gibsonton. They have become defacto bird sanctuaries especially for the American Oystercatcher, however increased shipping traffic is causing edge erosion. The project investigates techniques to promote erosion control and increase habitat attractiveness. The last project is an investigation into the treatment of stormwater from the Port. The creation of a GIS map of the Port with a comprehensive inventory of land uses and stormwater infrastructure is the first step. The aim of the project is to identify useable properties in close proximity to existing stormwater drainage systems. These can be used as remediation ponds for stormwater treatment and cleaning. All four projects demonstrate a sincere commitment to both understanding the environmental problems caused by industrial Port use and the use of innovative remediation technologies.

The development of remediation technologies is becoming increasing sophisticated. There is considerable research into the development of remediation technologies that are applicable for waterfront remediation. Two important fields are; the remediation of contaminated dredging sediments and the remediation of contaminated stormwater. The technologies for the remediation of contaminated marine sediment can be separated into two areas. One is in-situ treatment, that is, treating the contaminated sediment while it is still in the water, the other is ex-situ treatment, removing the sediment though dredging and then treating it through a number of methods. In-situ treatment can be divided into two main techniques, capping and containment. Capping sediment involves placing a layer of clean material over the contaminated sediment, effectively isolating it. Containing sediment means isolating the contaminated waterway with a physical barrier, such as sheet piling or a cofferdam. This zone can also then used for the disposal of other contaminated sediments. The area can capped or treated with a biological treatment, this action uses microbiological processes to promote processes that help transform contaminates in the sediment to less toxic forms. Two treatment methodologies are used to decontaminate ex situ sediment; physical chemical treatment and thermal treatment (Reis, 2007).

One specific organic technique that has found success with treating contaminated dredgings is phytoremediation. This method uses plants, normally salix species, placed into the contaminated sediment. As the plants grow contaminants are taken up into the leaves of the plants while the roots dewater the soil and allow oxygen penetration into the soil profile. This technique has been used successfully used with petroleum based contaminates (El-Gendy AS, 2009).

The remediation of contaminated stormwater is a better-known process. A constructed wetland is used as an organic machine to clean contaminated stormwater. The wetland remediation model has a three-stage process where contaminated stormwater moves through the three zones, gradually depositing rubbish and sediment through a series of vegetative screens. The cleaned stormwater is then released into natural watercourses, streams, or the sea (Shaver, 2000)

Solutions and Recommendations

The problems of contemporary waterfront development are clear, the construction of a new waterfront city from industrial port land often ignores or elides the contaminated nature of the site for a variety of reasons. Economic motives are important, site remediation is often expensive, involves the engagement of multiple agencies, and can becomes politically difficult while the generic real estate formula for the development of waterfronts is well known and well proven economically. However a growing consciousness of the global environmental crisis means that is no longer possible to make any sort of urban development without paying at least lip service to sustainability.

Within this milieu, the author will sketch out a tentative five-stage methodology for the development of a sustainable waterfront. The first step is to change the conventional dimensions of the waterfront development site, where the limits of the project are determined by property boundaries and the water is treated as a passive blue edge. The site needs to be reframed as a larger, connected landscape. Two operations must take place; the first is a scaling operation, the scale of the conventional waterfront plan must be enlarged. This operation could take place at a range of scales, nesting within each other, but each scale would have to show some specific landscape/environmental condition, for example, one scale map/plan would be at the watershed/catchment level, another would be at a regional level, showing conditions such as existing vegetation patches. The second operation is a change in representational technology towards a system that is able to take cognisance of the larger landscape and ecological forces. An obviously choice is GIS, a mapping technology with a known ability to represent territory at a number of scales and conditions. The aptitude of this software to combine conditions, to generate new maps, and to build maps of active environmental conditions such as hydrological states, is well recognized (Campagna, 2006; Easa, Chan, & American Society of Civil Engineers. Geographic Information Systems Committee., 2000; Maguire, Batty, & Goodchild, 2005; Wadsworth & Treweek, 1999).

The second step in this tentative hypothesis would be an attempt to recreate the indigenous waterfront. While the aboriginal condition have obviously been lost in many waterfronts, in some cases for centuries, an understanding of these original conditions is important as part of any future remediation strategy. Mapping the original waterfront, both terrestrial and marine is important as is determining the original hydrological conditions, both saltwater and fresh water. Understanding the original coastal flora and its relationship to the unique topography is equally important.

The third step is to understand the specific location of waterfront contamination both terrestrial and marine. Brownfield sites, contaminated dredging, contaminated stormwater emissions are al important sources of waterfront pollution. The specific types of contaminants, volumes, states, points of discharge, must be determined. Dynamic conditions such as stormwater discharge and sea level rise must also be mapped.

The fourth step is the canvassing of suitable remediation techniques for specifically identified environmental problems. For example, in a previous section, we have seen how contaminated stormwater can be remediated through the use of constructed wetlands and brownfield sites can be decontaminated through phytoremediation.

The last step is to represent the waterfront site at the new landscape scale and to privilege the environmental remediation operations, combining them where possible with native flora. An example of this technique would be using native plants in the building of a stormwater wetland remediation system or using native plants as part of the phytoremediation process.

The implications of this tentative hypothesis is the development of a process, that presents the waterfront at a scale that encompassed the larger landscape, that identifies areas of environmental contamination, that presents the spatial consequences of remediation operations, and demonstrates how folding in a native ecology would combine to present a radically refigured territory for the operation of waterfront development.

Auckland Waterfront

A specific design case study, The Queens Wharf redevelopment competition, will be put forward to test some of the ideas that have been raised in the design hypothesis. Queens Wharf is located on the Auckland waterfront, New Zealand. The waterfront lies on the southern edge of the Waitamata Harbour, it is approximately 4.5 km. long from Westhaven, a large commercial marina, in the west, to the, Ports of Auckland, in the east. Topographically the waterfront is an apron of reclaimed land to the north of the Auckland CBD. The central city lies over three valley/gullies systems that run north/south, at right angles to the waterfront. These gullies form three urban catchment systems (Bush, 1971).

The waterfront has developed very much in the manner described by the Hoyle model. The Port operations, begun in the 1840s, have buried the native shoreline with a large-scale incremental reclamation programme. In the 1960s, the whole waterfront was occupied by Port activities, divorced from the rest of the city by a security barrier known locally as the 'red fence'. Since the 1980s the Port has gradually withdrawn its commercial operations from the western sector of the waterfront while expanding its operations into a container terminal built in the eastern part of the

Port. There have been a number of commercial waterfront developments in the vacated land; the Viaduct Basin and now the Wynyard Quarter are two successful projects, that have closely followed the generic waterfront development prescription ("Auckland Waterfront Authority," 2011).

Like many waterfronts around the world, the Auckland waterfront has serious environmental problems. There are three major environmental issues; a large number of brownfield sites are present on the waterfront especially on the Wynyard Quarter. A tank farm for mainly petroleum products stood on this site for many years contaminated the surrounding soil from leaking tanks

The Ports of Auckland has an active dredging programme, both to remove the accumulation of contaminated sediment from stormwater discharge into the harbour and to increase the channel size for larger ships. The Port has traditionally disposed of the dredging in a site in the adjacent Hauraki Gulf; this dumping had incurred controversy with claims of the despoliation of valuable snapper spawning grounds. More recently the Port has used the dredging, combined with cement, to form a stabilized landfill for the expansion of the container port. ("Ports of Auckalnd," 2010) The third major environmental issue on the Auckland waterfront is the contamination of the Harbour from untreated stormwater. The three catchment zones of the Auckland CBD; Parnell, Queen Street, and Freeman Bay, are highly urbanized with mostly impermeable surfaces. This condition, combined with Auckland high rainfall and often sudden torrential down pours results in violent stormwater surges. Stormwater discharge points are located at a number of positions along the harbour edge. In some cases, storm bombs and single point outfalls have combined to produce large scale discharges resulting in polluted stormwater plumes extending into the harbour by as much as a kilometre.

The Queens Wharf Competition: Design Case Study (Bradbury. Matthew, 2009)

In 2009, the Auckland City Council held a design completion for a new public space to be developed on the Queens Wharf ("Queens Wharf Competition," 2009). The brief was to provide a 'party central' for visitors to the Rugby World Cup, (the final game is to be held in Auckland in September 2011). The brief was to transform the Queens Wharf, from a site occupied by two large corrugated iron cargo sheds used for imported car fumigation to a communal gathering space for Aucklanders and visitors alike. However the completion brief isolated Queens Wharf as a terrestrial site, an extension of the city, with a limited social and nonexistent environmental programme. The author decided to use the competition as an opportunity to develop a design project that connected to the larger environmental problem/issues of the Auckland waterfront and demonstrated how the use of environmental remediation technologies could lead to a radically reconfigured urban plan. The competition was a chance to test the design process hypothesis.

The first step in testing the design hypotheses was to build a digital terrain model of downtown Auckland and the waterfront (Figure 1). This model, stripped of buildings and civil infrastructure, reveals the underlying topography of the city, the ridge/gully catchments and reclaimed land. Drilling down to Queens Wharf, the site can be seen not simply as an extension of the immediate street network but as part of a larger catchment system defined by the Hobson Street, Karangahape Road and Symonds Street ridges. Queen Street, the traditional main street of Auckland, runs down the centre of the valley, the whole catchment, roughly a hectare in size, opens to the north and the harbour.

Figure 1. Digital terrain model of downtown Auckland and the waterfront

The second step of the process is to determine the native flora of the old waterfront. Based on research into similar sites on the Waitamata, the two most common environmental conditions that would have been found on the original site were, coastal cliff and coastal wetland ecotones. A typical transect of the coastal cliff ecotone reveals such species as; Pohutakawa, *Metrosideros excelsa*, Karo, *Pittosporum crassifolium*, Taupata, *Coprosma repens*, Houpara, *Pseudopanax lessonii* and *Astelia banksii.* A transect through the coastal wetland ecotone reveals a gradient of native species determined by water conditions; fresh water, brackish, or salt water. Common species are; Typha *orientalis, Baumea articulata,* and *B. rubiginosa* (Morton, 1993).

The next step is to determine the location of any contaminates. The Queens Wharf site presents two critical environmental issues, a contaminated seabed that requires regular dredging and polluted stormwater discharges. Stormwater from the Queen Street catchment is presently collected in a large tank under the Captain Cook wharf, adjacent to Queens Wharf, before being discharged untreated into the harbour.

The fourth step is to look for appropriate remediation techniques. Two green remediation technologies, as outlined in the previous section, would be used as possible solutions to the sites particular environmental issues. Contaminated stormwater would be remediated through a constructed wetland and the polluted seabed would be remediated through containment and phytoremediation. The most obvious problem with these solutions is the simple spatial requirements associated with the technical specifications for these treatments vs. the densely occupied site of downtown Auckland. This is a common issue on almost every waterfront project; there is no space available for remediation measures, especially when the typical waterfront real estate programme is initiated. This is where the design hypothesis sketched out in the previous section meets with the reality of the site and real estate intentions. But it is also where the larger conceptual framework of the hypothesis the connection of site to larger landscape can be drawn upon. By rescaling the site to not only represent the terrestrial water catchment but also the coastal condition of the littoral including the tidal rise and fall and the underwater topography a broader site for remediation operations are revealed.

The case study now makes a leap, conceptual and physical, to the water's edge, rather than accepting the water as a blue spectacle, the design proposal reforms the edge as an intermodal zone, a constructed gradient between freshwater and salt water. To form the new littoral between the two waters, an artificial wall made up of the contaminated dredging is proposed. As the contaminated dredging settles, the top layer can be planted with salix species, initiating a phytoremediation process to clean the pollutants from the dredging. As the process of decontamination proceeds, native species, drawn from the original coastal ecotones, can be substituted for the salix species. The proposed dredging barrier creates a kind of lagoon between the land and the harbour. This new topography creates an opportunity to construct an artificial wetland to treat contaminated stormwater by redirecting the existing stormwater flow from the Captain Cook wharf outfall to the new Queens Wharf lagoon. The technical requirement of the new wetland would derive from the amount of contaminated stormwater discharged from the Queen Street catchment. Within the constructed lagoon a typical three bay stormwater treatment system can be constructed, native species are introduced as part of the remediation process (Figure 2). The use of indigenous plant species in the stormwater cleaning process and the gradually replacement of exotic species with native species in the dredging remediation process will over time lead to the creation of an indigenous ecotones, a saltwater/freshwater littoral.

Figure 2. Queens Wharf Lagoon

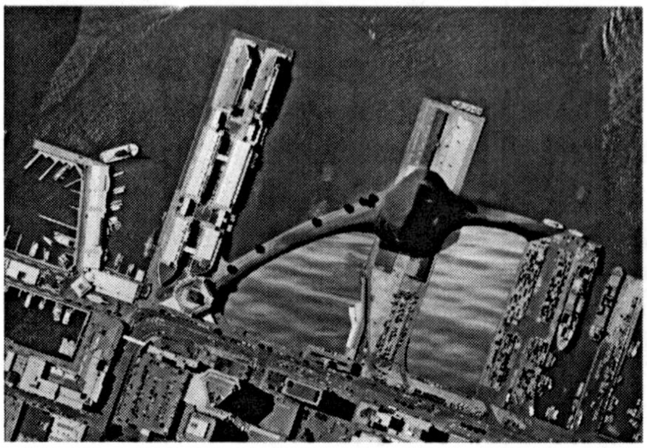

The Queens Wharf competition design case study helps to illustrate how two environmental remediation techniques; stormwater cleaning, and dredging remediation could be bought together to build an environmental infrastructure that simultaneously cleans stormwater and remediates polluted dredging. The design case study while illustrating a small and local design problem helps to point to a number of future scenarios in the development of a sustainable waterfront.

FUTURE RESEARCH DIRECTIONS

Two speculations about future research directions could be advanced through the results of this design case study. The first is to examine the implication of this process for the construction of the contemporary urban waterfront. The generic waterfront design model as outlined earlier in the chapter, directs the citizen as a passive spectator/consumer in the constructed waterfront spectacle. The particular spatial typologies; the promenade, the waterfront square, and the associated retail programme, all determine the way the contemporary urban waterfront is experienced by the citizen. The Queens Wharf design case study offers the citizen a radically reconfigured urban experience, one in which new kinds of urban encounters are engendered by a hybridized nature/city, a remediation machine (Figure 3).

The other speculative direction this design project leads to is the effect a fragment of native flora could have on the making of a new ecological structure for the contemporary city. The idea of the city as part of an urban ecology is not new, in the paper, Ecology and Urban Planning (Niemela, 1998). Jari Neimela, seeks to define the field of urban ecology and associated techniques. Neimela describes urban ecology as being 'ecological research in the urban setting' (Rebele. 1994). Neimela discusses how to incorporate ecological knowledge with urban planning regimes, what kind of nature exists in the city and

Figure 3. The Queens Wharf design and remediation

how can it be quantified. Neimela explores biotope-mapping techniques, information about location size and species compostion of biotope patches. This data can be measured and presented as maps and databases. Neimela discusses the knowledge of specific processes of human induced disturbance and the effects on succession. Lastly Neimela discusses specific eco system management regimes designed for urban nature to maintain or encourage high species diversity and habitat richness. An important goal of urban ecology is biodiversity; Neimela defines biodiversity in relationship to urban planning as both a mechanism by which ecological systems adapt to their environment and a buffer that ensures systems resilience against disturbance. This concept leads to a principal of maintaining biodiversity through a number of deliberate design strategies such as the provision of parks/green areas and the protection of existing natural areas in the city. Examining the concept of biodiversity in the city, the authors; Sandra Sweeny, Emrah Engindeniz and Sultan Gunduz discuss this idea in the paper, Ecological concepts necessary to the conservation of biodiversity in urban environments (Sweeney, 2007).The authors begin with a definition of biodiversity; 'Biodiversity is the diversity of life and processes in community's ecosystems and landscapes' (West 1993) Biodiversity critically operates over a range of scales. Biodiversity is made up of three elements compositional, structural, and functioning parts of an eco system (Noss, 1992) an example would be species (compositional) is linked to habitat diversity (structural), that is dependent on containing renewal through natural processes (functional). The concept of island biogeography, (MacArthur, 1967) is used to understand the location of nature within the urban environment. Two processes determine the viability of organisms within the 'island'; these are the habitat patch and the dispersal distance between patches, or the species relationship. The authors discuss the reasons how biodiversity can become lost, the term MVP or minimum viable population is introduced. Reasons why species populations can become extinct are discussed; reduced generic diversity, and changing relationship between the 'island' (sink) and the 'mainland' (source). With increasing urbanisation, mainlands themselves become islands, lost of linkages between islands is another source of species extinction. The tendency of city development to homogenise all activity fields including the biological is discussed, evidence in studies of the development planning and building of parks are examples of this tendency. The main actors/agents are; a growing proportion of non-native species becoming established, disturbance to resources, natural enemies, and the physical environment. The authors finish by outlining a way in which cities can become ecologically sustainable by connecting to a landscape structure that generates biodiversity through support of ecological processes. The techniques of attaining such goals are discussed by the authors Matthew Luck and Jianguo Wu, in the paper 'A gradient analysis of urban landscape pattern, a case study from the Phoenix metropolitan region, Arizona USA" (Luck, 2002). This study is an investigation into an attempt to relate the urban growth of Phoenix to ecological processes using gradient analysis and landscape pattern analysis. The author's uses a number of landscape metrics taken from McGargal and Marks (McGarigal, 1994) these are; Patch richness, this is based on a number of factors, the number of patch types in the landscape, a measure of diversity of patch types, class percent of landscape, the proportion of total area occupied by a particular patch type and a measure of dominance of patch types. Largest patch index, the proportion of total area occupied by the largest patch of a patch type. Patch density, the number of patches per 100ha. Mean patch size, the area occupied by a particular patch type divided by the mean patch size; a measure of relative variability, landscape shape index, the landscape boundary and total edge within the landscape divided by the total area, adjusted by a constant for a square standard area weighted mean shape index, a mean patch based shape index weighted by patch size.

The specific metrics that this research propose, could connect and extend the Queens Wharf project through a scalar telescoping of the site to include evidence of regional native vegetation. From this data the viable size and species diversity of the case study could be modified to enable it to perform as a viable patch, as part of the regional ecology.

CONCLUSION

This chapter explores a topical subject, the sustainable development of the waterfront through the lens of environmental remediation. This chapter has canvassed four fields of research; the way in which conventional waterfront developments are framed and developed, the serious environmental problems that many waterfront sites have inherited as part of their industrial heritage, the wide variety of environmental remediation technologies available, and lastly, the field of sustainable urbanism and its subset, urban ecology.

From these four fields, a speculative hypothesis is advanced that attempts to bring these sectors together in fruitful and productive work. The contemporary waterfront is rescaled as a landscape, a part of a larger urban environment. Studied as an ecology, many contemporary waterfronts have serious and long-term environmental damage, a broken part of a larger ecological system. An investigation into the ways in which this damage can be repaired using green remediation technology is advanced. A hypothesis is put forward that offers a way in which remediation processes could lead new urban design procedures. A design case study is advanced to test the adeptness of the hypothesis. The case study provides a concrete example of a way in which the hypothesis might be tested and at the same time gestures to a larger realm in which the waterfront might act as part of a large-scale urban ecological rehabilitation. The structure of the chapter moves between scales, from the larger theoretical concerns of urban sustainability, drilling down to the tangible interaction between environment remediation techniques and waterfront development, then telescopes out to show how this study can also gesture towards the larger realm of social and ecological restoration.

REFERENCES

Auckland Waterfront Authority. (2011). *Website*. Retrieved February 2011, from http://www.waterfrontauckland.co.nz/

Bradbury, M., Ison, N., Parlane, J., & White, J. (2009). *Queens Wharf competition*. Auckland, New Zealand: Auckland City Council.

Breen, A., & Rigby, D. (1996). *The new waterfront: A worldwide urban success story*. London, UK: Thames and Hudson.

Brugmann, J. (2010). *Welcome to the urban revolution: How cities are changing the world*. New York, NY: Bloomsbury Press.

Bunce, S. A. D. G. (2007). Introduction to political ecologies of urban waterfront transformations. *Cities (London, England)*, *24*(4), 251–258. doi:10.1016/j.cities.2007.02.001

Bush, G. W. A. (1971). *Decently and in order*. Auckland, New Zealand: Collins Bros. and Co.

Campagna, M. (2006). *GIS for sustainable development*. Boca Raton, FL: CRC Press.

Dale, A. (1947). *Fashionable Brighton, 1820-1860*. London, UK: Country Life.

Easa, S., & Chan, Y. (2000). *Urban planning and development applications of GIS*. Reston, VA: American Society of Civil Engineers.

El-Gendy, A. S. S. S., Brice, D., Garretson, J. H., & Schnoor, J. (2009). Assessments of the efficacy of a long-term application of a phytoremediation system using hybrid poplar trees at former oil tank farm sites. *Water Environment Research*, *81*(5), 486–498. doi:10.2175/106143008X357011

El Harim, M. (2007). *Development of an intergrated sustainability assessment toolkit*. Retrieved from http://download.sue-mot.org/Conference-2007/Papers/El-Haram.pdf

EPA. (2004). *Portfields. A coordinated approach to revitalizing ports*. Retrieved from http://www.response.restoration.noaa.gov/book_shelf/1656_0904port_report.pdf

FCA. (2005). *Knoxville South Waterfront redevelopment plan*. Retrieved February 2011, from http://www.Knoxville-South-Waterfront-Redevelopment-Feasibility-Study.html

Gastil, R. (2002). *Beyond the edge New York's new waterfront* (1st ed.). New York, NY: Princeton Architectural Press.

GCE. (2010). *Summary report and recommendations for phase 1 – Market analysis*. Retrieved December 2010, from http://www.mission.ca/wp-content/uploads/Phase-1-Market-Analysis-Working-Notes-July-29-10.pdf

Government of New Zealand. (2009). *Queens Wharf competition*. Retrieved October 2010, from http://www.arc.govt.nz/auckland/waterfront-development/queens-wharf/queens-wharf_home.cfm

Grant, J. (2006). *Planning the good community: New urbanism in theory and practice*. Oxford, UK: Taylor & Francis.

Hodson, M., & Simon, M. (2010). *World cities and climate change*. Maidenhead, UK: Open University Press.

Hoyle, B. H. D. (1981). *Sea port and development strategies*. London, UK: Wiley.

Kunstler, J. (1994). *The new urbanism: Toward an architecture of community*. New York, NY: Mcgraw-Hill.

Luck, M., & Jianguo, W. (2002). A gradient analysis of urban landscape pattern: A case study from the Phoenix metropolitan region, Arizona, USA. *Landscape Ecology*, *17*(4), 327–339. doi:10.1023/A:1020512723753

MacArthur, R. W. E. (1967). *The theory of island biogeography*. Princeton, NJ: Princeton University Press.

Maguire, D. J., Batty, M., & Goodchild, M. F. (2005). *GIS, spatial analysis, and modeling* (1st ed.). Redlands, CA: ESRI Press.

Martin, A. A. (2009). *Waterfront brownfield redevelopment study technical overview.* Retrieved February, 2011, from http://www.mission.ca/wp-content/uploads/28544-Waterfront-Brownfield-Redevelopment-Study-Technical-Overview-Sept-24-reduced1.pdf

McGarigal, K. M., & Barbara, J. (1994). *Fragstats. Spatial pattern analysis programme for quantifying landscape structures.* Retrieved December, 2010, from http://www.umass.edu/landeco/pubs/mcgarigal.marks.1995.pdf

Meyer, H. (1999). *City and port: Urban planning as a cultural venture in London, Barcelona, New York and Rotterdam: Changing relations between public urban space and large-scale infrastructure.* Utrecht, The Netherlands: International Books.

Morton, J. A. C. (1993). *Shore vegetation.* Auckland, New Zealand: David Bateman Ltd.

MVH. (2006). *PD mission.* Retrieved January 2011, from http://www.mvhinc.com/pd_mission.shtml

Niemela, J. (1998). Ecology and urban planning. *Biodiversity and Conservation, 8,* 119–131. doi:10.1023/A:1008817325994

Norris, D. (2003). *Build it, they will come.* M.E. Sharpe.

Noss, R. (1992). Indicators for monitoring biodiversity: A hierarchical approach. *Conservation Biology, 4,* 355–364. doi:10.1111/j.1523-1739.1990.tb00309.x

Ports of Auckland. (2010). *Commmunity environment: Dredging.* Retrieved January 2011, from http://www.poal.co.nz/community_environment/dredging.htm

Reis, E., Lodolo, A., & Stanislav, M. (2007). *Survey of sediment remediation technologies.* Retrieved February 2011, from www.ics.trieste.it/media/117892/df5310.pdf

Shane, A., Megan, G., & Thomas, E. (2000). Urban environmental sustaianability metrics: A provisional set. *Journal of Environmental Planning and Management, 43*(5), 643. doi:10.1080/713676586

Shaver, E. (2000). *Low impact stormwater management.* Auckland, New Zealand: Auckland Regional Authority.

Spiekerman, K., & Michael, W. (2003). Modelling urban sustainability. *International Journal of Urban Science, 7*(i), 47–64. doi:10.1080/12265934.2003.9693522

Sweeney, S. E., & Emrah. S. G. (2007). Ecological concepts necessary to the conservation of biodiversity in urban environments. *ITU A|Z, 4*(1), 56-72.

Tampa Port Authority. (2005). *Portfields project descriptions.* Retrieved from http://www.response.restoration.noaa.gov/book_shelf/1670_05tampa_portproj_descrip.pdf

UNCED. (1992). *Conventions on biological diversity.* New York, NY: Department of Public Information. United Nations.

Vallega, A. (2001). Urban waterfront facing integrated coastal management. *Ocean and Coastal Management, 44,* 371–410. doi:10.1016/S0964-5691(01)00056-4

Wadsworth, R., & Treweek, J. (1999). *GIS for ecology: An introduction.* Harlow, MA: Addison Wesley Longman.

WCED. (1987). *Our common future.* New York, NY: Oxford University Press.

West, N. (1993). Biodiversity of rangelands. *Journal of Range Management, 46,* 2–13. doi:10.2307/4002440

Williams, S. (1993). *Docklands.* London, UK: Phaidon Press.

This chapter was previously published in Green and Ecological Technologies for Urban Planning edited by Ozge Yalciner Ercoskun, pages 274-292, copyright 2012 by Information Science Reference (an imprint of IGI Global).

Chapter 94
Multifunctional Agriculture and the Green Economy

Andrei Jean-Vasile
Petroleum and Gas University of Ploiesti, Romania

ABSTRACT

Multifunctional agriculture represents a proactive approach of the role and effects of agriculture both on rural communities and traditions. The multifunctional agriculture passes beyond the classical functions of the agriculture bringing new gainful activities for the rural farmers. Understanding properly the multifunctional agriculture and farm diversity can offer all those involved new trends in valuing agricultural potential. This chapter presents a short but very comprehensive analysis regarding the multifunctional agricultural concept and farm diversity, in the context of the massive changes that have marked the actual evolution agrarian economy. It presents the evolution of multifunctional agriculture concept in the international approach and also analyzes the agriculture functions. In addition, it presents a short study case regarding the evolution of farm structure development in the Romanian agrarian sector.

THE CONCEPT OF MULTIFUNCTIONAL AGRICULTURE

In a general approach to the concept, multifunctional agriculture refers to all goods, products, and services resulting from agricultural activity. The transposition of the Rio Earth Summit documents (1992), in Agenda 21 of the EU has significantly contributed to the realization of this new concept, so the multifunctional agriculture obtained an increasingly important role in the scientific and political debate regarding the future of agriculture and rural development (Renting et al, 2009).

The concept of multifunctional agriculture is widely debated in scientific research and still rises living controversy at international level, nurturing over time with new meanings. Multifunctional agriculture is in the current economic development a defining element in the development of sustainable agriculture in the production relations

DOI: 10.4018/978-1-4666-4852-4.ch094

as a whole, encompassing both the economic, social and ecological functions but it especially is involved in reconciling different approaches to the role of agriculture in providing food security and the multivalent use of rural areas beyond its traditional functions. Referring to the term of multifunctionality, a group of Romanian authors (Luminiţa et al., 2003), claims that multifunctionality is the word that Europe uses to describe the fundamental link between sustainable agriculture, health security, natural balance, maintaining the quality of the landscape and of the environment, and all what is important for developing countries, particularly food security (Luminiţa et al., 2003).

The concept of multifunctional agriculture has recently appeared in literature, being defined for the first time in the Rio Earth Summit in 1992. The evolution of this concept has proved over time the adherence of agriculture to the changing demands imposed by practicing sustainable agriculture and to respect the multiple functions it is called to perform in terms of the role it plays in the national economy.

According to a British study (Marsden and Sonnino, 2008), the concept of multifunctional agriculture can be analyzed in terms of three approaches: as palliative to the productivist model and to reduce the price-cost ratio, as a form of regulation of consumption in rural areas and not least as part of the rural development process. (Marsden and Sonnino, 2008) The first approach reduces the term multifunctionality of agriculture to the notion of agricultural pluriactivity, as a survival strategy of farmers less competitive on the market and to reduce the harsh conditions imposed by the demands of a competitive market, being form as well as palliative against poverty in rural areas.

Multifunctional agriculture can be a form of regulation of consumption in rural areas, as a reflection of postproductivist paradigm, where the values are centered more on the character of diversification of land use. As outlined in the same study (Marsden and Sonnino, 2008), agriculture begins to lose its centrality in society, and nature is designed particularly in terms of landscape capitalization as a commodity.

Understood as part of the development of rural areas, multifunctional agriculture showcase in integrative form, both the productivist character of agriculture, as a branch of material production, and the positive externalities it generates. To benefit from the multifunctional attribute, agriculture, according to Terry Marsden must meet the three key conditions (Marsden and Sonnino, 2008), respectively: firstly it must generate income and employment opportunities for the agricultural sector, secondly it must contribute to the construction and development of a new agricultural sector to correspond to the needs and expectations of the whole society, and finally, it must involve radical redefinition and reconfiguration of rural resources, to different degrees and levels, especially of industrial agriculture.

Multifunctional agriculture defines a concept of particular significance that places agriculture at the confluence of the requirements imposed by the food security of the population and the multiple functions, social, environmental and cultural that it must realize at the level of rural communities by reconciling the conventional agricultural practices with the need for environmental protection and conservation of rural landscapes, imposing a new approach regarding the relation agriculture-environment-rural area.

Analyzing the concept of multifunctional agriculture, the British Professor Geoff Wilson states that this explains the multidimensional coexistence of the porductivist regime and therefore of the post-productivist one and thereby it comes to represent an accurate description of the multidimensional nature of the changes in rural environment and in agricultural ones (Wilson, 2001). In the evolution of the concept of multifunctional agriculture are worth noted three fundamental concepts promulgated by bodies such as FAO-WTO, OECD and the EU-27.

The FAO-WTO Multifunctionality Concept

In the view of the World Trading Organization (WTO), multifunctionality is a fad that must be removed, and agriculture should receive equal treatment, non-discriminatory, being treated under the same equality conditions with industry and the other economic branches. Starting from the idea that agriculture is distorted by applying various grants and subsidies as well as by the existence of trade barriers, affecting in this sense the free access to food, raw materials, and livelihood of the farmers around the world, it is necessary to reconsider these issues on agriculture and a fundamental reform of the economic sectors.

As advocated in literature, the arguments to justify the trade barriers and the preferential treatment that agriculture has undergone was never justified in terms of multifunctionality, because, multifunctionality never justified trade barriers or can justify the domestic production subsidies, the domestic support measures of agriculture vary by nation and the nations had to increase the importance of multifunctionality to disguise the internal measures of protection (Paarlberg et al, 2002). These arguments highlight the idea that, multifunctionality very often served as a tool for masking the protective measures and agricultural financing, amid deepening the global competitive conditions and trade liberalization. Multifunctionality providing in this sense the reason for continuing the subsidization of agricultural policies.

The agreement on Agriculture within the WTO is the result of the negotiations made during 1986-1994 within the Uruguay Round. During these negotiations, the states have agreed to remove the barriers affecting free trade in agricultural products, to reduce agricultural subsidies that distort the international trade, with some exceptions in the case of poor or underdeveloped countries. All these measures have made that the multifunctional theme shoul be the black sheep of multilateral trade negotiations between groups (Luminita et al, 2003).

Although the main objective of the Agreement is to reform the agricultural commerce with agricultural products and to imprint the agricultural policies a character more oriented towards market requirements, has not completely banned subsidizing the rural economies by imposing the condition that this financing is achieved through policies that not distort trade practices. Within WTO the reglementations on subsidies and grants paid to agricultural policies are structured in tapes, grouped by the degree in which they distort the commercial character and the measures of support designed for agriculture.

According to WTO practice, all internal measures to support agriculture which distort production and trade are grouped in the box orange (the amber box) for which there are set minims limits (5% for the developed countries and 10% for the developing states) and the states which allocates amounts greater than this limit are required to reduce them over time. The blue box group those measures which, while distorting the trade practices are estimated to be transition policies to the further reforms considering more the quota.

The measures which do not distort trade practices or have a minimal effect on them are included in the green box. Analyzed from the perspective of WTO, agricultural multifunctionality is a barrier to be removed for the full liberalization of trade, being in this accepts only a plea pleaded by numerous countries to additional finance agriculture and to circumvent the constraints of commercial regulations. On the other hand, Daniel Daianu claims that agriculture cannot be left solely to market forces. As market fundamentalism is harmful in finance, the more it must be fought in agriculture. This, of course, does not mean to neglect the logic of market functioning, efficiency considerations. (Daianu, 2009)

Multifunctional Agriculture in OECD Concept

A different perception of multifunctional agriculture is provided by the Organization for Economic Cooperation and European Development OECD, which in its documents has highlighted the need to express the multifunctional character of agriculture (Bruno, 2004), especially in terms of reducing agricultural subsidies and high food production and agricultural resources used (OECD, 2001a), within the Member States where was noticed a weak shift towards considering the principle of sustainable agriculture and especially imposing its productivist character.

The awareness of the impact that agriculture has on the development of rural communities, not only through the size of production but also through the strong character on modeling their evolution required a reconsideration of the role and place of agriculture in the macroeconomic context. As stated in OECD studies, compared to commodity production, agriculture produces a range of both positive and negative effects. Multifunctionality is neither specific nor exclusive to agriculture, but there are specific problems in this sector (OECD, 2003).

The OECD approach on multifunctionality is considering two fundamental premises on which modern agriculture starts from in carrying out its functions as an economic branch. Agriculture produces both goods, but also it generates intrinsic effects in the countryside. Viewed from this perspective, multifunctional agriculture makes a complex series of functions other than those specific to its classes. Multifunctionality is both a classic feature of the agricultural production process but through the multiple influences it exerts on rural environment and rural communities it has gained new facets, transforming itself into government policy.

The study write by OECD in 2001, defines multifunctionality in terms of the dual character of agriculture (OECD, 2001a). On one hand, it captures the productivist character of agriculture, given the existence of multiple freight and non-freight products produced together by agriculture, and on the other hand the fact that certain non-freight products take the shape of externalities or public goods, provided that there is no market for them, either they are malfunctioning (OECD, 2001a). The central idea of the definition given by OECD's research is the role that mixed goods and the externalities generated by agriculture have on rural communities and on the economy in general.

Multifunctionality in the concept of this organization shows the importance that the production of non-freight goods, public goods and externalities have on the classic function of agriculture, given that agriculture has ceased to be a central element of the national development in countries with developed economies. Agricultural markets and agricultural production continues to score fundamentally the rural economy in the case of developing states, and many rural communities are still dependent on the income from agriculture.

Externalities and public goods generated by agriculture are directly reflected on rural communities who see them as part of the component of the rural economy. The shift towards non-freight size of agriculture and reducing direct subsidies of production and the awareness of its social and environmental functions, sets a new agricultural model where the agricultural production respect the environmental criteria, maintaining the countryside and traditions.

Most times in the past agriculture was the engine for economic growth in rural areas also being the main source of income for the rural population and factor in labor employment in these areas. Approaching the concept of multifunctional agriculture, in OECD perspective reflects the multiple influences that agriculture has both on the national economy and especially on rural communities. Agriculture is both a vector in capitalizing the agricultural potential and an enhancer factor of the characteristics of rural communities in the particular context of the new trends regarding environmental protection, practicing the models of sustainable agriculture and capitalizing the specificity of rural communities.

Multifunctional Agriculture and the EU Approach

The interest in the effects of multifunctional agriculture in the European Union is a recent one following the steps of OECD. Thus, in the European literature (Antonio et al., 2007) multifunctional agriculture is recognized as having a crucial role in maintaining the viability of rural areas in Europe and has a particular significance in the conservation of resources and nature (Antonio et al., 2007).

Although the concept of multifunctional agriculture has been noted since the early 1980s in the many European documents such as the European Commission's report in 1988 entitled The future of rural society, in which it is mentioned the multiple contribution of agriculture in the economic development, in environmental management and in the viability of rural areas, the official confirmation of this concept, as highlighted in the literature (Wilson, 2009) was performed in 1993 at the European Council for Agricultural Law which has materialized the European efforts to harmonize agricultural legislation to ensure legal fundamentals for sustainable agriculture in Europe.

The community approach regarding multifunctionality is considering the multiple dimensions which agriculture makes in rural communities in Europe. A change in vision, which marked the evolution of the CAP, through moving from the mechanism of direct support of agricultural production to the one of direct support of farmers' income, is a true example of promoting multifunctionality policies at the level of the rural community.

Diminishing the role of agricultural production in the economy of rural communities as well as their orientation towards the practices of sustainable agriculture and promoting local values has contributed to the diversification of rural exploitation beyond the traditional constraints. Rural communities are no longer linked in an indestructible way to the farms. In these circumstances we must consider the effects that both cross compliance and decoupling the production payments had on the evolution of the rural economy and rural communities.

Promoting multifunctional agriculture in Europe was done within the European agricultural model, through two basic tools: cross compliance and decoupling the benefits and financial support. If we consider the first aspect, cross compliance defines the connecting elements between the actual farming and the environmental components, sustainable development and food quality of agricultural origin. Through this dimension it is recognized the defining role that agriculture holds in rural areas, as enhancer factor in the development of these communities. With a wide range of measures from which to choose, farmers can choose between different modes of production, they can choose their crops and guide their decision on production.

In the realization of agricultural multifunctionality, decoupling allows a greater freedom of the farmer in adopting the production measures, farms being forced to focus on grant activities they must transform themselves into competitivty poles. Although most of the time disconnecting means a sharp reduction in subsidies, the effect of these decisions is reflected on the orientation towards organic farming capable of generating replacement income for lost subsidies. Modulation involves reallocating a portion of the financial resources for direct payment schemes of the Pillar I of the CAP and their reorientation towards the coordinates of rural development contained in Pillar II.

In summary, multifunctional agriculture is promoted within the common agricultural policy and it captures the EU orientation towards a sustainable agriculture in harmony with the environment and the capitalization of the countryside, with its emphasis on the nonproductivist size of it and on the complex nature that rural communities play in the sustainable development of Community economy.

In essence, the European model of multifunctional agriculture is to ensure sufficient quantities of agricultural products, diversified in quality and affordable to most consumers, it allows a fair standard of living for the rural residents, it provides jobs in complex economic activities, environmental protection and the use of local renewable resources. The community concept on multifunctional agriculture is part of the trend definition provided by the OECD and comes to provide a dynamic understanding of the main agricultural model. Multifunctionality provides the European rural communities the possibility to capitalize the features, traditions, landscapes and other specific elements.

THE ROLE AND FUNCTION OF AGRICULTURE IN THE ACTUAL ECONOMY

When comes to the agriculture`s role and functions, in literature (Subic, 2007) is identified two main categories of functions that agriculture has to fulfill: the first group consists of agriculture primary functions or classical ones, and the second category refers especially to the complementary agriculture functions.

The first category of functions is represented mainly by the food security determinations, the role of economic growth factor, and the ones regarding the environmental protection. Among the complementary functions it is mentioned the role of raw materials supplier for the other industries and markets for industrial production, and also a role in reducing unemployment. According to (Vincze, 2002), the economic role of agriculture is described by the contribution of this economic sector in achieving some major issues as food and raw materials production, assuring a proper food supply chain for population, employing rural workforce or promoting rural industrialization.

The main function that agriculture is called to fulfill is to ensure the food resources for the population, by producing all the necessary means of subsistence, satisfying a primary human need through the participation in food security. Agriculture offers in this context the material support for development of human society through the important food resources that it provides for the population from capitalizing the natural potential. Having a well defined role in the whole economic structure, agriculture contributes to satisfying the food consumption demand of the population, also being a material support for other branches of the economy.

Ensuring adequate food supplies for the population and hence food security must be understood from the perspective of availability of food resources, the ability to be produced, as well as free access of the population to it. As (Rabantu, 2010) argues, whatever their nature, plant, animal or mixed, foods are essential for humans, since they provide the energy and basic nutrients, very necessary for the purposes of metabolic processes, growth and individual development (Rabantu, 2010, p. 161).

The concept of food security is very complex, and with multiple meanings. In a broad sense, food security is linked to the idea of free access and the right to food for the population. Regarding the definition of the right to food, this is, according to UN Resolution the right to have regular access, permanent and free, either directly or for a fee, to appropriate food in terms of quantity and quality and sufficiency, according to the cultural traditions of population of which the consumer is a part of and able to provide mental and physical life, individual and collective, free from fear, fulfilled and dignified (UN, 2001; Ziegler, 2001).

The international efforts to regulate and to ensure the right to food were multiple. Recognizing the role that agriculture has in delivering this right, now fundamental, the superior capitalization of the agricultural potential available is a viable solution to limit food insecurity and to ensure food self-sufficiency of the population.

In this regards, it is remarked the FAO efforts, which has adopted in November 2004, after the 127 Session of the FAO Council, a series of 19 basic guidelines (FAO Council, 2004) to provide the concept "right to food" an integrative understanding from the social, cultural and economic perspective. To FAO demarches are competed to UN initiative, materialized in four guidelines (Oliver de Schutter, 2009) that consider aspects regarding the trade which must fall within the objectives of sustainable development and to the liberalization impact which should be measured based on the needs of people in the state of food insecurity and not only based on aggregated indicators.

Food production remains, under these conditions, the main function that agriculture must fulfill, on the expense of others elements such as biofuels production, tourism, etc. Although ones tried to achieve an agricultural production that cover the food needs of the population, the phenomenon of food insecurity which is still manifesting globally. Given the issues of food insecurity, a FAO report highlights some worrying characteristics, such as: the food insecurity is increasing, and as a result of the present crisis and the degree of concentration is very high – and it also has a strong rural character (FAO, Rabontu, 2010).

The second function regards the complex role of agriculture which is not limited to matters relating to security and food self-sufficiency, but it actively contributes to the economic growth, having a significant role in capitalizing the economic potential of Romania. As advocated in literature (Popescu, 2001), the contribution of agriculture to the economy is massive because agriculture is the branch which traditionally transfers income to develop other sectors and activities of the national economy as a whole, through a number of channels (Popescu, 2001, p. 23).

In the case of Romania, agriculture became in the last twenty years, the main economic sector, being a driving element of economic growth through the implications it has in the income of the population by creating jobs, by modeling the development of rural communities, actively participating in the exploitation of local resources and their transfer to manufacturing branches. The existence of multiple relationships between agriculture and other industries makes the evolution to considerably influence the development of agricultural production growth and the existence of agricultural overpopulation may be a prerequisite of an active labor transfer to other branches of the economy and ensuring a high level of productivity in terms of human resources available in the industry.

The role of agriculture in providing economic growth is contemplated by the weight that it holds the GDP and GNI as well as in the structure of employment in Romania and in the structure of exports. As identified in the previous chapter the contribution of Romanian agriculture is about four times higher than in the other European economies (Romania 6.1% of GDP compared to 1.7% in the EU) and the share of employment in agriculture is 8%. If we consider only these issues, the role of agriculture in the national economy is crucial for its development.

Developing a competitive agriculture cannot be achieved without taking into account the environmental component and the compliance with the criteria of forming a sustainable agriculture, responsible towards the environment.

The third function is considering the symbiotic relationship between agriculture-environment which is crucial, the environment dramatically influence the agricultural development both by the quality of land available to agricultural practice, but especially through the influence of climatic factors, agriculture in its turn putting its mark. Although the concept of sustainable agriculture has recently emerged, following the FAO Council in 1998, it managed to impose itself in practice.

According to the FAO definition the sustainable agriculture represents the management and conservation of the natural basic resource and the orientation of technological and institutional exchange in such a way as to satisfy human needs for present and future generations (FAO Trainer's Manual).

Practicing sustainable agriculture requires complying with environmental conditions and its integration in a system of organic production to ensure the protection of natural resources, the conservation of traditional rural landscape and biodiversity conservation. The extensive process of Romanian agriculture transition to a market economy has created the premises for the practice of sustainable agriculture in Romania, not willingly, but rather by the lack of financial resources that have helped reducing the practice of applying chemical treatments and intensive agriculture. Under these conditions, the sustainable agriculture can be a viable alternative for Romanian rural development, of national economy overall through the high added value that these green products, sustainable made themselves.

As noted in some studies (Rusali, 2008), organic farming, although it had a growing trend in the recent years, is still a restricted sector in Romania, for the development of which substantial financial and political support is required through specific measures appropriate for the development given its real potential for supporting sustainable development of rural areas (Rusali, 2008, p. 424).

The development of sustainable agriculture in Romania may be, in this context, a functional solution of capitalizing the agricultural potential and of adapting the farming practices to the demands imposed by the economic integration in the EU, especially that the development and financing of this sector is included in Pillar II of the CAP. Agro-environment payments mobilize significant financial community resources that the Romanian farmers can benefit from, they are encouraged to promote environmentally friendly agricultural practices and conserve biodiversity.

Sustainable agriculture can be analyzed from the perspective of the ecological footprint, which synthetically express the pressure that mankind exerts on the biosphere, through consuming (Stanciu, 2009). Viewed in terms of this indicator, Romania has the lowest carbon footprint of all EU Member States, only 2.4 gha per person, being an ecological debtor country with a biocapacity of 2.3 gha per person. The ecological footprint of a country includes all farmland, pastures, forests, and fishing grounds required to produce food, fiber and timber to consume and absorb the waste they emit (Global Footprint Network, 2007).

Agriculture is the economic branch that proves the biggest addiction to the natural factor, climatic conditions they recover, or even modify by its influence, decisively participating to create favorable conditions for environmental protection by preserving and increasing carbon sinks present in soil, humus, and biomass production setting. The complex role that agriculture plays in human society involves the observance of environmental practices and promotes a healthy environment by restoring traditional links between nature and agriculture.

The fourth function highlights the role of source of raw material that agriculture has for manufacturing industries such as light industry and food industry. Agriculture provides these industries the needed raw material production - flax and hemp for fiber, tobacco, rapeseed oil and biodiesel, medicinal and aromatic plants, sugar beet, vegetables, cereals and so on, making an important social and economic function by sustaining their supply. Correspondingly, agriculture is also market for industrial production. Thus, we see interdependence between agriculture and other branches of the national economy, confirming the role that agriculture plays in mobilizing national productive forces. To all these features mentioned above is added the role that agriculture plays in increasing employment and reducing unemployment.

Agriculture is in the current conditions a viable job provider with vital importance at least for the rural communities. Agriculture's capacity to absorb the surplus of agricultural labor force is quite high, but not fully capitalized. For arguing the role of job provider and unemployment leverage, in Table 1 is presented the structure of labor force in agriculture, in some European countries in 2007. As a general view, in EU-27, about 11,693 thousands of annual work unit were used in 2007 and percentage of 78% represented the family farm labor force used in European agricultural holdings.

The role of agriculture is more complex than it was believed by now. It mobilizes a very important labor force for the rural economies and bonds tightly the family relations. The regular farm labor force represents 92% of total labor force at the EU-27 countries, and holds a very significant percentage for all the countries. The agricultural sector became in this context a major employment agent for the EU economy. If we take into account the agricultural holders being a natural person we have the complex imagine of the role of the agricultural in the current EU economy.

Countries as Italy, Poland, and Romania stand out with a strong farming sector dominated by individual farms. The age structure of the holding`s owners is a conditioned factor in developing new farm`s dimensions in economy. For example, in Romania, the agricultural holders aged >65

Table 1. The farm labor force in some EU-27 counties in 2007

	Total Farm Labor Force (1 000 AWU)	Regular Farm Labor Force (% of total)	Full-Time Regular Farm Labor Force (% of total)	Female Regular Farm Labor Force (% of total)	Family Farm Labor Force (% of total)	Agric. Holders Being A Natural Person (1 000)	Agric. Holders <35 Years Old (1 000)	Agric. Holders >=65 Years Old (1 000)
EU-27	11.693	92	34	34	78	13.441	823	4.584
Belgium	66	95	71	29	79	44	3	9
Bulgaria	491	95	38	39	85	490	15	222
Germany	609	91	50	28	69	365	28	27
Spain	968	82	42	20	65	988	44	361
France	805	89	67	25	47	428	34	66
Italy	1.302	90	37	30	84	1.664	49	741
Latvia	105	99	30	50	84	108	8	32
Lithuania	180	98	14	48	85	230	10	93
Hungary	403	97	25	37	77	619	47	172
Austria	163	97	53	41	88	161	16	18
Poland	2.263	97	34	42	95	2.387	294	388
Portugal	338	93	35	41	82	269	5	130
Romania	2.205	93	4	42	90	3.914	167	1.762
Slovenia	84	96	21	41	92	75	3	26
Slovakia	91	96	40	32	44	67	2	22
Finland	72	94	56	30	83	67	6	4
Sweden	65	97	42	26	76	68	4	15
UK	341	93	55	23	67	283	7	92

Source: Authors own processing based on Europe in figures – Eurostat yearbook, Eurostat (2011) years old hold the majority, and they are hard to be switched to the multifunctional dimensions of the agriculture. They are mainly trapped to the traditional agriculture methods.

Although labor force employed in agriculture has a significant share in the national rural economy, the low level of income of the rural population relies, as presented in literature (Rusali, 2008), in the need to develop a diversified rural economy based on alternative sources of income and a diversification of employment opportunities for employment in non-agricultural activities as a sustainable source of livelihood for the rural communities and the development of rural economy (Rusali, 2008).

In a brief presentation the agriculture functions are presented in Figure 1.

MULTIFUNCTIONAL FARMS AND FARM DIVERSIFICATION

The agrarian structures and the forms of organization in agriculture are the concentrated expression of the special conditions of development of the national agriculture over time, of measurements and historical conditions in an attempt to balance and highly capitalize the land, the experiences and practices of the national agriculture.

Farms are an important link in capitalizing the agricultural potential by better organizing the production factors in accordance with the requirements of national agricultural policy, to ensure sufficient agricultural production.

The practical experience has required over time three major types of agricultural structures (Bold and Dragici, 2010):

- The English system, organized on the basis of concentrating the land and the labor resources in large farms with high economic efficiency, able to realize and exploit the agricultural production obtained under the requirements of the market economy;
- The Danish system, centered around individual proprietary farm, understood as a structure capable of using the essential agricultural potential, has promoted the strengthening of cooperation system in agriculture as an effective tool in securing the capital circuit including a highly integrated with food industry for all farms, no matter the size;

Figure 1. Agriculture functions (source: own author design)

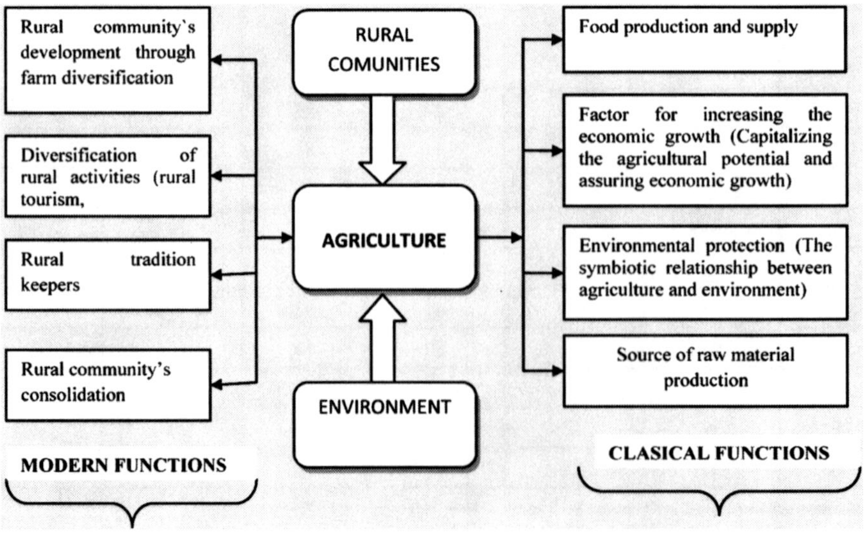

- The Russian system, cooperative or of state, has promoted the concentration of land, capital and livestock into large production structures with high degree of specialization and concentration in accordance with the industry but with diminished efficiency.

Over time Romanian agriculture, as claimed by the professors Ioan Bold and Avram Craciun, in a recent study, reflects three main forms of development: the traditional type, in peasant holdings – the household type, with a medium technical level in the agricultural associations and agricultural companies and – the industrial type, in compounds and resorts or specialized farms (Bold and Craciun, 2012).

The main form of exploiting the agricultural potential is the farm, which gathers and monetizes at a higher level the material, human and financial resources available in agriculture. The diversity of farms as well as their nature and dimensional structure has generated a wide range of types, leading to their classification according to several criteria.

Trying to provide a more comprehensive understanding of what means agricultural production structure, the literature offers a multitude of approaches to this notion. The evolution of agricultural structures after 1990 equally reflects the structural changes that took place in the Romanian agriculture as an immediate effect of the process of transition to a market economy in Romania, and the Romanian peasant desire to regain a lost good – the land. Returning to the private forms of exploitation of agricultural land and the agricultural orientation towards an open economy principles imposed reconsidering the role and place of the peasant in the economy as a whole.

Land restitution and the transformation of socialist forms of agricultural organization, existing until 1989, into holding structures, specific to the capitalist economy, has been a difficult process with profound implications on the Romanian rural environment. The decollectivization of agriculture has contributed to a variety of forms of exploitation, in which the small and medium exploitations are the most important, as well as transforming over 90% of the agricultural land in the country into private property.

Farm structures are equally the result of natural conditions and restrictions of land, hydrographic and the historical experiences of the rural communities in particular. Destructuring the socialist agricultural property and the emergence of private property, individual over the land were the essential steps in building a new farm system, in which the vast communist farm was quickly replaced by the small and very small farms with low economic efficiency. The farms' development during the period 1990-2000 highlights the changes that have occurred in the extensive process of privatization of Romanian agriculture and agricultural land restitution, land confiscated by the communist regime in the process of agriculture collectivization.

Holdings with anther gainful activity represent a good and very powerful indicator, reflecting the multifunctional character of the agriculture. By its manner, the agricultural holding is build on the consideration that it should provide a high degree of agricultural potential valuing by providing food from farm to tables. Developing another gainful activity besides agricultural function, for the agricultural holding is a progress in understanding their role and functions. In Table 2 is presented the evolution of Holdings with another gainful activity, in some EU-27 countries, including Romania, during 2000-2007.

In all the traditional and well development EU-27 countries, it can be noticed an increasement of the number of holdings with another gainful activity, except countries as Italy, Latvia, Hungary where these holdings has registered a decreasement. The agricultural holdings have to diversify their traditional activities and to respond to new and increasing environmental requirements and demands, and also, to capitalize the values of the cultural model. In this context, the multi-

Table 2. Holdings with another gainful activity, in some European countries during 2000-2007

Country	2000	2003	2005	2007
Belgium	1.920	2.090	2.140	1.910
Bulgaria	NA	27.040	10.970	10.310
Denmark	6.710	6.960	9.480	10.460
Germany	NA	80.040	87.870	80.590
Ireland	5.120	6.460	5.860	6.590
Greece	11.800	10.500	14.050	12.760
Spain	NA	25.810	35.030	37.470
France	NA	151.390	141.670	124.650
Italy	188.540	89.510	105.390	108.780
Hungary	45.160	86.640	36.140	31.830
Austria	33.250	32.550	36.500	35.580
Portugal	33.890	35.480	29.030	20.460
Romania	NA	186.570	942.090	617.690
Slovenia	3.990	2.870	3.150	3.120
Slovakia	2.440	2.810	1.610	3.150
Finland	17.340	18.960	20.460	18.870
UK	45.010	45.650	68.730	69.680

Note: NA= not available data
Source: Author own processing from Eurostat database (2012)

functional farm becomes a necessity. The majority of them had to become multifunctional if they want to beneficiate from the financial allotments through the Common Agricultural Policy mechanism.

Farm Development in Romania 2002-2010

The evolution of farm structures on the analyzed interval confirms the dual character of the Romanian agriculture. The extreme polarization of the categories of farms is maintained. Small and unincorporated farms dominate the national agricultural landscape by numbers, the agricultural land and labor force mobilized while large and very large farms capitalize 5.853 million ha of agricultural area. So, after a period of more than 10 years of transition, the dual evolution of the national agricultural structures can be confirmed on solid statistical basis obtained in extensive statistical research such as census or agricultural structure survey. Until the General Agricultural Census of 2002, Romania received an information basis consistent and uniform, allowing a thorough understanding of local farm situation.

As a general observation of the analysis of farms' development after 2000 is to reduce the total number of farms from 4,484,900 in 2002 to 3,856,200 in 2010. In this context, non-traditional agricultural production structures may prove to be fit well into this non-traditional agricultural environment (Sabates, 2002). In Table 3 is synthetically presented the evolution of the main indicators at national level regarding the farms.

In 2002 from the total holdings of 4,485,000 units, unincorporated farms numbered 4.462 thousand units, which represented 99.48%, and were using 7.709 thousand ha (55.3% of the total area) with an average size of the holding of 1.73 ha. In the same time incorporated farms represent 23000 units (0.5% of the total holdings) and were using 6.222 thousand ha (44.6% of the agricultural area), with an average farm size of 274.43 ha.

In 2005, there were 5.1% fewer farms in Romania than in 2002, respectively 4256 agricultural units. The number of the individual farms decreased 5.0% compared to 2002, reaching 4,238 farms, while the number of incorporated farms decreased by 19.4% compared to the same year, reaching 18,263 units.

Two years later, in 2007, when the *Structure survey in agriculture* was conducted, a rate of 99.5% of the total 3931 thousand farms, were individual farms, respectively 3.914 thousand farms, which owned 65% of the agricultural area used, and the 17 000 incorporated agricultural holdings, only 35% of the agricultural area used.

The number of farms registered in the year 2010 was of 3,856 thousand units, of which 3,825 thousand unincorporated farms and 31,000 incorporated farms. The number of unincorporated

Table 3. Evolution of the main national farm indicators, by the legal status

Indicators	Years	M.U.	Total Agricultural Areas	Unincorporated Farms	Incorporated Farms
Farms	2002	thousands	4485	4462	23
	2005	thousands	4256	4238	18
	2007	thousands	3931	3914	17
	2010	thousands	3856	3825	31
Total area	2002	thousands of ha	15708	8454	7254
	2005	thousands of ha	15442	9886	5556
	2007	thousands of ha	15265	9591	5674
	2010	thousands of ha	15867	8488	7379
Agricultural area used (AAU) (including idle arable land)	2002	thousands of ha	13931	7709	6222
	2005	thousands of ha	13907	9102	4805
	2007	thousands of ha	13753	8966	4787
	2010	thousands of ha	13298	7445	5853
Agricultural area used belonging to a farm	2002	ha	3,11	1,73	274,43
	2005	ha	3,27	2,15	263,08
	2007	ha	3,50	2,29	270,45
	2010	ha	3,45	1,95	190,84

Source: Author's own processing based on The General Agricultural Census 2010, INSE

farms decreased by 14.3% compared to 2002, while incorporated farms increased by 34.8% since 2002. In Table 4 is presented the evolution of the agricultural holdings and agricultural area used, by legal status, in Romania 2002-2010.

According to the data presented in Table 4, and in this range, individual farms continue to be the main form of exploitation of the national agricultural potential, with a share of more than 99.5% of the total Romanian agricultural holdings,

Table 4. The evolution of farms and agricultural area used, by legal status, 2002-2010

Farms' Legal Status	Farms – Total (thousands)				Farms Which Used Agricultural Area (thousands)			
	2002	2005	2007	2010	2002	2005	2007	2010
Total	4484,9	4256,2	3931,4	3856,2	4299,4	4121,2	3851,8	3721,9
Individual farms	4462,2	4237,9	3913,7	3820,4	4277,3	4103,4	3834,4	3686,7
Incorporated units	22,7	18,3	17,7	30,7	22,0	17,8	17,4	30,2
Societies/agricultural associations	2,3	1,6	1,5	1,4	2,2	1,6	1,5	1,4
Commercial societies	6,1	4,8	5,1	16,5	5,7	4,6	4,9	16,1
Cooperative units	0,1	0,1	0,1	0,1	0,1	0,1	0,1	0,1
Other types	8,5	6,9	6,8	9,4	8,4	6,8	6,8	9,4

Source: Author's own studies based on the RGA results (2002 şi 2010), The Structural Household Survey, 2007 and The Romanian Statistical Yearbook (2003-2011)

although it records a slight decrease in the number of 641 800 units, from 4,462,200 in 2002 to 3,820,400 in 2010. As shown in a specific study (Alexandru, 2005), in Western countries (we consider here, firstly, the EU and the U.S.), the best farms were the private-family ones self-managed or in lease. (Alexandru, 2005)

Romanian individual farms revolve mostly in the category of subsistence or semi-subsistence farms, and their great number, is a major problem in forming and strengthening a functioning agricultural economy, market-oriented, in Romania. The limited access of the members of this category of holdings, to financing sources and modern means of production has a significant impact on both the level of farms' profitability and farmers' income level, which is in this case reduced, affecting the quality of life of an important segment of the agricultural rural population. As opposed to these, the agricultural units with legal personality and the agricultural companies record a slight trend towards consolidation. The number of agricultural units with legal personality, increasing by about 8,000 units, from 22,700 in 2002 to 30,700 in 2010, while the number of companies increased from 6100 in 2002 to 16,5000 in 2010. In Table 5 is presented in detail the structure of farms and agricultural surface used, depending on the legal status of the holdings in the year 2010, after performing the RGA.

Analyzed in detail, the structure of the 3,856,200 Romanian agricultural exploitations shows strong agricultural exploitation individualization. Of the 3,825,600 unincorporated farms (with a share of 99.2% of the total exploitations) recorded in 2010, individual holdings were 3,820,400 (99.86%), while freelancers, individual companies or family businesses were gathering 5200 exploitations (0.14%).

In the same year 2010, incorporated farms were gathering 30 700 units of which 53.4% (16,400 units) were privately owned companies, municipalities and local councils had 8.8% (2,700 units), co-operatist unities 0.3% (0.100 units) and 1400 units (4.56%) were companies or individual associations. In Figure 2 is presented the evolution of total agricultural area used (thousands of hectares) of farms, depending on the legal status held in the period 2002-2010.

According to the representation in Figure 2, the total agricultural area used is approximately constant for the entire period, respectively around 13,000 hectares. Of the 13.931 thousand ha used in 2002, 7.709 thousand ha belonged to the individual holdings, 6.222 thousand ha to the incorporated units, 976 000 ha to the agricultural societies and associations, 2.169 thousand ha to the commercial agricultural companies and the exploiting cooperative agricultural units had approximately 2000 ha.

In 2010 from a total agricultural area used of 13.298 thousand ha, with 633,000 ha less than in 2002, 7,154,100 hectares were used by individual farms, 5,852,900 ha were used by incorporated units, 556,8000 ha were used by agricultural companies or associations, 3.173 thousand hectares were used by commercial agricultural companies and 8,200 ha by cooperative units. In fact, in 2010, 42.7% of the total agricultural area used was for unincorporated farms and 17.7% for incorporated holdings, as opposed to 2007, when 85.5% of the total agricultural area used was for unincorporated farms and 50.2% for incorporated units. In the Table 6 is presented the evolution of the distribution of the agricultural area used by a farm in Romania, according to legal status, in the period 2002-2010.

Regarding the evolution of the agricultural area used by a farm (medium size), in Romania, in the period 2002-2010, there is an almost insignificant increase of 0.34 ha of it, from 3.11 ha in 2002 to 3.45 ha in 2010. If in 2002 it was of 3.11 ha/unit, in 2005, it reaches to 3.27 ha/unit, and in 2007 to 3.50 ha/unit, for in 2010 we see a slight decrease, an average size reaching 3.45 ha/unit.

If we consider the development of the medium size category of holdings, one can see a dual tendency, of growth for individual farms, and the

Table 5. The farm's structure, the agricultural land used and average agricultural area used on a farm, by legal status, 2010

Farm's Legal Status	MU	Total Farms	Farms Which Uses Agricultural Area	Agricultural Area Used -ha-	Average Agricultural Area Used (ha) on a Farm
Unincorporated farms	thousands	3825,6	3691,7	3825,6	**2,0**
	%	99,2%	99,19%	99,2%	-
Individual farms	thousands	3820,4	3686,7	3820,4	1,9
	%	99,9%	99,9%	99,9%	-
Freelancers, individual companies, family businesses	thousands	5,2	5,0	5,2	56.18
	%	0,14%	0,14%	0,14%	-
Incorporated farms	thousands	30,7	30,2	30,7	**190.84**
	%	0,8%	0,81%	0,80%	-
Autonomous administration	thousands	0,1	0,1	0,1	323.41
	%	0,3%	0,3%	0,3%	-
Societies/Individual associations	thousands	1,4	1,4	1,4	400.57
	%	4,56%	4,64%	4,56%	
Commercial societies with capital mainly private	thousands	16,4	16	16,4	193.14
	%	53,4%	53,0%	53,4%	-
Commercial societies with capital mainly of state	thousands	0,1	0,1	0,1	49.36
	%	0,3%	0,3%	0,3%	-
Institutes, research stations, schools with agricultural sections	thousands	0,2	0,2	0,2	288.00
	%	0,7%	0,7%	0,7%	-
Local Councils/Municipalities	thousands	2,7	2,7	2,7	575.59
	%	8,8%	8,9%	8,8%	-
Cooperative units	thousands	0,1	0,1	0,1	120.24
	%	0,3%	0,3%	0,3%	-
Other types	thousands	9,4	9,4	9,4	47.63
	%	30,6%	31,1%	30,6%	-
Total	thousands	3856,2	3721,9	3856,2	**16497,0**

Source: Author's own studies based on the results of The General Agricultural Census, 2010, INSE

reduction of the size for the incorporated ones. The situation is noteworthy in the period under review, due in particular to the syncopated evolution of farm structures, reflecting the uncertainties of the proprietary relations in agriculture. If in 2005 a return to individual farms was of average 2.15 ha/unit, with more than 0.42 ha in 2002, when it was of only 1.73 ha/unit, in 2010 it reaches 1.87 ha/unit, with0.42 ha less than in 2007.

In the case of incorporated farms, the average agricultural area used, in 2010, was of 190.78 ha, compared to 274.43 ha in 2002. Cooperative units are the only categories of holdings which significantly strengthens its size. If in 2002 the average area used by them was of only 27.19 ha/unit, in 2010, the average agricultural area used by these increases to 120.2 ha/unit. Although the average size of a farm in Romania records a slight

Figure 2. The evolution of total agricultural area used (thousands of hectares) of farms, in Romania, 2002-2010 (source: author's own studies based on the RGA 2010 data)

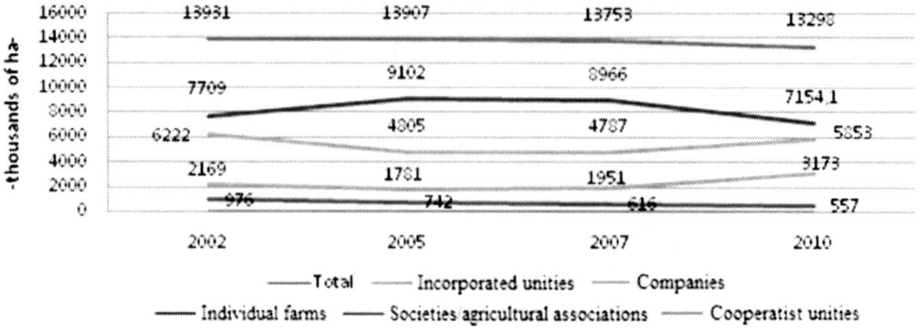

Table 6. The agricultural area used by a farm, according to farms' legal status, in the period 2002-2010

Farms' Legal Status	Agricultural Area Used, Average Distributed (hectares)							
	On a Farm				On a Farm Which Used The Agricultural Area			
	2002	2005	2007	2010	2002	2005	2007	2010
Total	3,11	3,27	3,50	3.45	3,24	3,37	3,57	3.57
Individual farms	1,73	2,15	2,29	1.87	1,80	2,22	2,34	1,94
Incorporated units	274,43	263,08	270,45	190.84	282,23	269,28	275,37	193.70
Cooperative units	27,19	30,06	212,51	120.24	30,72	36,48	215,54	122.03
Other types	24,49	22,35	48,68	47.63	24,68	22,54	49,11	47.82

Source: Author's own studies based on the results of The General Agricultural Census, 2010, INSE

increase, we are witnessing to the reduction of the average area of farms, by type of units, which is characteristic of the last period, highlighted by the General Agricultural Census data. Regarding the distribution of holdings by category of use of the agricultural area used, the situation is presented in Table 7.

From a total of 8.774 thousand ha of holdings which used land in 2002, 61.9% meaning (5,437,000 ha of farms) were unincorporated farms and the remaining 3.337 thousand ha (38.1%), incorporated farms. A total of 169 000 ha were farms formed as family gardens. Pastures and hayfields were recovered in the form of 4.644 thousand ha of farms, of which 40.4% incorporated and 59.6% unincorporated. Farms which had permanent crops were 344 000 ha of which 225 000 ha (65.4%) were incorporated holdings and 119 000 ha (34.6%) unincorporated. In 2010, the number of farms using arable land was reduced by 469 000 ha (-5.34%), reaching 8.305 thousand ha, of which 4.721 thousand ha of unincorporated farms (56.8%) and 3.584 thousand ha incorporated farms (43.2%), observing in the latter an increase of 247 000 ha compared to 2002. The holdings under the form of family gardens grow by 7.7% (13,000 ha), reaching 182 000 ha in 2010.

The total number of farms which capitalized pastures and hayfields as well as the ones who had permanent crops diminished in the period under study. In the first group, unincorporated farms represented 51.3% (2,306,000 ha) with 28% more than (428,000 ha) in 2002, while the number of incorporated farms has decreased by 20.9% respectively from 2.766 thousand ha in 2002 to 2.188 thousand ha in 2010.

Table 7. Farms' development in the category of use of the agricultural area used, 2002-2010, thousands of ha

Category of Use of the Agricultural Area Used	Years	Total Farms	Unincorporated Farms	Incorporated Farms
Arable land	2002	8774	5437	3337
	2005	8867	6285	2582
	2007	8691	6046	2645
	2010	8305	4721	3584
Family gardens	2002	169	169	-
	2005	171	171	-
	2007	178	178	-
	2010	182	182	-
Pastures and hayfields	2002	4644	1878	2766
	2005	4530	2385	2145
	2007	4540	2457	2083
	2010	4494	2306	2188
Permanent crops	2002	344	225	119
	2005	339	261	78
	2007	344	284	60
	2010	317	236	81

Source: Author's own processing based on the results of The General Agricultural Census, 2010, INSE

In the case of holdings which had permanent crops we see a decrease of 7.8%, respectively from 344,000 ha in 2002 to 317,000 ha in 2010. Unincorporated farms increased by 4.3%, respectively from 225,000 hectares in 2002 to 236,000 ha in 2010, and in this case the increasement is made at the expense of the incorporated ones, which are reduced by 32%, from 119,000 hectares in 2002 to 81,000 ha in 2010. In Table 8 is presented the detailed structure of the agricultural holdings, the agricultural area used by type of use, the total agricultural area and the unused one, depending on the legal status of the agricultural holdings in the year 2010.

Unincorporated agricultural holdings held in the year 2010, a proportion of 99.3% of the total arable land, fully family gardens, 99.0% pastures and hayfields and 97.1% permanent crops. Among them, there stand the individual farms holding more than 99% of the total agricultural area used, excepting the permanent crops where the proportion is of only 10%, while incorporated agricultural holdings represent less than 1% of the total agricultural area used. Of the 20 600 farms, 53% are privately owned companies, 4.6% individual companies or associations, 9% public administration and only 0.2% cooperative units.

CONCLUSION

Understanding the role and function of the agriculture represents a major challenge, not only from the theoretical point of view but both from the practice approach. Multifunctional agriculture represents a very actual concept not only in promoting the rural communities values, but also in developing a new dimension of the agriculture in a larger context of climate change, rural responsibility and rural development. During the years

Table 8. Farms, agricultural area used by type of use, unused agricultural land, total area by legal status of agricultural holdings

The Legal Status of the Agricultural Holding	MU	Agricultural Land Used					Farms' Unused Agricultural Land	Farms' Total Agricultural Land
		Arable Land	Family Gardens	Pastures and Hayfields	Permanent Crops	Total		
Unincorporated farms	Thousands	2729,5	2479,2	1497,4	11202,3	3691,7	314,6	3811,2
	%	99,3%	100,0%	99,0%	997,1%	99,2%	98,9%	99,2%
Individual farms	Thousands	2725,1	2477,1	1494,9	1119,2	3686,7	314,3	3806,1
	%	99,8%	99,9%	99,8%	10,0%	99,9%	99,9%	99,9%
Freelancers, individual companies, family businesses	Thousands	4,4	2,2	2,5	1,0	5,0	0,3	5,1
	%	0,16%	0,09%	0,17%	0,01%	0,13%	0,10%	0,13%
Incorporated farms	Thousands	20,6	-	14,8	3,2	30,2	3,6	30,8
	%	0,75%	-	0,98%	0,28%	0,81%	1,12%	0,80%
Autonomous administration	Thousands	0,04	-	0,04	0,03	0,05	0,02	0,05
	%	0,2%	-	0,3%	1,0%	0,2%	0,7%	0,2%
Societies/Individual associations	Thousands	1,0	-	0,5	0,1	1,4	0,1	1,4
	%	4,9%	-	3,5%	2,6%	4,6%	2,4%	4,5%
Commercial societies with capital mainly private	Thousands	12,5	-	5,0	1,6	16,0	1,7	16,3
	%	60,5%	-	33,7%	51,4%	53,0%	46,6%	53,1%
Commercial societies with capital mainly of state	Thousands	0,03	-	0,05	0,01	0,07	0,02	0,07
	%	0,2%	-	0,3%	0,4%	0,2%	0,5%	0,2%
Institutes, research stations, schools with agricultural sections	Thousands	0,2	-	0,1	0,1	0,2	0,0	0,2
	%	0,7%	-	0,6%	2,2%	0,6%	1,0%	0,6%
Local Councils/ Municipalities	Thousands	0,8	-	2,6	0,2	2,7	0,7	2,7
	%	4,0%	-	17,8%	6,1%	9,0%	18,5%	8,8%
Cooperatives	Thousands	0,02	-	0,05	0,00	0,07	0,00	0,07
	%	0,1%	-	0,3%	0,1%	0,2%	0,1%	0,2%
Other types	Thousands	5,8	-	6,2	1,1	9,4	1,0	9,4
	%	28,3%	-	42,2%	33,8%	31,1%	28,7%	30,6%
Total	Thousands	2750,1	2479,2	1512,2	1123,5	3721,9	318,2	3841,8

Source: Author's own studies based on the results of The General Agricultural Census, 2010, INSE

agriculture becomes a major economic sector, not only in Romania, but also at the European Union, too and it has diversified its area of action beyond the food production function and land exploitation.

Rural areas are now multifunctional as agriculture improved its role in developing rural community's perspectives in a changing business environment. There is less dependence form the agriculture and new areas of action have a raised. Farmers are now close to the environment by promoting sustainable technologies and they have returned to more traditional agricultural practices. When comes to multifunctional agriculture, we should agree that now agriculture, provides more goods and externalities than ever, and the useful range of services are tighter bond to rural communities and agriculture than food systems and raw materials suppliers before.

The multifunctional agriculture represents the future of the rural communities in a large and very competitive world economy. Diversifying the agriculture activities and increasing the role of farms in promoting sustainable development, finding new gainful activities for rural population will create a more competitive sector with a high impact on the national grad economy. Individual farming still represents a very active instrument in reducing unemployment and promoting feasible economic structures and it assures a massive weight in households consumption.

REFERENCES

Alexandru T. (2006). *Dimensiunea economică a exploataţiilor agricole*. Analele Ştiinţifice Ale Universităţii Alexandru Ioan Cuza Din Iaşi, Tomul LII/LII Ştiinţe Economice 2005/2006.

Antonio, G. S., & Alberto, G. G. (2007). A comprehensive assessment of multifunctional agricultural land-use systems in Spain using a multi-dimensional evaluative model. *Agriculture, Ecosystems & Environment*, (120): 82–91.

Bold, I., & Crăciun, A. (2012). *Organizarea teritoriului agricol: Concepte-tradiţii-istorie*. Timişoara: Editura Mirton.

Bold, I., & Drăghici, M. (2010). *Eseuri*. Iaşi: Editura Terra Nostra.

Bruno, L. (2004). Debating the multifunctionality of agriculture: From trade negotiation to development policies by the south. *Journal of Agrarian Change*, *4*(3), 336–360. doi:10.1111/j.1471-0366.2004.00082.x

Cecilia, R. (2010). Siguranţa alimentară şi rolul său în evolutia comertului cu produse alimentare. *Analele Universităţii Constantin Brâncuşi din Târgu Jiu*, (2), 161-172.

Council, F. A. O. (n.d.). *Voluntary guidelines to support the progressive realization of the right to adequate food in the context of national food security*. Retrieved 27.02.2012, from http://www.fao.org/docrep/meeting/009/y9825e/y9825e00.htm

Daianu, D. (2009). *Capitalismul încotro? Criza economică, mersul ideilor. instituţii*. Iaşi: Editura Polirom.

Dan, G. M., Daniela, G., Luminiţa, C., & Virginia, C. (2004). Ierarhizarea priorităţilor de dezvoltare agricolă şi rurală în România. Influenţele noii reforme a Politicii Agricole Comunitare, Institutul European din România – Studiul de impact (PAIS II) Nr. 11, Bucureşti.

de Schutter, O. (2009). *Mission to the world trade, (A/HRC/10/005/Add.2), 9 martie 2009*. Retrieved from www.ohchr.org

European Commission. (2011). *Europe in figures - Eurostat yearbook 2011*. Luxembourg: Publications Office of the European Union.

Eurostat (2011), Europe in figures – Eurostat yearbook, Publications Office of the European Union, Luxembourg, retreive 09.07.2011, from http://epp.eurostat.ec.europa.eu/cache/ITY_OFFPUB/KS-CD-11-001/EN/KS-CD-11-001-EN.PDF

Eurostat database (2012), Agriculture and fisheries section, Retrieved 09.07.2011, from http://epp.eurostat.ec.europa.eu/portal/page/portal/agriculture/introduction

FAO. (1995). FAO trainer's manual: Vol. 1. *Sustainability issues in agricultural and rural development policies*. Rome: FAO.

Global Footprint Network şi WWF. (2012). *Europa 2007: Produsul intern brut şi amprenta ecologică*. Retrieved 27.02.2012, from http://www.footprintnetwork.org/en/index.php/GFN/page/national_assessments/

Luminiţa, C., Constantin, C., & Valeriu, I. F. (2003). *Agricultura între restricţiile comerciale globale şi politicile comunitare. Centrul Român de Economie Comparată şi Consens*. Bucureşti: Academia Română, Editura Expert.

Marsden, T., & Sonnino, R. (2008). Rural development and the regional state: Denying multifunctional agriculture in the UK. *Journal of Rural Studies*, (24): 422–431. doi:10.1016/j.jrurstud.2008.04.001

National Institute of Statistics. (2008). *The structure survey in agriculture 2007*. Bucharest, Romania: INSE.

National Institute of Statistics. (2012). *The general agricultural census 2010*. Bucharest, Romania: INSE.

OECD. (2001a). *Policy commission on the future of farming and food, multifunctionality: Towards an analytical framework: Farming and food: A sustainable future*. London: DEFRA.

OECD. (2001b). *Multifunctionality: Towards an analytical framework*. Paris: OECD Publications.

OECD. (2003). *Multifunctionality: The policy implications*. Paris: OECD Publications.

OECD. (2006). *The new rural paradigm: Policies and governance*. Paris: OECD Rural Policy Reviews.

Paarlberg, P. L., Bredahl, M., & Lee, J. G. (2002). Multifunctionality and agricultural trade negotiations. *Review of Agricultural Economics*, *24*(2), 322–335. doi:10.1111/1467-9353.00022

Popescu, M. (2001). *Lecţiile tranziţiei: Agricultura 1990-2000*. Bucureşti: Editura Expert.

Renting, H., Rossing, Groot, Van der Ploeg, Laurent, Perraud, … Van Ittersum. (2009). Exploring multifunctional agriculture: A review of conceptual approaches and prospects for an integrative transitional framework. *Journal of Environmental Management*, (90): S112–S123. doi:10.1016/j.jenvman.2008.11.014 PMID:19121889

Rusali, M. (2008). Agricultura ecologică-dimensiuni şi implicaţii asupra dezvoltării durabile. *Economie Agrară şi Dezvoltare Rurală*, (4), 375-498.

Sabates, W. R. (2002). Farm strategy, self-selection and productivity: Can small farming groups offer production benefits to farmers in post-socialist Romania? *World Development*, *30*(10), 1737–1753. doi:10.1016/S0305-750X(02)00063-3

Stanciu, M. (2009). Amprenta ecologică a României - O nouă perspectivă asupra dezvoltării. *Calitatea Vieţii*, *20*(3–4), 271–288.

Wilson, G. (2001). From productivism to post-productivismyand back again? Exploring the (un)changed natural and mental landscapes of European agriculture. *Transactions of the Institute of British Geographers*, (26): 77–102. doi:10.1111/1475-5661.00007

Wilson, G. (2009). The spatiality of multifunctional agriculture: A human geography perspective. *Geoforum*, (40): 269–280. doi:10.1016/j.geoforum.2008.12.007

WTO. (2002). *Domestic support in agriculture: The boxes*. Retrieved 09.07.2011, from http://www.wto.org/english/tratop_e/agric_e/agric_e.htm

ADDITIONAL READING

Alene, A. D., & Manyong, V. (2007). The effects of education on agricultural productivity under traditional and improved technology in northern Nigeria: An endogenous switching regression analysis. *Empirical Economics*, *32*, 141–159. doi:10.1007/s00181-006-0076-3

Anne, M. H. (1996). Agricultural diversification into tourism: Evidence of a European community development programme. *Tourism Management*, *17*(2), 103–111. doi:10.1016/0261-5177(95)00113-1

Barrett, C., Moser, C., McHugh, O., & Barison, J. (2004). Better technology, better plots, or better farmers? Identifying changes in productivity and risk among Malagasy rice farmers. *American Journal of Agricultural Economics*, *86*, 869–888. doi:10.1111/j.0002-9092.2004.00640.x

Ciutacu, C., Chivu, L., & Carle, M. (2008). Étude comparative dans le secteur de l'agriculture de l'Union Européenne. *Romanian Journal of Economics*, *27*(2), 40–56.

Dimitri, D., & Dimitri, S. (1996). Farm business and the development of alternative farm enterprises: An empirical analysis in Greece. *Journal of Rural Studies*, *12*(3), 273–283. doi:10.1016/0743-0167(96)00017-4

Giller, K. E., Witter, E., Corbeels, M., & Tittonell, P. (2009). Conservation agriculture and smallholder farming in Africa: The heretics' view. *Field Crops Research*, *114*, 23–34. doi:10.1016/j.fcr.2009.06.017

Ingo, Z. (2011). Multifunctional peri-urban agriculture - A review of societal demands and the provision of goods and services by farming. *Land Use Policy*, *28*(4), 639–648. doi:10.1016/j.landusepol.2011.01.008

Istudor, N., Ion, R. A., & Turek, A. (2008). Consumers' perceptions about ethics in agro-food products' marketing. *Amfiteatru Economic*, *23*, 92–97.

John, D., Nicholas, M., & Alan, K. (1997). New perspectives on farm household incomes. *Journal of Rural Studies*, *13*(1), 57–64. doi:10.1016/S0743-0167(96)00047-2

Lapple, D., & van Rensburg, T. (2011). Adoption or organic farming: Are there differences between early and late adoption? *Ecological Economics*, *70*, 1406–1414. doi:10.1016/j.ecolecon.2011.03.002

Lazar, C., Lazar, M., & Dimian, G. (2010). The implications of the global economic crisis on the Romanian sustainable development. *Calitatea Acces la Succes*, *11*(113), 140–143.

Magnar, F., Markus, S., & Karin, G. (2012). Stability and variety – Products, organization and institutionalization in farm tourism. *Tourism Management*, *33*(4), 895–909. doi:10.1016/j.tourman.2011.08.015

Nicholas, J. E., & Brian, W. I. (1989). A conceptual framework for investigating farm-based accommodation and tourism. *Britain Journal of Rural Studies*, *5*(3), 257–266. doi:10.1016/0743-0167(89)90004-1

Peter, H. I., Colin, P., Steve, W., & Andrew, D. (2010). The future of small farms: Trajectories and policy priorities. *World Development*, *38*(10), 1349–1361. doi:10.1016/j.worlddev.2009.06.012

Rigby, D., & Cáceres, D. (2001). Organic farming and the sustainability of agricultural systems. *Agricultural Systems*, *68*, 21–40. doi:10.1016/S0308-521X(00)00060-3

Salvatore, D. F., Ivan, P., Aleksi, A., & Tom, M. R. (2010). Agrobiodiversity, farm profits and land fragmentation: Evidence from Bulgaria. *Land Use Policy*, *27*(3), 763–771. doi:10.1016/j.landusepol.2009.10.007

Shoshanah, M. I., & Jeff, S. S. (2012). Farm persistence and adaptation at the rural–urban interface: Succession and farm adjustment. *Journal of Rural Studies*, *28*(1), 107–117. doi:10.1016/j.jrurstud.2011.07.005

Steve, W., Johann, K., & Luis, L. (2010). The future of small farms. *World Development*, *38*(10), 1341–1348. doi:10.1016/j.worlddev.2009.06.013

Steven, H., Peter, H., & Thomas, R. (2010). The rural non-farm economy: Prospects for growth and poverty reduction. *World Development*, *38*(10), 1429–1441. doi:10.1016/j.worlddev.2009.06.008

Toderoiu, F. (2004). Romanian agrifood sector – Macroeconomic correlations of domestic competitiveness. *Journal for Economic Forecasting*, *1*(4), 87–105.

Turek, R. A. (Ed.). (2009). *Competitivitatea pe filiera cerealelor planificabile*. Editura Ars Acdemica.

Turner, J., & Taylor, M. (1998). *Applied farm management*. Oxford, UK: Blackwell Science.

This chapter was previously published in Sustainable Technologies, Policies, and Constraints in the Green Economy edited by Dorel Dusmanescu, Andrei Jean-Vasile, Turek Rahoveanu Adrian, and Jonel Subic, pages 290-311, copyright 2013 by Information Science Reference (an imprint of IGI Global).

Chapter 95
Sustainable Urbanism Revisited:
A Holistic Framework Based on Tradition and Contemporary Orientations

Derya Oktay
Eastern Mediterranean University, North Cyprus

ABSTRACT

At a time of uncontrolled globalization in which serious environmental problems are threatening cities and their inhabitants, as cultural integrity is constantly under attack and many cities lack socially inclusive and responsive environments, there is an urgent need for a radical shift towards a holistic strategy for sustainable urbanism combining ecological sustainability and socio-cultural sustainability. This calls for sensitivity to the traditional urbanism and impact of global ideas, practices, and technologies on local social and cultural practices. In line with these, this chapter aims to establish an environmentally sound and human friendly framework for sustainable urbanism. In this context, the study firstly provides a conceptual understanding of sustainable urbanism and a critical review of its philosophical and practical framework; secondly, it provides an assessment of contemporary approaches to sustainable urbanism; thirdly, the chapter analyses the traditional Turkish (Ottoman) city which provides valuable clues for sustainable development, and discusses possible research directions that could help promote the concept of sustainable urbanism.

INTRODUCTION

Changes that have taken place in the world over the past twenty years, including ecological disturbances and radical changes in traditional settlements have produced cities that are not just chaotic and monotonous in appearance, but have serious environmental problems threatening their inhabitants. Sustainable urbanism, on that ground, appears as a sound framework that draws attention to the immense opportunity to redesign the built environment in a manner that supports a higher quality of life and human health. In this context, when sustainable urbanism is characterised, what is usually addressed as the main concern is natural environment, and hence ecological sustainability, a

DOI: 10.4018/978-1-4666-4852-4.ch095

condition that could be explained with the climate change, the inevitable environmental crisis. Today's development practices consume enormous amounts of land and natural resources, damage ecosystems, produce a wide variety of pollutants and toxic chemicals, create ever-growing distances and inequities between groups of people, fuel global warming, and undermine local community and social values, economies, and quality of life. These incremental changes imply a more critical state in cities of traditional societies where transformations in the urban level are still visible.

What is questioned in this chapter is that, given our knowledge that environmental sustanability is a crucial need, are the contemporary approaches adequate for all settings? At a time of uncontrolled globalization in which sense of place, history and cultural distinctiveness is constantly under attack and many cities lack socially inclusive and responsive environments, do these approaches also integrate social-cultural dimensions? These call for a new understanding of traditional settlements as they represent good uses of local environmental and social values in their times.

The chapter first provides a theoretical underpinning of sustainable urbanism and a critical review of its philosophical and practical framework; second, assessing contemporary approaches to sustainable urbanism and analysing the traditional Turkish (Ottoman) City, proposes a holistic framework for sustainable urbanism that integrates environmental considerations with social-cultural sustainability.

BACKGROUND: A CRITICAL REVIEW OF CONTEMPORARY APPROACHES TO SUSTAINABLE URBANISM

The concept of "sustainability" in its modern sense emerged in the early 1970s in response to a dramatic growth in understanding that modern development practices were leading to worldwide environmental and social crises. During the seventies and eighties, the word "sustainability" used to be connected with the quotation from the Brundlant Report "development which meets present needs without compromising the ability of future generations to achieve their own needs and aspirations" (WCED, 1987). Over the decades, that definition has attracted a lot of discussions. Adams (1990) has criticized the Brundlant Commission's approach for being too accommodating to the interests of the industrialized nations and for not questioning the desirability of continued economic growth. Hence, the notion later developed to describe the goal of integrating concerns and analyses that join economic development and ecological health (Eid, 2003). Hereafter, the notion of endurance and continuity was thought to be the domain of natural science that studied environmental measures to ensure that controlled growth meant that we use the earth in a way that endowed the same rights for future generations. Falling beyond the realm of natural science, the city, the community and their concerns were treated as separate entities, rather than being incorporated into the sustainability context (Haughton & Hunter, 1994; Berg, Magilavy & Zuckerman, 1990). What is more, most of the literature viewed the city and urban living as detrimental to the natural environment and hence a challenge to sustainable development. Among these, the influential book *Design with Nature* by Scottish landscape architect Ian McHarg (1969) was the most influential[1]. On the other hand, since the city is an organic and dynamic entity and may take many different forms and meanings at different time intervals, we are bound to take the "time" factor into account. Sustainability, then, can be regarded as a perspective or paradigm in which we consider the three dimensions of society, economy and environment together, extending the fourth dimension of time."

Sustainable urbanism grows out of three late 20th Century reform movements that have transcended McHarg's antisocial environmentalism to highlight "sustainable development," that is a development which is non-damaging to the environment and which improves the long-term health of human and ecological systems (Wheeler, 2004, 24). The "New Urbanism," "Smart Growth" and "Green Architecture" movements provide the philosophical and practical framework of sustainable urbanism. Each of these movements, however, has revealed certain insularity.

Within architecture and urban design the movement known as the New Urbanism, which appeared in the early 1990s and has become a strong force for re-evaluating the physical layout of communities, cannot be considered efficient and urban, as its focus has been better-designed suburban development. New Urbanism cannot be considered new either, as it revives many ideas about the city and planning that was mainstream before the Modern Movement. Another criticism about New Urbanism is about the elitism within the movement (Kelbaugh, 2002). Indeed, the movement is open to criticism on a number of fronts – in particular for being focused on better-designed suburban development, often for upper income groups, rather than the creation of truly "urban" places, and for not incorporating green building design and landscaping. Furthermore, it can be considered a new type of "ideal vision" conceived, ordained and disseminated from above and not rooted in specific places or local cultures.

Just a few years later, in the mid-1990s, "Smart Growth" evolved as an effort to recast the policy debate over sprawl in a way that more directly linked the environment, the economy and daily life concerns in pursuit of a positive and sustainable urban growth as essential to the quality of the city and urban life (Tregoning, 2006; Wheeler 2006; Girardet, 2004). The movement focused especially on mechanisms to promote more compact, walkable[2] and economically efficient urban development.

Compact cities are argued to offer opportunities to reduce fuel consumption for traveling, as homes, work and leisure facilities are closer together. They are also favored by many in the field of urbanism because urban land can be re-used, while rural land beyond the urban edge is protected. Economic benefits, due to high concentrations of people supporting local economics and easier access to services and facilities, are also suggested. Compact cities with higher densities may also mean that people are more likely to meet each other on the street than in low density areas (Talen 1999; Duany, Plater-Zyberk and Speck, 2001), and people may have a stronger sense of attachment to place (Nelessen, 1994; Diamond and Richardson, 1997). Ultimately, a good quality of life is argued to be sustained, with high concentrations of people providing social conditions conducive to vibrancy, liveliness and cultural production and consumption.

However, there are many who insist that the case of the compact city is not proven, and there are many counter-arguments on its "negativity." The overriding problem with the compact city is that it requires us to ignore the causes and effects of decentralisation, and benefits it may bring" (Thomas and Cousins 1996, 56). As Gordon and Richardson (1997) argue, "sprawl may even be desirable in terms of increasing individual choice and mobility," leaving out the immense social and economic costs of this form of development, and the fact that sprawling cities often do not in fact give residents the choice to live in good, affordable housing in well-located and walkable neighbourhoods. However, the compact city concept can still be considered contradictory to the preference for suburban living where the advantages of concentration might change into disadvantages through congestion which would outweigh energy consumption benefits of the compact city. Indeed, empirical research verifies the preference for suburban living in many cities where the city cannot offer an ideal living environment in its central parts[3]. Tregoning (2006), concerning that contradiction between people's approach to sprawl and their appreciation of the environment, states that "high and middle-income suburbanites may be au fait with and increasingly concerned about traffic, rapid growth, health and other consequences, they remain largely and persistently happy with the quality of their immediate neighbourhoods in those 'hygienic enclaves'." These contradictions indicate a serious problem indeed and require a thorough understanding of determinants.

On the other hand, anti-sprawl strategies, which have obvious consequences for green and open space, have frequently lead to deadlocks in planning, especially concerning green space (Ståhle, 2010). Research supports the intuitive belief of a beneficial relationship between contact with nature and quality of life. A city with high-quality and generous green spaces symbolizes good planning and management, a healthy environment for humans, vegetation and wildlife populations, and bestows pride on its citizenry and government (Jim, 2004). On that ground, it can be stated that if green space is deprived, a compact city may become the antithesis of a green city.

Further, the compact city makes little sense for developing countries because the context is completely different from North American and European countries whose cities have experienced declining populations and deindustrialization. Cities of developing countries have much higher densities than their counterparts in developed countries, and they are not becoming significantly less compact in spite of decelerating population growth and the beginnings of decentralisation (Richardson, Bae, and Baxamusa, 2000; Angel, Sheppard, and Civco, 2005). Moreover, there are some other issues which necessitate developing country cities to be making realistic - yet minimal - plans for urban expansion. Rapid urbanization and higher densities, especially in some developing countries, have obvious consequences in terms of the choice of transportation modes, living conditions, congestion and pollution, and could compromise an environmentally sound planning. In most of these cities, city cannot be restructured into a compact sustainable city within the current planning framework that is limited to a two-dimensional thinking and the private land-owning interests, at the expense of long-term sustainability. Moreover, the increasing amount of people with anti-urban values (owing to the usual problems of the central parts of the city such as traffic congestion, air pollution, insufficient car parking, and lack of green spaces) are powerful elements supporting the low-density city.

Sustainability is most certainly concerned with extravagent use of finite resources and the efficient management of the ecosystem. It also addresses the need to ensure that what we do now does not negatively affect what future generations may wish to do. Murrain (1993) adapts this assertion to the role the urban designer can play: "Sustainability is about structuring town form such that the individual has choice but never at the expense of the citizens as possible to successfully determine the outcome of their daily lives in so far as the layout of the town and the location of uses can exist." On that ground, what we need is "good mixed-use" or "fine grain mixed-use" (Jacobs, 1961; Bentley et al., 1985), not just in relation to the inner city but equally for the urban edge and new settlements.

Our growing sensitivity for environmental issues has taught us that buildings are one of the principal users of energy and materials, and the major causes for damaging nature. Accordingly, new approaches are needed in the conception, theorization and implementation of architectural practices, which will generate architectural designs responsive to the new conception of the relations between nature and the humans where the latter are part of the former depending upon and determining it in dynamic and parametric relationship. In this context, "green architecture" or "sustainable building" emerges as the practice of creating structures and using processes that are environmentally responsible and resource-efficient throughout a building's life-cycle. Although new technologies are constantly being developed to complement current practices in creating greener structures, the common objective is that green buildings are designed to reduce the overall impact of the built environment on human health and the natural environment by efficiently using energy, water, and other resources, protecting occupant health and improving employee productivity, and reducing waste, pollution and environmental degradation.

What is disregarded in all these approaches is that cities also have social-cultural aspects. As stated by Rees (1997),

Cities are the engines of economic growth, the centres of social discourse and the living repositories of human cultural achievement, but also nodes of pure consumption and entropic black holes of industrial society.

Jane Jacobs (1961), who strongly blamed the built environment of modern times for all sorts of social dysfunction, revitalized the inhuman kind of thinking of urban planning and design. Indeed, most urban and suburban development during the past 50 years has been relatively generic, with little sense of place, history, or cultural distinctiveness. Many critics decry low-density, car-oriented, suburban style development, which they label socially isolating, segregating and alienating, calling instead for widespread use of higher density, mixed-use planning principles that lessen reliance on the automobile and increase social interactions. All these factors lead a long-term decline in the extent to which citizens participate in community groups and social institutions, and this decline of community participation is at least partly related to the physical nature of our cities and towns (Ehrenhalt, 1995; Moe and Wilkie, 1997)[4].

The acceleration of globalization has initiated a process of urban transformation, posing some serious threats and challenges to the public spaces of cities, among others. As cities have grown larger and spread wider, urban functions have disintegrated and public spaces, which are important to a democratic and inclusive society, have lost much of their significance in urban life. They became "empty spaces," a space of abstract freedom but no enduring human connection (Sennett, 1994, 375). Economic liberalization and social polarization and fragmentation have turned public spaces into subjects of contestation, reflected in neglect and decline or privatization and exclusion. Public realm, in this context, is shrinking and losing its meaning in people's life. For a long time, owing to the affects of the Modern Movement in architecture, it has been common practice in the development of new districts to prioritise the buildings themselves, then, if possible the public life. The results are deserted city centres and deserted neighbourhoods and urban spaces, where one gets the impression that the city is for cars, not for people.

Gehl (2006) contributes to the idea of the quality of experience in the public realm and points to the differences between the "welcoming environment" and the price we pay for moving fast and thus creating an "unwelcoming environment." The unwelcoming environment is exemplified by a development that presents us with blank walls and turns its back on its surroundings. The welcoming environment is the modern meeting place, the space enclosed and defined by buildings with activites that spill out into the space and create the vitality that makes the space pleasant and memorable. Gehl's concern with urban spaces also mirror the principles of the Italy-based "Cittáslow" (Slow City) movement[5], where the aim is to create an environment that resists the pressure to do everything faster, and therefore to improve the quality of life.

Almost 50 years ago Melvin Webber's renowned paper *The urban place and the nonplace urban realm* (1963) was attempting to persuade people that the traditional role of urban places as the setting for interaction and exchange were no longer necessary. From this perspective, advances in communication technology from video phones to rapid transport systems would stand for "urban place." In the last decade, explosions of information technology have caught up with Webber's forecast with the prediction of millions of people eventually working from home, and electronic media, like e-mail and the internet, are allowing groups to plan and organise events and open space use much more readily than before. Those who advocate the low density suburban developments use this phenomenon as an additional supportive point, positing that once the obligation for commuting is taken away, the arguments against "sprawl" diminish. In contrast to this and much more convincing, is the argument that if people work at home then there is even greater need for a range of facilities and diverse settings in close proximity to minimise the increased isolationism resulting from the loss of urban experience (Murrain, 1993).

Face-to-face human interactions in the public realm, indeed, are intensely relevant for supporting livability, safety and control, economic development, participation, and identity (Jacobs, 1961; Goffman, 1967). On the other hand, research observing people in real-life situations determines how the built environment impacts social wellness (Newman, 1973; Gehl, 1987/2006; Whyte, 1988). Electronic media are no substitute of face-to-face contact as proved by Andi Harris (1991), formerly of Apple Computers and the founder of Telemorphix, whose research into the organisation and potential of telecommuting among the teleworkers indicate that "the more high-tech we become, the greater our need to come together." Since it is urban public spaces that provide the opportunity to meet and watch others (strangers), we should unquestionably use their potential to the full through enhancing their quality and their accessibility by all.

One issue many cities are faced today is that privately owned, controlled spaces of modern urban commerce and design are isolating people from the city spaces which are important to a democratic and inclusive society. The shift from the traditional commercial strip to the sanitised shopping mall has a devastating effect on the city centre with significant reduction in city-centre trade, deteriorated atmosphere and weakened identity. In fact, people do not have freedom in these sanitised spaces as their behaviour and their acts are under control (i.e. they cannot organise a protest meeting or speak loudly), as the aim of shopping centres is not meeting the requirements for a decent everyday life, but encourage spending. Generic in form, replicated in cities all over the world, the shopping mall is a standardized provision of standardized experiences: consumption over production, convenience over difficulty, passivity over activity, homogeneity over heterogeneity (Evans, Taylor and Fraser, 1996). Moreover, as observed by Augé (1995), in the shopping malls, there is an automatic assumption of guilt about anybody not engaged in purposeful consumer behaviour.

King (2009) and Littig and Griessler (2005) suggest social sustainability means the satisfaction of basic human needs, the continual reproduction of humans and the subsequent continuation of culture. Mc Kenzie (2004, 120) more efficiently defines social sustainability as "a life-enhancing condition with communities, and a process within communities that can achieve that condition." In this understanding, social sustainability is a system of cultural relations in which the positive aspects of disparate cultures are valued and promoted and there is widespread participation of citizens not only politically but also socially in all areas of urban life environment.

The recent efforts towards more sustainable urban environments have revealed that, in order for sustainable urbanism to move forward and gain traction, it is essential that it be seen by citizens as playing an integral role in addressing the key issues of our times. The shift to a more sustainable lifestyle necessitates the communities to integrate individualised and privatised environmental action into everyday life and to achieve resource savings in a more extensive context using less water, less energy, less fuel for transportation and leading to less CO_2 emmissions.

To this point, we have to ask ourselves what specific measures need to be taken to create sustainable urban environments, and how environmental and social concerns can be brought together into one convincing scenario, in which everyone benefits. In this context, it is important to understand that the idea of sustainability is not new, and the traditional cities are excellent examples to learn from regarding various dimensions of sustainable urbanism. Sensitivity to tradition allows us to excavate the sophisticated repository of knowledge embedded in planning and design principles and processes linked to the ecological and socio-economic contexts of times past. However, factors such as rapid population growth, an unbalanced population movement due to shifts from rural to urban areas, the possible integration of the country to the capitalist world economy and significant changes in expectations and life styles all combine, in their various ways, to erode the viability of traditional approaches to shelter provision. This means that whilst there are some aspects of traditional approaches which still work well, other aspects may have become inefficient or unworkable, or generally unsustainable. On that ground, this chapter will focus on the Ottoman (Turkish) city, which teaches many lessons that can contribute to meeting contemporary and future planning and design needs provided that their viability are checked for each case and in a time-based perspective.

LESSONS FROM THE OTTOMAN CITY

The Ottoman city, built collaboratively by various cultures on a geographical setting extending from Middle Asia to Anatolia, from Mediterranean to Balkans, demonstrates sensitivity to local topography, Islamic and Christian philosophies about the natural world, and local habits and traditions built from a multitude of human values over centuries (Cerasi,1999). As Günay highlighted (2005, 22),

Tradition, custom and religion have produced a philosophy of life based on contentment with very little. The people were very economical. They were not given to luxury. There was simplicity in everything...[6] *But, in spite of this simplicity, there was a real plenty. There was abondence and great variety of food. There were many rooms and they were large*[7]*. It was a healthy, contended community.*

From an urban and social point of view, the main characteristic of the Ottoman city was its compartmentalization by *mahalles* (neighbourhoods), the outcome of ethnic particularities and religious differences. The *mahalle* was a geographical entity as well as a homogeneous community providing social and economic collaboration among neighbours (Figures 1 and 2). Each *mahalle* had its own characteristics and provided an indicative, unique social environment for their inhabitants. However, spatial segregation that was based on ethnicity and profession leaded separate lives within each minority, and therefore indicated a negative aspect from the perspective of contemporary sustainable urbanism. The *mahalle* was self-sufficient as well through the presence of a variety of functions including a religious-social center, small local market, fountains, *imaret* (open kitchen) and at times, workshops (Oktay 2004). As a result of the closed economy, every household produced their own foodstuffs, i.e. vegetables, fruits and a variety of seasonal produce that can be preserved and stored.

The efforts of numerous private builders (masters) in residential areas were guided only by a few simple rules of civility, assuring individuality within the neighborhood as well as community identity apart from the works of government. It is a remarkable lesson that every house in the Ottoman city was different, even as there is an overall unity and consistency in building technique, scale and character (Eldem,1987). As such, despite the lack of an organising development plan at the governmental level, that is a must in today's development practices, the respect to local environmental and social values made the Ottoman City a sustainable settlement regarding many points.

Figure 1. A typical layout of mahalle

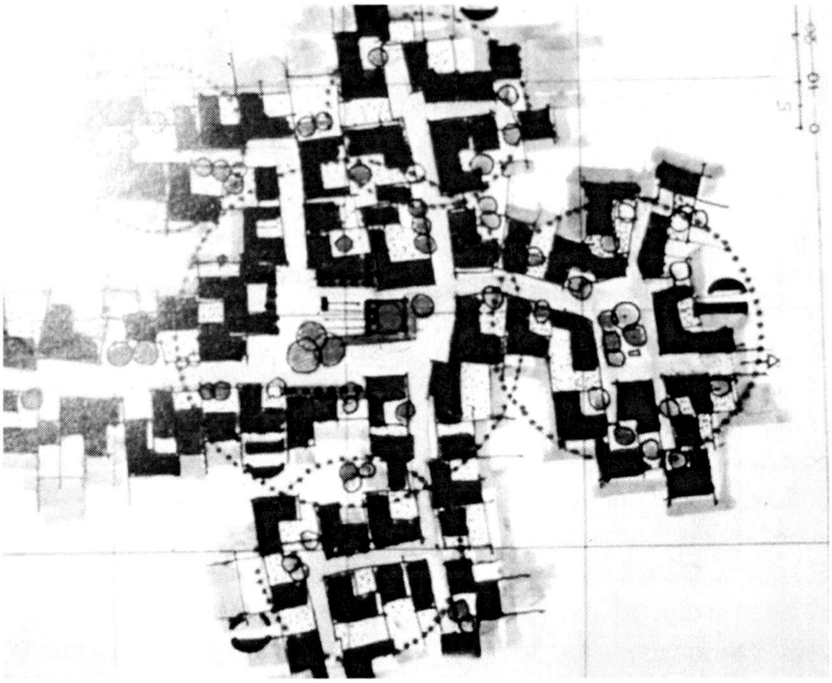

Figure 2. A view to the traditional townscape in Safranbolu

The space of the traditional (Ottoman) city was, at a functional level, clearly divided into public and private realms. The public realm, often in the town center, contained all the collective activities of the town, such as trade and commerce, religion, education, administration, and urban facilities, resulted in a fine-grain mixed-use character. The main public node and the representation of people's power were bestowed to the citadel, the Friday mosque and its courtyard, and the bazaar. One of these elements, the main - often covered - street or streets of the city, the bazaar or *arasta*, functioned also as a communication channel, connecting these to each other and top the less important activities such as public baths, water storages, and educational centers, hence creating a vivid public realm in a spatial continuum (Figure 3). This space was the meeting place of the local people with each other, with the political, religious, end economic hierarchies, and with the outside world. However, owing to the cultural codes and realities of the time which were very different from those of the modern Republic of Turkey, most of the the public facilities were perceived as the territory of men as the traditional role of women necessiated them to spend the majority of their time in the house and in its environs, which constituted the private realm.

The street system in residential areas was mostly pedestrian and had a hierarchical order: from the main streets spread out narrower streets

Figure 3. A representation of Arasta

that themselves had dead-end branches that lead to individual houses. In this system, only the main through-fares separated the urban fabric. This system was achieved through a process of organic growth in which the street pattern was gradually adjusted and changed according to the peculiarities of the land and needs of the local people, where there was no need for wider streets and a low level of accessibility was required. Despite the criticism of the street system from the viewpoint of accessibility and vehicular traffic, a conservationist principle is said to exist in this organic growth that concentrates on the minimum space required (Madanipour, 2009). Moreover, the hierarchical pattern of streets with dead-end branches serving a group of houses created privacy for the dwellers (especially for women - as a significant need at the time) and helped create a strong sense of belonging to their negihbourhood. From an urbanistic point of view, this organic character of the street, in the state of continuous becoming, produces an effect of great expressiveness, and therefore, enhances the character in the Ottoman city. The street also bore a potential for social activities. Children of similar ages played together and identified themselves with the street they lived in. Fountains of running water were found at many street corners where women had the chance to meet their neighbours and have a chat whilst getting water every morning and evening.

On the other hand, *avlu*, the courtyard of each house, an isolated environment that is well defined and well protected, served a variety of uses including social gathering, such as wedding and circumcision parties, women's preparing winter food together, or just spending time together, and helped create a more cohesive community in the *mahalle.*

Owing to the fact that Ottoman urbanism was never based on the kind of strong formalism characteristic of western cultures, a generally informal character was dominant in cities. In this context, there were no formal public open spaces, i.e. well-defined squares, or monumental axes to be found in the cityscape. However, despite having no planned squares and the lack of an active use of *meydan* by people, there was a social and psychological tendency towards meeting and gathering in open spaces of natural character (Figure 4) (Eldem 1987; Cerasi 1999).

The Ottoman city possessed various attributes that generated an ecologically sustainable environment. Regional climatic characteristics were reflected on the patterns of settlements, and accordingly every region produced its own characteristic urban fabric and architecture. For instance, in Safranbolu, one of the most characteristic towns in the northwestern Black Sea region of Anatolia (Turkey), hard winters forced the people to settle in sheltered valleys (Günay, 2005, 21). The pre-existing topographic character of the

Figure 4. Socializing in natural open space

site was apparent at the urban scale even in intense built-up areas. The green gardens, i.e. vegetable gardens and patches (*bostan*), orchards, and so forth, implied a green belt dividing the quarters (sometimes occupying a natural creek as observed in Antep, a city in the southeastern region of Anatolia) and bounded the town[8] (Aru 1998, 12), and contributed to the self-sufficiency in general. The small squares at the intersection of streets with trees created opportunity for access to nature in the public realm as well. The streets that were defined by high walls of the residential courtyards provided a protected and comfortable space, and being divided into two by a 10-14 cm water canal running through the middle, helped distribute water to gardens, and prevented the rainwater from flowing into the courtyards (Figure 5).

The presence of a variety of house plans all with a courtyard, *avlu,* or garden in every region of Anatolia reveals the fact that there was a natural relationship between such a layout and the Anatolian life-style (Kuban, 1983). With its fruit trees, flowers and small kitchen garden, the *avlu*, separated from the street by a wall, was the closest relation the house has to nature; and thus it also provided the inhabitant with direct access to nature, and enhanced both the building ecology and self-sufficiency of the house.

Figure 5. A street in the old Nicosia

All these peculiarities, on the contrary to many newly developed urban environments in Turkey and around the world, make the Ottoman city an ideal model for ecologically and socially sustainable cities despite its shortcomings in terms of viability of certain aspects (i.e. women's limited use of the public realm) for today's cities and urban life. Since sustainability needs to be assessed considering the cultural codes and realities of the time, as discussed in the background section of this chapter, these shortcomings may be tolerated within the larger, holistic context provided that requirements for every aspect of life are satisfied in today's urban planning and urban design.

REDEFINING ESSENTIALS FOR SUSTAINABLE URBANISM

Based on our critical review of contemporary approaches to sustainable urbanism and our analysis of the Ottoman city as an ideal model for sustainable urbanism, I would advocate that new urban planning and design endeavours should comprise a human dimension and demonstrate respect to regional characteristics. In this context, the following aspects are considered essential: Context-sensitive compactness and de-fragmentation; Completeness: good mixed-use; Connectedness: integrated transportation and land use; Ecological sensitivity; A focus on place and public spaces; Social-cultural sustainability and cohesive neighbourhood; and Sustainable life style.

Context-Sensitive Compactness and De-Fragmentation

Assuming that urban sprawl is mostly a negative phonemenon characterized by a low density development pattern and presenting as many problems the traffic congestion, massive absorption of green space, inadequate infrastructures and services, low quality developments, as well as bureaucratic dysfunctions and lack of financial resources (Downs, 1999), a proactive management of urban growth containing sprawl can be considered essential. Urban design of compact cities can obviously contribute to a more sustainable way of life, particularly in industrialised societies. However, as revealed through the ideas exemplified by the traditional Ottoman city that comply with regional characteristics, it cannot be expected that cities should all fit the same formula. What is needed is not a radical set of measures, but by a complete diagnosis of the territory, identifying local characteristics (i.e. climate, landscape, identity, culture and traditions), specificities, demands and dynamics, and an estimating and evaluation of the urban development processes, through comparing the demand and the offer for urban growth, and consideration of the issues of "where" and "how" the urban settlement grow.

Inspired by the Ottoman city and *mahalle,* the contemporary city could be thought as an entity made up cohesive and identifiable districts, and smaller towns of functional diversity could be created in the vicinity of the city rather than reaching unacceptable levels of density and population. In this context, density should be related to design in such a way that its advantages and disadvantages are investigated by considering local social dynamics (need for privacy, degree of privacy, neighbourly relations, and so forth) and environmental values (green infrastructure, made of wetlands, forests, groundwater recharge zones, and so forth), and new scenarios for "defragmentation" where open growth may find its placement.

Completeness: Good Mixed-Use

Fine-grain mixed-use is sought in urban expansion in order for those environments to be lively, safe, sensorily rich, choice laden, economically and spatially efficient and ecologically diverse; sustainable as far as the built environment can believably be. When these objectives are applied to what we have been producing in contemporary expansion and new settlement proposals we are nowhere near achieving them. On that ground, Murrain's definition (1993) of good mixed-use as "a finely grained mix of primary land uses, namely a variety of dwellings and workplaces with housing predominant, closely integrated with all other support services, within convenient distance of the majority of the homes" is useful. This good mixed-use was an important component of the public realm in the Ottoman city. Containing all the collective activities (i.e. trade and commerce, religion, education, administration, and urban facilities), the central parts of the city revealed a fine-grain mixed-use character and helped the local people meet with each other (despite the limited frequency by women owing to the cultural codes of the time) and with the outside world. The main street and the bazaar or *arasta* in the Ottoman city, functioned as a communication channel, connecting the main activities to each other and top the less important activities (i.e. public baths, water storages, and educational centers), and created a vivid public realm in a spatial continuum. These characteristics can be re-interpreted as a model when planning and/or re-designing our cities whose central parts are deteriorating owing to the lack of diversity of main functions (business, commerce, housing, recreation) and the effects of privately owned, intraverted spaces of modern urban commerce and design.

Connectedness: Integrated Transportation and Land Use

In a sustainable urban environment, people should have abundant opportunities to walk, ride (if possible), bike, (if necessary) use a wheelchair around the neighbourhoods, as well as having access to good public transport. These varied transportation options would increase access to services and facilites, help reduce car dependency and thus congestion and pollution, achieve a reduction of energy consumption and help maintain a high-level of energy-efficient and environment-friendly mobility inside the city or city region. The findings of empirical surveys (i.e. Lund, 2002; Kim and Kaplan, 2004; Khandokar, 2009) suggest that the social and physical contexts of the walkable neighborhood enhance casual interactions and social participations and are likely to contribute to the sense of community. In the Ottoman city, the walkability of the streets (at a time of the unavailability of motor cars but other means of transport such as horses and donkeys) was enhanced by human scale, physical convenience (protection from sun, rain, etc.) due to the narrow and winding streets following the natural contours of the land, and pleasant continuity of the outer walls of the houses and courtyards. From these, one important lesson for the contemporary city is designing the city streets first for people taking into account the functional and aesthetic needs of people rather than complying with cars only.

Ecological Sensitivity

As observed in the Ottoman settlements which reveal an ideal integration with the natural environment and climate, sustainable urbanism seeks to connect people to nature and natural systems, even in dense urban environments. In this context, an attempt at integrating such features as edible landscapes of fruit trees and large vegetable patches (allotments) into the city would be beneficial for dwellers in terms of lower heating and cooling bills, lower food costs, and reduced risk of flooding and landslide damage[9]. Trees with canopies can be used for their shadowing effect, and for the definition of spaces both in streets and courtyards. When a more flexible design is possible, the traditional concept of courtyard can be reinterpreted and modified in the new housing developments, and walk-up type housing blocks can be arranged around a semi-private courtyard space in some areas in a diversed typological pattern. In order to eliminate safety problems and to enhance the sense of place, the design of the residential complex should be based on the principles of responsive urban design by providing active edges (mixed-use if possible) along the streets and encouraging active use of the courtyards by residents. At the building scale, other important aspects to ecological sensitivity are the use of local and regional materials of natural character, conformity of the building to its environs and in particular to the climate, the flexibility to adapt to changing conditions over time, and the rich variety of spaces extending from interior spaces to open spaces through various types of semi-open spaces.

A Focus on Place and Public Spaces

Although public spaces form a crucial feature of sustainable and liveable cities, contemporary urban environments frequently lack enough space kept aside for them, and most of those spaces which are introduced as "public spaces" miss spatial, ecological and social qualities, and cannot be considered "places for people." Inspired by the Ottoman city, new urban areas could be planned and designed around a hierarchy of spaces for different purposes, the idea of main shopping strip could be revived in order to prevent the shopping malls to be the norm, and the street pattern could be organized in a way that each street has an identity through the continuity, design and functional layout of buildings. In the contemporary city, streets, squares and public parks are the only places where people truly meet as equals, and a high-quality public realm may help create a sense of belonging and collective identity.

Social Sustainability

Social sustainability is a system of social-cultural relations in which the positive aspects of disparate cultures are valued and promoted and there is widespread participation of citizens not only politically but also socially in all areas of urban life environment. Its success depends on the level of people's expectations, behaviour, value systems, transparency and accountability in both public and private decision-making. As the most appealing aspect of sustainable urbanism is to be the sustainable neighbourhood with its societal benefits, we must widen our definition of the sustainable urban neighbourhood to include social as well as environmental concerns as reflected in *mahalle*, the cohesive neighbourhood unit in the Ottoman city. However, we should not ignore the great changes that happened in the daily life of people, i.e. significant increase in percentage of working women, women's equal participation in almost all aspects of life, and so forth.

Sustainable Lifestyle

Everything we do as professionals and as human beings in the name of sustainability means very little if we don't actually change environmental behaviour of consumers, companies, communities and governments. Adopting sustainable lifestyles require incorporating a range of behavioural responses from energy saving and water conservation, to waste recycling and green consumption, and these would influence the urban quality of life without damaging the planet for the future. In the Ottoman city, in the early Ottoman and Seljuk periods in particular, owing to the preferred simplicity in every aspect of life and self-sufficiency in many senses, people generally adopted a sustainable lifestyle, and it was a healthy and contended community. In today's cities, what is needed for sustainable lifestyle is "education for sustainable development" and hence "ecological citizenship," that would enable urban residents to develop the knowledge, values and skills to participate in decisions about the ways they do things individually and collectively, both locally and globally.

FUTURE RESEARCH DIRECTIONS

In the beginning of the 21st century, cities appear as "catalysts" of deep-reaching negotiation process between tradition and innovation, between stagnation and continuity, and efforts to achieve urban sustainability may require overcoming many other complexities in addition to the presented framework.

As highlighted by Rowland (2006, 187), the new agenda for sustainable development is uncertain as the ground rules are changing:

That uncertainty is an interesting issue. Whereas the concept of branding is about creating certainty and uniformity, the dissection of the layers of the city and their interdependencies promote uncertainty, and the need for flexibility. So we can see a divergence appearing between a more dynamic approach to urban development, which explores issues of change, and one which creates the status and sameness we see in many of the masterplans currently being produced.

Sustainability often seems to be grounded in the future, and can seem difficult to achieve because the future is unpredictable, and the needs of future generations unknown. Part of the challenge for sustainable urbanism therefore is to plan for uncertainty and to find ways of accommodating dynamism in the city. Rather than being too obsessive about predicting the future, a thorough understanding of the concept of "resilient city" is needed through further research.

One challenge for the sustainable urbanism paradigm is to resolve the issues between the increasing demand for environmentally responsible development and traditional townscape. Zero-footprint development and high eco-scores demand new morphologies and typologies. Establishing the balance to create a new townscape without isolating the "green" buildings from their urban or regional contexts will therefore be a new future of sustainable urbanism.

Finding ways to be able to respond to the impact of globalisation in favor of public interest is another critical issue as we witness a weakening, a shrinking public sphere and an expansion of the space reserved for the private interest and the market forces. As Keleş highlights (2007), urban sustainability cannot be left to market forces alone, because the market tends to make decisions based on short-term perspective whereas sustainability is a matter of long-term concern.

Another timely research direction may tap a culture of inclusion in today's multi-cultural cities. Since it is acknowledged by a growing body of professionals and academics and highlighted by Wood (2010) that diversity is a resource and not necessarily a problem, strategies should be sought after in order to promote the concept of "intercultural city" that aims at acquiring a pluralistic identity. Today, in many "multi-cultural" cities, what looks like diversity to the outsider (tourist) is in fact a number of isolated areas with sharp division lines of non-interaction (Sennett, 1999). To make diversity a positive value and difference that really encourages and aids sociability in the city, it is not enough to find communal places where people feel good together, and have a nice time. Since the places where social interaction appears in the city may not be places of leisure or places of consumption, but they may be places where people meet about issues of conflict (such as schools), experiencing diversity must be taken more seriously.

CONCLUSION

As we live in environments that have often been very damaged, in ecological, social and cultural terms, there is an urgent need for a radical shift towards a holistic approach to sustainable urban planning/design, combining ecological and social-cultural sustainability. This calls for sensitivity to traditional urbanism and impact of global ideas, practices and technologies on local social and cultural practices. In that sense, the Ottoman city, in the early Ottoman and Seljuk periods in particular, possesses various characteristics that can inform modern planning and urban design.

Urban design of compact cities can obviously contribute to a more sustainable way of life, particularly in industrialised societies. However, since cities are all different in form and structure owing to a host of place-specific factors, it cannot be expected that they should all fit the same formula when it comes to the question of a sustainable urban form. The degree of compactness and/or defragmentation should therefore be context-sensitive. Inspired by the Ottoman city and *mahalle* that comply with local environmental and social-cultural values of the time, the contemporary city could be reconsidered as an entity made up of cohesive districts, and smaller towns of functional diversity could be created in the vicinity of the city rather than reaching unacceptable levels of density and population.

A sustainable community endeavours to promote multi-functional rather than mono-functional settlement patterns by providing compact urban centres, with a broad range of services and amenities in close proximity. This reduces the need for vehicular and public transport, thereby decreasing demands on infrastructure and energy resources, while promoting pedestrian accessibility and community. The fine-grain mixed-use in the public realm of the Ottoman city can be re-interpreted as a model when planning and/or re-designing our cities whose central parts are deteriorating owing to the lack of diversity of main functions and the negative effects of privately owned, intraverted spaces of modern urban commerce and design.

In the course of environmental transition, cities could attempt to keep as many as possible of the environment-sustainability ingredients, including green spaces. In that sense, an attempt at integrating such features as edible landscapes and directing some of the efforts of greening towards streets would be beneficial.

What matters in terms of "green architecture" or "sustainable buildings" is that the concept of the relationship between nature and the architecture as a design philosophy be restored, without resorting to superficial mimicry. It is worrying that so-called contemporary green buildings are often considered in isolation from their urban or regional contexts. It should be accepted that a city is not a simple collection of buildings, and green or "zero-energy" buildings alone do not create a sustainable city. What are important to green architecture are the use of local and regional materials, conformity of the building to its environs and in particular to the climate, the flexibility to adapt to changing conditions over time, and the rich variety of spaces extending from interior spaces to open spaces through a variety of semi-open spaces.

High-tech innovation and new sustainable technologies undoubtedly have an important role to play, but in an energy-depleted world, cities that can de-link from their dependence on these are likely to be more resilient.

Inspired by the Ottoman city, new urban areas could be planned and designed around a hierarchy of spaces for different purposes, the idea of shopping strip could be revived in order to prevent the shopping malls to be the norm, and the street pattern could be organized in a way that each street has an identity through the continuity, design and functional layout of buildings. We can move towards more inclusive urban design approach that not only views the public realm as an outside room with equitable access, but also as a welcoming place where a variety of users benefit from it and place a value on it as they interact with other people and their own prior experiences.

We must widen our definition of the sustainable urban neighbourhood to include social as well as environmental concerns as reflected in *mahalle*, the social-spatial unit in the Ottoman city, without ignoring the great changes that happened in the daily life of people. In the new settlements, there must be places that foster special rituals where all residents come together in common pursuit and observance as used to be done in the streets and courtyards. There should be places, which support multiple public activities, recreation, and settings arranged to encourage safe, and everyday, personal exchanges among people who might otherwise remain strangers.

Naturally these ideas and principles will not achieve their objective without an appropriate application strategy. Urban planning and design is a shared responsibility and putting aims into practice depends on evaluations within a far broader political-economic context. For this reason, a common vision shared by every strand of society needs to be determined and for this to materialize in the long run, uncompromising efforts must be made that do not resort to low quality and cheap solutions. Finally, policy-makers need to become a little more subversive in how change towards a more sustainable environment is sold, and governmental pressure on individuals to engage with environmental practices is strongly needed.

ACKNOWLEDGMENT

The author thanks the anonymous referees who provided useful feedback on an early version of this chapter, and Ms. Güneş Özdural for her valuable comments on the Ottoman city and her proof-reading.

REFERENCES

Adams, W. M. (1990). *Green development: Environment and sustainability in the third world*. New York, NY: Routledge.

Angel, S., Sheppard, S. C., & Civco, D. L. (2005). *The dynamics of global urban expansion*. Washington, DC: The World Bank Transport and Urban Development Department.

Aru, K. A. (1998). Türk kenti. Istanbul, Turkey: Building-Industry Center (YEM).

Augé, M. (1995). *Non-places: Introduction to the anthropology of supermodernity*. London, UK: Verso.

Bentley, I., Alcock, A., Murrain, P., McGlynn, S., & Smith, P. (1985). *Responsive environments*. London, UK: Architectural Press.

Berg, P., Magilavy, B., & Zuckerman, S. (1990). *A green city program for the San Francisco Bay Area and beyond*. San Francisco, CA: Planet Drum Foundation and Wingbow Press.

Cerasi, M. M. (1999). *Osmanlı kenti: Osmanlı kentinde 18. ve 19. yüzyıllarda kent uygarlığı ve mimarisi*. Istanbul, Turkey: Yapı Kredi Yayınları.

Diamond, H. L., & Noonan, P. F. (1996). *Land use in America*. Washington, DC: Island Press.

Dillard, J., Dujon, V., & King, M. C. (2009). Reflection and directions for the future. In J. Dillard, V. Dujon, & M. C. King (Eds.), *Understanding the social dimension of sustainability* (pp. 285–290). New York, NY: Routledge.

Dobson, A. (2009). Ecological citizenship. In M. Reynolds, C. Blackmore & M. J. Smith (Eds.). The environmental responsibility reader (256-270). London, UK: Zed Books.

Downs, A. (1999). Some realities about sprawl and urban decline. *Housing Policy Debate*, *10*(4), 955–974.

Duany, A., Plater-Zyberk, E., & Speck, J. (2001). *Suburban nation: The rise of sprawl and the decline of the American dream*. New York, NY: North Point Press.

Ehrenhalt, A. (1995). *The lost city: The forgotten virtues of community in America*. New York, NY: Basic Books.

Eid, Y. Y. (2003). Sustainable urban communities: History defying cultural conflict. In G. Moser, E. Pol, Y. Bernard, M. Bonnes, J. A. Corralizia, & M. V. Giuliani (Eds.), *People, places and sustainability* (pp. 83–93). Seattle, WA: Hogrefe & Huber Publishers.

Eldem, S. H. (1987). *Turkish houses: Ottoman period* (Vol. 3). Istanbul, Turkey: T.A.Ç.

Evans, K., Taylor, I., & Fraser, P. (1996). *A tale of two cities: Global change, local feeling and everyday life in the north of England*. New York, NY: Taylor & Francis.

Farr, D. (2008). *Sustainable urbanism: Urban design with nature*. New Jersey: Wiley.

Gehl, G. (2006). Life, spaces, buildings: And in said order please. In M. Moor, & J. Rowland (Eds.), *Urban design futures* (pp. 70–75). Oxon, UK: Routledge.

Gehl, J. (1987). *Life between buildings: Using public space*. New York, NY: Van Nostrand Reinhold.

Girardet, H. (2004). *Cities, people, planet*. Chichester, UK: Wiley-Academy.

Goffman, E. (1967). *Interaction ritual: Essays in face-to-face behavior*. Chicago, IL: Aldine.

Gordon, P., & Richardson, R. (1997). Are compact cities a desirable planning goal? *Journal of the American Planning Association. American Planning Association*, *63*(1), 95–107. doi:10.1080/01944369708975727

Gore, A. (2006). *An inconvenient truth*. Emmaus, PA: Rodale Press.

Günay, R. (2005). *Safranbolu houses*. Istanbul, Turkey: Yapı Yayın.

Harris, A. (1991). Telecommuting: A growing transport alternative for office commuters. *Proceedings of Technological and Societal Alternatives in Transportation 1991*. Report sponsored by the IEEE Society on Social Implications of Technology.

Haughton, G., & Hunter, C. (1994). *Sustainable cities*. London, UK: Jessica Kinsley Publishers.

Jacobs, J. (1961). *The death and life of great American cities*. New York, NY: Random House.

Kelbaugh, D. S. (2002). *Repairing the American metropolis: Common place revisited*. Seattle, WA: University of Washington Press.

Keleş, R. (2007). Sustainable urban development under unsustainable conditions. In D. Oktay (Ed.), *Inquiry into urban environment: Issues concerning urban, housing and the built environment* (pp. 27–38). Famagusta, Cyprus: EMU Press.

Khandokar, F. (2009). Briefing: User-perspectives on walkable neighbourhoods. *Proceedings of ICE: Urban Design and Planning*, *162*(DP4), 155–158. doi:10.1680/udap.2009.162.4.155

Kim, J., & Kaplan, R. (2004). Physical and psychological factors in sense of community: New urbanist Kentlands and nearby Orchard Village. *Environment and Behavior*, *36*(3), 313–340. doi:10.1177/0013916503260236

Kuban, D. (1986). *Turkish culture and arts*. Istanbul, UK: BBA.

Lund, H. (2002). Pedestrian environments and sense of community. *Journal of Planning Education and Research*, *21*, 301–312. doi:10.1177/0739456X0202100307

Madanipour, A. (1999). Why are the design and development of public spaces significant for cities? *Environment and Planning. B, Planning & Design*, *26*(6), 879–891. doi:10.1068/b260879

Madanipour, A. (2009). *Whose public space? International case studies in urban design and development*. New York, NY: Routledge.

Mitchell, D. (2003). *The right to the city: Social justice and the fight for public space*. New York, NY: Guilford Press.

Moe, R., & Wilkie, C. (1997). *Changing places: Rebuilding community in the age of sprawl*. New York, NY: Henry Holt & Co.

Murrain, P. (1993). Urban expansion: Look back and learn. In R. Hayward, & S. McGlynn (Eds.), *Making better places: Urban design now* (pp. 83–94). Oxford, UK: Butterworth.

Nelessen, A. C. (1994). *Visions for a new American dream: Process, principles and an ordinance to plan and design small urban communities*. Chicago, IL: American Planning Association.

Oktay, D. (2001). *Planning housing environments for sustainability: Evaluations in Cypriot settlements*. Istanbul, Turkey: YEM.

Oktay, D. (2004). Urban design for sustainability: A study on the Turkish city. *International Journal for Sustainable Development and World Ecology*, *11*(1), 24–35. doi:10.1080/13504500409469808

Oktay, D. (2007). *Interview with Lucien Kroll*. Brussel, March 2004/Famagusta, April 2007.

Oktay, D. (2010). *Gazimağusa'da kentsel yaşam kalitesi: Araştırma, planlama ve yönetim için göstergeler (Quality of urban life in Famagusta: Implications for research, planning and policy)*. Famagusta, Cyprus: EMU Press.

Oktay, D., & Conteh, F. (2007). Towards sustainable urban growth in Famagusta. *Proceedings of ENHR Conference: Sustainable Urban Areas*, Rotterdam.

Oktay, D., & Rüstemli, A. (2011). The quality of urban life and neighbourhood satisfaction in Famagusta. In R. W. Marans, & R. Stimson (Eds.), *Investigating quality of urban life: Theory, methods and empirical research* (pp. 232–249). New York, NY: Springer. doi:10.1007/978-94-007-1742-8_10

Rees, W. E. (1997). Urban ecosystems: The human dimension. *Urban Ecosystems*, *1*(1), 63–75. doi:10.1023/A:1014380105620

Richardson, H. W., Bae, C. C., & Baxamusa, M. (2000). Compact cities in developing countries. In R. Burgess, & M. Jenks (Eds.), *Compact cities: Sustainable urban forms for developing countries* (pp. 25–36). London, UK: Spon Press.

Sennett, R. (1994). *Flesh and stone: The body and the city in western civilization*. New York, NY: Norton.

Sennett, R. (1999). The challenges of urban diversity. In L. Nyström (Ed.), *City and culture: Cultural processes and urban sustainability* (pp. 128–134). Stockholm, Sweden: The Swedish Urban Environmental Council.

Ståhle, A. (2010). More green space in a denser city: Critical relations between user experience and urban form. *URBAN DESIGN International*, *15*(1), 47–67. doi:10.1057/udi.2009.27

Talen, E. (1999). Sense of community and neighbourhood form: An assessment of the social doctrine of new urbanism. *Urban Studies (Edinburgh, Scotland)*, *36*, 1361–1379. doi:10.1080/0042098993033

Thomas, L., & Cousins, W. (1996). The compact city: A successful, desirable and achievable urban form? In M. Jenks, E. Burton, & K. Williams (Eds.), *The compact city: A sustainable urban form?* (pp. 53–65). London, UK: E & FN Spon.

(1994). Towards a global ecological citizen. In Van Steenbergen, B. (Ed.), *The condition of citizenship*. London, UK: Sage.

Tregoning, H. (2006). It's sprawl, but it's my sprawl. In M. Moor, & J. Rowland (Eds.), *Urban design futures* (pp. 38–45). New York, NY: Routledge.

Van der Ryn, S., & Cowan, S. (1996). *Ecological design*. Washington, DC: Island Press.

WCED (World Commission On Environment and Development). (1987). *Our common future (Brundtland Report)*. Oxford, UK: Oxford University Press.

Wheeler, S. M. (2006). *Planning for sustainability: Creating livable, equitable and ecological communities*. New York, NY: Routledge.

Wood, P. (2010). *Planning for the intercultural city*. Introductory speech, NECE Workshop on City and Diversity – Challenges for Citizenship Education, 24-26 June, Barcelona.

ENDNOTES

1. The book tells the story of McHarg's harsh reaction as a young man against the pollution, ugliness, and a lack of vegetation in his native Glasgow, which instilled in him, and in many of his generation and later generations (i.e. Saunders, 1981) a lifelong bias against cities pointing to a link between cities and pathology.
2. Southwork (2005) defines walkability as the extent to which the built environment supports and encourages walking by providing for pedestrians comfort and safety, connecting people with varied destinations within a reasonable amount of time and effort, and offering visual interest in journeys through the network."

3 One recent example is a quality of life survey carried out by the author in Famagusta (North Cyprus), where uncontrollable sprawl and rapid urban development characterize the urban development in the last 15 years. Findings of this comprehensive survey reveal that, in response to the question about a desirable place to live, more than a third (38%) of 400 respondents preferred a neighbourhood comprising single family garden houses with access to natural areas, where a car was an absolute necessity (owing to the lack of public transportation) (Oktay, 2010; Oktay and Rüstemli, 2011).

4 Ehrenhalt (1995), in that tradition, observed three communities in the Chicago area as they were forty years ago, and as they are today. The result is a challenging work of social criticism that vividly evokes a time and place as follows: "What has vanished is the neighborhood-centered life that once buzzed around them - the shops where merchants and customers knew one another; the comings and goings between adjacent back yards; the activity in the streets that was monitored during all the waking hours of the day by the informal law enforcement system of the neighborhood, the at-home mothers." Moe and Wilkie (1997), on the other hand, blamed the policies favoring suburban living which in tandem had the effect of directing massive amounts of investment to the suburbs to the detriment of inner cities. Traditional neighbourhoods as a consequence were labeled as "bad" and the stage was set for the rise of the development industry.

5 Cittaslow (literally *Slow City*), inspired by the Slow Food organization, is a movement founded in Italy in 1999 and aims to improve the quality of urban life while resisting the "fast-lane," homogenized world seen in modern cities, celebrating and supporting diversity of culture and the specialties of a town and its hinterland (http://www.slow-movement.com).

6 In the earlier Ottoman and Selçuk periods in particular, during which the power was bestowed to local people equally.

7 This could be explained with the tradition of the extended family structure that exerted a strong influence on the design of the Ottoman houses.

8 Walled cities obviously need to be excluded from that generalization.

9 In that sense, a good example is the Soviet experiment where each family was alloted around 100 m^2 within sensible walking distance. This experiment still continues in most Russian cities where regular bus service carries people to the vegetable gardens. Another experiment is "popular gardens" of Havana (Cuba) that are the small parcels of state-owned land in the city, cultivated by individuals or community groups in response to ongoing food shortages.

This chapter was previously published in Green and Ecological Technologies for Urban Planning edited by Ozge Yalciner Ercoskun, pages 17-36, copyright 2012 by Information Science Reference (an imprint of IGI Global).

Chapter 96
Towards Zero Energy Buildings (ZEB):
The Role of Environmental Technologies

Paris A. Fokaides
RD Hydraulis Ltd, Cyprus

ABSTRACT

In 2009, European Union (EU) member states forged a long-awaited compromise on the recast buildings directive, agreeing that all new buildings would have to comply with high energy-performance standards by the end of 2020. The recast Energy Performance of Buildings Directive, which was finally announced in May 2010, requires the public sector to take the lead by owning buildings with "nearly zero" energy consumption standards by the end of 2018, which is two years in advance of the private sector. The objective of this chapter is to discuss both the range of potential consequences to European cities resulting from widespread implementation of zero energy buildings (ZEBs) and the relevant environmental technologies in accordance with the national goals set by the EU Member States. As EU member states are moving ahead with their targets and strategies for ZEBs, this chapter presents the most possible scenarios for the implementation of the EU recast buildings directive regarding ZEBs by 2020. A detailed review regarding the existing EU member states' definitions and policies on low energy buildings and ZEBs, and the current status of RES technologies for ZEBs is also presented. Finally, some first thoughts are provided regarding the minimisation of energy consumption in the building sector and the green city goal, as energy is considered to be one of the most important chapters when evaluating a green community. The next step for the integration of green buildings would be the adoption of principles resulting from ZEB analyses and descriptions in existing green building models.

DOI: 10.4018/978-1-4666-4852-4.ch096

INTRODUCTION

High energy prices and global climate changes force developed societies to reconsider their energy consumption habits. In the building sector, energy consumption continues to increase, primarily because new buildings are constructed faster than old ones are demolished. Because traditional building use consumes almost 40% of the total fossil fuel energy in the US and European Union, energy efficient buildings are becoming more interesting. A zero energy building (ZEB), or net zero energy building, is a general term applied to a building when it has zero net energy consumption and zero carbon emissions annually. This term implies that the ZEB energy demand for heat and electrical power is reduced, and this reduced demand is met on an annual basis from on-site renewable energy technologies. The zero energy building design rationale is a progression from passive sustainable building design. However, the object of a ZEB goes even further; not only should the energy consumption of the building be minimised by means of passive design methods, but the building itself should be able to balance its energy requirements with active techniques and renewable technologies. Zero energy buildings (ZEBs) constitute a growing part of the European building stock because two-thirds of European countries already include them in national roadmaps as a goal for future new buildings. Sustainable building is a rapidly growing practice in new construction development and is constantly gaining importance and popularity.

ZEBs are built with significant energy-saving features. The heating and cooling loads are lowered by using highly efficient equipment, natural ventilation, and other techniques. ZEBs are normally designed to use passive solar heat gain and shading combined with thermal mass to stabilise diurnal temperature variations throughout the day. To reduce the net consumed energy for heating and cooling purposes, these advanced buildings will need significantly higher insulation levels than conventional buildings. ZEBs employ super-insulation to significantly decrease the heat transfer through the walls, roof, floor, and windows. There are countless different insulation products available (many of which are natural) that can be used to provide the required high thermal resistance values (R-Values). The elimination of thermal bridging in ZEBs is taken into consideration not only because of energy loss, but also to avoid condensation that could result in damage to the building fabric. ZEBs also require the use of onsite renewable energy technologies to offset the building's primary energy use.

There are several definitions of what ZEB means, with a particular difference in usage between North America and Europe. According to the interim conclusions of the European concerted action for the building energy performance implementation directive, currently 23 different terms are identified across Europe for defining high performance buildings. The variation exists not only in terms of the absolute level of energy consumption in a low energy building, but also in terms of the deviation from the minimum requirements as stated in the national building regulations (Concerted Action, 2008). Further, the national calculation methods vary from country to country, which makes it rather complicated to compare the absolute values of the energy requirements. For instance, energy performance can be calculated by conditioned floor area, habitable area, or gross floor area based on primary or consumed energy (EN 15217:2007). Additionally, different energy consuming processes are included in the different calculations of the energy performance. These different standards in calculation inputs and assumptions can easily result in important deviations.

The development of modern ZEBs became possible not only through the progress made in new construction technologies and techniques but also by academic research on existing and experimental buildings through the collection of precise energy performance feedback. Today's advanced models can precisely demonstrate the efficacy of engineering design solutions. Advance projections of a ZEB even during the design stage are possible with the advancements in simulation tools (Parker, 2009). Simulation tools, now possible with the recent advancements, can perform with a range of design variables, such as building orientation, insulation type and values of the building elements, overhang depth, air tightness, heating and cooling efficiency, lighting and other equipment, and local climate. These tools can guide designers and engineers in their predictions of how a building will perform prior to its construction and enable them to follow the correct solution in terms of economic feasibility and lifecycle assessment.

ZEBs have many advantageous features, such as the following:

- Improved thermal comfort because of more-uniform interior temperatures,
- Being autarkic under certain conditions – a feature which is highly important, particularly during blackouts or periods of increased energy prices,
- Environmental sustainability (in terms of energy and pollution minimisation),
- Reduced total net monthly cost of living and higher resale values.

A ZEB will, however, have a higher initial cost. Additional costs for ZEBs cannot be predicted with precision, and in all cases they depend on specific conditions. Up to 10% extra upfront investment costs are reported; however, this initial cost is on a clearly declining trend (Sartori and Hestnes, 2007). The cost of energy efficient buildings is generally higher because of the extra costs associated with the improved insulation of all building components, such as windows. Another reason is that most entrepreneurs are not used to the new technologies, which increases both the invested time and resources in planning and the cost of education and quality assurance. Additionally, the lack of designer know-how increases the risk of design failure and poor performance of ZEB features. These issues only help contribute to the idea that energy-efficient buildings are expensive.

Public acceptance must be recognised as an important issue in shaping the widespread implementation of ZEBs and in achieving energy policy targets.

EXISTING STATUS OF ZEBs AND LOW ENERGY BUILDINGS IN EUROPE

Historical Background

Passive and vernacular architecture have been linked for centuries. In ancient times, people used to deal with the problem of achieving thermal comfort by choosing the appropriate construction solutions for their buildings (Zhai and Previtali, 2010). The question was much more simplified at that time: how to make a building warm during winter and cold during summer. In 2800 BC, the sun's direction had already become a basic design precondition for the fanes in ancient Egypt. In ancient China, people utilised light and heat from the sun in a spontaneous way, such as a north-south layout, large-size openable windows, and deep cornices (Tian and Qin, 2007). Socrates, a Greek philosopher, studied this problem almost 2500 years ago and derived a concept that is known in the early literature as the "Socratic House." According to Xenophon's Memorabilia, the essence of Socrates' studies was the influence of the sun's movement and the wind direction on the shape, form, and construction of a building. The "Socratic House's base plan was trapezoidal, with its base oriented to the south and its roof falling toward the north to avoid wind blasts. Archaeological evidence from Greece shows that the standard house plan evolved during the fifth century BC so that every house could make maximum use of the sun's warm rays during winter. By designing the streets in a checkerboard pattern running east–west and north–south, the ancient Greeks planned entire cities in Greece and Asia Minor, such as Colophon, Olynthus, and Priene, to allow every homeowner access to sunlight during winter to warm their homes (Perlin, 2004).

Existing Terminology

The most significant terms describing a ZEB are the following (Torcellini et al., 2006):

- Net zero site energy use
- Net zero source energy use
- Net zero energy emissions
- Net zero cost
- Off the grid
- Energy-plus

The main difference between a net zero site and a net zero source energy use building, is that in the latter case, the ZEB generates the energy it consumes (including the energy losses due to transport) through using on-site renewable energy sources. The off-the-grid ZEBs require distributed renewable energy generation and energy storage capability; this allows them to be autarkic on an hourly or even smaller basis. As far as the net zero cost building is concerned, the cost of purchasing energy should be balanced by the income from selling electricity generated on-site back to the grid. Outside the United States and Canada, a ZEB is usually defined as a building with zero net energy emissions, also known as a zero carbon building or zero emissions building. Under this definition, the carbon emissions generated from on-site or off-site fossil fuel use should be balanced by the amount of on-site renewable energy production. An energy positive building, also named energy plus building, is a building that on average produces more energy from renewable energy sources on-site than it consumes from the grid. This is achieved by using a combination of small-scale power generators and low-energy building techniques, such as passive solar building design, insulation and careful site selection, design, and orientation of the building.

ZEB definitions do not include emissions that are produced during the construction of the building or the embodied energy of the structure. Embodied energy is the energy that was used during the production process of a product, including raw material extraction, transport, manufacture, and deconstruction. However, it is possible that the consumed and embodied energy during the construction of a new building could dwarf the operational energy savings over its useful life (Ramesha et al., 2010).

European Union: Directive 2010/31 on the Energy Performance of Buildings

In 2009, European Union (EU) member states forged a long-awaited compromise on the recast buildings directive, agreeing that all new buildings would have to comply with high energy-performance standards after the end of 2020. The recast Energy Performance of Buildings Directive, which was finally announced in May 2010, requires the public sector to take the lead by owning buildings with "nearly zero" energy standards by the end of 2018, which is two years in advance of the private sector. This directive emphasises the necessity to implement more concrete actions to achieve the great unrealised potential for energy savings in buildings and to reduce the large differences between the Member States' results in this sector. In addition to the Energy Performance of Buildings directive, an eco-design and energy labelling directive was also revised within the framework of the EU sustainable consumption and production policies. These legal frameworks are expected to attain a large share of the energy savings in the building sector.

According to the recast of the European directive on the energy performance of buildings, a nearly ZEB means a building that has a very high energy performance, determined through a general calculation framework. Given the varying climatic and regulatory conditions across Europe, it is difficult to define exactly the concept and specific requirements for the entire EU. Therefore, the directive promotes the improvement of building energy performance within the Union, taking into account outdoor climatic and local conditions as well as indoor climate requirements and cost-effectiveness. The nearly zero amount of required energy is expected to be covered by energy from renewable sources produced on-site or nearby. Based on the recast, all member states are obliged to draw up national plans for increasing the number of nearly ZEBs. The national plans must include specific targets differentiated according to the category of building. In addition, all European member states must develop policies and take measures such as the setting of targets to stimulate the transformation of buildings that are refurbished into nearly ZEBs. The national plans should include at least the following elements:

1. Putting into practice the detailed applications of the definition of nearly ZEBs, reflecting national, regional and local conditions and including a numerical indicator of primary energy use.
2. Intermediate targets for improving the energy performance of new buildings by 2015.
3. Information on the policies and financial or other measures adopted.

Moreover, the European Commission has proposed additional EU financial instruments:

1. At least a 15% increase in the share of European Regional Development Fund money available for energy-efficiency measures.
2. Establishment of an Energy Efficiency Fund by 2014 with the goal to promote private and public investment in projects related to the improvement of buildings energy efficiency.
3. Value Added Tax (VAT) reductions for goods and services related to energy efficiency and renewable energy.

Starting December 31, 2012, and every three years thereafter, the European commission will publish a report on the progress of member states in increasing the number of nearly ZEBs. On the basis of the mentioned report, the Commission will develop an action plan and, if necessary, propose measures to increase the number of those buildings and encourage best practices related to the cost-effective transformation of existing buildings into nearly ZEBs (see Figure 1).

The definition of ZEBs across Europe is an interesting subject considering the pursuit of low energy buildings in the EU. Currently, only seven member states have defined low energy buildings: Germany, Austria, France, UK, Denmark, Finland and the Czech Republic (Thomsen and Wittchen,

Figure 1. The European roadmap to ZEB (2010-2020)

2008). Almost all of these definitions exclude existing or renovated buildings and are intended for new buildings; they also apply in almost all cases to both residential and non-residential buildings. Additionally, in some EU countries, such as France and the UK, labels have been introduced that help consumers understand the nationally standardised low energy buildings. The EU-27 residential building stock offers, however, a high potential for energy efficiency gains. Because the ZEB concept is expected to be a progressive evolution from other low-energy building designs, a more detailed review on the existing low energy building schemes in EU is presented in the following sub-sections.

EUROPEAN STANDARDS ON LOW ENERGY BUILDINGS

Germany: The Passivhaus

In Germany, the Passivhaus standard is currently applied. The Passivhaus standard was developed through a number of research projects conducted by Professor Wolfgang Feist of the Institute for Housing and the Environment of the Darmstadt University in Germany *(PassivHaus Institut, 2010)*. As of August 2010, there were approximately 25,000 certified structures of all types in Europe, and this standard is considered to be the most internationally influential. The vast majority of Passivhaus structures have been built in Germany, Austria, and Scandinavia. According to the Passivhaus standard for central Europe, the building must be designed to have an annual energy demand of not more than 15 kWh/m²/year for heating and 15 kWh/m²/year for cooling. Also, the buildings' energy consumption must not exceed a peak heat load of 10 W/m². Finally, the total primary energy consumption should not exceed 120 kWh/m²/year.

To achieve the major decrease in heating energy consumption that is required by the Passivhaus standard, a shift in the current approach to building design and construction is necessary. Therefore, the "Passivhaus Planning Package" (PHPP) simulation tool will need to be used for the design of the building. A dwelling that achieves the Passivhaus standard typically includes high levels of insulation with minimal thermal bridges and increased utilisation of solar and internal gains for heating purposes. Passivhaus dwellings have also excellent levels of air tightness and indoor air quality, which is provided by a whole building mechanical ventilation system that is equipped with highly efficient heat recovery systems. By specifying these features, the design for the heating load is limited to the load that can be transported by the minimum required ventilation air. Thus, a Passivhaus does not require any traditional heating system or active cooling to achieve comfort. The reduced heating demand can be typically met by using a compact services unit that integrates heating, hot water, and ventilation in a single unit. The realisation of passive houses and buildings places high demands on the components used (PassivHaus Institut, 2010):

1. Thermal insulation: U-values under 0.1 W/m²/K as a rule.
2. Construction without thermal bridges to a large extent.
3. Proven excellent air tightness using pressure tests.
4. Glazing with Ug values below 0.8 W/m²/K at high total solar energy transmittance.
5. Windows with Uw values below 0.8 W/m²/K.
6. Highly-efficient heat recovery from ventilation system (>75%) at low electricity consumption.
7. Lowest heat losses when producing and distributing domestic hot water.
8. Highly-efficient use of household electricity.

However, the initial applications of the standard Passivhaus concept had some major limitations for southern climates, where the problem of household energy use does not only refer to providing warm buildings in winter but also, and in some cases more importantly, to providing cool buildings in summer. The Passivhaus standard has been recently revised to account for passive design requirements such as cooling loads and other energy consumptions in the Mediterranean region.

France: BBC-EFFINERGIE

In France, five levels of regulatory requirements for the energy performance of buildings are defined; BBC in this case stands for "Low Energy Consumption Building" (Bâtiment Basse Consommation). The BBC-Effinergie label was created jointly with the French Ministry of Housing and the Effinergie association. The Effinergie association is an initiative taken by regional authorities that aims to promote low energy buildings in France. Annual requirements in primary energy for heating, cooling, ventilation, hot water, and lighting to obtain the BBC-Effinergie label for new dwellings must be lower than 50 kWh/m²/year. The calculation is made according to the French "Thermal Regulation 2005," in which the diversity of climates is taken into account by a coefficient of climate harshness. As a consequence, the values of this requirement fluctuate between 40 and 65 kWh/m²/year according to the region. By 2020, all new buildings in France are expected to be energy positive, thus they should produce more energy than they consume.

UK: Code for Sustainable Homes

The British Code for Sustainable Homes is an environmental impact rating system for housing in the UK, setting new standards for energy efficiency and sustainability that remain, however, not mandatory under current building regulations. It comprises an assessment system aimed at improving the environmental impact of new homes by introducing minimum standards in nine areas, including the use of energy, carbon dioxide emissions, and water. This Code uses a sustainability rating system that is indicated by stars to communicate the overall sustainability performance of a building. A building can achieve a sustainability rating from one to six stars depending on the extent to which it has achieved the Code's standards. One star is the entry level and above the level of the Building Regulations; six stars is the highest level. Based on this code, a building will have to have zero carbon emissions to achieve the highest sustainable level (Level 6).

In December 2006, UK government made the first proposals that all new homes would be "zero carbon" after 2016. At that time, zero-carbon status will require that, over the course of a year, a building does not emit any CO_2 resulting from energy needed for heating, hot water, lighting, or appliances. The minimum requirements for new buildings are expected to increase from Level 3 in 2010 to Level 6 in 2016. All of the carbon reductions are expected to be achieved using local "off-site" technologies connected to the development by a direct physical connection.

Denmark: Low Energy Class 1 and 2

Denmark's current building regulations define two low energy classes:: Low Energy Class 2 and Low Energy Class 1. These two classes are defined as having a calculated energy performance of 75% and 50%, respectively, better than the minimum energy performance of a new building. According to this scheme, the minimum requirement for residential buildings is 70 + 2200/A kWh/m²/year, and for other buildings, 95 + 2200/A kWh/m²/year (A is the conventional building area). The calculated energy performance of a building includes energy for heating, ventilation, cooling, and domestic hot water. Auxiliary energy consumption (e.g., pumps, fans, etc.) is included by a factor of 2.5. In the case where a building's indoor temperatures are too high (>26°C), a penalty related to the energy performance is also included. This fictive amount of energy is calculated as the energy needed to decrease the indoor temperature to 26 °C using a mechanical cooling system with a low coefficient of performance (COP=2). In other non-residential buildings, electricity for artificial lighting is included in the energy performance as well. By 2020, the energy requirements in Denmark are expected to be reduced by 75% compared with the minimum level from 2006 (Dyrbøl et al., 2009).

Other European Countries

According to the Austrian standard klima:aktiv, low energy buildings (such as the case of a single family house) should not exceed 60 kWh/m²a. The rationale used for determining a building's energy performance is very similar to the one used in Germany (Passivhaus). However, in Austria, the indicator of 15 kWh/m²/year in certain cases refers to the useful area and, in others, to the conditioned space, whereas according to the German standard, this space is the total area of the building.

In the Czech Republic, recommendations based on the German Passivhaus are also valid. In the CSN 730540-2, Czech standard low energy buildings and passive buildings are defined according to Professor Feist's limits, 15 kWh/m²/year. In the case of Low energy or Passive buildings, U-values should be lower than 66% of the currently demanded U-values. Also, according to this standard, in the case of a passive building, thermal losses should not exceed 0.3 W/m²/K on average, and the U-values of the roof and windows should be 0.12 and 0.8 W/m²/K, respectively.

Finally, in Finland, a description of low energy buildings is included in the Finnish building code. When designing a Low Energy Building, the calculated heat loss should not exceed 60% of the heat loss calculated according to reference values stated in the building regulations.

ZEBs AND ENVIRONMENTAL TECHNOLOGIES

The pyramid of energy sustainability and the role it plays in buildings consists of three levels: energy conservation, energy efficiency and, renewable energy sources (Figure 2). The promotion of sustainable environmental technologies is directly related to all these aspects and is expected to play a crucial role in the realisation of ZEBs. Based on the existing knowledge and European low energy buildings standards, the technological applications, which will certainly participate in the composition of a ZEB, are as follows:

- Thermal insulating materials and buildings techniques – Energy conservation
- Building appliances and lighting – Energy efficiency
- On-site installations of alternative technologies for the production of electricity, heating, and cooling – Renewable energy sources

Figure 2. The pyramid of sustainable energy development

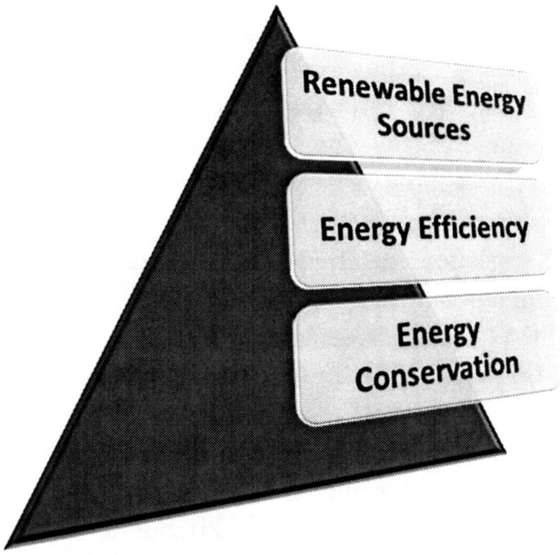

There is a general consensus that building energy use and environmental impact must be significantly reduced in order to reach the ZEB goals; however, there have been many discussions in the recent years as to the proper technologies and applications to achieve this. In the following sub-sections, the state-of-the-art status regarding the above mentioned building parameters are presented in detail.

Thermal Insulating Materials and Buildings Techniques

The most cost-effective way to optimise the energy performance of a building is to ensure that the energy demand of its conditioned space is minimised. The building envelope plays a crucial role in determining the building's energy demand because it is the primary enclosure for the conditioned space used to fulfil the function of the building. High performing and renewable heating systems in a poorly insulated building cannot be sufficiently defended on environmental or economic grounds. Therefore, thermal insulation of buildings is the simplest way to prevent heat loss and achieve economy in energy usage.

Selecting appropriate insulation materials is the key to design and construct a ZEB. Conventional insulation materials are often opaque and porous and can be classified into fibrous, cellular, granular, and reflecting types (Kaushika and Sumathy, 2003). The thermal characteristic of these insulation materials is specified in terms of thermal conductivity. The European market of insulation materials is characterised by the domination of two groups of products: inorganic fibrous materials and organic foamy materials (EURIMA 2010). In terms of insulating capabilities, almost all thermal insulating materials feature similar performance. Some commonly used thermal insulation materials, such as glass fibre (thermal conductivity: 0.03), alumina silicate (0.035), and mineral wool (0.04) have low thermal conductivity, which depends on the number of air cells packed at the cores of solid media. However, as far as other parameters are concerned, such as the reaction to fire and moisture or mechanical properties, insulation materials present some important differences. Furthermore, environmental and public health aspects play an increasing role, both in the search for 'optimum' materials for given applications and in the aims set by the industry for future developments (Papadopoulos). According to Humm (Humm, 2005), the thermal insulation of a low-energy house should be at least 20 cm thick. Phase changing material in the walls and vacuum insulation are considered to be the cutting edge building technology in this area. Depending on the type of construction, the complete external component can be between 25 and 60 cm thick in total and avoid thermal bridges. The problem of thermal bridges occurs wherever the insulated building envelope is penetrated by components that allow the passage of heat from inside the building to the outside environment. Many buildings lose more heat via avoidable thermal bridges than over the entire uninterrupted wall. Transitions and junctions require special care.

In concert with the opaque construction sections, windows are considered to be the most important component in mitigating the exterior environment because they are the weakest link in the control of thermal and mass transport through the exterior envelope. The thermal characteristics of the window frames, the seals between glazing and frame, and the durability of the gasket seals on operable units are all absolutely critical to the proper performance of the unit. ZEBs have a minimum requirement of high-quality sealants between the frame and the exterior wall in addition to the use of double- and triple-glazed insulated window units (Fernandez 2010). Insulated double-glazing significantly improves the performance of a window unit. Compared to a single-glazed unit, a double-glazed window can save up to three times as much heating energy; low-emission (low-e) double-glazed windows can save up to four times as much heating energy. Modern triple-glazed windows achieve U values of 0.6–0.8 W/m²/K. The use of gas filler between panes also improves performance; these windows have two low-e layers and are filled with argon or krypton (Wong et al., 2007). Aluminium and PVC window frames can certainly increase performance over their timber and steel frame counterparts.

Synthetic building materials with insulating properties enable the desire for simple buildability. Over the past few years, the industry has been offering a wide selection of building materials that provide both load-bearing and insulating functions. The load-bearing material is so permeated with air-filled pores that it exhibits barely-sufficient load-carrying capacity, while the air captured in the pores, with its poor conduction, provides an insulating effect. The insulating function weakens the load-bearing material, thus causing the ratio of strength to insulation to be determined for each case. Therefore, there exists a blurred dividing line between a load-bearing material with insulating properties and a load-bearing insulation material (Humm, 2007).

Another important area of development in the field of thermal insulating materials is related to nanotechnology. Bulk insulating materials, such as bulk nanoporous silica compounds and aerogel, are considered to be the next generation in the area of buildings thermal insulation. NanoPore Insulation LLC has developed some novel bulk nanoporous silica compounds with embedded organic molecules that were found to perform up to 10 times better than conventional insulating materials (Zhu et al., 2004). The superior insulation characteristics of these low-density, highly porous solids are a result of their unique shape and their small size, which is between 10 and 100 nm depending on the number of pores. These new insulating compounds have been tested in a wide variety of insulation applications that require excellent thermal performance, optimum energy efficiency, or minimum insulation thickness with excellent results.

The environmental impact of the thermal insulating materials building techniques may be performed by means of a life cycle assessment. Life cycle assessment analysis is an approach that accounts for all energy inputs to a material (Ramesh et al., 2010). The system boundaries of this analysis include the energy use of the following phases: manufacture, use, and demolition. In Sartori and Hestnes (2007), the life cycle energy consumption of conventional, low energy and zero energy buildings was reviewed. For this purpose, six versions of a building, one conventional, one ZEB, and four low energy alternatives, were analysed in a German context. This study surprisingly revealed that the life cycle energy of the ZEB was increased compared to some of its low energy versions. In the case of low energy buildings, the increase in embodied energy due to energy saving measures was minor compared to the relevant decrease in the operating energy; hence, their life cycle energy was significantly reduced. In contrast, in the case of a ZEB, although the operating energy was considered to be zero, the building's embodied energy was so high that the life cycle energy of some of the low energy cases was exceeded. The findings of this study strongly indicate that it is possible that a ZEB may not be, under certain conditions, the optimum solution in terms of net energy consumption. Hence, the performance of a life cycle analysis, when designing a ZEB, is a necessity and should therefore be carried out in all cases. Similar conclusions were also drawn in Winther and Hestens (1999).

Onsite Energy Production by Renewable Energy Sources

Interest in the use of renewable energy has increased during the past few decades beginning with the first energy crisis of the early 70s. Renewable energy is used to produce both electricity at central grid-connected generators and dispersed locations and heat in the residential, commercial, and industrial sectors. Producing energy from renewable energy sources is essential for a sustainable energy future. Renewable energy is a promising alternative solution because it is clean and environmentally safe. Compared to the fossil fuel energy sources they replace, the conversion of renewable energy sources also does not produce any greenhouse gases or other pollutants. Renewable energy technologies are maturing at different rates. Some have already penetrated the market, while others are just demonstrating their potential. A number of renewable energy technologies and related approaches and systems are available to supplement and even supplant conventional fuel sources in cities.

To evaluate the options for significant commercial renewable energy source utilisation, the costs of energy production, compared to conventional energy supply, are of outstanding importance. The outcome defines which renewable energy source technologies are feasible under given circumstances and what measures should be taken to achieve cost competitiveness in a specific time frame. It is certainly expected that the cost differences between renewable energy sources and conventional energies will constantly decrease, mainly due to further technical development and growing production rates of renewable energy technologies. Thus, a growing share of the technical renewable energy sources' potentials can be exploited economically in the long term.

Reductions in overall energy consumption levels by employment of advanced renewable energy technologies strategies would without any doubt bear considerable benefits within the buildings sector. However, rigorous renewable energy applications are perceived to impose restricted aesthetics on building design. Studies related to public attitudes towards novel energy technologies have shown that public acceptability is the most important issue in shaping the widespread implementation of renewable energy technologies and the achievement of energy policy targets. Furthermore, it is commonly assumed that public attitudes need to change to make more radical scenarios regarding the implementation of renewable energy technologies feasible (Devine-Wright, 2007).

In technologically advanced renewable economies, energy supplies will no longer exclusively depend on large, centralised supply models but will unfold in a more diverse and differentiated manner. The renewable energy technologies that are currently considered as applicable for onsite energy production (and thus may in the short term contribute in achieving the ZEB goal) are solar energy and biomass. Although technical applications related to wind, geothermal, and hydro energy are considered reliable, specific limitations with regard to the energy potential, especially in dense urban environments, significantly reduce their perspectives in small-scale applications and therefore their integration in the building. In a broader sense, large-scale installations could potentially participate in the energy balance of a ZEB in terms of a trading mechanism between large producers and individuals. However, a necessary precondition to realise such a mechanism would be the existence of a scheme that could guarantee the origin of electricity, heating, and cooling produced from renewable energy sources (see Figure 3). The implementation of the directive on the promotion of the use of energy from renewable sources is an issue that is currently being discussed in the EU.

Solar Energy

Aside from other renewable energy sources systems, which are based on solar energy in a broader sense (wind energy, biomass, and hydroelectric energy), two basic categories of technologies are usually applied to convert solar energy into other useful forms of end-energy. First, solar-photovoltaic (PV) modules convert sunlight directly into electricity. Second, solar-thermal power systems use solar radiation to convert solar energy to thermal energy, which is then used for space heating and cooling purposes as well as for domestic hot water.

Figure 3. Suitable renewable energy sources technologies for ZEBs

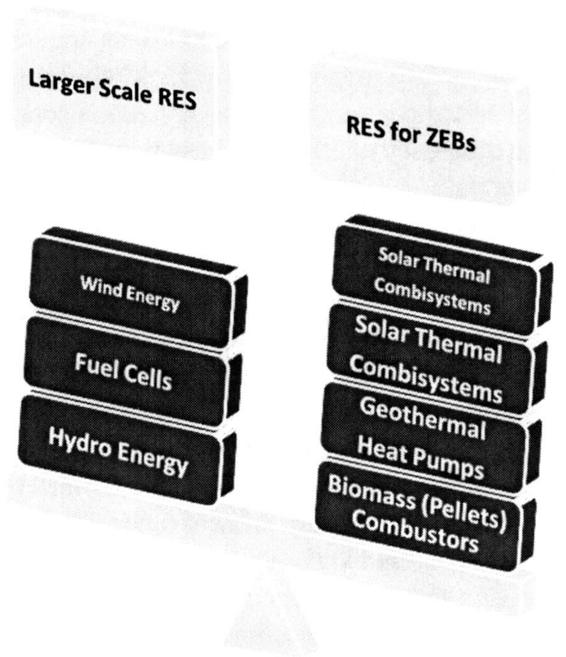

Photovoltaics

There are a vast number of applications for photovoltaic (PV) technology, covering all types of autonomous systems and grid connections. The power range is extremely wide due to the modularity of PV power, varying from milliwatts to megawatts. PVs are attractive because of their high reliability, long lifetime, modularity, and low maintenance costs. Solar cells presently have practical efficiencies of 15–20%. PV electricity generation has a low energy density per surface, and its efficiency does not depend on the size of the plant. PV modules in buildings are usually placed on the roof and sometimes also on south-facing building walls. They provide enough electric energy to cover a part of the building's needs. The roof-top modules are placed at a distance from the roof in order to allow rejection of the absorbed solar radiation that is not transformed to electricity. The incorporation of photovoltaic modules in the structure of a building is expected to be one of the best possible solutions in terms of achieving the ZEB goals. PV Systems are gradually becoming, on all levels of construction and design, an integral part of the overall building concept.

A diverse array of technical solutions needs to be applied and developed so that PV electricity will become cheap enough to compete with grid electricity. Advances in PV technology enable the use of PV-modules, in addition to electricity production, for several other purposes on a building, such as a roof-light, wall-cladding, or a window-pane. Integrating photovoltaic panels into building facades also represents a significant step forward in the application PVs in buildings (BI-PV). Such facades may serve not only as a renewable source of electricity but also as a source of heat for building heating and cooling. One technique for reducing the cost of solar electricity is to use a reflector for focusing radiation onto the PV cells, thus allowing expensive PV cells to be replaced by considerably cheaper reflector material (Davidsson et al., 2010). These collectors, also known as photovoltaic/thermal (PV/T) hybrid collectors, can produce electricity and thermal energy simultaneously and are expected to significantly change the cost effectiveness of the collectors.

Solar Thermal Power Systems

Solar home heating is usually achieved with passive solar systems that are less expensive. However, it is instructive to consider the more elegant active solar systems that pump solar heated water to store in hot water tanks for distribution to radiators. In Europe and worldwide, the largest sector of the solar thermal market is domestic water heating. A natural extension, however, of a solar domestic water heating system is adding capability to offset a portion of the space heating load in the same building, hence the name "combisystem." Most combisystems intended for residential applications treat domestic water heating as the primary load and thus take advantage of high solar energy availability in the summer. Beyond their domestic hot water "base load," combisystems typically use greater collector areas and larger storage tanks to capture and contain additional energy that can offset a portion of the building's space heating load (Caleffi, 2010).

Solar-assisted cooling also appears to be a promising alternative to conventional electrically driven air-conditioning units. Heat-driven solar-assisted chillers, utilising the sorption phenomenon, seem in this sense likely to have an excellent potential in the space for realisation of ZEB targets, especially in summer dominant weather conditions. When using the sorption phenomena (heating, cooling, and heat storage), a whole range of refrigerant fluids is available as working pairs (sorbate/sorbent) that can be useful for a wide range of applications. The pairs of interest in the field of sorption for building applications are adsorption (e.g., water/silica gel), absorption (e.g., water/lithium bromide), and chemical reaction (e.g., ammonia/inorganic salts). Nevertheless, solar-assisted cooling technologies have not yet reached an acceptable level of technological sophistication and are still considered to be under development. There is a need, for example, for cost-efficient and operationally friendly heat storage as well as a backup system (Oxizidis and Papadopoulos, 2008).

The building integration of present solar thermal systems is generally characterised by a low level of architectural quality. Most often solar collectors are considered only as technical elements and are confined to the roof top, be it flat or pitched. The architectural quality of most existing building integrated solar thermal systems (BIST) is quite poor, which often discourages potential new users (Probst and Roecker, 2007). Different BIST roofing systems can deliver different levels of thermal energy to match the varying needs of building occupants. The roof focusing system can deliver heat to the building at up to 400 °C. The transpired collector can deliver preheated air that is 15-30 °C warmer than ambient air. At optimum air flows, the system can operate at a thermal efficiency on the order of 80%. The solar roofing tile can deliver air heated to medium temperatures of 40-95 °C, or approximately 55 °C above the ambient temperature, and is projected to operate at an efficiency of 30-40% (Kalogirou, 2009).

Biomass

Biomass is currently the most used renewable energy source and will continue to be so in the foreseeable future. For biomass resources, several potential sources may be used. Biomass resources are agricultural and forest residues, algae and grasses, animal manure, organic wastes, biomaterials, and energy crops. In the near future, biomass and related technologies will have the potential to provide a cost-effective and sustainable supply of energy while at the same time aiding countries in meeting their GHG reduction targets.

Biomass combustion is a mature technology that is, in many cases, competitive with fossil fuels. Biomass heat is most habitually used in countries with good resource availability and particularly where district heating systems are utilised (Seyboth et al., 2009). One biomass fuel that has gained national attention as a result of the rising fossil fuel prices is wood pellets. Wood pellets are compressed by-products from the forest products industry, such as woodchips and sawdust. Wood pellets are a locally available and are a cost-effective heating fuel with several advantages over other types of biomass. Nevertheless, emissions of harmful gases are important parameters in addition to the efficiency and thermal performance of pellet heating units. The national building codes and emission regulations include limits of allowable emissions of noxious gases for wood pellet boilers.

Energy from biomass needs to achieve a significant cost reduction by a factor of at least 2 to become competitive with other conventional fuels. Research needs to further focus on technical and non-technical barriers of the most promising biomass supply-use chains. The issues that still need to be addressed are combustion technologies in terms of necessary biomass pre-treatment for minimisation of pollutant emissions, gasification technologies such as biogas and syngas production, and biofuel production techniques.

Other Renewable Energy Source Applications for Buildings

Recurring energy sources also include wind, geothermal, and hydroelectric energy. Although the applicability of these renewable energy sources in small-scale applications is limited, geothermal energy can be directly used for space heating and cooling. Currently in Europe, about 95,000 dwellings are heated by geothermal energy (Mega, 05) by employing ground-source heat pumps, also known as geothermal heat pumps. Ground source heat pumps use electricity to move rather than generate heat and are recognised to be outstanding heating, cooling, and water-heating systems. The primary benefit of ground source heat pumps compared to other systems is the increase in operating efficiency, which translates to reduced heating and cooling costs.

Appliances and Lighting

Domestic appliances, such as refrigerators, freezers, ovens, stoves, clothes washers and dryers, and dishwashers consume together with lighting as much as 20% of a building's total energy consumption. Therefore, the ZEB target will be hard to achieve if specific measures are not taken to reduce these consumptions. Enormous energy savings are also possible by using energy efficient equipment, effective controls, and careful design.

Household energy-consumption is transient, and rates of energy use vary dramatically with time of day and time of year. Although one domestic appliance may use less than 1 kWh per day, appliance usage in general results in large demands for electricity at peak times (Wood and Newborough, 2003). Promoting energy-conscious behaviour effectively throughout the population requires a better understanding of the interface between people and the equipment they use. Energy labelling of domestic appliances has already been hugely successful since its EU introduction in 1992. In the same year, the US Environmental Protection Agency introduced the voluntary ENERGY STAR labelling program designed to identify and promote energy-efficient products. Today, 90% of appliances sold in Europe are in class "A" – the best existing class. Current energy efficient domestic appliances in Europe use 40% less energy than conventional models did in 2001. Following this, if all households changed their ten-year-old (or older) appliances for new ones, 20 billion kWh of electricity would be saved annually in Europe (Ecosavings – Electrolux). However, new technologies are further reducing the energy consumption of an A-class product by as much as 50%. The EU has therefore decided to extend the scale upwards with three new classes: "A+," "A++," and "A+++" for products with a better than A performance.

Using less electric lighting reduces heat gain, thus saving air-conditioning energy and improving thermal comfort. Electric lighting design also strongly affects visual performance and visual comfort by aiming to maintain adequate and appropriate illumination while controlling reflection and glare. There are several types of energy efficient lighting and affordable lighting technologies. The electricity used over the lifetime of a single incandescent bulb costs 5 to 10 times the original purchase price of the bulb itself. Compact Fluorescent Lights (CFL) and Light Emitting Diode (LED) bulbs have revolutionised energy-efficient lighting. CFLs are simply miniature versions of full-sized fluorescents. They screw into standard lamp sockets and produce light that looks just like the common incandescent bulbs - not like the fluorescent lighting we associate with factories and schools. LEDs are small, solid light bulbs that are extremely energy-efficient. New LED bulbs are grouped in clusters with diffuser lenses that have broadened the applications for LED use in the home (Eartheasy).

Building energy management systems are also expected to play a significant role towards a ZEB because these systems can contribute to the continuous energy management and therefore to the achievement of the possible energy and cost savings. A building energy management system is a computer-based control system installed in buildings, most commonly in a large building where it controls and monitors the building's mechanical and electrical equipment, such as ventilation, lighting, power systems, fire systems, and security systems. Building energy management systems consist of software and hardware and are generally applied to the control of active systems. The performance of the building energy management systems is directly related to the amount of energy consumed in buildings in addition to the thermal comfort of the buildings' occupants. The majority of recent developments in building energy management systems have followed the advances made in computer technology, telecommunications, and information technology. In this context, a number of modern techniques and methods have been proposed in the international literature for improving specific systems' controls (Doukas et al., 2007).

Energy consumption in domestic buildings is not only determined by the type and number of electrical appliances but also by the use of these appliances by the occupants of the building. Therefore, the ZEB is not completed with the design and construction of the building; a successful ZEB must also take into account how the occupants' behaviours have an important impact on the energy performance of the building. Therefore, the ZEB occupant must be conscious of daily habits and patterns that affect the proper maintenance of the appliances. Understanding the way certain energy efficiency features of the building work is therefore highly essential. Changing furnace filters, having heating and cooling systems cleaned regularly, periodically checking the operation of solar systems, and maintaining exterior caulking and painting are only a few examples of ways in which a homeowner can assure a long-lived, high-performance building (NAHB Research Centre).

CONCLUSION

Sustainable building is a rapidly growing practice in new construction development and is constantly gaining importance and popularity. Following the European Commission's proposal in 2008 for an update of the 2002 Energy Performance of Buildings Directive, the recast was adopted by the European Parliament and the Council of the European Union on 19 May 2010. For new buildings, the recast fixes 2020 as the deadline for all new buildings to be "nearly zero energy." The promotion of sustainable environmental technologies will therefore be directly related to all the aspects of sustainable building, and is expected to play a crucial role in the realisation of ZEBs. Based on the existing knowledge and European low energy buildings standards, the technological applications, which will certainly participate in the composition of a ZEB, will be the energy conservation, the efficient use of energy and the promotion of renewable energy technologies. This paper considers the aspects of various low energy standards and illustrates the most possible scenarios regarding the promotion of eco technologies for the achievement of the zero energy goals.

Once the target of ZEB in Europe by 2020 is achieved, the next step will need to be identified. It is more than obvious that by the time the building sector will be capable of covering its own energy demands, discussions will be held on possible ways to satisfy parts of other energy sectors needs. The transportation sector for example, which is expected to grow more rapidly than any other up to 2020 and beyond, faces currently more hurdles than any other energy consuming sector in promoting renewable energy sources technologies. The European target of 20% by 2020 for the penetration of renewable energy sources in all other energy sectors is limited to the modest 10% for transport biofuels. Therefore, the excess of renewable energy in buildings could be transferred to transportation through electricity storage for electric or hybrid vehicles. A necessary precondition for this will be the establishment of the energy plus buildings. The Heliotrope in Freiburg, designed by Rolf Disch, a German architect, is the first building in the world to create more energy than it uses. Since 2006, Disch's "Solar Settlement" in Freiburg, Germany, has been an on-going project with 59 energy plus buildings (see Figure 4). France has also announced that by 2020 all new buildings will be energy plus.

It should be expected that the European ZEB target for 2020 will soon be extended to other fields of sustainable development and that the environmental technologies will soon be employed in a broader area of applications. The goal of green building and sustainable architecture is to use resources more efficiently and reduce a building's negative impact on the environment. ZEBs, at their current status, achieve only one key green-building goal, the minimisation of energy use. Therefore, ZEBs may not be considered sustainable in all fields of interest. By the time existing green building certification schemes such as BREEAM[1] and LEED[2] upgrade their requirements and endorse the zero energy consumption, a approach of a green, sustainable building will be achieved (see Figure 5). Under these schemes, not only energy but also other sustainability fields, such as water efficiency, atmospheric environment, materials, and resources, will be taken into consideration. Under these conditions, the lower ecological impact over the life of the building will be achieved and the vision for sustainable, ecological friendly buildings will be fulfilled.

Figure 4. Freiburg's "Solar Settlement" of Rolf Disch

Figure 5. Green sustainable buildings' holistic approach

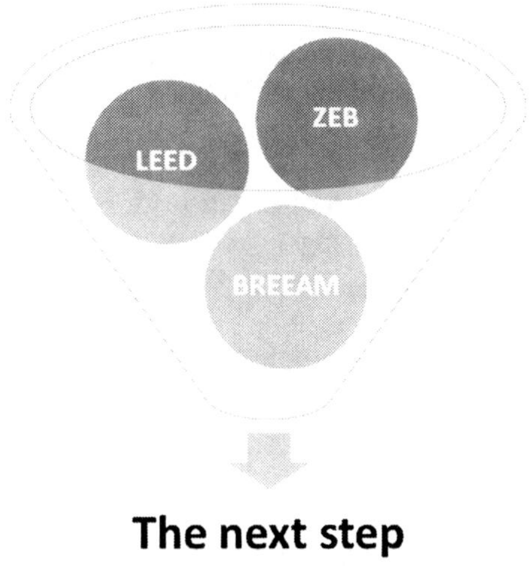

REFERENCES

Caleffi. (2010). *Solar thermal combisystems*. Retrieved October 3, 2010, from http://www.caleffi.us

Concerted Action for the implementation of EPBD. (2008). *Implementation of the Energy Performance of Buildings Directive Country reports 2008*. Retrieved October 8, 2010, from http://www.buildup.eu/publications/1916

Concerted action for the implementation of EPBD. (2010). *Executive summary report on the interim conclusions of the concerted action.* Retrieved April 1, 2010, from http://www.epbd-ca.org/Medias/Pdf/CA_Summary%20report_Feb2010.pdf

Davidsson, H., Perers, B., & Karlsson, B. (2010). Performance of a multifunctional PV/T hybrid solar window. *Solar Energy*, *84*, 365–372. doi:10.1016/j.solener.2009.11.006

Department for Communities and Local Government. UK. (2006). Building a greener future: Towards zero carbon development.

Devine-Wright, P. (2007). *Reconsidering public attitudes and public acceptance of renewable energy technologies: A critical review.* Working paper of the research project "Beyond Nimbyism: A multidisciplinary investigation of public engagement with renewable energy technologies." Retrieved August 31, 2010, from http://www.sed.manchester.ac.uk/research/beyond_nimbyism/

Doukas, H., Patlitzianas, K. D., Iatropoulos, K., & Psarras, J. (2007). Intelligent building energy management system using rule sets. *Building and Environment*, *42*, 3562–3569. doi:10.1016/j.buildenv.2006.10.024

Dyrbøl, S., Eriksen, K. E., Thomsen, K. E., Wittchen, K. M. (2009). *Energy and carbon impact of very low energy building. ECEEE summer study: Act! Innovate! Deliver! Reducing energy demand sustainably.*

Eartheasy. (2010). *Solutions for sustainable leaving: Lighting*. Retrieved September 25, 2010, from http://eartheasy.com/live_energyeff_lighting.htm

Ecosavings. (2010). *Electrolux*. Retrieved July 16, 2010 from http://www.electrolux.com/ecosaving

Effinergie. (2010). *Home*. Retrieved September 17, 2010, from http://www.effinergie.org/site/Effinergie/

EN 15217. (2007). Energy performance of buildings - Methods for expressing energy performance and for energy certification of buildings.

EURIMA. (2010). *Home*. Retrieved September 26, 2010, from http://www.eurima.org/

European Parliament. (2002). *Directive 2002/91/EC on the energy performance of buildings.*

European Parliament. (2010). *Directive 2010/31/EC on the energy performance of buildings* (recast).

Fernandez, J. (2010). Materials and construction for low-energy buildings in China. In L. A. Glicksman, & J. Lin (Eds.), *Sustainable urban housing in China* (pp. 44–73). Dordrecht, The Netherlands: Springer.

Humm, O. (2005). Sustainability fundamentals of architecture. In A. Deplazes (Ed.), Constructing architecture. Materials, processes, structure: A handbook (pp. 278-301). Staufen bei Freiburg, Germany: Birkhauser.

Kalogirou, S. (2009). *Solar energy engineering, processes and systems*. Oxford, UK: Academic Press.

Kaushika, N., & Sumathy, K. (2003). Solar transparent insulation materials: A review. *Renewable & Sustainable Energy Reviews*, *7*(4), 317–351. doi:10.1016/S1364-0321(03)00067-4

Klima:aktiv. (2010). *Home*. Retrieved September 19, 2010, from http://www.klimaaktiv.at/

Mega, V. (2005). *Sustainable development, energy, and the city: A civilisation of visions and actions*. New York, NY: Springer-Verlag.

NAHB Research Centre. (2010). *Seven steps to a ZEH*. Retrieved September 15, 2010, from http://www.toolbase.org/Home-Building-Topics/zero-energy-homes/seven-steps-zeh

Oxizidis, S., & Papadopoulos, A. M. (2008). Solar air conditioning: A review of technological and market perspectives. *Advances in Building Energy Research*, *2*, 123–157.

Papadopoulos, M. A. (2005). State of the art in thermal insulation materials and aims for future developments. *Energy and Building*, *37*, 77–86. doi:10.1016/j.enbuild.2004.05.006

Parker, D. (2009). Very low energy homes in the United States: Perspectives on performance from measured data. *Energy and Building*, *41*, 512–520. doi:10.1016/j.enbuild.2008.11.017

PassivHaus Institut. (2010). *Home*. Retrieved September 17, 2010, from http://www.passiv.de/index_phi.html

Perlin, J. (2004). Solar energy. In *History of encyclopedia of energy, 5*. Elsevier. doi:10.1016/B0-12-176480-X/00035-8

Probst, M. C. M., & Roecker, C. (2007). Towards an improved architectural quality of building integrated solar thermal systems (BIST). *Solar Energy*, *81*, 1104–1116. doi:10.1016/j.solener.2007.02.009

Ramesh, T., Prakash, R., & Shukla, K. K. (2010). Life cycle energy analysis of buildings: An overview. *Energy and Building*, *42*, 1592–1600. doi:10.1016/j.enbuild.2010.05.007

Sartori, I., & Hestnes, A. G. (2007). Energy use in the life cycle of conventional and low energy buildings: A review article. *Energy and Building*, *39*, 249–257. doi:10.1016/j.enbuild.2006.07.001

Seyboth, K., Beurskens, L., & Frasch, F. (2009). Renewable heating and cooling: The key to a large contribution of renewable energy. *Proceedings of ISES World Congress 2007* (vol. I – vol. V), *11*, 2854-2858.

Thomsen, K. E., & Wittchen, K. M. (2008). *European national strategies to move towards very low energy buildings*. Danish Building Research Institute, Aalborg University.

Tian, L., & Qin, Y. (2007). Utilization of renewable energy in architectural design. *Frontiers of Architecture and Civil Engineering in China*, *1*(1), 114–122. doi:10.1007/s11709-007-0012-4

Torcellini, P., Pless, S., Deru, M., & Crawley, D. (2006). *Zero energy buildings: A critical look at the definition*. Conference Paper, NREL/CP-550-39833.

Winther, B. N., & Hestnes, A. G. (1999). Solar versus green: The analysis of a Norwegian row house. *Solar Energy*, *66*(6), 387–393. doi:10.1016/S0038-092X(99)00037-7

Wong, I. L., Eames, P. C., & Perera, R. S. (2007). A review of transparent insulation systems and the evaluation of payback period for building applications. *Solar Energy*, *81*, 1058–1071. doi:10.1016/j.solener.2007.04.004

Wood, G., & Newborough, N. (2003). Dynamic energy-consumption indicators for domestic appliances: Environment, behaviour and design. *Energy and Building*, *35*(8), 821–841. doi:10.1016/S0378-7788(02)00241-4

Zhai, Z., & Previtali, J. M. (2010). Ancient vernacular architecture: Characteristics categorization and energy performance evaluation. *Energy and Building*, *42*, 357–365. doi:10.1016/j.enbuild.2009.10.002

Zhu, W., Bartos, P. J. M., & Porro, A. (2004). Application of nanotechnology in construction: Summary of a state-of-the-art report. *Materials and Structures*, *37*, 649–658. doi:10.1007/BF02483294

ENDNOTES

1 BRE Environmental Assessment Method (BREEAM) is a voluntary measurement rating for green buildings that was established in the UK by the Building Research Establishment (BRE).

2 Leadership in Energy & Environmental Design (LEED) is an internationally recognised green building certification system, developed by the U.S. Green Building Council, providing third-party verification that a building or community was designed and built using sustainable strategies.

This chapter was previously published in Green and Ecological Technologies for Urban Planning edited by Ozge Yalciner Ercoskun, pages 93-111, copyright 2012 by Information Science Reference (an imprint of IGI Global).

Index

A

abatement costs 77, 736, 743
Academic Social Responsibility (ASR) 1173
adult education 1150-1151, 1158, 1160-1161, 1481-1483, 1485-1491
Advanced Configuration and Power Interface (ACPI) 512
Advisory Group on Energy and Climate Change (AGECC) 305
aeronautics innovation 885
African Renewable Energy Alliance (AREA) 311
agricultural census
general 1712-1713, 1715-1718, 1720
alternatively-fuelled goods 1297, 1310
American Rainwater Catchment Systems Association (ARCS) 1671
analog operational-amplifier 348, 364
Analytical Hierarchy Process (AHP) 242, 953
analytical network protocol 581
Asian Institute of Technology (AIT) 1437
axiomatic design 242-244, 247, 251, 254-259

B

barriers to development 1546, 1549, 1552, 1557
Baseboard Management Controller (BMC) 663
Best Management Practice (BMP) 799
biogas technology 229, 1519, 1527
biomass combustion 1529, 1538, 1754
blended shore education 1478-1484, 1488-1490
boost converter 319-322, 324-326, 330, 332-333
Bristol Urban Consolidation Centre 1306
Brundtland Commission 170, 189, 847, 944, 992, 1314, 1455-1456, 1471, 1686
buck converter 319-322, 324, 330, 333
built environment 61, 186, 189-190, 197-198, 200, 517, 547, 619, 637, 1184-1186, 1200, 1463, 1687, 1723, 1726-1728, 1733, 1739-1740
bulk heterojunction 368-370, 373, 379-382
bulk nanoporous 1751
business model 88, 217, 762, 904, 906-907, 910, 912, 1120, 1172, 1546, 1548, 1552, 1554, 1560, 1574

C

Capability Maturity Model Integrated (CMMI) 128
capacity building 42, 50, 76, 205, 775, 817, 851, 1057, 1059, 1063, 1440, 1603, 1607, 1676
cap and trade 859
carbon accounting 734-735
Carbon Dioxide (CO2) 23, 65, 85, 249, 1282
carbon footprint 88, 97, 165, 283, 495-496, 498-499, 676, 678, 683, 723, 732, 735-736, 739-740, 742-743, 745-747, 792, 931, 1291, 1708
product-level 732, 735
carbon neutral growth 889, 897, 901
Caribbean Environmental Health Institute (CEHI) 1671, 1679
cellular communications 515
Center for Training and Demonstration of Appropriate Technologies (CEDECAP) 1054
certificate of entitlement 682, 692
change detection 152-153, 608, 612-613, 619-620
circuit switching 468-470, 478
circular economy 1218, 1228-1230
civic engagement 155, 547, 1124, 1129-1130, 1176, 1444-1445, 1447, 1451-1452
clean technologies 227, 384-386, 388, 1109
Climate Prediction and Application Center (ICPAC) 807
climate-smart agriculture 20-21, 44-45, 48
Cloud computing 279-283, 298-301, 390, 463-465, 476, 478, 538, 558-560, 564, 943
Cloud federation 279-282, 284-285, 287, 297-298, 301
cluster computing 661-662

coastal ecotones 1696
cob block 536
collaborative learning 1003, 1006, 1463, 1490, 1492-1495, 1499, 1503-1504
common agricultural policies 157
Communities of Interests (CoI) 852
Communities of Practice (CoP) 988
Community Risk Assessment (CRA) 1420
Compact Fluorescent Lights (CFL) 1755
conjugated molecules 365
conjugated polymers 365-366, 368, 379-380, 382
conservation management 849, 854-855, 865-866, 1318
Consumer Generated Energy (CoGE) 304-305, 312, 316
contaminated stormwater 1688, 1690, 1692-1693, 1695-1696
contemporary waterfront 1684, 1688, 1692, 1698
continuous monitoring 733
Convective Heat Transfer Coefficient (CHTC) 1031
Corporate Social Responsibility (CSR) 105, 111-112, 117, 154, 160, 205-206, 217-218, 393, 749, 757-758, 762, 765, 795-796, 849, 865, 896, 957, 1118-1119, 1173, 1175, 1180, 1307, 1466, 1602
corporate sustainability 183, 197, 201-206, 212-214, 217-219, 765, 901, 1106-1107, 1111, 1117, 1119-1120, 1169, 1314, 1316-1317, 1324, 1655
corporate sustainable development 219
county regional economy 1218, 1230
creative economy 1218-1219, 1221, 1225, 1228-1229
cropping calendar 1426
cultural codes 1730, 1732-1733
cultural diversity 1, 3, 13-14, 182
cultural norms 1232-1233, 1243

D

data aggregation 501, 512-514, 1339, 1356
dc-dc converters 318-321, 323-324, 327-328, 330-333
dc motor 348-357, 359, 361-364, 1371
Decision Support Systems (DSS) 568, 572, 803, 811
deficit model 1014-1015, 1025
Delphi method 1560, 1564, 1574
desalinated water 681, 692
diesel generator 413-416, 419-422, 424-425
digital activism 1121
digital cities 537, 541-542, 548-549
digital control 323, 330-332, 334
digital controller 323, 331, 335, 338, 346
Digital Ecosystem (DE) 551, 554, 564
Digital Elevation Model (DEM) 1030
Digital Habitat Ecosystem Architecture (DHEA) 551-552, 555, 558
Digital Habitat Ecosystem (DHE) 551-552, 556, 564
direct expansion GSHPs 261
discrete monitoring 733
Doubly Fed Induction Generator (DFIG) 1576
dynamic alignment 981, 986
Dynamic Power Management (DPM) 661

E

Earth bermed 536
eco-awareness 105-106
eco-hydrology 807-809, 812
eco-innovation 388
ecological accounting 1647, 1656
ecological and economical monitoring 1656
ecological footprint 4, 6, 17, 161, 163, 165-166, 171, 180, 182, 184, 789, 794, 1294, 1687, 1708
ecological impacts 799
ecological literacy 1445, 1452
ecological sensitivity 56, 1732, 1734
Ecological Tax Reform (ETR) 1246-1247
economic cost 413, 419, 422, 527, 1545
economic growth 9, 16, 68, 119-120, 149, 158, 166, 304-305, 384-385, 496, 681, 695-696, 709, 717, 723, 750, 777, 785, 977, 979, 987, 1048, 1186, 1215, 1222, 1228, 1234-1235, 1248, 1325, 1381, 1383, 1395, 1534, 1575, 1648, 1704, 1706-1707, 1724, 1727
economic value 4-5, 9, 11, 630, 751, 819-821, 830-831, 833-834, 839, 842, 1168, 1171, 1549, 1560-1563, 1565, 1567-1572, 1574
ecosystem approach 21, 48
educational alignment 999-1000
educational development 1454-1455, 1466-1467, 1470
educational engagement 999
educational imagination 999
education for Sustainability 988-989, 993, 1007-1008, 1452
effective operations 889, 902
efficient infrastructure 889, 902
efficient pricing 1246-1248, 1254, 1256-1257, 1264
efficient resource use 551
electrical generation 1638

electrical vehicle 1637-1638
Electronic Road Pricing (ERP) 684
emergent strategy 219
End-of-Life (EOL) product 945
End of Life Vehicles (ELV) 945
energy awareness 656
energy efficient residential blocks 1032
energy harvesting 497, 505-506, 512-514, 516
energy management 226, 230, 241, 245-246, 279, 390, 392, 397, 411, 430, 479-483, 493, 564, 672, 692, 709, 714, 730, 792, 857-859, 864-865, 1026, 1105, 1264, 1358, 1377-1378, 1497, 1614, 1630, 1637, 1639, 1756, 1758
energy optimization 260, 278, 347, 413, 426, 428, 445, 1590
energy policy 61, 82-83, 152, 155, 222, 236, 238, 241, 257-259, 304, 316, 564, 704, 836, 838, 901, 1009-1011, 1015, 1021, 1023-1025, 1048, 1066, 1084, 1086-1088, 1091-1093, 1104, 1244-1245, 1258-1260, 1262, 1264, 1376, 1381, 1397-1398, 1545, 1547, 1555-1556, 1573-1574, 1609, 1636, 1744, 1752
energy provision 1060-1061
energy-saving daemon 658-659, 669, 672-673
energy saving modes
    evaluation of 664-665
energy security 148, 225, 410, 713-714, 718, 728, 730, 891, 896, 1009-1010, 1047, 1073, 1240, 1394, 1530-1531, 1545
energy simulation 413
Engineers without Borders 1057
environmental assets 1157, 1163, 1656
environmental education 14, 150, 1005, 1008, 1444-1446, 1452, 1456, 1492-1493, 1495-1496, 1504
Environmental, Health and Safety (EHS) 945, 949
Environmental Impact Assessment (EIA) 800, 808, 811
Environmental Management System (EMS) 124, 131, 134-135, 889
environmental policy 66, 68, 73, 75, 79, 82-83, 126-129, 131-133, 136, 617, 691, 769, 780, 784-785, 794, 869, 900, 956, 1212, 1541, 1653
environmental rationality 1-3, 7, 9-16
environmental thinking 1, 125, 128, 389
environment protection costs 1656
EU-GCC clean energy cooperation 223, 241
European Modular System (EMS) 1266
experiential learning 1154-1155, 1161-1163, 1175-1177, 1450, 1463
explanatory style 963-965, 967, 970, 973
explorative learning 1492-1495
extended enterprise 103, 111
External Quantum Efficiency (EQE) 367
eye tracking 479, 482, 491-492, 494

F

farm business 1072-1079, 1081, 1084-1086, 1721
farm to fork 788, 797
FAWN approach 661
Feed-in-Tariff (FIT) 1547, 1557
Feed-in-Tariff (FIT) Programs 1547, 1557
field-based training 1151, 1155, 1161, 1163
Field-Programmable Gate Array (FPGA) 462
fiscal measures 677, 681, 688, 1246-1248, 1251-1257, 1264
fixed solar panel 335
flooding Optimization Algorithms 445
Focus Group Discussion (FGD) 1414
food security 20, 45, 48, 304, 311, 567-568, 571, 783, 808, 834, 836, 841, 1204, 1380-1384, 1388, 1390, 1396, 1432, 1443, 1675, 1702, 1706, 1719
food supply chain 745, 786, 797, 1706
Free Prior and Informed Consent (FPIC) 851
frequency control 1576-1581, 1584-1590, 1638
fuel cells 77, 80, 228, 233, 236-237, 240, 891, 1093, 1357-1359, 1371-1376, 1379, 1529, 1578, 1589
fuzzy sets 175, 242-243, 250, 257, 259-260, 879-880

G

game-based approach 1492
generating capacity 227, 337, 1105, 1625, 1633
geographical web services 617
Geographic Information Science (GIScience) 539, 603
Geographic Information System (GIS) 824, 1413
    spatial analysis 606
    urban environmental applications 603-604
global education 1153-1154, 1160, 1163
global governance gap 990
globally harmonized agreement 885-886, 890
Global Mining Initiative (GMI) 848
Global Reporting Initiative (GRI) 905, 1316-1317, 1330
global warming 8, 10, 18, 31-32, 41, 44, 46, 52, 70, 118, 120, 160, 247, 249, 257, 275, 303, 384-385, 389, 411, 430, 480, 518, 552-553, 640-642, 656, 735, 745, 749, 841, 869, 962, 974, 976-977, 985, 1282, 1358, 1530, 1724

Green500 list 658-659
Green architecture 1725-1726, 1737
Green computing 282, 285, 301, 386, 388-389, 448-449, 460, 480, 492, 494, 497, 507, 510-512, 642, 657, 674, 1379
Green consumers 752
Green economy 19, 46, 48, 55, 85, 140, 148, 151, 153-154, 304, 769-770, 772-774, 777, 783-785, 1217, 1296, 1438, 1656, 1662, 1676, 1678, 1701, 1722
Green empowerment 1057
Green energy 55, 226, 280, 285, 304, 399-400, 411, 1378, 1550
Green GDP 161, 166, 184
Greenhouse Gas (GHG) 148, 305, 726, 734, 849, 858, 887, 1247, 1520
 emissions 23, 39, 48, 66-70, 72-81, 83, 85, 225, 228-230, 305, 726, 734, 736, 787, 849, 857-859, 866, 887, 1247, 1250-1251, 1390, 1520, 1536-1537
Green initiatives 86-87, 89, 91-92, 94-99, 102
Green IS 386, 389, 391
Green IT 280, 301, 386, 388-391, 394, 943
Green logistics 87-89, 92, 95-102, 789, 797, 870-871, 873-874, 876
Green product 867
Green project management 121, 123, 128, 133, 915, 927, 929, 942
Green technologies 388
Greenthink 121-126, 129, 131-137, 139
Green vehicles 685, 690, 692
Green-Yellow Belts 623, 626-627, 633
Gross Domestic Product (GDP) 166, 721, 1252
ground source 261, 268-269, 271, 274-277, 1755

H

Heathrow Airport Limited (HAL) 1305
hidden Markov model 645, 657
High-Density Polyethylene (HDPE) 1664
Highest Occupied Molecular Orbital (HOMO) 378
High Performance Computing (HPC) 658
home environments 551-552, 556, 560
Hoyles development model 1685
human computer interface 479
hybrid cars 685, 906

I

IATA 4 pillars strategies 901
image differencing 612
image rationing 612
image regression 612
Independent Energy Producers (IPP) 312
industrial coupling 1218
industrial symbiosis 1218-1219, 1221-1226, 1230-1231
Information and Communication Technologies for Development (ICT4D) 1050, 1596
Information Communication Technology (ICT)
 business engendered 975, 986
 socially engendered 975, 987
information system 109, 145, 149, 229, 399, 405, 569, 571, 576, 580, 604, 624, 657, 823-824, 1137, 1139, 1147, 1214, 1404, 1413
information transparency 745
Infrastructure as a Service (IaaS) 301, 390
Infrastructure Planning Commission (IPC) 1018, 1022, 1026
ink jet printing 376
institutional assets 1157, 1163
instructional design 1490
integrated change control 125
integrated design 103, 108, 427, 546, 1033, 1043
Integrated Sustainable Supply Chain (ISSC) index 953
Integrated Water Resource Management (IWRM) 810
Intellectual Property Rights (IPR) 373
Intelligent Chassis Management Bus (ICMB) 663
Intelligent Platform Management Bus (IPMB) 663
Intelligent Platform Management Interface (IPMI) 663, 674
Intergovernmental Authority on Development (IGAD) 807
international development partnerships 1068, 1591
International Energy Agency (IEA) 226, 306, 714, 745, 1565
International/Intercultural Education 1490
International Organization for Standardization (ISO) 122-123, 139, 948
International Project Management Association (IPMA) 904
 world congress 218, 913, 915
International Rainwater Harvesting Alliance (IRHA) 1671
interoperable clouds 279, 301
investment intentions 1072, 1074-1076, 1079-1083

K

Kerio Valley Development Authority (KVDA) 811
Key Performance Indicator (KPI) 1185

key resources 157, 562
knife-over-edge coating 374, 377-378, 381

L

lag compensator 349, 360-364
Lake Victoria Development Authority (LBDA) 811
Lake Victoria Environmental Management Programme (LVEMP) 811
land cover 140, 145, 152, 608-609, 612-613, 625, 839-840, 843, 1145, 1400, 1402-1404, 1407, 1409-1411, 1413
land degradation 166, 773, 808, 833-834, 836, 1235, 1400-1405, 1408-1413
land management 20, 23, 27, 539, 633, 842, 1400, 1402-1403, 1406, 1408-1410, 1412-1413
LAR-1 algorithm 445
Latin America and the Caribbean (LAC) 1658
lean thinking 933-936, 942-943
legal interregnum 1212
Life Cycle Analysis (LCA) 195, 271, 387, 947, 1106
life cycle assessment 183, 185-186, 189-191, 389, 717, 795-797, 914, 932, 943, 952, 955-956, 1118-1119, 1228, 1390, 1398, 1751
lifecycle eco-compliance 103, 105, 107-108
Light Emitting Diode (LED) 1755
Lighting and Thermal (LT) energy model 1030
lignocellulosic biomass 1519, 1529, 1531
load balancing 434, 512, 659-660, 674, 1144, 1333-1335, 1338-1341, 1350-1351, 1354-1356
load consolidation 1265, 1267-1268, 1270-1271, 1273, 1276, 1300, 1302, 1309
Load Frequency Control (LFC) 1576, 1578
London Construction Consolidation Centre (LCCC) 1303
London Heathrow Retail Consolidation Centre 1305
London Micro-Consolidation Centre 1307
Longer and Heavier Vehicles (LHV) 1265
long-term thinking 941, 943
Lowest Unoccupied Molecular Orbital (LUMO) 378
low-income market 1120
ludic system 985-986

M

machine learning 640-645, 649, 653-654, 657
Magic Packet Technology 663, 675
market conditions 108, 1549, 1595, 1614, 1639
material handling industry 121-123, 138
mathematical model 321, 417, 420, 867, 952
maturity model 128, 904-905, 907-908, 913, 915-916
mean square error 1371, 1379
mental models 185, 187-188, 193, 195-200
Millennium Development Goals (MDG) 179, 303-304, 306, 308, 1048, 1164-1166, 1173, 1181, 1658
Mining Association of Canada (MAC) 846, 850
Mining Watch Canada 848, 866
modeling analysis 1563, 1572, 1575
morphogenic change 974, 980, 984, 986
morphostatic change 974, 980, 986
mud and stud 536
Multi-Criteria Decision Analysis (MCDA) 604
multifunctional agriculture 1701-1702, 1704-1706, 1717-1720
municipal solid waste 868, 879-880, 1092, 1516, 1519, 1523, 1526, 1536-1538, 1540-1543

N

national action plan 771, 778-779, 783-785
National Environmental Management Authority (NEMA) 798, 800
National Innovation System (NIS) 1050
national policies 711, 1294
Natura 2000 network 140, 150-152, 157, 630, 634
natural potentials 785
natural resources conservation 149, 155, 621, 623, 625, 631-634
Near-Infrared Reflectance (NIR) 609
neguentropic productivity 1, 11
net generation 1102, 1105
net metering 313, 1095-1096, 1098-1099, 1101, 1105
network lifetime 434, 436, 445, 1332, 1339, 1342, 1350-1352, 1356
Newater 681, 692
New Public Management (NPM) 989
Nomenclature of Territorial Units for Statistics (NUTS) 1206
Not In My Back Yard (NIMBY) 1009, 1026
NRM problems 816-819, 831
number of retransmissions 431, 438-439, 445-446
nutritional assets 1157, 1163

O

off-grid communities 1049, 1051-1052, 1059-1060, 1064, 1068, 1591-1593, 1597-1598, 1604, 1608
oil sands 713-720, 722-730, 847

on-site renewable energy 1622, 1632-1633, 1639, 1743, 1745
Operation, Maintenance, and Surveillance (OMS) 856
optical interconnects 463, 467-468, 470, 474, 476-478
organic agriculture 38, 153, 769-771, 773-775, 777-785, 978, 1135
Organic Light-Emitting Diodes (OLED) 367
oxygen generating surfaces 621, 623, 625-626, 631-632, 634

P

Pacific Small Island Developing States (PSIDS) 1671
packet switching 468-470, 478
paradigm shifts 4, 974
Participatory Rural Appraisal (PRA) 1414
Passivhaus Planning Package (PHPP) 1747
Passivhaus standard, the 1747-1748
payload neutrality 1270, 1275
PD controller 349, 355-358, 360, 363-364, 1577
People for Rainwater (PR) 1671
Petite Martinique 1659, 1662-1664, 1676
photoresistor cell 364
Photovoltaic Devices (PVD) 366
photovoltaic modules 349, 515, 1362-1364, 1375, 1377, 1379, 1753
photovoltaic panel 343, 348-349
photovoltaic (PV) 228, 327, 336, 350, 414, 696, 1048, 1357-1358, 1374-1375, 1753
photovoltaic (PV) technology 336, 1048, 1753
photovoltaic source 1357, 1359
photovoltaic systems 294, 296, 301, 319, 330, 333, 345, 1096, 1098-1100, 1374-1375
photovoltaic/thermal (PV/T) hybrid collectors 1753
phytoremediation process 1693, 1696
Planning Act 2008 1018, 1020, 1026
Platform as a Service (PaaS) 390
policy Instruments 676-678, 687, 690, 692, 1073, 1088-1090, 1102-1104, 1252, 1378, 1410
politics of difference 1, 7, 1125
poly(3-hexylthiophene) (P3HT) 366, 382
polyaniline (PANI) 367
poly(ethylene terephthalate) (PET) 367
polymer cells 373
polyvinyl chloride (PVC) 1664
Pond-Sand Filer (PSF) 1430
Portfields project 1691, 1700
positioning analysis 1560, 1572, 1575
positive economic measures 889, 897, 899, 902
positive emotions 958-962, 965, 967-970, 973
Positive Psychological Capital (PsyCap) 973
positive psychology 958-960, 962-963, 967-968, 970, 972-973
postproductivist paradigm 1702
poverty reduction 151, 204, 316, 700, 710-711, 728, 772-773, 832, 1066, 1111, 1164-1166, 1171, 1173, 1175, 1177, 1181, 1204, 1233, 1235, 1418, 1610, 1722
power conservation factor 429, 431, 442, 446
Power Purchase Agreements (PPA) 1557
power's management and control device 1379
practical action 1047-1049, 1053-1055, 1057-1060, 1062-1065, 1068-1069, 1239, 1243-1244
pragmatic energy-mix 1243
Principles for Responsible Management Education (PRME) 1179, 1181
probabilistic algorithm 440, 442-443, 446
product innovation 217, 1107, 1120, 1219
Productive Use of Renewable Energy (PURE) 303, 305-306
product lifecycle assessment 107
product lifecycle management 106, 394
product service systems 111, 119, 1106-1110, 1112, 1116-1120
project autonomy 201, 212, 214, 219
project charter 125, 928
Project Management Body of Knowledge (PMBOK) 1187
Project Management Institute (PMI) 1187
Project Management Office (PMO) 134
project stakeholder management 211, 219
project strategy 201-202, 209-212, 214, 216, 218-219
project sustainable development strategy 220
Proportional-Integral-Derivative (PID) Control 334
Proton Exchange Membrane (PEM) 1359, 1372, 1379
PSR model 162-163, 184
public and stakeholder engagement 1015, 1026
public engagement 581, 1009, 1015, 1019, 1021-1022, 1758
Public Infrastructure Renewal (PIR) 616
pure flooding 431, 439-440, 446

Q

quantitative approaches 161, 816-818, 830-831
query processing 1332-1334, 1355-1356

R

Radio Frequency Identification (RFID) 732
Rainwater Harvesting (RWH) 807, 1657-1658, 1671, 1680
rammed Earth 518, 520-536
Reduced Instruction Set Computer (RISC) 462
reductionist and systems approach 200
remote sensing 145-146, 151, 153, 155-156, 550, 602-604, 606, 608-610, 618-620, 624, 635, 843-844, 1401, 1412
Renewable electricity production tax credit 1092, 1102, 1104-1105
Renewable Energy (RE) 693, 1073, 1544, 1557
  cooperatives 1544-1546, 1548, 1551-1554, 1557
  investment strategies 1234-1236, 1242
  powered lighting technologies 1068, 1591
  resources 227, 236, 240, 244, 335, 414, 1055, 1105, 1497, 1592
Renewable Energy (RE) sources
  energy production 1751
Renewable Energy Technology (RET) 313
renewable portfolio standard 1095, 1097-1098, 1100, 1103-1105
renewable technologies 414, 1091-1092, 1232, 1743
Requests For Proposal (RFP) 133
resilience indicators 54
resource allocation 245-246, 259, 282, 495, 497, 503, 507, 509-512, 516, 604, 828, 956, 1218, 1223, 1336, 1356
resource-based view 206, 1560, 1562, 1573-1575
Resource Management Systems (RMS) 658-659
resources conservation 141, 145, 147-151, 155, 157, 621, 623-625, 631-634, 814
reverse logistic network design 867-868, 873
reverse logistics 103, 109, 218, 792, 867, 870-871, 874, 876-880, 956, 1505-1506, 1509-1510, 1515, 1517
review of evidence 816
risk management 21-22, 51, 61, 82, 125, 131, 212, 404, 411, 796, 799, 843, 848, 860, 927, 942, 958, 971
roll-to-roll techniques 376-377
Rooftop Rainwater Harvesting (RTRWH) 1677
root locus 323, 331, 354-356, 359, 361-362, 364
route discovery 430-431, 435, 439-440, 442-443, 446
routing protocol 446
Royal Commission on Environmental Pollution (RCEP) 1185, 1201
rural development 20-22, 38, 47-48, 140-141, 147-151, 154-157, 570-571, 580-581, 710, 773, 777, 779-780, 783-785, 1047, 1054, 1084-1085, 1203-1217, 1390, 1407, 1441, 1701-1702, 1705, 1708, 1717, 1719-1720
rural development theory 1203
rural economy 43, 147, 1211-1212, 1704-1705, 1710
rural electrical services enterprise 1058
rural electrification 239-240, 1047-1048, 1050, 1059, 1067, 1069, 1381, 1602, 1610, 1612
rural environment 141, 157, 1702, 1704, 1711

S

saline water 1423, 1426, 1436
salix species 1692, 1696
screen printing 375
secondary reserve 1576, 1578, 1586-1587
sectoral policy 1213
sensor-based condition monitoring 732, 743, 746
Sensor Data Record (SDR) Repository 663
Server Load Balancing (SLB) 659
shutdown of unused hardware 659
Sigma Sustainability Marketing Guide 751, 767
Single-Wall Carbon Nanotubes (SWNT) 372
Sky View Factor (SVF) 1041
slot die coating 377
Small Island Countries (SIC) 1415
Smart Grid 397, 403, 406-407, 409-411, 561-562, 756, 1614, 1616, 1621-1622, 1636-1639
Smart growth 58, 1725
SMART objective 1330
Smart space/environment 657
Smart systems 555-556, 560-561, 564-565
social assets 1157, 1163
social cohesion 141, 989, 991, 1005
social engagement 1181
social innovation 1000, 1121-1124, 1132, 1134
socialist agricultural property 1711
social justice-based training 1155, 1161, 1163
Social Network Analysis (SNA) 192, 196, 200
social participation 992, 995, 1004
social responsibility 105, 111-112, 117, 154, 160, 205-206, 217-218, 393, 749, 757-758, 762, 765, 772, 790, 795-796, 849, 865, 896, 914, 948, 951, 955, 957, 1118-1119, 1164-1166, 1168-1181, 1224, 1307, 1466, 1469, 1556, 1602
social shaping of technology 974, 979-980, 986
social sustainability 808, 905, 1108, 1316, 1320, 1330, 1728, 1734

socio-political-environmental changes 1235
Software as a Service (SaaS) 390
soil compaction 977, 1403-1404, 1406, 1409, 1413
solar absorption 230, 335
solar cells array 1358, 1379
solar photovoltaic servo tracking controlled system 348, 364
solar thermal power systems 1753
solar thermal technologies 696, 698-700, 712
solar tracker 332-333, 335, 337, 343, 347, 349
solar tracking control 348, 363
solid waste management 537, 542-545, 690, 825, 837, 869, 879-880, 1516, 1520, 1525, 1542
spatial information 145, 569, 625, 1137-1139, 1147, 1404
special nature reserves 34, 1640, 1644-1645, 1653, 1656
spectrum of Green 943
stabilized Earth 524, 531, 536
stormwater drainage systems 1691
stormwater treatment and cleaning 1691
strategic execution framework 929-930, 943
strategic fit 986-987
Strength, Weakness, Opportunities, and Threats (SWOT) 811
strong air traffic growth 902
structural change 948, 981, 986-987
success criteria 928, 1323, 1327, 1330, 1605
supervisory committee 1057-1058, 1063, 1065
Supervisory Control and Data Acquisition (SCADA) Systems 404, 412
Supply Chain Management (SCM)
    sustainable 88, 102, 788, 794, 796-797, 948, 955
support educators 1454, 1470
sustainability dimension 166, 1330
sustainability education and leadership 1454-1456, 1459, 1464, 1466, 1471
sustainability indicators 54, 158-159, 170, 176, 180, 182, 184-185, 188-189, 200, 259, 582, 585, 596, 789, 795, 846, 951-952, 956, 1313-1317, 1319-1320, 1325, 1327-1329, 1686
Sustainability-Oriented Innovation (SOI) 1106, 1108, 1120
Sustainability Reporting Guidelines (SRG) 905
sustainable adaptation 1414-1415, 1417-1419, 1432-1440
sustainable agriculture 18, 20, 36, 38, 42, 44-45, 48, 570, 773, 783-784, 836, 1411, 1701-1702, 1704-1705, 1707-1708
Sustainable Agriculture and Rural Development (SARD) 20, 48
sustainable aviation 885, 890, 892, 895, 898, 901
sustainable business models 976
Sustainable Development (SD) 201-202, 241, 386, 1183
    assessment 951
    literature review 945
    stakeholders 948
    three dimensions 944
sustainable education 1150, 1161, 1163, 1181
sustainable energy planning 246, 259
sustainable innovations 388
sustainable lifestyle 1728, 1735
sustainable livelihood 696, 804, 1414-1415, 1418, 1426, 1432
sustainable manufacturing 387
sustainable marketing 751-753, 766
sustainable procurement 797
Sustainable production and consumption 765, 772-773, 974, 984, 987
Sustainable system innovation 902
sustainable urbanism 62, 1686, 1698, 1723-1725, 1728-1729, 1732, 1734-1735, 1738
sustainable waste management 867-870, 879, 1505-1506
synthetic indicators 1207-1208, 1413
System Event Log (SEL) 663
systems approach 159, 190, 192, 200, 236, 677, 689, 692, 713, 951, 1087
systems thinking 197, 549, 677, 690, 692, 1183-1184, 1189-1191, 1193, 1201

T

taking social action 1152-1153, 1155, 1157, 1163
Tana and Athi Development Authority (TARDA) 811
technology adoption 179, 1438, 1557
technology diffusion 698, 1050, 1064, 1419, 1439, 1591, 1595
territorial network 1206
Third Party Logistics (3PL) companines 102
time response 348, 355, 363-364
tipping point 196, 200, 1166, 1450
TORQUE/Maui Adapter 671
Total Suspended Solids (TSS) 1667
tourism industry 1220, 1225, 1228-1229, 1446
Towards Sustainable Mining (TSM) 846, 849, 1318
Transformed Difference Vegetation Index (TDVI) 610, 618

transition management 885
transport and logistics service 98-100, 102
transport externalities 1246, 1259, 1261, 1264
transport logistics 1277
travel sustainable plan 1290, 1296
tri-dimensional integration 1184, 1190, 1201
Triple Bottom Line (TBL or 3BL) 188, 190, 195, 206, 386, 751, 765, 846-847, 865, 905, 914, 927, 941, 944, 951, 955, 1108, 1117, 1120, 1187, 1443-1444, 1455
tri-sector partnerships 715, 730, 1049, 1594

U

United Nations Environmental Programme (UNEP) 1671
United Nations Global Compact (UNGC) 1164-1166, 1169, 1180-1181
Urban Canopy Layer (UCL) 1030
Urban Consolidation Centres (UCC) 1297
urban design strategies 49, 51, 60
urban environment 61-62, 584, 604, 608-611, 614, 617, 625, 631, 1027, 1030-1031, 1044-1045, 1137-1141, 1143-1147, 1285, 1292, 1697-1698, 1733, 1739
urban planning 50, 62-64, 151, 153, 156, 546, 583, 600, 615, 617-621, 623-624, 631-632, 634-639, 843, 1028, 1046, 1132, 1140, 1147, 1282, 1295, 1437, 1689, 1696-1697, 1699-1700, 1727, 1732, 1736-1737, 1741, 1760
urban resilience 49-50, 55, 57-58, 60, 62, 64
urban sustainability 52, 581, 597, 599, 603, 1282, 1684, 1686-1687, 1698, 1700, 1735-1736, 1740

V

Value Added Tax (VAT) 1746
value of nature reserves 1656
vegetation indices 608-611, 618
virtual infrastructure manager 301
virtualization of clusters 659
Vulnerability and Capacity Assessment (VCA) 1420
vulnerability indicators 24, 1413

W

war of currents 398, 412
waste to energy routes 1519
waterfront development 582, 584, 586, 588-589, 598, 1683-1686, 1688-1689, 1691-1694, 1698
waterfront developments 1684, 1691, 1694, 1698
Water Resources Management Authority (WRMA) 798, 800
water scarcity 808, 811, 1659
Watershed Management (WSM) 798-800, 805, 809-813, 815, 1687
Wattle and Daub 519, 526, 536
Wavelength-Selective Switches (WSS) 478
WebGis 1137-1140, 1142-1146
wells-to-wheels analysis 730
Wide Field Sensor (WIFS) 610
wind turbine 346, 415, 929, 1011, 1025, 1103, 1546-1548, 1550, 1555, 1577-1578, 1580, 1588-1590
wireless distributed computing 495, 497, 503, 506-508, 510-513, 516
wood-based energy 1560-1561, 1563, 1572, 1574-1575
Work Breakdown Structure (WBS) 126
workforce management 397, 405, 407, 412
world benefit 1164-1167, 1173, 1178-1179, 1181
World Meteorological Organization (WMO) 807
World Summit on Sustainable Development (WSSD) 306
World Trading Organization (WTO) 1703

Z

Zero Energy Buildings (ZEB) 1742-1743
  describing terms 1745
  environmental technologies 1749

CPSIA information can be obtained at www.ICGtesting.com
Printed in the USA
BVOW06*1923060414
349698BV00004B/61/P